TELEVISION
THEORY AND SERVICING

Charles G. Buscombe

Reston Publishing Company, Inc.
A Prentice-Hall Company
Reston, Virginia 22090

To my wife Beryl and our sons
Brian, Renny, Gary and Dale.

Library of Congress Cataloging in Publication Data

Buscombe, Charles G.
   Television theory and servicing.
   Includes index.
   1. Television—Receivers and reception. 2. Television—Repairing. I. Title.
TK6653.B87   1984   621.388'87   83-11139
ISBN 0-8359-7544-4

© 1984 by Reston Publishing Company, Inc.
A *Prentice-Hall Company*
Reston, Virginia 22090

All rights reserved. No part of this book may
be reproduced, in any way, or by any means,
without permission in writing from the publisher.

10   9   8   7   6   5   4   3   2   1

Interior design and production: Jack Zibulsky

Printed in the United States of America

CONTENTS

PREFACE, ix

1 TELEVISION FUNDAMENTALS: The Station

1-1 Television as a System, 1 • 1-2 Bandpass Considerations and Channel Allocations, 2 • 1-3 Vestigal Sideband Transmission, 2 • 1-4 Compatability, 7 • 1-5 Monochrome System, 7 • 1-6 Camera, 7 • 1-7 Scanning and Synchronization, 9 • 1-8 Basic Camera Operation, 11 • 1-9 Composite Video Signal, 11 • 1-10 Test Patterns, 11 • 1-11 The Wonderful World of Color, 11 • 1-12 Color Resolution and Mixing Principles, 12 • 1-13 Color Station Functions, 13 • 1-14 Tricolor Camera, 13 • 1-15 Developing the Luminance or Y Signal, 13 • 1-16 Color Difference Signals, 15 • 1-17 The 3.58-MHz Subcarrier, 15 • 1-18 Modulating the Subcarrier, 15 • 1-19 Color Sync, 16 • 1-20 Interleaving and Total Composite Color Signal, 16 • 1-21 Special-Purpose Test Signals, 16 • 1-22 The Sound Signal, 17 • 1-23 Modulating the Sound and Video Transmitters, 17 • 1-24 Summary, 18 • Review Questions, 20

2 TELEVISION FUNDAMENTALS: The Receiver

2-1 Monochrome Receiver, 23 • 2-2 Chassis Configuration and Stage Recognition, 28 • 2-3 Controls, 28 • 2-4 Color Receiver, 29 • 2-5 Color CRT Fundamentals and Color Mixing, 31 • 2-6 Color Sync Stages, 32 • 2-7 Miscellaneous Circuits, 33 • 2-8 Color Picture, 33 • 2-9 Shadow Mask Color Tube, 33 • 2-10 Aperture Grill (in-line) Color Tubes, 34 • 2-11 Purity and Convergence, 34 • 2-12 Developing the White Raster, 34 • 2-13 Developing a BW Picture, 34 • 2-14 Gray-Scale Tracking, 35 • 2-15 Developing a Color Picture, 35 • 2-16 Degaussing, 35 • 2-17 Color Set Controls, 36 • 2-18 Summary, 36 • Review Questions, 39

3 TELEVISION TEST EQUIPMENT and Servicing Aids

3-1 Introduction, 43 • 3-2 Common Test Instruments, 45 • 3-3 Meter, 45 • 3-4 Basic Signal Generators, 47 • 3-5 Signal Tracers, 49 • 3-6 Cathode-Ray Oscilloscope, 50 • 3-7 Grid Dip Oscillator, 52 • 3-8 Tube Checker, 53 • 3-9 C/R Bridge (capacity analyzer), 53 • 3-10 Special Television Test Instruments, 53 • 3-11 Digital Volmeter and Digital Multimeters, 53 • 3-12 Special Purpose Signal Generators, 56 • 3-13 TV Stage Analyzers, 60 • 3-14 Counters, 60 • 3-15 Dual Trace, Triggered Sweep Oscilloscope, 61 • 3-16 Vectorscope, 63 • 3-17 Diode and Transistor Testers, 64 • 3-18 Flyback and Yoke Tester, 66 • 3-19 Picture Tube Tester and Rejuvenator, 68 • 3-20 Service Aids and Accessories, 69 • 3-21 Maintenance, Calibration, and Repair of Common Test Equipment, 73 • Summary, 76 • Review Questions, 79 • Troubleshooting Questionnaire, 82

iv CONTENTS

4 The TROUBLESHOOTING APPROACH

4-1 Introduction, 83 • 4-2 Software in Servicing, 84 • 4-3 Trouble Symptoms and Analytical Reasoning, 84 • 4-4 Other Considerations, 85 • 4-5 Control Adjustments, 86 • 4-6 Safety, 86 • 4-7 Developing a Plan of Action, 88 • 4-8 Inspection, 90 • 4-9 Localizing Trouble to the Stage or Module Level, 91 • 4-10 Signal Flow and Stage Checking, 92 • 4-11 Stage Checking by Injection, 93 • 4-12 Signal Tracing the Station Signals, 96 • 4-13 Summary, 100 • Review Questions, 104 • Troubleshooting Questionnaire, 104

5 FAULT ISOLATION

5-1 Isolating the Defect, 107 • 5-2 Adjustments, 109 • 5-3 Waveform Analysis, 109 • 5-4 Voltage Testing, 110 • 5-5 Resistance Checking, 114 • 5-6 Current Testing, 116 • 5-7 Is the Pix Tube Suspect?, 117 • 5-8 Component Defects, Substitution, and Testing, 117 • 5-9 Testing Solid-State (Semiconductor) Devices, 123 • 5-10 Checking Modules, 133 • 5-11 Working with Printed Circuit Boards (PCBs), 134 • 5-12 Replacing Components, 136 • 5-13 Intermittents and Other Problems, 137 • 5-14 Alignment, 140 • 5-15 Air Checking the Repaired Set, 141 • 5-16 Summary, 142 • Review Questions, 145 • Troubleshooting Questionnaire, 148

6 LOW-VOLTAGE POWER SUPPLY

6-1 LVPS Module, 151 • 6-2 Types of Power Supplies, 153 • 6-3 The AC Input Circuits, 153 • 6-4 Basic LVPS Circuits, 156 • 6-5 Filters, 161 • 6-6 DC Voltage Distribution, 162 • 6-7 Voltage Regulators, 163 • 6-8 Overload Protection, 167 • 6-9 Battery-Operated Portables, 167 • 6-10 The Typical LVPS, 167 • 6-11 Troubleshooting the LVPS, 168 • 6-12 Troubleshooting the AC Power Circuits, 170 • 6-13 Troubleshooting the DC Circuits, 175 • 6-14 Air Checking the Repaired Set, 179 • 6-15 Summary, 180 • Review Questions, 183 • Troubleshooting Questionnaire, 187

7 The VERTICAL SWEEP SECTION

7-1 Vertical Sweep Requirements, 191 • 7-2 Vertical Sweep Stages, 191 • 7-3 Vertical Oscillator, 192 • 7-4 Sweep-Dependent Circuits, 192 • 7-5 Vertical Sweep Circuit Fundamentals, 193 • 7-6 Generating the Sawtooth Wave, 193 • 7-7 Waveshaping, 194 • 7-8 Typical Vertical Oscillator Circuits, 195 • 7-9 Digital Countdown Sweep-Generating System, 200 • 7-10 Synchronization, 200 • 7-11 Buffer and Driver Amplifiers, 203 • 7-12 Vertical Output Stage, 204 • 7-13 Two-Transistor Output Stage, 208 • 7-14 Vertical Sweep-Dependent Circuits, 210 • 7-15 Raster Centering, 216 • 7-16 Complete Vertical Sweep Section, 216 • 7-17 Troubleshooting the Vertical Sweep Section, 217 • 7-18 Localizing the Problem Area, 223 • 7-19 Isolating the Defect, 224 • 7-20 Troubleshooting the Repair of Driver Stage(s), 228 • 7-21 Troubleshooting the Output Stage, 229 • 7-22 Troubleshooting the Complementary Symmetry Output Stage, 231 • 7-23 Raster Centering Problems, 233 • 7-24 Pincushioning Problems, 234 • 7-25 Dynamic Convergence Problems, 235 • 7-26 Troubleshooting for IC Defects, 235 • 7-27 Intermittents and Other Problems, 236 • 7-28 Air Checking the Repaired Set, 236 • 7-29 Summary, 236 • Review Questions, 239 • Troubleshooting Questionnaire, 241

8 HORIZONTAL SWEEP SECTION: Oscillator, AFC, and Driver Stages

8-1 Introduction, 245 • 8-2 Functional Description of Horizontal Stages, 247 • 8-3 Horizontal Sweep Circuit Fundamentals, 248 • 8-4 Typical Horizontal Oscillator Circuits, 249 • 8-5 Oscillator Synchronization, 251 • 8-6 Basic Phase Detector, 254 • 8-7 Oscillator Control, 257 • 8-8 Buffer-Driver Amplifiers, 258 • 8-9 Typical Countdown Sweep Circuit, 260 • 8-10 Troubleshooting the Horizontal Oscillator, AFC, and Driver Stages, 260 • 8-11 Localizing the Problem, 266 • 8-12 Isolating the Defect, 267 • 8-13 Troubleshooting the Oscillator, 268 • 8-14 Troubleshooting the AFC, 271 • 8-15 Reactance Control Circuit, 272 • 8-16 Oscillator and AFC Adjustments, 273 • 8-17 Troubleshooting the Driver Stage(s), 273 • 8-18 Intermittents, 274 • 8-19 Air Checking the Repaired Set, 274 • 8-20 Summary, 275 • Review Questions, 277 • Troubleshooting Questionnaire, 278

9 HORIZONTAL OUTPUT and HV Section

9-1 Introduction, 281 • 9-2 Horizontal Output Stage, 281 • 9-3 SCR Sweep System, 285 • 9-4 Horizontal Sweep-Dependent Circuits, 288 • 9-5 Horizontal Retrace Blanking, 289 • 9-6 Pincushion Correction, 289 • 9-7 Dynamic Convergence, 291 • 9-8 Horizontal Output Control Circuits, 292 • 9-9

Sweep-Dependent Power Supplies, 292 • 9-10 HV and Focus Voltage Power Supplies, 295 • 9-11 Developing the HV, 296 • 9-12 HV Regulator Circuits, 296 • 9-13 Overvoltage Protective Circuits, 297 • 9-14 Focus Voltage Circuits, 298 • 9-15 HV Arcing and Corona Discharge, 299 • 9-16 Typical Horizontal Output and HV Circuit, 300 • 9-17 Troubleshooting the Horizontal Output, HV, and Related Circuits, 301 • 9-18 Localizing the Problem, 302 • 9-19 Isolating the Defect, 310 • 9-20 Troubleshooting the Output Amplifier, 311 • 9-21 Air Checking the Repaired Set, 314 • 9-22 Summary, 315 • Review Questions, 317 • Troubleshooting Questionnaire, 318

10 SYNC SECTION

10-1 Introduction, 321 • 10-2 Sync Separator, 322 • 10-3 Noise Immunity, 324 • 10-4 Sync Amplifier, 327 • 10-5 Integrator and Vertical Synchronization, 328 • 10-6 Differentiating Network and Horizontal (AFC) Control, 330 • 10-7 Typical Sync Circuit, 330 • 10-8 Troubleshooting Sync Stages, 332 • 10-9 Localizing the Fault, 332 • 10-10 Loss of Vertical and Horizontal Sync, 332 • 10-11 Poor Vertical Sync, 341 • 10-12 Poor Horizontal Sync, 341 • 10-13 Isolating the Defect, 341 • 10-14 Intermittent Sync, 344 • 10-15 Air Checking the Repaired Set, 344 • 10-16 Summary, 344 • Review Questions, 346 • Troubleshooting Questionnaire, 348

11 AUTOMATIC GAIN CONTROL

11-1 Basic AGC Circuits, 351 • 11-2 Feeding the Controlled Stages, 353 • 11-3 Controlling Transistor Gain, 353 • 11-4 Controlling a FET, 355 • 11-5 Keyed AGC, 356 • 11-6 Keyer Input Signals, 357 • 11-7 Keyer Circuit Variations, 359 • 11-8 Amplified AGC, 359 • 11-9 Controlling the IF Stages, 362 • 11-10 Controlling the RF Stage(s), 365 • 11-11 Integrated-Circuit AGC, 369 • 11-12 Typical AGC System, 369 • 11-13 Troubleshooting and Repair, 370 • 11-14 Analyzing the Trouble Symptoms, 370 • 11-15 Localizing the Trouble, 372 • 11-16 Complexities of AGC Troubleshooting, 374 • 11-17 AGC Testing with a Signal Received, 380 • 11-18 AGC Clamping with a Bias Box, 380 • 11-19 Keyer Problems, 381 • 11-20 Checking AGC Delay Action, 381 • 11-21 Control Adjustment Procedures, 382 • 11-22 Localizing AGC Troubles to an IC, 382 • 11-23 Isolating the Defect, 382 • 11-24 AGC Amplifier, 384 • 11-25 AGC Bus Problems, 385 • 11-26 Troubleshooting the Delay Circuit, 385 • 11-27 Troubleshooting AGC Voltage Regulators, 386 • 11-28 AGC Component Replacement, 386 • 11-29 Intermittent AGC Troubles, 387 • 11-30 Multiple Troubles, 387 • 11-31 Air Checking the Repaired Set, 387 • 11-32 Summary, 387 • Review Questions, 392 • Troubleshooting Questionnaire, 395

12 CHANNEL SELECTORS, TUNING SYSTEMS, and Remote Control

12-1 Introduction, 401 • 12-2 General Physical Considerations, 401 • 12-3 Basic Tuner Functions, 403 • 12-4 Tuned Circuit Fundamentals, 405 • 12-5 RF-Mixer Tuned Circuits of a VHF Tuner, 409 • 12-6 RF-Mixer Tuned Circuits of a UHF Tuner, 409 • 12-7 VHF and UHF Oscillator Tuned Circuits, 410 • 12-8 Functional Operation of Tuner Stages, 411 • 12-9 Stage Functions, 413 • 12-10 VHF Tuner Types, 415 • 12-11 UHF Tuners, 418 • 12-12 Automatic Fine Tuning (AFT), 422 • 12-13 Phase-Locked-Loop (PLL) Tuning Systems, 426 • 12-14 PLL Fundamentals, 426 • 12-15 Typical PLL Tuning System, 429 • 12-16 Star System, 431 • 12-17 Remote-Control Systems, 431 • 12-18 Another Front-End PLL Tuning System, 435 • 12-19 Troubleshooting Channel Selectors and Tuning Systems, 436 • 12-20 Consider the Trouble Symptoms, 436 • 12-21 Which Tuner is Suspect?, 436 • 12-22 Faulty Tuner or Chassis Problems?, 437 • 12-23 Two Options if Tuner is Defective, 455 • 12-24 Tool Requirements, 455 • 12-25 Inspection of Faulty Tuner, 455 • 12-26 Troubleshooting and Repair of VHF Tuners, 455 • 12-27 Cleaning and Repair of Channel Selector Contacts, 455 • 12-28 Localizing VHF Tuner Troubles, 456 • 12-29 Isolating VHF Tuner Defects, 457 • 12-30 Alignment of the RF-Mixer Stages, 459 • 12-31 Checking the RF-Mixer Response, 459 • 12-32 RF-Mixer Alignment Procedures, 460 • 12-33 Troubleshooting the Oscillator Stage, 462 • 12-34 Troubleshooting and Repair of UHF Tuners, 465 • 12-35 Mechanical Problems with VHF/UHF Tuners, 467 • 12-36 Tuner Replacement, 468 • 12-37 AFT Troubleshooting and Alignment, 468 • 12-38 Troubleshooting PLL Tuning Systems, 470 • 12-39 Troubleshooting Remote-Control Systems, 472 • 12-40 Transmitter Testing, 473 • 12-41 Testing Remote Receivers, 474 • 12-42 Intermittent Tuner or Related Systems, 475 • 12-43 Air Checking the Repaired Set, 476 • 12-44 Summary, 476 • Review Questions, 479 • Troubleshooting Questionnaire, 482

vi CONTENTS

13 VIDEO IF and Detector Stages

13-1 Introduction, 487 • 13-2 Video Frequencies and Bandwidth Requirements, 487 • 13-3 IF Amplification, 488 • 13-4 Sound Takeoff, 489 • 13-5 IF Response and Bandpass Considerations, 489 • 13-6 Normal and Abnormal Marker Locations, 491 • 13-7 Color Set Response Curves, 492 • 13-8 Interstage Coupling Methods, 493 • 13-9 Wave Traps, 495 • 13-10 Trap Circuits, 497 • 13-11 Preamplification Tuning, 497 • 13-12 Feedback, Oscillations, and Neutralizing, 497 • 13-13 Typical IF Circuits, 502 • 13-14 Basic Video Detector, 503 • 13-15 Composite Video Signal, 505 • 13-16 Synchronous Detector, 506 • 13-17 Typical IF Strip Using ICs, 508 • 13-18 Troubleshooting the IF Strip and Detector, 510 • 13-19 Symptoms Analysis, 510 • 13-20 Inspection, 516 • 13-21 Localizing the Fault, 516 • 13-22 Video Overload, 518 • 13-23 Localizing Trouble by Signal Injection, 518 • 13-24 Interference Problems, 520 • 13-25 Regeneration and IF Oscillations, 522 • 13-26 Isolating Trouble in the IF Strip, 522 • 13-27 Troubleshooting the Basic Detector, 523 • 13-28 Troubleshooting the Synchronous Detector, 524 • 13-29 IF Alignment, 525 • 13-30 Alignment Evaluation, 525 • 13-31 Preliminary Remarks and Procedures, 526 • 13-32 Alignment Evaluation with Signal Generator and EVM, 527 • 13-33 Alignment Evaluation Using Sweep Equipment, 529 • 13-34 Alignment Equipment Particulars, 529 • 13-35 Sweep Fundamentals, 530 • 13-36 Obtaining a Response Curve, 531 • 13-37 Alignment Checking with Marker Signals, 532 • 13-38 Alignment Preliminaries, 535 • 13-39 Alignment Procedures, 536 • 13-40 Alignment with Generator and EVM, 536 • 13-41 Overall Alignment Using Sweep Equipment, 537 • 13-42 Alignment of Lumped Tuned Circuits, 539 • 13-43 Bar Sweep Alignment, 539 • 13-44 Synchronous Detector Alignment, 540 • 13-45 Fringe-Area Problems, 540 • 13-46 Intermittents, 541 • 13-47 Air Checking the Repaired Set, 541 • 13-48 Summary, 541 • Review Questions, 544 • Troubleshooting Questionnaire, 547

14 The VIDEO AMPLIFIER

14-1 Video Amplifier, 554 • 14-2 DC Restoration, 555 • 14-3 Frequency Response, 555 • 14-4 Contrast Control, 559 • 14-5 Brightness Control, 559 • 14-6 DC Control Stage, 562 • 14-7 Automatic Brightness Control, 562 • 14-8 Spot Killer, 562 • 14-9 Spark Gaps (SG) and HV Arcing, 564 • 14-10 Retrace Blanking, 564 • 14-11 Typical BW Video Amplifier Circuits, 565 • 14-12 Color Sets, 569 • 14-13 Comb Filter, 569 • 14-14 Delay Line, 573 • 14-15 Color Receiver Brightness Control, 573 • 14-16 Chroma Considerations, 578 • 14-17 Matrixing, 578 • 14-18 Service Setup Switch, 580 • 14-19 Video Amplifier of a Typical Modern Color Set, 582 • 14-20 Another Receiver Circuit, 583 • 14-21 Troubleshooting the Video Amplifier Section, 583 • 14-22 Analyzing the Symptoms, 583 • 14-23 Inspection, 586 • 14-24 Localizing the Fault, 586 • 14-25 Isolating the Defect, 600 • 14-26 Troubleshooting BW Sets, 600 • 14-27 Raster Troubles, 600 • 14-28 Troubleshooting CRT Bias Problems, 601 • 14-29 LDR Brightness Control Problems, 602 • 14-30 Blooming, 602 • 14-31 Arcing, 603 • 14-32 Hum Bars, 603 • 14-33 Raster and Pix Shading, 603 • 14-34 Pix Troubles (BW Sets), 604 • 14-35 No Pix or Weak Pix, 604 • 14-36 Excessive Contrast, 605 • 14-37 Video Overload, 605 • 14-38 Video Ringing, 605 • 14-39 Hum Bars on Pix Only, 605 • 14-40 Pix Smearing, 606 • 14-41 Loss of Fine Detail, 606 • 14-42 Interference, 606 • 14-43 Troubleshooting Color Sets, 607 • 14-44 Raster Troubles, 607 • 14-45 Bias Checking, 607 • 14-46 No Raster or Dim Raster, 608 • 14-47 Excessive Brightness, 608 • 14-48 ABL Troubleshooting, 608 • 14-49 Blooming, 609 • 14-50 Arcing, 609 • 14-51 Service Setup Switch Problems, 610 • 14-52 Color-Tinted Raster, 610 • 14-53 Pix Troubles, 611 • 14-54 No Pix or Weak Pix, 611 • 14-55 Color Pix (No BW Pix), 611 • 14-56 No Color or Weak Color, 611 • 14-57 BW Pix Ok, Poor Color Pix, 612 • 14-58 Troubleshooting the Comb Filter, 612 • 14-59 Checking Frequency Response (Fig. 14-3), 613 • 14-60 Delay Line (DL) Problems, 613 • 14-61 Interference, 614 • 14-62 IC Problems, 614 • 14-63 Intermittents, 614 • 14-64 Air Checking the Repaired Set, 614 • 14-65 Summary, 615 • Review Questions, 618 • Troubleshooting Questionnaire, 621

15 PIX TUBES and Associated Circuits

15-1 The Black-and-White (BW) CRT, 625 • 15-2 The Bowl, 625 • 15-3 Deflection Angles, 626 • 15-4 Implosion Protection, 626 • 15-5 The Screen, 626 • 15-6 Ion Burn Prevention, 626 • 15-7 The Electron Gun, 627 • 15-8 The Heater/Cathode, 628 • 15-9 The Control Grid (Grid No. 1), 628 • 15-10 The First Anode (Grid No. 2), 628 • 15-11 The Focus Electrode (Grid No. 3), 628 • 15-12 Grids No. 4 and No. 5, 629 • 15-13 Base/Socket Connections, 629 • 15-14 CRT Operating Voltages, 629 • 15-15 Beam Current, 629 • 15-16 Spark Gaps, 629 • 15-17 External Gun Components, 630 • 15-18 The Deflection Yoke, 630 • 15-19 The Raster Centering Ring, 631 • 15-20 Pincushion Adjusters, 631 • 15-21 Beam Deflection, 631 • 15-22 Raster Scanning, 632 • 15-23 Deflection Sensitivity, 634 • 15-24 Deflection Angles, 634 • 15-25   The Signal Input Circuit, 634 • 15-26   Producing Bar and Line Patterns, 635 • 15-27   The Picture, 637 • 15-28   Developing the Pix, 640 • 15-29   Color, 643 • 15-30   The Color Pix Tube, 644 • 15-31   The Delta (Shadow-Mask) CRT, 646 • 15-32   The Screen, 646 • 15-33   The Shadow Mask, 646 • 15-34   The Electron Guns, 647 • 15-35   Internal Gun Elements, 647 • 15-36   Base/Socket Connections, 647 • 15-37   Operating Voltages, 647 • 15-38   Beam Current, 647 • 15-39   HV Control and the X-Ray Hazard, 648 • 15-40   External Components, 648 • 15-41   The Deflection Yoke, 648 • 15-42   The Degaussing Coil, 648 • 15-43   The Purity Ring, 649 • 15-44   The Blue Lateral Adjuster, 649 • 15-45   Convergence and Beam Landing, 649 • 15-46   The Convergence Assembly, 650 • 15-47   Static Convergence, 650 • 15-48   The Blue Lateral Adjuster Functions, 652 • 15-49   Dynamic Convergence, 653 • 15-50   Pincushion Correction, 653 • 15-51   CRT Operating Controls, 654 • 15-52   The In-Line Pix Tube, 655 • 15-53   The 3-Gun In-Line CRT, 655 • 15-54   The Screen, 657 • 15-55   The Aperture Grill, 657 • 15-56   The Electron Guns, 657 • 15-57   The Trinitron, 659 • 15-58   External Gun Components, 660 • 15-59   Signal Input Circuits, 662 • 15-60   Developing the White Raster, 662 • 15-61   Developing a BW Pix, 662 • 15-62   Producing a Color Pix, 663 • 15-63   Producing Color-Bar Patterns, 665 • 15-64   Projection-Type Receivers, 667 • 15-65   Troubleshooting the CRT and Associated Circuits, 669 • 15-66   Use of a CRT Test JIG, 669 • 15-67   Analyzing the Symptoms, 669 • 15-68   Inspection, 669 • 15-69   First-Impression Suspects, 669 • 15-70   Symptoms and Troubleshooting Procedures, 672 • 15-71   No Raster or Dim Raster (Sound OK), 672 • 15-72   Checking CRT Operating Voltages, 676 • 15-73   Small Raster, 677 • 15-74   Raster Brightens During Warm-up, 677 • 15-75   Distorted Raster, 678 • 15-76   Excessive Brightness, 678 • 15-77   Neck Shadows, 678 • 15-78   Tilted Raster, 678 • 15-79   Raster not Centered, 678 • 15-80   Black Horizontal Bars, 678 • 15-81   Ripple on Raster Edges, 678 • 15-82   Poor Focus, 678 • 15-83   Blooming, 678 • 15-84   Bright Horizontal Flashes on Screen, 679 • 15-85   Arcing in CRT, 679 • 15-86   Ion Burns, 679 • 15-87   Arcing External to the CRT, 679 • 15-88   CRT Socket Defects, 679 • 15-89   Yoke Defects, 680 • 15-90   Raster Width, Horizontal Linearity, or Centering Problems, 680 • 15-91   Raster Height, Vertical Linearity, or Centering Problems, 680 • 15-92   No Pix Highlights, 680 • 15-93   Screen Goes Dark, 680 • 15-94   Raster Slow Coming On, 680 • 15-95   Pix Inverted and/or Transposed, 680 • 15-96   Symptoms and Troubleshooting Procedures (Color Sets Only), 680 • 15-97   Raster is Red, Green, or Blue, 681 • 15-98   Raster is Yellow, Magenta, or Cyan, 681 • 15-99   Color-Tinted Raster, 681 • 15-100   Gray-Scale Tracking Problem, 681 • 15-101   Jumbled Whirls of Color, 681 • 15-102   Need to Frequently Degauss CRT, 681 • 15-103   Color-Spotted Raster, 681 • 15-104   Color Contamination, 681 • 15-105   Color Outlines on BW Pix, 682 • 15-106   No BW Pix Reception, 682 • 15-107   No Color Pix (BW Pix OK), 682 • 15-108   Blooming, 682 • 15-109   Pix Tube Defects and Trouble Symptoms, 682 • 15-110   Pix Tube Testing, 682 • 15-111   Pix Tube Rejuvenation, 683 • 15-112   Flashing to Remove Shorts, 684 • 15-113   Replacing Pix Tubes, 684 • 15-114   Replacing a BW CRT, 684 • 15-115   Replacing a Color CRT, 684 • 15-116   Pix Tube Disposal, 685 • 15-117   BW Receiver Set-up Adjustments, 686 • 15-118   Color Receiver Set-up Adjustments (General), 686 • 15-119   Preliminary Adjustments, 686 • 15-120   Purity and Convergence, 687 • 15-121   Delta CRT Set-up Procedures, 689 • 15-122   Adjustment Procedures, 691 • 15-123   In-Line CRT and Trinitron Set-up Procedures, 696 • 15-124   Purity and Convergence Problems (General), 701 • 15-125   Troubleshooting Projection Receivers, 701 • 15-126   Intermittents, 702 • 15-127   Air-Checking the Repaired Set, 702 • 15-128   Summary, 703 • Review Questions, 707 • Troubleshooting Questionnaire, 710

16   The CHROMA SECTION

16-1   Signal Development at the Station, 715 • 16-2   The Receiver, 719 • 16-3   The Chroma Signal Stages, 721 • 16-4   Color Sync, 723 • 16-5   Chroma Switching, 724 • 16-6   The Chroma Circuits, 725 • 16-7   Bandpass Amplifier (BPA), 725 • 16-8   Typical BPA Circuit, 728 • 16-9   Automatic Color Control (ACC), 730 • 16-10   Chroma-Switching Circuits, 730 • 16-11   The Chroma Demodulators, 732 • 16-12   Demodulator Systems, 738 • 16-13   Typical Demodulator Circuits, 741 • 16-14   Automatic Tint Control (ATC), 743 • 16-15   The Chroma Output Circuits, 744 • 16-16   The Subcarrier Oscillator, 747 • 16-17   Subcarrier Phase-Shifting Networks, 749 • 16-18   Oscillator Control Circuits, 749 • 16-19   Chroma ICs, 750 • 16-20   Troubleshooting the Chroma Section, 752 • 16-21   Symptoms Analysis, 752 • 16-22   Inspection, 761 • 16-23   Localizing the Problem, 761 • 16-24   Trouble Symptoms and Stage Checking, 762 • 16-25   No Color or Weak Color, 762 • 16-26   Excessive Color, 763 • 16-27   Loss of ACC Function, 763 • 16-28   No BW Pix, Poor Color Pix, 763 • 16-29   No Pix (BW or Color), 763 • 16-30   Weak or Missing Colors, 764 • 16-31   Wrong Colors, 764 • 16-32   Erratic Colors, 764 • 16-33   Color Smear, 764 • 16-34   Tint Problems, 765 • 16-35   No Color Pix; Color Background to BW Pic, 765 • 16-36   Loss of Color Sync, 765 • 16-37   APC Alignment Procedures, 766 • 16-38   Hum Bars, 768 • 16-39   Troubleshooting IC Circuits, 768 • 16-40   Intermittent Chroma Troubles, 768 • 16-41

Killer Threshold Adjustment, 768 • 16-42 Bandpass Amplifier Alignment, 769 • 16-43 VSM Alignment, 770 • 16-44 Air-Checking the Repaired Set, 771 • 16-45 Summary, 772 • Review Questions, 775 • Troubleshooting Questionnaire, 776

17 The SOUND SECTION

17-1 The FM Sound Signal, 779 • 17-2 Receiver Functions, 779 • 17-3 The Sound Takeoff, 782 • 17-4 The Sound IF Detector, 782 • 17-5 The Sound IF Amplifier, 783 • 17-6 Sound IF Limiters, 783 • 17-7 FM Detectors, 783 • 17-8 The Audio Amplifiers, 789 • 17-9 The AF Voltage Amplifier, 789 • 17-10 Driver Amplifier, 789 • 17-11 The Output Stage, 790 • 17-12 The Speaker, 794 • 17-13 Volume and Tone Controls, 795 • 17-14 Negative Feedback, 795 • 17-15 Sound-Section ICs, 796 • 17-16 Troubleshooting the Sound Section, 796 • 17-17 Inspection and Symptoms Analysis, 796 • 17-18 Localizing the Problem, 804 • 17-19 Isolating the Defect, 805 • 17-20 Troubleshooting the Sound IF Detector, 805 • 17-21 Troubleshooting the IF/Limiter Stages, 805 • 17-22 Troubleshooting FM Detectors, 806 • 17-23 Alignment, 806 • 17-24 Troubleshooting the AF Stages, 808 • 17-25 Speaker Testing and Repairs, 812 • 17-26 Troubleshooting Sound ICs, 813 • 17-27 Intermittent Sound, 813 • 17-28 Air-Checking the Repaired Set, 813 • 17-29 Summary, 814 • Review Questions, 816 • Troubleshooting Questionnaire, 818

SAMPLE CIRCUITS of a Typical Modular Color Set (GE)

Sample Circuit A, 820-821 • Sample Circuit B, 822-824 • Sample Circuit C, 825-826 • Sample Circuit D, 827

INDEX, 829

PREFACE

We've come a long way in fifty short years since radio broadcasting was in its infancy—a time when it was considered sheer magic to sit before a box-like affair with its loud speaker, to twiddle the dials and, between shrieks and squeals, marvel at an age when you could listen to music, the news, or weather reports, seemingly from out of nowhere and merely at the touch of a button.

Radio is still with us, but our more blasé and sophisticated society has long since ceased to wonder at this marvelous creation, taking it for granted, even as its video counterpart, television.

Television. A means of communication that has influenced the lives and cultures of people the world over. With an ever-expanding market for home entertainment and enlightenment, the television industry has been hard-pressed to keep pace with technology and a growing list of applications—industry, aerospace, education, and news media, to name a few.

In the past few years we have witnessed great progress in receiver design, especially the conversion from vacuum tubes to semiconductors and the development of that wonderfully versatile device, the microchip. Then, there are the television off-shoots—cable TV, satellite ground stations, large-screen projection sets, electronic channel selection, and other microprocessor-controlled functions. In the offing, stereo TV sound, 1200-line high resolution pictures, the flat picture tube, and 3-D are taking shape.

This book is written for the student who will one day be called upon to service and maintain this equipment. Unlike most texts, theory and practice are treated with equal importance, and every effort has been made to keep it current with the latest trends in receiver design and troubleshooting techniques. Up-to-date circuits are described and analyzed along with most troubles that occur and the troubleshooting procedures for each. The beginning student is often overwhelmed by too much detail, but this book treats all subjects in a logical and straight-forward manner progressing gradually from the simple to the complex. This text possesses a number of unique features and contains information others treat superficially or not at all. At the end of each chapter, important points are condensed and summarized along with numerous thought-provoking review questions. Most chapters also include a troubleshooting questionnaire to stimulate analytical reasoning. Many of the troubleshooting flow drawings and trouble symptoms charts included list practically every conceivable complaint with step-by-step servicing procedures.

Prerequisites to the course of study include knowledge of electronic fundamentals and solid-state devices, but some review of these basics is included where necessary. While not essential, some practical background in working with radio, tube-type TVs, or other electronic equipment is also desirable.

The contents of each chapter are summarized as follows:

Chapter 1 provides a brief introduction to TV basics and what takes place at the station, with emphasis on the generation and processing of signals that are used by the receiver.

Chapter 2 deals with receiver fundamentals, both black and white, and color. The subject is treated lightly at this point, with appropriate details reserved for later chapters.

Chapter 3 contains information on all types of test equipment and accessories, from the most basic and inexpensive instruments used by the novice, to the most up-to-date equipment found in modern service shops. Included is a section on the repair and maintenance of such equipment.

Chapter 4 covers troubleshooting procedures for localizing faults to any section of the receiver. The treatment here is of a general nature with details covered in subsequent chapters.

Chapter 5 is a follow-up to Chapter 4. Here, procedures for pinpointing defects or other abnormal conditions that have previously been localized are covered. Also included are component and circuit testing procedures.

Chapters 6 through 17 treat the various stages and sections of receivers individually. Each chapter is in two parts—description and operational theory, followed by troubleshooting procedures. Included are symptoms charts and flow drawings.

This textbook is designed to serve the needs of student and teacher alike. With its many practical servicing hints and short-cut procedures, it should also prove useful to the technician in the field.

In conclusion, I wish to thank all those who contributed their experiences and suggestions in the preparation of this book, in particular, my colleagues in the teaching profession and associates in the field of TV repair.

I also wish to acknowledge the following for their kind permission to use circuit drawings, photos of test equipment and other diagrams in this book: B & K Precision Dynascan Corp., General Electric-Television Service Division, GTE Sylvania, Intertec Publishing Co. (Electronic Servicing Magazine), Heath Co., Hickok Electrical Instrument Co., RCA Consumer Electronics Division, Sencore News, Triplett Corp., Winegard Co., Zenith Radio Corp. Also, Bobbs-Merrill Co., Inc., Howard W. Sams & Co., Leader Instrument Corp., McGraw-Hill, Inc., Motorola, Inc., Prentice-Hall, Inc., and Sprague Co.

C.G.B.

Chapter 1

TELEVISION FUNDAMENTALS
THE STATION

Fig. 1-1 Zeroing in on the defect.

the shop. Two specialties each with its own qualifications. With a small operation, one person may handle both jobs.

To attain the enviable position of a skilled technician one must pay the price; time devoted to study, and practical application, preferably an apprenticeship in a busy repair shop working with professionals willing to share their knowledge and expertise with others. One must also keep abreast of advances in the field. (See Fig. 1-1.)

1-1 TELEVISION AS A SYSTEM

Television repair. The branch of electronics with an ever-growing need for skilled-technicians. A natural progression from the repair of radios and other electronic equipment. A profession that is interesting, challenging and rewarding all at the same time. A profession for the technical-minded individual with an eye to the future.

There are two kinds of technicians. The "outside" technician who makes house calls, and the "bench" technician who repairs sets in

Learning about television receivers requires some understanding of what takes place at a TV *station*, particularly the nature of the signals that are generated and transmitted. As shown in Fig. 1-2, television operates as a *system,* the station, and the numerous receivers within range of its signals. The station produces two kinds of signals, *video,* originating with the camera or tape recording, and *sound,* from a microphone or other source. Each of the two signals is generated, processed, and transmitted separately,

1

2 TELEVISION FUNDAMENTALS: THE STATION

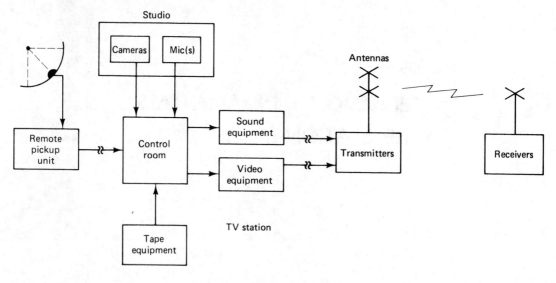

Fig. 1-2 Television as a system.

video as an AM (amplitude-modulated) signal, and sound as an FM (frequency modulated) signal.

1-2 BANDPASS CONSIDERATIONS AND CHANNEL ALLOCATIONS

Considering such things as the distance between stations, radiated power, geographic location, and topography, each station is assigned a particular channel to reduce the possibility of interchannel interference. A station may operate in either of two TV bands: VHF (very high frequency) (channels 2 through 13), extending from 54 to 216 megahertz (MHz), or UHF (ultra high frequency) (channels 14 through 83), extending from 470 to 890 MHz (see Table 1-1). Note that each channel occupies 6 MHz of the spectrum, a very *wide* bandpass, which can best be appreciated by realizing that nearly six radio broadcast bands would be required to accommodate even one TV station. The frequency allocations are in numerical sequence except for a couple of gaps where frequencies were previously assigned to other services. Also note that the video and sound RF carrier frequencies are always 4.5 MHz apart.

Figure 1-3a shows several channels with the relative location of their carriers, and the sidebands where the sound and picture information is contained, including color. In Fig. 1-3b, a comparison is made with an AM broadcast station. Note that for RF the *sound* is always on the *high* side, and the *picture* (pix) is on the *low* side. Also note the bandpass for the sound, 500 kilohertz (kHz), as compared to 10 kHz for radio and 150 kHz for FM broadcasting.

In Fig. 1-3c, color sidebands (designated I and Q) are shown added within the same bandpass as the monochrome video.

1-3 VESTIGIAL SIDEBAND TRANSMISSION

With radio, *all* the upper and lower sidebands are transmitted equally, but with TV it is not feasible because of limitations of available frequencies. Some communication systems use

Table 1-1 Television Channel Frequency Allocations

Channel Number	Frequency Limits (MHz)	Picture Carrier (MHz)	Sound Carrier (MHz)
VHF Television Station Frequencies			
2	54–60	55.25	59.75
3	60–66	61.25	65.75
4	66–72	67.25	71.75
5	76–82	77.25	81.75
6	82–88	83.25	87.75
7	174–180	175.25	179.75
8	180–186	181.25	185.75
9	186–192	187.25	191.75
10	192–198	193.25	197.75
11	198–204	199.25	203.75
12	204–210	205.25	209.75
13	210–216	211.25	215.75

UHF Television Station Frequencies

Channel Number	Frequency Range MHz	Picture Carrier MHz	Sound Carrier MHz
14	470–476	471.25	475.75
15	476–482	477.25	481.75
16	482–488	483.25	487.75
17	488–494	489.25	493.75
18	494–500	495.25	499.75
19	500–506	501.25	505.75
20	506–512	507.25	511.75
21	512–518	513.25	517.75
22	518–524	519.25	523.75
23	524–530	525.25	529.75
24	530–536	531.25	535.75
25	536–542	537.25	541.75
26	542–548	543.25	547.75
27	548–554	549.25	553.75
28	554–560	555.25	559.75
29	560–566	561.25	565.75
30	566–572	567.25	571.75
31	572–578	573.25	577.75
32	578–584	579.25	583.75
33	584–590	585.25	589.75

(*cont.*)

Table 1-1 (cont'd)

UHF Television Station Frequencies

Channel Number	Frequency Range MHz	Picture Carrier MHz	Sound Carrier MHz
34	590-596	591.25	595.75
35	596-602	597.25	601.75
36	602-608	603.25	607.75
37	608-614	609.25	613.75
38	614-620	615.25	619.75
39	620-626	621.25	625.75
40	626-632	627.25	631.75
41	632-638	633.25	637.75
42	638-644	639.25	643.75
43	644-650	645.25	649.75
44	650-656	651.25	655.75
45	656-662	657.25	661.75
46	662-668	663.25	667.75
47	668-674	669.25	673.75
48	674-680	675.25	679.75
49	680-686	681.25	685.75
50	686-692	687.25	691.75
51	692-698	693.25	697.75
52	698-704	699.25	703.75
53	704-710	705.25	709.75
54	710-716	711.25	715.75
55	716-722	717.25	721.75
56	722-728	723.25	727.75
57	728-734	729.25	733.75
58	734-740	735.25	739.75
59	740-746	741.25	745.75
60	746-752	747.25	751.75
61	752-758	753.25	757.75
62	758-764	759.25	763.75
63	764-770	765.25	769.75
64	770-776	771.25	775.75
65	776-782	777.25	781.75
66	782-788	783.25	787.75
67	788-794	789.25	793.75
68	794-800	795.25	799.75
69	800-806	801.25	805.75
70	806-812	807.25	811.75

Table 1-1 (cont'd)

UHF Television Station Frequencies

Channel Number	Frequency Range MHz	Picture Carrier MHz	Sound Carrier MHz
71	812–818	813.25	817.75
72	818–824	819.25	823.75
73	824–830	825.25	829.75
74	830–836	831.25	835.75
75	836–842	837.25	841.75
76	842–848	843.25	847.75
77	848–854	849.25	853.75
78	854–860	855.25	859.75
79	860–866	861.25	865.75
80	866–872	867.25	871.75
81	872–878	873.25	877.75
82	878–884	879.25	883.75
83	884–890	885.25	889.75

single sideband (SSB) transmission (where only one sideband is transmitted), which is all that is really necessary considering the duplication of intelligence in the sidebands. With television, *all* the *upper* sideband is transmitted as shown in Fig. 1-3b, with only a portion (vestige) of the *lower* sideband.

The bandwidth of the video is approximately 4 MHz before it is attenuated where the curve drops off. In Fig. 1-3c, note the chroma (I signal) is also vestigial, but the Q signal has *double sidebands*.

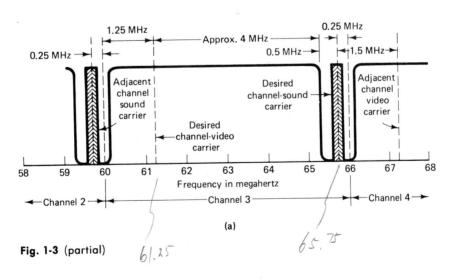

Fig. 1-3 (partial)

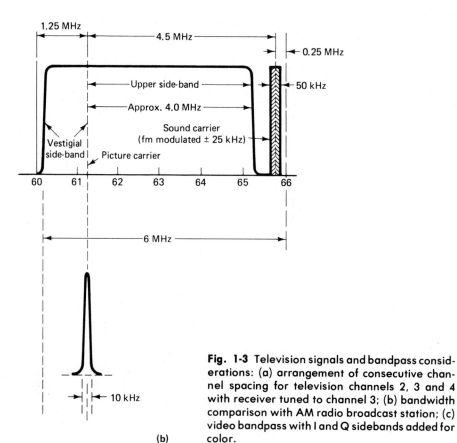

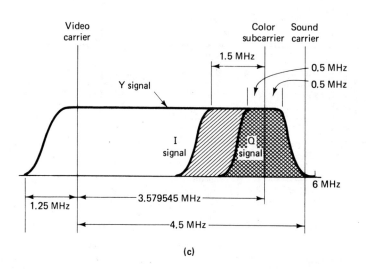

Fig. 1-3 Television signals and bandpass considerations: (a) arrangement of consecutive channel spacing for television channels 2, 3 and 4 with receiver tuned to channel 3; (b) bandwidth comparison with AM radio broadcast station; (c) video bandpass with I and Q sidebands added for color.

1-4 COMPATIBILITY

In the early planning days of color TV, many problems had to be resolved before it could become a reality. These problems were mostly in the area of *compatibility,* that is, how to develop a system that would not conflict with existing monochrome standards, which had long been in use. Four major criteria had to be satisfied:

1. Existing monochrome receivers were to receive good-quality pictures in black and white from a color telecast.
2. Both black and white (BW) and color sets were to receive good BW pictures from a monochrome telecast.
3. Color sets were to produce pictures of better than acceptable quality in full color.
4. All the information needed to produce a color picture had to be transmitted within the same bandpass as the monochrome system.

The system adopted and currently in use in this country is called the NTSC system; these standards have been named for the agency responsible for their adoption, the National Television Systems Committee.

1-5 MONOCHROME SYSTEM

Color and monochrome TV are compatible. Since the principles of monochrome (black and white) are fundamental to both, it will be considered first.

• **Picture Makeup and Resolution**

A picture that is photographically reproduced consists of a myriad of tiny dots or picture *elements.* The dots that make up a coarse-grained newspaper picture (halftone) can be easily seen with the aid of a magnifying glass. Due to the limitations of the human eye, when viewed at a distance, these picture elements cannot be seen and the picture quality is considered acceptable. With black and white, the variations in shading are determined by the number, density, and spacing of the picture elements. The greater the number of dots, the smaller is their size; and the closer the spacing, the greater is the detail. For example, a quality photo taken with a good camera and fine-grained film has good detail because the dots are so small as to be indistinguishable (unless the photo is blown up, in which case the enlargement becomes grainy). The individual dots of a picture all have the same shade. What may appear as a black area is simply a greater concentration of dots. White in a picture is the complete absence of dots. Various shades of gray are represented by variations in the dispersal of the dots.

These same relations also hold true for a black and white image on a television screen, which consists of over a quarter-million elements, regardless of picture size. The picture on a small screen has considerable detail. The picture on a large screen, however, must be viewed at a distance to preserve the illusion.

1-6 CAMERA

In practically all television stations the monochrome camera has been replaced by the color camera. However, monochrome cameras are still used to some extent in industry, closed-circuit security systems, and the classroom.

All video signals originate with the *camera,* which converts visual images into electrical impulses. A basic camera consists of a *camera tube* and its associated *beam-deflection* system, an *optical* system, a *monitor* or *viewing tube,* and the necessary *amplifiers* and *control circuitry.*

The main camera item is the cathode ray tube (CRT) of which there are several types in

8 TELEVISION FUNDAMENTALS: THE STATION

current use: the *image orthicon* (now rapidly becoming obsolete), the *vidicon,* and the *plumbicon.* Each tube type has characteristics specially suitable for particular applications. An important property of a camera tube is a high light sensitivity (for televising scenes in dim light). A small hand-held camera used extensively for on-the-spot newscasting is the *Minicam.*

The typical camera, (Fig. 1-4), basically consists of the following:

1. A light-sensitive *target plate* (mosaic) with a million or so photosensitive particles deposited on one surface of the plate.

2. A lens system for focusing the image or scene being televised onto the target, where each photosensitive particle corresponds with one picture element.

3. An *electron gun* that sequentially *scans* each picture element.

4. A *beam-deflection* system. Some tubes (as in an oscilloscope) have built-in electrostatic deflection plates. Most, however (like a receiver picture tube), employ electromagnetic deflection, which requires an external deflection *yoke.*

5. A built-in monitor screen (view finder).

(a)

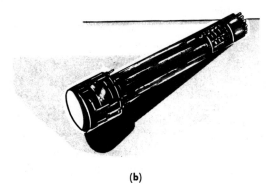

(b)

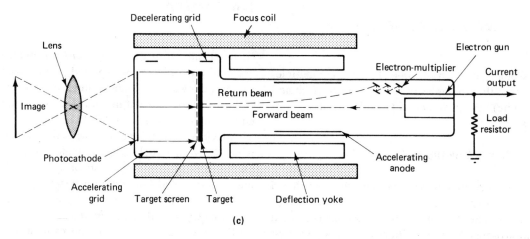

(c)

Fig. 1-4 (partial)

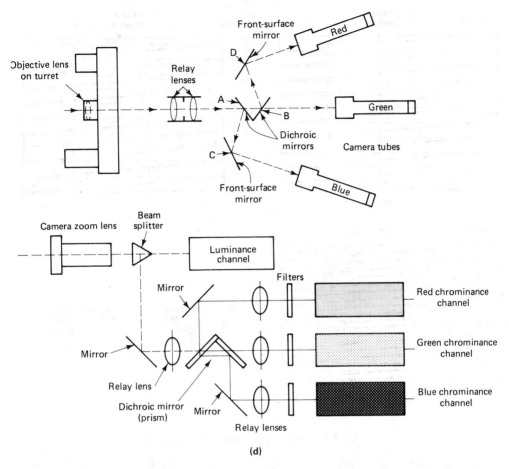

Fig. 1-4 The television camera: (a) typical color camera; (b) typical camera tube; (c) in a tri-color camera, light from the image is split three ways; (d) a four-tube color camera.

1-7 SCANNING AND SYNCHRONIZATION

In scanning, the electron beam is made to move both vertically and horizontally over the face of the target in a precisely controlled manner. The beam movement (as shown in Fig. 1-5) is known as *interlaced scanning* and is described briefly as follows: Using the upper-left corner as an arbitrary starting point, the beam moves to the right under the influence of the horizontal deflection system. At the same time, the vertical deflection is pulling the beam downward. Thus the beam travel is slightly downward to the right, as shown. This constitutes one scanning line, which may be called line 1. At the proper moment the horizontal deflection influence is reversed, and the beam is rapidly deflected back to the left side. This is called *horizontal retrace*, and the beam (still under the influence of the vertical deflection) travels downward to the left, as shown by broken lines. Once again the horizontal deflection is reversed and the action is re-

10 TELEVISION FUNDAMENTALS: THE STATION

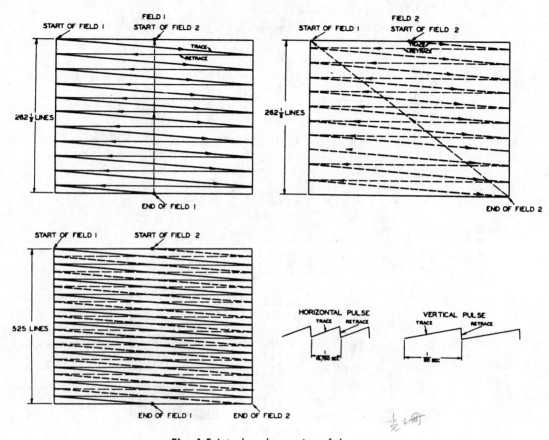

Fig. 1-5 Interlaced scanning of the raster.

peated to trace out another line (line 3) and others, until the beam reaches the bottom of the target at the completion of approximately 262½ lines. This ends the scanning of a complete *field*.

Now the *vertical* deflection is *reversed*, and the beam moves rapidly upward in what is called *vertical retrace*. The return to the top is in a zigzag manner, since the beam is also under the influence of the horizontal deflection. Although not normally visible, a number of horizontal *lines* are traced out during this retrace period.

Reaching top center, the beam again starts downward, this time tracing out *even numbered lines* that are *interlaced* with the first scanning. After scanning another 262½ lines (approximately), we have completed a second field, and the process is repeated. Two such fields or *scan-*

nings represent a *frame*, the scanning of virtually every picture element on the target plate.

With the American system (BW), 15,750 lines are scanned every second, 60 fields or 30 complete frames per second, a rate more than fast enough to provide the illusion of motion in a picture with absolutely no flicker. For color TV, the sweep rates are slightly lower. A frame is considered to be made up of 525 lines of picture information. In practice, however, there are only about 500, the rest being "used up" (and blanked out) during the *vertical retrace* interval.

Both vertical and horizontal sweep rates are precisely controlled by *synchronizing* (sync) *pulses* generated at the station, along with *blanking pulses* whose function is to cut off the CRT beam current during vertical and horizontal retrace

1-8 BASIC CAMERA OPERATION

As each picture element on the target is sequentially scanned, a voltage is developed by *photoemission* (or *photoconduction*, depending on the tube type). The amount of voltage generated by this process depends on the brightness of the picture element being scanned at that instant. For a black element, the voltage is zero. For the brightest elements the voltage is maximum. Thus the camera output is constantly fluctuating as the beam scans each and every image dot on the target. This information is transmitted, and the synchronized beam travel on a receiver picture tube reproduces the dots with the same levels of brightness (and at the same locations) as on the original camera tube.

1-9 COMPOSITE VIDEO SIGNAL

The lines of video information generated by the camera are broken up by the insertion of the relatively short duration blanking and sync pulses. However, the small amounts of picture that are lost during the brief vertical and horizontal retrace periods are not noticeable to a viewer at the receiving end. The composite video signal, as shown in Fig. 1-6, consists of a complex series of waveforms generated at the station. In the drawing, this series of waveforms represents *one line* of pix information. When transmitted, it contains all the information necessary to produce a monochrome picture on a receiver screen. The same signal is transmitted by a color station when broadcasting in monochrome. At the extreme left is a *horizontal blanking pulse* (or pedestal) to *cut off* the camera beam (and the beam of the picture tube) during the horizontal *retrace* interval. A horizontal sync pulse (which controls the horizontal beam trace) rides on the blanking pedestal. The voltage variations immediately following are the video generated in the scanning (in this case) of the last line of a field at the bottom of the target.

With the beam ready to start vertical retrace, a vertical blanking pulse of relatively long duration is introduced. Note the time intervals in the figure. As stated earlier, the horizontal sweep is still operating, and controlled, during vertical retrace. Some of the pulses on the vertical blanking initiate vertical retrace, and some are for controlling the horizontal.

When the beam reaches the top of the target and starts downward, video for the top line of the picture is produced as shown at the right. Blanking and sync pulses are continuous and repetitious, but the video of course is always changing in accordance with the image being scanned.

1-10 TEST PATTERNS

Once familiar to the home viewer, the monochrome test pattern is no longer regularly transmitted as an aid in the setup and adjusting of receivers. However, it is still transmitted at times between programming schedules for making camera and transmitter adjustments.

The test pattern (which occurs in many forms) provides a great deal of information relating to sweep linearity, interlacing, video frequency response, picture resolution, and the like. A typical test pattern is shown in Fig. 15-7.

1-11 THE WONDERFUL WORLD OF COLOR

It is hard to imagine a world without color, with everything seen only in drab and lifeless shades

12 TELEVISION FUNDAMENTALS: THE STATION

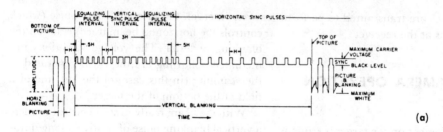

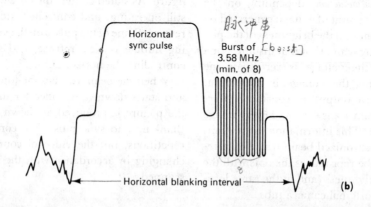

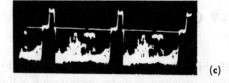

Fig. 1-6 (a) composite video signal; (b) the color burst; (c) pulses and two lines of video as observed on a CRO.

of black and white. But thankfully, the beauties of nature and of material things as well can be viewed as they are, in every conceivable hue. And so it is with television. The advent of color, with its rich enhancement of what we see on our screens, adds another dimension to our viewing pleasure.

1-12 COLOR RESOLUTION AND MIXING PRINCIPLES

In studying light, one finds there are three primary colors from which all others are derived:
red, green, and *blue.* As Isaac Newton discovered some 300 years ago, white light, to the human eye and brain, exists as a mixture of these three primaries in the proper proportions. Stated differently, *white light* is actually a combination of *all* colors of the spectrum.

Similarly, various shades of gray are represented as a blending of the same three primaries but at different levels of brightness. Other colors (the *complementaries of the primaries*) exist as proportionate mixtures, for example, magenta, a blending of red and blue, or yellow, a mix of red and green. With light, the colors are said to be *additive* (unlike *paint* primaries, which are different and *subtractive*).

1-15 Developing the Luminance or Y Signal   13

All colors are considered to have three distinct and different characteristics: *hue*, *saturation*, and *brightness*. Hue refers to the *identity* of a color, whether red, yellow, violet, and so on. The terms hue and color are synonymous and may be used interchangeably.

Saturation is the *strength or depth* of a color. For example, blue, while still retaining its identity, may be deep, vivid, or pale, depending on the amount it is *diluted* by white light. In the absence of white, the color is said to be *pure* or *fully saturated*.

Brightness relates to the amount of light energy emitted by a color. Although related, it is not to be confused with saturation, which is a separate and distinct condition. For example, a color may possess two extremes of brightness, extremely bright or barely visible. To help understand the difference between the two, consider that a monochrome picture consists of only brightness changes of the elements that make up the image, whereas a color picutre can have variations of both. Color TV takes into account all three of these characteristics.

1-13 COLOR STATION FUNCTIONS

The generating and processing of color signals by the station is shown in the basic block diagram of Fig. 1-7. The color signals originate from the cameras, whose outputs are identified with the three primary colors. In general, two kinds of signals are produced: (1) a *luminance* or Y signal that is comparable to the output of a monochrome camera, and (2) color or *chroma* signals. Initially, each signal follows a separate route, as shown in Fig. 1-7. After processing by numerous chroma stages, the signals are recombined (multiplexed) and transmitted together. A brief description of what takes place in each functional block follows. In Fig. 1-7, note in particular the names applied to the various signals, since the same terminology is used in a color *receiver*. Also note the relative amplitudes (from the three camera tubes that combine to produce black and white).

1-14 TRICOLOR CAMERA

The typical color camera differs from a monochrome camera in that it contains not one, but either three or four separate tubes (CRTs), one for each of the three primary colors, and a separate tube (when used) to develop the Y signal. The most recent cameras use only *one* CRT (called the *spectroflex*).

Light from the image being viewed is projected onto the photosensitive target of all three tubes, as shown in Fig. 1-4. An elaborate system of mirrors and optical filters, however, selectively permits only one color in the image to reach the target of the corresponding color tube. In other words, each tube "sees" only that color contained in the original image with which it is identified, that is, only that portion of that color contained in any of the complementary colors being viewed. An output voltage is produced from each tube and is identified with that particular tube. The amplitude of the output voltage varies in accordance with the intensity of the color being scanned at any given moment. The outputs of the three tubes are adjusted in the proportions of 30% red, 59% green, and 11% blue to produce a composite voltage corresponding to their proportions in white light. As stated earlier, some cameras have a separate CRT to develop the BW signal *directly*.

1-15 DEVELOPING THE LUMINANCE OR Y SIGNAL

This signal is comparable to the video output from a monochrome camera and represents

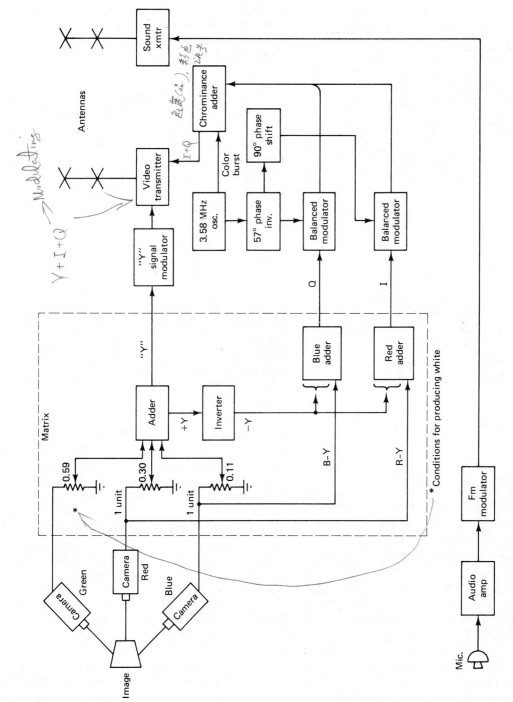

Fig. 1-7 Color TV station block diagram.

1-17 THE 3.58–MHz SUBCARRIER

To combine (or interleave) the color information with the luminance signal in the same bandpass, it becomes necessary to convert it into sidebands of a carrier (or subcarrier to be more exact). Since there are two color difference signals, two subcarriers are required. In practice they both come from the same source, but the phase of one is displaced from the other (in quadrature) by an angle of 90°.

The frequency of the subcarrier is 3.579545 MHz, roughly rounded off to 3.58 MHz. It is an unmodulated continuous-wave (CW) signal, sometimes referred to as a *reference signal* (which it is, considering its function). This subcarrier, along with the R − Y color difference signal, is fed to a modulator as shown in Fig. 1-7. The out-of-phase subcarrier (the quadrature signal) along with the B − Y color difference signal feeds the other modulator.

1-18 MODULATING THE SUBCARRIER

The system, sometimes called a *divided-carrier* system, uses two *balanced* modulators, or *encoders*. The purpose of the modulators is to *phase modulate* the two subcarriers with the two color difference signals. Modulation of the R − Y (I) signal takes place in one modulator, and of the B − Y (Q) signal in the other. Although the modulation sidebands of both are combined at the modulator outputs, they still exist as separate signals since each has its own subcarrier to which it relates.

Inasmuch as the subcarriers are out of phase by 90°, it follows that the chroma sidebands are also out of phase by 90°. The phase relationship

changes in brightness only. Its purpose is twofold: to produce a BW picture on a monochrome set and to produce the *fine* picture detail that is sharp and crisp on a color set.

With a three-tube system, the outputs of the three camera tubes are combined, then divided into the proper proportions for white, in a *matrix*, (see Fig. 1-7), which is simply a resistive voltage-divider network, and the three outputs are individually adjustable as previously explained. When combined as one voltage, they constitute the luminance or Y signal, which conveys brightness information only.

In the absence of a particular color in the image scanned, the output from that tube will of course be zero. A *black* image is the result of zero output from *all three* camera tubes.

1-16 COLOR DIFFERENCE SIGNALS

This term applies to color signals from which the luminance or Y component has been removed. They are generally developed in two *adder* stages, as shown in Fig. 1-7. The two adders receive an input from the cameras plus an *inverted-Y signal*. The adder outputs are the color difference signals, (R − Y) and (B − Y). These two signals convey all the information necessary to produce a picture in full color at the receiver.

Why is a green signal not provided? Green can be obtained from the proper proportions of the red and blue color *difference* signals at the receiver. Therefore, to develop and transmit a green signal is not only unnecessary, but would add considerably to the complexity of the system.

Bear in mind that the two color difference signals (in the absence of the brightness or Y component) represent pure and fully saturated colors of *all* hues.

between each chroma signal and its subcarrier establishes the *hue*. Amplitude variations at the modulator outputs correspond with the degree of *saturation*. Prior to transmission, the 3.58 MHz subcarrier is intentionally held back to reduce the possibility of *beat* interference at the receiver.

To summarize, *phasing* equates with *hue*, and instantaneous *voltage levels* with *brightness*; and when the camera is scanning white areas of an image, the voltage drops to zero and only the Y signal is transmitted.

1-19 COLOR SYNC

The circuit that generates the subcarrier reference signal must be extremely stable both in frequency and phase, since the slightest phase shift represents a change in hue. Inasmuch as the receiver has its own 3.58-MHz generator, this poses a problem. To ensure that the receiver subcarrier is locked in phase with its counterpart at the station, a color sync signal (called the *color burst signal*) is transmitted to control the receiver oscillator. The burst consists of an 8- to 10-cycle sampling of the station subcarrier generator. During a color transmission, it is introduced onto the *back porch* of each and every horizontal pulse. See Fig. 1-6. At the receiver, the frequency and phase of its subcarrier are compared with the burst signal once for each line being scanned. Where a shift occurs, the subcarrier is automatically brought into step with the one at the station to ensure that the correct hues are produced at all times.

1-20 INTERLEAVING AND TOTAL COMPOSITE COLOR SIGNAL

The NTSC standards require that the chrominance information be contained within the same bandpass as the monochrome video signal. This is made possible by a proper choice of sweep frequencies relative to the frequency of the subcarrier reference signal. Under these conditions, the chroma and video information is able to coexist as alternate *clusters* of energy side by side within the bandpass without interfering. This takes place in an *adder* stage immediately prior to modulation of the station carrier, as shown in Fig. 1-7. The output of the adder is called the *total composite color signal,* which is essentially the same as the monochrome composite signal of Fig. 1-6a, with the addition of the chroma information and color burst. When a color TV station is broadcasting *monochrome* only, the chroma and burst are *not* present and the signal is the same as shown in Fig. 1-6a.

1-21 SPECIAL-PURPOSE TEST SIGNALS

A number of special test signals are generated by the station for making adjustments, monitoring, and verifying the integrity of the equipment. The most common test signals are as follows:

1. *Color bar pattern:* This consists of *ten* vertical color bars ranging from yellow on the left, through various hues, to green at the extreme right. It is used by the station prior to going on the air to adjust cameras and other equipment. Sometimes the color pattern is transmitted to enable a check of chroma conditions at the receiver.

2. *Vertical interval test signals (VITS):* These test signals are often transmitted on lines 17, 18, or 20 during each vertical retrace interval (the first 21 lines of a field occur during retrace and are normally not seen by the viewer). The VITS signals are not visible because they take place when the screen is blanked out. At the station the VITS signals are used to monitor transmitting equipment and for checking automatic color control circuits. At the receiver, a VITS display permits rapid evaluation of intermediate-frequency (IF) amplifier response and operation of video amplifiers.

3. *Vertical interval reference signal (VIR):* This is a special signal now being transmitted by most stations during a color broadcast to ensure good and consistent color fidelity in their programming. Prior to the use of VIR, a set owner had to periodically readjust the controls whenever channels were switched or whenever the station changed from one camera to another. There are many possible reasons for the changes in color fidelity, involving either the station, the receiver, or both. Regardless of the cause, VIR helps greatly to maintain true and consistent color.

The VIR signal is inserted, along with the color burst, on the back porch of line 19 horizontal pulse during vertical retrace. Its presence is continuously monitored and analyzed by the station. Any deviation from normal conditions, such as an error in the chroma reference level, is immediately recognized and corrected. VIR signals differ from VITS in that they monitor the parameters of the *program material* being transmitted.

As of this writing, at least one brand of receiver utilizes the VIR signal to automatically make corrections in the *receiver* itself. This makes it possible to dispense with all user color controls.

1-22 THE SOUND SIGNAL

Little has been said thus far about the sound signal, partly because with TV it understandably takes second place to the video signal. There are several reasons why FM is used (over AM) for TV sound: better fidelity, less chance of interaction with the AM video signal, and reduced pickup of noise and other kinds of AM interference.

With FM sound transmission, a system of *preemphasis* and *deemphasis* is used. At the station, the higher (treble) frequencies are overemphasized compared to the lower (bass) audio frequencies, using preemphasis. At the receiver, the highs are attenuated (using deemphasis) to their normal level. The advantage of this technique is that static and background noise impulses (which are mostly high pitched) are also attenuated.

Figure 1-7 shows the audio signal path from the microphone, tape, or other source, through the various amplifiers, from where it is routed to the FM sound transmitter.

1-23 MODULATING THE SOUND AND VIDEO TRANSMITTERS

As stated earlier, there are actually two transmitters, an AM transmitter for the video (see Fig. 1-7) and an FM transmitter for the audio. In the video transmitter, an RF filter removes part of the low-frequency sideband (vestigial sideband).

• **Frequency Modulation**

Figure 1-8a shows an *unmodulated* carrier, and Fig. 1-8b shows the carrier when it is (frequency) modulated. For comparison, an AM (radio) wave train is shown in Fig. 1-8c. It's obvious that the two methods of modulation (AM and FM) differ in a number of ways. For example, with FM the *amplitude* of the carrier remains *constant* with or without modulation. In the FM modulation process, the carrier is made to shift both higher and lower than its *resting frequency* (the assigned carrier frequency). The *rate* of frequency changes corresponds with the audio frequencies modulating the carrier. Rapid rates of change relate to the high-pitched audio signals, while slow rates of change equate with low audio frequencies.

The *extent* of the actual frequency changes above and below the carrier is called the *deviation* or *sweep* and corresponds to the volume level; small amounts of shift occur with a low volume, and greater frequency shifts represent higher volume levels. Television stations are permitted to sweep up to ±25 kHz, a total of 50 kHz for 100% modulation. This compares with ±5 kHz for AM radio. It follows that circuits handling TV sound must have a proportionately wider bandwidth.

Leaving the transmitters, the AM and FM carriers are multiplexed (combined) to feed the transmitter antenna system, which radiates the two carrier signals in the form of electromagnetic waves.

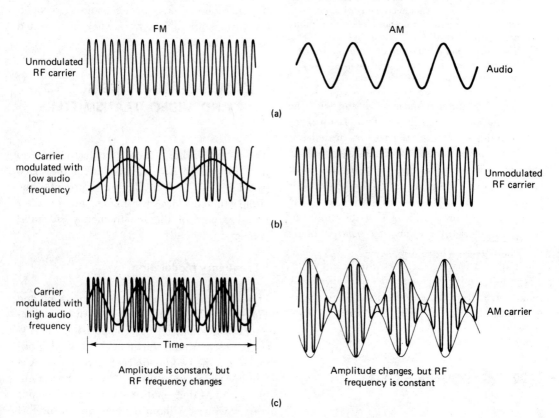

Fig. 1-8 Frequency modulation and amplitude modulation comparisons.

1-24 SUMMARY

The important points presented in this chapter are now summarized:

1. A TV station puts out two signals; a frequency-modulated (FM) *sound* carrier and an amplitude-modulated (AM) *video* carrier. The *useful* information is contained in the sidebands, which together occupy 6 MHz of the spectrum. The video and sound carriers have a separation of 4.5 MHz. Video sidebands extend up to 4 MHz for BW and up to 4.2 MHz for color. The color sidebands are transmitted within the same bandpass as the BW signal.

2. A BW camera generates voltages whose amplitudes vary in accordance with brightness levels of the pix elements being scanned. This *video information,* along with sync and blanking pulses, is transmitted as *sidebands* of the station carrier.

3. The U.S. system is based on a 525-line picture produced by *two* successive, interlaced scannings. Each scanning (of 262½ lines) is called a *field. Two* fields represent a *frame* or complete pix. In the absence of a pix, this is called a *raster.* The scanning rates (for BW) are 60 hertz (Hz) for the *vertical* sweep and 15,750 Hz for the *horizontal.* For color, the rates are slightly lower. Raster scanning is the same at the receiver as at the station. Receiver scanning is synchronized with that of the camera tube (or tubes) by *sync pulses* included as part of the transmitted signal.

4. A system of *vestigial sideband transmission* is used for both BW and color. Although one sideband extends the full amount (from 0 to 4 MHz), the other sideband only goes from 0 to 1.5 MHz. The object is to limit the bandpass requirements. If *double-sideband* transmission were used (as with radio), instead of 6 MHz, each station would require about 9 MHz of the "crowded" RF spectrum.

5. Except for two frequency gaps, *VHF* channels (2 through 13) extend from 54 to 216 MHz. To prevent interference, the channel allocation of a station depends on its geographic location and other factors. Except for the higher carrier frequencies assigned, *UHF* TV is identical to VHF TV. The UHF band (channels 14 through 83) extends from 470 to 890 MHz.

6. The quality of TV sound is better than for *radio* because of the wider bandpass allocation (50 kHz compared to 10 kHz). Another difference is the use of *frequency* modulation (FM) instead of AM. With FM, the *volume* of sound is determined by the *amount* of frequency sweep (or deviation) from the carrier frequency. The *rate* of the sweep corresponds with the *audible frequencies* in the sidebands. A system of *preemphasis* and *deemphasis* is used to combat static and other high-pitched interference.

7. Color has *three* distinct characteristics that are taken into account in the transmission and reception of color. *Hue* is the *identity* of the color, be it red, blue, or whatever. *Saturation* is the *intensity* of the color. And *brightness* is the amount of light emitted by a color. The three *primary* colors are *red, green,* and *blue.* White light is produced by a mixture of 30% red, 59% green, and 11% blue. The same proportions, but of varying intensities, create different shades of gray. Different proportions of the primaries produce all the complementary colors, yellow, violet, magenta, and so on.

8. The tri-color camera at the station contains three camera tubes, each identified with one of the three primary colors. Each tube sees only its "own color" in the image being televised. The outputs of the three tubes are combined in the proportions equivalent to white light. The combined signal is known as the luminance or *Y* signal. It corresponds with the video signal developed by a monochrome station. The Y signal is used to produce the BW picture on a monochrome set, to produce the fine detail on a color picture, and to determine the average brightness level for the colors.

9. *Color difference* signals represent pure, fully saturated colors after the *brightness* component has been *removed.* At the station, outputs from the red and blue cameras are combined with an *inverted-Y signal* to produce two color difference signals (R - Y) and (B -

Y). These two signals are recovered at the receiver to feed the picture tube, which uses them to reproduce images in all colors. (Green is not transmitted since it is developed at the receiver by the proper proportions of red and blue.)

10. At the station, the two color difference signals are used to phase modulate two sub-carriers. The subcarrier is a CW sine-wave with a frequency of approximately 3.58 MHz, which falls within the bandpass of the regular monochrome signal. This signal, along with the R - Y (I) signal, is fed to one modulator. An out-of-phase sampling of the subcarrier along with the B - Y (Q) signal is fed to the other modulator. In the modulation process the subcarriers are canceled out, and only the phase-modulated sidebands appear at the modulator outputs. Combined, they still retain their identity, since each relates to a separate subcarrier. The phase difference between the subcarriers is a fixed 90°; hence the two outputs from the modulators are always 90° apart. However, the phasing between each and its own subcarrier is always changing in accordance with the hue being scanned at that instant. Remember that hue corresponds with changes in phase, and that the amplitude variations determine the degree of saturation of the color. After modulation, the chroma sidebands are combined with the Y signal prior to transmission.

11. Short bursts (color burst) of the 3.58 MHz subcarrier generated at the station are transmitted as part of the composite video signal during a color broadcast. Approximately *8 cycles of this signal is superimposed on the back porch of each horizontal blanking pulse*. The receiver uses this signal as a standard reference for synchronizing the demodulation of the color signals. In this way the receiver becomes a *slave* of the station.

REVIEW QUESTIONS

1. (a.) What is meant by *vestigial* sideband transmission?
(b). Why is it used?

2. What is meant by *compatible* color TV?

3. Where at the station is the following produced: (a) the carrier, (b) the sidebands? State the purpose of each.

4. Explain the purpose of (a) the test pattern; (b) a color bar display pattern.

5. Why does the video occupy so much more of the allocated 6-MHz bandwidth than does the sound?

6. (a) Why is geographic location a factor in assigning station channels?
(b) State other considerations.

7. (a) Why is 4 MHz considered the upper limit for video?
(b) Why is 1.5 MHz the upper limit for chroma information?

8. What is accomplished by the chroma modulators at the station?

9. Where at the station are the following kinds of modulation used: (a) AM, (b) FM, and (c) phase modulation?

10. How is the Y signal developed by the station?

11. (a) What is a fully saturated color?
(b) What signal varies the degree of saturation?

12. (a) What symbols represent the color signals?
(b) How are they produced?

13. Explain the basic differences between an AM and FM signal.

14. At the station, where are the luminance and chroma signals (a) separated and (b) recombined

15. (a) What is the purpose of the VIR signal?
(b) How is it transmitted?

16. Why is the green signal not transmitted?

17. Why is the subcarrier not transmitted?

18. (a) What is a matrix? (b) An adder?

19. Explain the purpose of the following camera items: (a) the electron gun(s), (b) the target, (c) the arrangement of mirrors and lenses, (d) the viewing screen?

20. How is it possible to transmit two different signals (the video or Y signal and the chroma signal) all within the same passband?

21. (a) Is a *complete* pix focused onto the target of each camera tube?
(b) What *color(s)* appear on the target of each tube?

22. What is meant by a *color difference* signal?

Chapter 2

TELEVISION FUNDAMENTALS THE RECEIVER

2-1 MONOCHROME RECEIVER

Figure 2-1 shows an overall block diagram of a typical *monochrome* receiver. Some blocks represent a single stage; other blocks may represent two or more stages. Signal flow is indicated by arrows. Some of the *signals* referred to in Fig. 2-1 are generated internally within the receiver. Others come from the station (via the antenna) as part of the received signal. The blocks shown in Fig. 2-1 also apply to a color set (where their function is the same). A color set has additional stages, which will be considered later. The purpose and function of the various stages common to both is briefly explained in the following paragraphs.

- **Low-Voltage Power Supply**

This is the one common denominator of all stages. It furnishes the dc power (*B* voltage) required by tubes and solid-state devices in performing their various functions. It also supplies the low ac heater voltage for the picture tube (and other tubes in the case of older sets). The *B* supply of most solid-state receivers is *voltage regulated*.

- **Vertical Sweep Section**

This consists of a vertical sweep *oscillator, amplifier(s)*, and the vertical windings of a *deflection yoke*, which is mounted on the neck of the picture tube. The vertical sweep (as explained earlier for a camera tube) deflects the CRT electron beam vertically over the face of the picture tube. The sweep oscillator is the source of the sweep voltage and operates at approximately 60 Hz. the amplifier(s) develops the necessary sweep *power* required by the yoke to produce a strong magnetic field.

- **Horizontal Sweep Section**

This section performs a number of functions. In addition to horizontal beam deflection, it's the source of high voltage (HV) and focus voltage for the picture tube, "scan-derived" dc power for various circuits, and "keying" voltages for a number of circuits.

As with the vertical sweep, there is a horizontal oscillator that generates a horizontal sweep voltage (at a frequency of about 15, 750 Hz), a driver amplifier, and a power amplifier to excite the horizontal yoke coils of the picture tube.

23

24 TELEVISION FUNDAMENTALS: THE RECEIVER

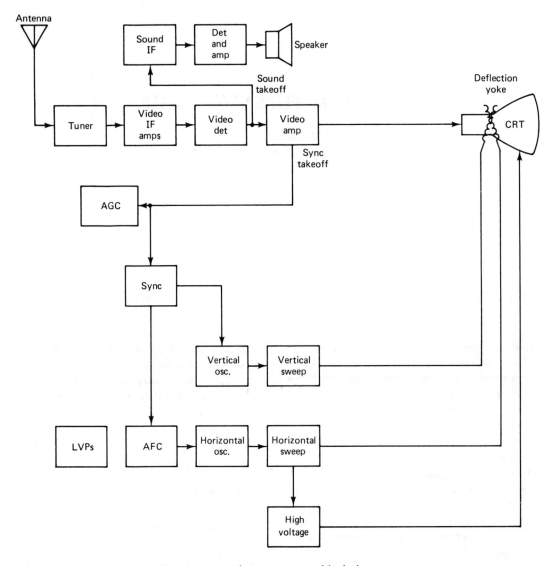

Fig. 2-1 Monochrome receiver block diagram.

- **High Voltage and Focus Voltage Supply**

A picture tube requires an anode supply of many thousands of volts dc, and a focus voltage around a thousand volts or so. Both come from the high-voltage (HV) *flyback transformer* in the horizontal output stage.

- **Sync Stages**

Here, vertical and horizontal sync pulses are stripped from the composite video signal and used to control the two sweep oscillators. Often a *noise immunity* circuit prevents strong noise bursts from inadvertently triggering the sweep.

2-1 Monochrome Receiver

- **Integrator Network**

This is a low-pass *RC* filter at the input of the vertical oscillator. Its job is to reject the horizontal sync pulses while building up a cumulative pulse for direct triggering of the vertical sweep.

- **Horizontal Automatic Frequency Control**

(AFC) This is essentially an *error-sensing* circuit that recognizes and automatically corrects for any changes in frequency and phase of the horizontal oscillator. It is not to be confused with *color AFC* of a color set, although they both operate in the same manner.

- **Signal Path: The Antenna**

The antenna intercepts the electromagnetic waves sent out by the station and converts them into electrical signal currents to feed the tuner. Antenna elements resonate within the VHF and UHF frequency bands.

- **Tuner or Channel Selector(s)**

Two selectors are required, one for VHF and one for UHF. Each tuner selects the desired channel signal from the multitude of different frequencies coming from the antenna. Its tuned circuits must pass the video and sound sidebands that occupy 6 MHz, as shown originally in Fig. 1-3. Conversion from RF to IF takes place in the tuner (as with radio) except that there are *two* carriers. Hence the tuner output consists of two signals, a video IF carrier with its sidebands and the modulated FM sound IF carrier.

- **Video or Picture IF Section**

This consists of three or four amplifier stages pretuned to embrace all frequencies within a 3.5- to 4-MHz passband, depending on the set. Still retaining their 4.5-MHz separation, the video and sound signals are both amplified as *one* in this section.

- **Video Detector**

Here the video IF carrier is *demodulated* to extract the useful picture information contained in its sidebands. Also in this stage (of a BW set) a new and lower sound IF is produced by the *beat* between the picture and sound carriers. As might be expected, the new sound IF is 4.5 MHz, the difference between the two.

- **Video Amplifier(s)**

Usually these consist of two stages to boost the video signal to a level that will adequately drive the picture tube for good picture contrast. The overall response must be "flat," from about 30 Hz up to 3.5 MHz or better.

- **Black and White (BW) Picture Tube and Associated Components**

The pix tube (CRT) is the end of the line (or output) for the video signal. A typical CRT, shown in Fig. 2-2, primarily consists of an *electron gun* and a *screen*. The screen is a thin film of phosphor deposited on the inner surface of the glass face. When struck by electrons, the phosphor particles emit light. Each particle glows momentarily after impact. The CRT gun (which "shoots" the beam of concentrated electrons at the screen) contains all the elements found in most small vacuum tubes; and although physically different, their functions are essentially the same. The gun contains (1) a *heated cathode* (the source or emitter of electrons), (2) a *control grid* for regulating and varying the beam current, and (3) an *anode* for accelerating the electrons to the screen. There is also a *focus grid* for converging the beam to a fine pinpoint where it strikes the screen phosphor surface.

A *deflection yoke* (Fig. 2-2) is mounted on the tube neck close to the tube flare. Its purpose, in

26 TELEVISION FUNDAMENTALS: THE RECEIVER

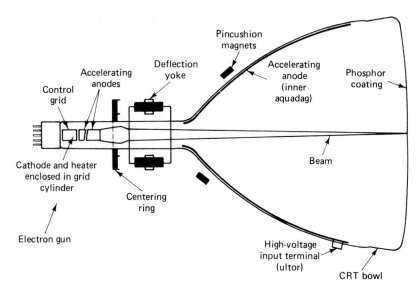

Fig. 2-2 Black-and-white pix tube fuctional drawing.

conjunction with the sweep circuits (as explained earlier), is to sweep the beam both vertically and horizontally over the screen surface. Beam deflection in a receiver pix tube is the same as with a camera tube at the station. The sweeps are synchronized by virtue of the vertical and horizontal sync pulses as transmitted. The light we see on the screen is called a *raster*. Mounted on the tube neck is a *centering ring*, a magnetic device for centering the raster on the screen.

Except for the alternating current supplied to the heater (filament), all the elements of the tube are supplied with dc *operating voltages*. Several thousand volts are required for the anode to accelerate the electrons over the great distance from the cathode to the screen.

The amplified video signal is fed either to the cathode or the control grid (depending on the set) to *intensity modulate* the beam current. Variations of the video signal cause the beam to increase or decrease, producing corresponding changes in brightness as it strikes the screen.

Picture Development. A picture is produced by varying the brightness of the scanned phosphor dots in accordance with amplitude variations of the video signal. White dots are produced by a high beam current. Black dots are the result of beam current cutoff. Different shades of gray are developed by in-between variations. Figure 2-3 shows the relationship between video variations and one line of a picture.

Note the signal *polarity*, an important consideration. The polarity must be *positive going* when fed to the cathode or *negative* when driving the grid. If polarity is reversed, the result is a *negative* picture, as in photography.

• **Automatic Gain Control (AGC) Stage**

The AGC circuitry automatically regulates the gain or *sensitivity* of a set, that is, its ability to respond to either weak or strong signals, depending on local receiving conditions. AGC is similar to AVC (automatic volume control)

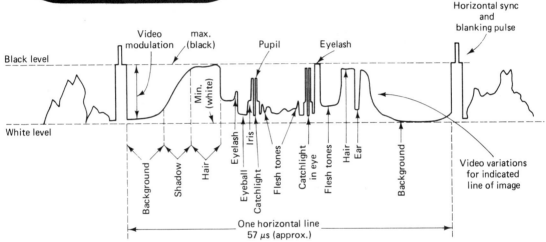

Fig. 2-3 Picture development.

From: *Television Service Manual*, ©1977, by Robert G. Middleton, used with permission of the publisher, The Bobbs-Merrill Company, Inc.

on a radio. If the AFC action is defeated or not working properly, the symptoms may be overloading on strong signals, complete loss of picture and sound, or poor sync. Most sets have an AGC or *threshold* control that should be adjusted for the location where the set is to be used.

- **Sound IF Amplifier**

The 4.5-MHz frequency-modulated (FM) sound signal is fed to the sound IF amplifier(s) from a *sound takeoff* point. This may be anywhere between the video detector and the CRT input. (With a color set, the sound takeoff is from the video *IF* strip.) The FM sound IF bandpass is understandably narrower than the video IF bandpass.

- **Sound Detector**

This circuit *demodulates* or extracts the AF sidebands from the FM carrier. The detector may be a transistor or a matched pair of diodes. Most solid-state receivers now combine the functions of IF, detection, and amplification in a single integrated circuit (IC) chip.

- **Audio Amplifier**

As in a conventional AM or FM radio, this consists of two stages, a *voltage* amplifier (to obtain sufficient gain), followed by a *power* amplifier to drive the speaker.

- **Speaker**

The speaker reproduces the original sound developed at the station.

2-2 CHASSIS CONFIGURATION AND STAGE RECOGNITION

Unfortunately, the stages and components as laid out do not conform with the orderly presentation of a block or pictorial diagram. Finding your way around a new and unfamiliar chassis is not always easy, especially with solid-state circuit designs and where service data are not readily available. However, certain components are basic to all sets, with careful observation and by learning to recognize certain parts and landmarks, you can usually identify all the major circuits.

Modern sets use one or more printed circuit boards (PCBs). A typical PCB contains all (or most) of the components for one or more stages. The components are mounted on one side of the PCB, with the wiring on the other. Most PCBs have markings (road mapping) on one or both sides of the board as an aid in identifying parts and their connections.

2-3 CONTROLS

There are two categories: *user* controls, familiar to every set owner, and *service*-type adjusters. User controls and the more familiar service controls are described next. Service adjusters are described in later chapters as appropriate. For color set controls, see Sec. 2-17.

1. *ON–OFF switch and volume control:* These are usually combined on a single control.

2. *Tone Control:* Some sets have tone control and some do not.

3. *Channel selector(s) or tuners:* All modern sets have both a VHF tuner and UHF tuner. Sometimes the tuners share a common dial and indicator assembly; sometimes they are completely separate. Channel numbers may be engraved or etched or have a translucent window illuminated from behind. More expensive sets may use light-emitting diodes (LEDs) or a digital readout that appears directly on the screen.

4. *Fine tuner control:* This control operates in conjunction with the channel selectors for precise tuning of the oscillator. It may be a separate control or part of the channel selector assembly. VHF and UHF fine tuners may be either separate or combined.

5. *Automatic fine tuner disabling switch:* When switched ON, AFT automatically controls the tuner oscillator, eliminating the need for manual fine tuning. Fine tuning is in the manual mode when the switch is OFF.

6. *Contrast (picture) control:* Connected in the video amplifier circuit, this front panel control regulates picture contrast by varying the gain of the video signals before they reach the CRT input.

7. *Brightness control:* Connected to the CRT grid-cathode input (or in a video stage when direct coupling is used), this control functions as a bias control to vary the average intensity of the CRT beam current.

8. *Focus control:* This control is usually located at the rear of the set. It regulates voltage applied to the CRT focus grid for sharpest scanning lines and picture detail.

9. *Height or vertical size:* This control is located on the rear apron. It controls the amount of vertical deflection by varying the amplitude of the vertical sweep voltage.

10. *Vertical linearity:* This *wave-shaping* control is connected in the vertical section and located on the rear apron. It is adjusted for uniform spacing of the scanning lines to achieve the proper vertical proportions of picture objects.

11. *Vertical hold:* This control, located at either the front or rear of the set, varies the frequency of

the vertical oscillator for synchronizing with the station. If misadjusted, the picture may roll vertically and fail to lock into sync.

12. *Horizontal hold:* This control may be at the front or rear of the set. It varies the frequency of the horizontal oscillator for synchronization with the station.

13. *Width or horizontal size:* A rear control that varies the amplitude of the horizontal sweep voltage for proper amount of horizontal deflection.

14. *AGC threshold control:* A rear control that establishes receiver sensitivity for both weak and strong signals.

2-4 COLOR RECEIVER

Figure 2-4 shows a typical color receiver. Figure 2-5 shows an overall block diagram of a typical color set. The color-related stages are in heavy outlines, superimposed over the monochrome diagram of Fig. 2-1. The other (monochrome) blocks are common to *both* types of receivers. Certain differences and peculiarities may apply to color sets; these will be covered in later chapters.

Color-related stages may be divided into two categories: (1) *chroma stages,* which process the color signals from the station, and (2) *color sync stages,* sometimes referred to as *AFPC* (automatic frequency and phase control) circuits, whose job is to maintain color sync with the station. Blocks representing the color stages are briefly described next.

• **Bandpass Amplifier and Color Killer**

The total composite color signal, including the color burst, goes from the takeoff point in the video amplifier to the bandpass amplifier, as shown in the diagram. The amplifier is controlled (switched ON and OFF) by a *killer* stage. During a monochrome program the killer senses the absense of a color burst (and therefore no color information) and automatically disables the bandpass amplifier, preventing the passage of any noise on to the CRT, and the picture is developed from the Y signal only. When color signals are present, the killer stage (by sensing

Fig. 2-4 Typical small-screen color set with in-line CRT (chassis extended for servicing).

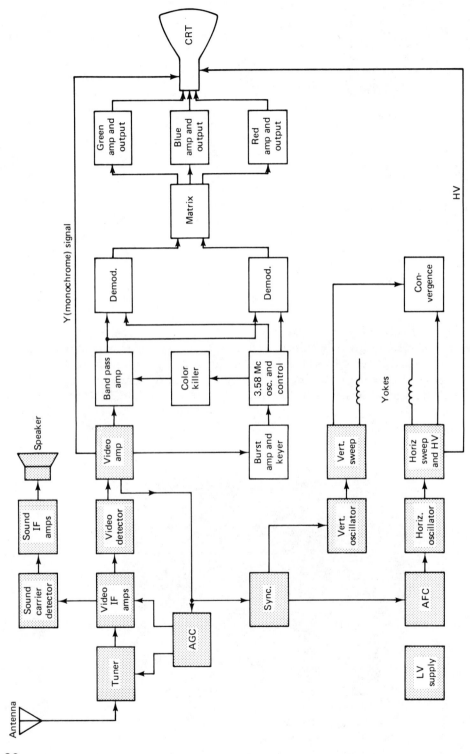

Fig. 2-5 Color set block diagram. (Note: Shaded blocks are also found in a black-and-white set.)

the presence of the burst) removes the inhibit on the amplifier, permitting it to amplify the chroma signals and pass them on to the next stage. The *threshold level* at which the stage is *gated* is established by the setting of a *killer control* on the back of the set.

Without this circuitry, *a monochrome set* ignores the burst signal and the color information being broadcast. It responds only to the Y signal, which corresponds to the regular video of a monochrome telecast.

- **Chroma Demodulators**

The purpose of the demodulators (sometimes called *synchronous* or *product detectors*) is exactly opposite to that of the modulators at the station; they decode and recover the chroma sidebands from the phase-modulated subcarrier(s). Since the modulation process required two subcarriers, for the two color difference signals (R - Y) and (B - Y), the same is true for demodulation. The phase relationship between the demodulated signals will be the same as the difference between the subcarriers.

There are three basic systems of demodulation. The original system (and the one providing the best color, but now obsolete) is called the *I and Q system*. It used an I demodulator to process the I sidebands to recover the R - Y color difference signal. The Q demodulator was used for the Q sidebands to recover the B - Y color difference signal. This required two locally generated CW reference signals (subcarriers) 90° apart in phase to take the place of the subcarriers that were not transmitted.

One of the most popular systems now in use has an *X and Z demodulator,* as shown in Fig. 2-5. It, too, is a *synchronous demodulator* in which the two circuit transistors are made to conduct on alternate cycles of the applied 3.58-MHz sinewave. This X and Z system is simpler and more economical to produce than the I and Q system, but it does not utilize the full 1.5-MHz capability of the latter. Economically, the I and Q system is not justified considering the slight benefits in color rendition obtained at greater cost.

Another system, called the *(R - Y) (B - Y) system,* uses a phase difference of 57.5° between the two subcarriers. With this system the G - Y green signal is usually derived in the demodulators rather than later as with the other systems.

REMINDER: _____

The *phase difference* between the chroma signals and that of the subcarriers establishes the *hue* that is reproduced at that instant of time, and the *amplitude* of the demodulated color difference signals is responsible for the degree of *saturation* or *intensity* of the color.

- **Color Difference Amplifiers**

Sometimes these are classed as *matrix* systems, of which there are a number of possible configurations (see Chapter 14). An amplifier uses *active* devices such as transistors, but matrixes use *passive* devices (diodes). Other matrixes are simply a network of voltage-dividing resistors. In some cases the green (G - Y) voltage is developed in the demodulators; in other cases it may be produced in the amplifiers as a proportionate sampling of the R - Y and B - Y voltages. Alternately, it may occur in the *pix* tube itself.

2-5 COLOR CRT FUNDAMENTALS AND COLOR MIXING

A *conventional* color picture tube has three guns each identified with one of the three primary colors: *red, green,* and *blue*. The screen is coated with phosphors sensitive to these three primaries. When red phosphors are impacted by the electron stream, we see a red image, for the green phospors, a green image, and so with the blue. The colors we can view however are not limited to these three. Fortunately, our sense of sight enables us to see "something that isn't there." Although the phosphor dots are too small to be seen individually, we see larger areas of an image as a *blending* of the primaries. For example a mixture in the proportions of 30% red, 59% green, and 11% blue will appear as white. As previously stated, grays are produced by "dull" primaries. The less saturated the color the lighter will be the shade of gray, and vice versa.

And the other colors are seen in the mind's eye as a result of mixing. Every conceivable hue is possible according to the proportions in the mix. Yellow, for example, is produced as a blending of red and green phosphors. To blend, it is not necessary for the colors to overlap; they need only be in close proximity when viewed at a distance.

One shortcoming of the human eye is its inability to discern color on extremely small picture elements. Because of this, only the *prominent* details of a picture are seen in color. The fine detail is seen only in black and white, something of which we may be completely unaware.

As shown in Fig. 2-5, each of the CRT guns is driven by discrete signals from the color difference amplifiers and the three cathodes, in parallel, with a Y signal.

2-6 COLOR SYNC STAGES

Color sync is *not* to be confused with synchronization of the CRT beam-scanning process. The block in Fig. 2-5 marked AFPC (automatic frequency and phase control) contains all the stages responsible for synchronizing color reproduction at the receiver with that being televised. As previously stated, demodulation of the color signals requires the generation of a 3.58-MHz subcarrier to substitute for that which was suppressed at the station. Loss of color sync results in developing the wrong hues or a complete loss of color.

• **The 3.58-MHz Subcarrier Stage**

This oscillator stage is *crystal controlled* for frequency stability, which is very important in developing the correct *hues*. Strictly, hue is a function of phase, but the oscillator must also be stable in *frequency* before it can be *phased* by the burst signal. This stage has two outputs 90° out of phase (or in quadrature). These two outputs are fed to their respective demodulators as shown in Fig. 2-5.

REMINDER: _____

Although the phase relationship between *chroma* and the two subcarriers is always changing, the phase difference between the two subcarriers must remain *constant*.

• **Burst Amplifier and Keyer**

The burst signal is a sampling of the station subcarrier and is transmitted immediately following each horizontal sync pulse, as shown in Fig. 2-5. It is compared (in phase) with the locally generated subcarrier, and if any discrepancy exists, a correction is made.

The composite signal feeds the burst amplifier as shown in Fig. 2-5. The amplifier is *keyed to conduct* (by a pulse from the horizontal sweep circuit) only for the duration of horizontal blanking. Thus the video (which could result in inadvertent triggering) is rejected and only the burst gets through. Sometimes a separate *keyer stage* is used.

• **Color Phase Detector and Reactance Stage**

Here the signal generated by the local subcarrier oscillator is compared with the incoming burst signal. Should a phase difference occur, a *correction* or *error* voltage is developed.

The reactance stage is a *simulated capacitive reactance*, in modern sets usually a *varactor diode*, connected across the oscillator tuned circuit. Variations in the developed error voltage across the varactor diode create changes in its (simulated) capacity to bring the oscillator back in step and restore normal operation.

The foregoing deals with the major functional stages of a set. There are also *incidental* stages or miscellaneous circuits such as those described next (refer to Fig. 2-5).

2-7 MISCELLANEOUS CIRCUITS

• Automatic Color Control (ACC)

This is a means of automatically controlling the gain of the bandpass amplifier in accordance with the strength of the received signal. It's a *closed-loop* circuit operating much like AGC.

• DC Restorer

Where a coupling capacitor is used between the video detector and/or the chroma demodulators and the CRT inputs, the *dc level* representing the average brightness and/or saturation levels of the televised scene are lost. The average dc level can be restored with a diode connected at the CRT input(s). Most solid-state sets use *direct coupling*, so there is no need for additional restoration.

• Blanker

This circuit operates in conjunction with the color killer to furnish a *gating pulse* for disabling the bandpass amplifier during BW reception.

• Pincushioning

This is a sweep problem associated with large-screen rectangular CRTs. It causes the raster to bow inward at top and bottom and/or at the sides. With BW sets, correction is made with small magnets. With color sets, antipincushioning circuits are required to correct the problem.

2-8 COLOR PICTURE

The physical characteristics and functional descriptions of color pix tubes, their associated components, and how BW and color pictures are reproduced is described briefly in the following paragraphs. For further details, see chapter 16.

2-9 SHADOW MASK COLOR TUBE

The heart of a color set of course is the tricolor picture tube, of which there are two basic types in common use. The original type, still popular, is called a *shadow mask tube*. Phosphors of the three primary colors are deposited on the inner side of the CRT faceplate as tiny dots. The dots are arranged in triangular groups of three called *triads*. (see Fig. 15-9). About ½ inch behind the screen is a gauzelike perforated plate called a *shadow mask*. It has approximately 500,000 tiny holes, equivalent to the number of triads on the screen. The shadow mask is precisely positioned so that each hole is exactly opposite the center of each triad.

The tube has three identical electron guns, each named for one of the primary colors. When the proper setup adjustments have been made, the beams from the three guns converge in the mask holes sequentially during the scanning process. Emerging, the beam of each gun strikes its proper color dot in the triad beyond, as shown in Fig. 15-9. This calls for a very critical and precise alignment of the mask during manufacture, and proper adjustments when the tube is put to use. Any misalignment and the beams will strike the wrong color dots or none at all, resulting in a degraded picture.

Under normal conditions, however, each gun will faithfully produce images of its own color, and when properly mixed, as explained earlier, the right proportions of the three primaries produces white, grays, and all the complementary hues.

2-10 APERTURE GRILL (IN-LINE) COLOR TUBES

These more recent tubes have their three electron guns positioned *in-line* instead of in a triangular (delta) pattern. The screen phosphors are arranged in *vertical strips* rather than triads, and the mask has *grills* with slots that correspond to the screen phosphors. Convergence of the beams takes place within the slot areas. Although beam trace is *horizontal,* the pix appears to be made up of *vertical* lines, as compared to the horizontal scanning lines of a delta tube. Details of a typical in-line tube are shown in Fig. 15-14. The tube has a *short* neck for increased beam current and brightness, and a *narrow* neck that brings the yoke field closer to the beam for increased beam deflection sensitivity. Convergence is no great problem with this tube, and the conventional convergence assembly and its associated panel are no longer needed.

One type of aperture grill tube is the *Trinitron,* a Japanese (Sony) import. Other than the yoke, the only item on the tube neck is either a small adjustable coil called a *twist coil* or a PM-type adjuster. The tube has built-in *convergence plates.* There is only one convergence adjustment, called the *vertical static control.* This is a pot to regulate the amount and direction of dc current through the twist coil, if there is one.

2-11 PURITY AND CONVERGENCE

These two interrelated functions exert control over the three beams, individually and together, until they impact their associated phospor dots (or stripes) on the screen. Any tube imperfections, problems with associated assemblies, misadjustments, or abnormal voltages and the beams may strike the wrong color dots or stripes, . . . or none at all, creating wrong hues, impure colors, and a degraded picture. Purity and convergence is controlled with external components on the tube neck, built-in tube elements, and various controls.

2-12 DEVELOPING THE WHITE RASTER

The strength or intensity of each of the three beams is controlled by individual, preset, screen-grid controls, in the proper proportions of 30% red, 59% green, and 11% blue that are necessary to produce white. However, as a prerequisite for obtaining a *pure white* raster, it is necessary to first develop *pure* rasters of the three primary colors as previously explained. To go a step further and produce the *whitest whites* possible over a wide range of operating conditions requires a technique called *gray-scale tracking.*

2-13 DEVELOPING A BW PICTURE

Just as a pure raster in each of the three primary colors is a prerequisite to a white raster, a good

BW picture is needed before you can get a good color pix. Assuming a white raster, and with the color control turned down, a BW picture is produced by the Y signal only. Any color fringes (color outlines) on a BW picture call for further adjustment of purity and convergence.

2-14 GRAY-SCALE TRACKING

Without this adjustment, what should be white areas of a BW or color *pix* may appear *grayish* for different settings of the brightness control and for different signal levels. This is due mostly to variations in the three guns and the relative brightness emitted by their screen phosphors. Most sets compensate by providing color *drive controls*. Usually there are two (for blue and green), but sometimes a red drive is also included. Whereas the brightness control varies the bias of all three guns simultaneously, the color drive controls regulate the guns individually (as well as the amount of drive each receives from the Y signal). These drive controls are adjusted on a BW *pix* for the whitest whites (highlights) consistent with different settings of the brightness control.

2-15 DEVELOPING A COLOR PICTURE

Assuming a good BW *pix* (which also means the raster is normal), all that is needed to produce a good color picture is to feed each gun with their respective R - Y, G - Y, and B - Y color difference signals. Color images are reproduced as follows:

When the beam is scanning a picture element, image, or area of the screen that should appear red, for example, only the red gun is operating and the green and blue guns are momentarily cut off by the signal. Similarly, only the green gun is operating to produce green, and the blue gun to produce blue. For white, all three guns are in use, their beams illuminating all three phosphors in the triads (or strips) in the proper proportions, as determined by the incoming signal(s). The same applies to the gray areas; the same proportions of the three primaries are present but with variations of intensity. For the complementary colors, only the proportions of the primaries are changed in accordance with the voltage levels of the ever-changing color difference signals. For black areas, all three beams are cut off, as with a monochrome tube. Finally, the smallest pix objects (the fine detail) produced by the highest video frequencies appear only in black and white, of which our eyes are not aware.

2-16 DEGAUSSING

A shadow mask or aperture grill may become magnetized by external influences such as vacuum cleaners, other electrical appliances, and soldering guns. Even the earth's magnetic field may magnetize the tube. If magnetized by the slightest amount, it will adversely affect the CRT beams, creating purity and convergence problems. To prevent this, one or several large coils are mounted around and near the front of the tube in close proximity to the mask or grill. The coil is energized from the 110-V, 60-Hz supply line. Whether for a watch, a CRT, or a battleship, the principle of degaussing is the same, that is, the use of an ac field to demagnetize steel or other ferrous metals.

Older sets had a switch to energize the degaussing coil manually whenever required. With modern sets, degaussing is *automatic,* and a strong ac field is produced *momentarily* every time the set is turned on. In stubborn cases (if the degaussing coil is not strong enough

2-17 COLOR SET CONTROLS

As with a monochrome set, a distinction is made between *user* controls and *service-type* adjusters. Only the former are considered here; the others are covered in subsequent chapters, as appropriate. Also omitted are those user controls found on a BW set and previously covered in Sec. 2-3. Alternate names are included in some cases.

1. *Color/intensity/chroma control:* This control is usually found in the bandpass amplifier stage. It controls the strength of the chroma signal to obtain the desired amount of color saturation to suit the individual.

2. *Hue/tint:* This control is used to establish the proper color identity or hue. Since hue changes with phase, the control may be found in any of the following circuits: the subcarrier oscillator, color AFC stage, or the bandpass amplifier.

3. *Raster tint:* (known by various trade names): Connected in the CRT cathode circuits, this provides manual control over the relative bias of the guns to obtain an optional *background tint*. It is not found on all sets.

4. *Preset color controls* (e.g., Instamatic): A set equipped with this feature obviates the need for frequent adjustment of certain controls such as color intensity, hue, contrast, and so on.

5. *Remote controls:* This is a popular option found on many of the more expensive receivers. It consists of two main items, a remote hand-held sending unit and a receiving unit with its associated control mechanisms. Many different systems are in use. With one system the sending unit is a battery-operated miniature transmitter that transmits different *supersonic* signals (in the 30- to 50-kHz range). The transmitter of one popular system is entirely mechanical, developing supersonic *sound* signals to command the various receiver functions. Another system uses *infrared* as the control medium.

At the receiver, a supersonic microphone (or other pickup device) responds to the signals and initiates the desired response. The basic functions are channel selection, power ON-OFF control, and volume. The more complex systems can also control color intensity and hue. In most cases, the tuner and other controls are driven by small reversible motors. For details on remote control systems see Sec. 12-16.

2-18 SUMMARY

1. The receiver picture tube has an *electron gun* that "shoots" a beam of electrons at the phosphor-coated screen. Visible light is produced by the impact. The beam is constantly deflected vertically and horizontally to produce a *raster*. The received video signal is *demodulated* and then fed to the CRT. Amplitude variations of the video signal vary the intensity of the beam and therefore the amount of brightness from the impact at any given instant. Hence, each pix element on the screen reproduces images ranging from white to black in accordance with the image on the camera tube at the station. Although scanning is *sequential,* we see a complete image due to the high scanning rate and *persistence of vision* of our eyes.

2. The *highest* video frequencies (near the upper limit of 4 MHz) are responsible for the *smallest* pix details. The *low* frequencies (those close to the carrier) produce the *larger* objects (on a horizontal plane). When viewed at a distance, we are unaware of the scanning lines that make up the pix.

Summary

3. Picture elements are reproduced at the precise correct locations on the screen because (a) the received video information modulates the CRT beam intensity at practically the same instant it is developed at the camera, and (b) receiver scanning is synchronized *exactly* with the scanning of the camera tube(s) by virtue of the sync pulses received for that purpose.

4. White picture elements are produced by the *highest* beam intensity. *Grays* correspond to a *lower* intensity. *Black* occurs when the CRT beam is completely cut off.

5. Briefly, the functions performed by the various *signal stages* of a BW receiver are as follows: (a) The tuner receives and selects the desired channel with its total band of frequencies, including both sound and pix signals. Both pix and sound carriers (with their modulation sidebands) are converted to *lower* (IF) frequencies; (b) the video and sound IF signals are amplified, demodulated, and amplified again; the video signal is fed to the CRT, the sound signal to the speaker; (c) at some point, the *sync signal* is separated from the other signals, amplified, and processed to control the vertical and horizontal sweep circuits.

6. A color set has a *subcarrier oscillator* to produce a color subcarrier frequency of (3.58 MHz) to take the place of the subcarrier signal suppressed at the station. This subcarrier is necessary for the demodulation of the chroma signals. For good *frequency stability,* this oscillator is *crystal* controlled. Different colors (hues) are produced by the varying phase relationship between the subcarrier signal and the chroma signals at any given instant. This requires that the subcarrier be *locked* precisely in phase with its counterpart at the station. This is accomplished by the 3.58-MHz *color burst* received during a color program. *Manual* control over phasing (and the hue of pix objects) is provided by a receiver control.

7. At the station, the chroma signals were developed by the modulation process as explained earlier. At the color receiver, the reverse takes place; that is, the chroma signal is *demodulated* with the aid of the locally generated subcarrier. The subcarrier oscillator has *two outputs* that are out of phase by 90° (or some other amount depending on the demodulator circuit used). The demodulators are supplied with two kinds of signals: a subcarrier signal and the chroma sidebands of the *received* signal. In the demodulation process, the same *color difference* signals are recovered as those produced at the station.

8. Most sets have a hue control. It is normally required because phasing is so *critical*. It is adjusted to obtain familiar flesh tones. Most sets also have a control for adjusting color intensity or saturation. The control simply regulates the *strength* of the chroma signals.

9. Most stations transmit a *VIR* (vertical interval reference) signal for accurately controlling hue, saturation, and other critical aspects of the color signal. The VIR signal is used in two ways: by station personnel to ensure the transmission of a good color signal, and at the receiver, where special circuits monitor and control color fidelity and stability. Receivers having this feature have no need for the manual color controls found on other sets.

10. *All* pix elements are not reproduced in color. The human eye cannot discern color on *very* small details, so the *highest* chroma signal frequency is only 1.5 MHz. Thus the smallest pix detail (produced by frequencies from 1.5 to 4 MHz) are viewed only in black and white.

11. Presently, the most popular color picture tube is the *shadow mask (delta)* type. It has three separate electron guns, each related to one of the three primary colors. The screen is coated with three color phosphor dots arranged in triads. A perforated shadow mask is positioned behind the screen with its holes exactly corresponding with the triads. When properly set up, the beam from each gun is able to go through each and every hole to strike the proper color dot beyond. This is quite a feat, considering that the beam is constantly being deflected with an ever-changing angle. Initially, a receiver is adjusted to produce a white raster, just as with a monochrome set. Although all phosphor dots of the three colors are illuminated, as seen up close, at a distance the eye sees the mixture as *white*. Similarly, with the proper proportions of the primary colors, we are able to see any and all the complementary colors. The *intensity* of any color being viewed is determined by the instantaneous amplitude of the color difference signal and the average brightness level, and by the Y signal, which simultaneously feeds the three guns.

12. To prevent noise and the like from reaching the picture tube via the chroma stages during BW reception, a *color killer* is used to disable the bandpass amplifier when it cannot sense the presence of a color burst.

13. The *aperture grill* or *in-line* type of color pix tube is used in many modern receivers. As compared to a *delta-* or *triad*-type color tube, the three electron guns are positioned in line rather than in the conventional triangle arrangement. The three-color phosphors on the screen are arranged in *vertical strips* (instead of *triads*). In place of a shadow mask the tube has an *aperture grill* with vertical openings that correspond to the screen phosphors. Unlike the delta-type color tube, an in-line tube is much easier to converge. Instead of a three-section convergence yoke, only a small coil is required, mounted on the neck or built into the deflection yoke. Some sets dispense with a coil in favor of a system of adjustable magnetic rings.

14. *In addition to* all the stages, components, and circuitry of a BW set, a color receiver has the following: (a) A color pix tube with its associated circuits and adjustments for developing *pure fields* of the three primary colors. (b) *Chrominance stages,* where the received color information is demodulated, mixed, and amplified to drive the three guns of the CRT. (c) The video amplifier or Y channel (not much different from a BW set). The Y signal is necessary in the mixing process to establish the correct brightness level of the colors. Since the maximum frequency handled by the chroma stages is 1.5 MHz, *higher* frequencies (those that reproduce the finest pix detail) must be furnished by the video section. (d) *Color sync* stages (sometimes called the AFPC (automatic frequency and phase control) section. Here the recevied *color burst* is used to synchronize the subcarrier oscillator to the station. Colors of the correct hue are produced *only* when the color sync is *precisely correct*.

15. Conditions required for obtaining a good color pix include (a) a *white raster*, where there is no *tinting* (background color) or *color contamination,* (b) a good *BW* pix, where there is no color *fringing* (color around edges of pix objects), (c) proper chroma and video (Y) channel signals being fed to the CRT, and (d) proper convergence of the three electron beams of the CRT.

16. To obtain a white raster, the CRT must be capable of developing *pure fields* of the three primary colors. Then the relative beam intensities of the three guns are adjusted (with screen grid controls) to illuminate the color fields in their proper proportions: 30% red, 59% green, and 11% blue. This is determined *visually*. In addition, *drive controls* are provided to obtain the *brightest possible highlights* (gray-scale tracking) on a BW pix.

17. When a station is broadcasting in BW *only,* the absence of a *color burst* is sensed by a *color killer* stage (in a color set) which turns off the chroma stages. The pix tube is then driven only by signals from the video amplifier fed simultaneously to all three guns.

18. A BW set cannot respond to color signals. To reproduce a color broadcast in BW, the receiver simply ignores the color burst and the chroma sidebands, using only the Y signal that was transmitted.

19. Any ferrous materials in and around a color pix tube may become magnetized by such things as appliances, speaker magnets, and even the earth's magnetic field. When this happens, proper convergence of the CRT beams becomes difficult or impossible. To prevent this, all sets have a built-in *degaussing coil* (or set of coils) that is momentarily supplied with 60-Hz line current every time the set is turned on. Where this is not completely effective, the tube can be *thoroughly demagnetized* with an external degaussing coil as part of the servicing setup procedure.

20. The conditions for developing a color pix can be briefly described as follows: (a) The three guns must be adjusted (in their proper proportions) to obtain a white raster. The brightness control simultaneously regulates the beam current of *all three guns*. (b) Static and dynamic convergence is adjusted for a precise overlap of all three color fields, so there is no color visible on a BW pix. (c) The R − Y, G − Y, and B − Y *color difference signals* individually control the beam currents of their respective guns. For example, to produce *red* at a given instant, the R − Y signal excites the red gun and the other two guns are turned off. Similarly, to produce *green,* the G − Y signal activates the green gun and the other two are disabled. For *blue,* the B − Y signal activates the blue gun and the other two are disabled. To produce any of the *complementary colors,* the guns are completely (or partially) energized in their proper proportions. The complementary colors of course are only an illusion in the mind's eye, since the CRT can *only* produce the three *primary* colors. To produce *white,* all three guns are simultaneously energized in the proportions of 30% red, 59% green, and 11% blue. To produce *black,* the three beams are driven to *cutoff*.

REVIEW QUESTIONS

1. What relative size of pix objects are produced by the (a) *high* video frequencies, and (b) *low* video frequencies? Explain.

TELEVISION FUNDAMENTALS: THE RECEIVER

2. What is the purpose of the electron gun in a camera tube, and a receiver pix tube?

3. Explain how intensity variations of the video signal modulate the CRT beam to produce pix elements ranging from white to black.

4. State the purpose of (a) the sweep circuits, and (b) the deflection yoke.

5. What is meant by *sequential scanning* of picture elements?

6. Name four oscillator stages found in a color receiver.

7. (a) Explain the composite video signal transmitted during a broadcast in BW.
(b) In what ways is this signal changed for a color broadcast?
(c) Explain the purpose of all pulses that make up the total signal.

8. Why is the bandpass requirement for video so much greater than with AM radio?

9. Explain how the brightness control works.

10. (a) Explain the importance of linear beam deflection.
(b) How is a pix affected by nonlinearity in the vertical and horizontal sweep circuits?

11. (a) What is a raster? (b) How is it developed? (c) Is it possible to have a pix without a raster, and vice versa?

12. What is the numerical relationship between the number of pix lines and the vertical and horizontal sweep rates?

13. Why are the vertical and horizontal retrace lines not visible under normal conditions?

14. (a) Explain the purpose of all the pulses of the composite video signal that are transmitted during the vertical retrace interval.
(b) Are all equalizing pulses used for each field?

15. Explain why various pix elements are produced at the same locations on a pix tube as on the camera tube target.

16. Which receiver controls regulate the vertical and horizontal sweep rates?

17. (a) What gives the illusion of motion on a TV pix?
(b) Compare the *flicker rate* with motion picture film.

18. Why are the sync and blanking pulses not normally seen on the screen?

19. (a) Explain the blanking level of the composite video signal. What receiver control establishes this level?
(b) What levels of the video produces white, gray, and black?
(c) What is the *blacker-than-black* region of the signal?

20. Why does a weak signal at the CRT input result in loss of pix contrast?

21. (a) Explain the importance of signal polarity at the input to the pix tube.
(b) What is the effect if reversed?
(c) Why is polarity of no importance in stages ahead of the detector?

22. A *takeoff point* in a receiver is where certain signals become separated for rerouting. Where in a set would you expect to find the following: (a) sound takeoff, (b) sync takeoff, (c) chroma takeoff, (d) color burst takeoff?

23. Although there are only three different color phosphors on a color pix tube, we are able to see all other colors including white. Explain.

24. How is the Y signal used by (a) a monochrome set, and (b) a color set?

25. Why is the luminance signal fed to all guns of a color tube and the color difference signals to each gun individually?

26. What two kinds of information are conveyed by the chroma signal? What information comes from the Y signal?

27. (a) What is the delay line and why is it required?
(b) How would the pix be affected if it were omitted?

28. What is meant by degaussing and how is it accomplished?

29. What is accomplished by the chroma demodulators in the receiver?

30. Why are *two* subcarriers required for modulation and demodulation of a color signal?

31. State two functions of the color burst.

32. (a) What is meant by gray-scale tracking?
(b) What controls are used?
(c) Is this operation performed on a raster, a BW pix, or a color pix?

33. Explain the basic differences between the delta-type color tube and trinitron (in-line) tubes.

34. (a) Which receiver control varies the amplitude of the chroma signal?
(b) The phase difference?
(c) What effect has each on the picture?

35. What is meant by AFPC and how is it accomplished?

36. (a) Why is a good raster prerequisite to obtaining a good BW pix?
(b) Why is a good BW pix essential to obtaining a good color pix?

37. Explain the purpose of the VIR signal now transmitted by most stations.

38. Explain the electromagnetic deflection of a CRT beam.

39. What is meant by a field and a frame?
(b) How much time is required for each?

40. What is meant by interlaced scanning?

41. Why is a pix composed of *fewer* than 525 lines?

42. With movie film, *complete* pictures are projected onto the screen. In what way does the development of a complete TV pix differ?

42 TELEVISION FUNDAMENTALS: THE RECEIVER

43. What is the effect on the quality of a pix when the high and/or low video frequencies are attenuated? Explain why.

44. What would be the effect on (a) the raster and (b) the pix if one or two guns on a color tube become weak?

45. What is meant by R - Y, B - Y, and G - Y?

46. (a) What are the I and Q signals? (b) State the frequency range of each.

47. How is it possible for two chroma signals to coexist and retain their identity when sharing a common output at the station modulators and a common input at the receiver demodulators?

48. (a) How is the phase difference between the two subcarrier outputs established?
(b) What receiver component establishes the frequency of the subcarrier?
(c) How is the phase controlled?

49. (a) What receiver stage handles the burst signal?
(b) Why is it keyed or gated by the horizontal sweep?

50. (a) What stage is responsible for enabling and disabling the bandpass amplifier?
(b) Why is it done?

51. (a) Where in the receiver are the luminance and chroma signals separated and (b) then recombined?

52. The Y signal (representing brightness levels) is separated at the station in order to develop *fully saturated* color signals. When and where is the Y signal restored?

53. How is the color phasing of the receiver synchronized with the station?

54. Where and how is the 4.5-MHz sound IF carrier developed in (a) a BW set?
(b) a color set?

55. What is meant by a complementary color?

56. Explain the difference between beam synchronization and color sync.

57. Any color has three characteristics, brightness, saturation, and hue. Define each.

58. Why are the finest details of a pix not reproduced in color?

59. The video (Y) signal is required when receiving a BW program on a color set. Why is this signal required for a *color* pix?

60. Explain why *more than one* color will be missing from a pix if *one* of the guns is dead.

61. In what way is the quality of a BW pix degraded if the video bandpass of the receiver is considerably less than 3.5 or 4 MHz? Explain why.

62. A color CRT can *only* reproduce the three primary colors, yet we see *many* hues on a color pix. Explain.

63. What is the purpose of a shadow mask or aperture grill?

64. The amplitude of a modulated FM sound carrier remains *constant*.
(a) How are variations in audio frequencies conveyed?
(b) What determines the volume level of the audio signal?

Chapter 3

TELEVISION TEST EQUIPMENT AND SERVICING AIDS

3-1 INTRODUCTION

In the absence of a completely automated checker (the long-dreamed-of instrument that can instantly identify a faulty stage or component), fast and efficient TV repair still depends on the knowledge and skills of the individual technician. This in turn depends on the tools and test equipment available. Occasionally, an experienced troubleshooter can get by with just an intuitive, perceptive sixth sense and an analytical mind (desirable attributes worth cultivating). But for the most part, attempting to repair a complex modern television without at least a few essential pieces of testing equipment can be frustrating, time consuming, and usually unproductive. We have reached an age where precision testing is *in* and guesswork is *out*.

Some of the equipment used in the repair of radio and other electronic devices can be put to good use in TV servicing. But when it comes to the testing and alignment of more complex circuitry, additional and more sophisticated items are needed. For the beginner, starting out on a small scale with modest means, this of course poses a problem.

But it needn't be. Since many service problems are straightforward and relatively simple, for a time at least, one can get by with only the essentials. For the more difficult problems, and for those aspiring to become full-time professionals, more and better equipment can gradually be acquired. High-quality instruments are expensive. Fortunately, many are available in kit form for the do-it-yourselfer at a considerable saving. See Fig. 3-1.

44 TELEVISION TEST EQUIPMENT AND SERVICING AIDS

Fig. 3-1 Most instruments are available in kit form for the do-it-yourselfer.

Fig. 3-2 Basic meter types.

3-2 COMMON TEST INSTRUMENTS

A brief descriptive review follows of the more common and conventional instruments familiar to most readers. Represented are such items as meters, signal generators and tracers, the oscilloscope, and so on. See Fig. 3-2. Subsequent paragraphs deal with instruments specifically designed for TV servicing, including a number of commonly used aids and accessories. Details on the *use* of each item of test equipment are given in later chapters where appropriate.

3-3 METER

This instrument (for measuring voltage, current, and resistance) is absolutely indispensable for troubleshooting any kind of electronic equipment. Meters have evolved dramatically over the years. First was the basic 1-milliampere (mA) "thousand-ohms-per-volt" instrument, which is fine for checking batteries, power supplies, and the like, but is practically useless for voltage checking in high-resistance circuits.

Then came the vacuum tube voltmeter (VTVM) and its solid-state counterpart the electronic voltmeter (EVM), both still popular because of their high sensitivity, negligible loading of circuits under test, and other desirable features.

And we now have the digital voltmeter (DVM) and the digital multimeter (DMM) with their numeric digital readouts, all the best features of the EVM, and more besides.

• **Basic Current Meter**

A meter *movement* (the moving coil assembly), whether used as a voltmeter, ohmmeter, or whatever, is basically a *current*-indicating device. To measure current, a circuit must be "broken into" with the meter inserted *in series with a load*. Because of the inconvenience of having to unsolder or disconnect circuits, current is usually determined by computation, using measured values of voltage and resistance. There are times, however, when actual measurements are called for.

The *lowest* current range (microamperes, milliamperes, or amperes) of a meter is determined by the sensitivity of the basic movement. Higher current ranges require the use of *shunts*. Since the current through the movement is very small, any additional current must be bypassed around it (through the shunt). The lower the shunt resistance, the more current it will pass, increasing the current *range* of the combination. For measuring relatively high currents, the shunt is usually a small fraction of an ohm.

A shunt must never be opened when taking readings. For measuring current, always start with the highest range and work down. Finally, *never* connect a current meter across a voltage source, no matter how low.

• **DC Voltmeter**

A basic voltmeter is simply a sensitive current meter with a series-connected *multiplier* resistor of a relatively high value. Instead of units of current, the scale is calibrated in volts.

Most meters have several *ranges*, as selected by a switch, or by using pin jacks, and each range has its own multiplier resistor (precision units with a 1% tolerance or better). The higher the voltage range and/or the more sensitive the movement, the higher will be the value of the multiplier that is required. Most general-purpose instruments have ranges from 0 to 1 V to 0 to 1 kV. For solid-state testing, where small increments of a volt are to be measured, a *lower* range is necessary. Different ranges make use of a *multiplying factor* ($\times 1$, $\times 10$, $\times 100$, etc.) for interpreting the scale markings.

An unavoidable error is created when measuring voltages in high-resistance circuits. Since the meter current must pass through the resistance of the circuit under test, an *IR* voltage drop is created, and the reading obtained will be less than the true voltage by the amount of this IR drop. The higher the circuit resistance and/or the less sensitive the meter (the higher its current drain), the greater will be the error. This is known as *circuit loading*. For example, a 0 to 1 mA, low-sensitivity meter is fine for measuring current, but not for voltage, since its accuracy is unacceptable in most instances. On the other hand, a voltmeter using a 50-microampere (μA) movement is practical for many uses, and a 20-μA instrument is even better.

• AC Voltmeter

A basic dc voltmeter can be made to measure ac by adding a rectifier. The rectifier may consist of several low-current diodes connected in a *bridge* circuit. Most meters have separate multiplier resistors and ac and dc, so the same scale markings can be used. Scale calibration in root mean square (rms) is accurate only for sinusoidal waveforms. For other waveforms, the readings will be in error (higher for square waves and lower for sawtooth waves). The accuracy of such a meter drops off rapidly for frequencies higher than audio.

• Ohmmeter

By adding a battery and a current-limiting resistor, any basic meter movement can be used as a *continuity checker* or calibrated directly in ohms. The maximum resistance that can be measured is determined mainly by the sensitivity of the movement: the greater the sensitivity, the higher the resistance that can be measured, using a given amount of battery voltage.

The ohmmeter has many uses: circuit continuity testing; measuring resistance of coils, resistors, and other components; checking capacitors, diodes, and transistors for leakage, shorts, or opens, to name a few. To avoid damage, certain precautions must be observed. As with a current meter, never attempt readings on a *live* circuit where voltage is present. Never try to test a *charged* capacitor. Some ohmmeters (on the lowest and highest ranges) have enough voltage and/or current at the leads to destroy transistors or ICs being tested. It is important to *know* your instrument. Finally, take heed when the zero adjust potentiometer must be set full on. The battery needs replacing and, if ignored, may leak and create serious corrosion damage.

• Volt-Ohmmeter

Sometimes called a voltohmist or multimeter, the Volt-ohmmeter (VOM), is a combination instrument, a multirange current meter, ac–dc voltmeter, and ohmmeter. Pocket-sized thousand-ohms-per-volt meters are available for just a few dollars. The more expensive quality instruments have a movement sensitivity ranging from 20 kilohms/volt (kΩ/V) to 100 kΩ/V. Accuracy for most VOMs is around 2% for dc volts and 3% for ac and ohms. A typical VOM has a zero adjust control, a mode switch, and a range switch, and provides for measuring all values of *E, I*, and *R* normally encountered in electronic circuits. Quality meters are protected against overload; for other less expensive meters, an improper setting of the function or range switch can spell disaster.

• Overload Protection

The most critical and expensive part of a VOM is the movement itself. Unfortunately, movements can be easily destroyed by an inadvertent overload. Modern instruments are protected in different ways against this hazard. The simplest (but least reliable) method is fusing. Others employ fast-acting circuit breakers that can be reset. Another means is by shunting the movement with two zener diodes having reversed polarity.

• **VTVM/EVM**

By far the most popular instrument (and for good reasons) is the VTVM (vacuum tube voltmeter) or its solid-state equivalent the EVM (electronic voltmeter). Because of its high input impedance (around 10 MΩ), there is negligible loading on a circuit under test, providing a high degree of accuracy when voltage checking high-resistance circuits. Having a greater sensitivity than a VOM, it can measure resistance (and leakage) up to hundreds of megohms. In addition, the meter movement is self-protected against overload. Although most VOMs provide for current ranges, VTVMs do not, lessening the chances of burnout.

On ac, some meters are calibrated for rms and others for peak-to-peak (p-p) voltage, a most useful feature in TV servicing. Some have both. An advantage of a solid-state EVM over a VTVM is that there is no warm-up time. Most EVMs use one or more field effect transistors (FETs) for increased sensitivity and reduced loading.

For dc volts, a VTVM/EVM has a *polarity reversing switch* so that the common lead will always be at ground potential. The dc lead is a shielded cable with an isolation resistor (about 1 MΩ) built into the probe. Usually there is a *zero center* scale to facilitate the alignment of certain TV circuits.

For measuring or monitoring the modulation component of an RF signal, a *demodulator probe* may be used when switched to a dc voltage range. A *high-voltage probe* enables the meter to read up to about 25 kV for checking the HV applied to a receiver picture tube.

The accuracy of a typical VTVM/EVM is between 1.0 and 3.0%. In addition to the tests performed by a VOM, this instrument has other important uses: signal tracing and gain checking receiver circuits, alignment, and so on.

Another type of meter (to be described later) is the DVM/DMM. A distinct departure from other types, it operates using computer logic circuits and has a digital readout instead of a direct reading meter movement.

3-4 BASIC SIGNAL GENERATORS

Basically, a signal generator is an instrument that provides a convenient source of signal (for test purposes) to simulate those signals put out by the station or those normally generated within the set. Depending on the type, a generator may develop either RF or AF signals. Radio frequencies are used for working on the high-frequency stages of a receiver, and AF for the low-frequency stages.

• **Signal Injection (with a generator)**

Signal injection is one good method for locating a bad stage. Other uses of a generator are for stage gain checking, frequency response and distortion checking, the alignment of tuned circuits, and temporary substitution of certain CW oscillators in the receiver. Three generator types found on most service benches are briefly described next. Refer to Fig. 3-3. (Special generators used in TV servicing are described in Sec. 3-10.)

Fig. 3-3 (partial)

48 TELEVISION TEST EQUIPMENT AND SERVICING AIDS

Fig. 3-3 Basic AF-RF signal generators and pencil-type noise generator.

• **RF Signal Generator**

A typical RF generator produces signals ranging from about 100 kHz to around 50 MHz (on fundamentals) and much higher on harmonics. This includes those frequencies required in the servicing of both radio and television equipment. Specifically, for the latter, this includes all VHF and UHF channel frequencies, sound and video IFs, and chroma-related signals. The generator has provision for two kinds of signals: unmodulated RF (CW), and modulated RF using a built-in 400-Hz oscillator. This low-level AF signal is also available for other uses. The RF is delivered to the set being tested via a shielded cable. The strength of the signal is relatively low (measured in microvolts) to correspond with station signal levels normally encountered in a receiver. The amplitude is adjustable with two *attenuator* or gain controls. One is a step-type coarse adjustment and the other a continuously variable fine control. With some instruments the controls are calibrated in microvolts. Others may use a direct reading meter.

The most important consideration of an RF generator is frequency stability and calibration accuracy of the main tuning dial. For certain TV frequencies (like 4.5 and 3.58 MHz), where extreme accuracy is essential, some instruments have built-in crystals or a *calibrator*. The output of a generator should be fairly uniform over all ranges, but can be expected to drop off somewhat at the highest frequencies.

• **AF Signal Generator**

A typical AF generator will develop both sine- and square-wave signals from 20 Hz to 100 kHz or better. Although the 400-Hz modulation voltage from an RF generator is suitable for signal tracing low-level *voltage amplifiers* in this frequency range, a stronger and continuously variable signal as produced by an AF generator can be useful (and sometimes necessary) when working on high-level output stages. The output from most AF generators is between 10 and 30 V and is adjustable with coarse and fine attenuator controls. The most important features of an AF generator are undistorted waveshape for both sine- and square-wave outputs, uniform output over the entire frequency range, and good frequency stability.

In addition to stage checking by injection, other uses are for frequency response checking of AF, video, and chroma amplifiers; substitution checking of vertical and horizontal sweep oscillators; stage-gain testing; amplifier distortion testing; square-wave *ringing* tests on yokes and transformers; and for producing bar patterns for linearity testing of vertical and horizontal sweep circuits.

- **Noise Generators**

Although frowned on by some technicians, this simple and inexpensive instrument can substitute for an RF-AF generator when such instruments are not available, for example on house calls. Its main function is to locate a bad stage. The signal generated is a pulse, rich in harmonics, that can *shock excite* its way through RF, IF, AF, and video stages. It is heard as a characteristic *hash* sound. Most noise generators are available in the form of pencil-like probes. Some include a built-in amplifier for use as a signal *tracer* for listening tests.

3-5 SIGNAL TRACERS

As opposed to stage checking by *injection*, this type of instrument is used to follow the progress of either a station signal or a generator signal through the various stages of a set. The most basic signal tracer is a pair of headphones, although their sensitivity to weak signals is limited. Other instruments such as an oscilloscope (CRO) and EVM (on ac) can also be used. Instruments specifically known as tracers are basically nothing more than high-gain voltage amplifiers. In fact, any amplifier will work well, such as the audio stages of a radio or TV set. For working with RF and IF stages a *demodulator probe* is required. The indication may be *aural* (using phones or a built-in speaker) or visual (a *tuning eye*, self-contained meter, or an externally connected EVM or a CRO) See Fig. 3-4.

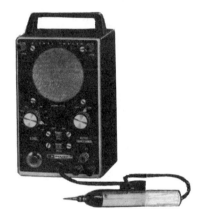
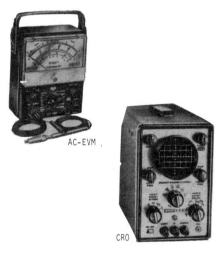

AC-EVM

CRO

Fig. 3-4 Aural and visual signal tracers.

3-6 CATHODE-RAY OSCILLOSCOPE

The cathode-ray oscilloscope (CRO) or *scope*, as it is commonly called, is one of the more sophisticated and versatile instruments available to the electronic technician. In the hands of a capable operator, it is an indispensable tool. A CRO eliminates much of the quesswork in TV troubleshooting and contributes greatly to a technician's efficiency. Unfortunately, the novice, for one reason or another, is sometimes disinclined to exploit its full potential, a grave mistake. A CRO has many uses that cannot be duplicated by any other instrument. With its visual display of a recurrent signal under test, *instant evaluation* is made possible; this includes waveform analysis and, if the CRO is calibrated, a peak-to-peak voltage reading. In signal tracing, this is particularly important, since most schematics indicate the normal or expected waveforms at key circuit points, along with their p-p amplitudes. Other uses of the CRO include stage-gain checking, frequency response tests, and the alignment of tuned circuits.

Modern wide-band CROs with special features are quite expensive; but for the most part, even the simplest basic instrument is satisfactory for many troubleshooting tasks. Along with the ability to interpret what is viewed on the CRO screen, a good understanding of its operation can be helpful. This can best be described with the aid of a block diagram (see Fig. 3-5).

The basic functions of a CRO are similar in many ways to those of a TV receiver. In fact, you can actually get a picture on a CRO by simply feeding it with a video signal and the vertical sweep voltage from a receiver. The main part of any scope is the CRT. As with the picture tube in a receiver, there is an electron gun with all the conventional elements. One difference, however, between a scope CRT and a TV pix tube is the manner of deflection. Instead of electromagnetic, a scope uses electrostatic (ES) deflection with built-in deflection plates, of which there are two pairs.

The CRT screen is coated with a green phosphor, which provides good illumination of the trace. Brightness is adjustable with an *intensity control*. A *focus control* varies the sharpness of the trace. Vertical and horizontal *positioning controls* allow for centering the trace on the screen.

The horizontal trace of the beam is developed by an internal sweep oscillator, as shown in Fig. 3-5. It produces a sawtooth-shaped wave for linear deflection. The resulting left-to-right trace represents units of time, a *time base* for the signal being displayed. With some CROs the sawtooth is fed directly to the horizontal deflection plates of the CRT. In Fig. 3-5, a buffer amplifier is shown for added sensitivity; this is desirable when there is a need to expand a display on the horizontal (X) axis.

The frequency (and phasing) of the oscillator is adjustable with two controls: a step-type *coarse frequency* adjustment and a fine-frequency or *vernier* control. The frequency range is typically from near zero to about 200 kHz. Depending on the setting of a selector switch, the horizontal trace may originate from either the internal oscillator or some external source using the horizontal input terminal provided.

Vertical deflection of the beam is the result of amplitude variations of the signal under test, which is applied to *vertical input* terminals. A shielded cable is used to prevent power line hum pickup and other interference. A vertical amplifier ensures adequate deflection when checking low-level signals. One or two vertical gain controls are provided; in some cases they are calibrated directly in millivolts and volts.

To display a waveform, the frequency of the sweep oscillator is adjusted in accordance with that of the signal under test. Its frequency must be the same as (or a submultiple of) the signal to make the display appear stationary. Where the frequencies are equal, one wave is produced. To

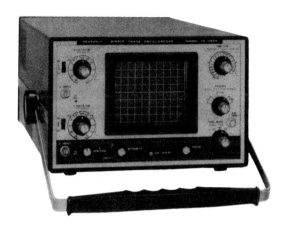

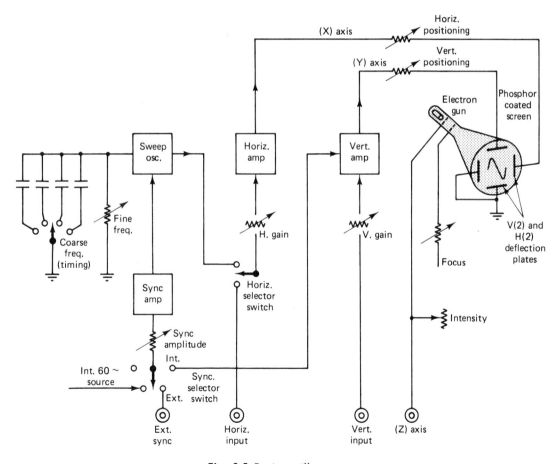

Fig. 3-5 Basic oscilloscope.

obtain two waves, the oscillator frequency must be one-half the signal frequency. For three waves, the ratio is 1:3, and so on. In practice, two or three waveforms are usually displayed. Other oscillator settings produce a random and meaningless display of lines. Only *recurrent waves* can be displayed (such as TV sweep voltages and sync and blanking pulses). Other signals, like video (when there is motion in the picture), appear as a blurred and ever-changing pattern. As the beam reaches the right side of the CRT after completing each trace, the rapid dropoff of the sweep sawtooth results in an almost instantaneous retrace to the left side, and the process is repeated. Some scopes blank out the retrace so that it will not be visible at low timing speeds.

As with television, some means of synchronization is necessary to stabilize the display. As shown in Fig. 3-5, a *sync selector switch* permits the selection of alternate sync sources to control the oscillator. When set for *internal sync*, the oscillator is self-synchronized by a sampling of the signal under test. When set for *line* (or 60 Hz), the power line may be used in some cases. For *external sync*, other suitable sources may be used by connecting to the *external sync terminal* provided. In Fig. 3-5, a sync amplifier is used. Precise adjustment of the *sync amplitude control* is necessary for optimum stability of the display.

Since deflection is linear, the scope may be calibrated to indicate the peak-to-peak value of any signal under test. A ruled transparent grid (graticule) is positioned in front of the tube face for this purpose, and the reading is expressed in volts per centimeter of vertical deflection. As a convenience for calibration, many CROs have a built-in accurate voltage standard. Calibration procedures are explained in Sec. 3-21.

For observing the modulation of RF signals (as in the RF-IF stages of a receiver), a demodulator probe must be used to demodulate the signal before it is applied to the vertical input terminals. They and other special-purpose probes are described in Sec. 3-20.

As stated earlier, the most basic and inexpensive oscilloscope can perform many servicing tasks, but (aside from the special features incorporated in better scopes) its utility is governed by certain important factors. Also, the choice of a scope should depend on its intended use. In low-frequency work, for example, a response up to 100 kHz might be adequate, but for the chroma circuits in color TV it must go much higher. Some considerations regarding a general-purpose scope are as follows:

1. *Sensitivity:* There must be sufficient gain to produce adequate vertical deflection for weak signals normally encountered in the early stages of a receiver. A typical deflection factor is about 10 millivolts/centimeter (mV/cm). Multiplying factors of a step-type attenuator extends the range to several hundred volts.

2. *Frequency response:* The bandwidth of the vertical section must be sufficient for the intended use. A response to about 1.0 MHz is typical for a general-purpose instrument. To observe display details or higher frequencies (the 3.58-MHz color burst for example), the response must be good to 4 or 5 MHz.

3. *Horizontal sweep:* Must be linear. The upper frequency should be at least 100 kHz with good stability. Some scopes have two preset positions on the control for 60 and 15,750 Hz, the frequencies most often used in TV and specified on the waveforms shown on most schematics.

4. *Vertical input impedance:* Should be high to prevent undue loading of the circuit under test. One megohm is typical.

Precision scopes with special features are described in Sec. 3-15.

3-7 GRID DIP OSCILLATOR

The grid dip oscillator (GDO) (Fig. 13-13) is an inexpensive instrument that has many uses: rough alignment and resonant frequency checking of tuned circuits; as an RF signal generator (of limited accuracy unless equipped with plug-in crystals); for measuring unknown inductance and capacity values; and as a signal monitor (especially useful in identifying RF interference).

A GDO is basically a high-frequency oscillator with a meter or tuning eye indicator. It uses plug-in coils covering all useful frequencies. When its coil is inductively coupled to a tuned circuit, resonance is indicated by a dip of the meter with the frequency read on a calibrated dial. For greatest accuracy, the least possible coupling should be used.

3-8 TUBE CHECKER

Although not used too much in an age of solid state, a tube checker does have a place on the service bench (or on house calls), where it can be a time saver under certain circumstances. Its indications are not always accurate or conclusive, however, and substitution becomes the best recourse. A tube tester checks for emission, shorts, leakage, gas, and an open heater. Most modern testers have an extension harness and adapter plugs and sockets for checking picture tubes (see Figure 3-14).

3-9 C/R BRIDGE (CAPACITY ANALYZER)

The C/R bridge (Fig. 3-13) uses a *bridge circuit* to measure capacitors and resistors with a high degree of accuracy. With a large calibrated scale it covers a wide range of values (capacitors from just a few picofarads to several thousand microfarads and resistors from near zero ohms to several hundred megohms). Some instruments also measure inductance. The instrument also checks for leakage, shorts or opens in solid dielectic capacitors, and the quality (power factor) of electrolytics. Some testers use a tuning eye as a *null* indicator and a meter to indicate the voltage applied to a capacitor under test.

3-10 SPECIAL TELEVISION TEST INSTRUMENTS

Although all the instruments just described can be used for TV repair, other special-purpose types are needed to keep pace with advancing TV technology. Such instruments are described in the following paragraphs; details on their use are given in subsequent chapters as appropriate.

3-11 DIGITAL VOLTMETER AND DIGITAL MULTIMETERS

Although the conventional VOM and EVM still have their place, digital meters are rapidly taking over, and for very good reasons. Compared to a meter movement its numeric display is easier and faster to read and with greater accuracy (no problems of interpretation or parallax). Besides, the DMM is more sensitive, which means reduced loading and greatly expanded ranges, particularly on the low side—a *must* for solid-state troubleshooting. For one accustomed to other meter types, a certain amount of familiarity with and understanding of such things as conversion factors (converting from basic units to higher or lower equivalents), decimals, and the calibration units (which vary with different instruments) is required.

Most DMMs have the same basic panel controls as a VOM or EVM: function switch and range switch, which are sometimes combined. Some meters have a zero adjust; others do not. For comparison, DMMs are judged mainly by the number of *digits* in their readout; the more the digits, the greater is the range and accuracy on all functions, but, as might be expected, the higher the cost.

Each digit of a display is able to display all numerals from 0 through 9. An exception (for most meters) is the digit at the extreme left, which only indicates a 1. It is called a *half-digit*. A basic and least expensive DMM usually has 2½ digits. The use of the half-digit can be confusing: enough for now that it permits readings beyond whatever range is selected, whether it be volts, amperes, or ohms. A 2½-digit meter, which can read up to two decimal places, is sufficient for most applications where extreme accuracy is not required. Laboratory-type DMMs often have 4 or 5 digits. Considering utility and cost, the most practical meter for servicing has 3½ digits.

The maximum display for a 3½-digit instrument is 1999, the numeral 1 representing the *half-digit*. The lowest reading is 0.001. For a reading to be meaningful, the placement of the decimal point is all important (the decimal is not considered a digit). The *most significant* figure in a readout is the one at the left, which always precedes the decimal. The *least significant* figure is the one at the extreme right, which follows the decimal. Changing ranges changes the position of the decimal.

Not all meters use the same calibration units in their readout. Some are in volts or millivolts, others in amperes or milliamperes, ohms or kilohms, and so on. For example, if calibration is in volts, a readout of 0.0003 represents 300 μV; if in amperes, a display of 0.001 would be 1.0 mA and 0.010 is 10 mA. If calibration is in kilohms (which is often the case), a display of 0.001 represents 1Ω, 10.5 is 10,000 Ω, and 10,000 is 10,000 kΩ or 10 MΩ. With some meters, the nonsignificant zeros are displayed; with others they may be blanked out to avoid confusion.

A DMM makes use of computer-type logic for its operation, where the values of voltage, current, or resistance under test are continuously sampled, converting amplitude variations into digital values for the readout display. For current and resistance measuring, the logic circuitry sees each as a voltage. Rectifiers are used for the ac functions. Two basic types of DMMs are available: the "bench-type" ac or battery-operated meter; and portable battery-operated instruments for field use, some of which are extremely compact, about the size of a pocket calculator. Battery-operated models often use rechargable nickel-cadmium batteries and a built-in charger. Some instruments incorporate a variety of features, for example the capability for measuring capacitor values, although this is not typical. Two modern DMM's are illustrated in Fig. 3-6.

Photo courtesy of Hewlett-Packard.

Fig. 3-6 Digital multimeters are rapidly replacing the conventional EVM.

A DMM differs in many ways from other meter types. As an aid in the understanding and evaluation of this type of instrument, its specific and unique features are described next.

3-11 Digital Voltmeter and Digital Multimeters

• Types of Readout Displays

The earliest bench-type DMMs used Nixie Lights, gas-discharge tubes containing all numerals from 0 through 9, which when individually energized became visible with an orangish glow. One lamp is required for each digit in the display. Currently popular instruments employ either light-emitting diodes (LEDs) or liquid-crystal displays (LCDs). Some LEDs form the numerals with illuminated dots (one dot per LED), but most have a *7-segment display*. The desired numerals are illuminated by activating the segments in a particular manner. LCDs also use a 7-segment display. Both types produce a red glow.

Some DMMs include a conventional meter movement that can be useful in indicating *trends*, that is, where a reading is continually changing as when peaking tuned circuits during alignment.

• Polarity Indication

All DMMs automatically indicate the polarity of a dc voltage or current, and there is no need to transpose test leads or operate a reversing switch. With each dc reading a (+) or (−) sign appears at the extreme left of the display. Some meters use only a minus sign, and positive polarity is *assumed* in the absence of a sign.

• Zero Adjusting

A zero-adjust control, when provided, is adjusted for a display of all zeros with test leads shorted. Many DMMs have automatic zeroing, which obviates the need for a control.

• Sensitivity and Resolution

Sensitivity is depicted as the lowest reading that can be displayed on the lowest range, as indicated by the least significant digit of the readout. For example, with a 4-digit DMM on the (lowest) 100-mV range, the last digit represents a sensitivity of 10 μV.

Resolution is usually stated as a percentage. A 4-digit meter for example can resolve one digit or count in 10,000s for a resolution of 0.01%.

• Overranging

Practically all DMMs possess this desirable feature, which enables them to read beyond the normal limits of any selected range and for all functions. This makes it unnecessary to switch to the next higher range when a value being measured exceeds that indicated by the range switch. For example, a 3½-digit DMM on a 1-V range (any function) can actually read up to 1.999, and on a 1000 range up to 1999, an almost 100% increase in each case. This is more than a convenience since the accuracy of a reading increases as you approach the upper limits of a range and *beyond*. The reason for this is the greater number of digits that follow the decimal for the higher readings, compared with readings near the low side of any range. A numeral 1 at the left of a display is sometimes called the *overrange digit*. Overranging is usually stated as a percentage (e.g., 100% as stated previously for the 3½-digit meter). The less the maximum count (display), the lower will be the percentage of overranging, and vice versa, before the need to switch ranges.

Different methods are used to indicate when overranging takes place. With some meters the display blinks on and off or a light comes on. Dashed lines may appear or the display may be blanked out. In some cases the overrange digit flashes on.

• Autoranging

With this feature, range switching is automatic and there is no need for a range switch. Switching may occur either *up* or *down* depending on the value of the last reading. Autoranging automatically selects a range that provides the greatest accuracy. Where a measured value rises above a range limit, high accuracy is assured by a reading in the overrange region. Where a value is near the lower limits of a range, switching is downward for a reading in the upper region of the lower range. Sometimes, even when a value being measured is stable, the least significant digit of the readout will change from one number to another. This is called *bobble*, and with most DMMs it is considered normal.

- **Typical Meter Ranges**

There are usually four or five ranges for each function. With overranging, measurements are possible beyond the highest stated range and, as with any meter, well below the lowest range. Typical ranges are as follows:

- DC volts: 200 μV to 1200 V full scale.
- AC volts: 200 μV to 1200 V over a range of 50 to approximately 500 Hz.
- Current (both ac and dc): 200 μA to 2 A.
- Resistance: 50 mΩ to 10 to 20 MΩ.

The dc may be extended to about 50 kV with a high voltage probe and to ac frequency with a demodulator probe.

- **Accuracy**

This is stated as a percentage of the actual readout rather than full-scale deflection as with a conventional meter. A typical DMM may be off by only one or two counts as indicated by the least significant digit of the readout. Accuracy differs for different functions and, as previously explained, is greatest at the high end of a given range. The accuracy for dc volts is usually between 0.1 and 0.5%. For ac volts, it may be somewhat lower, for current, 1%, and for resistance, about 0.25%. For ac, accuracy is based on a sine-wave and falls off rapidly beyond several thousand hertz. Some meters provide adjustments for calibrating when necessary.

- **Input Resistance**

Input resistance is usually 10 MΩ, the same as for an EVM.

- **Resistance Power**

This is an indication of how much voltage and/or current is present at the test leads when checking resistance, an important consideration when checking semiconductors, since they can easily be destroyed by excessive power from a test instrument. With DMMs it usually runs about 10 V or less.

- **Meter Protection**

Since a serious overload can destroy a DMM, sometimes beyond repair, most are protected against overload on all ranges and functions. Either fuses or circuit breakers are used. Maximum voltage limits should never be exceeded.

3-12 SPECIAL-PURPOSE SIGNAL GENERATORS

The basic generators previously described can be used to some extent, but others are needed for certain specific TV troubleshooting procedures. The following paragraphs deal with those instruments specially designed for TV, both BW and color.

- **Sweep Generator.**

Whereas a conventional RF signal generator develops either an unmodulated RF (CW) signal or an *amplitude-modulated* (AM) signal, a sweep generator by comparison puts out a *frequency-modulated* (FM) signal. It is used primarily in the alignment of tuned circuits in the tuner, IF, and chroma stages of a set, usually in conjunction with an oscilloscope. This method of alignment, known as *sweep* or *visual alignment*, produces a *response curve* on the scope, such as shown in Fig. 13-18 representing the amplitude variations for all frequencies that fall within the *bandpass* of the tuned stage or stages under test.

Alignment requires the use of three instruments, a sweep generator, a *marker* generator, and a scope. They may be separate items or a combination instrument, such as shown in Fig. 3-7.

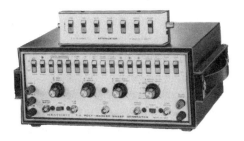

Fig. 3-7 Typical sweep-marker generator as used with a CRO for TV alignment.

The main tuning dial of the sweep generator is calibrated for all frequencies used in television. Accuracy of calibration is not too important, as will be explained later. With the sweep modulation turned off, the output of the generator is a steady unmodulated RF *carrier* at a frequency determined by the dial setting. A *sweep width* or *deviation control* controls the (FM) modulation. It is calibrated from zero to about 30 MHz and its setting establishes the *amount* of *frequency deviation* either side of the center carrier frequency. For example, if set for a sweep of 10 MHz with the tuning dial set for 50 MHz, the RF output continuously sweeps ±5 MHz between 45 and 55 MHz. If the sweep is increased to 20 MHz, the output will vary from 40 to 60 MHz. In use, the bandwidth of the stages being investigated or aligned dictates the amount of sweep required. For example, an IF section with a bandpass of 4 MHz requires at least that amount and more, and the tuner RF stages, 8 to 10 MHz or better. If the amount of sweep used is too low, only a (center) portion of a curve will be displayed by the CRO. (See Fig 13-17l) Where too much sweep is used, frequencies beyond the limits of a curve will be swept, compressing the response and making it difficult to interpret (Fig. 13-17m).

The *rate* of frequency change is established at 60 Hz and is sinusoidal. In other words, regardless of the *amount* of sweep, the RF varies above and below the carrier frequency at 60 times a second.

In tracing out a response curve, a sampling of this 60-Hz sweep voltage is used by the scope (in lieu of its internal sweep oscillator) for horizontal deflection. The interconnections for sweep alignment are shown in Fig. 12-23. The RF signal from the sweep generator is fed to the input of the tuned stages of the receiver as shown. The amplitude of the RF is adjusted with an *output level* control on the generator. The output of the receiver (from some suitable point) feeds the vertical input of the scope. Supplied from the same source, the horizontal movement of the scope beam follows and coincides with the generator sweep. At the same time the beam is deflected vertically, in response to voltage variations according to the alignment of the tuned circuits. The result is a *response curve* representing the bandpass of the set.

To trace out a single curve, it is necessary that the horizontal CRO trace be in step or in phase with the generator sweep. This is accomplished with a *phasing control* on the generator. If improperly adjusted, you get two curves (Fig. 13-17 i and j), one for normal trace and one during right-to-left retrace of the beam. They are normally adjusted to overlap. The alternative is to *blank out* the retrace with a *blanking control* on either the generator or the CRO.

• **Marker Signal.**

A response curve alone has little or no significance unless you can identify particular frequencies that correspond with various points on the curve. This is done with a *marker*, usually provided by a *marker generator* that puts out an *accurate* unmodulated RF signal. Any generator can be used provided it is *accurately calibrated* and has a frequency range covering all TV tuned circuits. The combination instrument shown in Fig. 3-7 has a built-in marker generator with crystal control for frequencies commonly used. The marker signal is loosely coupled to the receiver stages along with the sweep signal, as shown in Fig. 12-23. It appears as a *blip* on the response curve, from the *beat* between the two signals.

For a better understanding, imagine both generators connected directly to the CRO without the receiver stages in between. You get a horizontal trace whose width is determined by the amount of 60-Hz sweep voltage from the generator and the horizontal gain setting of the scope. Marker pips can be made to appear at any point on the line; their frequencies are determined by the setting of the marker generator dial. For a 50-MHz carrier with a sweep of 10 MHz, a marker of 50 MHz will appear at the center of the line (provided the calibration of the two generators agrees) with markers for 45 and 55 MHz at the two extremes. With a sweep of 20 MHz the ends of the line will represent 40 and 60 MHz, respectively. Only one marker appears at a time, but it can be moved at will by tuning the generator.

With the receiver included in the hookup (Fig. 12-23) and a response curve traced on the scope, the markers can be positioned anywhere on the curve for frequency identification and bandwidth determination. In practice, the strength of the marker can be somewhat critical: if too weak, it is apt to be obscured; if too strong, it may distort the curve. The strength of the marker can be controlled by varying the coupling or from a level control on the generator. Sometimes two or more markers are desired simultaneously, which calls for separate signal sources. An alternative is a *passive marker* consisting of a calibrated wave trap connected series with the sweep generator output. It shows up as a slight dip or notch on the curve.

Details of sweep-visual alignment are covered in Sec. 12-30 and 13-35.

• **Function Generator**

This is a versatile, general-purpose precision generator. It can be very useful for tracing audio, video, chroma, sweep, and sync stages, and for frequency response and distortion checking of most amplifiers. A typical function generator is shown in Fig. 3-3. Different output waveforms are provided: sine, square, and triangular. Some models also include pulses. A typical generator covers from near 0 Hz up to 1 to 5 MHz. Output is adjustable with a step-type attenuator from zero to about 10 V p-p. Some generators have sweep provisions.

Requirements, as for any generator, are calibration accuracy, frequency stability, adequate output with precise control, and minimum harmonic distortion.

• **Crosshatch Generator**

This type of generator produces signals that when fed to the video section of a set develop a number of stationary vertical or horizontal bars on the screen, or a *crosshatch* pattern of both. It can be used for evaluating and adjusting sweep *linearity*, *pincushioning*, and *convergence*. Although a convenience, this generator is not essential since most generators can produce bar patterns. Its just a matter of frequency. For example, a 60-Hz voltage develops one *horizontal* bar; 120 Hz, two bars; and so on for greater numbers, provided the frequency is an even multiple of 60, the vertical sweep rate. A voltage at 15,750 Hz, the horizontal sweep rate, produces one *vertical* bar, and a correspondingly greater number for even multiples of 15,750. For linearity checking, a great many closely spaced bars are needed. The amount of voltage determines the darkness or contrast of the bars. A square wave gives a more sharply defined bar than does a sine wave. A typical crosshatch generator uses switching to produce a fixed number of bars.

- **Dot, Crosshatch, Color-Bar (Pattern) Generators**

At one time, separate generators were required for these three functions, long considered essential for fast and efficient color TV servicing. Now practically all models feature them in one combination instrument. Figure 3-8 shows a compact, pocket-sized version used mainly for house calls.

In the *dot mode*, the generator develops signals that produces a pattern of uniformly spaced white dots on the picture tube screen; in the *crosshatch mode*, a pattern of white lines is produced. Either may be used for checking and adjusting the beam convergence of a set. The generator alternately turns the three beams of the CRT on and off a number of times during each vertical and horizontal scan. If the triads of the three fields exactly overlap while the beams are on (i.e., perfect convergence), the dots will be white. Where there is misconvergence on portions of the screen, the dots in those areas will show color. The same holds true for a white crosshatch pattern. Convergence adjustments are made as needed until there is no color visible on either pattern.

Most generators of this type provide a variety of patterns that can be used, including a single dot that can be positioned anywhere on the face of the tube. Some have a control to vary spot size. Convergence procedures are described in Sec. 15-22.

A color-bar generator develops signals that display a pattern of vertical color bars on the screen. There are usually ten bars of different hues, from yellow-orange on the left, through orange, red, magenta, reddish blue, blue, greenish blue, cyan, bluish green, to green on the right. Each color corresponds to a specific phase angle. The pattern is used for checking chroma stages, evaluating color reproduction, and making adjustments. If a receiver is operating properly, all hues will be displayed in their proper sequence and with the correct degree of saturation. Troubles are indicated when one or more colors are weak, missing, or out of sequence. There are three types of color-bar generators:

1. *Rainbow generator:* This is the simplest and least expensive type. It develops its own subcar-

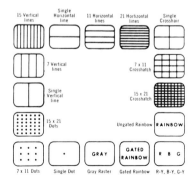

Fig. 3-8 A dot–crosshatch–color-bar generator can develop a variety of patterns.

rier (of 3.56 MHz), which is constantly changing phase to produce a rainbow of colors. Although the bars do not accurately reflect true values of hue or saturation, they do serve their purpose within limits. There can be problems in signal tracing with this type of generator.

2. *Keyed-rainbow generator:* This is the choice of most technicians, for reasons of economy and utility. The generator is keyed on and off, which creates black dividing lines between the color bars. All bars are firmly locked in sync. These conditions make it easier to assess color reproduction and signal trace chroma stages with a scope. Unfortunately, since the *simulated chroma* signals as generated all have the same amplitude, the relative brightness of the various colors is *not true* as compared to programmed station material. For further detail see Sec. 15-63.

3. *NTSC color-bar generator:* This type of generator develops *fully saturated* bars of the precisely correct hues in accordance with NTSC standards; and different signal amplitudes ensure identical brightness levels for all colors. White and black is also included in the display. As an option, some generators can also display each color bar individually.

Most dot-bar generators have two kinds of outputs: (1) a video signal for feeding video amplifier stages direct, and (2) a modulated RF signal that can be fed to the VHF antenna terminals with the tuner set for a specific channel. Some even provide a modulated IF signal for injecting into those stages. Some instruments have additional features, such as a sound carrier so the receiver fine tuner can be adjusted correctly to reject the *920-kHz beat* that can be a problem. One model includes disabling switches for the CRT guns. Besides a pattern selector, all instruments have a *chroma level control* to vary the intensity of the display according to ambient light conditions.

3-13 TV STAGE ANALYZERS

Analyzers, under various trade names, have been around for many years and are popular with technicians who like a number of different functions in one instrument. Basically, an analyzer is a multipurpose signal generator and tracer for all stages of a TV. A sophisticated version will generate all types of signals, RF, IF, video, sound, chroma, sync pulses, color burst, and signals to produce dot, crosshatch, and color-bar patterns. As a signal tracer it can trace and simultaneously monitor signals in all stages. Some models also include a CRO.

Figure 3-9a shows a popular multi-purpose instrument of this type called a video analyzer (Sencore). It is an extremely accurate and versatile instrument that can reduce servicing time to a minimum. Some of its functions include: a peak reading voltmeter; alignment generator; color-bar generator; dot/crosshatch generator; a ringer tester for shorted yokes and flyback transformers; *bar sweep* patterns up to 3.56 MHz; chroma bar sweep; tuner substitute; an accurate standard 1 V p-p video signal; and a variable dc supply.

3-14 COUNTERS

Frequency counters (Fig. 3-9b) are now found in many service shops for checking or monitoring the frequency of receiver oscillators, calibrating precision test equipment, and servicing video tape recorders (VTRs). A typical counter is able to measure frequencies from 1.0 Hz to 50 MHz or more, with an accuracy ranging from 1 to 10 parts per million. Most counters also provide a count on *intervals* and *events* when those modes are selected.

Counters use computer-type logic circuits and have a LED digital display like on a DMM (see Sec. 3-11). As with a DMM, they may also have *overrange* capabilities and feature *autoranging* that dispenses with a range switch. There are few panel controls: a mode switch, count reset, and an attenuator for the input signal under test. Pocket-sized counters are available with

3-15 Dual Trace, Triggered Sweep Oscilloscope

Fig. 3-9 (a) Video analyzer; (b) frequency counter.

many of the features of the larger and more expensive models.

3-15 DUAL TRACE, TRIGGERED SWEEP OSCILLOSCOPE

Although a reasonably good general-purpose oscilloscope as described in Sec. 3-6 is satisfactory for most troubleshooting problems, this CRO (see Fig. 3-10) is able to do a better job under certain circumstances e.g. viewing VIR and VITS signals and the color burst. The two features, dual trace and triggered sweep, are separate and independent functions. Some CROs include one or the other, but most quality instruments incorporate both, along with other refinements such as magnetic tape storage. Kits are available for converting an ordinary CRO to dual trace.

The dual trace capability makes it possible to display two separate signals simultaneously (usually for comparison), which otherwise requires the use of two scopes. Some uses for dual trace in TV servicing include the following:

1. Input-output gain checks on amplifiers and level comparison checks on similar or related stages.

2. For localizing intermittent conditions, where one trace monitors the input and the other the output of a stage.

3. Simultaneous comparison of two signals for such things as distortion, proper waveforms, relative phase displacement, and amplitude levels of the R - Y, G - Y, and B - Y signals at the CRT input. Dual trace makes it possible to instantly check the time delay introduced by the *delay line* in a video amplifier, and the phase relationships between horizontal sweep *feedback pulses* relative to other signals in *triggered* or *gated* circuits such as horizontal AFC, burst stage, killer stage, keyed AGC, and so on. Still another use is phase checking the two outputs of a 3.58-MHz subcarrier oscillator.

4. For checking and *measuring* the timing and duration of pulses, for both periodic and random events.

5. For quickly locating the stage where signal overdrive or pulse clipping occurs.

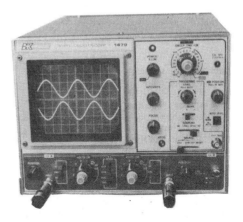

Fig. 3-10 Dual trace, triggered sweep CRO is a good investment.

Triggered sweep is most desirable in a scope. Besides being able to resolve and display more details, it greatly improves the *locking* (sync), particularly when observing rapid pulses at a low repetition rate, something that limits the usefulness of an ordinary CRO. It also makes possible good locking on the lowest-level signals and at high frequencies. Besides the stable locking and the ability to display complex waveforms in great detail, the triggering feature provides for the precise calibration of time intervals, the locking of signals that periodically change frequency, and the observation and measurement of portions of wave trains, including the smallest portions of individual cycles.

• Dual Trace: Functional Operation

There are two identical and completely independent vertical channels including separate input terminals and matching *gain* or *level controls*. The typical frequency response of each vertical amplifier ranges from dc (zero cycles) to at least 10 MHz and higher, depending on the quality of the CRO. Each channel has two gain controls, a step-type switch and a variable pot. Calibration is usually in volts per centimeter (V/cm) in reference to the squared-off *graticule* positioned in front of the CRT. A scope must be able to display a wide range of signal levels, from as low as about 1 mV to as high as 20 to 60 V or so. For example, when the vertical deflection from a signal covers 3 squares (centimeters) and the range switch is on the × 10 position, the voltage is 30 p-p. Calibration accuracy is typically between 3% and 5%.

The two input signals under test are fed to the vertical deflection plates of the CRT via an *electronic switch* whose function is to alternately switch from one input to the other. Due to rapid switching and the *afterglow* of the screen phosphor, there is the illusion that the two displays are simultaneous. There is little if any flicker except at very low sweep rates. Adapters are available for converting conventional scopes to dual trace.

• Triggered Sweep: Functional Operation

With this function, the horizontal sweep is initiated (triggered) by the signal under test. In the conventional scope, the sweep (as produced by the internal sweep oscillator) is *recurrent*, that is, continuous without interruption at a frequency determined by the setting of the sweep frequency controls; and the sync (the stationary locking of the display) is dependent on sources that are not always effective, making it difficult or impossible to observe details and make measurement of certain signals being displayed.

With triggered sweep, the frequency controls have a different name and are usually referred to as the *time/div* (division) or *time base* control; and the time base (horizontal trace) can be accurately calibrated into units or divisions of time. Typical frequency range is 100 nanoseconds (ns)/cm to about 0.3 s/cm. With triggered sweep, the sweep is (normally) *not* recurrent. In the absence of a signal input (signal under test), when in the triggering mode, the CRT beam is blanked out and the horizontal sweep is inhibited. A signal, when applied, initiates the start of the sweep and the beam is *unblanked* for the duration of the trace. During the sweep, a *lockout circuit* prevents noise bursts or other extraneous signals from influencing the sync. This accounts for the extremely stable locking capability. The signal under test must be of sufficient amplitude and of the correct polarity before triggering can occur. When these conditions are met, (in accordance with the setting of certain panel controls), *one* horizontal sweep is produced, after which the beam retraces to the left side to await the next cycle to be triggered. Some scopes provide for a horizontal trace even in the absence of a signal, since without any visible display there may be some uncertainty as to whether the scope is operating.

Some additional features included in quality scopes are as follows:

1. *AC-DC inputs:* Normally, a scope has a blocking capacitor at the vertical input. It passes an ac signal, at the same time blocking any dc that might otherwise shift the base line of the display, sometimes shifting it clear off the screen. But at times there is a need to measure the dc component. This calls for direct coupling between the input terminal(s) and the CRT. The amount of deflection for dc is the same as for p-p values of an ac voltage.

2. *Sweep magnification:* This provides a /5 or /10 magnification (widening) of the horizontal trace when portions of a signal need expanding for a closer look and detailed observation, for example, in viewing individual cycles of a color burst or VIR signal, which are otherwise difficult to resolve.

3. *Polarity reversal switch:* This makes it possible to view a display either upright or inverted, regardless of polarity.

4. *Provisions for calibration:* Many scopes provide an accurate voltage source for checking and calibrating vertical deflection.

5. *TVV and TVH sweep rates:* A selector provides either a 60-Hz sweep or a 15.75-kHz sweep, the two most commonly used sweep frequencies.

3-16 VECTORSCOPE

The circular display pattern produced by this type of CRO can be used instead of (or in addition to) the color-bar pattern on the screen of the receiver pix tube, since it provides essentially the same kind of information. It is called a *vectorgram*, or more commonly, a *daisy-petal* display, which it resembles (see Fig. 3–11). Each petal corresponds to a color of the bar pattern. By analyzing the display you can instantly determine and evaluate the chroma conditions in a set, for example, alignment, and the proper functioning of the chroma demodulators, matrixes, color difference amplifiers, and color sync circuits.

Fig. 3-11 Vectorscope with its daisy-petal display speeds up chroma troubleshooting.

A vectorscope is really not too different from a conventional oscilloscope; in fact, any good general-purpose scope can be used for this purpose, with certain modifications. To obtain a meaningful display, the response of the horizontal amplifier must be flat to at least 1 MHz. And both vertical and horizontal sections must be able to handle signals up to 100 V p-p or better. (This requirement can be waived when working on high-level stages where signals can be fed directly to the vertical and horizontal deflection plates of the scope CRT, if access terminals are provided.) The vertical and horizontal inputs must be isolated with no common grounding. One must also disable the horizontal sweep. Some conventional CROs provide the necessary switching for use as a vectorscope.

Since a color-bar generator is also needed for this method, many vectorscopes include it. Some also feature such conveniences as gun-

64 TELEVISION TEST EQUIPMENT AND SERVICING AIDS

disabling switches for the receiver pix tube. With the exception of the usual sweep controls (which are not required), a vectorscope has the same basic controls as on any basic scope. Its operation is explained as follows.

- **Lissajous Figures and Rotating Vectors**

If equal-amplitude sine waves (as from an AF signal generator or the 60-Hz power line) are fed to the horizontal and vertical deflection plates of a scope, you get a perfect and stationary circle if the waves are exactly in phase. If there is a slight phase shift, the pattern becomes an ellipse where the amount of slope corresponds with the amount of shift. By connecting the scope inputs to the R - Y and B - Y outputs of the receiver demodulators, you also get a Lissajous pattern, in this case the *daisy-petal* display. Once for every sweep of the 15,734 Hz of the receiver, a 360° phase shift takes place between the subcarrier oscillator and the chroma signals, as supplied by the color-bar generator. This produces *all* the hues of the spectrum, ten of which are displayed as bars on the pix tube and petals on the scope display. Thus each petal can be considered a *vector* representing a definite phase relationship and color.

The various colors can be identified by markings on a graticule placed in front of the CRT. See Fig. 3-11. Going clockwise from the *burst reference point* (at 9 o'clock) are numbers from 1 to 10 that correspond with the numbered bars reading from left to right on a pix tube display. A petal representing yellowish-orange for example should be at the 1 o'clock position, bright red straight up at 12 o'clock, and so on, over the 360 degrees of the circle. Although not indicated on the graticule, each color is phase related. For example, the bluish-green vector is at the 6 o'clock or 270° point, and red is at 90°. Bear in mind that the vectors represent fully saturated colors since the Y or brightness component is not present at the demodulator outputs. The length of each petal as it extends outward from the center of the display represents the relative *amplitude* of that particular color difference signal. The roundness of each petal provides a relative indication of chroma bandpass. The pattern should normally remain stationary, but can be made to *rotate* if the hue control of the set is turned. When a pattern continues to rotate, it indicates a sync problem in the receiver.

3-17 DIODE AND TRANSISTOR TESTERS

Most technicians know how to check a diode or a transistor with nothing more than an ohmmeter. And more often than not such basic tests will show up leakage or an open or shorted junction, the most common troubles associated with these devices. However, such methods are time consuming and the results are not always valid or conclusive. A proper instrument designed for the purpose can take a lot of guesswork out of the testing, can do it faster, and can perform other tests required for a true and complete evaluation of the condition of the unit.

The simplest and least expensive testers (called "quick testers") are usually limited to basic go-no-go tests (as performed by an ohmmeter) and in identifying transistor types (NPN or PNP) and the terminal leads and base connections. Better instruments are able to test diodes and transistors both in and out of a circuit and can measure *current gain* of *bipolar* transistors and the GM (transconductance) of *FETs*. Typical testers of this type are shown in Fig. 3-12.

The testing and replacement of a diode or transistor (if found to be defective) usually follows a definite sequence. The various tests (in a logical sequence) are explained next.

3-17 Diode and Transistor Testers 65

Fig. 3-12 Typical transistor testers. Note the short-preventing miniclips.

- **Testing Diodes**

The testing of diodes is relatively simple. If not known, the first step is identifying the anode and cathode terminal leads using appropriate switches on the tester. Then *forward* and *reverse* current readings are taken, which correspond with the *front-to-back resistance ratio* of the diode. The higher this ratio is, consistent with a low forward resistance, the better the diode. This test also indicates an *open* or a *direct short*.

Initial testing is normally done *in circuit* without removal and with the receiver power turned off. Avoid unnecessary removal whenever possible since this increases the risk of heat and strain damage to the unit. Depending on test results, the unit can then be removed if necessary for further out-of-circuit verification testing and possible replacement. But before replacement, always make circuit voltage tests for conditions that may damage or destroy the new part. It is also wise to check the new diode before installing and to *double-check* its polarity.

Some instruments are able to check special diode types, such as SCRs for example. Others can even check the *avalanche* (breakdown point) of zeners.

- **Testing Bipolar Transistors:**

Conventional Types Both NPN and PNP transistors with the normal connections, base (B), collector (C), and emitter (E), may be tested either in or out of a circuit. In-circuit testing requires three probes with suitable clips that will not readily short out closely spaced PC connections. Sockets are provided for out-of-circuit testing. The various tests in their proper sequence are explained next:

Transistor and Lead Identification. All testers have a *type* switch for determining whether a PNP or NPN type and switch must be in the proper position for all subsequent testing. Another switch is provided for identifying unmarked connecting leads. This switch goes through all possible interconnecting combinations until the right one is found as indicated by a light coming on. With some instruments this searching operation is performed automatically using computer logic.

Go-No-Go Test. As with diode checking, a test is made for opens, shorts, and leakage for each junction. Leakage current is indicated on the meter scale. As explained for diodes, the test is normally performed in circuit, followed by an out-of-circuit test to verify.

There are two important leakage tests. The *ICBO test* checks for any *reversed voltage* developed between the collector and base with the emitter open. In operation, such leakage has the effect of increasing the *forward bias* on the transistor. The *ICES test* checks for leakage between collector and emitter with the base shorted to the emitter. During such tests, voltages applied to the transistor are limited to a safe value to prevent any damage due to overload. A switch enables separate testing of small (low power) transistors and high-power units rated at 1 watt (W) or more.

DC Beta Test. The term *beta* is used to indicate the relative *current gain* of a transistor under normal operating conditions. It is expressed as a *ratio* comparing the amount of *E* to *C* current developed by a given amount of *E* to *B* current. For example if 1 mA of *E* to *B* current produces an *E* to *C* current of 50 mA, the beta is 50. A meaningful beta test cannot be made if the transistor is defective in any way or if there is a low circuit resistance shunting the junctions. In the latter case, testing must be done after removal from the circuit. Some testers have two or more beta *ranges*; there is also a CAL (calibrate) button and a *test button*.

AC Beta Test. Whereas dc beta is checked under *static* conditions, this test is *dynamic*, with suitable ac voltages applied to the transistor. AC beta is expressed as a change in collector current over a change in base current (I_c/I_b), with the collector voltage held constant. AC beta normally decreases at high frequencies; at low frequencies, ac and dc beta are about equal.

REMINDER: _____

When tests are inconclusive, remove the transistor from the circuit and recheck. Make sure there are no circuit defects that could damage a new unit when making a replacement. Check the new transistor prior to installation.

• **Curve Tracers**

This method checks dynamic characteristics by displaying a series of *signature patterns* on a CRO. It is useful for accurate comparison tests.

• **Field-Effect Transistors**

These units call for a different mode of checking since their operation more closely resembles a vacuum tube than a transistor. For testing FETs, a meter is calibrated in microsiemens (formerly micromhos), the unit of transconductance, the same as for tubes. To avoid damage, certain precautions (to be described later) must be taken. The units are so sensitive in fact that even with normal handling some damage invariably results.

The first tests are to identify the connecting leads, the *source* (S), the *gate* (G), and the *drain* (D). Tests are then made for leakage between the different electrodes. Then a *zero bias drain current test* (similar to cathode emission of a tube) is made followed by the GM test.

3-18 FLYBACK AND YOKE TESTER

A number of symptoms and clues can lead you to suspect a bad flyback (horizontal output transformer) and deflection yoke (both prone to breakdowns because of the high pulsed voltages that are present). Unfortunately, other defects can produce the same symptoms. To unnecessarily replace such major components, particularly on a color set, can be quite costly in both time and dollars.

Some conditions, like an open winding or the visible arcing that leaves a telltale carbonized leakage path, can be easily verified. But the big problem, and one that is not always easy to diagnose with absolute certainty, is shorted windings. The reason is that even a few shorted turns may not show any significant or measurable change in resistance. To this end, a flyback and yoke tester (Fig. 3-13) can often be helpful. Two types of testers are available to the technician, each operating on a different principle.

- **Dip-Meter Type of Flyback and Yoke Tester**

Shorted windings in any coil or transformer reduce its inductive reactance, which in turn raises its natural resonant frequency. The more severe the short is, the greater the increase. This type of tester consists basically of a power oscillator (operating around 1500 Hz) with a sensitive meter as an indicator. It operates much like a GDO (Sec. 3-7). If connected to a good flyback or yoke, the meter will read accordingly. If there are shorted windings, the change in frequency plus the loading of the oscillator causes the meter to read in the "bad" region. The instrument will also check for direct shorts or an open circuit. The instrument has one shortcoming. It cannot discriminate between a low Q coil that is all right and a high Q coil with a few shorted turns.

- **Ringer Type of Flyback and Yoke Tester**

Unless defective, any coil can be *shock excited* into *ringing* (oscillating) at its natural resonant frequency, with the *damped oscillations* indicating on a meter or scope. This instrument generates low-frequency, sharply spiked pulses for this purpose. When connected to a good flyback or yoke, the decay is gradual, producing anywhere from 10 to 100 cycles as indicated by the meter. With shorted windings, the ringing will rapidly diminish and read accordingly. The more turns that are shorted, the faster the decay. As with the previously described tester, the results are not always conclusive, however, depending on Q and other factors.

Ringing can also be done using an AF signal generator and a CRO (see Sec. 7-21).

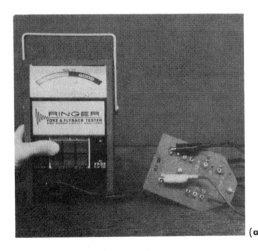

Fig. 3-13 Component testers: (a) flyback transformer being checked with a flyback and yoke tester; (b) a useful instrument for checking and accurately measuring R and C values.

3-19 PICTURE TUBE TESTER AND REJUVENATOR

The single most expensive item in a TV set is the picture tube. Not to mention the difficult and time-consuming job of replacement and setup, every effort should be made to ensure a correct diagnosis before considering replacement. Many defects involving seemingly unrelated stages and circuits can masquerade as a bad CRT, and where the condition of the tube is not known, *every possibility* should be checked out before condemning the tube. This includes such things as the HV, focus voltage, and other operating voltages, brilliance and screen control circuits, setup adjustments, the possibility of socket troubles, and so on. Where such things are found, chances are the CRT is all right. Given sufficient time and proper techniques, the condition of a CRT (as with any tube) can be determined without the aid of a proper tester. But a CRT tester, if not an essential instrument, can be a real time saver.

A typical checker, such as shown in Fig. 3-14, will show up most of the defects common to pix tubes, both BW and color: opens, shorts, leakage, gas, and loss of emission. It can even predict probable life expectancy by checking for emission dropoff at low heater voltage. Leakage and short tests indicate which tube elements are involved, as indicated by neon glow lamps. Some instruments have one meter, and some have three meters to simultaneously measure the cathode emission of the three guns of a color CRT. Many regular tube testers provide for checking CRTs, with optional harness and connector assemblies to handle all CRTs including trinitron and other in-line types.

But there are some conditions a checker will not show up, such as focus and other problems due to gun defects or purity and convergence problems caused by a warped or misaligned shadow mask.

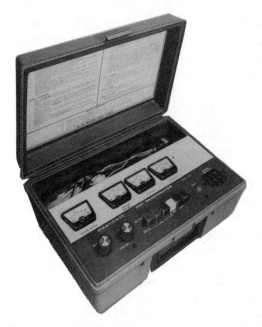

Fig. 3-14 Typical pix tube tester and rejuvenator.

• **Rejuvenation**

More often than not a weak tube can be restored, at least temporarily. This is done by applying a higher than normal voltage to the grids, which removes any *contamination buildup* on the cathode and exposes new coated areas for increased emission. This must be done carefully and judiciously, however; otherwise complete and irreversible *stripping* of the cathode occurs and the tube must be replaced. To prevent this, the *cleaning process* is *timed*. Because of the possibility of damaging a good tube, the procedure should not be indiscriminately used without first checking the tube's condition. Emission can also be restored with a heater brightener or *booster* (see Sec. 3-20). After rejuvenation, a tube may be good for several years or fail within a few hours. There is no guarantee.

Shorts or leakage due to a buildup between elements can usually be remedied by *flashing* with several hundred volts of dc. Because of the risk to the tube, short and leakage tests should be frequently made between flashing. In some cases a *weld* can be made to restore an internal open circuit. Detailed procedures for testing and restoring picture tubes are described in Chapter 15.

3-20 SERVICE AIDS AND ACCESSORIES

Besides the regular test instruments described so far, there are a great many miscellaneous items (see Fig. 3-15) that can be of great help in servicing TVs. Some of the following might be considered gadgets, useful on occasion but hardly essential. Others, like the *tuner subber* and *CRT test jig*, are almost a must. Some make good do-it-yourself projects. Many good books are available on the construction of such items.

• RC Substitution (Decade) Box

A *sub box* is simply an assortment of resistors and/or capacitors covering most values in common use. It can be useful for *temporary* substitution or in determining the approximate value of a component when a resistor becomes burned beyond recognition or the markings on a capacitor are obscured and no schematic is available. Resistors and capacitors may be contained in separate substitution boxes or combined in one box. A typical sub box has either terminals or clip leads with a selector switch. When substituting, make sure you are not exceeding the wattage rating of the resistors or the voltage of the capacitors.

• Bias Box

This is a source of low-voltage dc that is variable from zero to 15 V or so. It is used for *clamping* AGC circuits at some predetermined voltage as an aid in troubleshooting and for experimentally adjusting the bias on critical transistors. It may be a battery or a *well-filtered* ac-powered supply. For use on transistor circuits it must have both (+) and (−) outputs or a reversing switch. Ideally, it should have its own accurately calibrated voltmeter.

• Variable DC Power Supply

Unlike a bias box (for which it can also be used), this *supply* must be able to deliver considerable current if called upon to do so, as when supplying the dc power for a single stage or even substituting for the LVPS of a receiver. It should be capable of delivering up to 2 A or more with good voltage regulation (1.0% or better). A typical supply of this type has a dc output up to 20 V or so with a ripple no greater than 10 mV. The unit should have its own voltmeter and current meter, especially the latter for monitoring current drain of a circuit under test. Most supplies have some sort of overload protection, preferably a resettable circuit breaker.

• CRT Test Jig

As a temporary substitute for the color TV pix tube while troubleshooting a set at the shop, this unit can be a great convenience and time saver. On a house call it makes it possible to remove only the chassis and leave the heavy cabinet and pix tube behind. In most cases this also eliminates the need for reconvergence after the set is returned. The jig can also be used to isolate a trouble to either the CRT or the chassis without a lot of testing. Normally, it *cannot* help when troubles involve items left intact and not taken to the shop, such as the deflection yoke, convergence assembly, degausser coils, and the CRT.

A typical test jig (see Fig. 3-15) consists of a color pix tube along with the essential assemblies mounted on its neck: a deflection yoke,

70 TELEVISION TEST EQUIPMENT AND SERVICING AIDS

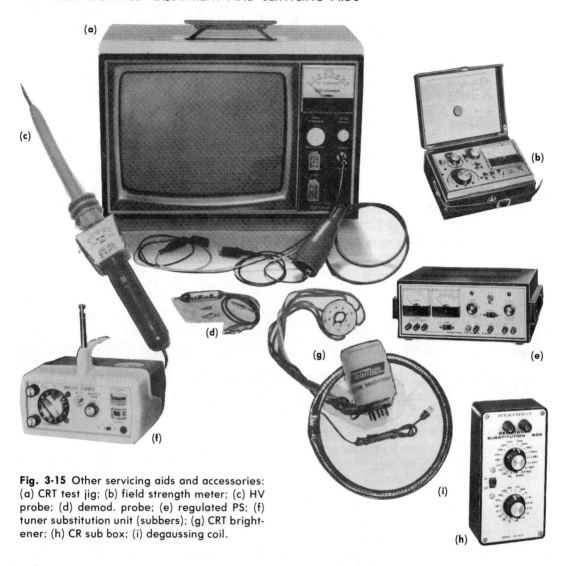

Fig. 3-15 Other servicing aids and accessories: (a) CRT test jig; (b) field strength meter; (c) HV probe; (d) demod. probe; (e) regulated PS; (f) tuner substitution unit (subbers); (g) CRT brightener; (h) CR sub box; (i) degaussing coil.

color purity adjusters, static convergence assembly, and in a few models also the dynamic convergence unit, although normally this is not considered essential. All voltages, including HV and deflection (sweep) voltages, are supplied from the set being tested. Some models have a built-in meter for monitoring the HV and in some cases a speaker to substitute for the one left in the cabinet.

Because of the many makes and models of sets, each with different cable connectors, CRT basing configurations, an so on, a variety of *adapters* is required, as well as *extender* cables, for different hookups.

Typical interconnections between a TV chassis and a test jig include a long anode lead to feed HV from the set to the test jig CRT. Normally, focus voltage is furnished by the receiver's focus circuit. In cases where it is too low to produce a raster, it can be taken from a voltage divider off the HV as shown. An alternative is to use the voltage from another set.

Impedance matching between the jig yoke and the flyback transformer in the set can be a problem, producing some undesirable effects like yoke *ringing* and difficulties with raster *size*. Some jigs employ a universal yoke, others a matching device, as shown.

Most jigs only have provision for *center* (static) convergence. This calls for a *simulated convergence* load, which plugs into the chassis connector. Where the jig has a dynamic convergence panel and assembly, the chassis connects to the jig via a suitable cable.

If necessary, a jumper can sometimes be placed across the degaussing coil connector on the chassis to complete the circuit, in lieu of the missing coil.

Despite inconveniences and having to keep up to date with new adapters, a test jig is a big plus in the long run. Some technicians even make up their own jig, although fabricating extender cables and adapters poses quite a problem.

• Tuner Substitution Units (Subbers)

The most common trouble symptoms for which the TV tuner (channel selector) may or may not be responsible are inability to receive one or more channels, weak reception with *snow*, and mechanical problems (which are usually obvious). But it is not always easy to decide whether the fault lies with the tuner, the antenna, or the "chassis." One troubleshooting approach calls for signal tracing with a generator and then checking all voltages applied to the tuner. If still in doubt, you can place the receiver and a good set back to back while substituting (swapping) tuners. A proper tuner substitution unit makes the job much easier and faster (see Fig. 3-15).

A substitution unit is basically an all-channel TV tuner with the necessary terminals and switching to facilitate hookup while temporarily replacing the suspect tuner in the set. It has its own power source (usually batteries) and is not dependent on the set being tested. With an antenna connected to the unit, an IF *station signal* is developed for any desired channel to feed the IF input of the set. The regular shielded coaxial from the set may be disconnected from the receiver tuner and plugged into the unit, or the unit output may be fed to any IF stage via clip leads, which can be helpful in isolating a bad IF stage. Some units have a separate UHF tuner, others a phono jack for use with an external UHF tuner. The substitution unit has an RF gain control in lieu of the AGC normally provided by the set.

Some models have a built-in meter to measure signal strength. If accurately calibrated, it can double as a field-strength meter (FSM). Some models also develop sound and video signals for tracing those stages of a set.

A one- or two-channel alternative to a substitution unit is a dot and color-bar generator as described in Sec. 3-12. Or a substitution unit can be made up from almost any TV tuner having a 40-MHz output.

• Isolation Transformer

A receiver (or test instrument) that is "hot" (its chassis connected to one side of the power line) represents a potential safety and equipment hazard under certain circumstances, depending on which way the power plug is inserted and what is grounded. The problem is overcome with an isolation transformer, which is simply a transformer having a 1 : 1 ratio that is inserted between the line and the set or test instrument in question. Some units are adjustable up to 150 V, which can be helpful in compensating for low or high line voltage or when checking for intermittents. Note that some such items are *auto-*

transformers and do *not* provide isolation. In an emergency you can use two filament transformers back to back with their 6-V windings connected together.

• **Variable Autotransformer**

(Variac) As with a variable isolation transformer, this item is used to raise or lower the line voltage during certain test procedures. But it does *not* provide isolation. An ordinary filament or power transformer can be used for this purpose by connecting the low-voltage secondary coil or coils in series with the primary. Depending on the *phasing*, (which way the windings are connected), the arrangement may be series *aiding* or *opposing* to either raise or lower the output by the amount of the secondary voltage. An important use is in diagnosing intermittents (Sec. 5-13).

• **Field-Strength Meter**

This instrument is basically a small all-channel TV receiver with a meter to indicate signal strength. In fact, any portable set with a few changes can be made to serve the purpose. It is a useful aid for "one-man" antenna installations for locating the best antenna site, for checking and evaluating an installation, for MATV and CATV (community antenna systems) installations, and for antenna *orientation* in particular, except where there are ghosts or interference problems. It is almost a *must* for surveying weak signal areas.

A typical FSM is battery operated, with a sensitivity or range switch for signal levels between 10 μV and about 100 mV (see Fig. 3-15).

• **Pix Tube Brighteners**

(Filament Boosters) This is a small transformer that connects between the CRT base and the socket connector. By raising the heater voltage by about 25%, it can often extend the useful life of a weak tube. It also helps in checking the *condition* of a CRT. If adding the booster results in a brighter raster (increased emission) with improved picture quality, then the CRT is presumed to be weak.

Most brighteners are *autotransformers*. Where a tube has a cathode-to-heater short, an *isolation*-type booster must be used. Every CRT requires a particular type of booster, considering its socket configuration. The current rating of units for large color tubes is considerably greater than for smaller BW tubes. Boosters used with older tube-type sets often had provisions for series or parallel heater arrangements.

• **Picture Tube Degausser**

The purity and convergence of a color set are adversely affected by even the slightest magnetic influence, as from a partially magnetized CRT shadow mask or the nearby chassis. All modern sets overcome this problem with a built-in *degausser coil*. It is located inside the set close to the front of the tube. *Demagnetization* occurs each time the set is turned on when the coil is momentarily energized with ac. Sometimes, however, the built-in degausser is not completely effective and an external degausser must be used (see Fig. 3-15).

A typical service-type degausser consists of a large coil about 12 in. in diameter. It is equipped with a switch and a long cord for plugging into a 110-V outlet. A suitable substitute is the degausser coils removed from an old set. Or a degausser can be made up using about 500 turns of no. 20 wire wrapped with tape. Use of the degausser is explained in Sec. 15-119.

• **Color CRT Gun Disabler**

When evaluating single color fields and making purity adjustments during setup procedures, it is necessary to individually disable the three guns. One method is by adjusting the three screen controls. A better and faster method is with a switching arrangement that selectively disables the guns individually (by overbiasing them) with 100-kΩ resistors connected from control grids to ground. A typical *gun killer* can be made up using three switches and resistors, and clip leads for connecting to the CRT grids. This arrangement is built into many color-bar generators as a convenience.

- **Probes, Leads, and Cables**

Those commonly used for interconnecting equipment, making measurements, and probing circuits are described next.

Clip Leads or Jumpers. It seems you never have enough of these short insulated leads with alligator clips attached, when it comes to making temporary hookups. They are certainly convenient, but never forget the risks involved, the damage that can be done from inadvertent shorts, especially when clipped to PC boards or transistor or IC terminals. Probes fitted with *mini-clips* should always be used when connecting meters and other instruments to closely-spaced circuit contacts.

CRT Socket Extender Cables. With most sets, the wire bundle connecting to the CRT is too short to permit working on the set under certain conditions, and an extender cable can make the task easier. Separate cables (or adapters) are needed with the proper plugs and sockets for all types of tubes. If excessively long or if the control grid wires are too close to the others, you can expect the fine pix detail to be smeared or degraded to some extent.

High-Voltage Probe. This is a *must* for measuring or monitoring the high-voltage dc applied to a pix tube anode, which should always be maintained at the level specified for each receiver. Basically, the probe consists of a *multiplier* resistance of very high value (several thousand megohms) for extending the range of a voltmeter (usually to about 50 kV). The special type of multiplier is housed in a well-insulated probe and handle (see Fig. 3-15), with a ground wire attached. Such a probe can be used with any meter (VTVM or EVM) with an input impedance of 10 MΩ or more. Some probes have their own built-in meter.

Demodulator Probe. This probe functions as a *detector* to demodulate RF-IF carriers so the modulation signals can be indicated on a scope or a meter. It consists of a diode, load resistor, and blocking capacitor connected to the instrument via a shielded cable.

Signal Tracing Probes. A variety of probes fall in this catagory, each for a specific task. All employ a shielded-grounded coaxial with connectors to match the instrument. A basic probe provides a straight-through, *direct* connection. Some probes have a built-in *isolation* resistor so that they will not load the circuit under test. For high frequencies, a *high-impedance probe* containing a low-value capacitor is generally used. For low frequencies, a *low-impedance probe* having a larger capacitor is used. For extremely high frequencies of low level, a probe may have its own preamp. For additional attenuation of a signal, a probe may have a two- or three-position *range multiplier switch*. The use of probes is explained in subsequent chapters as appropriate.

3-21 MAINTENANCE, CALIBRATION, AND REPAIR OF COMMON TEST EQUIPMENT

The TV technician often encounters new and unfamiliar sets, and is constantly beset with problems he or she has never experienced before. The technician soon learns to take them in stride. Yet, unaccountably, when his own *test* equipment breaks down, he shies away and usually ends up sending it out for repairs. There is no valid reason for this. After all, a test instrument is just another electronic device and no more difficult to service than the sets one works on everyday. The following information is intended as a *guide*, an assist for those who wish to attempt their own repair of certain instruments, including maintenance and calibration.

• **Meters**

The VOM. The most common troubles are a damaged meter movement (burned out, bent pointer, fused or distorted torsion springs) due to overload; one or more functions not working; one or more ranges not working or in error; problems involving the switching or zero adjustments; and the test leads and probes, which are subject to considerable abuse.

Where the voltage readings on one range are off, it is usually a bad multiplier resistor, which can easily be replaced with a precision resistor of 1% tolerance or better. Trouble with ac volts may be a bad multiplier, or if *all* ac ranges are affected, a defective rectifier unit. Older meters used selenium stacks, which can be replaced with individual solid-state diodes.

Trouble with the ohmmeter ranges usually means a run-down battery, which makes it impossible to zero adjust. The battery should be checked periodically (under load). A weak or dead battery should be removed immediately to prevent internal corrosion, which can quickly destroy the instrument.

One or more current ranges not working probably mean an open shunt. A replacement can be made using wire from a discarded wire-wound potentiometer or high-wattage resistor shunt. Values are critical and should be wired directly to the switch terminals since their lead resistance is part of the shunt.

Erratic switches are often due to dirty, corroded, or distorted contacts.

Repairs to the meter movement itself can be a tricky business. Some defects may be minor; others, more serious, may call for replacement. A bent pointer can be straightened; but be careful, it may break off or result in other damage. Make sure the pointer does not drag on the glass or scale card. If a pointer hangs up at some point, it could be a damaged jewelled bearing, a speck of dust, but, more likely, one of the two torsion springs that carries the current to the coil. Often these springs develop overlapping loops which can sometimes be corrected by gently blowing. Probing, if necessary, requires good eye-hand coordination, lots of patience, a toothpick or the equivalent, and a sensitive touch. An open coil or spring is generally not repairable.

To check dc calibration, a new flashlight cell measures close to 1.6-V; a car battery a little over 12 V. Readings taken with another *accurate* meter in parallel may be used for voltage calibration.

For ac volts, you can check against a 6.3-V heater circuit or the 120-V line, although both should be considered approximate. The resistance (ohms) ranges can be checked by measuring known (banded) precision resistors, making allowance for their tolerance limits. To check current ranges, connect a test meter in series with another *standard* meter in a suitable test circuit.

The VTVM and EVM. This type of meter uses either a dual triode tube or a pair of transistors in a *balanced-bridge* circuit. The bridge output feeds a dc amplifier, which operates the meter movement. At the input of the instrument is the usual range switch and ac-dc multiplier resistors. It has a high (10 MΩ) input resistance. Most modern meters use a FET for increased sensitivity. A solid-state rectifier is used for ac measurements.

3-21 Maintenance, Calibration, and Repair

When a meter breaks down (which is rarely, with normal usage), a *typical* circuit can be helpful, although the *proper schematic* is always desirable. If a meter is completely inoperative, look for a blown fuse or a bad circuit breaker (assuming it will not reset). Check the power source, whether line operated or batteries. Unlike a VOM, the battery is required for *all* functions, not just the ohmmeter.

REMINDER: _____

Check and replace the battery periodically to avoid damage and possible destruction from leakage and corrosion.

Where every function is in trouble, check circuits common to all, including the FET and transistors.

CAUTION: _____

Extreme care is necessary in the handling and testing of a sensitive and *easily damaged* FET.

Where a faulty meter movement is suspected, proceed as described previously. However, this condition is rare since the movement is *self-protected* against overload.

Where dc ranges only are affected, check the same things as for a VOM: the mode and function switches, the dc multipliers, and the dc cable and probe with its built-in limiting resistor. Where ac ranges are affected, check the switches, the ac multipliers, and the rectifier unit. The same tests as for dc also apply to the ohmmeter function.

Where some or all readings appear inaccurate, check calibration as for a VOM. Check transistors, particularly for leakage if the bridge will not balance. If an older instrument, check the tubes. Adjust calibration adjusters if necessary.

The DVM and DMM. Unless qualified in checking logic circuits, avoid becoming too involved except for the most obvious defects, such as bad test lead, battery, switches, or power supply troubles. If you have reason to suspect a particular IC chip, it can easily be replaced. One good test is for current drain, with a milliammeter connected in series with the dc supply source. Excessive current usually means a shorted or leaky transistor; too little current means an open circuit or that one or more transistors are not conducting.

A *counter*, if available, can be used to check the timing functions. The readout LEDs can be tested with an ohmmeter. The *forward* resistance is normally about 100 kΩ, the reversed reading close to infinity. They can also be tested by applying voltage (of the correct polarity) to see if they light. Most LEDs operate on 1.6 V dc.

• **Signal Generators**

The most common troubles are as follows: for an RF generator, no output or low output, calibration errors, or frequency drift; for AF, low or no output or a distorted waveform, and erratic operation of controls. Consider the nature of the problem and the probable circuits involved, and concentrate your efforts in that area. Start by checking the output cable for an open or ground.

Check the operation of the oscillator stage, the source of the signal. If not working, it could be that the tube or transistor is at fault. If necessary, check applied operating voltages and associated components. If off calibration, verify that the dial is not slipping. Excessive heat and/or poor ventilation are the likely causes of frequency drift, especially with tube-type instruments. Calibrate against an accurate standard (WWV, a radio, or communications receiver). If the oscillator is working, signal trace through to the output. Some generators do not have a blocking capacitor, so look for an output attenuator or control that may be burned.

For *sweep* troubles with a sweep generator, check the *sweep oscillator* and the reactance control circuits.

For loss of AF output, use the same general approach as for an RF signal. Use a CRO to check distortion of both square- and sine-wave outputs.

Crosshatch and dot and color-bar generators can be tested using the same general procedures. If reproduced color bars are of the wrong hue, out of sequence, or drifting, check out the crystal oscillator stages and their AFC control circuits. Most instruments have one or more adjusters to correct for calibration errors.

• **Oscilloscope**

Some of the more common troubles that can develop in a CRO are as follows:

1. *No visible trace:* Check for an open in the power cord or plug or a bad ON–OFF switch. Check the CRT for a possible burned-out heater. Check all CRT operating voltages. Compare voltage readings at the four deflection plates. Where one is lower than the other (and trace is off the screen), suspect a leaky HV coupling capacitor and/or its associated high-value resistor in the output stage.

2. *Dim trace:* May be caused by a weak CRT (loss of emission) or low operating voltages. Check the brilliance (intensity) control circuit.

3. *Trace off center:* As for no trace, it could be a leaky HV coupling capacitor at the deflection plates. Check operation of the V and H centering controls.

4. *No horizontal trace or loss of width:* Check operation of the sweep oscillator and the driver amplifier when used. Check operation of the width control. If triggered sweep, check triggering pulses for proper amplitude and proper operation of the sweep enabling circuits.

5. *Poor horizontal linearity* (which distorts the waveform being tested): Check linearity of sweep circuit waveform with another scope. Troubleshoot the sweep oscillator and amplifier stages.

6. *Horizontal sweep calibration errors:* Look for a bad tube, transistor, or other component in the sweep oscillator stage.

7. *Time/div calibration error:* Same as for frequency errors. Also check the horizontal driver stage.

8. *No vertical trace or low amplitude:* Vertical amplifier is probably at fault. Signal trace from vertical input terminal up to the CRT deflection plates. Look for damaged components at vertical input including gain and sensitivity controls, as caused by checking signal voltages greater than specified. If dual channel, and one channel only is at fault, check out *both* channels, comparing one channel against the other.

9. *Poor vertical linearity or off calibration:* Test as in step 8.

• **Vertical P-P Voltage Calibration Procedures**

Make sure the graticule is properly in place in front of the CRT screen. If the CRO has an internal voltage source for this purpose, use it to check the amount of deflection at the proper setting of the vertical gain control(s). If a calibration error exists, determine a new setting for the controls; record the setting or make a mark on the panel as appropriate. Future use in making p-p voltage tests requires the controls to be set at these markings. Any *external* source of ac can also be used for calibration. If the meter used for checking reads rms, then compute (i.e., p-p = rms x 2.8).

3-22 SUMMARY

1. Most test instruments used for radio repair can also be used (with certain limitations) for TV. Additional instruments are also needed (especially for color TV servicing). Some are expensive. But to the full-time professional, their cost is repaid many times over.

Summary

2. A quality type VOM (one with a basic sensitivity of 50 μA or less) is a desirable adjunct or backup to the regular shop EVM. And even a less sensitive instrument can be useful for general-purpose testing, such as voltage testing batteries, power supplies, and low-impedance circuits; continuity testing; and current measuring. But remember, such meters are easily damaged or destroyed. Unless protected by a fuse or some other device, even a momentary overload can be very costly. The most common mistake is touching the probes to some voltage point while the meter is set to read current or resistance.

3. The basic rules in the use of a meter (particularly a VOM) are as follows: (a) When a voltage is to be measured, *double-check* that the selector is not set for some current or resistance range. (b) Make sure a set is turned OFF before making resistance tests. (c) When checking *current*, make sure the meter is in *series* with the load and *never across a voltage source*. (d) When measuring voltage or current, verify the polarity of the test leads and *start with the highest range*. If leads are reversed, a slight backward deflection may go unnoticed and the temptation is to switch to lower ranges, creating an overload. (e) When current checking, make the connections (and later disconnect them) only with the set turned off. (f) Do not trust the accuracy of resistance readings near either extremes of the scale. Select a range that gives a reading near mid-scale.

4. A VTVM or EVM (its solid-state conterpart) has two great advantages over a VOM or multimeter: (a) It has a high input impedance and draws very little current from the circuit under test. This gives high accuracy when voltage checking high-resistance circuits and extends the resistance ranges while using only a nominal amount of battery voltage. (b) The meter movement is self-protected against burnout. As part of an electronic bridge circuit, any current overload does not go through the movement.

5. The amout of current in a circuit is usually *computed* from the voltage developed across a known value of resistance. Where current is to be measured directly, take all necessary precautions to prevent meter damage. Make sure the set is *off* prior to making connections. *Open* the circuit under test, and observing the correct lead polarity, connect meter *series with the load*. Avoid ''haywire'' hookups where the test leads may accidently ground out. Keep one hand on the ac line cord, ready to remove the power at the first indications of trouble. Select a range at least equal to the anticipated current flow. Avoid switching ranges with the power on. Turn the set *off* before disconnecting the meter.

6. The current range of a basic meter movement can be extended using a suitable *shunt*. Since the shunt must pass all the current (except the small amount required by the movement), it must be of low resistance with a high power rating. Where there is a need to extend the current range of a meter (in an emergency-type situation), the shunt should be connected directly across the movement terminals since the resistance of hookup wire and switch contacts is often greater than the shunt itself. A further word of caution: *Never* open the shunt circuit with current flowing. Should this happen, the movement will be instantly damaged or destroyed.

7. When an extra voltmeter is required, any sensitive current meter can be used by simply adding a series multiplier resistor. An existing voltmeter can be made to read higher voltages with an additional multiplier. The resistance of a multiplier is computed using values of meter current and the desired voltage range. Choose a range where the same scale markings can be used with a suitable multiplying factor.

8. Most progressive service shops now favor the DVM or DMM, and for good reason. With an instant digital readout, it is faster to use than a conventional meter. There is no need to interpret a measurement as is necessary with the calibrated scale on a regular meter. It reads fractional voltages, particularly important in solid-state servicing. There is no need to switch leads for a reversed polarity. The feature called *autoranging* eliminates the need for a range switch. It is self-protected against overload.

9. Many types of signal generators are available; some are designed for a particular task, others are multipurpose. Those commonly used for TV servicing are the following: (a) *AF generator:* It has a fairly high output and covers all audio frequencies. It has many uses: injection testing audio, video, and chroma stages; for gain checking these stages; frequency response and distortion checking; and so on. (b) *RF generator:* These instruments have a fairly *low* output (consistent with signal levels of the stages under test). A typical RF generator covers all RF and IF frequencies, including TV channel frequencies of the VHF–UHF bands. They usually provide a single AF frequency, which can be used externally or for modulating the RF carrier. Such a generator can be used for signal injection testing all high-frequency stages such as the tuner, sound, and video IF strips; gain checking these stages; alignment of tuned circuits; and so on. Some special-purpose generators include the *sweep generator* used for alignment, the *crosshatch generator* for checking linearity and pincushion problems, the *dot generator* for checking and making convergence adjustments, and the *color-bar* generator for checking and evaluating chroma stages. (c) *Signal tracer:* This instrument is used to follow the passage of a station signal (or any internally generated signal) from one stage to another through the set. The *readout* or indicator may be aural (from a speaker) or a meter. Actually, any audio amplifier can be used for tracing AF stages. For tracing RF and IF stages, a typical tracer has several *tuned* amplifier stages. (d) *Oscilloscope* (CRO): Like the meter, this is an indispensable tool for modern-day servicing. It provides an instant display of any recurrent ac signal voltage for observing waveforms, measuring p-p amplitude, and frequency checking. It is particularly useful in signal tracing practically all stages of a receiver. Three special-purpose-type CROs are the *dual trace* CRO, the *triggered sweep* CRO, and the vectorscope. These features are often combined in one instrument. (e) *Diode and transistor testers:* Although most diode or transistor defects can be found with an ohmmeter, a proper tester can make the job easier and faster. Most such instruments can make *in-circuit* tests, although the results are not always conclusive. A simple go-no-go (or quick check) is usually all that is necessary, but some instruments also have the capability for measuring beta and for checking FETs and thyristors. Another use for these instruments is identifying unknown transistor types and identifying unmarked connecting leads. Some instruments called *curve tracers* evaluate transistors by a series of *signature patterns* on a CRO. (f) *Pix tube tester and rejuvenator:* These are two distinct functions often found in one instrument. Pix tube testing is basically the same as for any other type of vacuum tube. An added feature is a life-expectancy test, which can be quite useful. Unless the cathode(s) or a CRT are completely *stripped* and incapable of produc-

ing electrons, emission can often be restored (and the life of the tube extended) by momentarily applying a positive dc voltage to the control grid(s). Internal shorts (which are quite common) can often be burned out by flashing with HV.

10. Many technicians are either too busy or feel unqualified to service their own test equipment. This is unfortunate since most repairs are quite simple and require only the same basic know-how as for any electronic equipment. At the very least, every technician should *maintain* his equipment by repairing cables as needed, replacing batteries, and the like.

REVIEW QUESTIONS

1. Why is it *essential* that a meter used for servicing solid-state equipment have *low* and *accurate* voltage ranges?

2. Explain why a VTVM or EVM gives more accurate voltage readings than a VOM when checking high-resistance circuits.

3. State two reasons why most EVMs and VOMs do not have a current-measuring capability.

4. Although the dc voltage range of a meter may be more than adequate, measurements should not be taken where high-amplitude pulse voltages are present, such as at the output of the horizontal output transistor. Explain why.

5. (a) Compute the value of a shunt resistor required to extend the range of a 50-μA (100-Ω) meter movement to read 1 A full scale.
(b) What type of resistor would be required?
(c) Explain why.

6. (a) It is desired to convert a 1-mA, 50-Ω meter movement to a voltmeter with a range of 0 to 100 V. Compute the *exact* value of the series multiplier that is required.
(b) Would this be a suitable instrument for checking high-resistance circuits?
(c) Why? Explain *why* a more sensitive meter would do a better job.

7. (a) Without actually checking it, how would you know when the battery in a meter is getting weak?
(b) What *serious consequence* may result if it is not replaced?

8. For checking diodes and transistors with an ohmmeter, it is important to know the polarity and amount of voltage at the test leads. Explain why.

9. Name some desirable features of a DMM as compared to a VOM or EVM.

10. Many signal generators do not have a capacitor in series with the output. If not (or when in doubt), add one externally before injecting a signal into a solid-state receiver. Explain this precaution and the possible consequence if it's not being observed.

11. (a) Which has the greater output voltage, an RF or an AF signal generator? (b) Why?

80 TELEVISION TEST EQUIPMENT AND SERVICING AIDS

12. What kind of signal is obtained from an RF generator with the AF, (a) switched on; (b) switched off?

13. What is a signal tracer and how is it used?

14. A demodulator probe is required when signal tracing high-frequency stages with a meter or CRO. Explain why. (b) How does it function?

15. (a) Give four important uses of a CRO.
(b) Where is a dual trace CRO used to advantage?
(c) What are the advantages of a triggered sweep CRO over a basic CRO?

16. What is a noise generator? For what purpose can it be used? What are its limitations?

17. Briefly state what the following are used for: (a) grid dip oscillator (GDO); (b) C/R bridge; (c) sweep generator; (d) function generator; (e) TV stage analyzers; (f) vectorscope; (g) ringers; (h) subber; (i) CRT test jig.

18. Name six tests or functions that can be performed by a diode and transistor tester.

19. Explain the operating principle of a flyback and yoke tester.

20. (a) How does a pix tube tester differ from an ordinary vacuum tube tester?
(b) How is a CRT rejuvenated?
(c) How are internal CRT shorts removed?

21. Explain the use of an isolation transformer.

22. All electronic voltmeters have a very high input impedance. Explain why this is desirable.

23. What CRO circuits would you check for the following problems? (a) No light on the screen. (b) A bright horizontal line but no vertical deflection. (c) Insufficient horizontal trace. (d) An off-center trace that cannot be corrected. (e) A very dim trace but all other functions normal. (f) The CRO is dead and the indicator does not light. (g) A display can be had but it will not lock. (h) Only one display obtained on a dual trace CRO. (i) All displays show indications of hum voltage even with the vertical gain turned down.

24. How would you calibrate a CRO to read p-p voltages?

25. (a) What is a *Lissajous figure*?
(b) How is it produced?
(c) Give a practical use.

26. (a) What is the frequency of the oscillator in a CRO if you see three cycles of a 5-kH signal? Explain why.
(b) What frequency is required in order to view 10 waves of the 60-Hz line voltage?

27. Why is it not possible to view a nonrecurrent waveform on a CRO?

28. Why must the horizontal sweep of a CRO be *linear*?

29. (a) Why are meter shunts such *low* resistance?
(b) Why are voltage multipliers such *high* resistance?

30. What are the *x, y,* and *z* axes of a CRO display?

31. Explain the purpose of the three instruments required for aligning tuned circuits using the seeep alignment method.

32. Calibration accuracy of a sweep generator is not too critical, but a marker generator must be very accurate. Explain why.

33. Prior to aligning tuned circuits, a signal generator should be warmed up for 20 minutes or so. Explain why.

34. What advantages has a dual trace CRO or a triggered sweep CRO over an ordinary CRO?

35. What is the rationale of the life-expectancy test on a CRT checker?

36. A typical meter can measure up to about 1 kV; higher voltages require an external HV probe with a built-in multiplier. Why is this necessary?

37. What two factors of a CRO determine the highest frequency that can be displayed?

38. Why is a current meter so susceptible to damage?

39. What are the essentials of a good CRO?

40. Explain the desirability of having the following features on a DVM: (a) autoranging; (b) overranging.

41. A CRO is calibrated for 2 V per division of the squared graticule.
(a) What is the *peak* value of a display that covers 4 divisions?
(b) How would you make the CRO read 10 V per division?

42. (a) What is meant by ringing a coil or transformer?
(b) What trouble condition shows up with such a test?

43. What is the purpose of electronic switching in a dual trace CRO?

44. Explain the use and limitations of a CRT test jig.

45. Give the advantages of an EVM over a VOM.

46. Explain *how* the CRO traces out a waveform display.

47. (a) How would you use a CRO to determine the approximate frequency of an ac voltage? (b) to accurately determine the frequency with the aid of a signal generator?

48. Give four uses for an RF signal generator in servicing TV.

49. (a) To view a greater number of cycles on a CRO, do you increase or decrease the sweep frequency?
(b) Explain why.

50. What is the purpose of a (a) color-bar generator, (b) dot generator, (c) crosshatch generator?

51. What is the main difference between a triggered sweep CRO and an ordinary CRO?

52. Explain the purpose of the following CRO controls: (a) frequency or time base con-

trol, (b) vernier, (c) polarity, (d) sync selector, (e) horizontal phasing, (f) trigger level, (g) trigger mode, (h) volts/cm.

TROUBLESHOOTING QUESTIONNAIRE

1. An ohmmeter will not zero adjust. State several possible causes.

2. The output attenuator of a signal generator is burned open. State the probable cause.

3. One voltage range on a meter is inaccurate. State the probable cause.

4. All ac ranges of a VOM are inoperative. State the possible causes.

5. Give several possible causes when the display on a CRO is completely off the screen.

6. A CRT with 6 V on its heater has low emission. Emission improves when voltage is increased to 7 or 8 V. What is wrong and what can be done?

7. After several cycles of rejuvenation, a CRT becomes progressively weaker. Why did this happen and what is the prognosis?

Chapter 4

THE TROUBLESHOOTING APPROACH

4-1 INTRODUCTION

Anyone with an understanding of electronic fundamentals and who has become reasonably proficient in the servicing of radio and other electronic devices should have no difficulties troubleshooting TV. Comparing the two types, monochrome (BW) sets are the simpler, but once the fundamentals are mastered, color receivers may also be taken in stride.

A typical modern TV may contain some 50 to 100 transistors, about half that number of diodes, and up to a dozen or so ICs. But despite its complexity, TV in some respects can be easier to service than radio. This is because the pix tube serves as an indicator for most of the set's functions, in addition to providing a visual display of trouble symptoms that arise.

What can be said of the technician called upon to service this equipment? Besides a good understanding of circuit functions and test procedures, he or she must possess or develop many unique qualities and skills. First and foremost is an analytical mind capable of accurately assessing problems, with powers of concentration for more difficult and perplexing ones (see Fig.

Fig. 4-1 Concentration.

4-1). A good technician must also have the tenacity to follow through in a logical manner when the way seems clear and the flexibility to make an about face when on the wrong track. The technician must also be observant, ever alert to spot the slightest clue, have skills of manipulation and dexterity when handling microminiature components, have the ability to make decisions, and so on.

How simple or how difficult any given servicing problem may be depends on the experience

84 THE TROUBLESHOOTING APPROACH

and expertise of the technician. Problems that are relatively simple and straightforward may be overwhelming to the novice and require a great deal of time and effort to discover and correct. The same troubles, to one of greater experience, will be handled with ease and in a fraction of the time. But such skills must be acquired. And the key to that is knowledge and application. The learning of circuit functions, troubleshooting procedures, and techniques is learning by *doing,* by working on sets at every opportunity. And as time goes on, as the beginner develops more and more confidence in his abilities he will find those problems that once seemed difficult have become simplified, even commonplace.

This chapter (along with chapter 5) deals mostly with *generalities;* . . . troubleshooting procedures that apply to the *overall* receiver. Because of the diversity of subject matter, and lest the reader be overwhelmed at this point by the rather large volume of material covered, it should be read *once* at this time, for a grasp of *general* procedures; and used *later* for reference as needed.

4-2 SOFTWARE IN SERVICING

Although a schematic is not required for each and every repair job, a proper schematic along with other pertinent service data is usually a big help. For many troubles, such information is absolutely necessary. Most shops consider schematics to be part of the stock-in-trade, and equally as important as their test equipment. Diagrams and servicing literature are available from a number of sources. Most popular are *folders,* packages that contain, in great detail, all the essential information on all makes and models of sets. In addition to the schematic, this includes photos and drawings showing the location of all components, resistance and voltage charts, CRT setup procedures, and alignment information. The schematics also show normal waveforms and p-p voltages that should be obtained at all key circuit test points.

Circuits and other information are also available from libraries and the receiver manufacturers. If an exact circuit is not available, you can often get by with another (preferably of the same make) since there are often similarities between different models produced by the same manufacturer within a one- or two-year span.

Lacking such information, you are "on your own." It may take longer, but you can always make up your own partial schematic diagram as you proceed. As an apprentice, you can consider it an invaluable part of the learning process. Actually, it is not too difficult since most modern receivers have the circuit printed directly on the PCBs.

4-3 TROUBLE SYMPTOMS AND ANALYTICAL REASONING

The first step for any troublesome set is to consider the customer complaints. Observe the trouble symptoms, *define the problem,* and make a tentative judgment. Although sometimes vague and misleading, do not discount the nontechnical remarks and observations of the set owner, which often provide valuable time-saving clues.

When first turning the set on to observe and consider the complaint, be careful and alert, ready to pull the plug at the first signs of smoke, sparking, or overheating that can result in further damage. If the screen does not light up in a reasonable amount of time, switch it off and consider the possible causes for failure. See if the CRT filament is lit. Look for indications of an overloaded horizontal output stage, including the flyback transformer and other parts in the HV section. Be thorough. Bear in mind that (more often than not) there are *multiple* troubles to be noted.

In deciding on the most likely areas of the set involved for any particular troubles, a knowledge and understanding of the overall block diagram (Figs. 2-1 and 2-5) is absolutely *essential*. Until memorized, it's a good idea to keep such a block diagram constantly at hand.

A word of caution at the start. Although one should cultivate and learn to trust one's judgment, do not be hasty in jumping to conclusions or too anxious to begin testing. The experienced technician takes time to consider all possibilities. Clear thinking and logical reasoning at this point can often save time later and prevent getting off on the wrong foot. Troubleshooting involves extremely complex thought processes, involving the brain and the stimuli it receives from the various sensors (sight, touch, smell, etc.), which in turn suggest the appropriate action for any given circumstance.

Start by asking yourself questions that relate to your problem. Such as: Is there a raster? If so, we at least know we have both LV and HV and also that the pix tube is functioning. Is there sufficient brightness? If not, the HV or focus voltage may be low, the bias too high, or the CRT itself is defective.

Does the raster have enough height and width, with good linearity? If not, one or the other sweep circuits are probably at fault. If both, then the LVPS. Remember that serious troubles in the horizontal output kill the HV and therefore the raster.

Critically examine the picture on the screen. Is there adequate contrast, indicating the video amplifiers and all previous stages are doing their job. Does the pix have good detail and resolution with freedom from smearing and other problems?

And what of the sync, both vertical and horizontal, especially during warmup and when switching channels? There are many possible causes for trouble: the sweep oscillators, sync stages, the video amplifiers (even when there is good pix contrast). One way to judge the relative strength of blanking and sync pulses (up to the pix tube input) is to observe how the blanking and sync *bars* appear on the screen (See Fig. 10-8.).

And color? If color is absent, *all* chroma stages (including the killer stage and the subcarrier oscillator) are suspect. Where there is color, are all primaries and complementary hues present in their proper proportions and with adequate saturation? Does the tint control cover its full range with good flesh tones obtained at its midpoint? For color *sync* problems, the automatic frequency and phase control (AFPC) circuits may be in trouble or in need of alignment.

4-4 OTHER CONSIDERATIONS

There are many other things to consider when evaluating trouble conditions as they relate to the overall block diagram. Remember that some stages handle only one kind of signal; other stages carry several, such as the tuner and IF strip, which simultaneously pass *all* signals coming from the station, that is, sound, video, sync, and color information.

• Multiple Troubles

This is usually a problem with older sets. There are two kinds: multiple *faults* or *defects*, and multiple *symptoms*. The distinction between the two should be clearly understood.

Multiple *defects*, for example, may exist in two or more completely unrelated stages, each resulting in a different *symptom*. With multiple defects, each must be treated *separately*. The logical order of priority is to first take care of any *raster problems;* then the *BW pix;* then *sync;* and, finally any *color* difficulties, *in that order.*

86 THE TROUBLESHOOTING APPROACH

Multiple *symptoms,* on the other hand may result from *one* specific defect. In this case, replacement of that one bad part, making adjustments, or whatever will clear up *all* the symptoms. A classic example is an open filter in the LVPS (the common denominator of all stages). Any problems with the B voltage can create a *variety of symptoms,* such as a poor raster, weak pix, unstable sync, color problems, and so on.

Whether or not multiple symptoms are the result of *one* or *more* faults is not easy to determine without further testing.

4-5 CONTROL ADJUSTMENTS

During analysis (and at later times in the troubleshooting process) it is sometimes necessary to adjust the various controls and adjusters in order to reach certain conclusions. Such adjustments should be considered preliminary and tentative at this time. One must know what to expect as each control is adjusted. Practice beforehand on a set that is in proper operating condition. Controls and adjusters are identified in Sec. 2-3 through 2-10. Detailed adjustment procedures are described in subsequent chapters as appropriate. At this point, a few rules and guidelines are in order:

1. Don't tamper with adjustments that have not been identified.

2. Don't try adjusting (aligning) tuned circuits this early in the procedure.

3. Indiscriminately adjusting *several* controls or adjusters can compound your problems, mask the original symptoms, and add to the problem of diagnosing.

4. For symptoms indicating a weak CRT, try adjusting KINE BIAS (master brightness control).

5. Make sure the service setup switch is in the NORMAL position.

6. If there is excessive contrast with sync instability or no pix or weak pix with excessive snow, try adjusting the AGC threshold control.

7. If there is no color or weak color, try adjusting the killer threshold control.

8. Expect some overlapping interaction when adjusting the vertical controls (height, linearity, and vertical hold). The *critical* adjustment of *all three* has a direct bearing on sync stability.

9. Good focus may not be possible until all abnormal conditions of the horizontal sweep, the HV, and CRT operation (including the HV adjustment) have been corrected.

10. If there is no raster or excessive *blooming,* back off on the brightness control and/or the three screen controls until the trouble has been found and corrected.

11. When a control does not act normally (erratic, ineffective, best results are obtained at one extreme setting, etc.), the control itself may be at fault, or it may provide a clue as to *which circuit* is in trouble.

4-6 SAFETY

This is a subject too often ignored or lightly dismissed until one is faced with some unpleasant, even dangerous situation. It is smart to be safety conscious and aware of common hazards and to have a knowledge of first aid procedures. There are two kinds of hazards, those regarding *personal* safety, and those involving damage to the TV receiver components.

• **Safety Hazards and Precautions**

The wise and responsible technician always observes safe and sensible practices as they apply to himself and his co-workers, with concern for the safety of the user of the equipment he is repairing. Whenever possible, avoid working alone and in seclusion, especially when engaged in potentially hazardous tasks, such as pix tube handling or working with the HV. Never allow small children near sets being serviced. Other hazards to be aware of are discussed in the following.

Electrical Shock. Avoid physical contact with the HV, especially in color sets where there is 25 kV or more. Don't put too much trust in the insulation of the HV anode lead; it could be cracked. A pix tube can retain a high voltage charge for long periods. Always discharge the tube before handling. The danger here is from a shock-produced reflex reaction, such as dropping the tube.

Even the lower, 120-V ac line voltage and the dc B voltage of a set (which may be as much as 400 V or so) can also give you a bad shock. One precaution is to always keep one hand behind you when taking measurements in a live chassis with the danger of inadvertently touching some "hot spot." When working on problems that do not require a raster, it is a good idea to disable the HV. A quick method of doing this is to remove the screws securing the case (collector) of the horizontal output transistor to break the circuit.

A particular hazard is the power line if working near water pipes or grounded appliances or while standing on cement floors in damp weather. Metal-topped benches can also be bad. Good practice dictates the use of an isolation transformer, especially where a chassis or the test equipment is electrically hot.

Avoid "haywire hookups." The good technician seldom gets a shock. If you find yourself being shocked too often, then you know you are being careless and taking chances. It is time to practice safety.

For the safety of the set owner, always check for a possible ground or leakage between the 120-V input and the chassis.

Working on antenna installations can be dangerous. A potentially lethal situation is where a mast may accidentally contact nearby power lines.

Pix Tube Breakage. The danger of a pix tube *implosion* and injury from flying glass is ever present and very real. Fortunately, if proper precautions are observed, such breakage is rare. Avoid tapping or accidentally striking the CRT when working on the chassis or replacing the tube. The greatest danger is breakage of the face plate.

Burns. Avoid grasping high-wattage resistors, power transistors, and supposedly unplugged soldering irons and the like. When soldering, take care that molten solder does not fall on the skin or splatter into your eyes.

Cleaning Solvents. Use only approved chemicals (*never* use carbon tetrachloride which is highly toxic). Ensure good ventilation when using aerosol spray cans and don't breathe the vapors. If a chemical enters the eyes, flush immediately with water and call a doctor.

Fire Hazard. Don't leave an exposed chassis operating unattended for long periods, especially near combustibles. Poor workmanship in soldering and wiring, sparking at a bad power line interlock, a bad ON-OFF switch, and faulty "instant-on" sets are the cause of many fires. Never use an overrated fuse or make a permanent jumper across a fuse, circuit breaker, or a heater circuit fuse link.

X-Radiation. X-radiation is potentially hazardous. Make sure the HV is not excessive and that all shields and baffles are in place.

Physical Strains. Avoid lifting and manhandling heavy chassis and pix tubes by yourself. When lifting, keep object close to the chest, using *thigh* instead of back muscles.

- **Damage to Components**

Components, especially solid state, can be easily damaged or destroyed.

88 THE TROUBLESHOOTING APPROACH

1. Never spark the HV as is commonly done with tube-type sets. If there are signs of arcing, turn the set off *immediately;* *then* consider the problem.

2. Don't probe a live circuit with a metallic screwdriver. When taking voltage readings, be careful the probe itself does not create a short.

3. Don't indiscriminately bridge substitution capacitors across live circuit points. A charged capacitor (especially electrolytics) may instantly burn out one or more small transistors or ICs. Always discharge a capacitor first and connect only with the set turned *off.*

4. Don't operate the set unless all heat sinks are in place and the chassis has adequate ventilation.

5. Don't check transistors with an ohmmeter unless you know your meter. The meter's internal battery may destroy low-power transistors and *FETs* in particular.

6. A strong signal from a signal generator can destroy transistors, ICs, and FETs. Keep the generator output at the lowest possible level.

7. Connect the signal generator with the set turned *off.* Connect the ground lead *first* and remove it *last.*

8. Avoid circuit *disturbance tests* and replacing components with the set turned on.

9. Use great care in handling delicate components. Be careful working around certain fragile components like exposed coils. Pix tube base pins are also vulnerable. Be careful when probing transistors, diodes, ICs, and other easily damaged components.

10. Make sure automatic shutdown circuits are operating properly.

11. When turning on a set during testing, keep one hand on or near the *cheater cord.* Its the quickest way to disable the set if things go wrong.

12. Breakage of the CRT can have serious consequences. Be careful not to strike the glass with any metallic object. Breakage of the tube bell or faceplate can produce a devastating implosion. For proper handling of CRT, see Chapter 15.

13. With transformerless sets and test equipment where the chassis is "hot," use a line isolation transformer whenever possible.

4-7 DEVELOPING A PLAN OF ACTION

Costly and sophisticated test equipment is fine, and the human brain has no equal, but no amount of reasoning alone will restore the operation of a troublesome receiver. The reasoning must be converted into some form of activity, such as performing various tests and the like. Sometimes the initial tests lead directly to the trouble, sometimes not. But each test provides new information and stimulates fresh ideas for additional testing, a new approach, or a change in direction, which ultimately leads to a solution.

Every troubleshooting problem represents a distinct and different set of conditions and no two technicians will approach a problem in exactly the same manner. Depending on circumstances and other factors, it is always important to have a general *overall* plan of attack, that is, a *troubleshooting procedure.* A suggested plan is shown in Fig. 4-2 in which, depending on the symptoms, specific actions are taken, step-by-step in a logical sequence.

In general, the main steps in troubleshooting are (1) analyze the trouble symptom(s), (2) inspect, (3) make tests to isolate the troublesome stage or section, and (4) making additional tests to pinpoint the defective component(s) or circuits. This chapter deals only with the first three, to localize the trouble to one particular area of the set. Chapter 5 continues from there to zero in on the precise cause and make the necessary repairs (see Fig. 5-2).

The procedures as described in the following paragraphs are *general,* as applicable to all stages of a set, and are intended only as an introduction to the fundamentals of troubleshooting the overall receiver. *Specific* and detailed procedures and troubles peculiar to individual stages are contained in later chapters as appropriate.

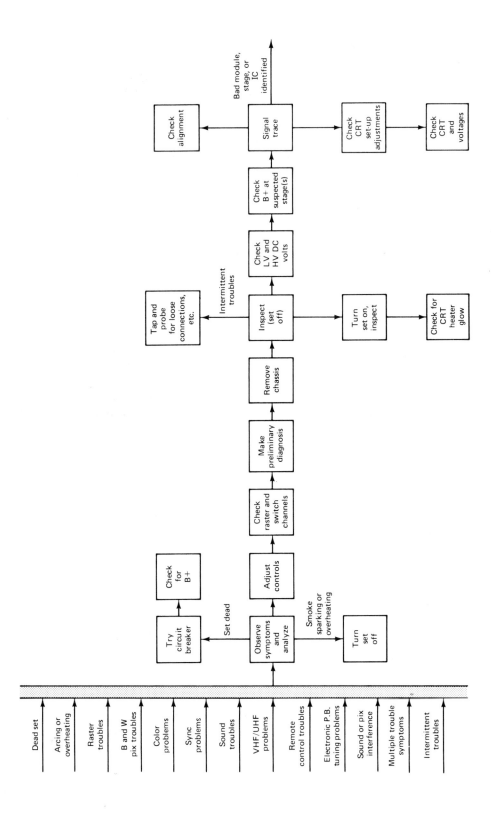

Fig. 4-2 Localizing the problem.

4-8 INSPECTION

The first logical step in the troubleshooting sequence (after analyzing the symptoms) is to make a thorough *inspection*. Such an inspection often brings to light the more obvious conditions that can lead directly to the *cause*, eliminating the need for much unnecessary testing. Inspection is normally performed in several ways: first, without removing the chassis and with the set turned off, and then with the set turned on. Finally, the chassis is removed if that becomes necessary.

Each inspection involves full use of four human senses: sight, sound, smell, and touch. Make written notes on any abnormal conditions found, rather than trusting them to memory. While inspecting, familiarize yourself with the chassis layout, noting if possible the location of stages and the identity of major components for future reference. Brush or blow away any accumulation of dust that may hinder the inspection.

Where chassis removal is necessary, first make sure that all necessary plugs and cables are disconnected (including the speaker) and that all front panel knobs have been removed. Be careful not to strike the picture tube or disturb the critical setup adjustments on the neck. During inspection, keep all four senses alert for the conditions described in the following.

• **Sight**

Your eyes, your greatest resource, can often spot things, sometimes by sheer accident, that can lead you straight to the trouble. For example, a broken or charred resistor is one obvious result of a current overload. Scorched insulation and carbonized deposits may be caused by arcing, which may be visible with the set turned on. Look for melted wax deposits, especially under transformers, as a clue to overheating. If suspected, check for HV arcing or corona around the flyback transformer and HV components with the room darkened. Also look for visible signs of smoke when the set is on.

Inspect PCBs for cracks and other damage. Resolder any questionable soldered connections, especially at terminal points and component tabs that are subject to stress. When the symptom is no raster, see if the CRT heater(s) lights up.

• **Sound**

Listen for the crackling, popping sounds of arcing or the telltale *hiss* of corona. Turn off all lights to localize the source of arcing in a darkened room. With the volume turned up, note any mechanical vibrations, which may be caused by loose cabinet or chassis items or a bad speaker. Unusual sounds from the speaker (like hum and sync buzz) can be indicative of certain troubles.

• **Smell**

Be alert to the acrid odors of burning insulation. Learn to recognize the characteristic odors given off by certain overheated components, such as transformers, electrolytic capacitors, and resistors. The sweetish smell of ozone is a sure indication of corona discharge, which may be all but invisible.

• **Touch**

Become familiar with the *normal* operating temperature of various components. A light touch with the hand or finger is sufficient to detect overheating. After operating for a time, power transformers, power transistors, and high-wattage resistors normally dissipate a fair amount of heat, but watch for overheating—as well as the reverse. Common clues to trouble are a high-wattage resistor or a power transistor that runs *cold* or a low-wattage resistor or small transistor that gets *hot*. A capacitor that runs warm is a sign of trouble (capacitors dissipate very little energy); frequently this is caused by internal arcing. Sections of a yoke or flyback that overheat often indicate shorted windings. Of course, such touch testing of HV components should be made *after* the set is turned *off*.

4-9 LOCALIZING TROUBLE TO THE STAGE OR MODULE LEVEL

Usually, after considering the symptoms (as they relate to the overall block diagram), a decision can be made with reasonable certainty that localizes the trouble to a specific section of the receiver. Before a *final* diagnosis is possible, however, certain tests must be performed, first to verify (or disprove) such reasoning, and then to further localize the defect to a particular stage or module. In some cases, only a few quick checks may be necessary for verification. At other times such testing may become quite involved and time consuming, especially where there are multiple troubles or the symptoms are confusing. And there will be times when you find yourself on the wrong track, when test results fail to back up your reasoning; it then becomes necessary to call a halt and reassess the problem from another angle.

It must be emphasized that the object at this time is *not* to pinpoint the exact cause, the defect itself, but rather to *localize* the fault to a particular stage or area. Sometimes this can be done with voltage, current, and resistance tests, or by signal tracing, or by a combination of these.

A good place to start (especially if the set is dead) is with the LVPS. A couple of preliminary measurements will establish whether or not there is B voltage. And if there is no raster, check for HV at the CRT. When the raster is good but there are pix problems, *signal tracing* may be called for.

Signal tracing (as described in Sec. 4-10) is a means of localizing troubles to a particular stage where the normal signal flow is either interrupted, attenuated, or distorted. There are two ways of doing this: (1) by *injection,* where a signal from a generator or other suitable signal source is fed to the input and output of the various stages, or (2) where a CRO or other type of indicating instrument is used to follow the stage-to-stage progress of *station signals* or those generated *within the set.*

Normally it is not too difficult to locate a *dead* stage where a signal is *lost*. The greater problem is with a *weak* stage or when the signal has been degraded or distorted. In other words, the less obvious the symptoms or the *closer the test results to the norm, the more difficult the servicing problem.* Under these conditions, testing must be more precise, with greater reliance on detail and slight deviations that otherwise might be considered insignificant.

Another thing to remember is that the defect itself may not necessarily be found in the stage where troubles are first observed. An inoperative IF strip, for example, may not be working because the AGC is at fault. Similarly, when direct coupling is used, where the operation of one stage depends on another or a series of stages, the defect may turn up far afield from the starting point.

For localizing troubles, the functional areas of a TV receiver, in general, may be categorized as follows: the *power supplies* (both LV and HV), *signal circuits* (including *those received* and *those generated within the set*), the pix *tube* with its associated components and related circuitry. For the purpose of localization, the important tests are briefly described next. *Detailed procedures* for *specific stages* and *functional areas* are included in subsequent chapters as appropriate.

REMINDER: _____

If there is no need to observe the pix tube when making certain tests, it is often best to disable the HV. And remember the following sequence of priorities: first the raster, then the BW pix, sync, sound, and the color pix, in that order.

• **Power Supplies**

Where the LVPS is suspect, check for ac at various points between the power plug and the rectifiers. Check for dc at each output including the regulator circuit. Compare dc voltage readings with those values shown on the schematic. Watch for signs of overload. For details, see Chapter 6.

Where the HV is suspect, check for HV dc at the CRT anode. Check focus voltage at the CRT socket. Compare readings with those specified.

• **Pix Tube**

Where there is a raster problem, verify its condition. Check HV and all operating voltages at the socket. Adjust HV, kine bias, and focus voltage as necessary. If it is a color CRT, compare measurements and adjustments for each gun. Make setup adjustments as required.

REMINDER: _____

Many conditions can emulate a bad CRT, so don't be hasty in condemning it. It is often possible to use the HV and/or the CRT of one set to check another by placing and interconnecting them back-to-back.

• **Signal Stages**

Localize the faulty stage by signal tracing by injection using one of the methods described in the following paragraphs. Circuits that handle the station signals are the tuner, all video and sound stages, chroma, and sync stages. Internally generated signals include the sweep voltages in the deflection circuits, the subcarrier signal(s) of a color set, and keying or gating pulses for control circuits such as the color killer, AGC, and AFC.

• **Oscillator Stages**

Common test methods include (1) dc voltage checks, (2) checking for a generated signal using an ac EVM or a CRO, and (3) signal substitution from the same stage of another set or a suitable CW signal from a generator.

4-10 SIGNAL FLOW AND STAGE CHECKING

The object here is to localize trouble to a particular stage or section. The four basic signal tracing methods are as follows. The method chosen under any given circumstance depends in part on which section of the set is involved, the kind of equipment available, and personal preference.

    1. By *injection:* a signal from a generator (or other suitable signal source) is introduced at various circuit points.

    2. Tracing the progress of *station* signals or locally generated signals using a suitable indicator such as a CRO or an ac meter.

    3. A variation of methods 1 and 2, using an indicator to follow a generator signal from stage to stage.

    4. What is sometimes referred to as *signal snatching:* signals from one section of the same (or another) set are used to check other stages. The speaker and pix tube are used as indicators.

4-11 STAGE CHECKING BY INJECTION

This method (see Fig. 4-3) requires a signal generator and an indicator. The kind of signal used in each instance depends on which section of the set is involved, each stage requiring a particular type of signal to *simulate* waveforms normally handled by that stage. In general, an RF signal (of the proper frequency range) is required for tracing high-frequency tuned stages; and an AF signal (of any convenient frequency) is used for most other stages. A noise generator (Sec. 3-4) can be used for *all* stages (RF, IF, AF, chroma, etc.). There must also be some means of indicating *if* (and *how well*) a signal is getting through a particular stage. A CRO or ac EVM is commonly used for this purpose. Other indicators that can be used are an aural type signal tracer (Sec. 3-5), TV analyzers (Sec. 3-13), or

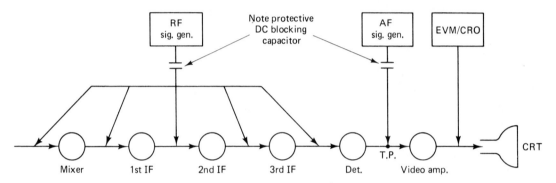

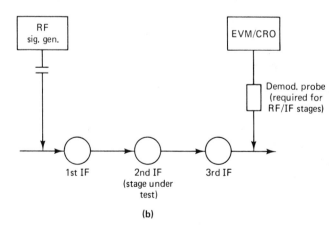

Fig. 4-3 Localizing trouble by signal injection: (a) overall gain checking; (b) single-stage gain checking (typical).

the speaker and/or pix tube of the receiver itself. The CRO or EVM when used as an indicator, may be connected at the output of the stage (or stages) under test or somewhere beyond. With a signal fed to the input of a stage and the indicator (at the output of the stage) showing a normal response, the stage can be considered trouble free. No indication or a weak indication means the signal is lost or attenuated at some point within the stage. We can then assume that the trouble is localized to this particular stage.

There are two methods of signal tracing by injection: (1) The generator may be connected to the receiver input (or the input to a stage or series of stages) while the CRO or EVM (which indicates the generator response) is moved from point to point. (2) The indicator is connected at the output of a stage or stages or some convenient point downstream, and the generator signal is alternately injected at various points.

Since some sections of a set (such as an IF strip) have two or more stages, the question is where and how to begin. It is often a matter of preference. If we start by introducing a signal at the receiver (or stage) input and working *forward* (in the direction of signal flow), the indicated response can be expected to get progressively *weaker* as we approach the output, since there are fewer stages involved. Conversely, when working backward (toward the receiver input) with an injected signal, the response should get stronger. When there is trouble, and *no* response, the faulty stage will be the last one checked prior to obtaining a normal indication. As opposed to the preceding, when the point of signal injection and only the indicator are moved, working forward with the indicator should show an increasing response, and working backward (toward the set input), the response will become weaker.

As an example, consider the IF strip shown in Fig. 4-3. If there is no indication when signal is fed to the first and second stages (with the indicator at the output), but the response is normal when injection is made to the third stage, then the second stage is presumed to be bad. If injection to the second stage gives a response, but there is no response with injection to the first stage, then the latter is the faulty stage.

The normal amount of gain or voltage amplification to expect from a given stage depends on its function and other factors. Such information is sometimes shown on a schematic. In general, the small, high-gain RF and video transistors in a set can ber expected to develop a much higher amplification than the power output amplifiers.

The voltage gain of any stage is the amount of signal at its output compared to that fed to its input as measured with an ac EVM or a calibrated CRO. With signal injection, signal *overdrive* can create problems, such as distortion, reduced gain, and other misleading indications, as well as the risk of damaging transistors. Consequently, use the *least* amount of signal that produces an indication. The strength of signal can be varied with the generator attenuator control(s) or with changes in coupling between the generator and the stage input. Where possible, use the lowest (most sensitive) range on the indicator (the EVM or CRO), except where there are problems of hum pickup and instability. This ensures that the generator output will be kept to a minimum. In general, *weak* signals are used for the *low-level* high-gain RF and IF stages, and *stronger* signals for injection at power output stages and at the pix tube input.

In conclusion, a few reminders: All connections are made with the set and the signal generator turned *off*. To avoid damage, always use a blocking capacitor series with the generator output cable. Always connect the ground lead *first* and remove it *last*. Be careful that bare test probes and alligator clips do not create shorts, especially with PCBs and ICs. When an IC is used, make sure you connect the generator and the indicator to the proper ter-

4-11 Stage Checking by Injection

minals When testing by signal injection, the antenna leads should be disconnected to prevent station signals or interference from influencing the readings.

The kinds of signals used in tracing the different stages of a receiver are described next in Sec. 4-10.

• **RF Generator**

Included in this category are the general-purpose generator (see Sec. 3-4), the function generator (Sec. 3-12), the sweep generator (Sec. 3-12), and the marker generator. An RF generator is used for tracing the tuner, video and sound IFs, and the chroma AFPC stages.

For tracing the RF/mixer stages of a tuner, the generator frequency is determined by the channel selected. Because of the wide bandpass, choose a frequency approximately midway between the limits, for example 57 MHz for channel 2. The signal must be modulated using either the internal AF modulation or some external source. The CRO or EVM (indicator) is preferably connected at some point beyond the video detector; if ahead of the detector, a demodulator probe must be used. With the generator attenuated to a *very low* level, the signal is injected to various circuit points such as (when working forward) the antenna terminals, input of the RF stage, and the mixer input.

When tracing the video IF stages, tune the generator to the approximate center of the IF bandpass (around 43 MHz, depending on the set). The indicator is connected at the detector output or beyond. If desired, the bars produced on the pix tube by the generator modulation can be used as a relative indicator; the *darker* the bars are, the *stronger* the indication, and vice versa. The number of bars as determined by the modulation frequency has no significance for this purpose. To recognize variations in the response, the bars must be made to appear *gray* prior to each test by adjusting the brightness and contrast controls as required. When signal tracing forward in the direction of signal flow, inject the signal first at the mixer and then at the input of each successive IF stage. Unless very low level signals are used, the AGC should be disabled by *clamping* with a *bias box* or by adjusting the *threshold control*.

When tracing the sound IF section, tune the generator to 4.5 MHz with modulation turned on. The indicator may be connected anywhere between the detector output and the speaker. The speaker response to the generator tone can also be used as a relative indicator if desired. The receiver volume control should be turned up.

When tracing the AFPC section of a color set, tune the generator to 3.58 MHz. For the bandpass amplifier(s), the indicator may be connected to the demodulator outputs or the control grids of the pix tube. The bars on the CRT may also be used as a relative indicator, as previously explained. When checking the color burst amplifier, connect the indicator to the color phase detector.

REMINDER:

To pass a signal, the bandpass amplifiers must be *turned on* by the killer stage or by adjusting the killer control.

• **AF Signal Generator**

An AF signal is used for tracing the following stages: audio amplifiers, video amplifiers, chroma amplifiers, sync stages, and the vertical and horizontal sweep circuits. The injection signal may come from a number of sources, including a variable frequency AF generator, or the AF modulation signal of an RF generator. Since the former may have a fairly high output, keep the attenuator turned down to avoid overloading and damage to transistors. The frequency used is not important.

For tracing AF stages, inject the signal at

various points between the sound detector and the speaker. The indicator may be connected to either the output stage or the speaker itself. The generator tone from the speaker provides a relative indication of the response. (Signal *strength only* can be determined with this test. *Distortion* testing involves waveform checking with a CRO, or listening to music).

For tracing video amplifiers, inject the signal at various points between the video detector and the CRT input, where the indicator is connected. If desired, the bar pattern on the screen can be used for a relative indication. The receiver contrast control should be turned full on and the brightness set to get *gray* bars.

For tracing color-difference amplifiers and matrix circuits, inject the signal at various points between the demodulator outputs and the CRT control grids, where the indicator is connected. The bar pattern can also be used.

For tracing sweep circuits, inject the signal at various points between the sweep oscillators and either the vertical yoke or the input of the horizontal output stage, as applicable, where the indicators are connected.

For tracing sync stages, a pulse generator is preferable, but a sine- or square-wave signal will serve the purpose. Inject the signal at various points between the point of sync takeoff in the video amplifier and the vertical integrator and/or the horizontal AFC, where the indicator is connected.

NOTE: _____

It may be necessary to disable the sweep oscillator(s) to prevent their signals from influencing the response.

- **Dot, Crosshatch, Color-Bar Generators**

Under certain circumstances a crosshatch generator (see Sec. 3-12) can be used in lieu of an AF generator. A dot and color-bar generator (Sec. 3-12) (although primarily for checking convergence and evaluating the chroma stages) can also be used for signal tracing. The indicator may be the pix tube bar pattern or a vectorscope display, as desired.

- **Other Signal Sources**

Other kinds of signals can often be used in the absence of a regular signal generator, for example, the *noise generator* (Sec. 3-4), whose broadband signal can go through *any* stage. Substituting for an AF signal, a 60-Hz signal voltage is available from the hot side of a tube filament (heater) circuit or the 120-V line or from the vertical sweep circuit, provided a blocking capacitor is used in series with the test lead.

Any convenient AF amplifier can be readily converted into a test oscillator by simply adding a small positive-feedback capacitor from the output stage back to the input. Signals of different amplitudes can be obtained by tapping in at various points of the test oscillator. As before, a series blocking capacitor is essential.

4-12 SIGNAL TRACING THE STATION SIGNALS

No signal generator is required with this method. The antenna is connected to the set, and a suitable indicator is used to follow the progress of the *station signals* step by step through the various stages (see Fig. 4-4). As with the injection method, the indicator may be a CRO, an ac EVM, a signal tracer, or any convenient AF amplifier. A CRO is the preferred instrument since it provides a visual indication of the shape of waveforms in addition to amplitude levels. (The indicator must have high sensitivity when tracing *low-level* input stages. This is no problem in the output stages where signal levels are high.) Most schematics show the *normal* waveforms to expect at various circuit points under normal conditions. Signal amplitudes

4-12 Signal Tracing the Station Signals   97

Fig. 4-4 Signal tracing to localize trouble.

cannot be stated, however, since they depend on local receiving conditions and other factors. A demodulator probe must be used in conjunction with each type of indicator when tracing high-frequency stages.

As with the injection method, signal tracing may be *forward* (moving the indicator through the various stages in the direction of normal signal *flow*) *or the reverse.* If working forward, the indication should get stronger as you approach the output stages; when working backward (toward the receiver input), the responses will be weaker. Any malfunctioning stage in the signal path will disrupt or degrade the signal in some manner.

Using the IF strip again as an example (Fig. 4-5), if when moving the indicator from left to right a response is obtained at points *A* and *B* but not at *C,* then the third stage can be considered defective. If, when moving the indicator from right to left there is no response at *C* or *B,* but there is at *A,* then the second stage is at fault. As stated earlier, the gain of an amplifier is its output voltage compared to the voltage at its input.

Signal tracing of the station signals through the different stages is given next.

• Video Stages

This includes the tuner, the IF strip and detector, and the video amplifiers. A *demodulator probe* is required when tracing those stages *ahead* of the detector. When an aural-type tracer is used, the expected response is a *raspy sound* as

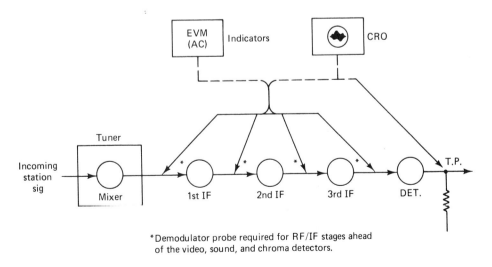

*Demodulator probe required for RF/IF stages ahead of the video, sound, and chroma detectors.

Fig. 4-5 Signal tracing IF and detector stages.

produced by the low video frequencies and the vertical pulses. When using a CRO, portions of the composite video signal (Fig. 1-6) should be observed. As explained in Sec. 3-6, the frequency or time base controls of the CRO are adjusted in accordance with the frequency of the recurrent waveform being observed. Usually it is either a multiple of 60 or 15,750 for observing about three pulses plus the video. Since video frequencies are seldom recurrent, the video display is constantly in motion and has a blurred appearance.

Proper interpretation of a video response on a CRO is of the utmost importance, and one must learn to work with patterns that appear either upright or inverted. When tracing the video amplifier stages following the detector, watch for signs of signal overload, pulse clipping, or compression (see Chapter 14). Verify that proper inversion is taking place and that the signal is of the correct polarity at each point up to the CRT input.

REMINDER: _____

The response should be strongest at the CRT, with a negative-going polarity if fed to the grid(s), or positive at the cathode(s). Each common-emitter stage normally inverts the signal between input and output. If it does not, the result is a *negative pix* and sometimes poor sync (depending on the point of sync takeoff).

- **Sound Stages**

This includes all stages from the *sound takeoff* point to the speaker, that is, the IFs, detector, and AF amplifiers. Using an aural-type tracer as an indicator, you hear the station sound. With a CRO you see the audio fluctuations. (An ac meter can also be used.) For tracing the IF stage(s), a *demodulator* probe is required. A *low-impedance* probe is best for audio stages *following* the detector.

- **Sync Stages**

Any of the foregoing indicators can be used to trace the sync pulses from the point of *sync takeoff* to the input of the vertical sweep oscillator and/or the horizontal AFC. A *high*-impedance probe usually gives the best results. Using an aural tracer, the raspy sound of the pulses is heard. With a CRO, a clean display of the pulses is desired (with little or no video) for points following the sync separator. No amplification should be expected from a clipper–separator stage.

NOTE: _____

When using an ac meter as an indicator, it is often difficult to distinguish a pulse reading from that caused by video, hum, the vertical oscillator, or other kinds of interference.

Poor sync is often the result of either sync clipping or compression in a video stage prior to sync takeoff. A quick way to determine this is to observe the *blanking and sync bars* on the CRT screen and compare their contrast with the darkest pix objects (see Figure 10-8).

- **Chroma Stages**

This includes the bandpass amplifier(s), color demodulators, matrixes, and color difference amplifiers. A triggered-sweep CRO is the preferred indicator. When tracing the burst and chroma stages, the receiver color control should be trurned full *on* and the killer threshold adjusted (usually counterclockwise,) to enable the bandpass amplifiers.

For loss of color, follow the chrominance signal through all stages from the point of chroma takeoff (in the video amplifier) to all three grids of the color CRT, where the signal should be strongest. For color sync problems, trace the burst signal from point of takeoff, through the burst amplifier, to the color phase detector.

REMINDER: _____

A pulse from the flyback transformer is required to key the burst amplifier into conduction.

- **Signal Tracing Sweep and Color Subcarrier Circuits**

As compared with station signals, these signal voltages are self-generated *within the set*. To prevent interference while signal tracing, the antenna input leads should be disconnected and/or the set tuned to a blank channel.

Any of the indicators described can be used, but a CRO is preferable for these circuits because the *shape* of waveforms must be considered as well as amplitude. Most schematics indicate what waveforms to expect.

1. Trace the *vertical* sweep voltage from source to load: begin at the oscillator output, and trace through the driver amplifier(s), and the vertical output stage to the yoke. If the trouble symptom is loss of sweep, the *amplitude* of the response is the main concern. If a vertical linearity problem exists, *waveshapes* are important. If an aural-type tracer is used, you hear the raspy sound of the 60-Hz sweep, which changes in pitch as the vertical hold control is adjusted.

2. Trace the *horizontal* sweep voltage from the horizontal oscillator through the driver(s) to the *input* of the output stage. Horizontal waveforms should be checked against schematic for shape and amplitude.

CAUTION: _____

It is unwise to check voltages at the output or at the yoke because the high-amplitude pulsed voltage can cause instrument damage.

3. Trace the 3.58 MHz generated by the color subcarrier oscillator, through the phase shifting networks to each of the color demodulators,

where amplitudes should be equal. Either a good-quality CRO or an ac EVM will serve as an indicator for this purpose.

- **Signal Checking at Control Stages**

Included are such circuits as AGC, ACC, horizontal AFC, color AFPC, tuner AFT, blanker, and color killer. The object of signal checking here is to verify the *presence* of the proper applied signals that contribute to the function of each control stage. Both *waveforms* and *amplitude* must be verified. In most instances, a CRO is used. Procedures for each stage are described in subsequent chapters as appropriate.

- **Signal Snatching**

The following emergency-type procedures can be used to locate a bad stage if and when regular test instruments are *not* available. The technique is to check one section of the set by transposing appropriate signals from other stages (or from another set) using the pix tube and speaker as the indicators. For example, using the video amplifier section to check the AF stages and vice versa.

Using the Audio Section as an Aural Tracer. A lead with a series blocking capacitor is connected to the volume control at the amplifier input. When tracing video and sync stages, a raspy buzz sound is heard where the signal exists. For tracing the vertical sweep section, a sweep buzz is heard that changes in pitch with the vertical hold control.

Pix Tube as an Indicator. A lead with a series blocking capacitor is connected to the video amplifier input. When used for tracing the sweep stages, a bar pattern provides the indication. When tracing sound stages, the bar pattern fluctuates with the audio. The chroma stages can also be traced in this manner.

Subcarrier Oscillator as a Signal Source. A lead with a series blocking capacitor is connected to the oscillator output, and the 3.58-MHz voltage is used to signal trace the burst and bandpass amplifiers. A suitable indicator is connected at the demodulators or beyond.

- **RF Generator as a Substitute for Receiver Oscillators**

The *unmodulated* CW RF (at the proper frequency and amplitude) is injected into the set as a temporary substitute for any receiver oscillator whose condition is suspect. This is one of several methods to determine if an oscillator is functioning.

4-13 SUMMARY

1. Skills in TV servicing come with practice. The enterprising beginner jumps at every opportunity to work on a set. Each one is a challenge adding to experience. And don't be turned off by the difficult; there is great satisfaction in solving problems that have stumped others.

2. Develop a general troubleshooting plan, one that is flexible enough to fit all circumstances. Perseverance and tenacity are fine if you are on the right track; if not, it can work against you. When things are not working out, be ready to change direction or try a fresh approach.

3. The troubleshooting sequence in most cases is as follows: consider the symptoms; inspect, making full use of all senses, ever alert for the slightest clue; localize the trouble area by signal tracing, voltage checking, or other means; Make *E, I,* and *R* tests in the faulty section to pinpoint the defect. But again, be flexible, ready to skip steps, take shortcuts, or alter the sequence when warranted.

4. Unquestionably, the proper schematic can be a big help, even essential at times. But learn to get along without it when you must. Diagrams of similar models and other makes of sets can be helpful, backed up by a good general knowledge of circuits. If necessary, make up your own partial sketch of the stage being worked on as testing progresses.

5. Don't be hasty by bringing out your equipment or digging into a chassis. Time spent observing symptoms and in analytical reasoning is not wasted. False starts and following the wrong leads can often be avoided by carefully planning your first moves.

6. All TV troubleshooting is based on a thorough understanding of the block diagram. Keep a mental picture of it before you at all times.

7. During initial inspection with the set on, watch for signs of sparking or overheating that can damage components. Avoid leaving a troublesome set operating unattended for long periods.

8. Multiple troubles are common enough. Learn to distinguish between multiple *symptoms* that may have *one* underlying *cause* and *multiple defects,* often in different and

unrelated sections of the set. A common example of the former is the LVPS, the stage common to all circuits, where a defect can produce a variety of symptoms with *one* underlying cause. For multiple symptoms, consider first things first: conditions affecting the *raster,* then *the BW pix,* the *sync,* the *color pix,* and then *color sync,* in that order.

9. Be conscious of safety hazards at all times. In particular, be aware of potentially dangerous voltages, the possibility of an imploding CRT unless treated with respect, physical problems resulting from lifting the heavy chassis, CRT, or the entire set, and danger associated with rooftop antenna installations and repairs.

10. Avoid practices that can damage or destroy components. Avoid straining or striking the CRT. Avoid spark testing the HV. Be careful about creating shorts when live-circuit probing a PCB. An isolation transformer should be used when working with "hot" test instruments and transformerless sets. When bridging a capacitor across circuit points, discharge it *each time* before making contact. Double-check connections for possible errors after making a replacement. Avoid stress and heat damage to solid-state components. Use a heat sink. Make sure power is *off* before making or breaking circuit connections. Always connect a generator ground lead *first* and remove it *last.* Make sure there is a blocking capacitor in the generator hot lead. Take care not to create shorts or grounds when voltage-checking transistor and IC terminals. Some ohmmeters can damage or destroy a transistor on the low or high resistance ranges; determine its output voltage and short-circuit current capability beforehand.

11. One way to determine if an oscillator is functioning is by substitution, by injecting a suitable CW signal from a generator or the oscillator stage of another set.

12. Misadjusted controls create problems, for example, loss of pix and sound if the AGC threshold control is turned off (fully ccw), or loss of color if the killer control is turned fully clockwise (cw). Blooming is a common problem when the brightness or color controls are turned up, particularly if the three color screen controls are set too high or there are problems with the HV regulation. Don't expect perfect focus within the range of the control when there are other problems such as too little or too much HV, abnormal CRT operating voltages, or a weak CRT.

13. Careless handling of a pix tube can be hazardous. Since it can hold a charge for long periods, always ground out the CRT HV *ultor* contact before handling. It is not so much the shock hazard as the *reflex action* that can result in injury when touching an HV point or dropping the tube.

14. Signals in any stage following a demodulator must have the correct polarity. This can be determined with a CRO, where (for a given instrument) the display may appear upright *or* inverted. Another way is with a dc voltmeter, where the pointer kicks either upward (or downward) on signal peaks.

15. Tuned circuits need to be *critically* aligned. Don't tamper with adjustments unless they are *known to be* off resonance; and *never* move adjustments aimlessly, hoping that the trouble will disappear.

16. It is easy to *misread* the color coding on a resistor that is scorched. Check the schematic for the proper value.

102 THE TROUBLESHOOTING APPROACH

REMINDER: _____

A high-wattage resistor normally runs warm; a *low-wattage* unit *never* does (unless there is an overload). The wattage rating of a replacement resistor must be at least as high as the original.

17. An overheating (or open) resistor is usually the *result of* a short, which should be investigated before replacing the resistor. Where a shorted component or wiring short is discovered check the schematic and look further for other components that may have been damaged by an overload.

18. Never cover up a carbonized leakage path with tape or an insulating aerosol spray. It just does not work. The carbon must be scraped or washed clean with solvent or the part replaced. When arcing has occurred in a flyback transformer or yoke, replacement is the only answer.

19. Open circuits due to hard-to-find hairline cracks in PCB soldered wiring are very common. Touch up all suspected areas with a low-wattage iron and a minimum of solder. Such a break may be bridged with solder or a short piece of wire soldered between connections. Another cause of intermittents occurs where large components (such as filters) break loose from the PC wiring, usually from accidental bumping or stress. Resolder as necessary.

20. Sustained arcing in the LVPS circuits will damage components and can start a fire. High-voltage arcing, evident by its loud crackling sounds, occurs at the HV rectifier, the HV anode lead, in a *spark gap,* and sometimes inside the CRT. *Corona* is a HV discharge (from sharp points usually) around the flyback transformer or the HV dc leads or connections. It gives off a characteristic *hissing* sound, the sweetish smell of ozone, and can be seen as a bluish glow in the dark.

21. Unless a defect is observed during inspection, the next logical step in a troubleshooting procedure is to *localize* the fault to a stage, group of stages, or some *functional* section. *After* it is localized, then it is time to zero in on the defect: a bad component, wiring problem, adjustments, and so on, the underlying *cause* of the problem.

22. Troubles are *localized* in signal stages by one of two methods: (a) tracing by *injection,* using a generator, or (b) by tracing the progress of a station signal or one generated in the set, using some form of *indicator,* usually a CRO, or an EVM set to measure ac.

23. For tracing by injection, the generator signal must conform with the kind of signal normally handled by the stages being checked, for example, an RF generator for high-frequency tuned stages and an AF generator for relatively low frequency circuits such as the sound, video, and chroma amplifiers.

24. To avoid distortion and misleading results from signal overdrive and possible damage to transistors or ICs, a generator signal level should never be stronger than that normally found in the stage under test. Lowest signal levels are in the *early stages* of a set; higher levels occur the closer you get to the outputs. To ensure a low generator output, always use the most sensitive ranges on the CRO or other indicator.

Summary

25. Schematics usually show expected waveforms and p-p amplitudes for signals generated *within* the set, but *not* for those coming from the station. In general, levels for the former are lower than found in tube-type sets because of the lower circuit impedances, and because transistors are basically *current* (not voltage) amplifiers. Since current gain is difficult to determine, stage gain is still expressed in volts as indicated by an ac EVM or a calibrated CRO.

26. When tracing by injection, a convenient indicator for relative signal strength is either the pix tube (bar patterns) for video and chroma stages or the speaker for the sound stages.

27. In the absence of a burst signal, the killer stage cannot trigger a chroma bandpass amplifier into conduction. So when signal tracing with a generator, disable the killer by turning the killer control CCW.

28. Sync pulses can be lost or attenuated in the video amplifier section. To determine if they are normal at the point of sync takeoff, examine the sync and blanking bars on the pix tube, comparing their contrast with black pix objects.

29. What you see on the pix tube can be a big help in deciding what section(s) of a set is in trouble. Analyze the off-channel raster. Judge the quality of the BW pix. If a color set, is it free of contamination? Remember a *good raster is prerequisite to a good pix.* Is there good vertical and horizontal sync stability? Critically evaluate the color pix and the color sync.

30. Where symptoms indicate a possible defective pix tube, it might be one of the first things to check. But don't be too anxious to replace it when test results are questionable. Many other conditions can emulate a bad tube.

31. When tracing AGC controlled stages, use either a low-level signal or *clamp* the AGC to prevent overloading and misleading results.

32. Most AF signal generators have sufficient output to damage or destroy transistors. When tracing, start with the lowest possible output and don't be tempted to turn the gain up full if there is no response from the stage under test. The signal level should not exceed the level normally present at a given circuit point.

33. Never operate a set for more than a few seconds without all transistor heat sinks in place. Unless the heat developed by a transistor junction is rapidly dissipated, the transistor will be destroyed or seriously damaged.

34. Remarks made by a set owner can often prove helpful in diagnosing a problem. When not offered, don't hesitate to ask questions. And bear in mind that analysis and decision making are an on-going process—testing, evaluating the results of such tests, one after another, until the problem is resolved.

35. Arcing (especially if low intensity or at some inaccessible location) can be difficult to pinpoint. One approach is to use a length of plastic or rubber tubing as a makeshift "stethoscope." Simply hold one end to your ear and probe with the other.

36. X-radiation is potentially harmful. It occurs because of excessive HV and/or improper shielding of HV components. Always measure the HV on large-screen color sets

104 THE TROUBLESHOOTING APPROACH

and never adjust higher than specified. Verify operation of the HV shutdown circuit (when used) and never defeat its function. Make sure all HV shields, containers, and baffles are in place and firmly secured.

REVIEW QUESTIONS

1. Explain how a system block diagram aids in troubleshooting.
2. Why are there *two* sound IF carriers, 41.25 MHz and 4.5 MHz?
3. Intermittent troubles are often the most difficult. Why?
4. For what trouble conditions might the HV be disabled, while working on a set? Explain why.
5. Explain the need for continuously monitoring a suspected stage when there are intermittent troubles.
6. What is the primary reason for signal tracing?
7. Give three characteristics of a signal that can be checked with a CRO.
8. Multiple troubles are normally considered in the following order: raster problems, pix and sound troubles, sync, then color. Explain why.
9. Inspection plays a major role in servicing. Cite examples, using each of the four senses that can lead you directly to the cause.
10. What is the primary purpose of any oscillator stage?
11. It is unwise to leave a faulty set operating unattended for an extended period. Give several reasons.
12. Improper procedures can damage or destroy solid-state components. Cite a number of examples.
13. Power supply problems have the highest priority. Why?
14. A proper schematic can be as important as your test equipment. What kinds of information can it provide?

TROUBLESHOOTING QUESTIONNAIRE

1. A color set has poor contrast for both BW and color programs. The raster is good. Name some stages and functions you would check.
2. A strong chroma signal is necessary for good color saturation. What stages would you check when a set has a good BW pix but weak color?
3. What stages might be responsible for *no color* if BW pix is ok?
4. Is it possible to have good color but a weak BW pix? Explain.

5. Name the stages to check when there is loss of color sync.

6. While air checking a set just prior to delivery, some minor trouble symptoms develop. What would you do?

7. An *overage* set has a bad CRT and a variety of trouble symptoms. What would you advise?

8. A set works perfect in the shop but not at the owner's home. What might be the cause, assuming no troubles have developed?

9. State three possible reasons when a component value in a set does not agree with the schematic.

10. Under what circumstances would you pull just the chassis, leaving the CRT in the customer's home?

11. What is the usual remedy when a tuner or other control is noisy and erratic?

12. A set owner complains of intermittent troubles, but extensive bench checking fails to find anything wrong. What would you do?

13. Smoke is observed coming from a chassis. How would you proceed?

14. One color is missing from the CRT raster, yet the CRT is good. Give several possibilities.

15. Where the LVPS is overloaded, explain how to quickly determine if this is due to a B voltage short or to a transistor drawing excessive current.

16. Neglecting to use a series blocking capacitor in a generator lead can damage a transistor or the generator. Explain why.

17. What is wrong if a voltage can be measured across the terminals of a fuse?

18. A transistor amplifier stage stops operating after making some stage gain tests with a signal generator. Cite four possible causes.

19. Signals at an amplifier are being monitored for an intermittent condition. When the set acts up, where would you look if (a) the signal remains normal at both the input and output? (b) the signal is normal at the input but not at the output? (c) no signal is present at the input *or* the output?

20. When there is no raster, would you normally suspect the CRT if (a) there is loss of HV? (b) the CRT heaters do not light up? (c) all CRT operating voltages check O.K.? (d) if there is also loss of sound? Explain your answers.

Chapter 5

FAULT ISOLATION

Chapter 4 described the procedures for *localizing* troubles to a stage or a functional *area* of the set. This chapter is concerned with isolating and correcting the *cause* of the defect, whether a bad component, an adjustment problem, or whatever. As with Chapter 4, the procedures are of a *general* nature, applicable to any stage and any kind of defect. Specific details for isolating faults in each particular stage are described in subsequent chapters.

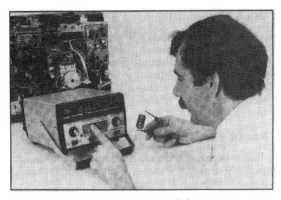

Fig. 5-1 Isolating the defect.

5-1 ISOLATING THE DEFECT

Assuming the trouble has been *localized* (with reasonable certainty) to a particular stage or section, the next step is to pinpoint the *cause*. This means concentrating your efforts only in the troublesome area, making various tests, measurements, and adjustments as shown in Fig. 5-2 (which is a follow-up of the troubleshooting flow diagram of Fig. 4-2).

For every complaint there are many possible causes. It may turn out to be a wiring short or open circuit, one or more defective components, misalignment, or the need for a simple adjustment. Usually, as you proceed, the results of each test or measurement determine which tests should follow, one after another in a logical manner until the problem is resolved. The results of each test add more information, bringing you closer to solving the problem.

Troubleshooting a *major* breakdown can be relatively easy. Seemingly "small" problems on the other hand may require considerable time and patience. Typical minor problems are a slight loss of height, borderline sync instability, color saturation that is slightly under par, flesh tones that are "not quite right," and so on. For

108 FAULT ISOLATION

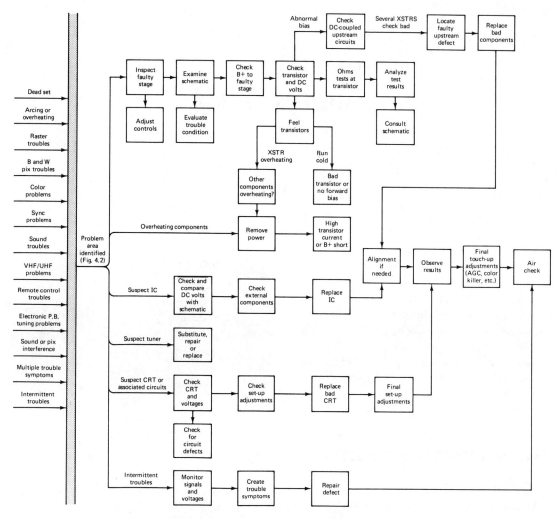

Fig. 5-2 Isolating defects and TV repair.

such problems, one must take a *closer look* at all measurements as well as components that check less than perfect.

Under these conditions some technicians elect to replace *all* small components even remotely related to the problem without bothering to test them (shotgun approach). Depending on circumstances, such an approach does have some justification when you weigh high labor costs against the relatively inexpensive parts. This should *not*, however, be considered normal practice.

• **Cause and Effect**

When a capacitor or a transistor shorts out, it often creates other secondary problems, like burned-out resistors, coils, or transformers, or in more serious cases, a chain reaction, where a whole string of transistors, domino fashion, may be wiped out in an instant. When a bad component is discovered, look for others that may have been damaged. It then becomes a question of which defect caused which breakdown. Was the shorted transistor, for example, the primary cause of other breakdowns? Or did the capacitor destroy the resistor, changing the transistor bias and causing its demise? The important thing is locating and replacing *all* the defective components, whether *cause* or *result*.

• **Testing**

Test operations and procedures for isolating a bad part or correcting a problem condition are shown in Fig. 5-2. Although the order shown is the most logical in most instances, the sequence may be altered or certain tests may be omitted, depending on circumstances, the nature of the problem, and individual preference. Some of the steps are similar to those performed in *localizing* the trouble. Unless circumstances indicate otherwise, all the tests are performed on the troublesome stage as localized previously in Chapter 4. The tests, described in the following paragraphs, are as follows: (1) adjustments, (2) waveform checking, (3) voltage tests, (4) resistance tests, and (5) current tests. These tests are followed by the testing of suspected components, including the pix tube.

Although even the simplest component or the most insignificant bit of wiring or soldered connection is crucial to the operation of any circuit, the single most central item in most stages is the *transistor*. It can be considered the functional focal point of the stage (or part of a stage) around which most of the troubleshooting tests are made. Proper operation of transistors largely determines the overall operation of the stage and the receiver as a whole. All tests (as just outlined and as described in the following)

are concentrated on the transistor and its associated components. See Fig. 5-3.

5-2 ADJUSTMENTS

While troubleshooting to isolate the *cause* of a malfunction (as with *localizing* troubles) (Sec. 4-5), it usually becomes necessary to make adjustments of controls and adjusters that might affect the operation of the troublesome stage. This includes both user (front panel controls) and service-type controls, as well as the CRT setup adjustments. It does *not* apply to the alignment of tuned circuits (which should *not* be adjusted at this time).

All such preliminary adjustments should be considered tentative. Final adjustments should be made later as described in subsequent chapters where appropriate.

5-3 WAVEFORM ANALYSIS

If not done previously, it may be worthwhile to use the CRO to display signals at various points in the faulty stage to help further isolate the trouble. Depending on the stage, this may include signals or pulses received or those generated within the set. Such signals should be analyzed for the proper p-p amplitude, polarity, and waveshape, as shown at test points on the schematic.

REMINDER: ────────────────

Signal *overload* can cause as many problems as too weak a signal. And CRO loading on critical and high-frequency stages can produce misleading results, such as distortion or a reduction in amplitude. Use the proper impedance matching probes (Sec. 3-20).

110 FAULT ISOLATION

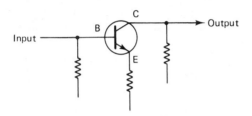

- Plug-in transistors makes servicing easier

- Scope check input/output signals
- Measure dc voltages
- Ohms tests
- Test transistor (in and out of circuit)
- Measure or compute current
- Test other components
- Substitute or replace transistor
- Replace other components as needed
- Observe results. Repeat voltage tests as necessary
- Other problems: PCB damage; wiring shorts or opens; transistor socket troubles; poor soldering, etc.

Fig. 5-3 The transistor, focal point for testing.

5-4 VOLTAGE TESTING

Most defects in a TV set create abnormal voltage conditions. The more serious problems usually result in drastic changes from the norm, which are relatively easy to recognize. Minor troubles, on the other hand, tend to produce *small* changes in voltage, a more difficult troubleshooting problem. In either case, an incorrect voltage reading is a good clue that often leads directly to a "fix."

When a set is operating at all, we know there must be *some* B voltage. Similarly, the presence of a *raster* shows there is HV. But the mere existence of voltage does not necessarily eliminate either circuit as a suspect. When a set is completely dead, the logical place to start checking is the LVPS.

• Voltage Checking the LVPS

The power supply used in inexpensive, small-screen BW sets is usually a simple unregulated supply with an output anywhere from about 12 to 30 V dc. More expensive, large-screen color sets use a more elaborate supply (actually several supplies in one) both regulated and unregulated, with outputs ranging from about 15 V or so up to several hundred volts in some cases. The higher voltages are for output stages (or vacuum tubes if it is a hybrid set). Some sets require a positive (+B) voltage; some a negative voltage; and some use both. Convenient test points to measure voltage are at the filter capacitor terminals, and at the "source" end of high-wattage resistors.

• **CRT Operating Voltages**

When there are raster problems (no raster, raster too dim or too bright, color contamination, etc.), concentrate testing in the pix tube area. In the absence of any clues, testing may begin by checking the condition of the tube (or making setup adjustments if needed), followed by voltage tests, or the reverse order may be used if preferred. Pix tube testing is described in Sec. 15-110.

All voltage measurements (excepting the HV) are made at the tube socket. Most of the readings depend on the settings of certain controls. With a color tube, it is customary to compare readings for each of the three guns (they should be the same) and to concentrate on any gun that is suspect. Observe normal precautions regarding the potential shock hazard and the danger of tube breakage, as described in Sec. 4-6. For detailed procedures on CRT voltage checking, see Sec. 15-72.

• **Voltage Testing at Transistor Terminals**

A transistor can only perform its intended function if supplied with the correct dc operating voltages. And since most defects (including the transistor itself) upset these voltages, voltage checking at the transistor terminals plays an important part in troubleshooting the faulty stage. A complete test calls for *five* separate measurements: from the base, emitter, and collector terminals, each in reference to ground; and from the base and collector terminals in respect to the emitter. The latter two readings are especially important since they represent the true bias conditions. Each measured value should be recorded for future reference and compared with those shown on the schematic.

With a good understanding of typical normal operating voltages, any gross discrepancy can be easily recognized. But *small* variations from the norm (which are often significant) can be easily overlooked. Hence the need for good servicing data and an accurate, sensitive low-range meter capable of measuring *fractional* voltages.

The kinds of defects that result in abnormal voltage conditions vary with each particular circuit and are too numerous to relate at this time. However, Table 5-1, in conjunction with the simplified diagram of Fig. 5-4, illustrates some common examples. Normal voltages are as indicated in Fig. 5-4.

In Fig. 5-4 the PNP transistor is used in a *common-emitter* configuration. Both base and collector require a *negative* voltage (relative to the emitter). Reminder: voltages indicated on schematics are *relative to ground.* Since the source voltage is *positive,* the emitter must operate at a higher positive potential than the other two elements. With this circuit, B/E bias is produced in two ways: from the voltage divider $R1/R2$, and the IR drop across the emitter resistor $R3$, due to emitter-collector current. The B/E bias is the *difference* between the B and E voltages, in this case, 0.2 V. Polarity is determined by the greater of the two, that is, the emitter, making it more positive than the base. Stated differently, the base is more negative than the emitter.

In the collector circuit, the current produces a 1 V drop across $L2$, and a 0.5 V drop across $R3$. Thus the true C/E operating voltage is -8.5 V. For an NPN transistor, the amounts would be the same but all polarities would be reversed.

For simplicity, Table 5-1 assumes there is only one of several possible causes for each abnormal voltage condition. In practice, there may be several defects resulting from a single cause, producing more than one abnormal voltage indication. Hence one abnormal reading by itself is seldom significant and can in fact be misleading. *All* voltages must be checked before reaching a conclusion, including measurements from B to E and C to E, the two readings of *greatest importance.* The defects indicated in the table are explained as follows:

Table 5-1 Transistor Voltage Testing and Troubleshooting Chart

Abnormal Voltage Condition (to Ground)	Possible Defects		
	Base	Emitter	Collector
Zero	Shorted $C1$, open $R1$, (B/C transistor short	Shorted $C2$, C/E short	Open $R3$, open transistor (C/E) open $R2$
Low	Leaky $C1$, $R1$ increased, B/C leakage	Leaky $C2$, C/E leakage	Shorted $L2$, increased $R3$, increased $R2$
High	$R2$ open or increased, open $L1$	Open $L2$, open transistor, Open $R2$	Open $L2$, shorted transistor

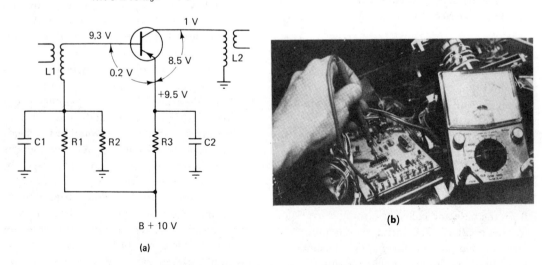

Fig. 5-4 Transistor and IC voltage checking: (a) transistor terminal voltages; (b) voltage checking at transistor and IC terminals requires great care to prevent shorts and one should always use miniclips or a needle-point probe.

5-4 Voltage Testing

1. *Zero base voltage:* The shorted $C1$ grounds out the voltage from the divider. An open $R1$ opens the circuit to the 10-V source. A shorted B/C junction practically grounds out the base via $L2$. The increased B/E bias causes high current and probable destruction of the transistor and $R3$.

2. *Low base voltage:* Same conditions as for zero base voltage, but to a lesser degree (see Table 5-1).

3. *High base voltage:* With $R2$ open there is a loss of voltage-divider action and the base voltage goes up. With $L1$ open, the meter reads the emitter voltage via the B/E junction. The loss of (B/E) bias turns off the transistor; no current means no drop across $R3$, full voltage at the emitter, and zero voltage on the collector.

4. *Zero emitter voltage:* A shorted $C2$ grounds out the emitter. A transistor short grounds the emitter via $L2$. $R3$ will overheat and burn out.

5. *Low emitter voltage:* Same as for zero emitter voltage but to a lesser degree.

6. *High emitter voltage:* With $L2$ open, there is no current flow and no drop across $R3$. With $R2$ open, the positive increase at the base represents reduced bias, causing low current, less drop across $R3$, and an increased emitter voltage.

7. *Zero collector voltage:* Transistor is cut off. Loss of current means no voltage developed across $L2$.

8. *Low collector voltage:* Current is reduced. Emitter voltage increases.

9. *High collector voltage:* No current flow, and meter reads the emitter voltage via the C/E junction.

- **Voltage Testing at FET Terminals**

Checking voltages at FET terminals follows the same general procedures as for other transistor types, despite the difference in the elements. These devices are described in Sec. 5-9, including operating parameters and precautions to prevent damage.

- **Voltage Testing at IC Terminals**

Since there is no practical method for "testing" an IC (as with other components), the procedure (after the trouble is localized and when the IC is suspect) is to make voltage checks at *all* terminals prior to removal and replacement. The object is to determine whether or not the fault is *external* to the IC. Compare the applied voltages with those shown on the schematic. Make certain no voltages appear where they should not be. Double-check all readings after removal and *before* replacement to prevent possible damage to the new unit. Further details are described in Sec. 5-9.

- **Direct Coupling Problems**

The use of coupling capacitors (except when shorted or leaky) effectively isolates one stage from another, as far as dc operating voltages are concerned. Unfortunately, this is not true when the output of one stage is coupled directly to the next, a common practice in modern sets. It can easily be seen that trouble in one stage (in one way or another) upsets the voltages and operating conditions of adjacent stages downstream. This poses a particular problem in troubleshooting when there are two or more stages directly coupled (all interdependent) and a defect is reflected all down the line. Troubles such as these can be a real challenge. The circuit shown in Fig. 5-5 serves to illustrate. If transistor $Q1$ or an associated component is defective, for example, it will affect the voltage across $R3$, upset the bias of $Q2$ and either kill or degrade its output in some manner, depending on the severity of the problem. Or working backward (upstream), a B/E or B/C short in $Q2$ will change the drop across $R3$ and affect the output of $Q1$. There are other possibilities, which will be described later. Thus when servicing such circuits, the *cause* of voltage discrepancies on one stage may be found far removed from where the trouble shows up. One approach is to isolate the suspected stage(s) by cutting across the PC wiring, (although this too in many cases will upset the voltages).

114 FAULT ISOLATION

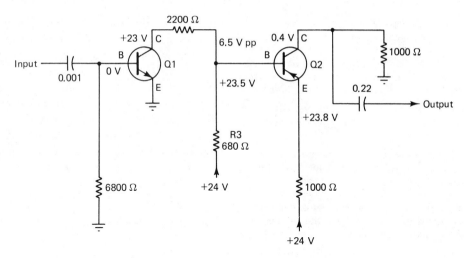

Fig. 5-5 Direct coupling and voltage distribution.

5-5 RESISTANCE CHECKING

Once a trouble is localized and voltage tests are made, resistance checking becomes the next logical step in isolating a component or circuit defect.

CAUTION: _____

Resistance checks are *always* performed with the power *off*. Make sure, prior to testing, by removing the line cord. A mistake can damage or destroy the multitester. And remember; capacitors can store a charge long after the set is turned off. Suspect this when the meter "pegs" (in either direction) or ohms readings differ with the leads reversed. Although the latter may be due to diode or transistor junctions. Always discharge filters and other capacitors before making ohmmeter tests. And one other point. With some meters, the internal battery can damage or destroy solid-state components. Know your meter. When in doubt, avoid both the lowest and the highest ranges.

- **Resistance Checking the LVPS and Circuit Wiring**

When there is little or no dc output from one or more of the PS sections, it could be a defect within the supply, or the dc feed to other stages could have high leakage or a direct short to ground. One way to determine the cause is by disconnecting the supply from the load and rechecking using voltage or resistance checks, or both. A direct short will measure *zero* ohms to ground. High leakage is caused by a *low* reading.

NOTE: _____

It is possible to have an overload even when there is a normally high ohms reading from the B voltage circuits. This is caused by excessive current drain of transistor(s) after the set is turned on.

An overheating component such as a resistor is a good place to start. If it is a series resistor in a dc power line, check the resistance from both ends to ground. The lowest reading represents the *load* side (the point closest to the defect) and the starting point for further testing.

In feeding voltage to the various stages, the components of the voltage distribution network range far and wide, which can create a lot of troubleshooting headaches. The main difficulty is the great number of branch circuits and possibilities when there is an overload. One approach is to *isolate* the fault by cutting wiring and/or components loose, one by one, while repeatedly taking measurements. A faster procedure is as follows.

Where there are numerous wires and components tied to a B voltage junction point, as in Fig. 5-6, for example, and there is a low or zero ohms reading to ground, the question is which of the components or branch circuits are responsible? The procedure is to measure the resistance (*to ground*) from the far end of each individual component, as follows.

NOTE: _____

The components are seldom conveniently bunched together as shown. They may be scattered throughout the chassis, which means following each section of wiring to its destination (at least on the schematic).

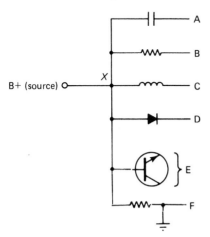

Fig. 5-6 Isolating a short circuit.

1. *Point A:* If reading is zero or lower than at point X (which is typically about 2 kΩ or more), the capacitor is either shorted or leaky. If the reading is high, this circuit is no longer suspect. Resume testing from point X.

2. *Points B and C:* If ohms reading (to ground) is lower than at X, the short or leakage is at these points or beyond.

3. *Points D and E:* A reading of zero ohms means there is a direct ground at these points. A low reading may mean there is trouble beyond, or the problem is due to the low forward resistance of the diode or transistor. Reverse meter leads and recheck.

When checking (in the same manner) beyond the points indicated in Fig. 5-6, the lower the reading, the closer you are to the *cause*.

EXCEPTION: _____

Where a component *connects* to ground (as at F), the reading at that point will naturally be zero ohms.

REMINDER: _____

Where the readings are normal (high) and there are indications of overload, the cause is usually high current drain of some transistor after power is applied.

• **Resistance Checking at Transistor Terminals**

Resistance measurements taken from transistor terminals to ground can be useful, especially when consulting the schematic for components that could relate to abnormal readings. The trouble is the transistor junctions are often in shunt with other components, which makes the measurements inconclusive until one or the other is cut loose and checked individually. This, and variations of junction resistances, the meter polarity, and other considerations, is why a schematic seldom provides this information.

• Resistance Checking at IC Terminals

Resistance checks (as with voltage tests) can help to determine whether there is a defect with the IC or its external circuitry. Prior to installing a new IC, comparison tests can also be made against the original unit.

5-6 CURRENT TESTING

When there is reason to suspect that the dc current in a circuit is either too high or too low, it can be measured directly or by computation. Sometimes there is a need to know the total current drain from the LVPS, for example, (which for a typical color set is about 2A). The meter (having a range at least as high as the amount to be measured) must be connected *in series,* which means *breaking the circuit* and inserting the meter between the source and the load. Branch circuits (where the amount of current is somewhat less) can be checked in the same manner.

CAUTION: _____

Make sure the meter is *not* connected *across a* circuit, as when checking voltage. Avoid haywire hookups. An inadvertent short can cost you a new meter. Make and break connections only with power *off.*

An alternative to *direct* measurement of current is by computation ($I = E/R$), using the measured voltage drop directly across a *known* value of resistance in the circuit.

When a meter of the proper range is not available or there is a risk of damaging it from an overload, a suitable lamp (e.g., a 12-V auto taillight) connected in series provides a visual and relative indication of current flow.

• Checking Transistor Current

Since current flow is basic to the operation of all circuits, it may seem strange that *direct measurement* of transistor current is seldom practiced. Reasons for not doing so are the time-consuming nuisance of having to break into the circuit and restore it later and the risk of meter damage, since current meters are especially vulnerable to overload from an accidental short. The preferred method is by computation, as explained previously.

REMINDER: _____

Excessive emitter or collector current is caused by internal leakage, or a direct short at one or more of the transistor junctions, or too much forward bias. If there is too little current, it may be a "weak" transistor with an open or high resistance junction or too little forward bias. The tolerance for collector current in most cases is about ± 20%. When in doubt, check the specifications.

A *power* transistor should normally feel warm to the touch. No sign of heating after several minutes of operation is an indication of trouble, as much so as the opposite condition of overheating. Most *small* transistors, unless they use a heat sink, should *never* get warm. Remember that what may appear as a normal reading of current does not necessarily mean a transistor is good. Instead of normal conduction, the current could be due to internal leakage. Perform the test explained earlier, by shorting the base to the emitter, this time observing the collector current. If transistor is good, the loss of bias should cause the current to drop to zero. For obscure troubles, it may be necessary to replace the unit and compare current readings with that of the original.

• Thermal Runaway

This is a condition (usually associated with *power output* transistors) where high collector current causes serious overheating and damage to the transistor and other components. It can be caused by internal leakage across the junctions or too much forward bias. When a transistor overheats, its breakdown voltage is reduced and its forward bias increases, resulting in a higher output current and even greater heat dissipation. Sometimes the condition goes unnoticed at first, then slowly worsens. (A factor sometimes ignored in transistor overheating is the efficiency of the heat sink. Use silicone grease as applicable and tighten all mounting screws.)

Although there are other causes, thermal runaway should be suspected when a transistor has an abnormally short life, necessitating frequent replacement.

5-7 IS THE PIX TUBE SUSPECT?

Many pix tubes are unnecessarily replaced because a technician failed to consider all the possibilities and make all the necessary tests and adjustments, not just on the tube itself but on all the related circuits and components. Or test results may be incorrectly interpreted. It is easy to go wrong, considering the many factors involved and the trouble conditions that can duplicate the symptoms of a bad tube. To further complicate matters, there is the great variety of possible CRT defects: a burned out heater (filament), internal shorts or open circuits, loss of emission (for one or more guns if its a color tube), misaligned gun elements or a warped shadow mask, gas, or an air leak, and so on.

In lieu of a CRT tester (which can be a useful, almost indispensable time saver), the evaluation of a CRT's condition depends on the results from a number of individual tests: checking for HV, checking *all* operating voltages at the CRT socket, checking for open or shorted elements (which are frequently intermittent), the setup adjustments, and so on.

For detailed procedures, see Chapter 15.

5-8 COMPONENT DEFECTS, SUBSTITUTION, AND TESTING

Although most readers are undoubtedly familiar with the common components found in any electronic circuit, test procedures are included at this time by way of review. Tests must be accurate and complete. Errors or inconclusive test results cause much wasted effort and sidetracking *away from* the defect. Since there is usually more than one method for checking a particular part, try several if there is reason for doubt.

- **Component Defects not Affecting E, I, and R Measurements**

Some defects have little or no effect on dc voltage or current readings and resistance measurements, which makes them more difficult to locate. An open capacitor is an example. Whenever opens are suspected, they can be tested or bridged with a suitable substitute while noting any change in set operation. The symptoms of an open filter capacitor are well known: a high ripple voltage and hum, reduced dc output, and so on.

An open coupling capacitor is often located by signal tracing (Sec. 4-10).

An open bypass capacitor will have considerable signal voltage across its terminals and may create degenerative effects, oscillation, or reduced stage gain.

NOTE: _____

Electrolytic-type capacitors are frequently used for coupling and bypassing and often become open or dried up.

Tuned circuit problems likewise have no effect on the operating voltages of a transistor. Examples are misalignment, shorted coil windings, and missing or broken core slugs, which can be difficult to spot.

118 FAULT ISOLATION

• Testing by Substitution

This is a time-honored method; the suspected component is cut loose (electrically isolated) and a suitable substitute connected in its place while observing the effect (if any) on the operation of a circuit and/or the overall operation of the set. *R/C substitution boxes* are commonly used for this purpose, although most items can be checked using this method, including yokes, flyback transformers, the CRT, and others. When substituting, certain factors must be considered:

Resistors: Values should be close to the original and the wattage rating at least as high.

Capacitors: Values should be close to (or greater in some cases) than the original. Voltage rating should be at least as high or greater than the original. For very small values, particularly in high-frequency stages, the leads should be short.

Yokes and transformers: A temporary substitute can often establish whether the part is defective. But don't expect 100% results unless an *exact* substitute is used.

Diodes: For PS rectifiers, the voltage and current rating must be at least as high as the original. For demodulators and AFC-type circuits, almost any small diode may be used.

Transistors: There are substitutes for most transistor types, but many factors must be considered. See Sec. 5-9.

ICs: The substitute *must be* an exact replacement.

CRTs: Most types can be made to work as a temporary substitute. Some things to consider are the type of socket and voltage requirements.

• Testing Resistors

Nearly all resistor failures are the result of a current overload. They may become open, changed in value, intermittent, or noisy. Resistor values are more critical in some circuits than in others. The usual method of testing is with an ohmmeter. For *precise* measurements, a CR bridge (as described in Sec. 3-9) is more accurate. Don't rely on the color coding.

In some cases a resistor can be checked in the circuit; at other times one end must be cut loose where the reading is influenced by other components such as coils, other resistors including controls, or solid-state devices. The rule is, if a resistor measures *higher* than its tolerance limits, it is definitely bad. If it measures *lower,* it must be disconnected from the circuit and rechecked.

REMINDER: _____

Even with the power off, an erroneous ohms reading is possible because of a residual voltage charge on capacitors, particularly filters. For in-circuit testing it is always a good idea to take *two* readings, one with the meter leads reversed. If there is a difference in readings, the resistor is either shunted by a diode, transistor junction, or an IC, or there is voltage in the circuit. Another clue is when the meter pointer kicks either upward or backward beyond the scale. Discharge the circuit before rechecking the resistor.

During inspection (Sec. 4-8) and again later when the troublesome stage has been localized, look for resistors that appear scorched, have changed color, or in extreme cases have become charred. It may be the result of a *previous* overload *or* related to the existing problem. Don't rely too much on the color bands, which tend to change appearance when heated; a red may appear brownish, for example, or a green becomes a blue. Such *overheated* resistors should *always* be replaced even when they check OK, because in some circuits they create noise and other disturbances. A little sideways pressure will often cause them to crumble.

High-value resistors, on the order of 50 kΩ or more, pass very little current, so there is no problem of overload. But they can still open up or change value for no discernible reason. Lower-value resistors (except wire-wound types) when subjected to overload can change value either up or down depending on the degree of charring.

Trouble symptoms that develop or change during receiver warm-up are sometimes caused

5-8 Component Defects, Substitution, and Testing

by a defective resistor. When suspected, heat the unit with a soldering gun while monitoring for resistance variations, usually upward. Value should return to normal when allowed to cool. When a resistor checks bad or overheats, the cause must be found and corrected before making a replacement. The *wattage* rating of a resistor must be at least as great as the original. In the case of a high-wattage unit, it should be kept clear of critical components and those susceptible to heat damage. Make sure there is adequate ventilation. If it is a flameproof resistor, replace with one of the same type.

When no schematic is available, determining the value or wattage of a defective resistor can be a problem, particularly if it is badly burned, broken, or charred. After correcting the cause, one approach is to experimentally bridge it with different substitute values (as in a *sub box*), while checking the stage function or the overall operation of the set. There is, however, a calculated risk to the method. Be alert to signs of overheating or other difficulties.

To determine the value of an open, unknown *wire-wound* resistor or potentiometer, simply add the ohmmeter measured values of each section on both sides of the "break."

• **Special Resistor Types**

    1. *Thermistor:* This unit undergoes an *intentional* resistance change with variations of heating created by current flow. Thermistors have a number of uses; one of the early types, called the *Globar* resistor was used to protect against tube burnout and other damage in *transformerless* sets. When cold, its resistance is high, limiting the current to a safe value. During warm-up, as the current increases, the resistance gradually decreases, reducing the *IR* drop to normal. The same basic units are frequently used in the vertical yoke circuit to maintain constant deflection to compensate for the heating of the yoke coils. They are also used in automatic degaussing circuits to shut off the ac flow after the first few seconds of set operation. The value of a thermistor is often too high when cold and/or fails to decrease sufficiently when heated. One test is to use an ohmmeter to monitor the changes from cold to hot when heated with a soldering iron or gun. Another method is by computation, using the measured *IR* voltage drops. An aerosol coolant spray can be useful here. Alternately heat and cool the thermistor while making voltage and resistance tests.

2. *Fusible resistors (fusistor):* This is a protective limiting resistor used in some LVPSs. Like a fuse, it opens the circuit when there is an overload.

3. *Light-dependent resistors (LDRs):* This type of resistor changes in value with variations of light intensity. It is used in automatic brightness control circuits to regulate screen brightness in accordance with changes in ambient room lighting.

4. *Voltage-dependent resistor (VDR or varistor):* The value of this unit varies with changes in applied voltage. One use is for transistor bias stabilization. Another is in automatic degaussing circuits. Resistance changes can be determined by computation using current and *IR* drop measurements.

• **Variable Resistors (Potentiometer Controls)**

Most "pots" are the low-wattage carbon strip or film type and are either rotary or *slide* operated. A number of miniaturized service-type pots (trimpots) are often located on PCBs inside the set for adjusting bias and other critical functions. High-wattage wire-wound (WW) pots are still used to some extent as raster centering adjusters and the like.

Potentiometer elements may become open, grounded, or noisy. Like a fixed resistor, it must be cut loose from the circuit to make a definitive test. To check total resistance, measure from one extreme lug to the other. To check *control action,* connect an ohmmeter from the center lug (wiper contact) to one outer terminal lug. Note resistance variation as control is varied from one extreme to the other. The changes should be smooth and gradual from the total value to near zero ohms. Repeat with meter leads from the center to the other outer terminal. Where variations are erratic or the control *cuts out,* chances

120 FAULT ISOLATION

are the element needs cleaning. Squirt a suitable solvent-lubricant in all cracks and openings. Adjust the control for a minute or so and it usually cleans itself. In extreme cases, the control may be disassembled for cleaning when replacement is not warranted.

When a replacement is called for, obtain an original part whenever possible; otherwise, a number of factors must be considered: physical style and dimensions, length and type of shaft, the resistance *taper,* and so on.

- **Testing Capacitors**

A capacitor may become open, shorted, leaky, or suffer a reduction in value (capacity). With solid dielectric types (mica, paper, ceramic, etc.) a reduction in value is not too common, but it is with electrolytics, which tend to dry up over a period of time. Capacitors of various types have a great many uses in television: for interstage coupling, bypassing, filtering, tuned circuits, and so on. Capacity values may vary from as little as a picofarad to several thousand microfarads. As with resistors, values may be critical in some circuits but not in others.

Under certain circumstances a capacitor may be tested *in circuit.* When in doubt, it's better to cut one end loose because of the shunting effect by other components. If not isolated, there is the same potential problems as when checking resistors, that is, charged capacitors in the circuit that either "pegs" the pointer or give an erroneous indication. Inspection seldom provides a clue regarding the condition of a capacitor, but sometimes the rolled paper types show evidence of bulging or seepage, and the *can*-type electrolytics may have accumulation of powdered electrolyte around their terminals.

The effectiveness of power supply filters can be determined by checking for ac *ripple voltage* using a CRO or EVM set for ac. An ac reading in excess of 1 or 2V at its terminal means the filter is not doing its job. The same applies to emitter and collector circuit bypass capacitors, assuming the stage is passing a signal. An open coupling capacitor is often found while signal tracing. When a generator signal gets through from one side but not from the other (or when a CRO shows signal on one side only), the unit is possibly open. To verify, bridge the suspect with a substitute of like value. Locating an open capacitor using the preceding method sounds simple. But generally, this kind of defect poses an elusive problem, mainly because it does not affect dc voltage readings.

A shorted or leaky capacitor, on the other hand, represents a different situation, since a *direct short* will kill any dc, and *leakage* will reduce the amount. In addition, there is usually overheating and other damaged components, resistors, and transistors as a result of the breakdown. When you encounter a burned resistor, look further for the *cause*—a shorted diode, capacitor, transistor, or whatever.

There are a number of alternative methods of checking capacitors, depending on circumstances and available equipment.

Ohmmeter Method. With capacitor removed or disconnected, when the meter leads are touched to a good capacitor the pointer deflects as the unit takes on a charge from the instrument battery. When fully charged, current ceases and the pointer returns to *infinity*. The time of charging depends on the amount of voltage and capacity; where they are both high, the time may be considerable, as with an electrolytic filter, for example. Conversely, a small value may charge so rapidly as to barely deflect the pointer.

A shorted capacitor is immediately apparent since the reading will be zero ohms. If leaky, the pointer will deflect toward zero and then move in the reverse direction as it should. But the final reading will be something *less than* infinity; the lower the ohms reading is, the greater the leakage, and vice versa.

5-8 Component Defects, Substitution, and Testing

If there is no deflection when touching with the probes (assuming a moderate-to-large capacity), then the unit is presumed to be open. But don't be fooled. Meter reaction to *small* capacitance values is often undetectable.

With this method, meter ranges should be changed as necessary. Start with a low ohms range, and even when a capacitor *seems* good, verify by switching to the highest (most sensitive) range. What appears as infinity on a low range may show up as leakage on the multi-megohm range.

One limitation of this ohmmeter method is the low voltage of the meter battery, since some capacitors only break down when subjected to higher voltages; an extreme example is HV filters of the ceramic type, which are normally checked by substitution.

Using a Voltmeter and a DC Power Source. This is a more sensitive method than the preceding since the voltage used may closely approach the rating of the unit under test. The dc can be supplied by either a separate power supply or from the set under test. *But it must not exceed the capacitor rating.* For out-of-the-circuit bench testing, the hookup is as shown in Fig. 5-7 a. Initially, select a meter range that will handle the supply voltage. When contact is made to a good capacitor, the pointer will deflect upward and then drop to zero in the time required to charge to the full value of the power source. If the capacitor is shorted, the meter will immediately indicate the full voltage. If leaky, the meter deflects upward but the pointer returns only part way toward the low end of the scale. The higher the final reading is, the greater the leakage, and vice versa. To detect small amounts of leakage, start with a high voltage range; then gradually drop to the lower ranges as capacitor assumes a charge. If the capacitor is open (or has a very small capacity), the meter will not react.

The test may be performed in circuit by cutting one end of the capacitor loose, as shown in

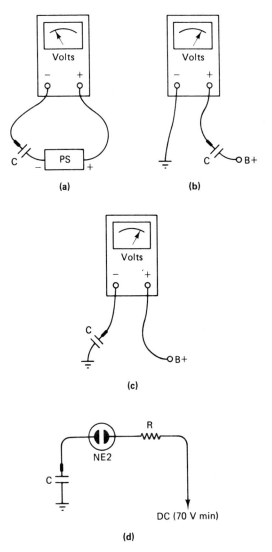

Fig. 5-7 Capacitor testing methods.

Fig. 5-7 b. Check for dc volts at both ends of the capacitor. Break the capacitor loose at the end where there is no voltage (or where the lower amount is if there is voltage at both ends). Test the capacitor as just described.

NOTE: _____

If in a signal circuit, a strong signal (or other ac

122 FAULT ISOLATION

component) can produce meter fluctuations that may be mistaken for leakage.

Another hookup where one end of a capacitor goes to ground but to no suitable voltage point is shown in Fig. 5-7c. The hot lead of the meter is tied to a convenient voltage point. Testing is the same as just described.

Using a Neon Glow Lamp and a DC Power Source. With this simple but effective hookup (Fig. 5-7 d), a glow lamp is used as the indicator. Since it requires at least 70 V to fire the gas, it cannot be used on lower voltages. Make sure the voltage used doesn't exceed the capacitor rating. A limiting resistor (R) protects the lamp. The testing procedure is the same as with a meter. With a good capacitor, the lamp will flash once and then go out. If shorted, there is a continuous dim glow or flicker to the lamp. If open, there is no indication. With this sensitive test, most electrolytics will show some leakage.

REMINDER: _____

Electrolytics are *polarized*.

Another simple check for electrolytics is to charge them with dc (not exceeding the capacitor rating). Wait a short time and then note the intensity of the spark when they are shorted out. A good unit will hold a charge for at least a half-minute or so. A continuous spark when charging the capacitor means it's either shorted or badly leaking.

NOTE: _____

Electrolytics not used for long periods may show some initial leakage until they have been "formed" at their rated voltage.

Testing by Substitution. This is a time-honored method where a sub box comes in handy. Make sure the substitution unit will handle the voltage. If an electrolytic, observe the correct polarity. In-circuit testing (by bridging) is only possible for a suspected *open* capacitor. If leaky or shorted, one end *must* be cut loose.

CAUTION: _____

To prevent destroying transistors, make sure the substitution unit is *discharged each time* before bridging into a circuit.

Testing with a CR Bridge. Besides *measuring* capacity, most of these instruments (see Sec. 3-9) will also check the capacitor's *condition*. They will also check the *power factor* of electrolytics, a test of quality or effectiveness. With the possible exception of electrolytics, capacitors showing the slightest leakage should *always* be replaced. Be sure the voltage rating is at least as high as the original.

• Coils and Transformers

Practically every section of a television set contains one or more coils or transformers of various types. There are the small, high-frequency coils used for coupling and tuned circuits of the channel selectors, the sound and video IFs, and the chroma and color burst stages. And there are the larger, *iron-core* types found in the LV and HV power supplies, the sweep circuit, and the yoke.

A number of factors determine the *inductance* of a coil, mainly, the number of turns, the physical size and shape, and the type of core, if any. Very small inductances as used in high-frequency stages may consist of only a few turns with a rating in microhenries. By comparison, the filter choke in the LVPS has many turns, an iron core, and a rating of 10 henries (H) or more. In general, small coils and transformers are found in stages where the signal frequency is high, and larger units in low-frequency stages, for example, the sweep and audio circuits. Coils in high-frequency tuned circuits usually have an adjustable iron core, or *slug*, for achieving *resonance*.

By definition, a coil consists of but one winding (of any number of turns). A *transformer* on the other hand has two or more coils inductively coupled. One winding, the *primary,* is energized from either an ac or a pdc (pulsating dc) source, such as the collector output of a transistor amplifier. The *secondary* winding(s) supplies ac power to a load. The output voltage of a transformer is determined by the *turns ratio.* It may be either a step-up or step-down transformer. Current and power output are determined by wire size and wattage rating. Since the power output of a device cannot exceed the input, if a transformer has a step-up *voltage* ratio, its *current* output is reduced proportionately. For a step-down voltage ratio, a higher current output is *possible.* The *actual* amount of current of course is governed by the load. Since transistors, unlike tubes, are *current* amplifiers, transformers used in solid-state receivers generally use heavier wire with lower ratios, by comparison. The highest *voltage* (25 kV or so) for the CRT is furnished by the many thousands of windings of the flyback transformer. But only fine wire is required for the coil since the *current* is negligible (the few microamperes represented by the CRT beam current). Transformers having the highest *power* rating are found in the LVPS, the sweep, and AF output stages. An *autotransformer* has its primary and secondary in series (e.g., a CRT heater brightener).

A simple continuity check will show up an open coil or transformer winding. Testing for shorted windings (a common defect) poses the greatest problem. The ohmic resistance values of coils is sometimes shown on a schematic. This is fine and can be helpful where many windings are shorted and the change in resistance is measurable. With only a few windings shorted, the test can be useless.

In power circuits, shorted windings result in overheating. But an *overload* produces the same effect. For an overheating power transformer in the LVPS, the secondaries can be individually cut loose as one conclusive test. If overheating continues, the transformer is bad; if not, it is a circuit overload problem.

A shorted flyback transformer will overheat, sometimes with visible arcing, which usually calls for replacement. One method of testing is by *ringing* with a flyback-yoke tester as described in Sec. 3-18. The only alternative is substitution (see Chapter 9).

Shorted windings in a tuned circuit coil are almost impossible to detect. A good clue, however, is when the coil will not tune or resonates at a higher frequency than normal. Before condemning a coil, first check for a broken or missing core slug.

5-9 TESTING SOLID-STATE (SEMICONDUCTOR) DEVICES

Diodes, transistors, FETs, and ICs have high reliability unless subjected to electrical stress or overload. Failures are commonplace. Frequently however, transistor breakdown is the result of a defect elsewhere.

Caution:

Never make a solid-state replacement without first determining if there is *another* defect (some primary *cause* of the breakdown), which if not corrected will result in damage to the replacement.

A solid-state device (like a capacitor) can become open, shorted, leaky, or intermittent, although the latter condition is not too common. When a unit is suspect, it can often be tested in circuit. This is desirable, to avoid damage due to removal. But where results are inconclusive, it must be isolated or removed for further testing. Many such devices, especially the miniature types, will not tolerate too much handling and must be treated with care. The technician must also contend with the inaccessibility and crowding of parts on a PCB. This calls for needle-point (or hook-type) probes for testing and small (low-wattage) soldering tips when making a replacement. To avoid damage when soldering, remember the four rules for good soldering: use a low-wattage iron, use a heat sink, do the job quickly, and avoid stress on pigtails.

When a unit is found defective, replace as described in Sec. 5-12.

• **Testing Diodes**

Diodes are used extensively in the modern television. There are silicon-type rectifiers found in the LVPS, which are tiny, efficient, and reliable. There are the "stick-type" HV rectifiers for the HV and focus voltage requirements of the CRT. There are the many subminiature varieties (both silicon and germanium types) with their numerous functions: signal detectors-demodulators, AGC clamping, AFC control circuits, isolation, the damping or suppression of transient oscillations. And there are many special-purpose types: zeners used as voltage regulators, thyristors (SCRs) used in some sweep circuits, and varactors (varicaps) used for controlling the tuner and other oscillators are a few.

A diode is the basic semiconductor. Its ability to conduct depends upon the polarity of an applied voltage. *When forward biased,* with negative on its cathode, and positive on the anode, it has high conductance (low resistance) to the flow of current. When *reverse biased,* with opposite polarity, it has low conductance (high resistance) to current flow and is essentially an open circuit.

A diode can become open, shorted, or leaky, although leakage is rare with modern silicon types. In most cases, an ohmmeter is used for testing. Sometimes a diode can be checked without removing it from the circuit. Under these conditions, the set must be turned *off* and the filters discharged. A more reliable and definitive test is with one lead disconnected or the unit completely removed.

When forward biased by the meter battery (negative test lead to cathode), a good diode will read very low, usually about 50 Ω or less depending on the type of diode, the meter, and the range selected. When reverse biased (with test leads transposed), a good diode should read very high, anywhere from 100 k Ω to several megohms; again, the actual amount depends on the factors mentioned. For ohmmeter testing small transistors, it is customary to use the $R \times 100$ range, and for power transistors, the $R \times 1$ range. In general, the higher the back-to-front resistance ratio, the better.

A shorted diode will read zero ohms in both directions. An open diode will read infinity in both directions. Leakage shows up as a low reading in both directions.

The leads (cathode-anode) of an *unmarked* diode can be identified while testing when you know the meter polarity. Another method is by comparing readings taken on another diode (any type) that *is* marked. Most quick-check types of diode-transistor testers can check a diode either in or out of the circuit. Such a unit is described in Sec. 3-17.

Substitution is another test method when an *open* diode is suspected. Simply bridge it while observing results.

CAUTION

Make *certain* the polarity is the same as for the original connections.

The testing of special purpose types of diodes commonly used in television sets is described next.

Zener Diode. This device is used when there is a need for a stabilized voltage, as in the voltage-regulated LVPS. Whereas a conventional diode breaks down when its rated inverse voltage is exceeded, a *zener* only functions in the breakdown

5-9 Testing Solid-State (Semiconductor) Devices

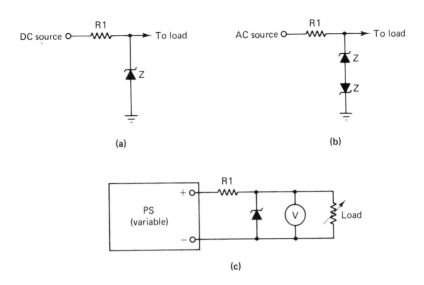

Fig. 5-8 Zener testing and PS regulation checking.

range. When connected in a circuit as in Fig. 5-8, it draws negligible current when the voltage across its terminals is below the predetermined breakdown value. If the source or load voltage increases beyond this point (the *zener* or *avalanche* rating of the unit), it begins to conduct. Its current drain produces a voltage drop across the series resistor $R1$, reducing the output voltage accordingly. The greater the increase in source voltage is, the higher the conductance and voltage drop to stabilize the output. Increases in *load current* are also compensated for in the same manner. Figure 5-8b shows two zeners connected back to back for regulating ac.

A zener, like any diode, can be tested for an open, short, or leakage using an ohmmeter as described. To check its regulating ability, it must be connected in a suitable circuit, as in Fig. 5-8c. A variable dc supply is used as the source, with a voltmeter connected across the zener. Gradually increasing the voltage from a low value will indicate at what level conduction takes place. A constant output when the source voltage or load current is made to vary (within limits) above and below this level indicates the unit is good.

Varactor Diode (Varicap). In effect, this is a voltage-sensitive variable capacitor or, to be more exact, an electronically *simulated* capacitor. Used in modern tuners it eliminates the need for mechanical switching. It is also used in AFT and other AFC circuits. Its simulated capacity varies inversely with the amount of reverse bias voltage applied to its terminals. In a tuner, a separate varactor is shunted across the tuned circuit of each stage, RF-mixer, and oscillator. Channel selection is accomplished by simultaneously applying a preset voltage (corresponding with the desired channel) to each of the units, usually by push-button control. Except for its control function, which is best tested in circuit by substitution, it is tested for open, short, or leakage as with other diodes.

Photosensitive Diodes. Used for automatic brightness control in some sets, this device develops a dc output that varies with the amount

126 FAULT ISOLATION

of ambient room lighting. It is tested for voltage output under actual operating conditions.

Thyristors. These are *switching* or *control*-type diodes having two or more junctions. There are two basic types, the *SCR* (silicon-controlled rectifier), and the *triac*. Unlike most diodes, they have three terminals, and are often stud-mounted for heat conduction. They are used in *switching* applications where relatively high currents can be controlled from small *triggering* pulses. Some typical uses are in horizontal sweep circuits where they replace driver-amplifier transistors; in regulated LV power supplies, and in some remote-control units.

The control (or switching) electrode on these devices is called the *gate*. With a forward bias applied to an SCR, it is normally nonconducting or in an *off* state. With a small forward bias applied to the gate, it switches to an *on* state where it conducts heavily. It can also be switched off by a reverse bias. A triac is essentially two SCRs parallel connected with reverse orientation. Unlike an SCR, it can be controlled by pulses of either polarity.

In high-power applications, these units require an efficient heat sink. Make sure the mounting stud nut is tight. When a unit overheats (thermal runaway), check for junction leakage or a circuit defect. Some diode-transistor testers will check thyristors but most will not. A simple test hookup is shown in Fig. 5-9. With switch closed, the indicator lamp should not light. With switch open, the gating action triggers conduction and the lamp should come on. If shorted, the lamp will be lit at all times. If open, the lamp will not light at either switch position. When ohmmeter testing an SCR, the reading from gate to cathode (in both directions) should be about 1 kΩ, and other readings should be close to infinity.

High-Voltage Diodes. These "stick-type" diodes are used as HV and focus voltage rectifiers

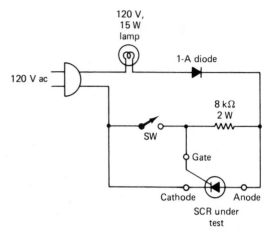

Fig. 5-9 SCR test circuit.

in all modern sets. Essentially, they consist of a great many diodes in series. They have a very high forward resistance, and regular ohmmeter testing is not conclusive, unless the device is completely shorted, which is very uncommon.

Testing by substitution is the preferred method. Another method is to connect the diode series with a power supply of several hundred volts, measuring the leakage resistance in both directions with a dc voltmeter. As with all diodes, double-check the polarity when making a replacement.

Light-Emitting Diodes (LEDs). These devices are being used more and more as function indicators in modern television sets. They give very little trouble. When they fail to light, usually the control circuitry is at fault. The testing of LEDs is described in Sec. 3-21.

- **Testing Transistors**

The typical TV receiver contains a great variety of transistors to meet the requirements of the different stage functions. Relatively *high power* transistors are used in the vertical, horizontal, and audio output stages. Numerous *low-power* transistors serve other functions: voltage

5-9 Testing Solid-State (Semiconductor) Devices   127

amplifiers, drivers, oscillators, and so on. Some transistors are general-purpose types; others are chosen for their high frequency characteristics or other considerations. Some transistors are special-purpose types. Some sets use either NPN or PNP types exclusively, but most sets use both. Germanium transistors are used mainly for high-power applications, whereas silicon types are used almost exclusively in modern sets. In general, testing procedures are the same for all transistor types.

A transistor can be considered as two diodes connected back-to-back, as shown in Fig. 5-10. Note the difference between the two types. With an NPN, the two *anodes* are common to the base. With a PNP, the *cathodes* junction with the base, and the anodes become the C and E electrodes. Because of this similarity to diodes, testing procedures are essentially the same, and their junctions are subject to the same kinds of breakdown: shorts, opens, or leakage.

There are several testing methods: using an ohmmeter, using a diode-transistor checker, or, in the final analysis, by substitution.

Testing With an Ohmmeter. Six separate measurements are required. Since the resistance depends on whether a junction is *forward* or *reverse* biased by the meter battery, each of the three junctions must be checked for both polarities. Although results are not always conclusive, initial testing should be made in circuit to save time and to avoid the component damage that often occurs during transistor removal.

Problems inherent with in-circuit testing are created by the shunting effect of other components (resistors, coils, and solid-state devices). For example, a junction shunted by a 100-Ω resistor makes testing impossible, since

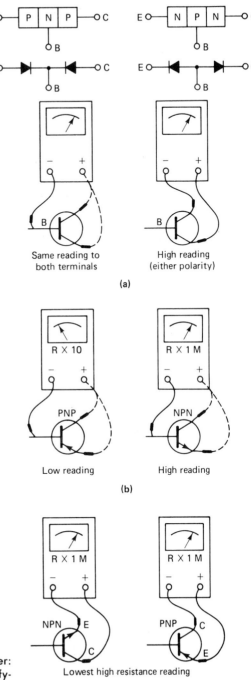

Fig. 5-10 Transistor testing with an ohmmeter: (a) identifying transistor terminals; (b) identifying transistor type; (c) identifying C and E terminals.

the reading, regardless of meter polarity, will always be 100 Ω or less.

But in-circuit tests *can* provide useful information. In the example, a reading of 0 Ω (in either direction) would mean the transistor was shorted, or if greater than 100 Ω, then the *resistor* is bad. If the reading is *infinity*, then both the junction *and* the resistor are open. By analyzing the results, more information can be gained by taking repeat readings with the meter leads reversed.

REMINDER: ─────────────────────────

As with diodes, a transistor junction should have a high back-to-front resistance, the higher the better.

To *accurately* assess the condition of a suspected defective transistor, it must be isolated from the circuit. This may be by complete removal or (unless it is a plug-in unit) by temporarily cutting across the PC wiring to any two of the three terminals. The terminals (for most sets) are identified with B, E, and C markings stamped on the board. Transistor leads are usually coded in some manner, by the arrangement of the leads, the location of a colored dot or tab, and so on. Terminal identification information is provided by transistor substitution guidebooks. Note that *bottom views* are shown, which must be reversed when looking at them from above. Step 1 of the following procedure shows how to identify unmarked transistor leads.

A word of caution. Certain ohmmeters can damage or destroy small sensitive transistors. Either the meter battery may exceed its breakdown potential or, when forward biased, create a *current* overload. These conditions vary according to the meter and the range selected. In particular, avoid taking *forward-biased* readings on small transistors using the $R \times 1$ (or the highest) ranges. With EVMs and DMMs, however, there is no risk and *any* range may be used.

And one thing more. The *actual* resistance value for any given transistor junction cannot be specified because of the variable factors involved: the type of transistor, the meter, the resistance range selected. Different ranges will be used at different times during the testing. In general, the $R \times 100$ range is most suitable for small transistors, and the $R \times 1$ range for power transistors. Testing is performed as follows.

Step 1. Identifying the Leads (if uncoded or unmarked): Make sure the set is turned *off* and the capacitors are discharged. Start by arbitrarily connecting the common meter lead to any terminal and probing the other two. If the readings (whether low or high) are approximately *equal,* the common lead connection is the *base,* (Fig. 5-10a). If they are not, move the common meter lead to another terminal and repeat the two measurements. If necessary, move the meter lead to the remaining terminal and repeat until two equal readings are found. An alternate method is to locate two terminals measuring a high value regardless of meter polarity. They are probably the E and the C, and the other is the base. See Fig. 5-10 a.

Step 2. Identifying the Transistor Type: If a schematic is not available, proceed as follows. If the two readings taken in step 1 are *low* (approximately 100 Ω or less), with the *negative* meter lead at the base, the transistor is a PNP type. If the readings are high (100 kΩ or more), it is an NPN. Conversely, if the + meter lead is at the base, a low reading indicates an NPN, and a high reading, a PNP. See Fig 5-10(b).

Step 3. Identifying the C and E Terminals: Although the readings between these terminals are *high* (regardless of lead polarity), one reading is slightly higher (or lower) than the other, making it possible to identify them *relative to each other*. For the lowest reading with polarity as indicated, Fig. 5-10c shows both transistor types.

Step 4. Testing the Transistor: From the foregoing tests, the condition of the transistor should already be known. A zero ohms reading between any two terminals means the junction is *shorted.* A reading of *infinity* means it is *open*. But don't be misled. The high reading on many silicon types is close to infinity. For a typical germanium transistor, it is usually between 100 kΩ and one or two MΩ. Another chance for error is not realizing what meter range you are using. *Reminder:* a low reading on a high range for example, may look like zero ohms; or a high reading on a low range may be mistaken for infinity, an open circuit. Finally, when a desired high reading checks *low*, it means the junction is leaky. Leakage is common with power output transistors and germanium types in general, but it is comparatively rare with small silicon transistors. Leakage does not necessarily disable a stage, but it will seriously impair its operation by upsetting the critical bias conditions and may ultimately result in a breakdown.

Other Useful Tests. An out-of-circuit, relative *gain check* can be made as follows: Connect the ohmmeter between the E and C terminals (negative lead at C for a PNP type and at E for an NPN). Note the resistance reading. Connect a 1-kΩ resistor between B and C. With a good transistor the reading should drop. The lower the reading is, the higher the gain.

An in-circuit test can be made on a transistor by either increasing or decreasing its bias while noting any change in the collector voltage or current. To remove the bias, simply connect a jumper between B and E. If transistor is good, the loss of bias should cause a reduction in current and the C voltage to become equal to the source.

CAUTION: ─────────────────────

Don't mistakenly connect the B to C or the transistor will be destroyed by the excessive forward bias.

The preceding test is not practical for a grounded collector configuration. In this case, *increase* the forward bias slightly by connecting a resistor of a few thousand ohms between B and C, and look for an increase in the emitter voltage and/or current.

To become familiar with the foregoing procedures, it is a good idea to practice with an assortment of unknown diodes and transistors. When there are doubts in identifying leads and determining a transistor type or its condition, take comparative readings on a good unit of the same type.

Darlington Transistor. This special type of transistor is actually two transistors in *cascade* with the output of one feeding the input of the other. Used as a preamp or driver, it has a very high gain. High-power Darlington transistors are often found in output stages. Separate transistors sometimes use the Darlington connection. Testing is the same as for other transistors, except the resistance readings are higher.

• **Using a Diode and Transistor Tester**

Testing with this instrument is more convenient than the ohmmeter method, which can be tedious and time consuming. Of particular value is the *go-no-go* tests, which can be performed on a unit either in or out of the circuit. Another advantage over the ohmmeter is *beta testing* of a transistor, although this information is not often required in servicing. Typical testers are shown in Fig. 3-12 and described in Sec. 3-17.

Testing procedures vary somewhat with different instruments; a general procedure is described here. For in-circuit testing, *the set is turned off* and the instrument test leads are attached to the diode or transistor terminals. For transistors, a panel switch is set for the type being tested, either NPN or PNP. Some instruments also have a range switch for currents corresponding with low-, medium-, and high-power units. For out-of-circuit bench testing (a more reliable test), either the test leads or the instrument plug-in sockets can be used as desired. A problem with in-circuit testing is the difficulty in attaching the clip leads without creating shorts to adjoining connections.

When type and terminal identification are unknown, most testers have a switch that when rotated checks out various combinations. A panel light indicates this information. Terminal identification is also made by comparing forward and reverse currents. The same test determines whether a junction is open, shorted, or leaky. Leakage current is usually indicated on a panel meter.

130 FAULT ISOLATION

Unless plug-in transistors are used, in-circuit testing is usually attempted first. Where results are uncertain (due to shunting by low-resistance circuit components), the unit must be either isolated or removed for further testing.

For checking *beta* (current gain), a function switch is set for beta and a range switch for the anticipated current, which is indicated on the meter. One useful purpose of the beta check is in matching and comparing transistors used in critical circuits.

• **Curve Tracers**

These instruments display a family of dynamic characteristic curves (signature patterns) on a CRO as another means of checking a transistor, especially for an accurate determination of its beta value. It is especially useful in *comparing* and *matching* transistors, where two or more transistors (either PNP or NPN types) having the same beta are required.

• **Testing FETs**

A FET combines the desirable features of a pentode vacuum tube and a transistor. Its main features are high input impedance and low input capacity, high gain and transconductance, a low noise level, and low internal feedback capacities, which make neutralization unneccessary, even at ultrahigh frequencies. These characteristics make it ideally suited to certain applications, such as in the RF-mixer stages in a TV tuner and in some test instruments.

Like any transistor, a FET has two or more junctions. The terminals of a FET are identified as follows:

1. *Source:* This corresponds to the *emitter* of a bipolar transistor and the *cathode* of a vacuum tube. It connects to the source of electrons, the negative side of the power supply.
2. *Drain:* This compares with the collector on a transistor and the plate of a tube. It is the *output* terminal of the FET. Like a tube, it connects to the positive side of the power supply.
3. *Gate(s):* There may be one or two. This serves as the input terminal(s), corresponding with the base of a transistor and the control grid of a tube.
4. *Substrate:* This is the foundation or base support on which the different materials are mounted. Sometimes it connects to the *case*, at other times the source terminal or ground.
5. *Channel:* This is not a terminal, but the path of current flow between the source and the drain. It appears on a FET symbol as a vertical line.

There is a great variety of FETs, as with other transistor types, each with different and distinctive characteristics. There are two main types: the *junction gate* type (JFET) in which the gate is in direct contact with the channel material, and the *metal oxide semiconductor* type (MOSFET), in which the gate or gates are close to but insulated from the channel. All FETs are further identified as N channel or P channel types. And there are two basic types of MOSFETs, *single* and *dual* gate. With the dual-gate type, each gate has control over the output. As an RF amplifier in a tuner for example, the signal is applied to one gate with AGC control on the other. As a mixer, one gate is used for oscillator injection. A typical FET amplifier is shown in Fig. 5-11.

Biasing. In the normal circuit configuration, a FET uses *forward* bias, that is, *negative* on the gate (relative to the source) and positive on the drain. As with a vacuum tube (and unlike other transistors), a high negative *reverse* bias results in current cutoff. Forward bias consists of making the gate bias less negative. With a JFET, there must be sufficient negative bias so that the signal drive cannot result in a forward bias and the resulting changes in current. As with other transistors, bias voltage is obtained from the drop across a resistor, by using a voltage-divider arrangement, or a combination of both.

5-9 Testing Solid-State (Semiconductor) Devices

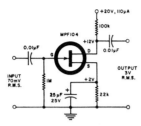

Fig. 5-11 Typical single-gate FET amplifier.

Troubleshooting a FET circuit is essentially the same as for other transistors: make voltage, current, and resistance tests as appropriate, followed by testing of the device when it is suspect. When making forward and reverse resistance checks on the junctions, make sure of the polarity of the meter leads.

NOTE:

The gate-to-source bias polarity requirements are opposite to what one might think; they are + for an N type and − for a P type.

Besides the ohmmeter method, a FET may be checked with a suitable tester; many diode and transistor testers have this capability.

Handling. Working with FETs entails certain risks that *must* be considered. As with all transistors, they can be damaged by heat, shock, moisture, and excessive voltage. The *gates* are particularly vulnerable to *static charges,* which can originate from many sources, especially in dry weather (e.g., testing or handling after combing your hair or walking on a nylon carpet). FETs in tuners are often destroyed from static discharges during electrical storms (even a minute charge can puncture the thin oxide film insulating the gate from the channel). Some FETs (IGFETs) have built-in gate protection, but most do not. Up to the moment of installation, a FET should be protected with a shorting ring or conductive foam available for the purpose. Avoid creating transients (as when bridging capacitors). Test the soldering iron and test equipment for leakage prior to use. As an added precaution the soldering iron should be *grounded.* With test equipment, always connect the ground lead first and remove it last. If possible, work on a grounded bench top and ground your body, particularly the hand that will contact the device.

• Checking ICs

An IC chip contains most of the small components that make up one or more complete stages, that is, the diodes, transistors, resistors, low-value capacitors, and the associated wiring. In fact, there is hardly a set made today that does not contain one or more of these versatile devices.

Physically, there are two kinds of ICs: (1) a flat, molded, dual-in-line (DIP) type with connector pins projecting on either side, and (2) a metal can T.O. type that resembles a transistor except for its many pigtails. See Fig. 5-12. The former may be soldered into a PCB but most often plugs into a socket to facilitate replacement. It is keyed for proper orientation when installing. The can type is usually wired in place.

An IC is *not repairable.* Although its *components* are not accessible for testing, a number of checks should be made external to the IC before considering replacement. Figure 5-12 shows the symbolic representation of a typical IC along with its internal circuitry. The latter is included here for the purpose of explanation but is seldom shown on a schematic diagram. The triangular-shaped symbol points in the direction of signal flow. For convenience and simplification of a diagram, sometimes a complex IC is shown as *two or more* symbols at various locations on the drawing. All pins of an IC are numbered for identification. Although knowing the internal circuitry is not a requirement for servicing, its purpose can be determined by its functional name and the interconnections with the rest of

132 FAULT ISOLATION

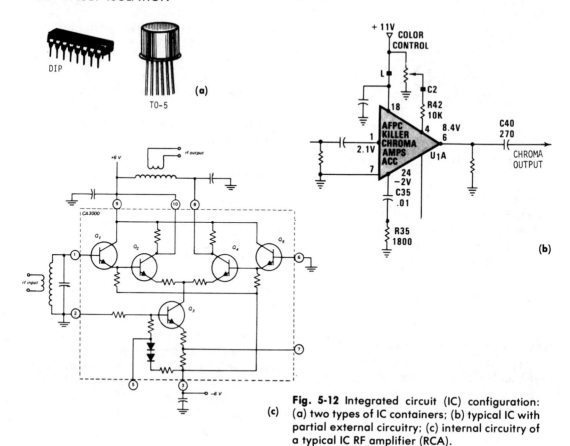

Fig. 5-12 Integrated circuit (IC) configuration: (a) two types of IC containers; (b) typical IC with partial external circuitry; (c) internal circuitry of a typical IC RF amplifier (RCA).

the circuit. In Fig. 5-12, note the connections to the numbered pins. Signals enter the IC on the left and emerge on the right. Other pins connect to the larger associated *external* components, such as coils, filters, and the like. Voltages are indicated when appropriate as an aid to troubleshooting.

Where the circuit of an IC is available (and it seldom is), it's sometimes possible to pinpoint an open, short, or other condition as a result of tests made at the terminal pins. Generally, where there's signal input but no output and all pin voltages check OK, the IC can be considered defective. As a precautionary measure prior to replacement, the following tests should be made.

1. *Signal tracing:* This is done by signal injection or by scoping the signal at the input and output connections. When a signal (of proper amplitude, waveform, polarity, etc.) gets through, then of course the IC is not suspect. When it does not, other testing at the terminals is required.

2. *Checking voltages:* Verify the amount and polarity of all voltages as shown. Loss of B voltage can be due to either internal or external defects. Remove the IC from its socket or cut the voltage line to the pin being checked. Use an ohmmeter to check for a ground (short) at the IC and the external circuit. Make sure there is no voltage at any socket pin unless it is the required voltage shown on the schematic. For abnormal voltages at any pin, locate and correct the cause *before* installing a new IC.

3. *Other tests:* Additional testing may be required depending on the IC stage functions. If the IC contains an oscillator, for example, use a CRO or EVM to check for the generated signal(s) at appropriate terminals. If there is a keyed circuit, check for the applied gating pulses. Here, again, the cause of trouble may be either internal or external to the IC.

When an IC replacement seems warranted, make a final precautionary check on socket voltages (see checking voltages, item 2) as a hedge against possible destruction of the new unit. In some cases you may go a step farther by making comparison ohmmeter checks on the suspect versus the new IC before installing. If both are the same, there is nothing wrong with the IC.

5-10 CHECKING MODULES

Some modern sets employ modular construction where all the components and circuitry of one or more stages or functional areas are contained in separate and discrete replaceable units (modules). At the time of writing, however, the trend is toward large PCBs containing many ICs. A typical module (as shown in Fig. 5-13) contains even the major components of a stage such as transformers, filters, coils, ICs, and in some cases even controls. A module is identified according to its function(s) (e.g., vertical module, chroma module, etc.). Some units (like the LVPS module) may exist as a small *subchassis*. Others (much like any PCB except they are removable) may slide into drawers. Modules are interconnected with other modules or the hand-wired sections of a chassis with the *quick-disconnect* type of plugs and cables, or edge-type contacts that engage when the module is in place.

Once a trouble has been localized, a suspected module is removed for inspection and further troubleshooting as necessary. For out-of-the-set voltage checking, the B voltage may be restored using an external power supply, of the proper value and polarity, connected directly to the module interface. In some cases, further signal tracing by *injection* is possible, as well as the scoping of waveforms.

Unless it's encapsulated, a module *can be repaired* by replacing defective components.

(a)

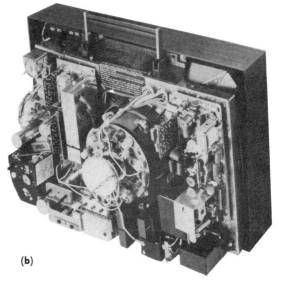

(b)

Fig. 5-13 (partial)

134 FAULT ISOLATION

(c)

(d)

Fig. 5-13 Replaceable modules: (a) modules plug into these edge connectors on the two main circuit boards; (b) rear view of chassis with all modules plugged in; (c) typical plug-in PCB modules; (d) replacing module in a slide-out drawer.

When troubles are intermittent or obscure or damage is extensive, it may be expedient to replace the entire module. It is impractical and uneconomical for a repair shop to stock the many different modules used by all sets; but they are not too expensive when purchased on an exchange basis. But before purchasing a module, be reasonably certain that replacement is *necessary*. A few guidelines and ground rules for installing modules are as follows:

Be careful not to bend or place undue stress on a PC-type module during removal or installation. Potential problems are breaks in the soldering to the larger components and cracks in the PC wiring, the same as with any PCB.

Never remove or install a module with power *on*.

After removal, inspect module for such things as poor soldering, dirty, corroded, or damaged interface contacts, cracks in PC board or wiring, discolored or burned resistors, carbonized leakage paths, and the like.

When loss of B voltage occurs at a module input, disconnect and check at the chassis side of the interface to determine if the trouble is internal or external to the module.

5-11 WORKING WITH PRINTED CIRCUIT BOARDS (PCBs)

Compared with the hand wired sets of the past, all modern receivers make use of PCBs that contain the parts and wiring for most, if not all, stages. A typical PCB (Fig. 5-14) consists of an insulating phenolic base or *substrate*. For the most part, the components are mounted on one side and the printed "wiring" on the other. A typical board also includes markings (roadmapping) showing component locations and identification and their interconnections to facilitate circuit tracing. Most substrates are translucent, and a high-intensity light enables viewing of the wiring from the component side.

A number of problems have long been associated with PCBs, such as the crowding of many parts into a small area adding to the difficulties in testing and replacement, problems with the printed circuit itself, and broken soldered connections and cracks in the wiring, which are often intermittent. When working with PCBs, certain techniques and precautions must be observed in the testing and handling of the easily damaged subminiature components. Special tools are required, such as needle-nosed pliers, small diagonal cutters, large and small

Fig. 5-14 Printed circuit boards (PCBs): (a) component side of a typical PCB; (b) wiring side of a typical PCB.

tweezers, a sharp pointed pick, a gooseneck lamp, preferably with a built-in magnifying lens; small magnifying lens or jeweller's loupe, small dentist-type mirror, low-wattage pencil-type soldering iron, surgical type hemostat for holding wires or components while soldering (optional), stiff-bristled toothbrush for removing molten solder or residue, suction bulb (solder sipper) and copper braid for removing molten solder, and a small solder pot (optional).

Good workmanship should be practiced at all times. This is of particular importance when it comes to soldering, which means just the right amount of solder and sufficient heat to make it *flow,* yet not enough to cause shorts, overheat or endanger components, or cause the wiring to peel and separate from the substrate. PCBs are somewhat fragile and subject to breakage. Avoid unecessary pressure, bending, or twisting that can create damage and cracks in the wiring, which is a very common and sometimes difficult problem.

Serious damage may call for replacement of the entire PCB as a unit. If the damage is minor,

136 FAULT ISOLATION

repairs can be made as follows. If board is cracked, restore its mechanical integrity using cement and reinforcing strips where necessary. Replace any damaged components. Rework all questionable soldered connections. Breaks in the wiring can be corrected either by resoldering across small cracks or bridging large gaps with hookup wire. Using an ohmmeter on its lowest range and a needle-point probe, check the continuity of all wiring in the area. This will locate hairline cracks that cannot be seen by inspection.

5-12 REPLACING COMPONENTS

To avoid errors, make careful note of the points of contact *before* removing the part to be replaced. Don't trust to memory. Make a sketch. For small transistors, where the B, E, and C terminals are not stamped on the board, it is a good idea to make your own markings. Note the polarity of diodes and small electrolytic capacitors. Self-induced troubles are always possible, so check and double-check everything. A wiring error can be costly and compounds your troubleshooting problems.

Components with pigtails are normally replaced by complete removal, feeding the leads of the new part through holes in the board and soldering to the PC wiring exactly as before. Because of crowding and inaccessibility, a simpler technique is to *cut* the original part loose, but leave its pigtails intact for connecting to the replacement. A caution here is against excessive and prolonged heating, which may melt the solder from the wiring side of the board. To avoid damage to small diodes and transistors, it is important to use a heat sink while soldering and unsoldering. Using long-nosed pliers, grasp each pigtail between the component and the point where heat is applied. Avoid prolonged heating. Use only a low-wattage iron with a small, well-tinned tip so that PC wiring will not peel away from the board. After installing, check the integrity of the soldering and lead separation and for wiring errors by a close-up inspection with a magnifying lens. Make sure there are no soldering splashes or foreign particles in the wiring.

FETs require *special* care in handling, since they can be damaged by even moderate heating and static charges. Most FETs are provided with a shorting device, which should be left in place until installation. Even a slight leakage in a soldering iron can do damage. Check for soldering iron leakage using the highest range of an ohmmeter and/or ground the tip during the soldering operation. Make sure the set is *unplugged* and the antenna disconnected.

ICs also rate special consideration. After removal of the suspected unit, check for voltage at *all* terminals and compare with the schematic. Excessive voltage (or voltage present where it should not be) will often destroy the replacement IC. Check for possible defective external components. Most ICs are coded. Make sure it is properly oriented before installing.

Transistor heat sinks are *important*. If a heat sink was used originally, be sure it is reinstalled after making a replacement. When replacing a power transistor that is insulated from the chassis, verify that its insulator is in place and coated on both sides with liberal amounts of heat-conducting silicone grease.

Following are some additional hints and precautions:

1. *Never* remove or replace parts with the set turned *on*. This also applies to soldering. Preferably, pull the plug. Besides the potential shock hazard, there is a good chance of creating inadvertent shorts that can destroy components.

2. The wattage rating of a replacement resistor must be *at least* equal to that of the original. Don't crowd high-wattage resistors close to combustible material, small transistors, or other components that may be damaged by heat or soldering that may melt.

3. The voltage rating of a replacement capacitor must be *at least* equal to that of the original. For electrolytics, make sure the polarity is correct. If in doubt, check the voltage beforehand. A replacement capacitor, especially electrolytics, can store enough energy to damage transistors. Always discharge capacitors before installing.

4. After making a replacement, verify the proper connections, diode polarity, transistor orientation, and the like, *before* the set is turned *on*. The simplest mistake can be costly. After applying power, watch closely for anything abnormal. Feel resistors, transistors, and other parts for signs of overheating.

5. Some parts, such as yokes, flyback, and power transformers, must be *exact* replacements.

6. Where conditions *change* after making a replacement or performing other operations, a *self-induced trouble* should not be ruled out, especially if the technician is interrupted in his task. Double back and recheck your last effort, verifying the accuracy of the connections.

7. When a component replacement does not result in the expected improvement in set operation, there are several possibilities: the replacement is defective or improperly installed, the part removed was not the cause of the trouble, the set has *multiple troubles*, it may be a critical circuit requiring some compensating readjustments, or the condition is intermittent.

8. When reworking PC wiring, use solder sparingly to prevent solder overlap to adjacent connections. Use a sipper or brush away any excess. Make sure large electrolytics, IF cans, and other large components are physically secure. Even a slight movement will break the soldering at their terminal lugs. Rework these *stress points* and other questionable terminal points. A little extra solder *here* is desirable.

9. One common result of arcing is the formation of a carbonized conducting path with potentially serious consequences. Sometimes it shows up by casual inspection with the set *off,* and at other times as a visible arc or burning with the set *on*. The charred area or conducting path *must be* interrupted or completely removed by scraping or, if necessary, by complete removal of the damaged portion of the board. Replacement is the alternative. Never apply tape or insulating compounds *over* the area.

10. The grounding of PCBs to the chassis is sometimes dependent on hold-down screws at strategic points. Hard-to-diagnose troubles can be caused by one poor ground. Tighten all such screws on a routine basis.

11. With a little practice, a solder pot can be a time saver in the removal of multicontact components such as IF cans and modules. *Caution:* To avoid damage, the operation must be performed *quickly.* Use extreme care to prevent injury from the molten solder or splashes in the eye.

12. To prevent the possibility of feedback or upsetting tuned circuits, a component should normally be replaced in the same manner and at exactly the same location as the original. Avoid disturbing *lead dress,* especially in high-frequency stages.

13. Wherever possible, use *identical* transistor replacements. When such are not available, refer to a *replacement guidebook* for a suitable substitute. There are many factors to be considered. The same applies to ICs, FETs, and certain types of diodes.

14. To avoid errors and save time when replacing components (like transformers, with their numerous leads), it is a good idea to *tag* all wires *as they are removed,* even though a schematic is available. Above all, don't trust to memory.

5-13 INTERMITTENTS AND OTHER PROBLEMS

One of the most exasperating problems a technician can face is that of the *intermittent,* a trouble condition that shows up (often *momentarily*) and then corrects itself for no apparent reason. Usually the intermittent condition is so "sensitized" that the on-again, off-again state can be triggered by the least disturbance: the slightest pressure on a PCB, switching channels, turning on a light, or walking across the room are examples. The main difficulty with intermittents is getting the set to remain in the trouble state long enough to make some tests and locate the fault. Even then the condition may be so critical that the mere contact of a meter probe is enough to restore normal operation.

The first thing is to learn the difference between a true intermittent and a trouble *symptom* that shows up occasionally because of the *marginal* operation of some circuit. With the former, it is the *cause* that is intermittent. With the latter, the malfunction, if there is one, is stable enough; it is the *symptoms* that come and go, for whatever the reason.

Since it is useless to make tests while a set is operating normally (except where measurements may be taken and recorded as a basis for future comparison), a technician may do one of two things: he may wait for the set to act up on its own (and hoping it will remain so long enough to be worked on), or he may take the initiative and *force* the trouble to show up, which is usually the best approach. In addition to noting changes in receiver operation, the suspected section or stages (if known) should be monitored using a CRO, EVM, or signal tracer. If the trouble area is not yet localized, monitor *major checkpoints* such as the signal input(s) to the CRT, the HV, chroma outputs, the speaker, and so on, as appropriate. A useful instrument is a *circuit analyzer* (see Sec. 3-13), which simultaneously monitors different stages and functions. A *dual-trace* CRO can also be used to advantage by simultaneously monitoring the input and output of a stage. AC and dc voltages at key points should also be monitored.

In general, there are two kinds of intermittents, those relating to *circuit* problems (opens or shorts in wiring, poor soldering etc.), and those caused by a *component defect*. The most common *intermittent-type* defects are as follows:

1. *Pix tube:* open or shorted elements, shadow mask warping when heated, intermittent base-to-socket contacts.

2. *Diodes and transistors:* intermittent contact between pigtails and junctions, poor contact between pigtails and sockets when used.

3. *ICs:* thermal opens and shorts, poor pin-to-socket contacts.

4. *Resistors:* may open or change in value.

5. *Capacitors:* intermittent opens between pigtail or lug and the internal connection.

6. *Coils and transformers:* open circuit at terminals, shorting windings, or grounding, intermittent arcing.

7. *Switches:* open or shorts due to dirty or damaged contacts.

8. *Circuit breakers:* intermittent open and latching problems.

9. *PC boards:* intermittent open or short in PC wiring.

Intermittents fall into one of the three major categories: (1) *thermal,* where intermittent conditions change with variations in temperature; (2) *electrical,* where changing conditions are related to voltage and/or current; and (3) *physical or mechanical,* where intermittent conditions are influenced by movement, tapping, vibration, and the like. When troubleshooting an intermittent, the first thing is to determine which of these categories apply, in order to decide on the appropriate action to *force* the set into its trouble mode at least long enough to permit testing. To repeat, the object is first to *create the problem.* Procedures are as follows.

Thermal Disturbances. Many intermittents show up only after a set has been operating for some time, and then behave normally when the chassis is removed and placed on the bench where it gets more cooling air. To determine if it is a thermal condition and to make it act up so that tests can be made, it is necessary to create a heat environment. One method is to cover the chassis with a large carton or blanket and, if necessary, to place a 40- or 60-W lamp inside.

NOTE: _____

This does not *localize* a problem since *all* components are heated together.

5-13 Intermittents and Other Problems

CAUTION:

Don't overheat. It can damage components.

Where *certain* parts are suspect, apply *localized* heat. Commonly used for this are a hair dryer, heat lamp, or a soldering gun applied to pigtails of resistors and capacitors, but never to solid-state devices. If possible, alternately heat and cool the part, using an aerosol *coolant spray*. When an intermittent does not respond to temperature changes, don't cook it to death. Try a different approach.

Physical or Mechanical Disturbances. Attempt to create the problem by tapping the *chassis* at various points using a solid object such as heavy pliers. When the set is susceptible to the slightest movement or vibration, try to localize the most sensitive area by moving the tapping point and reducing the intensity of the tapping. Once localized, *lightly* tap components using a pencil or plastic rod to zero in on the cause. Try *gentle* pressure and bending of the PCB. Gently probe suspected components, exerting some stress on pigtails and the like as appropriate. Separate any bare wires or contacts that could be shorting out. Be particularly alert for poor soldering, especially where large components connect with PC wiring. Watch for hairline cracks in the wiring or solder splashes bridging connections. Touch up questionable connections in the suspected trouble area and, if necessary, rework *all* soldering.

Intermittent contacts at a pix tube socket are fairly common, especially in the heater circuit where the current flow is greatest. The symptom is variations in heater brightness or when it goes out entirely. Try twisting the socket and moving the leads. To show up intermittent CRT opens and shorts that cause flashing on the screen, *lightly* tap the neck of the tube from all sides using a nonmetallic object.

Transistors and ICs sometimes develop poor contact with their sockets. Apply gentle pressure and slight movement from all angles and tap lightly. Try turning up the sound. If troubles show up only at high volume level, look for such things as a bad (scratchy) speaker, loose antenna leads or terminals, or intermittent pin-and-socket contacts on any module or PCB.

Electrical Disturbances. When the problem seems unrelated to either thermal or physical changes, it could be a condition affected by surges or transients, line voltage fluctuations, or any of the signals from the station or generated within the set. Once again, the object at this point is to make the set act up (by whatever means possible) before attempting to find the cause.

One approach is to increase all voltages and currents throughout the set by raising the line voltage. A *variac* or booster transformer is used as described in Sec. 3-20. If necessary, the voltage applied to the set can be increased to about 130 V, but no higher.

CAUTION:

Avoid a *prolonged* overload, which may damage or destroy components. Watch for any signs of overheating, especially in the power or flyback transformer or power transistors.

NOTE:

With some sets, the voltage-regulating circuit makes it impossible to raise voltages in this manner.

In some instances, troubles can be made to show up by *reducing* the line voltage. When a component is suspect, it can be disconnected and an overload applied *directly* without endangering other parts. Intermittent *arcing* is not uncommon. Usually it can be localized *visually*.

When conditions appear related to *signal* strength, try adjusting certain controls such as the AGC, contrast, chroma intensity, and color killer. At other times it may help to substitute a generator for a station signal as a means of controlling the level. Use an RF, AF, or color bar generator depending on which section of the set is suspect.

• **"Tough Dogs"**

This term aptly describes a set that is especially difficult to troubleshoot and repair, and pinpointing the cause, the elusive defective component (or whatever), can be a real challenge. Sometimes it involves multiple troubles, which compounds the problem. Or it may be an intermittent or strange unusual symptoms never before encountered.

As opposed to an intermittent, where changes are usually *abrupt,* some conditions are *gradual,* mostly during warm-up. It may be a slow buildup in height or width, or contrast or color, or a slow worsening of sync stability, for example. The cause is often leakage in a transistor junction, a resistor changing value, or thermal runaway in a power transistor. Another common trouble is a slow increase in raster brightness, which is usually due to a weak CRT.

Multiple *symptoms* to the layman can be especially confusing, for example, where *everything* seems affected, brightness, contrast, sweep, sync, color and even sound. Chances are there is but one cause (such as a dried-up filter in the LVPS that is common to *all* receiver functions). Then there are AGC problems, which can produce a variety of symptoms, such as loss of sync and a negative pix. Or consider the chroma output stages where a defect not only affects pix color, but may also be responsible for the absence of a raster or one of the three primary color fields.

Another possibility is wiring errors or wrong component values, especially if the set was recently worked on and there are obvious indications of poor workmanship. Under these circumstances the first step is to check the wiring in the trouble area against the schematic, while reworking any questionable soldered connections.

After agonizing over a problem for longer than you would care to admit, it may be time to consider replacing *all* the possible suspect components en masse, particularly the relatively inexpensive resistors and capacitors, and even the transistors unless they are *known* to be good. But use discretion. The "shotgun" approach should only be used as a last resort.

When you encounter a "tough dog," when nothing seems to work out, don't be ashamed to seek help. Check with others who may have had similar problems, a field service representative for your make of set if possible, since many defects tend to be repetitious. At least, don't give up. When the going gets rough, set it aside for a time to return to later with a clear mind and a fresh approach. Often that is all that is needed.

5-14 ALIGNMENT

Precise alignment of the various tuned stages of a set (the tuner, IFs, and chroma sections) is essential for proper operation, and misalignment can create problems that simulate those produced by component breakdowns or other defects. Fortunately, alignment is seldom if ever required except when the adjustments have been tampered with, when components have aged, over a period of time, when there are circuit disturbances from frequent servicing, or when components in a tuned circuit are replaced. Even then, it can be difficult to determine whether or not alignment *is* required, except *to go through the procedure.* Generally, it should be the *last thing* to suspect. The subject is treated in later chapters as appropriate.

NOTE: _____

Alignment *does not* include the adjustment of the

5-15 AIR CHECKING THE REPAIRED SET

There is nothing more disconcerting than after devoting much time to a repair job to receive a call soon after that it is not working properly. No matter how conscientious the effort, it is bound to happen occasionally, but if *too often,* the technician should look to *himself.*

To reduce the frequency of *callbacks,* a set should never leave the shop without an air check of at least several hours, depending on circumstances and the nature of the repair. This is a phase of TV servicing that is sometimes neglected or, at best, limited to a cursory checkout lasting but a few minutes. It is during those first few hours after turn-on when most problems show up. If nothing unusual happens, chances are the set is good. (Intermittent troubles of course rate *special* consideration.) Unless a definite cause has been found and corrected, don't take a chance. It is bound to reoccur, necessitating a repeat call. As insurance, take steps to trigger the condition (see Sec. 5-13) while the set is still on the bench.

Evaluating Receiver Performance. There are many things to watch for during the air check. And this is a good time to make final touch-up adjustments, previously overlooked, such as on height and linearity, HV, focus, raster centering, the AGC and killer threshold controls, and others. Make sure all user controls have sufficient range and that none operate at one extreme setting (which may be a clue to impending troubles), for example, vertical and horizontal hold controls that barely lock the pix or a focus control that hardly makes it. Make sure tint control covers the full range from magenta to green with good flesh tones at its approximate mid-setting. Check for noisy or erratic controls.

Critically evaluate all aspects of the set's operation, such as the following:

1. *Raster:* Is there adequate brightness and freedom from *blooming* at the extreme setting of control? Is there sufficient height and width, with raster properly centered with no tilt? Are the scanning lines uniformly spaced and in *perfect* focus? Is there any waviness to the lines or a need for pincushion correction at the edges?

2. *Pix:* Is there good contrast without smearing or streaking? Is there good resolution of fine detail, not just on large foreground objects and with consideration to screen size? Is there excessive snow, considering the reception area? Or signs of ghosting? What about vertical and horizontal linearity and subtle indications of foldover? Are there any sound bars or hum bars on or off channel? Is there good sync stability, both vertical and horizontal? If a color set, is there good color intensity on all channels? Are reproduced colors correct, especially skin tones? Is color sync stability good, particularly when first turned on and when switching channels?

3. *Tuner:* Are all VHF and UHF channels received well? Is the AFT properly adjusted? Is the tuner noisy or erratic when tapped or switching channels, indications that a cleaning is called for? The set owner will appreciate it.

4. *Sound:* Is there sufficient sound with good quality and freedom from hum, buzz, or other interference? Are there any acoustic vibrations at full volume? Is the control noisy?

A good idea is a *checklist,* as a reminder, posted in a conspicuous spot.

142 FAULT ISOLATION

5-16 SUMMARY

1. There are two types of transistors, NPN and PNP. The main differences from a servicing standpoint are the polarity of the applied voltages and the manner in which they are tested. In a normal circuit configuration, an NPN unit requires a positive bias and collector voltage; and a PNP type, negative voltages.

2. The emitter of a transistor corresponds with the cathode of a vacuum tube, the base compares with a control grid, and the collector with a tube's plate. The base is usually fed by the input signal and the output taken from the collector.

3. The B-to-E bias on a transistor is very low and very critical. *Forward* bias, (positive for an NPN and negative for a PNP) is required for conduction; the greater the voltage, the higher the collector current. A *reverse* bias reduces current, and if high enough, turns the transistor off.

4. A quick *in-circuit* test to determine if a transistor is working is to check for changes in collector voltage when shorting the base to the emitter. The resulting *loss of forward bias* turns the transistor *off,* and the collector voltage will increase to the value of the source (if the transistor is normal). If there is no change in collector voltage, the transistor is not operating. *Reminder: Never* short the B to C *directly.* As a test, use a 1- to 5-k Ω resistor to *increase* the forward bias.

5. Transistor current can be quickly determined by the voltage drop across either an emitter or collector circuit resistor, using Ohm's law. Unless there is leakage, base current is negligible.

6. A transistor, like a capacitor, can become open, shorted, or leaky. An *open* junction is frequently overlooked, especially between E and C.

7. A common problem with power output transistors is *thermal runaway.* The symptoms are overheating of the transistor and associated circuit resistors. The cause is usually B to C leakage, excessive forward bias, or an inefficient heat sink.

8. A FET has the desirable characteristics of both a transistor and a vacuum tube. Because of its high gain and input impedance, it is frequently used in low-level stages such as the tuner RF stage. A FET is very easily damaged.

9. All modern sets use PCBs. A typical PCB has the components mounted on one side of an insulating board or substrate and the printed circuit wiring on the other. The circuit consists of thin copper strips to which the components are soldered. The strips are easily broken by bending or other abuses. Because of the very close spacing, extreme care must be used in soldering to prevent shorts. Another problem is peeling of the wiring strips from excessive heating.

10. Some sets use *replaceable* modules or PCBs of the plug-in type. They can be easily removed for servicing or replacement as a unit. By comparison, an IC (because its components are not accessible) cannot be repaired.

11. Troubleshooting an IC involves (a) signal tracing or observation of input and output signals, (b) voltage checking at all terminal pins, (c) ohmmeter checking at the terminals, and (d) testing associated external components. Where such tests show no defects in the *external* circuit, but the stage is not functioning, the IC is presumed to be bad.

12. Most ICs are unique and there are no reliable substitutes except exact replacements. With transistors, when original parts are not available, substitution is often possible by referring to a substitution manual.

13. The type and terminals of an unknown transistor can be identified, as well as its condition, using either an ohmmeter or a diode and transistor tester. The sequence using an ohmmeter is to determine the base connection, which in turn establishes the type, whether NPN or PNP, by taking two sets of resistance readings. This is followed by determining which is the E and C and then testing to decide if the transistor is good or bad.

14. Even with a schematic, it is a good idea to make a sketch showing the orientation, connections, and so on, of transistors and other multicontact components *before* removal. Also double-check *after* replacement and before applying power.

15. Resistor values associated with the operation of a transistor are critical, particularly bias resistors in the base and emitter circuits.

16. Electrolytic capacitors are used *extensively* in solid-state receivers. Transistors are current devices that require low-impedance circuits (compared with tubes), which in turn calls for higher capacities for coupling and bypassing. Higher filter capacities in the LVPS are also needed because of the greater current requirements, and the low impedance of the load circuits.

17. Transistor voltages are critical, especially the bias. Use only an EVM or DMM (that draws negligible current), never a VOM. Don't trust readings near either end of the scale. Select a range that gives a mid-scale reading.

18. There are risks in measuring current. A wrong connection or an inadvertent short can destroy the instrument. Always make connections with the *power off.* Double-check connections before turning the set on. Make sure the meter range is high enough to handle the anticipated current.

19. When making ohmmeter tests, if the pointer kicks beyond its limits in either direction, it is a sure sign there is voltage present. To verify, make a voltage check. Be sure the set is turned off. Discharge capacitors, particularly electrolytics, before continuing with the test. Even the slightest voltage will cause an error in the ohms reading. Check by reversing the leads.

20. A diode or transistor junction can be considered bad if its *forward* resistance is *high* or if its *reverse* reading is *low.*

21. There are two questions that need answering if a transistor is not conducting: (a) is the transistor good, and (b) are there normal operating voltages at its terminals *enabling*

it to conduct? For incorrect voltages, remember the transistor itself can be responsible.

22. The basics of troubleshooting a transistor circuit can be summarized as follows: (a) When the transistor is not conducting because it is not *enabled,* don't blame the transistor. Check all voltages, particularly the B to E bias. (b) If the transistor is not conducting and the voltages are normal, chances are the transistor is defective. (c) If transistor is conducting too heavily and the bias checks *low* or *reversed,* it is *probably* leaky, although there *may* be circuit problems. (d) When transistor conduction is *low* and the bias is normal, the trouble is with the collector circuit. If there is zero C voltage, the C circuit may be open. If E/C voltage checks normal or a little *high,* the junction is probably open. (e) When transistor and all voltages are normal, yet the stage does not function properly, look for an open coupling capacitor or suspect misalignment.

23. A transistor may be either forward or reverse biased, regardless of polarity of the source, by making the B or the E either more + or more − relative to each other.

24. The true operating bias of a transistor is measured between B and E. This voltage is the *difference* between the B and E voltages as measured to *ground.*

25. Transistor operating parameters are interdependent. The bias affects the C voltage and current; and the latter, in turn, often affects the bias.

26. Changes in E circuit resistance have little effect on the E voltage, but greatly influence the C current.

27. When making a number of voltage or resistance measurements, it is considered good practice to record all readings when taken. Don't trust to memory (especially when dealing with fractional values). A record of such readings is necessary in analyzing a difficult problem, for later comparison as testing progresses, and for checking against the schematic.

28. There are two good reasons for elaborate testing prior to replacing an IC: (a) to ensure there are no external defects that might destroy the new IC, and (b) To determine whether it is the *IC* or some other component that is defective.

29. It is important to read the fine-print notes on a schematic. Failure to do so can result in wasted time and effort and misinterpreting test results. Such notes often contain useful servicing information.

30. "Roadmapping" on a PCB aids in circuit tracing and identifying and replacing componets.

31. As a general rule, an *open circuit* is often the *result* of some defect. A *short circuit* is usually the *cause* of problems.

32. The surge created by discharging filters with a jumper can destroy transistors and ICs. A better method is to *bleed* off the charge through a resistor of several thousand ohms. It takes a little longer, but it is *safer.*

33. To reduce the chance of creating shorts when voltage checking at closely spaced terminals on PCBs, transistors, and ICs, use probes having mini spring-loaded grips. Apply (and remove) the probes *only* with the set turned *off.*

34. In-circuit testing of transistors is not always reliable because of the shunting of the junctions by other circuit components. By the same token, some components cannot be tested in circuit because of the shunting by transistor junctions. In such cases, try reversing the meter leads. Instead of removal, disconnect two of the three leads.

35. In-circuit testing of transistors and other components must be done with the set turned *off*. If power is not removed, test results will be meaningless with possible damage to components and test equipment.

36. Three things should be known about an ohmmeter when used for checking transistors; (a) the polarity of the test leads, (b) the amount of voltage at the test leads for all range positions, and (c) the maximum current capability of the instrument as when the test leads are shorted together. High forward bias from the meter battery can (in some cases) damage or destroy transistors.

37. Transistor operating voltages (as shown on a schematic) are for normal operating conditions, and voltages taken with the transistor *removed* bear no relationship to those specified. There are times, however, when such readings can be useful, for example to verify the presence (or absence) of applied voltage and to determine if any abnormal voltages are being applied to the transistor because of some component or circuit defect.

38. Measuring the dc operating voltages at the transistor terminals should be one of the first steps in troubleshooting any stage. Normal voltage readings usually mean the circuit is operating properly, and any voltage discrepancies usually indicate a defect. Because of the low values, small departures from the norm are often overlooked.

REVIEW QUESTIONS

1. Why is a CRO such a desirable instrument for TV servicing?

2. State several kinds of receiver troubles that will not ordinarily show up with dc voltage checking.

3. Why is it often necessary to *disconnect* a diode, transistor, or resistor for testing?

4. Why is it desirable, when possible, to test some components without removing them?

5. What type of probes are most suitable for in-circuit transistor testing or checking operating voltages? Explain why.

6. Where resistance checks high to the E and C terminals of a transistor, with the negative lead at the base, is it a PNP or NPN type?

7. Explain the purpose of a zener diode and how it works in a circuit.

8. What is a varactor diode and where is it used?

9. What is a Darlington transistor?

146 FAULT ISOLATION

10. Name the elements of a MOSFET, comparing them with those of a vacuum tube and with a biopolar transistor.

11. In what circuits are FETs sometimes used to advantage? Explain why.

12. Define *forward* and *reverse* bias.

13. Why is an ohmmeter test of a HV-type rectifier diode not always reliable?

14. Explain why a reading of zero ohms on the highest range of an ohmmeter, or infinity on the lowest range, is almost meaningless.

15. Exactly *why* do the forward and reverse ohmmeter readings differ across a diode or transistor junction?

16. State the procedure in each case where an in-circuit resistor measures more or less than its coded value.

17. What precautions should be taken when bridging a suspected open capacitor with a substitute?

18. Under what circumstances might you replace all small, inexpensive components in a trouble area without bothering to check them?

19. What confusing results can be expected from making in-circuit ohmmeter tests at transistor terminals?

20. Give an example of how a transistor can have a bias of opposite polarity to that of the supply source.

21. Explain why an abnormal voltage condition on the first of several direct-coupled amplifiers shows up on all stages.

22. Why are actual current *measurements* seldom made on a transistor circuit? (b) How is current determined?

23. What is the result of inadvertently shorting the B to the C on a transistor. Explain your answer.

24. What determines whether a power transistor requires an insulator to separate it from its heat sink?

25. What precautions should be taken when handling, testing, or replacing FETs?

26. Why is it important that a set be turned off when connecting test instruments, soldering, or replacing components?

27. When voltage measured from B to ground on a transistor measures -1.2 V and the emitter measures -0.3 V, what is the amount and polarity of the operating bias?

28. If there is 18.5 V on the collector of a transistor and 1.5 V at the emitter, what is the true C to E operating voltage?

29. What is the purpose of silicone grease as applied to a power transistor insulator?

30. Explain the purpose and importance of a heat sink.

31. Why are direct-coupled stages often difficult to troubleshoot?

Review Questions **147**

32. How would you go about troubleshooting a multi-stage direct-coupled amplifier?

33. Give three reasons for temporarily connecting a lamp across a fuse or circuit breaker that constantly or intermittently opens up.

34. A badly scorched resistor should always be replaced. Why?

35. It is a good idea to *record* all voltages and other test results while troubleshooting. Why?

36. What is the fallacy of taping over a carbonized leakage path on a PCB, a component, or whatever?

37. Explain how and why it is possible to destroy ICs and transistors while bridging circuit points with a substitute capacitor.

38. When is an isolation transformer called for?

39. What is the difference between an *applied* and a *developed* voltage on a transistor?

40. What factors must be considered in choosing a substitute transistor?

41. What is the result of B to C leakage in a transistor?

42. State several possible causes when an incorrect bias voltage is measured.

43. How does each of the following defects affect the B, E, and C voltages of a transistor: (a) open E resistor, (b) open C resistor, (c) shorted E/C junction, (d) shorted B/E junction (e) shorted B/C junction, (f) open C/E junction?

44. One end of a 10-Ω resistor is connected to a -5 V source, the other end to a +15 V source. Compute *amount* and *direction* of current flow.

45. A circuit point that is bypassed with a capacitor is said to be at *ground potential*. Explain.

46. Certain defects can destroy a number of transistors domino fashion. Some may become open, others shorted. Explain why.

47. *Lead dress* and the physical location of components in high-frequency stages is often critical. Explain.

48. Explain the importance of *air checking* a set after repairs are completed.

49. With direct coupling, the output of one stage helps to establish the transistor bias of the following stage. Explain why.

50. Does the type of transistor or the polarity of its dc operating voltages relate to the polarity of the input or output signals?

51. When signal tracing with an aural tracer, what kind of sound will you hear from the following stages: (a) video amplifier? (b) AF amplifier? (c) sync stages? (d) chroma stages? (e) vertical sweep stages?

52. Under certain circumstances, it is a good idea to keep one hand on the cheater cord or plug when turning the set on. Explain why.

53. With intermittent troubles, the first step is to *create* the problem. Explain.

148 FAULT ISOLATION

54. Circuit tracing PCB wiring can be difficult and confusing. After visually tracing a circuit between two remote points, it is a good idea to verify with an ohmmeter check. Explain how and what meter range you would use.

55. Explain, with an example, how a dual-trace CRO can be used to advantage in isolating an intermittent problem.

56. Name three categories of intermittents and the steps you might take to create problems of each kind.

57. State the probable cause when full B voltage is measured at all three terminals of a transistor.

58. Describe two methods of identifying the anode and cathode terminals of an unmarked diode.

59. A resistance check of a coil or transformer will seldom reveal shorted windings. Why?

60. What trouble is indicated when the B to E or B to C readings of a transistor are (a) *high* with the ohmmeter connected either way? (b) *low* with the meter connected either way?

61. A transistor collector draws 0.1 mA when supplied from a 12-V source through a 5-kΩ load.
(a) What is the measured collector voltage using a meter that draws 20 μA?
(b) What is the *actual* operating voltage?

62. Even with a schematic available, it is a good idea to make up your own sketch (including appropriate notations) when replacing components. Explain why.

63. Explain how to use a piece of *braid* or a desoldering *sipper* for removing excess solder from a connection.

64. Small children should not be permitted in the area when servicing a set in the home. List a number of potential hazards.

65. What is meant by (a) transistor cut-off? (b) saturation? (c) How does bias determine each condition?

66. A PNP transistor is not conducting. The base voltage is +0.01 V, and the emitter voltage is +0.05 V. What is the problem?

TROUBLESHOOTING QUESTIONNAIRE

1. A capacitor, a 1-Ω coil and a 1-MΩ resistor are connected in parallel. State the probable cause if an ohmmeter across the combination measures (a) 1 Ω, (b) 1 M Ω, (c) 0 Ω.

2. State a logical procedure when there is no B voltage at an IC terminal.

3. Two PNP transistors are directly coupled. What would happen to the collector current of the second stage if the emitter resistor of the first stage opened up?

4. What is the probable result if a PNP transistor is mistakenly replaced with an NPN type?

5. The voltage at the ends of an overheating resistor measures 2 and 10 V, respectively. At which end would you look for the defect? Explain.

6. A power transistor may run too hot or too cold. Give possible causes in each case.

7. An IC is replaced and immediately goes bad. State possible reasons.

8. State four possible causes of excessive B/E bias on a transistor.

9. Where the measured current in an E circuit differs from the computed value, state the probable cause.

10. Your ohmmeter measures infinity between the C and E of a transistor. State two possibilities.

11. A diode or resistor checks open. Is it necessary to disconnect before bridging with a substitute? Explain. (b) What if it's *shorted*?

12. A voltage divider consists of a feed resistor $R1$ and a bleeder resistor $R2$ to ground. What will happen to the voltage at the junction if (a) $R1$ becomes open? (b) $R2$ becomes open? (c) a capacitor from the junction to ground becomes shorted?

13. Give three possible causes when a transistor has an abnormally high or low bias.

14. A set appears to operate normally, yet periodically pops the circuit breaker. State possible causes and how you would proceed.

15. A high-wattage dropping resistor from the LVPS runs cold. Which of the following apply? (a) It is a normal condition. (b) The circuit is grounded ahead of the resistor. (c) There is a ground after the resistor. (d) The resistor is open.

16. When making in-circuit tests, an ohmmeter pointer pegs beyond full scale, then reverses when the leads are transposed? State the cause and what to do about it.

17. An ohmmeter shows no continuity between two supposedly connected points on a PCB. Give two possibilities.

18. After touching up the soldering on a PCB, the set stops operating. What would you suspect?

19. State the possibilities when there is no improvement after replacing a suspected component.

20. A fluctuating resistance reading indicates a possible intermittent coil. An overvoltage is applied momentarily to the coil followed by a repeat resistance check. What conclusions would you draw if the reading is now (a) infinity? (b) steady? (c) the same as before?

21. An electrolytic capacitor is alternately bridged from one circuit point to another when the set suddenly goes dead. (a) State the probable cause. (b) Name two things wrong with the procedure as stated.

150 **FAULT ISOLATION**

22. *New* trouble symptoms are observed after replacing a bad component, resoldering connections, and so on. Name several possibilities.

23. A set may operate after replacing a blown fuse or resetting the CB, but this seldom represents a permanent fix. Give a follow-up procedure.

24. Small silicon transistors have very low leakage between the C/E junction, and an ohmmeter check often indicates *infinity* with the meter leads connected either way. How would you determine if the junction is *open?*

25. A resistor is badly scorched but still checks *OK*. Would you replace it? Why?

26. While soldering PCB wiring, the solder bridges several connections. How would you proceed?

27. An in-circuit check of a resistor indicates 10 Ω. With meter leads reversed, the reading is much higher.
(a) Explain the difference.
(b) Which reading is the more accurate of the two?

28. A receiver cuts out or operates erratically when a plug-in module is disturbed.
(a) What would you suspect?
(b) How would you proceed?

29. Mistakes are often made in visually tracing PC wiring and wires in cabling and congested areas. What simple test will verify the correct terminations?

30. Replacing a suspected IC without first testing at its terminals is foolhardy. Explain why.

31. It's unwise to repeatedly reset a circuit breaker that trips. Explain why.

32. What condition exists if (a) a low-wattage resistor runs hot? (b) a high-power resistor runs hot? (c) a high-power resistor runs cold?

Chapter 6

LOW-VOLTAGE POWER SUPPLY

6-1 LVPS MODULE

The LVPS supplies power for the entire set (see Figure 6-1). This includes the dc voltage and current requirements of all transistors, ac for the CRT filaments (and any other tube filaments if for a hybrid set), dial lights, and automatic deguassing if for a color set. Although it is the least complex of all TV circuits, it can be the site of many problems. As the *primary* source of dc voltage, the LVPS should not be confused with other power sources, such as the HV and focus voltage supplies that operate in conjunction with the horizontal sweep section. The horizontal sweep is also the source of secondary power supplies, known as *scan rectification*. Such power supplies, (PSs) in turn, are dependent on the LVPS, the common denominator of all stages.

The PSs of a solid-state TV receiver differ considerably from those in a tube-type set. For the latter, the requirements may be several hundred volts, with a current drain of usually less than about 200 mA. By comparison, transistors operate with much lower voltages, but the current demands are much greater, up to 2 A or more in many cases. In addition, the PSs (especially for color sets) must have extremely good voltage regulation, low ripple (be well filtered), and have adequate protection against overload.

Fig. 6-1 Typical regulated power supply module.

There are several variations in power supply circuits. Some use transformers, some do not. Half-wave, full-wave, and bridge rectifier circuits are used extensively, both separately and in combination, as are voltage multipliers (mostly doublers). The LVPS of most modern color sets consists of two or even three supplies in one unit, each designed to meet the requirements of specific stages of the set. Often both negative and positive dc outputs are provided. Practically all color sets use some form of voltage regulation.

A functional block diagram of a typical LVPS used in most solid-state color sets is shown in Fig. 6-2. By comparison, the PS for a BW set is much less complex, although the descriptions that follow, in general, apply to both types.

The first block (Fig. 6-2) represents the ac input circuits, all components and wiring that supply 60-Hz power to the rectifiers, and the CRT filament. The function of the rectifier block(s) is to change the ac to pulsating dc (pdc), with either a positive or negative polarity as required, and commonly referred to as either B+ or B-. *Silicon diodes* are used mostly, between one and four rectifiers, depending on the circuit. Silicon diodes are small and often difficult to locate on a chassis.

The purpose of the *filters* is to smooth out the pdc to produce a relatively pure dc. A typical *L/C* filter consists of an iron-core filter choke and two or more high-value capacitors. The choke opposes variations in current and the capacitors oppose changes in voltage. The capacitors act as reservoirs, alternately charging and discharging to fill in the gaps of the dc ripple component. Where the current is low, a resistor is sometimes used instead of a choke coil. Many modern receivers use an electronic ripple filter, commonly called an *active power filter*, or simply *APF*. Instead of a filter choke, a transistor is used that functions like a variable resistor between the source and the load. It operates much like a voltage regulator (to be described later) by lowering its conductance during peaks of the rectified dc and increasing its conductance during the gaps to smooth out the ripple.

Voltage Distribution. A complex network of resistors is used to feed the various stages of the set. Included are the *R/C decoupling networks* usually found close to the transistors and ICs. Resistance values used in the various *branch lines* depend on the voltage and current requirements

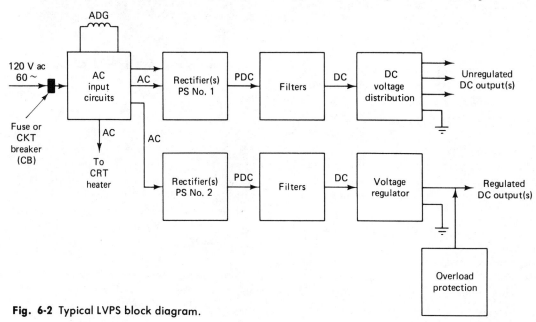

Fig. 6-2 Typical LVPS block diagram.

of the circuits being supplied. In addition to series dropping resistors, most transistor circuits also use voltage-divider arrangements, chiefly for bias. Capacitors are connected at all resistor junctions for additional filtering or decoupling.

Voltage Regulation. Ideally, the output of any supply source (generator, battery, power supply, or whatever) should remain *constant* regardless of variations in load current and (in the case of a PS) variations of the ac input voltage. This is not always easy to do. The problem is that every electrical source has some *internal resistance*. This has no effect on the open-circuit or no-load voltage measured at the output. But when current flows, an *IR* drop is developed across the internal resistance, reducing the output voltage accordingly. Where the internal resistance and/or the load current is low, the *IR* drop may be minimal and the output (the *difference* between *no-load* and *full-load* voltage) will be relatively low and constant. The opposite is true where internal resistance and load current is high. In a power supply, all components that carry dc contribute to the internal resistance. This includes the forward resistance of rectifiers and the resistance of choke and transformer coils.

Since the current drawn by every transistor in the set is fluctuating with its input signal, any variations in the dc output of the supply may result unless there is good regulation and filtering. The biggest problem occurs with power (output) transistors that draw heavy current. Such abnormal conditions create interaction between stage functions, especially if there is insufficient decoupling and isolation of the *B* supply lines. For example, if the PS regulation is poor and/or the drain of the *AF* output stage is abnormally high, audio signal variations will be present at the PS output. This, in turn, affects other stages powered from the same point. In the case of a video stage, for example, *sound bars* will appear on the screen. There are many more undesirable possibilities resulting from PS fluctuations.

Several kinds of *regulating* circuits are in common use, from the simplest, (using a passive zener diode) to more complex circuits using (active) transistors. These circuits are described in Sec. 6-7. They all serve the same function, to maintain the B voltage *constant* despite variations in load current or the ac power input. Some less-critical circuits do not require a regulated dc supply. Such stages are fed from some point *prior* to the *regulator,* as shown in Fig. 6-2.

Overload Protection. Basic protective devices are fuses and circuit breakers. Sometimes they are connected at the ac input, sometimes in the dc circuits. Their purpose is to protect the PS and other circuits against extensive damage by opening the circuit in the event of a short or other overload condition. Some modern sets use elaborate and sophisticated circuitry to sense impending overload problems.

6-2 TYPES OF POWER SUPPLIES

A number of factors determine the kind of LVPS used for a given set: quality and receiver cost, voltage and current requirements to be met, regulation, and so on. As a rule, the larger the screen size and the more costly the set, the greater the need for an elaborate PS and other associated circuitry. The basic power supplies commonly used in solid-state receivers are the half-wave transformerless PS, the half-wave voltage doubler, full-wave voltage doubler, voltage tripler, transformer-type full-wave supply, the bridge rectifier, and scan rectification.

6-3 THE AC INPUT CIRCUITS

This section includes all the components and wiring associated with the 60-Hz ac power.

154 LOW-VOLTAGE POWER SUPPLY

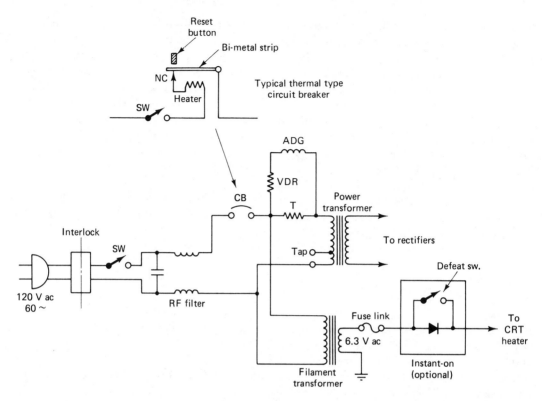

Fig. 6-3 Typical ac input circuit.

Generally, this section of the PS gives very little trouble. A partial diagram of a typical input circuit is shown in Fig. 6-3.

- **AC Power Line Circuit**

This comprises the *power cord*, the *ac interlock*, the *power-line noise filter*, the ON/OFF *switch*, the line *fuse* or *circuit breaker* (CB), *automatic degaussing* (ADG), and the *primary circuits of the power and filament transformers*. The power cord and interlock are both common sources of trouble. The interlock is a safety device to cut off the power when the rear panel is removed.

The ac power line filter usually consists of a small choke coil in each leg of the ac input and either one or two bypass capacitors. The coils are wound with heavy wire and are often connected directly to the interlock terminals. Sometimes they are inductively coupled using a *toroid*-type transformer. The purpose of the filter is to prevent noise pickup from the power line.

The ON/OFF power switch is usually on the volume control. When the set has the instant-on feature, it is often a *dual* switch.

Many older sets used a fuse in the ac input line as protection against an overload, but all modern sets (both BW and color) use a circuit breaker (CB) that can be reset after the overload condition has been corrected. Most CBs are of the *thermal* type with a *latching* device. When subjected to repeated overloads, these units sometimes become defective. For additional protection, some sets also use a filament circuit *fuse link*, consisting of a short length of no. 26 hookup wire.

The ADG coils around a color pix tube are also powered from the 60 Hz power.

From the input circuits, the 120-V ac feeds the primary winding of the power transformer (where used) and a separate filament transformer in some cases. With a "transformerless" supply, the 120 V ac goes directly to the rectifiers. Some power transformers have a tapped primary as a means of varying the turns to voltage ratio to compensate for low or high line voltage.

• **Power Transformers**

The basic features of a typical transformer are described in Sec. 5-8. Some TV transformers have a *single* secondary winding, some have several, either to step up voltage, step down voltage, or both. Many have a single multitapped winding when a number of separate dc outputs are required. Sometimes an autotransformer is used or a single coil shunted across the 120-V input. Autotransformers, however, *do not* provide power line *isolation* and are generally avoided. Where *ac* voltage regulation is used, a specially designed transformer is required that has loosely coupled windings and a saturable core.

• **Filament Transformer Circuits**

The heater (filament) of the pix tube (and other tubes if it is a hybrid set) is usually supplied from a 6.3-V winding on the power transformer or from a separate filament transformer. Low-voltage (6 V) dial lights also use the same source, although many modern sets use neon indicators, which require higher voltages (from 60 V to about 100 V). Some sets use a grounded heater circuit; others have one side of the circuit tied to B+ to reduce the potential difference between the cathode and heater of the CRT. This reduces the possibility of internal shorts. Another method is to operate the CRT heater from a winding on the flyback transformer. Because of its high current capability, a heater circuit represents a potential fire hazard in the event of a direct short. As protection, many heater circuits include a fuse link. If not provided in the design, it is a good idea to add one when servicing a set.

• **Instant On**

The instant on feature reduces the warm-up time of the pix tube by having its heater lit (at reduced voltage) *at all times,* even with the set turned off. There are several ways of doing this. Most sets have an *instant-on defeat switch* that either shorts out a series-connected diode or resistor or an extra heater voltage source, or functions as a voltage *selector.* Figure 6-3 shows a typical circuit.

• **Automatic Degaussing (ADG)**

The metal parts of a color pix tube (the shadow mask or aperture grill in particular) can become slightly magnetized from such things as a vacuum cleaner, a soldering gun, and even the earth's magnetic field. This degrades the pix by interfering with normal beam travel as required for good convergence. Degaussing is the method used to *demagnetize* the tube (as the jeweller does with a watch, or as the hull of a warship is degaussed).

A color pix tube has a large coil (or a series of smaller coils) mounted around the periphery of the tube, inside the cabinet, and close to the faceplate. The coil is momentarily supplied with ac from the 60-Hz power source. The earliest sets sometimes had a push-button panel switch to energize the coil when required. With modern sets, degaussing is automatic (ADG) and occurs for a few seconds each time the set is turned on, while the CRT is warming up.

Some ADG circuits are in series with the ac input, as in Fig. 6-4a. Sometimes it is shunted across either the 120-V ac line or a winding on

156 LOW-VOLTAGE POWER SUPPLY

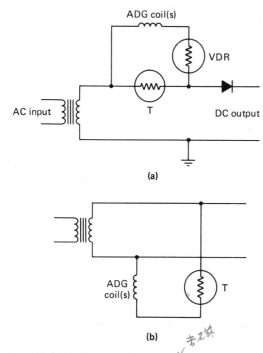

Fig. 6-4 Automatic degaussing circuits.

the power transformer. Figure 6-4a uses a *thermistor* and a voltage-dependent resistor (VDR or varistor) for automatic control. At the moment of turn on, thermistor resistance is high and varistor resistance is low, routing most of the ac through the degaussing coils. As more current is drawn by the set during warm-up, the resistance changes are reversed and less and less current goes through the coils until all of it flows through the thermistor to the set. Figure 4 a and b uses a thermistor with a positive temperature coefficient (PTC). When cold, its resistance is low and the coil is energized. During warm-up the resistance increases and the current drops to near zero.

• **Power Consumption**

The power required to operate a TV set is the product of the ac line voltage times the current drain. In early-model sets this consumed power was quite high (as much as 500 W for some large consoles). As might be expected, a color set with its greater number of components uses more power than a BW set. With concern for power wastage, modern sets are much more efficient; some large-screen consoles use as little as 100 W.

There are times when a technician may wish to check the power consumption as part of the troubleshooting procedure. For example, a set may appear to work properly, but with signs of overheating. Excessive current drain can be caused by such things as leaky filters, shorted transformer windings, leaky transistors with thermal runaway, or a consistently high line voltage.

6-4 BASIC LVPS CIRCUITS

• Half-Wave Transformerless LVPS

A transformerless PS has no power transformer and simply rectifies the 120-V ac input, as is done in most ac-operated radios. The circuit uses a single diode (Fig. 6-5) as a half-wave rectifier. The *no-load* output of such a supply is equal to the peak value of the ac, roughly 180 V. In practice, the operating (load voltage) is about the same as the rms value of the input, 120 V more or less depending on certain variables. Lower voltages are obtained by using suitable dropping resistors.

With half-wave rectification, ripple can be a problem, and high-value filter capacitors are required, especially if there is considerable current drain. In some cases, a resistor is used in place of the choke coil. The *fusible resistor R1* serves a dual function. It serves as a fuse in the event of an overload and also as a *limiting (surge) resistor* to limit the initial high charging current of the input capacitor $C1$ (which might otherwise destroy the diode rectifier). As with all power supplies, the input filter provides some filtering and has a great influence on the dc output. The output filter $C3$ has little effect on the dc; its main job is to filter out ripple.

6-4 Basic LVPS Circuits

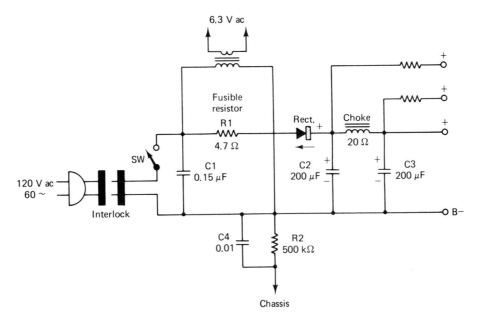

Fig. 6-5 Half-wave power supply.

With some sets (mostly BW), one side of the ac line (which is the B return) connects to the chassis. It is sometimes called a *hot* chassis. To protect the user, the cabinet and all knobs must be properly insulated. This configuration represents a potential shock and fire hazard. Sometimes the ac plug is polarized so it can only be inserted one way. When servicing such sets, it is advisable to use an isolation transformer (see Sec. 3-20.

In Fig. 6-5 (the preferred arrangement), the chassis is isolated from the power line. Resistor $R2$ and capacitor $C4$ provide the necessary path for stray ac and dc voltages. This circuit can be recognized in that the metal-can filters are *not* connected to the chassis. All voltage measurements are made relative to $B-$, instead of chassis ground.

• Half-Wave Voltage Doublers

Voltage multipliers have long been popular as a means of obtaining higher voltages without the need for an expensive power transformer, although in some cases a transformer is still used with these circuits. The circuit shown in Fig. 6-6 uses two rectifiers and three filter capacitors. Note that neither end of capacitor $C1$ is connected to ground. The operation is as follows: When the ac input polarity is as shown, $C1$ charges to its peak value (approximately 180 V) via rectifier diode $D1$ and the surge resistor $R1$. When the ac reverses, the line voltage in effect is *series aiding* with the voltage charge on $C1$, and the diode $D2$ conducts, charging input filter $C2$ to this value, about 360 V. This is under no-load conditions. In practice, with load current flowing, the output voltage drops to about 250 V. Because of the heavy drain and the low ripple frequency, very large filter capacities are needed, often as high as 2000 to 3000 μF. Their voltage ratings must be at least as high as the peak levels under no-load conditions. Note that the voltage rating of $C1$ is only half that of the other two.

Figure 6-7 shows another method used for doubling. It consists of two basic half-wave circuits with different polarity outputs. The volt-

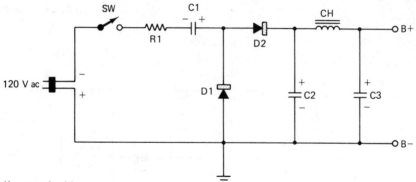

Fig. 6-6 Half-wave doubler circuit.

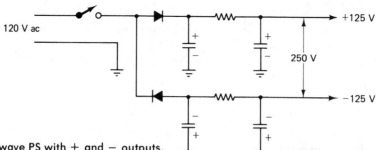

Fig. 6-7 Half-wave PS with + and − outputs.

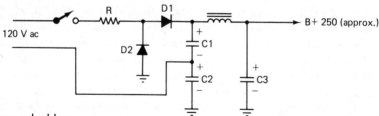

Fig. 6-8 Full-wave doubler.

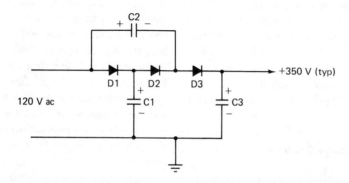

Fig. 6-9 Half-wave tripler.

age difference between the two outputs is about 250 V, with the -125-V line used as the negative return.

- **Full-Wave Voltage Doubler**

This circuit (Fig. 6-8) and its operation are much like the half-wave doubler (Fig. 6-6). Each diode conducts on alternate half-cycles of the ac, charging its associated capacitor. These two voltages are series aiding, and, once charged, their total represents the dc output of the supply, which, for a transformerless circuit as shown, is about 250 V under load. Note that neither end of $C1$ is grounded.

- **Voltage Tripler**

This circuit (Fig. 6-9) is sometimes used to obtain fairly high voltages when the current is low and good regulation is not essential. Its operation is as follows: When the ac input polarity is as shown, $D1$ conducts, charging $C1$ to the peak of the line voltage (under no-load conditions). When the ac reverses, the line voltage, series aiding with the charge on $C1$, causes $D2$ to conduct, charging $C2$ to twice the line voltage. On the next half-cycle, the high charge on $C2$, series aiding with the line voltage, causes $D3$ to conduct, charging $C3$ to about 480 V (no load). In operation with the set drawing current, the dc output is about 350 V. As with all multiplier circuits, the actual dc output is determined by the rate of discharging the capacitors compared with the replenishing of these charges.

- **Transformer-Type Full-Wave LVPS**

This circuit (Fig. 6-10) uses a center-tapped power transformer, and the two diodes conduct on alternate half-cycles. The output voltage is approximately the rms value of one half of the transformer secondary. Usually a step-down ratio is used to meet the low voltage require-

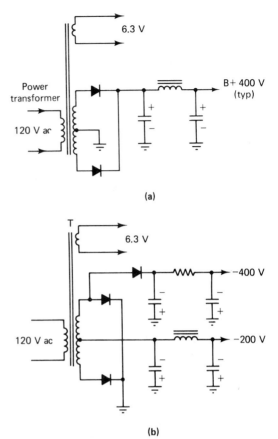

Fig. 6-10 Full-wave transformer-type power supplies: (a) (+) output PS; (b) (−) output PS.

ments of transistors. Another version of this circuit is shown in Fig. 6-10b. Here the diodes connect to ground and the transformer tap is the output, in this case, *negative*. Note that the only difference between a negative and a positive dc PS is the reversed connections of the rectifiers and filter capacitors. This circuit also uses the *total* transformer voltage to obtain increased voltage, requiring an extra half-wave rectifier as shown.

- **Bridge Rectifier**

This popular circuit (Fig. 6-11) uses *four* diode rectifiers in a bridge configuration. It is also a

160 LOW–VOLTAGE POWER SUPPLY

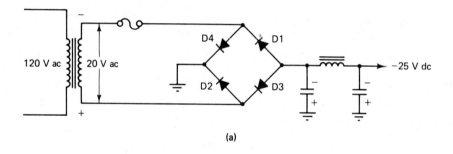

(a)

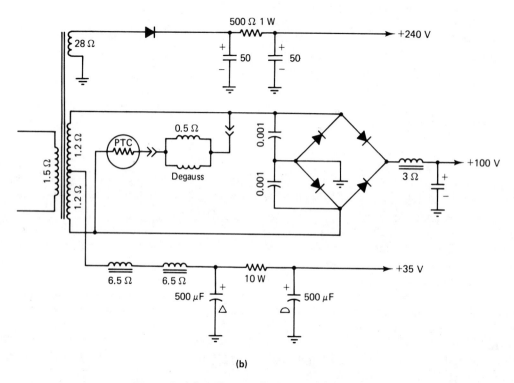

(b)

Fig. 6-11 (a) Full-wave bridge rectifier power supply; (b) dual output bridge type PS (note the low resistance of all coils).

full-wave rectifier, but it does not require a *tapped* transformer. The dc output is roughly equal to the total secondary voltage. Two rectifiers conduct in series for each alternate half-cycle of the ac. With the polarity shown, *D1* and *D2* are conducting to produce the output, with *D3* and *D4* conducting on the next half-cycle.

The four diodes may be either separate components or combined in an IC chip. For the latter, replacement of the entire package is necessary if one diode goes bad. Another version of a bridge circuit (Fig. 6–11 b) uses an off-center transformer tap to develop a *lower* voltage.

• **The Secondary PS (Scan Rectification)**

This system is used extensively in modern sets to supply LV to a number of stages. The voltage comes from a winding on the flyback transformer. Half-wave rectification is used and very little filtering is needed because of the high ripple frequency (15.75 kHz).

6-5 FILTERS

Although the trend is toward greater use of active power filters (APF), the majority of sets still use the conventional brute-force *L/C* pi-type filter. Filter capacities vary greatly, from about 20 µF or so to several thousand microfarads in circuits supplying heavy current. A general rule here is 1000 to 2000 µF per ampere of load current. The type of circuit is also a factor. Since a half-wave rectifier produces only one dc pulse for each cycle of ac, it is generally harder to filter than full wave where the ripple frequency is higher.

The capacitors are constantly being charged and discharged, taking on energy during the peaks of the dc pulses, and giving it up (to fill in the gaps between pulses and to supply the load) at other times. The greater the rate of discharge, the lower will be the average dc output. Remember that the input filter has the greatest influence on the output *voltage,* and the output filter carries most of the burden of filtering. Thus, when it comes to checking, some ac ripple voltage can normally be expected across the input filter, but across the output filter it should be quite low, ideally, near zero. Another point to remember is that at the moment of set turn on, the input filter acts initially as a direct short. Its high charging current can exceed the maximum rating of the rectifier(s), hence, the need for surge resistors in some circuits.

REMINDER: Electrolytic filters are polarized and must be wired correctly, and their voltage rating must be at least as high as the voltage peaks *under no-load conditions.*

Filter chokes are quite different from those found in tube-type receivers and they are not interchangeable. By comparison, they have less inductance and are wound with heavier wire to handle the increased current and to keep the IR drop to a minimum. Depending on the degree of filtering required and the load current, a TV choke may have an inductance of about 5 Henries and a dc resistance of 5-20 ohms.

• **The Active Power Filter (APF)**

A typical APT is shown in Fig. 6-12a. Note the absence of a filter choke (although capacitors are still required). The circuit uses two transistors, *Q1* (the power filter) and *Q2* (a filter driver). The operation is as follows: Any ripple voltage at the PS output is also present between the base and emitter of the driver *Q2* to be amplified, in-

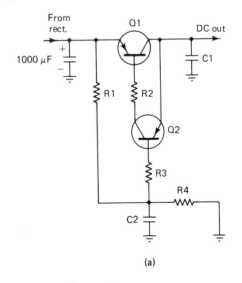

Fig. 6-12 (partial)

162 LOW-VOLTAGE POWER SUPPLY

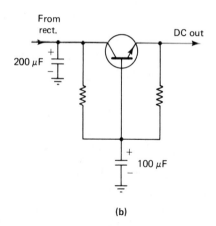

(b)

Fig. 6-12 Active power filters (APF).

verted, and applied to the base of $Q1$. The amplified ripple (of reverse phase) cancels out any ripple that may be present at the output of $Q1$. The circuit is also effective at audio and vertical and horizontal sweep frequencies, eliminating the possibility of the PS acting as a common coupling impedance between these stages.

Figure 6-12b represents a simplified APF circuit, which functions much like a zener regulator. The transistor is in series with the PS output. Any increase in the C/E current flow (due to ripple) develops a small *reverse* bias. This reduces the transistor conductance (increases its resistance), reducing the output for that instant. The reverse takes place during any *decrease* in ripple current, and the momentary increases in current fill in the ripple valleys.

6-6 DC VOLTAGE DISTRIBUTION

Because of the vast differences in the voltage and current requirements of the various stages, a complex network of both series dropping resistors and voltage dividers is required to feed the many transistors and ICs in a typical TV receiver. The *series resistors*, in addition to their main function of dropping voltage, also provide isolation (decoupling) between stages to prevent the interaction of signals. This requires filters and/or decoupling capacitors at each resistor junction. Depending on their relative R and C values, they also serve in some cases as either high- or low-frequency *filters*.

Voltage dividers, on the other hand, are used sparingly, because they waste power. In low-voltage circuits, however, this is usually minimal by using fairly high values of bleeder resistors to ground. Voltage dividers stabilize or improve the voltage regulation of critical circuits.

In dropping voltage, all resistors develop *heat* to a small or large degree. This heat must be safely dissipated to prevent damage to the resistor and other nearby components. In practice, the wattage *rating* of a resistor is usually at least 25% greater than the actual power dissipation. Even so, they can still burn out if an overload occurs.

Sometimes it is necessary to *compute* the value of a burned-up resistor, and a little exercise in figuring values is in order at this time. Figure 6-13 shows a *greatly simplified* hypothetical voltage distribution network to serve our purpose. Voltage and current requirements are as indicated, using round numbers for simplicity. Capacitors are also shown, although they have no effect on voltage and current distribution unless they become shorted or leaky.

Note the path of current flow for the various transistors. For $Q1$, the path is from the negative side of the PS, through the transistor from emitter to collector, through its load (the coil), the series dropping resistor $R1$, and back to the positive terminal of the PS. The current of $Q2$ goes from the negative side of the supply, through the transistor, $R3$, $R2$, and $R1$, then back to the supply. $Q3$ and $Q4$ are essentially in parallel. Their combined currents flow back to the PS, via $R6$, $R4$, $R2$, and $R1$. Note that the bleeder resistor $R5$ passes *only* bleeder current.

Using Ohm's law, the individual resistor values can be calculated as follows: $R1 = 40/1.11\ \Omega$, $R2 = 20/0.11\ \Omega$, $R3 = 20/0.1\ \Omega$, $R4 =$

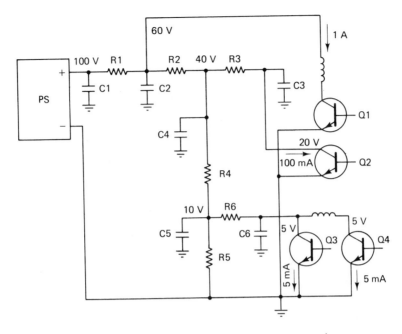

Fig. 6-13 Simplified voltage-distribution network.

$30/0.01\ \Omega$, and $R6 = 5/0.01\ \Omega$. In the example, the small bleeder current (which is seldom greater than filter leakage) is ignored. If, for instance, it were 0.5 mA, then $R5$ would be $10/0.0005$, or $20\ \text{k}\Omega$.

To summarize, observe that $R1$ handles the total current flow of *all* transistors. Resistor $R2$ passes the combined currents of $Q2$, $Q3$, and $Q4$. Resistor $R3$ passes only the current of $Q2$. Resistors $R4$ and $R6$ both pass the combined currents of $Q3$ and $Q4$. Using the preceding values, the power dissipated by each resistor is readily computed.

REMINDER:

The power rating of a resistor must be somewhat higher than the *actual* power dissipated.

- **Component Defects and Distribution Problems**

A defective component may be a shorted or leaky capacitor, a resistor that is open or has changed value, a bad IC or transistor, and so on. In some cases, such defects can upset the voltage and current distribution of the entire set. In other cases, a breakdown may affect only the stage or stages supplied from a single discrete branch line. Referring to Fig. 6-13, for example, if $C1$ or $C2$ becomes shorted, the result is a completely dead set. For a shorted $C5$ or $C6$, or an open $R6$, only $Q3$ and $Q4$ will be completely disabled. Further examples are described in Sec. 6-13. Noting which resistors run hot or cold and the voltages at the different circuit points are often prime indicators of the nature and location of a defect.

6-7 VOLTAGE REGULATORS

As explained earlier, variations in load current (the normal current drain of all transistors) create a fluctuating voltage drop across the internal resistance of a power supply, and the dc

164 LOW-VOLTAGE POWER SUPPLY

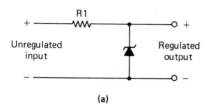

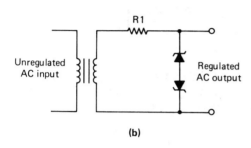

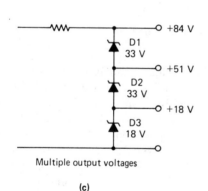

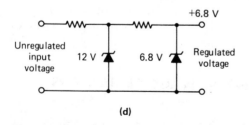

Fig. 6-14 Zener-diode voltage regulators: (a) simple dc regulator; (b) simple ac regulator; (c) stacked Zener diodes; (d) cascaded Zener-diode regulators.

output (under load with current flowing) tends to vary accordingly. With a well-designed PS, such changes are kept to a minimum. Voltage regulation is sometimes expressed as a percentage:

$$\frac{E \text{ (no load)} - E \text{ (full load)}}{E \text{ (full load)}} \times 100$$

The lower the percentage (the less the difference between no load and full load), the better the regulation is.

Let's take an example. The unloaded output of a PS is, say, 50 V, and its internal resistance, which is represented by the forward resistance of the rectifier(s), the filter choke, and power transformer, is 10 Ω. With a steady drain of 2 A, the internal drop is 20 V and the voltage output will drop to 30 V. The regulation in this case is

$$\frac{(50 - 30)}{30} \times 100$$

or nearly 70% (which is *poor*). Taking it further, if the load current *fluctuates*, say between 1.5 and 2.5 A, then the output will fluctuate between 35 and 25V, a total of 10 V p-p, creating many problems.

Certain critical TV circuits require a *stabilized* voltage using some form of regulator. Although very effective in controlling normal fluctuations, they cannot be expected to compensate for all *abnormal* conditions, for example, when one or more transistors draw excessive current or when ac line voltage fluctuations are extreme. Also bear in mind that where the PS output is consistently low or high, it is not necessarily a regulation problem. It may be caused by a low or high ac line voltage or insufficient or excessive load current caused by circuit defects. Different regulator circuits are described next.

• **The Zener-Diode Regulator**

This is the simplest and most commonly used regulator. A basic circuit is shown in Fig. 6-14a.

6-7 Voltage Regulators

As explained in Sec. 5-9, a zener, like any diode when reverse biased, has a very high resistance. When its *zener* or *avalanche* voltage is exceeded, it conducts; the higher the voltage is, the greater the conductance, and vice versa. When connected as shown, this current produces a changing voltage drop across the series resistor $R1$ and varies the output voltage accordingly. Hence the output is maintained reasonably constant regardless of variations of the ac input voltage or the dc output load current. Where the voltage tends to rise, it is suppressed by the increased zener current; any tendency to drop is compensated for by a reduction in zener current and the output rises to normal.

Zeners are available with different voltage and current ratings and can be used in a variety of ways. The circuit shown in Fig. 6-14a is for a positive dc PS; for a negative supply, the zener leads are simply reversed. For regulating *ac,* two zeners may be connected back to back as in Fig. 6-14b. For regulating voltages greater than the rating of a single zener, several may be connected in series, as in Fig. 6-14c. Sometimes taps are brought out from their junctions, as shown, to provide different regulated voltages. For improved regulation the circuit of Fig. 6-14d may be used with two or more zeners in *cascade.*

- **Transistor Voltage Regulators**

Transistors can be used for voltage regulation in a variety of ways, but basically they fall in one of two categories, *series* regulators or *shunt* regulators. Both NPN and PNP types are used depending on whether it is a − or a + dc PS. In many cases, the entire regulator circuit is contained in an IC chip.

The simple *series* regulator, shown in Fig. 6-15a, looks and operates much like a basic APF circuit (Fig. 6-12) except for the zener, which is present in most regulator circuits. The dc output is continuously compared with a reference voltage, in this case the input voltage between the B and C of the transistor. The base voltage is held constant by the zener. In essence, the transistor (in series with the output) acts as a variable resistor whose conductance changes with variations of either input or output voltage.

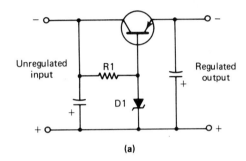

(a)

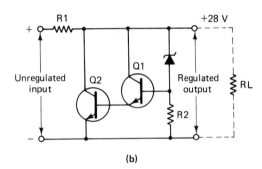

(b)

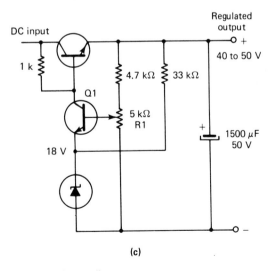

(c)

Fig. 6-15 (partial)

166 LOW-VOLTAGE POWER SUPPLY

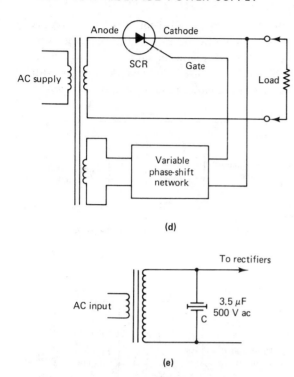

Figure 6-15b shows a basic *shunt*-type regulator. Compared to the previous circuit, here the transistor acts as a variable *load*. Transistor $Q1$ senses any changes in voltage and controls the conductance of $Q2$, whose current develops a voltage drop of the desired amount across series resistor $R1$ to change the output accordingly.

Another circuit variation is shown in Fig. 6-15c. This circuit has a trimpot ($R1$) to establish the desired voltage range over which the regulator operates. Variations in output as developed across the potentiometer circuit are fed to the error-sensing amplifier $Q1$ at any predetermined amplitude. A more elaborate circuit arrangement uses three transistors: an error sensor, a driver amplifier, followed by the regulating transistor, which has a much higher current rating than the others since it must handle the entire load current.

Fig. 6-15 Transistor, SCR, and transformer voltage regulators: (a) simple series voltage regulator; (b) schematic diagram, basic shunt regulator; (e) voltage regulating transformer.

• **Thyristor Voltage Regulators**

Compared to a transistor-type regulator, which provides a continuously variable *linear* control over voltage, an SCR or triac is a *switching*-type regulator. It simultaneously performs both the rectifier and regulator functions of the PS. A popular circuit configuration is shown in Fig. 6–

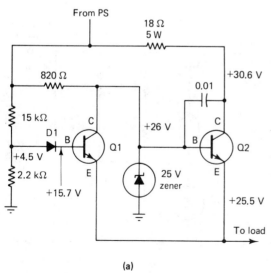

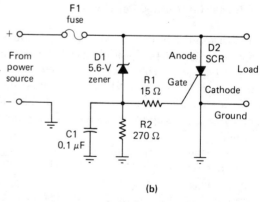

Fig. 6-16 Overload protection circuits.

15d. The phase-shift network is a timing circuit for the generation of *trigger pulses* for firing the SCR at precise times relative to the rectified pulses. The *duration* of SCR conduction determines whether the output is increased or decreased as required.

• **Voltage Regulating Transformer (VRT)**

The VRT (Fig. 6-15e) is a specially designed power transformer for regulating the amount of *ac* supplied to the rectifier(s). High output is obtained by *resonating* the transformer secondary winding to 60 Hz with a special *resonating capacitor* across the coil. The voltage applied to the rectifiers is a *square wave*. The resonating capacitor, (C1), usually about 3 to 4 μF, is an oil-filled paper type with a voltage rating high enough to withstand the total p-p voltage output of the transformer. Regulation is accomplished by operating the transformer core close to saturation. Normal variations of sine-wave input AC produces normal output voltage. But when the core becomes saturated and induction ceases, excessive changes of input voltage produce little if any voltage changes in the square-wave output. With this circuit, voltage measuring can be a problem when using rms reading meters, since a reading for square waves is considerably less than for sine waves.

6-8 OVERLOAD PROTECTION

Modern sets make good use of fuses (regular and slow-blow, fusisters and links), as well as circuit breakers in both ac and dc circuits. But all these devices are relatively slow acting and not completely reliable. Hence the need for additional protection, especially since transistors and ICs are so susceptible to overload. A number of circuits are in use to prevent catastrophic damage (like the all too frequent destruction of a whole series of components) from a momentary overvoltage or current overload.

A *current* overload is often the result of a B voltage short (a filter capacitor that has high leakage or becomes shorted, a shorted or leaky diode or transistor, etc.). A circuit that protects against current overload is shown in Fig. 6-16a. During normal operation, $D1$ is reverse biased and keeps $Q1$ biased to cutoff. If the output (load) voltage suddenly drops for any reason, $D1$ becomes forward biased into conduction, which in turn biases $Q1$ into saturation. When $Q1$ is conducting, $Q2$ (which passes all the load current) is biased to cutoff, effectively removing the load from the source.

An *overvoltage* condition often results from trouble in a voltage regulator circuit or when a load is removed. A simple protective circuit is shown in Fig. 6-16b. Normally, both the zener and the SCR are nonconducting. If the source voltage goes up, the zener conducts, producing a triggering pulse to fire the SCR into conduction. This constitutes almost a direct short on the PS and blows the fuse.

6-9 BATTERY-OPERATED PORTABLES

A typical battery portable will operate either from a 12-V car battery or 120-V, 60-Hz house current. Some have a built-in battery-charging circuit. For power line operation, the LVPS is conventional, developing 12 V dc, consistent with the battery voltage. In most cases the supply is regulated. Unlike most CRTs, the small screen pix tube has a directly heated 12-V (low-current) filament, which is supplied from the 12-V dc bus.

6-10 THE TYPICAL LVPS

The circuit of a late model color set is included at the back of this book. The example set uses modular construction. For simplicity, the different circuits drawings are arranged at the back of the book in the order of coverage in the text. The following description of the LVPS section refers to sample Circuit A.

The LVPS here is more or less conventional. A power transformer is used so there is no need to use an isolation transformer when servicing. A special self-regulating transformer is used (as described in Sec. 6-7) that operates in conjunction with a *resonating capacitor* (C 905) whose value is extremely critical. This is sometimes called a *ferroresonant* PS.

This LVPS is protected with two fuses, one in the 120 V AC input circuit, the other between the power transformer and the main PS rectifiers. The ac input circuit has the usual noise filter (L 900). The input power is switched via connector RL 91. Terminals 1, 2, and 3 connect either to remote control power turn-on circuitry if provided, or the front panel on/off switch. The ADG circuit is conventional. When cold, the VDR (R 902) provides a low resistance path for ac current through the degaussing coil. During receiver turn-on as R 902 heats up, its increasing resistance reduces the coil current to a very low value. Additional current reduction is created by bucking action from the transformer winding.

REMINDER: ───────────────

Any residual ADG current after warmup will degrade the pix.

The LVPS actually consists of five separate power supplies. The main power supply uses a bridge rectifer which is clearly evident. The rectifiers are supplied with ac from one winding of the power transformer and has a dc output of + 131 V. The 131 V supply is filtered with a choke coil, L 901. The other PSs use resistors for this function. Note the large value filter capacitors used in all PSs. Additional filtering is provided by R/C networks in the various voltage distribution circuits throughout the set. Note the small (0.001 mF capacitors shunted across all rec-

tifiers. Their function is to minimize RF radiation produced by shock excitation of the diodes.

Besides the bridge rectifier their are two additional full-wave PSs; one PS using diodes Y 905 and Y 906 develops + 23 V. The other PS uses diodes Y 907 and Y 908 to develop + 15 V. This PS is regulated by zener diodes Y 911 and Y 912.

There are also two half-wave PSs. One uses diodes Y 909 and Y 915 to produce + 9 V; the other uses diodes Y 913 and Y 914 to develop + 55 V, which is reduced to + 13 V by resistor R 907. Although one rectifier will do the job, each of these PSs uses two diodes in series as insurance against breakdown. Note the symbols used to facilitate identification of the different voltage points throughout the set. Also note the numerous taps on one winding of the transformer, with the off-center ground tap. One winding of the transformer supplies heater voltage for the CRT, with one side of the circuit connecting to B + to reduce the chance of internal shorts in the CRT.

Other dc voltages are developed by *scan rectification* in the horizontal output circuit as described later in Sec. 9-9.

6-11 TROUBLESHOOTING THE LVPS

Troubles normally associated with the LVPS include the following: completely dead set (no raster or sound); smoke or other evidence of overheating or burning; arcing; audible hum; hum bars on the pix tube; poor regulation where pix blinks or fluctuates with the sound; small raster (reduced width and height); multiple symptoms; blown fuse or tripped circuit breaker. Be guided by such symptoms in your troubleshooting approach.

Start by locating the PS components on the chassis. Note whether or not there is a power transformer. Study the schematic. If the PS has different dc outputs feeding different stages, try to relate the symptoms and stages to a particular dc source. Verify with a voltage check and com-

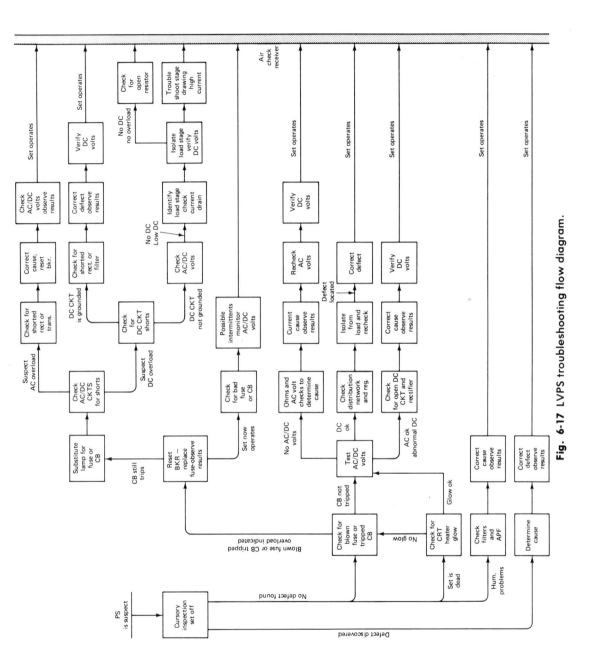

Fig. 6-17 LVPS troubleshooting flow diagram.

169

pare with the schematic. Where *all* stages are affected, the *main* sections of the PS are probably responsible.

Inspect the PS section for anything unusual. Make full use of all senses as described in Sec. 4-8. Be particularly alert for signs of overheating, smoke, and arcing. *Sniffing* and *touching* may provide good information. Feel all diodes, transistors, and *small* resistors in the area for signs of heating. If there are filtering problems, look for seepage or dried powder around electrolytic terminals. Where shorts or overheating is suspected, it is a good idea to monitor the PS output voltage during power turn on, rather than waiting for troubles to show up with the risk of component damage. Better still, make an ohmmeter check (at the filters) for a possible short *before* switching the set *ON*.

Don't be in a hurry to use your instruments. Some conclusions can be reached *without* a lot of unnecessary testing. For example, if the CRT heater and dial lights come on, we know there is *ac* power; and the presence of sound or raster means there is B voltage. Hum is usually due to a filtering problem. When components overheat, there is obviously an overload. When a high-wattage resistor runs cold, it is either open or is not passing current. This is often a good place to start. But sometimes testing is done in a reverse order, as when the first tests made show no voltage is reaching a transistor or IC of a particular stage, in which case testing is performed working backward toward the source.

Different technicians may approach a problem in different ways; but to avoid component damage power turn-on should be avoided until at least a few preliminary tests and observations have been made. A troubleshooting flow diagram for the LVPS is shown in Fig. 6-17.

REMINDER: _____

Observe common sense precautions against shock hazards and damage to components (see Sec. 4-6). Never spark B voltage circuits to ground, intentionally or otherwise.

For the purposes of troubleshooting, the PS can be represented as having two main functional areas: the *ac input circuits,* those between the power plug and the rectifier(s), ADG, and the CRT heater, and the *dc circuits,* from the rectifier output(s) to the receiver stages that use this voltage. In the following procedures, each of these two sections is considered separately. Where a component is suspect, test as described in Sec. 5-8.

6-12 TROUBLESHOOTING THE AC POWER CIRCUITS

Table 6-1 shows a quick-reference troubleshooting chart for the ac section of the PS. As a rule, fault finding and repairs are relatively simple and straightforward. Information presented in Table 6-1 is of a *general* nature and applicable to all types of PSs. Specific details are given next.

- **The 120-V ac Input Circuit**

One common trouble here is an intermittent or completely open line cord. Check continuity; or with set on, gently tug and twist at both ends near the plug and interlock, two points subject to the greatest abuse. Check for sprung or dirty contacts at the interlock and listen for signs of arcing. A suspected bad cord should be *replaced,* not repaired. Bad interlock contacts may be cleaned or tightened.

A defective ON/OFF power switch is not uncommon. Check with ohmmeter or by bridging with a jumper with power *on*. If the switch hangs up (latches) in either position, try injecting a solvent and work it repeatedly. Allow the residue to evaporate before turning set *on* to avoid the bad hum caused by leakage between the ac on the switch and the volume control to which it is attached. If there are any signs of heating or charring of the switch, it *must* be *replaced.* Dual

6-12 Troubleshooting the ac Power Circuits

Table 6-1 AC Circuits Troubleshooting Chart

Trouble Condition	Probable Cause	Remarks and Procedures
No ac or low ac at rectifier input (no overload indicated)	No power at 120-V ac outlet; faulty line cord, interlock, noise filter chokes, ac switch, ADG circuit open, tripped breaker or blown fuse, open electrolytic at doubler input, open transformer coil, open fusistor.	Make continuity checks or ac voltage checks to locate open circuit. CRT heater may or may not be lit. See if ADG coil is disconnected.
CRT heater or dial light not lit (no overload indicated)	Burned out tube or dial light. Blown fuse link, defective tube socket, faulty instant-on circuit.	Sound from the speaker indicates whether set is completely dead. *Note:* CRT heater may be supplied from flyback transformer.
Heater circuit overload	Defective filament transformer, grounded heater or dial light circuit, shorted heater circuit wiring.	Unless circuit is fused, wires will burn, creating a fire hazard.
Low or high line voltage	Inadequate house wiring, or problems with utility power source.	May occur only at certain times of day or night and in certain localities.
Low ac, low or no dc with indications of overheating	Shorted transformer windings, shorted rectifier(s), main dc bus is grounded, shorted or leaky filters.	Depending on location and extent of the overload, it may or may not blow fuses or trip the circuit breaker. Note which components are overheating.
ADG problems	Defective thermistor or varistor, open coil circuit.	May or may not affect PS output depending how coil is connected. Make sure coil is plugged in.
Blown fuse or circuit breaker tripped	Any drastic overload condition. Could be a faulty fuse or breaker if set works after resetting or replacement.	Reset breaker or replace fuse only *once*. If set operates, investigate further for possible intermittent condition.
Intermittent ac	Any open circuit or component as listed above.	Monitor ac volts at various points up to rectifier inputs. Make every effort to make the problem show up.
Arcing	Carbonized leakage path between connections. Charred components.	Increase contact spacing. Correct leakage condition, replace damaged components.

switches are notorious for this condition, especially those with an instant-on circuit.

A continuity check (with power *off*) quickly reveals a burned-out fuse or an open or erratic circuit breaker (CB). After resetting the breaker, check its latching by vigorously tapping. If intermittent, do not attempt a repair; replace the unit. Even when a new fuse or resetting the breaker restores set operation, look for the cause of the blown fuse or the tripped breaker. Make sure any replacement is of the correct current rating for the set. Never replace a *fusistor* with a fuse. The resistor is needed to limit the high charging current of the input filter when the set is turned on.

• **The ADG Circuit**

Where the degaussing coil is in series with the power input, a break in the circuit (as from a disconnected plug or connector) will disable the set. Where the ADG circuit is at fault, there may be partial or complete loss of demagnetization flux, leading to purity and convergence problems, or the circuit may be energized continuously, which shows up as a weaving mixture of colors on the screen. Where a bad thermistor or thyristor is suspected, test as described in Sec. 5-8.

• **The Power Transformer**

Voltage and continuity tests will show up an *open* winding. *Shorted* windings cause overheating and eventual burnout. Resistance readings are seldom useful because a *few* shorted windings cause so little change from the normal. Voltage (ac) measurements are more meaningful.

NOTE: _____

Don't expect the same resistance readings for both halves of a center-tapped secondary. The outermost windings (of greater diameter) require more wire and have a slightly higher resistance, although the *ac voltage* across each half should be the same. With multitapped transformers supplying different rectifier circuits, a variety of ac voltages can be expected. Refer to the schematic.

The power transformer of a typical large-screen color set normally runs quite hot. When there is internal arcing or shorted windings, it gives off the odor of burning insulation. To determine if overheating is caused by a transformer breakdown or is due to a current overload, try isolating the secondary winding(s) from their load. If overheating continues, the transformer must be bad.

Identifying the leads of a transformer can sometimes be a problem since color coding is not always standard. When making a replacement, first identify the primary winding, which usually measures less than 10 Ω. The low-voltage heater windings are much less than this and the HV coil much greater. Tentatively connect what appears to be the primary to 120 V ac using a fine wire "fuse" (in case of an error). Identify all secondary windings by measuring their ac outputs.

Transformers having an isolated primary are desirable from an electric shock and fire hazard standpoint, especially when servicing. Some sets use an *autotransformer* where primary and secondary windings are directly connected. This type does *not* provide isolation, and, when servicing, a 1 : 1 isolation transformer (Sec. 3-20) should be used.

Some transformers have a tapped primary to compensate for either low or high line voltage. Where the line voltage is normal or tends to be high, the entire winding is used. For consistently low line voltage, the tap may be used to boost the PS output. Sometimes a *variac* or booster transformer (Sec. 3-20) is used for this purpose. The cause of low line voltage may be inadequate house wiring or low voltage at the service entrance to the building. If the latter, it is the responsibility of the power company and they should be notified.

Some sets have a built-in circuit for auto-

6-12 Troubleshooting the ac Power Circuits

matically regulating the ac, either the line voltage or the ac applied to the rectifiers. They may use a triac or two zeners back to back, shunted across a winding, or a special voltage-regulating transformer (VRT) as described in Sec. 6-7. To check the regulation, vary the ac line voltage and/or the dc load current while monitoring the PS output.

- **Tube Heater(s) and Dial Light Supply Source**

The low-voltage ac for the CRT heater and dial lights may come from a winding on the power transformer, a separate filament transformer, or, in some cases, the flyback transformer. Because of the high current, a short in this circuit can have serious consequences, hence the use of *fuse links* as protection against overload. Never defeat their purpose with a permanent jumper. Checking this low-resistance circuit for a short or ground requires that the heater(s) be isolated from the source. Where an unfused heater circuit becomes shorted, scorched or burned-up wiring results. Look for the short at the end of the burned wires *farthest* from the source.

Some hybrid sets have their tube heaters (including the CRT) wired in series and powered from the 120-V line with a dropping resistor. When one tube goes out or the circuit is interrupted for any reason, none of them will light. The break may be found by testing the tubes individually or by a continuity test of the circuit. Another method is by checking for ac along the circuit and across each tube; the voltage across a good tube (without current) will be zero; across an open filament, you read the total line voltage.

- **Instant-on Problems**

When instant-on heater circuit troubles are suspected, observe the CRT heater, which should be at half-brilliance with the set plugged in but turned *off*. Heater brightness should increase to normal when the set is turned *on*. If in doubt, measure the voltage across the CRT heater contacts. If there is a problem, check the extra filament transformer, if there is one, the dual power switch, the instant-on defeat switch, and the series diode, when used. Poor contact between tube pins and socket is another cause of trouble. If the tube does not light, try wiggling the socket as a *test*.

- **Checking Power Consumption**

Sometimes a power check is warranted (e.g., where an overload is suspected or there is periodic fuse blowing). Few technicians possess a wattmeter, but there is an acceptable alternative method for checking power consumption. Simply connect a *low-value* resistor (1 Ω or less) between the 120-V line and the set. Compute the power using the measured ac voltage developed across the resistor. Compare the results with the wattage stamped on the set, making allowances for any difference in line voltage and the slight insertion loss created by the resistor.

- **Checking Power Line Isolation**

This should be a *must* for every transformerless set with a floating ground. It should be checked before it leaves the shop to ensure isolation. A short or leakage between the power line and the chassis is a potential shock and fire hazard to the set owner, particularly if the set is positioned near some grounded object such as a floor furnace, water or gas pipes, and certain appliances. An antenna grounded via the mast also increases the risk. The explanation is that one side of the 120-V, 60-Hz power is hot and the other side is grounded. This is why some sets use a *polarized* line plug that can only be inserted one way.

The isolation test is simple and requires but a few seconds. Remove the power plug from the outlet. With ohmmeter on the megohm range, check for leakage from each plug prong to

174 LOW–VOLTAGE POWER SUPPLY

Table 6–2 DC Circuits Troubleshooting Chart

Trouble Condition[a]	Probable Cause	Remarks and Procedures
No dc at rectifier output. No overload indicated.	Defective rectifier(s) (open)	An open rectifier is usually caused by an overload. Look for the cause before replacing.
Low dc at rectifier output. No overload indicated.	Open rectifier (where more than one is used), open or dried up input filter.	If a voltage doubler, all filters are suspect.
No or low dc at rectifier output, with overload indications.	Shorted or leaky rectifier(s), shorted or leaky filters, shorted power transistors, or zener diodes	Shorts or leakage in regulator or voltage-distribution network will lower the dc at rectifier outputs. May or may not result in a blown fuse or a tripped breaker.
No or low dc at output of regulated branch circuit (no overload indicated).	Series dropping resistor open or increased, open regulating transistor.	Assuming there is normal dc on the main dc bus.
No or low dc at regulator output (overload is indicated).	Leaky or shorted decoupling or regulator circuit capacitors, leaky/shorted zener or regulator transistor. Excessive current drain by load stage.	These troubles will reduce the dc at rectifier output. No dc regulation. If high load current suspected, isolate load and recheck.
Excessive dc voltage at regulator output (no overload is indicated).	Open zener. Shorted regulator transistor, insufficient current being drawn by load.	DC is not regulated under these conditions. If low load current, troubleshoot the faulty stage.
Hum troubles. (may be hum in sound, hum bars on pix, or both.)	Open or dried-up filters or APF circuit troubles. Open rectifier diode in doubling or bridge circuits.	If dc is low, suspect the input filter. If dc is OK, suspect output filter(s).
Arcing.	Damaged insulation of component, wiring, PCB, etc. Component or circuit connections almost touching, foreign matter in PCB or other wiring. Component defect. Carbonized leakage path on insulation parts.	Arcing may be intermittent or sustained. Use all senses to locate source.

6-13 Troubleshooting the dc Circuits

Table 6-2 (cont'd)

Trouble Condition[a]	Probable Cause	Remarks and Procedures
DC circuit fuse blows.	Any overload condition as listed above.	If no apparent cause, suspect intermittent condition. Take all necessary steps to recreate problem. Monitor fuse circuit current.
Intermittent dc (ac OK).	Any dc circuit defect as listed above.	Monitor dc at various circuit points. Try to recreate trouble condition.

[a]For all troubles listed, it's assumed there is ac at rectifier input(s)

chassis. Leakage should be at least 100 kΩ or better. If less than that or a direct ground, locate and correct the cause, which may be a wiring short, a shorted isolation capacitor, or, in the case of some doubler circuits, a shorted or leaky diode or filter in the PS.

6-13 TROUBLESHOOTING THE DC CIRCUITS

Assuming there is ac up to the rectifier(s), then any *dc* problems involve one or more of the following: the rectifier(s), filters, voltage-distribution networks, voltage regulators, *R/C* decoupling networks, or conditions associated with the *load*(s) (i.e., factors that influence the current drain of the various transistors and ICs). Table 6-2 shows a quick-reference troubleshooting chart with possible causes for the more common dc trouble conditions. This information is of a general nature as applicable to most types of PSs. Detailed troubleshooting procedures for each type of PS are described in the following paragraphs.

- **Low or No DC at Rectifier Output(s)**

If there are no signs of overheating, chances are one or more rectifier diodes are open, or there is a blown dc circuit fuse, if there is one. In either case, check further for the *cause,* which may be a shorted or leaky filter, probably the input filter. The reason no overload is apparent is because the open fuse or rectifier isolates the load from the source. Loss of dc across the output filter capacitor (if it checks high at the rectifier output) means the choke is probably open. When voltage is low at both ends of the choke and there is no overload, the input filter may have lost capacity or is open, or there is a rectifier problem; if several, only one may be open. Loss of voltage and components running hot signifies an overload. If one or more diodes overheat, check for a leaky or shorted input filter or a leaky or shorted APF transistor. If the choke also gets hot, it is probably the output filter or trouble further along the line. In either case, turn the set *off* (to avoid further component damage) while deciding on the next step.

When voltage checks high at the output filter, and there is partial or total loss of voltage on one branch line, and no overload is indicated, check for an open dropping resistor in that particular circuit. If the circuit is *regulated,* look for an open resistor, diode, or transistor in the regulator circuit. Such defects can also result in a *higher-than-normal* dc at the regulator output.

When *all* dc voltages check low with very low or zero dc on one particular branch line, with *in-*

dications of overload, the possibilities are: a leaky or shorted filter or decoupling capacitor on that line, a leaky or shorted zener or transistor if it is a regulated circuit, or there is excessive load current being drawn from that line. For the latter possibility, one approach is to break the circuit to isolate the load from the source to see if the dc is restored or increases appreciably. Conversely, if load current is too low for any reason, the voltage on that particular branch line will measure *high*. Typical causes for abnormally high or low current by transistors and ICs are discussed in Sec. 5-6.

• Troubleshooting DC Branch Lines

The cause for loss of voltage on a branched circuit (if there is *no* overload) is easy to locate by simply checking the various series-dropping resistors in the circuit. When there are signs of overload, it can be more difficult, since there are numerous possibilities. As previously explained, these fall in either of two categories: a short or high leakage to ground or a high load current. For the latter, this means troubleshooting whatever stage is supplied from that branch line.

There are two methods for locating a short: resistance checking and voltage checking. With resistance checking (as with the example shown in Fig. 5-6 and explained in Sec. 5-5), you take a number of readings, and the point of lowest resistance is the trouble site. With the voltage checking method, you look for the point of lowest voltage. In reference to Fig. 5-6, for example assuming there is voltage at the source, point X, but low or zero voltage at point A, then the capacitor is shorted or leaky. If A is higher than at X, then obviously A is supplied from a different source. For loss of voltage at B, if the resistor overheats, then the short or cause of high current drain is at B or beyond. If there is a low reading at C, then the short is at that point. If there is loss of voltage at D or E, then the trouble will be found at those junctions. For *excessive* voltage at D or E, either the diode or transistor is shorted, or these points are supplied from a different voltage source than point X. A zero reading at F can be expected, since that point is grounded.

REMINDER: _____

An overheating component is the best clue for which branch line to follow. Concentrate on the end of a resistor where voltage reading is lowest. Avoid getting on the wrong track by keeping in mind the association between the observed trouble symptoms, the stages most likely involved for such symptoms, and which particular branch lines supply those stages. To help locate the troublesome branch line, isolate the components by interrupting circuits at various points as appropriate, until the cause is identified. Table 6-3 shows how certain defects upset the normal voltage distribution of a typical circuit. (See Fig. 6-13)

• Reversed Polarity

When a reading of the wrong polarity is obtained from one of the PS outputs, check the following possibilities: reversed meter leads, a diode rectifier installed backward (it is very easy to do), or you are sidetracked onto the wrong circuit. Check with the schematic.

• Arcing

Arcing may be caused by either a breakdown in insulation (wiring, terminal strips, etc.) or inside a component. Usually, but not always, there is a good clue which shows up during inspection, like crackling or popping sounds, smoke, or visual indications (in a darkened room). When there is a charred leakage path, there may be prominent arcing, tiny pinpoints of light along the path, or actual combustion. Arcing inside a component is harder to find. If there are crackling sounds, it can sometimes be

6-13 Troubleshooting the dc Circuits

Table 6-3 DC Voltage Distribution Problems (see Fig. 6-13).

Shorted Capacitors (overload conditions)

C1	C2	C3	C4	C5	C6
No PS output. May trip CB or damage PS components.	No dc to all transistors. PS output reduced. R1 overheats.	No dc to Q2. R1, R2 and R3 overheat. PS output reduced.	No dc to Q2, or Q3/Q4. R1 and R2 overheat. PS output reduced.	No dc to Q3/Q4. Reduced output from PS and to Q1 and Q2. R1, R2, and R4 overheat.	No dc to Q3/Q4. Reduced output from PS and to Q1 and Q2. R1, R2, R4, and R6 overheat.

Open Resistors (No overload and resistors run cold)

R1	R2	R3	R4	R5	R6
High output from PS but no dc to all transistors.	High output from PS and to Q1. No dc to Q2, or Q3/Q4.	High output from PS and to Q1 and Q3/Q4. No dc to Q2.	High output from PS and to Q1 and Q2. No dc to Q3/Q4.	High dc to Q3/Q4.	High output from PS and to Q1 and Q2. No dc to Q3/Q4.

Excessive Transistor Current

Q1	Q2	Q3/Q4
Reduced output from PS and to all transistors. R1 overheats.	Reduced output from PS and to all transistors. R1, R2, and R3 overheat.	Reduced output from PS and to all transistors. R1, R2, R4, and R6 overheat.

localized using a length of rubber hose as a stethoscope.

Where arcing occurs between closely spaced points (as with PC wiring), try to separate the contacts. Examine carefully for any minute signs of charring, which *must* be removed by scraping. For extensive charring of terminal strips, PC boards, or components, replacement is the only reliable repair.

REMINDER:

Taping over or spraying a charred area does *no good;* it only covers up the problem.

• **Checking Voltage Regulation**

Some outputs from the LVPS are regulated, some are not. When the dc from a regulated supply is unstable, the possible causes are trouble with the regulator circuit, excessive line voltage variations, troubles with that portion of the PS, or excessive current demands by the load. Check the regulation as described in Sec. 5-9 using the hookup shown in Fig. 5-8. The test consists of monitoring the regulated output while making changes in ac input voltage and the dc output load current.

Regulator circuit problems are usually caused by a bad zener, SCR or transistor. If an IC, make precautionary voltage checks after removal and prior to replacement. Where line voltage fluctuations are suspected, monitor over a 24-hour period. If caused by inadequate house wiring, try another outlet. If it is the main service, complain to the power utility company.

REMINDER: _____

Consistently low or high line voltage can be corrected with a booster transformer, but this will not help the *regulation*.

Monitor the dc output of the PS *ahead of* the regulator. If unstable, look for the cause in the PS components as previously explained.

For suspected high load current, verify current drain either by computation (as explained earlier) or by actual measurement. Make sure the ammeter or milliammeter has sufficient range to handle the anticipated current (refer to the schematic) and is inserted *in series* after interrupting the circuit at some convenient point.

• **Hum and PS Ripple Checking**

With solid-state sets, the cause of hum is nearly always due to *poor filtering* of one or more outputs of the LVPS. Look for defective electrolytics with dried powder at their terminals as a quick visual check.

Different instruments can be used to locate a defective filter by testing: an EVM set to read *ac* voltage, a CRO, or some form of *aural* signal tracer. Simply probe each and every electrolytic capacitor in the PS section, looking for the one with the greatest amount of ac (ripple) voltage at its terminals. Good capacitors show a minimum of ripple. Next, try bridging the *suspect* with a substitute capacitor of the proper value and voltage rating. Make sure of correct polarity depending on whether it is a + or a − supply. If bridging a suspect capacitor reduces the amount of ripple voltage *significantly* and clears up the symptoms, then replace it. On older sets, several (or even *all*) the filters may require replacement. For a multisection electrolytic, replace *all* sections even if only one checks bad.

REMINDER: _____

Where there are ripple problems and the dc checks low, it is probably the *input* filter that is defective. If the dc checks OK, check the *output* filter and any capacitors beyond.

Where an APF circuit is used, check for an open zener or control transistor.

• **Secondary Power Supplies**

SCAN rectification is used in many modern sets. Voltage problems relate to the horizontal sweep circuit where they originate. For appropriate troubleshooting procedures, see Chapter 9.

• **Silicon Radiation**

Unless shunted with a small capacitor, silicon rectifiers sometimes develop a high-frequency radiation that is picked up by the tuner. Other remedies include: dress leads between antenna input and the tuner and keep clear of the LVPS, and lead dress of rectifier leads and ac line filter choke. Keep leads short and close to the chassis.

• **Troubleshooting AC and Battery-Operated TV Portables**

These sets operate from either the 120-V ac line or a 12-V battery. For battery operation, a selector switch connects the battery to the B voltage bus. For ac operation, the output of a conventional regulated supply is switched to the bus. Where a built-in charger is used, another switch position feeds the charger output to the battery.

As a first test in troubleshooting, check to see if the CRT heater is lit. If there is no dc on the 12-V bus for any switch position, look for trouble in the regulator circuit, the selector switch, or some stage supplied from the bus. If trouble only occurs for the ac position, check the PS using procedures previously described. But if trouble exists on the battery-operating position, a good possibility is that the battery is bad or requires charging.

REMINDER:

An open-circuit voltage check on a battery is *meaningless*. Check battery voltage *under load*. If the battery is defective or needs charging, the voltage will drop *drastically* when the set is turned on.

NOTE:

A common cause of low B voltage on these sets is excessive current drain by the directly heated CRT due to a partially shorted CRT heater. Check cold CRT filament resistance and compare with a CRT known to be good using an ohmmeter.

- **Troubleshooting Remote Control ON-OFF Switching Functions**

Power turn on for the set requires that the remote control unit be *on* at all times and the set plugged in. When the power ON or OFF button on the transmitter unit is pressed, the LVPS of the set is activated from a motor-driven switch or a stepping relay. Where remote control is not possible, look for the cause in the transmitter or the controlled devices. For further details, see Chapter 12.

- **Power Supply Intermittents**

Compared to other circuits, intermittent troubles in the PS are rather uncommon. For suspected loss of ac input voltage or dc output voltage, monitor voltages with a meter left connected as long as necessary. As explained in Sec. 5-13, intermittents can be made to show up in a number of ways. Intermittent fuse blowing or circuit breaker tripping is fairly common. Sometimes the trouble is a defective fuse or breaker, but, to be sure, operate the set for several hours while watching for a reoccurence of the trouble. In particular, monitor the temperature (by touching) of all power output transistors for possible thermal runaway; and listen for any signs of arcing or flashover, especially in the flyback and yoke areas.

Intermittent operation of the CRT heater is another trouble to monitor. If any visible changes in brilliance occur *after* warm-up, try wiggling the socket; dirty or corroded pins or socket contacts are a common cause of intermittent trouble. Try scraping and cleaning pins and sockets. If trouble persists, replace the socket.

6-14 AIR CHECKING THE REPAIRED SET

Never consider a set is repaired until it has been given a burn-in check for at least 30 minutes or more. After making repairs to PS circuits watch for signs of smoke, arcing or overheating components. Make sure the cheater makes good contact with the interlock. Some circuit breakers are intermittent. See if it opens up when tapped. Listen for hum and look for hum bars, a sure sign the filters are going bad. Make sure there is adequate width and height to the pix, a good indication of normal output from the PS. Check the operation of the off-on switch.

180 LOW-VOLTAGE POWER SUPPLY

6-15 SUMMARY

1. The LVPS develops *both* ac and dc voltages. Besides supplying the rectifier(s), the ac is used to heat the CRT filaments in some sets and to energize the ADG coils in a color set. After rectification, either + or – dc voltages of different amounts are supplied to all transistors, ICs, and the CRT, as required for their operation.

2. The dc output of any unregulated PS varies inversely with its internal resistance, represented by the rectifier(s), filter choke, and transformer, and the amount of current drain. For good regulation, a well-designed PS has *low* internal resistance. In the interest of efficiency and good regulation, the current demands of all transistors are kept to a minimum, consistent with their operation.

3. Because of its internal resistance, the dc output of a PS can be expected to drop if one or more transistors draw *excessive* current or if a filter or bypass capacitor is leaky. Conversely, the output voltage will *increase* if the receiver components (transistors and ICs) draw *less* than normal current.

4. The simplest voltage regulator (still used extensively) consists of only a zener diode and a series voltage dropping resistor. Within limits, any tendency of the dc output to change is compensated for by a change in zener conduction whose current develops an *IR* drop across the resistor, raising or lowering the output accordingly.

5. The most popular regulator circuit uses a zener and one or two transistors. The zener and/or one transistor senses any voltage changes and controls the conduction of a *regulating* transistor. Depending on the circuit, the regulating transistor is either in series with the load or shunted across the output, where it functions as a variable load.

6. An APF works much like the regulator described in paragraph 6. The conduction of a transistor connected in series with the output is made to compensate for the slightest voltage variations as represented by the ripple component. No filter choke is required by this circuit, and only nominal values of filter capacities are needed.

7. A simple half-wave PS uses a single diode to rectify either the 120 V ac directly, or the nominal voltage from a transformer winding. With a ripple frequency of only 60 Hz, it is harder to filter than 120-Hz full wave. It is used mostly to furnish medium, unregulated dc voltages where current requirements are low.

8. A full-wave transformer type PS requires a center-tapped winding and two rectifiers. *Each* half-cycle of the ac is alternately rectified to produce a ripple frequency of 120 Hz, which is relatively easier to filter.

9. Voltage multipliers (doublers and triplers) provide an inexpensive way of developing fairly high voltages without the need for a step-up transformer. Generally, the regulation of such a PS is not too good, since the load is supplied from the charging of one or more capacitors.

10. A bridge rectifier uses *four* rectifiers to produce a full-wave output. The full output of a transformer winding is used and does *not* require a tap. Two rectifiers conduct (in

series) for each alternation of the ac. The diodes may be *separate* units or *combined* in an IC package.

11. A power transformer may have a step-up or step-down ratio, or both, to meet the different voltage requirements of a set. Having a separate primary winding, it provides *isolation* from the 120-V ac power line. An autotransformer (with primary and secondary windings directly connected) does not provide this *desirable* isolation.

12. The heater voltage for the CRT and dial lights is furnished by a filament winding on the power transformer from a *separate* filament transformer or from a winding on the flyback transformer. The heater circuit should be fused to protect against possible fire damage from a short circuit. A typical fuse link consists of a 1-in. length of no. 26 wire.

13. With a set having the instant-on feature, the CRT filament is heated continuously at reduced brightness, even when the set is turned off. This shortens the warm-up period when the set is switched on. By reducing the initial surge, it can also prolong the life of the CRT filament.

14. The automatic degaussing coil(s) for demagnetizing a color CRT are energized from either the 120 V ac or from a winding on the power transformer. A thermistor and/or a varistor are generally used to control the momentary flow of demagnetizing current for a few seconds during receiver turn on.

15. Electrolytic filter capacitors frequently give trouble. They can develop excessive leakage, imposing an extra current drain on the PS. They also tend to dry up with age, losing capacity in the process. A completely dehydrated filter is the equivalent of an *open* capacitor.

16. Turn set *off before* bridging a suspected filter with a substitute prior to restoring power. Bridging *any live circuit* always entails the risk of creating an inadvertent short, and the surge from the charging current of a capacitor can endanger solid-state devices.

17. Voltages are distributed to the various stages of a set via a number of *branch lines* coming from the main bus. Many schematics identify these different voltage sources with coded symbols (e.g., a small square or triangle) to help in circuit tracing and analysis.

18. Two ways to reduce voltage between the output of the PS and the stage(s) being supplied are with series dropping resistors or voltage dividers. The voltage drop developed by a series resistor depends on its resistance and the amount of current flow. As expressed by Ohm's law, if the resistance and/or the current is high, the voltage drop will be high, and vice versa. The voltage on the load side of a resistor is the source voltage minus the amount of *IR* drop.

A *voltage divider* requires at least two separate resistors (or a tapped resistor). One resistor connects to the voltage source, the other to ground, with the load supplied from their junction. A voltage divider provides better voltage regulation than a series dropping resistor.

19. The output of a PS may be *regulated* in three ways: by regulating the 120-V ac input, regulating the dc using zeners and transistors, or ac regulation using a special voltage-regulating transformer (VRT). With a special design, the portion of the core associated with the winding supplying the B voltage is made to saturate during portions of the ac cycle. Since a saturated core cannot contribute to induction, any voltage or current variations during such periods are *not* passed on to the rectifiers, and from there to the output. Voltage applied to the rectifiers is a *square wave*.

20. Circuits are protected from overload in various ways. A regular fuse, as commonly used, blows from a current overload, opening the circuit almost instantly. To avoid fuse blowing from *momentary* surges, *slow-blow* fuses are employed. Practically all modern sets use a thermal-type circuit breaker that can be reset once the overload condition is corrected. Circuit breakers (CBs) are usually connected in the 120-V input circuit. Some sets employ special overload-sensing circuits that disable a set in a variety of ways.

21. In general, there are two steps in troubleshooting a PS: (1) check for *ac* up to the rectifiers (and the CRT heater), and (2) check for *dc* from the rectifiers, the filters, voltage regulator (if used), and the voltage-distribution network feeding the various stages.

22. One of the first things in troubleshooting a PS is to determine *which* is responsible, the PS itself or trouble with the load (the stages being supplied). If the voltage is *low*, for example, the PS itself may be defective, or conversely, the excess current drain may be caused by some defect in one or more receiver stages. One way to determine this is to *isolate* the load from the supply and measure the dc output. If the output goes *low* when the load is restored, the load is the cause.

23. Partial or complete *loss of voltage* on a *branch* line is a common problem. If accompained by a resistor overheating, look for a shorted or leaky capacitor, diode, or transistor. If there is *no* overheating, look for an open resistor or other component that may be open. Two methods for troubleshooting the distribution network are (1) use an ohmmeter, with set turned *off,* to locate an open or a short, and (2) take dc voltage readings at various points. The fault will usually be found at the point of lowest reading, whether ohms or volts.

24. An overheating resistor (or other component) is always a good clue in fault finding. Look for a short, leakage, or an excessive current drain working from the load side (the end farthest from the source).

25. A fusistor should *never* be replaced by an ordinary fuse or a resistor. A fuse alone will not limit the high charging current of the input filter. A resistor alone will not offer the protection of a fuse. However, the *dual* combination of a fuse *and* a proper-value resistor connected in series is a suitable substitute for a fusistor.

26. *Never repeatedly* replace a blown fuse or reset a circuit breaker without looking for the cause. It accomplishes nothing and causes further damage to other components.

27. As a precaution, an *isolation test* should be made on all transformerless sets, as well as certain other types, during or after being serviced. High leakage or a short from chassis to power line can be hazardous. When working on such transformerless sets, it is advisable to use an *isolation transformer*.

28. An open *input* filter results in *low* dc output and hum problems. An open *output* filter causes hum but has little or no effect on the dc.

29. A low-wattage resistor that runs hot signifies an overload. A high-wattage resistor that runs *cold* means it is either *open* or for some reason is not passing current.

30. *Arcing* can produce *serious* consequences. Symptoms of arcing are flashes on the screen, crackling sounds from the speaker, unstable sync, smoke, fire, and crackling or popping sounds from the site of the arc.

31. Poor voltage regulation and/or open filters or decoupling capacitors causes interaction between stage functions. Check for any ac ripple voltage on the B supply lines, using a CRO, ac voltmeter, or signal tracer.

32. For intermittent PS problems, monitor the ac and dc voltages at various key points. Intermittent fuse blowing or circuit-breaker (CB) tripping is a common trouble. Try bridging the fusistor or CB with a lamp as a temporary substitute to restore operation and to serve as an indicator for any changes in current. An ammeter may also be used, but it may be destroyed by an inadvertent short or an unexpected overload.

33. Incorrect voltages measured at transistor terminals may be due to a dc *source* voltage that is either too high or too low. For such incorrect voltages, always verify the source voltage when troubleshooting a stage. Voltages should be within 10% of the value specified on the schematic.

34. All stages are not necessarily supplied directly from the LVPS. Many modern receivers have one or more *secondary* power supplies, often called *scan rectification*. The dc is obtained by half-wave rectification of voltage supplied from a winding on the flyback transformer. With a high ripple frequency (15,750 Hz), very little filter capacity is required.

35. Many transformerless sets have a *floating ground*. When troubleshooting, this poses a hazard to components and test instruments, as well as the chance of receiving a shock. The risks can be reduced by using an *isolation transformer*.

36. All ac and dc voltages in a set depend on the value of the ac line voltage. Voltages specified on a schematic are usually for a line voltage of 120 V ac. To compensate for line voltage that is consistently high or low, some sets provide taps on the power transformer.

REVIEW QUESTIONS

1. Does the power consumption value stamped on a chassis represent the actual power consumed? Explain.

2. What is meant by the load on a PS? Explain *no load; full load*.

3. Exactly why does a PS output drop under load, and why does this voltage depend on the amount of load current?

184 LOW–VOLTAGE POWER SUPPLY

4. The current drain from a PS is usually somewhat greater than the combined drains of the transistors and ICs. Explain the difference.

5. What is a *fusistor* and why is it used?

6. State two functions of a *decoupling* network.

7. Why must the voltage rating of a *filter capacitor* be greater than the PS load voltage?

8. A PS generally uses rectifiers having the lowest possible forward resistance and transformers and filter chokes wound with the heaviest wire practical. Explain why.

9. Why are such large-value filter capacitors required in a solid-state TV PS?

10. Explain why a variac or booster-type transformer is sometimes used and how does it function?

11. What is the purpose of a tapped primary on a power transformer? b. Does the output increase or decrease when the tap is used? Explain why.

12. Why is the noise filter in the 120-V input line wound with such heavy wire?

13. State the purpose of the ac interlock and what trouble does it often develop?

14. After turning a set off the CRT heater may remain lit. Explain.

15. When does a zener diode start conducting and what does this do to the dc output voltage? Explain why.

16. What is SCAN rectification and why is it sometimes used?

17. How would you quickly determine (without a meter or a diagram) whether a PS develops a + or a – output?

18. What transistors in a set have the *highest* voltage and current requirements?

19. It is unwise to *repeatedly* reset a circuit breaker or replace a blown fuse. Why?

20. Name two main differences between power supplies having + or – outputs.

21. State the main advantage of full-wave over half-wave rectification.

22. Why is the input filter so important to the dc output of a PS?

23. Define *voltage regulation*, both good and poor.

24. Name two kinds of transistorized *voltage regulators* and explain how they differ.

25. Describe the operation of a typical ADG circuit.

26. A voltmeter of high *sensitivity* is *not* required for checking a PS. Why?

27. Why is a no-load voltage reading on a battery or PS almost meaningless?

28. State two reasons why a set should be turned *off* when bridging a filter with a substitute.

29. Explain how and why a disastrous short may develop by earth grounding a set having a transformerless PS.

30. Which components account for the internal resistance of a PS?

31. For what reasons might filters be connected (a) in series? (b) in parallel?

Review Questions

32. Why does the voltage at the junction of two voltage-dividing resistors *increase* if the bleeder resistor to ground increases or becomes open?

33. Give some examples of component defects that can result from (a) excessive voltage; (b) excessive current.

34. Is there any harm in using a resistor with a higher power rating or a capacitor with a higher voltage rating than specified?

35. What factors determine the (a) voltage output of a PS? (b) the current output?

36. What is the dc output voltage of an unloaded half-wave PS operating from 120 V ac?

37. What is a *brute-force* filter?

38. For intermittent overloads a fuse or circuit breaker may be temporarily shunted by a lamp. Give reasons for doing this.

39. What is the resultant capacity and voltage rating of two 20-μF, 100-V electrolytics connected back to back for use in an ac circuit?

40. In a voltage-regulating circuit, sometimes two transistors are connected in parallel. Why? What will probably happen if one becomes open?

41. What factors must be considered when choosing a rectifier replacement?

42. What advantages has an APF over a regular choke/capacitor-type filter?

43. An overheating resistor is a good place to start looking for trouble. Explain why.

44. What is the minimum voltage rating for a capacitor connected across the 120-V ac line?

45. When alternately bridging filters with a substitute, it *should be* discharged after and before each test. Why?

46. A typical CRT heater measures about 1 Ω, but *actual* current flow is much less than the calculated value. Explain.

47. A resistance check across a PS output shows 10 kΩ, but much less with the meter leads reversed. Is something wrong? Explain.

48. A 5Ω transformer primary supplied with 120 V ac would seem to draw a very high current. But such is not the case. Why?

49. How would you evaluate a resistor ($R1$) that is shunted by another, if (a) it reads much higher than the coded value? (b) less than coded value? Which reading calls for isolation and a recheck?

50. An electrolytic filter, mistakenly connected across the 120 V ac or a transformer secondary, may get hot, sizzle, and explode in seconds. Explain why.

51. Both a regular diode and a zener will avalanche at some predetermined voltage. A regular diode will be destroyed in the process, but a zener will not. Why?

52. As a rule of thumb, about 2000 μF of filter capacity is required for each ampere of expected load current. High current requires a high capacity. Why?

186 LOW–VOLTAGE POWER SUPPLY

53. A grounded dc branch line will not necessarily blow a fuse or trip a breaker. Why?

54. What is the effect if one diode in a bridge rectifier (a) becomes open? (b) shorted?

55. Two hum bars are displayed on the CRT screen. Is the PS half-wave or full-wave? Explain.

56. Why are resistors sometimes connected, (a) in series? (b) in parallel?

57. What is the probable result of creating an accidental short across the ac interlock?

58. During the servicing of a set, it is a good idea to wiggle the ac line cord at both ends. Explain why.

59. The effectiveness of a filter choke decreases as its core approaches saturation, as with excessive load current. Why?

60. How would you identify the uncoded leads of a power transformer?

61. The output from the LVPS may be a little higher or lower than specified by the schematic. Give two possible reasons.

62. The filter capacity requirements for a scan rectifier circuit are very low. Why?

63. Some sets have both + and − dc supplies. Why?

64. How would you connect a 6-V filament transformer to (a) boost the ac line voltage by 6 V? (b) reduce the line voltage by the same amount? (c) What is the principle involved?

65. In the instant turn-on mode a diode connected series with the CRT heater causes the filament to glow at half normal brilliance. Why? Why is a diode preferable to a resistor for this purpose?

66. Capacitors connected to the dc bus must be repeatedly discharged before an accurate resistance check can be made on the circuit. Explain why.

67. The 120-V power line is potentially more dangerous than a B supply of several hundred volts, or even the HV supply. Why?

68. Does a zener diode continue to regulate if the dc voltage across the load (a) drops below the zener level? (b) rises above the zener level? Explain.

69. What determines the size of wire to use as a conductor? (a) the amount of current? (b) the voltage? (c) or both?

70. What is meant by the power factor of an electrolytic capacitor and why is it more meaningful than a resistance check for leakage?

71. What is a floating ground?

72. What happens to the dc load voltage if a zener (a) becomes shorted? (b) open? (c) leaky? State reasons in each case.

73. In lieu of the proper 6 V battery, is it practical to use 4 flashlight cells in series to operate a battery-powered TV? Explain

74. Other than an inherent defect, what could cause a zener to short out?

75. What would cause a zener to pass more current than its normal rating, (a) an increase in load current? (b) an increase in source voltage? (c) either increase?

76. Under what conditions does a zener start drawing high current?

77. A PS delivers 1 A through a 50-Ω dropping resistor and a zener is used to regulate the 30-V output. (a) Compute the output voltage if the load circuit opens up. (b) What is the probable effect on the zener?

78. A large sewing needle makes a good tool for puncturing insulation when checking for an open circuit in a wire. Explain how you would pinpoint the break.

79. What is the result of installing a fuse whose rating is, (a) too low? (b) too high?

TROUBLESHOOTING QUESTIONNAIRE

1. What condition would you have if the zener diode in a simple regulator circuit (a) becomes open? (b) becomes shorted?

2. A set has hum problems and low dc supply voltage. Give three possible causes.

3. A high-power resistor runs cold. State several possible causes.

4. A carbonized leakage path has formed between two contacts, with signs of arcing. Would you (a) cover it with tape or some insulating compound or spray? (b) replace the part? (c) reroute the circuit?

5. What PS section would you check if there is (a) no ac reaching the rectifier(s)? (b) the ac is OK but there is no dc output? (c) there is some dc output but none on one of the branched circuits?

6. What is probably wrong if, when checking for a B + short across a filter, the ohmmeter (a) reads zero ohms? (b) less than 100 Ω? (c) infinity? (d) the pointer deflects beyond the scale in either direction?

7. A series-dropping resistor overheats. State several possible causes.

8. A set owner is troubled with periodic drops in line voltage. Would you make use of the tap on the power transformer, add a line booster, or neither? Explain your answer.

9. How can you quickly determine if a low or zero dc voltage reading is due to troubles at the source or the load?

10. A CRT heater either fails to come on or periodically changes in brilliance. What would you check for?

11. A resistor is running hot. Both resistance and voltage readings from one end to ground are *low*. Which end connects to the load and which way do you look for the cause?

12. What is wrong when a considerable amount of ac can be measured across a filter capacitor?

13. A power transformer overheats. After disconnecting all secondary wires, it continues to run hot. What is wrong? (b) what if it now runs cool?

14. What condition is indicated when the output of an unregulated PS is abnormally high?

15. A bridge-type PS has low dc output, a high ripple, no signs of overload, and substituting filters does not help. What would you suspect?

16. There is no dc output from a regulated PS. A jumper connected from C to E of the regulating transistor restores the voltage and the set operation. Name two possible causes.

17. A filament transformer is being used in the 120-V ac input as an autotransformer to boost the line voltage. Instead, the voltage is reduced, and reversing the plug in the outlet does not help. What is wrong?

18. One of the three heaters in a color CRT fails to light. Could the heater supply be at fault? Explain.

19. A PS has a negative voltage. What type of transistors are probably used in most of the stages?

20. A suspected leaky or shorted filter *must* be cut loose in order to test or substitute. Why?

21. The dc output from a branch line checks low until the load is removed. What can cause this besides a high load current?

22. Smoke is observed coming from a chassis, but the source is not immediately apparent. How would you proceed?

23. Bridging a rectifier with a substitute is an acceptable test procedure if it is open, but not if the rectifier is leaky or shorted. Explain.

24. A schematic shows two series-connected resistors to provide outputs of 20 V and 10 V from the 100-V PS. Where would you look for trouble if (a) PS output checks low and the other two points read zero? (b) if the source and the 20-V tap checks low and the 10-V point checks zero? (c) if zero volts at all three points?

25. There is voltage up to the heater pins on a CRT socket, yet the tube (which checks good) does not come on. What is wrong and how would you check it out?

26. A PTC resistor reads several hundred ohms when tested cold. Is the resistor *good* or *defective?* Explain.

27. The voltage measured across three series-connected resistors connected to a 100-V source measures 0 V, 100 V, and 0 V, respectively. What is wrong?

28. Which would you suspect, the input or output filter, if (a) there is low dc and a high ripple? (b) dc checks out but there are bad hum problems?

29. State the probable causes of a burned-out rectifier in a PS.

30. There is no dc output from a PS and no signs of an overload. Would you suspect an open or a short?

31. What happens to the voltage at a voltage-divider junction if (a) a capacitor at that point becomes open? (b) becomes shorted? (c) the bleeder section to ground opens up?

32. The series-dropping resistor in a simple zener regulator circuit is overheating. State possible causes.

33. Which filter, if leaky, would cause the filter choke to overheat?

34. What is the probable result of making wrong connections to the (square, half-moon, or triangle) coded terminals of a multisection electrolytic?

35. High-wattage resistors are supposed to get warm. State 3 possible causes where a resistor that checks ok, remains cold.

36. If a voltage can be measured across a fuse terminal, is the fuse good or bad?

37. A set is overheating with indications of a B voltage short. Would you (a) take voltage and current readings? (b) turn the set off and attempt to locate the cause by testing with an ohmmeter? Explain your answer.

38. There is no dc from a scan rectifier circuit. What would you check if (a) the raster is ok? (b) there is no raster, but sound is normal?

39. State the probable cause when the meter indicates a high ac voltage and a low dc voltage across the input filter.

40. A CRT shows filament continuity. There is voltage at the heater pins, and the tube neck gets warm. But there is no visible heater glow. What is wrong?

41. Using an ohmmeter and without the aid of a schematic, how would you locate a grounded component in a complex voltage-distribution network?

42. You discover three capacitors tied to a dc bus where voltage checks zero and there are indications of overload. How would you proceed?

43. The raster slowly gets smaller (both width and height) after set warmup. State the possible cause.

44. You suspect the PS is intermittent and wish to substitute using an external supply. What must be considered and how would you proceed?

45. You have just replaced numerous diodes, transistors, and ICs after a catastrophic failure. State the possible cause and precautionary checks before applying power.

46. You are bridging various filters one at a time with a substitute when the set goes dead. Subsequent checking shows one or more transistors destroyed. State the probable cause.

47. The voltage measured at the load end of a series-dropping resistor is the same as at the PS output. What condition is indicated?

48. After replacing a blown fuse or resetting the circuit breaker, a set appears to operate normally.
(a) State several possible reasons for the blown fuse or CB.
(b) Why should the set be given a thorough *air check* under these circumstances?

49. Some antennas are grounded via the mast and a vent pipe. If there is a flash, a blown fuse, or worse when the antenna is connected to the set, what is wrong?

50. Explain each abnormal condition created by defects listed in Table 6-3 (see Fig. 6-13).

51. What problems can be expected if a PS rectifier is wired in reverse?

52. A transformerless set has a low-resistance reading between floating ground, chassis, and the metal cabinet of the set. This represents a potential fire and shock hazard. What is wrong?

53. An electrolytic capacitor is to be connected between two circuit points where there are different amounts of voltage of the *same polarity*. (a) To whch point would you connect the positive terminal of the capacitor? (b) or does it matter?

54. A PS rectifier runs warm and has a short life. What would you suspect?

55. There are sounds of arcing and the set cuts off when the line cord is disturbed at the ac interlock. What would you do?

56. An ac line cord is suspected of being open or intermittent. How would you check it?

For the following questions refer to the Sample Circuits at the back of the book.

57. A 2.2-Ω fusistor series with one of the scan diodes is burned open. What would you suspect?

58. Would there be any *immediate* effect on the pix if the PTC resistor series with the degaussing coil opened up?

59. What is the effect on set operation if the SCR start-up thyristor opens up?

60. The emitter of regulating transistor Q 2903 measures *zero* volts. What would you suspect?

61. The +15 V dc from the regulated scan supply measures high and tends to fluctuate. What would you suspect?

62. No ac or dc voltages can be measured on the start-up SCR. Give two possible causes.

63. What would you check if there is excessive HV but the set appears to operate normally?

Chapter 7

THE VERTICAL SWEEP SECTION

7-1 VERTICAL SWEEP REQUIREMENTS

A raster is produced on both the camera tube and the receiver pix tube by simultaneously deflecting the CRT beam both vertically and horizontally. The function of the *vertical* sweep section (as explained in Sec. 1-7) is to deflect the beam *up and down* at a constant rate. For a BW set, the sweep rate is 60 Hz; for a color set it is a little slower. The *horizontal* sweep section (Chapter 9) moves the beam *sideways* at a *much faster* rate. Although *beam deflection* is the primary purpose of both sweep circuits, a number of other receiver functions depend on them for their operation. The vertical sweep dependent functions are *vertical retrace blanking, pincushion correction,* and, for color sets, *dynamic convergence.*

The vertical sweep section consists of a number of interconnected stages, as shown in the block diagram (Fig. 7-1). Some sets are more complex than others. A color set usually has more stages than a BW set and may use anywhere from three or four to six transistors, particularly if it is a large-screen color console that requires considerable sweep power. With some sets, all the components of the vertical sweep section are contained on a *separate vertical module,* which is identified as such. In modern sets, ICs are used extensively and may contain all but the larger components such as transformers, electrolytics, and power output transistors. The stages of a typical vertical sweep section are briefly described as follows.

7-2 VERTICAL SWEEP STAGES

The stages responsible for *beam deflection* are the oscillator, driver amplifier(s), and the output stage, which includes the deflection yoke. The purpose of the *oscillator* is to *generate* the vertical sweep voltage. Of primary importance is its *frequency stability*.

The purpose of the driver stage(s) (anywhere from one to three stages) is to amplify the sweep voltage generated by the oscillator. Driver amplifiers also serve as buffer stages to prevent

192 THE VERTICAL SWEEP SECTION

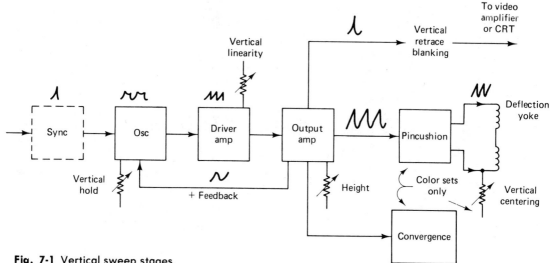

Fig. 7-1 Vertical sweep stages.

loading of the critical oscillator and to provide the necessary sweep *power* to drive the output stage.

The purpose of the output amplifier is to develop enough *sweep power* to drive the deflection yoke. The yoke converts sweep current into a magnetic field for vertical deflection of the CRT beam.

Various methods are used to couple the sweep voltage from one stage to another. This includes direct coupling, *R/C* coupling, and the use of transformers. Transformer coupling is most often used between the driver and output stage and between the output amplifier and the yoke.

7-3 VERTICAL OSCILLATOR

The vertical oscillator must be synchronized with its counterpart at the TV station. The oscillator is controlled in two ways: *manually,* with the vertical hold control, and *automatically,* by the vertical sync pulses coming from the station. The block in Fig. 7-1 marked *integrator* is an *RC* network whose job is to integrate the vertical sync pulses (coming from the sync stages) before they are applied to the oscillator.

7-4 SWEEP-DEPENDENT CIRCUITS

Interfacing with the output stage are the three sweep-dependent circuits mentioned earlier. With *vertical retrace blanking,* a high-amplitude vertical pulse is fed to the pix tube to ensure complete blanking of the screen during vertical retrace intervals.

Pincushioning is a form of raster distortion that occurs with all large-screen rectangular pix tubes unless corrections are made. Color receivers use *circuitry* to perform this function. To prevent *vertical* pincushioning, the vertical sweep waveform is modified (prior to feeding the yoke) to purposely introduce some nonlinearity to the vertical beam trace. *Horizontal* pincushioning is described in Chapter 9.

Convergence, of course, is associated only with color receivers. For *dynamic convergence* of *triad-type* pix tubes, modified sweep waveforms are applied to a *convergence assembly* on the neck of the tube. The object is to ensure good convergence of the three beams near the outer edges of the screen. Other simpler methods are employed for convergence of the more recent *in-line*-type pix tubes.

7-5 VERTICAL SWEEP CIRCUIT FUNDAMENTALS

Unlike most stages in a receiver (amplifiers and the like), an oscillator generates its own signal. The vertical oscillator signal is the *source* of the sweep voltage that results in deflection of the CRT beam. There are three important characteristics of the signal voltage generated by the vertical oscillator: *amplitude, frequency,* and *waveshape*.

Amplitude. Sweep *voltage* relates to the *amount* of sweep or beam deflection; the greater the voltage is, the greater the vertical deflection or raster *height*.

Frequency. For a pix to be locked *vertically in sync,* the oscillator frequency must be exactly 60 Hz for a BW set and 59.94 Hz for a color set as established by the setting of the hold control. When properly adjusted, the oscillator is *locked into sync* by incoming sync pulses. But unless the *free-running* frequency is correct, a number of abnormal picture displays are possible. For example, if the oscillator is running at precisely half the normal rate (30 Hz), you see two complete pictures, one above the other and separated by a *blanking/sync bar*. The explanation is that the reduced rate of beam travel coincides with the reception of *two* video and pulse wave trains, in other words, two fields instead of one. If frequency is one-third normal (20 Hz), you get three complete pictures, and so on. For exact multiples of 60 Hz the pictures will also be locked in sync. For frequencies *higher* than 60, by an *even* multiple, say 120 Hz, only one-half a picture is reproduced in the shortened time of vertical trace. But if the frequency is an *odd multiple* of 60, it will be out of sync, rolling up or down depending on whether frequency is greater or less than normal.

Waveshape. Under the influence of both vertical and horizontal sweep circuits, the CRT beam must scan the screen at a constant and uniform rate. This is accomplished, in part, by generating specially shaped (sawtooth) waveforms as shown in Fig. 7-2. For the vertical sweep, the constantly increasing voltage deflects the beam downward, creating uniformly spaced *scannning lines* and pictures in proper proportions. When the sawtooth voltage drops (reverses direction), the beam retraces back to the top of the screen. By comparison with the vertical trace period (which occupies most of the alloted 1/60th of a second), vertical *retrace* must take place in less than 1200 μs or 1.2 ms.

Where an improper waveshape or other condition produces a *nonlinear* sweep, the scanning lines will *not* be uniformly spaced, producing pix objects out of proportion (e.g., people with flattened or elongated heads, short or long legs, etc.). If the alloted retrace time is exceeded, the result is vertical *foldover*.

7-6 GENERATING THE SAWTOOTH WAVE

Sawtooth waveforms are produced by alternately charging and discharging a capacitor in the oscillator circuit. The peak-to-peak voltage thus developed is the sweep signal that drives the following stage. The charge and discharge cycles are produced by the switching action of the oscillator (or switching) transistor. A partial circuit is shown in Fig. 7-2. The charge and discharge currents of capacitor C take different paths. When the transistor is nonconducting, the capacitor charges (via the resistor R) from the dc

194 THE VERTICAL SWEEP SECTION

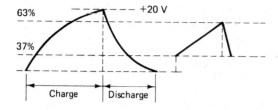

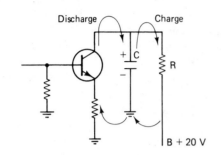

Fig. 7-2 Developing the sawtooth wave.

source. When the transistor is turned on, it provides a low-resistance path for the discharge of the capacitor.

Factors involved in the charging of the capacitor are its value (capacity), the amount of applied voltage, and the amount of resistance in the charging circuit. For example, a *small* capacitor becomes charged to a higher level and in a shorter period of time than a larger unit. Maximum sweep voltage would be developed if it were allowed to charge to the full value of the source voltage and then discharge to zero. In practice, this is not done because these extremes result in an *exponential* curve rather than the linear *sawtooth-shaped wave* that is desired. Instead, the capacitor is permitted to charge only to 63.2% of the applied voltage and to discharge only to the 37% level, as shown in Fig. 7-2b. The time required to charge a capacitor to the 63% level is expressed as TC (time constant in seconds), which is the product *RC,* where *R* is in ohms and *C* is in farads. For example, a 0.01 μF capacitor in series with a 100-kΩ resistor takes 0.001 s or 1 ms to charge to the 63.2% level.

The amount of time allowed for the charge

and discharge of the capacitor is established by the time intervals of the composite video signal (Fig. 1-7). And to stay within these limits, the *RC* values in the circuit are quite critical. Also, the junction resistance of the transistor (when conducting) must be very low to ensure a rapid discharge in the 1.2 ms allowed for vertical retrace. To further reduce the discharge time, the vertical deflection yoke is encouraged to ring or oscillate for a fraction of a cycle (more on this in Sec. 7-12.

7-7 WAVESHAPING

There are different ways to achieve linear deflection. Some sets develop the sawtooth in the oscillator, as just explained, and then preserve its shape by operating all subsequent stages as distortionless linear amplifiers. In most cases, however, the wave developed by the oscillator is modified by waveshaping circuits to satisfy the requirements of each particular set. This is usually accomplished by the vertical linearity control. Some sets also have a vertical *peaking control* to alter the waveform. The present trend is to dispense with both controls by using selective *negative feedback* over several or all vertical sweep stages.

Figure 7-3 shows how different waveshapes affect the linearity of beam deflection. *Normal* sweep is shown in part (a), where all scanning lines are uniformly spaced. In part (b), the initial fast rise *spreads* the lines at the *top* of the screen, and the slower rate of increase near the peak of the wave *compresses* the lines near the *bottom* of the screen. In part (c) the condition is reversed, where the voltage buildup starts slowly and then accelerates, compressing the top lines and spreading those at the bottom. The explanation is that the slower the beam travel, the greater the number of horizontal sweeps that can take place in that amount of time. The reverse holds true for a speedup of the vertical deflection.

7-8 Typical Vertical Oscillator Circuits

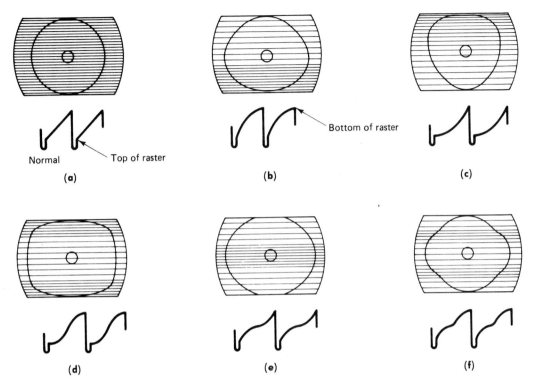

Fig. 7-3 Waveshapes and vertical linearity conditions.

The rest of the examples should be self-explanatory. In Fig. 7-3d the pix will be compressed at top and bottom and stretched in the middle. With part (e) it is the reverse, stretched at top and bottom, and crowded in the middle. Part (f) shows how a 60- or 120-Hz ripple voltage superimposed on the waveform creates alternate areas of crowding and compression that will slowly drift upward with phase changes. This effect is called breathing.

7-8 TYPICAL VERTICAL OSCILLATOR CIRCUITS

There are two basic oscillator circuits in common use, the *multivibrator* (MV) type and the *blocking oscillator*. Another, more recent innovation is known as the *countdown* system. With it, there are no sweep oscillators per se; instead, counters and digital logic are used to develop sweep voltages (see Sec. 7-9).

• The Multivibrator (MV)

The circuit of a MV-type oscillator looks much like any two-stage *RC*-coupled *amplifier;* in fact, such an amplifier becomes an oscillator simply by adding a *feedback loop* from the output of the second stage back to the input of the first stage. The feedback signal (known alternatively as *positive* or *regenerative feedback*) is in phase with and reinforces the signal at the input. This signal is the one *generated* that originates from an initial

196 THE VERTICAL SWEEP SECTION

impulse when power is first applied. This must not be confused with *degenerative* or *negative feedback,* where the signals are *out of phase.*

Depending on the circuit, the feedback loop may be a straight-through connection, or an *RC*-coupling network. Sometimes the oscillator stage is comprised of two separate transistors, one of which may be called the *switching* transistor. Other MV oscillator circuits use only a single transistor and obtain the necessary feedback from some following stage, either a buffer-driver amplifier or the power output stage. A variation of the two-transistor oscillator uses an unbypassed resistor common to both transistors, to provide the feedback.

With the typical MV using two transistors of the same type, oscillations are produced as they alternately drive each other between saturation and cutoff. The frequency is determined by the time constant of the *RC* component values and to some extent by the characteristics of the transistors and the applied operating voltages. Some circuits use both an NPN and a PNP transistor. With this configuration, both transistors are alternately turned off and driven into saturation.

The characteristic waveform of a MV is a modified square wave, and not the desired sawtooth required for beam deflection. This is not important, since, as previously stated, the oscillator merely functions as an electronic switch for the charging and discharging of a capacitor where the sawtooth wave is developed. The amplitude of the sawtooth waveform is determined by the amount the capacitor can become charged and discharged during the allotted time for each trace–retrace cycle.

Figure 7-4a shows a typical MV-type oscillator. Three transistors are shown: *Q1* and *Q2* produce the oscillations; *Q3* is a waveshaping or switching transistor for developing the sawtooth wave. Note that direct coupling is used between stages and in the feedback loop from the collector of *Q2* back to the base of *Q1*. Also note the waveforms and their polarity at the various circuit points. The waveforms are mostly sharply spiked pulses, except for the sawtooth waveform at the output of *Q3*. The *RC* network at the sync input to *Q1* is the *integrator*. At the output of *Q3* is *C2*, the sawtooth-forming capacitor. Both *Q2* and *Q3* act as switching transistors. The operation of the circuit is as follows:

1. At the start, *Q1* is reverse biased by the + dc from the junction of the voltage-dividing resistors *R1/R2.*

2. After *C1* becomes charged, *Q1* becomes forward biased into conduction. This is aided by the

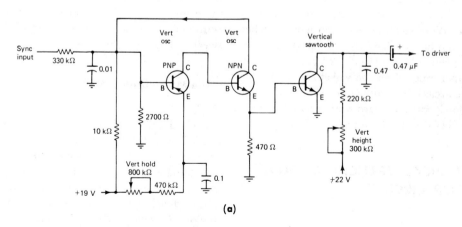

Fig. 7-4 (partial)

7-8 Typical Vertical Oscillator Circuits

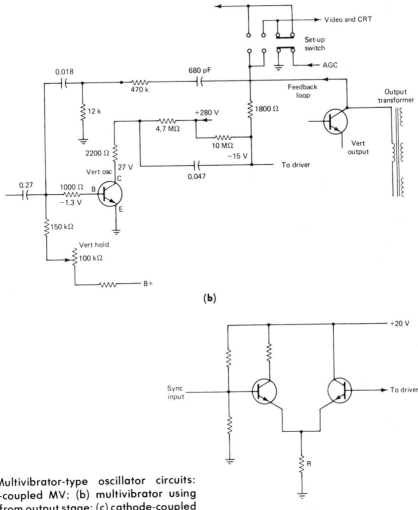

Fig. 7-4 Multivibrator-type oscillator circuits: (a) direct-coupled MV; (b) multivibrator using feedback from output stage; (c) cathode-coupled MV.

negative-going feedback pulse from Q2 and by the negative-going integrated sync pulse as received, which initiates vertical retrace. Conduction is via the B/E junction of Q2, which is also forward biased.

3. Conduction gradually depletes the charge on C1; Q1 becomes reverse biased and is turned off, completing the cycle. The bias of Q1 is also established by the critical setting of the hold control, to make the oscillator run *slightly slower* than 60 Hz so that the incoming sync pulse can maintain control by prematurely triggering Q1 into con-

duction sooner than it would in a free-running state.

4. While Q1 and Q2 are nonconducting, the waveshaping transistor Q3 is turned off and the sawtooth-forming capacitor C2 becomes charged (by the source voltage) via the height control and its series resistor R7. This produces the trace portion of the sawtooth as shown.

5. During conduction of Q1/Q2 (as initiated by the sync pulses), Q3 is turned on (by the forward bias developed across R5) and C2 is rapidly dis-

198 THE VERTICAL SWEEP SECTION

charged via the low resistance C/E junction of *Q3*. This dropoff of sawtooth voltage results in vertical retrace of the CRT beam.

6. To summarize: From the moment of triggering, the polarity of the various sweep pulses are negative-going at the base of *Q1*, positive-going from the collector of *Q1* to the base of *Q2*, negative-going feedback from the collector of *Q2* back to *Q1*, positive-going from the emitter follower *Q2* to the base of *Q3*, and a negative-going sawtooth from the collector of *Q3* to the next stage via coupling capacitor *C3*. Note that these signal polarities are *not* related to the polarities of the dc operating voltages as shown.

• Single-Transistor Multivibrator

Figure 7-4b shows how one of the oscillator transistors is eliminated in many receivers. The regenerative feedback (which is required to sustain oscillations) is obtained from the power output stage. The feedback must be in phase with the signal at the oscillator input. With this circuit configuration (where oscillator operation depends on the proper functioning of the output stage), troubleshooting becomes somewhat more complicated.

• The Blocking Oscillator

This circuit is distinguished from a MV in that it uses a small transformer for feedback and requires only one transistor. It also has greater frequency stability. Some circuits use a separate switching transistor for sawtooth development, as with the MV just described.

A simplified blocking oscillator is shown in Fig. 7-5. The transistor is a PNP type operating from a negative supply source. Resistor *R1* and the hold control act as a voltage divider to supply the bias. The blocking oscillator transformer (T) has a slight step-up ratio. Negative-going sync pulses from the integrator are fed to the base via the transformer secondary winding. Capacitor *C2* is the sawtooth-forming capacitor. Circuit operation is as follows:

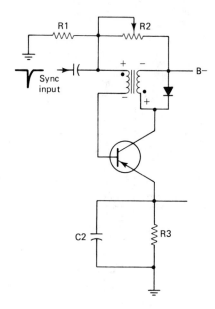

Fig. 7-5 Blocking oscillator.

1. At the instant power is applied, the transistor is forward biased from the voltage divider *R1/R2*. With the transistor conducting, the *IR* drop across *R3* charges *C2* to produce the sawtooth rise as shown. The increasing current and *IR* drop cause further increases in forward bias and still more current. With the increasing current, a voltage is induced into the transformer secondary having the polarity as shown, which produces still further increases in current. This continues until saturation is reached.

2. At the moment of saturation (where the current levels off), transformer action ceases and some of the forward bias is lost. With the loss of the transformer voltage, the charge on *C2* overcomes the dc from the divider, and the transistor becomes reverse biased into cutoff.

3. The collapsing field of the transformer induces a voltage of opposite polarity into the secondary, making the base even more positive to maintain cutoff. This blocked condition exists until the charge on *C2* becomes dissipated (via *R3*). This dropoff represents the retrace part of the cycle and continues until the transistor once again becomes forward biased and the action is repeated.

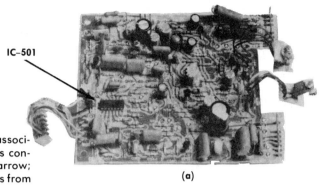

Fig. 7-6 Typical countdown system and associated circuits: (a) The countdown system is contained in a single IC (IC-501) indicated by arrow; (b) vertical and horizontal sweep originates from a single oscillator.

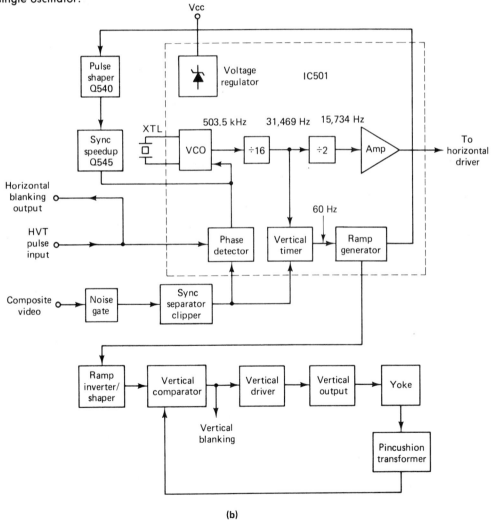

4. As with the MV, the oscillator is adjusted to free run a little more slowly than 60 Hz, and depends on the incoming sync pulses to initiate conduction.

7-9 DIGITAL COUNTDOWN SWEEP-GENERATING SYSTEM

This relatively new system replaces the conventional sweep oscillators (in a color TV set) by using digital countdown logic to develop both vertical and horizontal sweep signals. The block diagram of Fig. 7-6 shows a typical system, where most of the major functions take place in a single IC. A brief description follows.

A crystal-controlled oscillator (VCO) produces a signal of 503.5 kHz, which goes through two *divider circuits* to produce frequencies of 31,469 and 15,734 Hz. The 15,734-Hz signal is amplified to drive the horizontal sweep circuit.

REMINDER:

Both vertical and horizontal sweep frequencies for a color set are slightly lower than for BW sets.

The 31,469-Hz signal is further divided by a *vertical timer* to produce the 59.94-Hz signal for vertical deflection. With crystal control, the oscillator is extremely stable, as are all signals downstream. Precise phasing of the oscillator and the vertical timer is controlled by the station sync pulses, which enter the IC from an external sync separator stage as shown. Some sets include the sync, video amplifier, and AGC functions in this same IC.

From the vertical timer, two blocks, the *ramp generator* and the *inverter-shaper*, convert the 59.94-Hz signal into the desired sawtooth-shaped waveform. A *comparator* circuit accepts feedback from the vertical output stage to correct any nonlinearity of the wave. From the comparator, the signal leaves the IC to feed a conventional driver stage. Note the low-level takeoff points for the vertical and horizontal blanking pulses. Because of its accuracy, sets using this system do not require the usual vertical and horizontal hold controls.

7-10 SYNCHRONIZATION

To obtain an intelligible picture, raster scanning at the receiver must be *exactly* in step with the scanning of the camera tube at the station. To accomplish this, the station sends out vertical and horizontal sync pulses as part of the composite video signal (Fig. 1-6). The *vertical* sweep oscillator (when adjusted to free-run at the correct frequency) is *directly* controlled (triggered) by the vertical pulses. (By comparison, the *horizontal* oscillator is *indirectly* controlled using some form of AFC.) After each line of pix information in the scanning of a raster, the screen is momentarily blanked out for horizontal retrace by a horizontal blanking pulse (see Fig. 1-6). Shortly after the start of the retrace, a sharply spiked horizontal sync pulse is received for controlling the horizontal oscillator.

After scanning the last line of pix information (for each field), with the CRT beam near the bottom of the screen, the screen is blanked out by a vertical blanking pulse, for about 1.2 ms, the time alloted for vertical retrace. Riding above the vertical blanking level is a train of sync pulses, as shown. Some of them control the vertical oscillator, some the horizontal, and some do both. The vertical pulses initiate the vertical retrace interval, and the horizontal pulses maintain horizontal control during vertical retrace. All this occurs while the screen is blanked out.

• Integrator Network

This is a low-pass *RC* filter inserted between the sync stage(s) and the vertical oscillator (see Fig.

10-4). The vertical retrace portion of the composite video signal is a pulse train that contains not one, but *six* vertical sync pulses. These are integrated into a single, sharply spiked pulse for triggering the oscillator. As the pulses arrive, sequentially, one after the other, an accumulating pulse voltage is developed across the resistor *R1* (Fig. 10-4) in the oscillator input circuit. For a strong signal, the oscillator may be triggered on the first or second pulse. A weak signal may require all six to do the job. If the pulse level is *too weak*, the oscillator will be free-running and the picture will roll.

• **Oscillator Triggering**

With the oscillator transistor turned off (nonconducting), the sawtooth-forming capacitor is being charged to produce the active (downward trace) of the CRT beam. When a pulse arrives, the oscillator is turned on to discharge the capacitor and initiate beam retrace. Triggering action is shown in Fig. 7-7 with the (time) relationship between the generated sweep voltage and the incoming integrated sync pulses. Figure 7-7a shows the condition for normal sync. The frequency (and phase) of the oscillator is such that the pulse reaches a level that alters the transistor bias to turn the transistor on sooner than it normally would in a free-running state. For stable sync, the oscillator is adjusted (with the hold control) to operate a little slower than the pulse repetition rate.

In Fig. 7-7b, the oscillator is much too slow and the pulse arrives too late for triggering, with a resulting loss of sync. In Fig. 7-7c, the oscillator is too fast (pulse is premature) and too low on the curve to perform its function. Here again the oscillator is free-running with complete loss of sync. The sequence of events of raster scanning are briefly summarized as follows:

1. After the last line of pix information is scanned near the bottom of the screen, the pulse train arrives. The beam continues downward, scanning about six unmodulated lines. During this time, the horizontal oscillator is being controlled by alternate even and odd line *equalizing pulses* (see Fig. 1-6), for interlaced scanning of the even and odd line fields. The pulses are at half-line intervals, the even fields making use of pulses 2, 4, and 6. For the odd-line fields, the horizontal oscillator is controlled by pulses 1, 3, and 5. Until the vertical oscillator is triggered by the following series of vertical pulses, the CRT beam is still moving downward, with the *horizontal* oscillator controlled by the serrated *vertical* pulse.

2. With the integration of the vertical pulses and the triggering of the vertical oscillator as explained, the beam starts moving *up,* and will continue so for the duration of the vertical retrace interval. Six additional equalizing pulses are received following the vertical pulses; their purpose is to maintain horizontal sync for alternate even and odd line fields. Following the equalizing pulses are anywhere from 10 to 14 regular horizontal pulses. At some time during these pulses (or coincident with the arrival of video for the next field), vertical retrace is completed as the

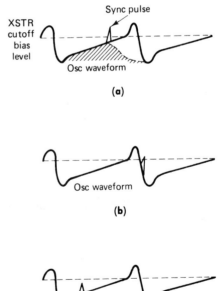

Fig. 7-7 Oscillator triggering: (a) normal sync. where pulse initiates retrace; (b) oscillator too slow (pulse arrives late), oscillator is free-running; (c) oscillator free-running (frequency too fast and pulse arrives early).

202 THE VERTICAL SWEEP SECTION

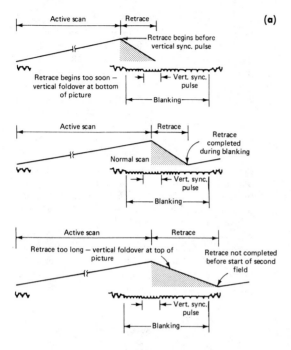

vertical oscillator completes its cycle, and the beam starts moving downward for the scanning of the next field.

3. As the beam starts down, if the first line of video has not yet arrived, there will be several unused scanning lines at the top of the picture. These will normally be blanked out until the vertical blanking pulse is ended and video starts coming through. As the beam continues down during the vertical trace period, 262½ lines are scanned for each field, a total of 525 for a *frame*, a complete picture.

4. With the beam reaching the bottom of the screen, the sequence is again repeated with the scanning of the last line of video and the arrival of another pulse train.

Figure 7-8a shows the time relationship between sawtooth formation and the pulses during the vertical retrace interval. Note that excessive retrace time creates an overlap where the beam has yet to reach the top of the screen when video for the next field is received and part of the pix is folded back on itself.

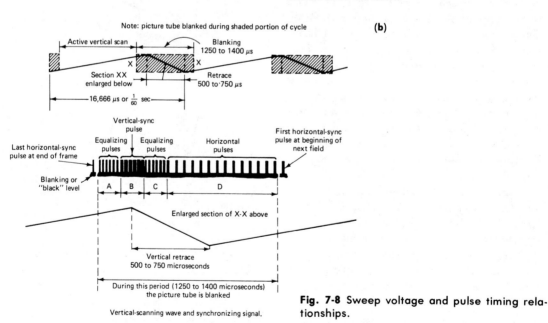

Fig. 7-8 Sweep voltage and pulse timing relationships.

From:
Television Service Manual, ©1977, by Robert G. Middleton, used with permission of the publisher, The Bobbs-Merrill Company, Inc.

Synchronizing Typical Oscillators

The polarity of sync pulses for controlling an oscillator depends on the type of transistor; to turn on a PNP-type transistor requires a negative-going pulse, and for an NPN type, a positive-going pulse. For the MV oscillator shown in Fig. 7-4a, the negative-going pulse (approximately 20 V p-p) reaches the base of $Q1$ via the integrator resistors. The pulse overcomes the normal B-to-E bias, forcing the transistor to conduct.

With the blocking oscillator circuit shown in Fig. 7-5, pulses from the integrator feed the transistor base via a coupling capacitor and one winding of the blocking oscillator transformer. The oscillator is blocked until the arrival of a pulse, which turns the transistor on.

For sync control of a typical countdown system, refer to the foldout schematic at the back of the book (and Fig. 7-6). Upon entering the IC, the pulses control a vertical *timer* whose function is to divide the frequency.

7-11 BUFFER AND DRIVER AMPLIFIERS

Some sets (mostly small-screen BW models) feed the vertical output stage directly from the oscillator. Most, however, use one, two, and sometimes three intermediate stages. These are known by different names, according to their function: *predriver, driver, buffer, inverter,* or simply *sweep amplifiers.* One purpose of these stages is to provide *isolation* to prevent the output stage from loading the critical waveshaping circuits of the oscillator. Another purpose, even more important, is to ensure that the output stage gets sufficient drive, either *voltage, power,* or *both,* depending on the circuit. A transistor using the emitter-follower configuration is often used because it has a high input impedance that does not load the oscillator and a low output impedance to match the output stage. Such a circuit, however, has *no voltage* gain and *does not invert* the signal. Where the power output stage uses *two* transistors, usually it is preceded by an inverter to drive one of them with the proper signal *polarity.* Where high amplification is needed, a *Darlington* transistor may be used.

Some sets use capacitive coupling between stages, but most use *direct coupling,* where their dc voltages and operating conditions are dependent on one another (which of course makes the servicing of them somewhat more difficult). In some cases, most of the oscillator and driver components are contained in an IC.

The height and linearity controls are often found in a driver stage, and *negative feedback* is commonly used from the sweep output back to the driver input as a means of reducing distortion. Sometimes this is so effective it eliminates the need for a linearity control. A degenerative or *negative* feedback loop must not be confused with the *positive* feedback loop required for MV operation.

A simplified version of a single driver stage is shown in Fig. 7-9. Note the linearity control that influences the waveshape by regulating the amount of negative feedback in the loop. The height control in this case also determines the amount of feedback (and sweep amplitude), as well as the gain of the output stage. As an emitter-follower, the stage does not contribute to voltage amplification; in fact, there is a slight loss as indicated. Input and output waveforms are of the same polarity and essentially have the same shape.

Also note the *service setup switch,* a portion of which ties in with the vertical sweep. In the position shown, it has no effect. When making CRT setup adjustments, it disables the vertical by grounding out the sawtooth waveform. It may tie in with *other* sweep stages and in some cases opens the circuit at some appropriate point.

With direct coupling, the bias for the output transistor is established by the *IR* drop across the driver emitter resistor $R3$ and the setting of

204 THE VERTICAL SWEEP SECTION

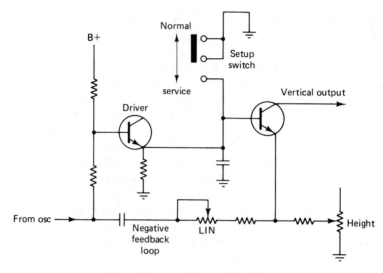

Fig. 7-9 Typical driver amplifier.

the height control. With this circuit there is considerable interaction between controls.

During the trace portion of the cycle, the positive-going increase in sawtooth voltage at the driver base turns the transistor on and produces a corresponding increase at its emitter and the base input of the output transistor. For the retrace period, the negative-going drop-off of sawtooth voltage drives both transistors momentarily to cutoff.

7-12 VERTICAL OUTPUT STAGE

A fairly high yoke *current* (approximately 0.5 A) is needed to produce the required amount of beam deflection, hence the need for a *power amplifier*. For a small-screen BW set, sufficient deflection can be obtained from an oscillator driving a one-transistor (single-ended) output amplifier. However, most large-screen color sets use *two* power transistors to meet the combined demands of the sweep and its related circuits.

For *beam deflection,* the main items in the output stage are the power transistor(s) and the *load,* which is the yoke. To obtain the required amount of sweep, the following conditions must be satisfied:

1. The output transistor(s) must have an adequate amount of drive from the previous stage.

2. The output stage must develop sufficient sweep power.

3. The sweep power must be efficiently transferred from the output transistor(s) to the yoke.

• **Single-ended Output Stage**

Figure 7-10 shows a basic *single-ended* output stage. The circuit is quite simple and conventional. Operation is usually class A and the transistor (in the common-emitter configuration) is directly coupled from a driver stage that supplies its bias. *Direct coupling* is used to drive the yoke. The purpose of the inductance (L1) that in effect shunts the yoke is to prevent raster centering problems. Raster centering on a BW set is done with a magnetic centering ring mounted on the

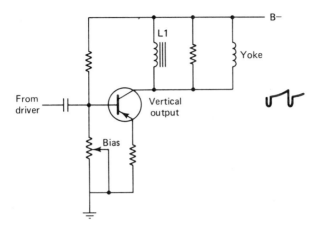

Fig. 7-10 Single-ended output stage.

neck of the CRT. This is not practical with color tubes, and centering is accomplished by passing dc through the yoke. This current, depending on its direction (polarity), produces flux that moves the beam either up or down by a small amount.

In Fig. 7-10, the dc component from the transistor tends to upset the raster centering in the same manner, except for the action of the choke coil. The choke has a high impedance to the sweep voltage, so most of the *sweep* current goes through the yoke as intended. But the coil has a very low *ohmic* resistance, shunting most of the *dc* around, rather than through, the yoke.

Note the shape, amplitude, and polarity of the waveforms. With a positive-going sawtooth on the base, the transistor inverts it to a negative polarity at the collector. With an input signal of 1 V and an output of 50 V, the voltage gain is 50. However, about two-thirds of this amount represents the sharp spike of the retrace interval and does not contribute to the amount of active sweep. Because of this spike, the waveform is trapezoidal rather than a true and linear sawtooth, as might be expected. The reason for this is as follows.

A linear (sawtooth) voltage applied to a *pure resistance* produces a linear current flow. This is not true for an inductance, and the yoke requires this trapezoidal waveform to produce a true sawtooth *current* and a linear sweep. The *spike* is the result of yoke excitation during the retrace period. The waveform shown is for a typical BW set. With a color set, the wave has a *butterfly-wings* appearance at this point because of the antipincushioning circuit.

• **The Deflection Yoke**

Beam deflection is the result of interactions between the magnetic fields around the yoke and the field around the CRT beam current. By definition, it is the same action that produces the torque in an electric motor. As with any current-carrying conductor, the beam current is surrounded by a circular magnetic field, which reacts against the field of the yoke. Depending on the direction of current, the fields either attract or repel, and the beam is deflected either up or down on the screen. Naturally, the larger the beam current and/or sweep current, the greater the flux and the amount of deflection.

The yoke has two pairs of coils, one pair for horizontal, the other for vertical deflection. To efficiently develop a strong magnetic field, it has a powdered-iron core. The vertical coils are on the sides and the horizontal coils are at top and bottom. The magnetic flux from each coil is at right angles to the winding. The inner coils (those next to the CRT neck and closest to the electron stream) are horizontal coils, since more horizontal than vertical deflection is required. All coils work almost independently from each other; the vertical coil on the left provides vertical sweep for the left side of the screen, with very little influence on the right side. Similarly, the coil on the right provides for right-side deflection. This is because of the loose inductive coupling that exists between the two coils. The only coupling between vertical and horizontal coils is the small amount of unavoidable capacitive coupling. If excessive, this may be a problem, creating crosstalk between the circuits that shows up as vertical jitter or poor interlacing. To prevent this, a capacitive ground or leakage path may be provided, or in some cases a small capacitor across the vertical coils bypasses any horizontal voltage that may be present.

Because of their wide variety, yokes (like transformers) are seldom interchangeable. For example, the opening in the yoke must physically match the diameter of the CRT neck since a snug fit (minimum air gap) is needed for maximum flux concentration. All modern yokes have a *cosine* type winding where the coils flare out and conform with the curvature of the CRT bell to produce uniform focus over the entire screen area. The vertical generally has more windings than the horizontal coils, hence a slightly higher vertical coil resistance. As with most coils involving high current and flux density, the actual number of windings is a trade-off between the opposing factors involved; more *ampere turns* means a stronger flux, but more windings means a higher resistance, which tends to reduce the current.

Yokes used on BW sets are somewhat smaller than for color sets. Figures 7–11 and Fig. 15–2 show typical yokes of each type for comparison.

The resistance of the yoke windings tends to increase with heat. This in turn reduces beam

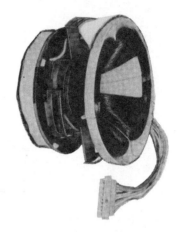

Magnet adjusters for purity and convergence

Fig. 7-11 Typical deflection yokes, and in-line CRT with yoke permanently attached (bonded).

deflection during warmup. To prevent this and to stabilize the height, some BW yokes have a built-in thermistor in series with the coils. As the coils warm up, so does the thermistor, and its gradually *reduced* resistance *increases* the deflection current. Another way to stabilize sweep is with a VDR (voltage-dependent resistor) shunted across the yoke or output transformer. Any *increase* in sweep amplitude reduces the VDR resistance to *decrease* the amount of sweep. Detailed yoke functions are described in Sec.'s 15–18, and 15–21.

7-12 Vertical Output Stage

• **Yoke Coupling Methods**

There are several alternative ways of feeding sweep power to the yoke. Figure 7-12a makes use of direct coupling with a high-inductance, low-resistance choke coil carrying most of the transistor current to prevent problems with raster centering. Compared with tube-type sets, such direct coupling (without the need for a transformer) is possible because the yoke impedance closely approximates that of the power output transistor. For the best possible impedance match, the yoke coils may be connected either *series* or *parallel,* as appropriate.

In Fig. 7-12b, a coupling capacitor has been added to pass the sweep signal while blocking the dc component. To pass sufficient power at the low 60-Hz frequency, the capacitor must have a large value (1000 to 2000 µF). Polarity must be observed since it is an electrolytic type. With this arrangement, some means must be provided for raster centering.

The circuit in Fig. 7-12c uses an output transformer to pass the sweep signal while blocking the dc, as well as impedance matching between the transistor and the yoke. It usually has a 1 : 1 ratio. In some cases an autotransformer is used, which may or may not require a coupling capacitor to block the dc. Some sets have a *multicoil* output transformer, with separate windings to supply the convergence and pincushion circuits, for raster centering and for feedback.

• **Yoke Ringing**

A coil such as a yoke when *shock excited* by a pulsed voltage produces a series of damped oscillations. Both the vertical and the horizontal yokes are encouraged to *ring* or oscillate for *one half-cycle.* This is to ensure a rapid drop-off of the sawtooth and a rapid retrace of the CRT beam. Prolonged ringing results in undesirable effects. For the vertical, this is prevented with *damping resistors* shunted across the yoke coils. The smaller their value is, the greater the damping, and vice versa. Too little damping creates problems at the top of the screen. Too much damping means a long retrace time, with the possibility of vertical foldover. Excessive ringing is sometimes prevented by using a *diode* shunted across the yoke or output transformer. The short-duration ringing produces the sharp spike (Fig. 7-10) for the required trapezoidal wave pattern.

• **Feedback Loops**

Feedback is used extensively in vertical sweep circuits. There are two kinds: *regenerative* (also known as *positive* or *in-phase*) feedback; and the opposite condition, *degenerative* (also referred to as *negative* or *out-of-phase* inverse) feedback.

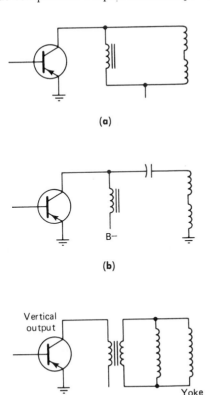

Fig. 7-12 Yoke-coupling methods.

Depending on the circuit, either type of loop may be *direct* or contain a coupling capacitor where it is necessary to block the dc. Usually there is an *RC* network, as a capacitive or resistive voltage divider, to regulate the amount of feedback.

Positive feedback is used to *sustain* oscillations when a MV-type oscillator is used. Some of the signal from the output is fed back, in phase, to reinforce the signal at the oscillator input, as explained in Sec. 7-8. Feedback may be from the second half of a two-stage oscillator or, as is more common, from the power output stage. The amount of feedback is fairly critical. If too little, the oscillator may not start up under adverse conditions like low line voltage. Also, the weakened oscillations may result in a reduction of raster height. On the other hand, *excessive* feedback can create problems like oscillator *squegging* (double triggering), where several frequencies are produced simultaneously. In some cases the vertical hold control is in the feedback loop.

In addition to the positive feedback loop for oscillations, some sets have *several negative* feedback loops. The purpose of negative feedback is to improve the response of all stages included in the loop to reduce distortion and obtain the best possible linearity of the vertical beam trace. Any abnormalities in the output waveform (as a result of nonlinear operation of some stage) are canceled out; in the process, however, negative feedback results in an unavoidable reduction in stage gain. But this can be overcome by adding another amplifier stage. A set may use a single negative feedback loop embracing most or all the amplifier stages; or there may be a separate feedback loop for each stage. In general, the greater the amount of feedback is, the better the response, but the greater the loss of gain. The use of feedback also tends to *stabilize* a circuit, making it less susceptible to changes from component aging, line voltage fluctuations, and the like. With some sets, the linearity control is in the feedback loop; with others, the response is so good that the control is *omitted*.

7-13 TWO-TRANSISTOR OUTPUT STAGE

There are two basic circuits in common use: the *complementary symmetry* circuit, and a similar version called the *quasi-* (or totem pole) complementary symmetry circuit. Both circuits use two power transistors operating either class AB or B. They operate in a push-pull fashion, conducting alternately, with each transistor responsible for scanning one half of the raster. The transistor outputs are combined to drive the yoke, either directly, or via a coupling capacitor or transformer.

- **Complementary Symmetry Circuit**

Two basic circuit configurations are shown in Fig. 7-13. Both circuits use *unlike* transistors, an NPN, and a PNP, *both* operating as emitter followers. In Fig. 7-13a, the yoke is fed through a coupling capacitor (cc) and only *one* source of B voltage is required. The voltage is split equally between the two transistors. In Fig. 7-13b, the yoke drive is *direct* (no cc), and the circuit requires *two* separate power sources of equal amounts but opposite polarities. With this arrangement, the junction of the transistors is at *zero* dc potential in reference to ground.

The operation of the two circuits is almost identical. For the conditions as shown, *Q1* is conducting with a small forward bias and *Q2* is cut off with a high reverse bias. The positive-going signal at the base of *Q1* causes increased conduction to supply sweep power to the yoke for scanning the top half of the screen. As the increasing signal amplitude overcomes the bias on *Q2*, it conducts to supply the trace for the bottom half of the screen. The outputs of the two transistors combine at their common emitter junction to feed the yoke. One problem peculiar to this circuit is *crossover distortion*, which may show up as a *notch* midway on the sawtooth wave. It occurs where neither transistor is con-

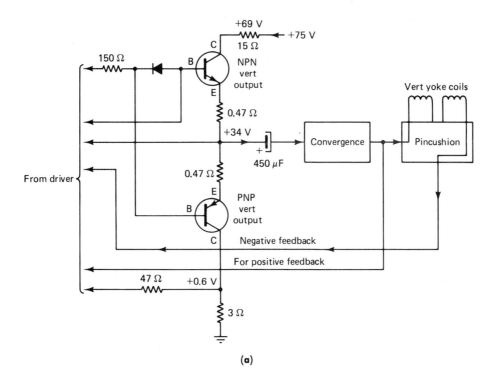

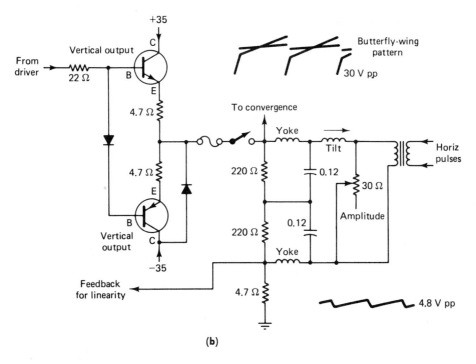

Fig. 7-13 Complementary symmetry output circuits.

210 THE VERTICAL SWEEP SECTION

ducting for an instant when one transistor takes over where the other leaves off. It is prevented by offsetting the bias on the transistors with a diode between their base connections and by using a fair amount of negative feedback. The two low-value emitter resistors *R1* and *R2* are for current limiting and are also protective devices that fuse under overload conditions (as when a transistor shorts out). In the circuit of Fig. 7-13a, resistor *R3* also serves as a fuse, in addition to providing a feedback voltage.

Resistor *R4* is a trimpot. It is a bias adjuster for the two output transistors. Bias voltage is critical. Too little forward bias means reduced output and a corresponding reduction in raster height. Too much bias can result in thermal runaway, which destroys transistors and other components. Bias is normally adjusted after replacing one or both transistors. Where an adjuster is not provided, some form of *bias stabilization* is generally used.

• **Quasi-Complementary Symmetry Circuit**

Two basic circuit configurations are shown in Fig. 7-14. Both circuits use a pair of *identical* transistors, one operating as an *emitter follower*, the other as a *common-emitter amplifier*. Because of this, the latter must be driven by a *phase inverter*. In Fig. 7-14a (as in Fig. 7-13a), the two transistors are supplied (in series) from one dc source, and a coupling capacitor passes sweep power to the yoke while blocking the dc. At the junction of *Q1/Q2* the dc voltage is one-half the supply source. In Fig. 7-14b (as in Fig. 7-13b), the amplifiers are *directly* coupled to the yoke and each transistor is supplied from a *separate dc source* of *opposite polarity*. If everything is in balance, the voltage at the junction feeding the yoke will be *zero*.

Except for the added inverter, the operation of both circuits is not too different from those in Fig. 7-13. In Fig. 7-14a, for example, the positive-going signal from a driver stage directly feeds the base of the output transistor *Q1*. Since *Q1* is an emitter follower (which does not invert the signal) and *Q2* is a common-emitter amplifier (which does invert), *Q2* is driven by an inverter *(Q3)*. With different amounts of bias, the two output transistors are alternately driven into conduction, first, *Q2,* to provide the trace for the top half of the screen, then *Q1,* to sweep the bottom half. Then both transistors are turned off and the yoke kickback provides energy for beam retrace.

It is essential that *Q1* and *Q2* receive equal drive and that their outputs be the same. This is complicated by the difference in their configurations, where one *(Q2)* is a voltage amplifier and the other is not. In practice, the gain of *Q2* is reduced to *unity* with negative feedback back to the inverter, and because the *Q2* load impedance (the yoke) is very low.

REMINDER: _____

Negative feedback reduces the gain of a stage by *cancellation;* the greater the feedback is, the lower the net gain.

For both circuits, the takeoff point for *retrace blanking* is the yoke. Other circuits get the pulses from some prior stage. Typical waveforms are shown at inputs and outputs; note their relative amplitude, shape, and polarity.

NOTE: _____

If the foregoing circuits are used with a *color set,* additional circuitry is required for raster centering, convergence, and pincushioning.

7-14 VERTICAL SWEEP-DEPENDENT CIRCUITS

Besides beam deflection, there are *other* vertical sweep related functions associated with the output stage as described next. See Fig. 7-15.

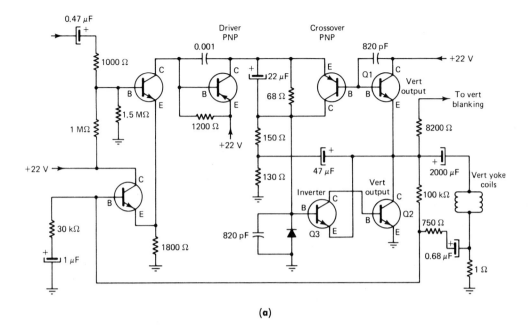

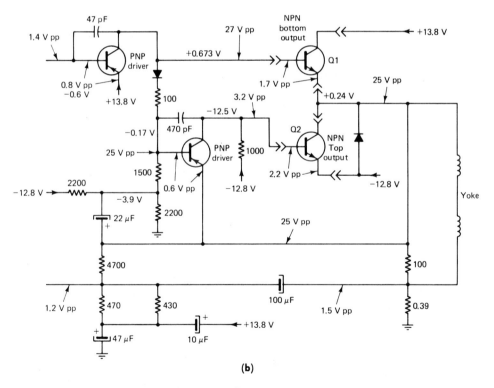

Fig. 7-14 Quasi-complementary symmetry circuits.

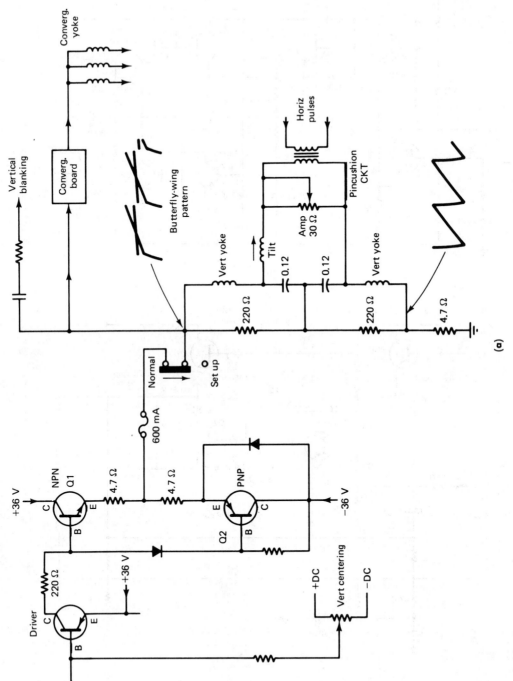

Fig. 7-15 (partial)

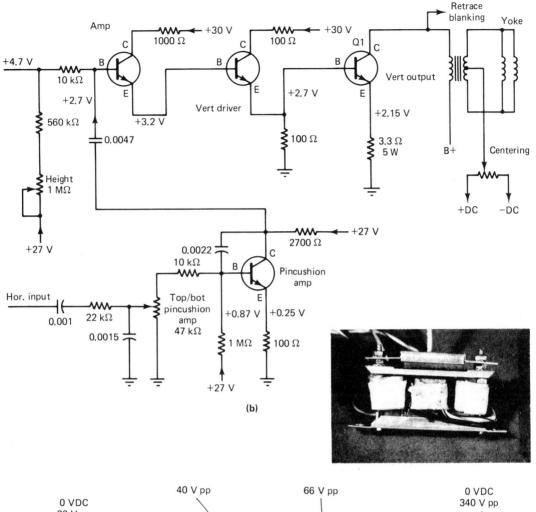

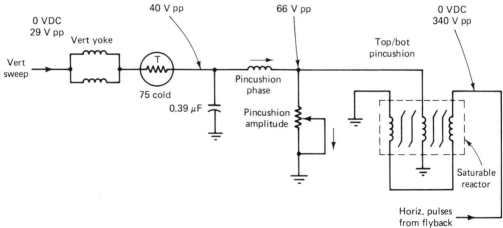

Fig. 7-15 Vertical sweep-dependent circuits.

214 THE VERTICAL SWEEP SECTION

• Vertical Retrace Blanking

Ideally, under good receiving conditions, the blanking pulses (which are part of the received signal) are strong enough to overbias the CRT and darken the screen during retrace intervals. Unfortunately, beam cutoff does not always occur, especially when the brightness control is turned up. Under these conditions, visible *retrace lines* appear on the pix, a series of widely spaced white lines sloping diagonally upward to the right.

All modern sets have a special circuit that largely overcomes this problem. *Sharply peaked* vertical sweep *pulses of short duration* are used to overbias the CRT at the proper times (i.e., at the beginning of each vertical retrace period after scanning a field). Depending on its polarity, the pulse may be applied to either the CRT grid, or its cathode, or to some suitable point in the video amplifier section. For added gain and/or a phase reversal, a separate *blanker transistor* is sometimes used. Most receivers also have *horizontal retrace blanking* (see Sec. 9-5). Vertical and horizontal blanking are separate functions. With some sets they are treated separately; with others, *both* pulses are routed together and handled as *one*.

Vertical blanking pulses may be taken from a driver stage or, what is more common, from the output amplifier, or a winding on the output transformer. The pulses must be of high amplitude (at least as great as the CRT cutoff bias). They must have the proper *polarity*, positive-going if they ultimately drive the CRT cathode or negative-going to drive the grid. Pulse *duration* is extremely important. It is established by the time constant of an *RC differentiating network* (see Fig. 7-15), whose function is to shape the sweep waveform into sharply spiked pulses as required.

The blanking pulse is developed at the *peak* of the vertical sawtooth and becomes effective from the beginning of the retrace interval. Certain defects create problems. For example, a pulse that is too weak cannot do its job and retrace lines will appear. If the pulse duration is too short, the lines may be visible on the upper portion of the screen. If the pulse duration is too long, blanking will go *beyond* vertical retrace and the top of the screen will be in shadow or completely blanked out.

Retrace blanking is not to be confused with *sync blanking*, or *raster blanking*, a feature of some modern sets where the CRT is blanked out to prevent the thin bright line from burning the screen in the event of sweep failure, or as a failsafe measure if the HV becomes excessive.

• Pincushion Correction

Pincushioning is a condition peculiar to large-screen, wide-angle rectangular tubes, both BW and color. Because of the flatness of the CRT faceplate, the nonconcentricity between the CRT guns (delta-type color tubes) and the yoke opening, and other factors, the four sides tend to bow inward. With BW tubes it is easily corrected with small permanent magnets placed at strategic points near the tube or built into the yoke assembly. With color tubes, special *antipincushioning* circuits are required where *top and bottom* and *side distortion* is treated separately. Side correction is described in Sec. 9-6.

For top and bottom correction, the vertical sweep must be increased for the *mid-portions* of each horizontal scan. This is done by superimposing a sampling of *horizontal* sweep voltage on the vertical sweep current flowing through the yoke. The phase of the correction voltage must be opposite for top and bottom areas of the screen, since the horizontal scanning lines must be bowed *upward* at the top and *downward* at the bottom. Too much correction, however, results in a barrel-shaped raster where the 4 sides bow outward.

The *horizontal* sweep may be introduced into the vertical yoke in various ways. Some sets do it

7-14 Vertical Sweep-Dependent Circuits

using a separate transformer; others use taps on the vertical output transformer. See Fig. 7-15.

The problem of top-bottom correction is complicated by the fact that correction voltages *of opposite phase* (polarity) are required for correction above and below the center line (the raster must be bowed upward at the top and downward at the bottom). One method feeds a *horizontal* parabolic wave (from the horizontal sweep or yoke circuit) via top-bottom correction circuitry to the vertical yoke. Another method develops a phased parabola from a sampling of the horizontal sweep voltage. It is fed to the input of the vertical sweep amplifier where it combines with the sawteeth to be amplified to drive the yoke. The result is greater vertical deflection (during the middle of the horizontal trace) to bow the raster *outward* at top and bottom to match the deflection at the corners.

Another popular circuit is shown in Fig. 7-15b. It uses two *saturable reactors* (sometimes called modulators), one for side correction, the other for top-bottom correction. Some sets perform both functions with a single reactor. A saturable reactor is a form of magnetic amplifier (Fig. 7-15c). The device has a core that is readily *saturated* when a specific amount of dc is passed through one winding. A PM-type magnetic shunt establishes the saturation point.

For top-bottom correction, horizontal pulses (from the flyback transformer) go through two of the coils of the reactor transformer $T2$, and the degree of saturation is determined by the vertical sweep current through the center winding. With core saturation there can be no induction; but when the core becomes unsaturated, horizontal pulses are induced into the center coil to modify the vertical deflection current at the correct times. The result is increased vertical sweep at the center of the horizontal scanning. The amount of correction is adjustable with the control pot $R1$. Phasing, for opposite polarity correction at top and bottom, is done with the adjustable coil $L1$.

• **Convergence**

Convergence (as described in Sec. 2-11) is where the three primary color fields are superimposed on each other to produce a BW raster. With modern in-line pix tubes, it is no great problem, but with *triad* (delta type) tubes, perfect convergence, especially near the outer edges of the tube, is not easy to achieve or maintain. *Static convergence* (for the central areas of the screen) is taken care of with small adjustable permanent magnets (one for each gun) mounted on the *convergence* yoke on the CRT neck. Adjacent to each of the three PM *static adjusters* on the yoke is a small electromagnet whose pole pieces are positioned directly over corresponding pole pieces in each of the three guns, as shown in Fig. 15-12. Each of the three electromagnets has two sets of coils. One coil is supplied with a sampling of the vertical sweep output; the other is fed from the horizontal sweep. Thus, in addition to normal vertical and horizontal deflection, the three beams are also influenced (to a small degree) by magnetic fields developed by the convergence coils. This is called *dynamic convergence*.

To obtain the desired result, the sweep voltages fed to the coils are modified (as with pincushioning) from a sawtooth to a *parabolic* shape, as shown in Fig. 15-12. Thus the amount of influence on each of the three beams will vary according to amplitude changes in the waveform relative to their landing positions on the screen at any given instant. Waveshape and amplitude adjusters are provided on a *convergence panel*, as shown in Fig. 15-11.

In some sets, convergence, pincushioning, and the yoke circuits are all interconnected in series, and the sweep current has only one path. With this arrangement, an open in convergence or pincushion coils will also result in no vertical deflection.

For in-line pix tubes, convergence is much simpler. In place of a convergence yoke with its complex circuitry and adjustments, there may

216 THE VERTICAL SWEEP SECTION

be only a small coil built into the deflection yoke or a twist coil on the CRT neck with only two adjusters. Some late-model sets have only adjustable magnetic rings on the tube neck and no electronic circuitry at all. For further details on convergence, see Chapters 9 and 15.

7-15 RASTER CENTERING

Any *dc* current in a yoke coil will produce a polarized magnetic field that tends to move the raster off center. Current in one direction through the horizontal yoke will shift the beam (and raster) to the left. Current in a reverse direction will move it to the right. Similarly, dc in the vertical yoke will move it either up or down. Where sweep voltage feeds the yoke via a coupling capacitor or transformer, the dc component is lost and there is no problem. With this arrangement, there is no need for a centering control. But many sets use direct coupling. Because of misadjustment, component aging, and other factors, some means must be provided to center the raster in both directions. With a BW CRT, this is accomplished with a *centering ring* on the tube neck. With a color tube it is done by feeding each yoke with a small amount of dc. The direction and polarity of the dc is adjusted by V and H centering potentiometers.

One method used for vertical centering is shown in Fig. 7-15a. The centering control is connected across one secondary winding of the vertical output transformer, and the control arm and the coil center tap connect to the yoke in a bridge circuit. The *net* dc voltage applied is about 1 V, the difference between two slightly different voltage sources. With the control arm at mid-setting, the voltage is zero. At the extreme settings, a small amount of + or − dc is impressed on the yoke.

A control method used with a complementary symmetry output stage is shown in Fig. 7-14. As described in Sec. 7-13, the two output transistors are normally balanced so there is zero dc at their junctions that drive the yoke. If there is an *unbalance* (as when one transistor conducts more or less than the other), the yoke receives a small amount of either + or − dc. The centering control does this deliberately by altering the bias on one driver amplifier, as shown.

7-16 COMPLETE VERTICAL SWEEP SECTION

Following is a brief functional description of the vertical sweep section of the color set schematic at the back of this book (see Sample Circuit B).

The plug-in sweep module of this particular set contains all the vertical sweep components and the sync separator and signal conditioning circuits. It also contains the circuits for generating the horizontal sweep signal. All the circuits for generating and controlling both sweep signals are contained in a single IC, as shown in block diagram form in Fig. 7-6. Also note the vertical sweep blocks external to the IC. A brief description of the operations performed by the IC is given in Sec. 7-9. Since the IC is not repairable, a detailed description is of little concern to the technician and will *not* be covered at this time.

Leaving the IC, the positive-going (1.5 V p-p) vertical sawtooth signal feeds the predriver amplifier (Q601) operating in the common emitter configuration. The height (vertical size) control varies the sweep amplitude by controlling the bias on this transistor. The inverted sweep output of Q 601 drives another amplifier, Q 603. Note the slight signal *loss* in Q601 due to the relatively high value emitter resistor, which is unbypassed. Another reason is the negative feedback (from the vertical output stage) applied to its collector.

Transistor Q 603 has a gain with an 11-V p-p signal at its output. Its waveform is modified by a negative feedback signal applied to its emitter circuit. The output of Q 603 drives a third and final stage, Q 605. This is an emitter-follower stage with a positive-going output of only 6 V at its emitter. Note the vertical blanking pulse that is taken from its input circuit to feed the video amplifier section of the set.

The dual-transistor output stage is a quasi-complementary symmetry amplifier. The two transistors (Q 607 and Q 609) are driven *out of phase* by the two output signals from Q 605. To obtain a balanced output from the two circuits, Q 609 operating as a common emitter amplifier has considerable gain (112-V p-p output with only 0.6-V p-p input). With a high drive signal (112 V p-p), Q 607, as an emitter follower with considerable feedback, has *no voltage gain,* so its output will match that of Q 609. The two output transistors (of the same type and driven by signals of opposite polarity) conduct alternately, Q 609 providing the sweep for the top half of the screen and Q 607 for the lower half. Both transistors are turned off by the high-amplitude feedback pulse from the yoke during vertical retrace.

The yoke is driven via a coupling capacitor C 619. For raster centering, a small amount of dc is introduced into the yoke via the centering control R 650. For top–bottom pincushioning, the set uses an uncomplicated system where horizontal sweep current is series fed from the horizontal yoke to one winding of a simple pincushion transformer (T 1810). The secondary coil is connected series with the vertical yoke. No pincushion adjustments are provided.

Note the absence of any dynamic convergence circuitry. Instead, the set employs a simple and unique self-converging system making use of a number of adjustable magnetic rings on the CRT neck. The adjustment procedure is described in Chapter 15.

7-17 TROUBLESHOOTING THE VERTICAL SWEEP SECTION

For most of the following procedures, it is assumed that the trouble has been localized to this section of the receiver. If not, be guided by the symptoms and the procedures described in Chapters 4 and 5. The most common trouble symptoms associated with the vertical sweep are: partial or complete loss of vertical deflection (raster height), poor vertical linearity, raster distortion, horizontal hum bars, and loss of sync. If it is an unfamiliar set, the first thing is to locate and identify the vertical module (or components) on the chassis. Study the schematic. Inspect, first with the set off, then with power applied. Make note of anything unusual or suspicious. Use all senses as explained in Sec. 4-8. Be especially alert for signs of overheating, burned or damaged components, poor soldering, and the like. Try adjusting the vertical controls, noting anything unusual in their operation. Analyze the symptoms and try to decide where to start, for example, the oscillator if it is a sync problem, driver-amplifier stages where there is reduced height but the sync is OK, or the yoke where there is keystoning. Be prepared to spend more time with intermittents, and small problems (e.g., barely enough height, less than perfect linearity, marginal sync).

A troubleshooting flow diagram is shown in Fig. 7-16. Table 7-1 lists a number of common vertical troubles, procedures, and possible causes. Use them as a guide if the trouble has not yet been localized to some specific stage or area of the sweep section.

The first step (as shown in the flow diagram) is to *localize* the fault to a particular stage or subsection and, after that, to *isolate* the defect. Don't neglect the inspection at the start, first with the set off, then again after power turn on. It is an *important* part of the procedure that can often

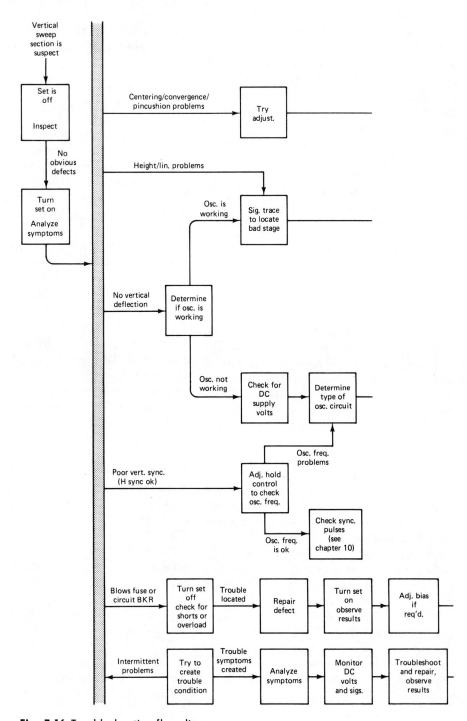

Fig. 7-16 Troubleshooting flow diagram.

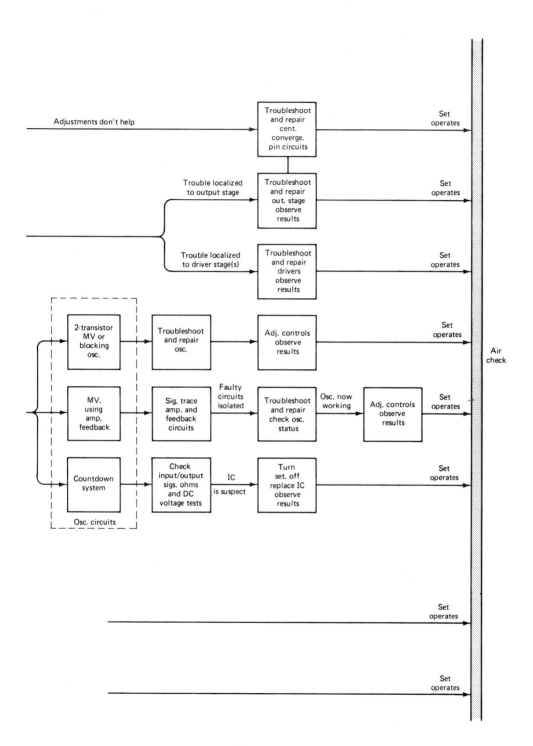

Table 7-1 Vertical Sweep Troubleshooting Chart

Trouble Condition	Probable Cause	Remarks and Procedures
No vertical deflection; thin horiz. line only	Loss of B voltage. Dead oscillator, driver, or output stage. Defective countdown IC.	Turn down brightness so will not burn CRT screen. Make voltage check and signal trace to localize trouble. Check components.
Insufficient height or poor linearity	Low B voltage. Weak oscillator, driver, or output stage. Negative feedback problems.	Adjust vertical controls as required. Check B voltage. Check input-output signal amplitude of each stage. Scope check waveforms. Check voltages and components in faulty stage.
No raster (screen is dark); sound ok	*Probable* cause is loss of HV, a defective CRT or abnormal CRT operating conditions.	May be loss of sweep and beam completely off screen at top or bottom. May also be due to CRT disabling from loss of vertical sweep (some sets only).
Vertical centering problems	Color set: defective centering control. DC voltage problems. Leaky or shorted yoke coupling capacitor. Unbalanced dual output transistors.	If BW set, adjust centering ring. For color set, check for dc yoke current. Check for unbalanced dual output stage.
Black area at top or bottom of screen	See remarks for the numerous possibilities.	May be a centering problem, insufficient height, poor linearity (compression), a retrace blanking problem, vertical foldover, or unbalanced sweep from dual output stage.
Vertical foldover, top or bottom of screen (portions of pix inverted)	Defective transistor, diode, or *R/C* components in sawtooth-forming circuit. Improper operation of vertical output stage. Open or leaky anti-ringing diode or VDR in output stage.	Check waveforms of all stages, particularly the trapezoidal wave at vertical output. Output transistor may be saturating, or operating near cutoff. May be a prolonged retrace. Check for excessive yoke damping.

Table 7-1 (cont'd)

Trouble Condition	Probable Cause	Remarks and Procedures
Vertical deflection for top *or* bottom half of screen only	Defective output transistor(s) or other defect affecting their operation.	Check for unbalance of dual output transistor circuits. Check dc input(s) and at transistor outputs.
Gradual loss of height	Thermal runaway of output transistor(s). Defective thermistor in yoke. Some resistor changing value during warm-up. Defective VDR in output stage.	Usually occurs within first half-hour of operation.
Blows fuses or circuit breaker opens up	Output transistor(s) shorted or drawing excessive current. Shorted or leaky filter on vertical dc bus. Defective fuse or circuit breaker.	Check for a B voltage short. Try replacing fuse or resetting breaker *once* only.
Breathing (Raster expands and contracts rhythmically.)	Open filters on vertical dc bus.	*Caution: Don't* bridge filters with set turned *on*.
Keystone raster shape (not rectangular)	Shorted or open yoke coil, open pincushion coil, open damping resistor across one yoke coil.	The narrow part of the keystone wedge points to the defective coil if it is a yoke problem.
Crossover distortion (poor linearity, center of screen)	Defective transistor in dual-output stage. Unbalance in outputs. Output bias problem. Misadjusted bias control.	For this problem a notch is seen near center of the active trace portion of the waveform.
Overheating components	If a transistor, it may itself be bad or too much forward bias. Overheating resistors are usually caused by a shorted capacitor or transistor or excessive current drain of a transistor.	Observe which components overheat, then *quickly* turn set off. Consider the possible causes and make tests as appropriate. Periodically check temperature of transistors by *feel*.
Pincushion distortion (raster edges bow inward)	Misadjustment or defect in antipincushion circuit. If BW set, may be a weak or an improperly positioned PM magnet.	Try pincushion adjustments. Scope check for butterfly-wings pattern at vertical output and parabolic waveform at pincushion circuit.

Table 7-1 (cont'd)

Trouble Condition	Probable Cause	Remarks and Procedures
Visible retrace lines (widely-spaced white diagonal lines)	Brightness set too high. Defect in retrace blanking circuit. Defective blanker transistor.	Lines should not be visible for normal setting of brightness control. Try touch-up adjustment of vertical hold control. If a color set, and excessive brightness, back off on kine bias control.
Poor vertical convergence	Improper parabolic waveform. Open vertical coil(s) on convergence yoke. Defect in convergence panel.	If unable to converge, check for parabolic waveform reaching converge yoke coils.
Poor interlacing or vertical jitter (non-uniform spacing of scanning lines)	Poor lead dress or insufficient filtering of the dc bus. Weak or excessively strong sync pulses. Signal overdrive.	Critically adjust vertical hold control. Verify there is good vertical sync stability, and no indications of signal overdrive. Scope check for horizontal sweep voltage in the vertical circuit (from poor lead dress, inductive or capacitive coupling, yoke leakage, or insufficient decoupling on the dc bus).
Loss of vertical sync (horizontal sync is OK) *Note:* Vertical sync problems include a variety of symptoms; multiple pix's, rolling, jitter, oscillator drift, etc.	Anything abnormal in the oscillator circuit will usually affect its frequency, e.g., capacitor leakage, resistors changed value, transistor leakage, etc. (For weak or missing pulses, check the integrator and sync stages per Chapter 10.)	Using hold control, check ability of oscillator to generate the correct frequency. If unable to obtain one pix within range of control, the oscillator stage is at fault. If one pix is obtained but unable to *lock,* sync pulses are probably weak or missing.
Intermittents	According to the nature of the problem. Any of the foregoing conditions can be intermittent.	Do whatever is necessary to create the trouble symptoms long enough to permit testing.

save time by eliminating much step-by-step testing.

7-18 LOCALIZING THE PROBLEM AREA

Before getting involved, make sure you are aware of all safety precautions as described in Sec. 4-6, especially the proper procedures for avoiding damage to components.

• DC Voltage Check

Depending on the symptoms, start by verifying the presence of B voltage at one or more key points as indicated on the schematic. If there is no voltage or the voltage checks low (with component overheating or other signs of overload), follow the procedures described in Secs. 6-6 and 6-13. If the dc is supplied from a secondary source (SCAN rectification), See Sec. 9-9.

• Signal Tracing

The purpose of signal tracing is to localize a fault to a specific stage or functional area. There are various methods, as described in Sec. 4-10. Further details (relating to the vertical section in particular) are given next.

Signal Tracing with a CRO or EVM. The CRO is the preferred method since it not only indicates the presence or absence of sweep voltage at various circuit points, but also provides an instant readout of p-p amplitudes, polarity, and waveshapes. For relative indications of signals, however, an EVM (on some ac range) is quite suitable.

Usually, the first step is to establish whether or not the oscillator is working. Look for a sweep voltage at its output or the input to the first driver stage. If no signal is being generated, check the schematic, noting the type of oscillator circuit used.

REMINDER:

Most MV-type oscillators depend on feedback from a driver or output stage (to sustain oscillations) and all stages *within the feedback loop are suspect*. Start with the output stage working back toward the oscillator, using *signal injection* (as described in Section 4-10.).

If the oscillator is working, trace the signal through the various stages working toward the yoke. What you are looking for depends on the nature of the problem. If there is complete loss of sweep, you determine at what point the signal is lost. For partial loss of height, the problem is more difficult, and the signal amplitude at each check point must be carefully considered as compared with those on the schematic. Where *linearity* is the problem, emphasis is on *waveshape*. Here, again, check with the schematic. Note the typical waveforms, their amplitudes and polarity, on the Sample circuits at the back of this book.

REMINDER:

Don't expect any voltage gain or phase reversal from an emitter-follower circuit. Also, remember to adjust the height and linearity controls while testing.

Open capacitors can often be discovered while signal tracing. For example, if the sweep signal is modulated by hum voltage, scope check at dc voltage points for excessive ripple. This often indicates an open filter or decoupling capacitor. Open electrolytics are a very common problem. Also, be on the watch for open coupling capacitors, which are most easily discovered while signal tracing. Where a coupling capacitor is used, look for attenuation or complete loss of signal between the output of one stage and the input of the next.

224 THE VERTICAL SWEEP SECTION

If troubles are *intermittent,* monitor the signal at each stage and observe any changes when the set acts up. Common sweep circuit intermittents are an oscillator that is slow in starting up or erratically cuts out, a sudden decrease in raster height, abrupt changes in linearity, and periodic loss of vertical sync.

Signal Injection. With this method, a 60-Hz signal from a test generator is injected at various points while noting the response at the power output stage or yoke. The indicator may be a CRO, an EVM, or the pix tube itself. When there is complete loss of vertical deflection, it is customary to start by feeding the signal to the output stage, working backward toward the oscillator.

REMINDER: _____

Make sure there is a blocking capacitor series with the generator lead. Keep generator output attenuated to a low level, especially when feeding predriver stages where the transistor might easily be destroyed by signal overdrive. When progressing from the output to the input of a common-emitter amplifier, look for some voltage gain. If it is an emitter follower, no gain can be expected.

Lacking a generator, other sources of signal can be used in an emergency for tracing by injection (see Sec. 4-11). They should be used with care to avoid damage to components.

7-19 ISOLATING THE DEFECT

Once the trouble has been localized to the stage or module level, make tests (as described in Chapter 5) to identify the *cause,* whether it is an abnormal voltage, a misadjustment, a bad component, or whatever. The various tests are generally performed in a logical sequence as follows; depending on circumstances, however, this sequence need not be rigidly adhered to.

1. Inspect for anything unusual, concentrating on the faulty stage.

2. Measure the dc supplied to the stage. Compare with the schematic. If there is no dc or it checks low, there are two possibilities: a short or high current drain by the faulty stage, or trouble at the source. For the latter, proceed as described in Sec. 6-13. If the source is a secondary power supply (SCAN rectification), troubleshoot as in Sec. 9-20.

3. Measure dc at transistor terminals and compare with the schematic. Be precise. Even fractional voltage discrepancies can make the difference between a transistor conducting or not conducting. Where direct coupling is used between stages, dc measurements are made difficult, since practically any circuit disturbance upsets the normal voltage distribution from that point onward. *Reminder:* Some component defects do not show up during normal E, I, and R testing (see Sec. 5-8.).

4. Resistance tests: Make resistance checks as appropriate from circuit points, transistor terminals, and all suspected components in an effort to identify a bad part (see Sec. 5-5).

5. Test suspected component(s) as in Sec. 5-8. Replace as necessary and observe results.

6. When an IC is suspect (and prior to considering replacement) perform the following: Scope check input–output signals. Check for proper dc applied voltages as specified on the schematic. Make resistance checks to determine possible shorts *external* to the IC. Such tests may show the IC is not at fault and may point up a defect that if not corrected may destroy the replacement IC (see Sec. 5-9).

Specifics on troubleshooting and testing the individual stages are described in the following paragraphs.

- **Troubleshooting and Repair of the Vertical Oscillator**

Three kinds of troubles can be attributed to the vertical oscillator: partial or complete loss of vertical deflection, poor linearity (where there are no drivers feeding the output stage), and frequency-sync problems.

No Vertical Deflection. The absence of any scope display (or ac reading on an EVM) at the oscillator output indicates a dead oscillator. Other ways to check the go-no-go status of an oscillator (as explained in Sec. 5-4) are to note small changes in the dc operating voltages at the transistor terminals when the circuit is disabled as by grounding with a capacitor. With some sets the oscillator output is very low (less than 2 V p-p), which can mislead one into thinking that no signal is being generated. Turn up the CRO gain or use a low-meter range before jumping to conclusions.

Testing of a troublesome oscillator is determined in part by the type of circuit. For example, the feedback necessary to sustain oscillations in a blocking oscillator takes place in a small transformer, which can go bad. With a two-transistor multivibrator, a breakdown of components in either circuit can kill the oscillations. Where the positive feedback comes from a driver or output stage, all components within the feedback loop become suspect.

As with any stage, once the trouble is localized, it is customary to start by checking and evaluating dc voltages at the transistor terminals (see Sec. 5-4). However, since most defects upset the readings, it becomes a problem deciding whether a wrong voltage is the *cause* or the *result* of the stage not oscillating. One common cause of a dead oscillator is loading by some circuit defect downstream, for example, a shorted driver transistor (even when stages are not directly coupled) or a shorted or leaky coupling capacitor. For this possibility, try isolating the oscillator by removing the driver transistor or breaking the connection to its base input. Loading (by some circuit disorder upstream) can also kill the oscillator, for example, a shorted sync transistor or capacitor in the integrator network. Here, again, isolation is one way to find out. Such defects will also upset the normal operating voltages on the oscillator, and the cause may also be found by following *these* clues.

As stated earlier, a number of stages (and numerous components) are suspect where the oscillator gets its feedback from a driver or output stage. If the amplifiers check out (by the signal injection method) and no defect can be found in the oscillator stage, the feedback loop may be at fault. Look for an open coupling capacitor or a poorly soldered connection. A shorted or leaky capacitor will also upset the oscillator voltages, as will trouble in the output stage if the loop is directly coupled.

For the testing of individual components, see Sec. 5-8. Look for open, shorted, or leaky capacitors, open or changed resistors, defective transistors or diodes, or a defective blocking oscillator transformer. A very common problem is high-value, low-wattage resistors that increase in value.

Intermittent Oscillator Operation. This is a fairly common problem, which usually occurs shortly after set turn on. Since other stages can produce the same loss of deflection, monitor the on-again, off-again sweep voltage at each stage. One common cause of an intermittent oscillator is poor junction contacts inside a transistor. While monitoring the signal, try coolant spray on the suspect transistor, alternating, if necessary, with the application of localized heat (of short duration). Also check for intermittent operation of the service setup switch.

Another condition is when an oscillator is slow in starting up, and the cause can be very elusive. Typical causes of a reluctant oscillator are a leaky transistor or capacitor, a resistor slightly off value, or excessive loading by some downstream stage. As a *temporary* expedient, try increasing the amount of positive feedback with a slightly larger capacitor in the feedback loop or a higher voltage-dividing resistor to ground; or increasing the B voltage to the stage. This is not a cure but a means of pinpointing the problem.

CAUTION:

Avoid making changes that alter the operating voltages of a transistor, especially the bias.

Insufficient Raster Height. Assuming it is not due to misadjusted controls, this problem indicates there is not enough sweep current flowing in the yoke, which in turn is caused by some defect upstream from that point. The loss of height may be severe or, where the loss is moderate, a problem that can be much more difficult to correct. If after signal tracing the stages and carefully evaluating the sweep amplitude at different points it is decided there is *low output* from the oscillator, start by checking the dc supply voltage. Low dc voltage is a common cause of reduced raster size.

In general, the same tests are performed as for a complete sweep loss. The difference is usually one of degree (i.e., the severity of the breakdown), for example, a capacitor or transistor with *leakage,* rather than a direct short, or a resistor *slightly* off value. Also look for capacitors that are open or nearly so, in the case of electrolytics. When in doubt, try substitution.

REMINDER:

Loading caused by defects in interfacing circuits can reduce the strength of oscillations and, therefore, the amount of oscillator output. If there is a feedback loop from an amplifier stage, verify the strength of the feedback pulse and make further tests as appropriate.

Vertical Linearity Problems. Here again the cause may be found in the oscillator or some subsequent stage. Examine the waveform at the oscillator output and compare with that shown on the schematic. Try adjusting the height, linearity, and peaking controls as appropriate.

REMINDER:

Where there are driver amplifier stages, linearity of sweep is usually established in these stages and the oscillator waveform is of little consequence. Where the oscillator *is* a possibility, check the transistor and the *RC* components in the waveshaping network. Verify the operation of the linearity and/or peaking control and its associated circuitry, and any negative feedback loop from a later stage. The linearity control normally has the greatest effect on the *top* portions of the screen.

Frequency and/or Sync Problems. Two conditions must be satisfied to have a pix locked in sync: (1) the oscillator must be *free-running* at the correct frequency, and (2) it must be supplied with repetitive sync pulses. To do its job of triggering the oscillator, the pulses must have sufficient amplitude and be of the proper shape and polarity.

Oscillator Frequency. The frequency of oscillations is determined primarily by the time constant of the transistor B/E circuits and is variable over a small range with the hold control. Almost any change or defect in the oscillator circuit (or interfacing circuits) will affect the frequency. Minor changes can be compensated for with the hold control. For more severe problems the frequency change may be *great* and *beyond* the range of the control to synchronize the pix.

Vertical roll is the usual symptom for loss of sync. Where a single pix is stationary (even momentarily), it means the frequency is correct. If the roll is *upward,* the oscillator is running *slow;* if roll is *downward,* it is running *fast.* Thus, with the pix tube as a convenient indicator, there is seldom a need to actually *measure* the frequency.

Another important consideration is the oscillator must have good frequency *stability.* Where there is frequency drift, the hold control must be periodically adjusted to maintain lock-in. When it does occur, try replacing the transistor even though it checks OK, in addition to the *R/C* components in the *critical* frequency-determining circuits.

When faced with a sync problem, the first step is to decide if it is an *oscillator or sync pulse* problem. It should be possible to obtain a single pix at the approximate mid-setting of the control and to roll the pix in both directions. If not, the oscillator stage is at fault. If pix locks at one *extreme* of the control, that too spells trouble. Check and replace all components that are at all questionable.

NOTE: _____

Due to the interaction between controls, sometimes the frequency problem may be corrected by a slight readjustment of the height and linearity controls.

Where the free-running oscillator frequency is off and the cause is obscure, in extreme cases (as a temporary expedient) it may be possible to obtain the correct frequency with a slight change in *R/C* component values. Decide whether the oscillator is fast or slow and alter the values accordingly. Avoid making drastic changes that will adversely affect height and linearity or the transistor bias.

A set using the countdown system for developing the vertical and horizontal sweep signals poses a special problem. Since there are no hold controls, nothing can be done when the frequency is incorrect except to replace the digital frequency-determining ICs. First make the necessary resistance and voltage checks as described in Sec. 5-9 and 7-26.

Oscillator Triggering. Once again, eliminate the oscillator as a suspect by making sure a pix can be rolled *both* upward and downward within range of the hold control. If the oscillator is operating normally, roll the pix *slowly* downward, then upward, at which point it should snap *abruptly* into sync. Where the upward roll continues *without hesitation,* it means a complete *loss of sync* and the oscillator is *not* being triggered because there are *no pulses* reaching the oscillator input. Where a pix *momentarily* stops rolling and/or occasionally loses sync or flips frames (as when switching channels), the sync is considered marginal and the cause is probably *weak* pulses.

NOTE: _____

The hold control should not be too critical; that is, pix should lock over at least 25% of its range. A critical control is another indication of weak pulses.

A CRO is normally used to check for pulses at the oscillator input. With the tuner set for a reasonably strong channel, connect the CRO to the base (or emitter, as applicable) of the oscillator transistor. If a pulse is present, note its amplitude, which normally runs between 2 and 10 V p-p or about 40% of the oscillator waveform. The precise amplitude cannot be specified on the schematic because it depends on the strength of the received signal. While viewing the pulse, note any hash, video, or hum voltage that can cause erratic or premature triggering.

Where the CRO shows no pulses (or weak pulses) at the oscillator input, scope trace back through the integrator network and, if necessary, the sync stages until a normal pulse is found (see Chapter 10). If pulses are lost or attenuated by the integrator, check the individual resistors and capacitors, unless they are contained in an IC, in which case it must be replaced as a *unit.*

If a set has poor or marginal pix contrast as well as unstable sync, troubleshoot the former condition and the sync may take care of itself. Another common problem is *signal overdrive* (often caused by a malfunctioning AGC circuit), where the contrast may be OK, but the peaks of the signal (the sync pulses) are either stripped or compressed, resulting in no or weak pulses at the oscillator input.

For poor vertical sync *and* unstable *horizontal* sync, the cause might be found near the point where the two kinds of pulses separate or in some stage ahead. Less likely, but still possible, is *two separate troubles* in both the vertical and horizontal circuits.

Poor *interlacing* and vertical *jitter* are two kinds of sync instability. For this problem, check for horizontal sweep voltage getting into the vertical by way of the B supply, coupling from poor lead dress, or *crosstalk* in the yoke. It can also be caused by weak (or overly strong) pulses, especially if there are problems with signal overload.

7-20 TROUBLESHOOTING AND REPAIR OF DRIVER STAGE(S)

Where total or partial loss of sweep or poor linearity has been localized to a driver stage, follow up with resistance, voltage, and component testing as described in Chapter 5. Since the conditions of height and linearity are closely related, possible causes of trouble and the procedures are essentially the same for both. But don't misread the symptoms. For example, what may at first appear to be loss of height may turn out to be a linearity problem, once the controls are adjusted. Or the reverse may be true.

When there is a black border at top or bottom, it could be either a height or linearity problem, an off-center raster, or foldover (where scanning lines are compressed or overlapped). A *shadowy bar* at the top is often caused by retrace blanking problems. When there is *some* deflection, start by adjusting both height and linearity controls.

NOTE:

Always adjust for a small amount of *overscan* (to more than fill the screen). This extra amount of height is needed because of possible drops in line voltage and variations of circuit tolerances, and also because a *picture* is *smaller than an off-channel raster,* because of the blanking pulses, which produce a blanked-out *border* on all four sides of the pix.

It is not advisable to adjust linearity by watching a pix. If convenient, use a crosshatch or bar generator, adjusting for uniform spacing of the bars. Another method is to use the raster scanning lines, although on an in-line CRT or a small screen they are difficult to resolve, especially if the focus is less than perfect. If a control is erratic or cuts out, try cleaning or replace, as necessary.

For complete loss of sweep, see if the transistor is conducting (either by checking the *IR* drop across an emitter or collector resistor or by jumping base to emitter, as explained in Sec. 5-9).

REMINDER:

Some drivers (especially predrivers) operate in the common-emitter configuration and provide a voltage gain; others are emitter followers where there is no gain. When a Darlington transistor checks bad, it must be replaced with one of the same type and beta characteristics.

When looking for the cause of incorrect operating voltages and the stages are directly coupled, remember the cause may be with an interfacing circuit either upstream or downstream. If there are *no dc* voltages at the transistor, check for dc input to the *stage* and, if necessary, follow back to the source.

For partial loss of sweep where the raster does not fill the screen, look for conditions that can *reduce* the sweep amplitude. Make sure the dc supply voltage is up to par. If the transistor check shows leakage, replace it. Check for troubles in negative feedback loops, either by testing the individual components or by opening the loop (which should normally result in an increase in stage gain). With the loop open, check dc and sweep voltages at both ends, and then once again (for comparison) when the circuit is restored.

For linearity or foldover problems, compare input and output waveforms with those indicated on the schematic. Make sure the pix is locked in sync, since frequency often has a bearing on waveshape. Check loops for loss or attenuation of the feedback signal. Unless caused by excessive retrace time, vertical foldover is usually the result of a bad transistor or incorrect operating voltages.

When scanning lines are alternately compressed and expanded over different portions of the screen, the cause is probably inadequate filtering of the dc supply. This is sometimes mistaken for hum bars. Frequently, the compressing and expanding of the lines is continuous, with a bar movement slowly upward, a condition called *breathing*. Scope check the dc bus and try substitute filters.

For intermittent troubles in a driver stage, simultaneously monitor the signal at input and output. Check for an erratic CRT setup switch and clean or replace as required. Try probing the wiring and wiggling the pigtails of all suspected components. Try heating and cooling the transistor(s).

7-21 TROUBLESHOOTING THE OUTPUT STAGE

As with the drivers, this stage can be responsible for complete or partial loss of sweep, poor linearity, or retrace blanking problems. For a color set there can also be problems with convergence, pincushioning, and raster centering. The defects most frequently encountered are shorted transistors, resistors damaged by the overload, or a faulty yoke.

If there is little or *no* vertical deflection, check the dc input to the stage and the dc operating voltages at the transistor terminals. Check to see if the transistor is conducting (by checking the IR drop across an emitter resistor or by shorting base to emitter). If it is not conducting and the voltages appear OK, the transistor is suspect. Check its temperature with your finger. As with any power output transistor, after 5 to 10 min. of operation it should be quite warm. If cold, it is not conducting (although the transistor itself may not be the cause). If too hot, check for a shorted or leaky junction or bias problems, two common causes of thermal runaway. Check the transistor, preferably out of the circuit.

REMINDER: ⎯⎯⎯⎯⎯⎯⎯⎯⎯⎯⎯⎯⎯⎯⎯

Before replacing a bad transistor, check for and correct any other circuit disorders. When a transistor has shorted out, it invariably damages the emitter resistor and, in some cases, other components, and causes possible fuse blowing. A replacement emitter resistor *must be* the exact value. When installing a transistor, don't forget the mica insulator (if called for). Use only *one* thickness and apply silicone grease liberally to both sides. Make sure there is no grit or burrs that might puncture the insulator when the mounting screws are tightened.

Where a bias control is provided, it should be adjusted after replacing a transistor, especially if there are signs of overheating. This is important, since improper bias can cause reduced height, poor linearity, and the ultimate destruction of the new transistor and even the output transformer or yoke. The procedure is simple: Connect a current meter in the emitter or collector circuit (or measure current indirectly with a voltmeter across the emitter resistor). Allow a 5 to 10 min. warm-up for the current to stabilize. While observing the screen, adjust trimpot to obtain the correct amount of current, as specified for the transistor type.

CAUTION: ⎯⎯⎯⎯⎯⎯⎯⎯⎯⎯⎯⎯⎯⎯⎯

The *idling current* must not exceed the rating as

specified. While making the adjustment, it is a good idea to monitor the sweep waveform to make sure no distortion takes place.

When there is no deflection and the transistor circuit checks OK, the cause may be with the coupling to the yoke or the yoke itself. Check for an open coupling capacitor or a winding on the output transformer as applicable. A keystone-shaped raster is usually caused by a shorted or open yoke. Other possible yoke defects include an open coil, leakage between the vertical and horizontal coils, leakage or shorts to ground via the core, arcing, or a defective built-in thermistor.

• **Yoke Testing**

When the yoke is suspect, proceed as follows: Remove the yoke cover for access to the connections. Inspect. Check continuity and resistance of each vertical coil and compare.

NOTE:

The vertical coil connections are those shunted by two equal-value resistors (typically 200–500 Ω); *horizontal* coils are sometimes shunted by one or more capacitors. Depending on the set, the vertical coils may be connected in series *or* parallel, and sometimes the resistors are omitted.

For a simple resistance check, there is no need to disconnect the resistors. A reading of about 10 Ω means the coil is good. If the reading is 200 Ω or so, the coil is open but the resistor is good. The resistance of the two coils must be *equal,* although this may be difficult to determine with an ohmmeter.

Shorted windings in a yoke coil are perhaps the most common trouble. Even one or two shorted turns is sufficient to cause problems of reduced height, keystoning, and possible burnout of the yoke. An ohmmeter will seldom detect the condition. There are several ways to check for shorted windings. After operating the set for 5 to 10 min, remove the power and feel the windings. Both vertical coils should be equally warm. When one is shorted, it will run hotter than the other.

Another test, where the coils are in series, is to connect the combination across a 6 to 10-V ac source and measure the voltage across each coil. They should be the same. A more conclusive test is by ringing the yoke. An instrument for this purpose is described in Sec. 3-18. This involves *shock exciting* each coil individually to produce damped oscillations viewed on a CRO. Shorted windings are indicated whenever the oscillatory waveform *rapidly* decays. The ringing test is made by exciting the yoke with a square wave from an AF signal generator and viewing the display on a CRO. Each vertical coil is tested separately for comparison. For this test, the yoke must be isolated from the shunting resistors and the rest of the circuit.

An open damping resistor (which is very uncommon since it is protected by the low resistance coil) results in excessive ringing. The effect on set operation is to alter the trace and retrace intervals, resulting in foldover or one or more horizontal bars at the top of the screen.

Arcing in a yoke calls for careful inspection, which usually means removing it from the CRT and looking for signs of arcing and overheating with the yoke connected and the set turned on.

CAUTION:

Avoid touching the horizontal connections and turn down the brightness to avoid spot burning of the screen. If possible, avoid removing the yoke in a *color set,* since this entails reconverging the set.

During inspection, look for the telltale signs of bared coil wires, carbonized deposits, and charred insulation. If charring is extensive, the yoke *must be replaced.* Look for crossed coil wires, especially between the vertical and the horizontal sections. This, and also problems involving broken wires and poor soldering, are often repairable.

The *thermistor*, often built into the yoke of a BW set, sometimes goes bad. If open, the result is no deflection. If its value increases, it causes reduced height. Test as described in Sec. 5-8.

When in doubt, a yoke may be tested by substitution. But this is only to establish its go-no-go condition, so don't expect to obtain a full and linear raster.

• **Feedback Loop Problems**

Certain defects in positive or negative feedback loops (that tie in with the output stage) can reduce or kill the sweep voltage. Where such troubles are suspected, troubleshoot the circuit as explained in Sec. 7-19. When replacing an output transformer having a feedback winding, be careful not to reverse coil connections that can create phasing problems. Common feedback problems are loss of feedback from a leaky or shorted capacitor to ground, too much feedback if it is open, or troubles at the receiving end of the loop if there is direct coupling.

• **Vertical Retrace Blanking Problems**

This is another interfacing circuit where defects can upset the operation of the vertical sweep or produce other symptoms, for example, problems with retrace lines on the pix (even at normal settings of the brightness control), or shading at the top of the screen.

First determine from the schematic, or by circuit tracing, which element(s) of the CRT (grid or cathode) is being fed with the blanking pulses. This in turn establishes the pulse polarity. (In some sets, the pulses feed the *video amplifier*). Most schematics show the p-p amplitude of the pulses, a very important consideration. Check for pulses at the CRT. If visible retrace lines are the problem, the pulses may be weak or missing. Either a peak reading EVM or a CRO may be used, although the latter is preferable since it also displays the *waveform*. If necessary, trace pulses back through the *R/C* network to the takeoff point in the sweep circuit. If there is sweep circuit loading, look for a shorted or leaky capacitor in the network; if not, it may be an open coupling capacitor or series resistor.

Shading at the top of the pix indicates the CRT blanking duration is too long. Changed resistor or capacitor values are the usual cause, although troubles with a blanking amplifier (when used) can also do the same thing.

7-22 TROUBLESHOOTING THE COMPLEMENTARY SYMMETRY OUTPUT STAGE

When operating properly, the circuit is balanced with both transistors contributing equally to the output. The transistors conduct alternately, one to produce the top half of the raster, the other, the bottom half. This can be a help in troubleshooting once it is decided which half of the circuit is responsible when scanning is non-uniform or limited to just a portion of the screen. The usual cause is a faulty transistor. Depending on the trouble symptoms, a good starting point is to check the dc supply(s) to this stage. Typically, it may be anywhere from about 12 to 100 V (consult the schematic for the exact amount). Compared to the other vertical stages (which are often supplied from *scan* rectification), the output stage usually gets its power from the main LVPS.

If it is a circuit where each transistor is supplied *from a separate dc source* (one positive, the other negative), check for dc at the junction of the two transistors (see Fig. 7-13b). If they are conducting equally, the two equal and opposite voltages should cancel out and the reading should be zero (or near zero, depending on the vertical centering adjustment in some cases). More on this later. If reading is greater than about 1 V (of either polarity), check further for the cause of the imbalance.

NOTE:

Where direct drive is used to the yoke, such an imbalance will *create* centering problems.

When there is only *one* dc source supplying both transistors, the dc voltage is (or should be) divided equally between them. Check the dc at their junction. The normal reading is approximately one-half the source voltage. Any drastic difference is an indication of trouble. For example, (as with any voltage divider), if the transistor connected to the source is *open* (or the one connected to ground is *shorted*), the reading at the junction will be zero. For the reverse condition, the junction voltage will equal the source. When both transistors are conducting, but one conducts more than the other, the anomaly will be correspondingly smaller.

When there is an imbalance for either type of circuit, check the dc bias on each transistor. To determine true operating conditions, make *three* measurements on each transistor, base to ground, emitter to ground, and base to emitter (the latter measurement represents the *actual* bias). Compare with voltages specified on the schematic. Where there is a discrepancy, look for the cause, which may be a bad transistor or some associated component. If there is a bias control (trimpot), don't forget to check its adjustment.

Any *dc* imbalance between the two transistors will affect their outputs, and the scanning of either the top or bottom half of the screen will be reduced; raster centering may also be affected in some cases.

The CRO can also supply useful information. Check the sweep signal at the base input of each transistor. They may or may not be equal depending on the circuit. Consult the schematic.

REMINDER:

The quasi circuit requires an inverter so that the *phase* of the signals driving each transistor will be *opposite*. If the signal drive to one or both circuits is not what it should be, go back and double-check the driver-inverter stages, or check for a shorted output transistor.

If the input signals appear to be good, scope the output signal at the junction of the two transistors. The amplitude of the output signal varies greatly with different sets and may be anywhere from about 25 to 100 V p-p. Check with the schematic. If it is a color set, the waveshape at this point will have the characteristic butterfly-wings appearance, as well as the sharp retrace spike (see Fig. 7-13). Since both circuits are emitter followers, no voltage gain (or phase reversal) can be expected between input and output.

Where only one transistor in a dual output stage checks bad, it is wise to replace them *both* to ensure the balanced condition, and because the other one may have been damaged by the overload. Generally, the low-value emitter resistors also have to be replaced at the same time; but don't use a higher *wattage* rating, because these units also function as protective *fuses*.

REMINDER:

After replacing transistors, don't forget to adjust the bias if a control is provided (see Sec. 7-13), and watch for signs of thermal overload.

Crossover Distortion. Unless there is a smooth transition where the current of one output transistor takes over where the other leaves off, a small notch will appear near the center of the scan portion of the output waveform. This produces a thin white streak across the center of the pix. For this condition, adjust the bias trimpot or check all bias stabilizing components in the base circuits, a diode, thermistor, or VDR, as applicable.

If there is no vertical deflection and the sweep amplitude at the transistor outputs is *excessive*, chances are there is an open in the load circuit. Check for a blown fuse if there is one, an open coupling capacitor, an open transformer coil in the coupling circuit, an open yoke, or an open pincushion or convergence coil if it is a color set.

CAUTION: _____

Where there is direct coupling, don't jumper a blown fuse or use one of a higher current rating; it can result in a burned-out yoke.

The usual cause of a blown fuse is a shorted transistor. When the complaint is *reduced* raster height and there is a coupling capacitor, it may have lost much of its capacity. Check by bridging with a suitable substitute. For troubles of this nature, it is a good idea (even when full sweep has been restored) to verify proper operation of the two transistors by double-checking the dc at their junction and comparing the *IR* drops across the two emitter resistors, which should be equal, assuming their values are identical.

7-23 RASTER CENTERING PROBLEMS

For a BW set, adjust the centering ring on the neck of the CRT. Twisting the tabs on the two magnetized washers relative to each other moves the beam (and the raster) up, down, and sideways. Make sure there is the same amount of overscan on all sides. If there is a problem with "neck shadows," check to see that the yoke is snug against the tube flare.

With a color set, centering problems involves the *circuitry*. With a two-transistor output stage (where zero voltage is developed at their junction, and the yoke is coupled *directly*), raster centering is accomplished by having the junction voltage vary a little in either a positive or negative direction. This is done by altering the bias of a driver stage via the control pot. With direct coupling to the output stage, this in turn causes one transistor to conduct more or less than the other to create a dc unbalance.

When an output transformer is used, some sets have the control shunted across a secondary winding, with the control arm feeding the yoke (as in Fig. 7-15) and the outer terminals connected to separate + and − dc sources.

For centering problems, the most common causes (depending on the type of circuit) are unbalanced conduction of dual-output transistors, a leaky coupling capacitor feeding the yoke, or a bad control.

REMINDER: _____

Don't be misled by symptoms. What appears to be an off-center raster may be a linearity problem (the crowding of scanning lines at top or bottom may leave a black border). It can also be caused by reduced *sweep* output from one of the power transistors.

Testing is simple. Check the dc at the wiper contact on the control or at the yoke. Vary the control. If the control is connected to + and − dc sources, a reading of zero dc should be obtained near mid-setting. If not, either the control is bad, one of the dc sources checks too high or too low, or dc is reaching the yoke in some other way because of a component or circuit defect. For the latter possibility, or as a check for coupling capacitor leakage, simply break the circuit. There will be an immediate loss of deflection, so turn down the brightness to prevent burning the screen. Note the position of the horizontal line. It should be exactly centered.

A control that is erratic or cuts out is not uncommon and should be replaced. When suspect, disconnect and test as described in Sec. 5-8.

Where a control is used to offset the bias of one of the output transistors and voltage varia-

234 THE VERTICAL SWEEP SECTION

tions at the control appear to be normal, check the components between the control and the stage being controlled.

7-24 PINCUSHIONING PROBLEMS

If it is a BW set, there may be two magnets mounted on brackets, one on each side of the pix tube. If necessary, the magnets can be moved a small amount in any direction (by bending the supports) to straighten out the raster.

For pincushion correction of a color set, first make sure there are no problems with the vertical or horizontal sweep. This is an *essential* first step since top-bottom correction requires a sampling of the horizontal sweep voltage, and side correction depends on the vertical sweep. If there is reason to believe the pincushion adjustments have been tampered with, try readjusting; otherwise, assume there is some component or circuit defect that must be corrected. Normally, adjustments are called for *only* after replacing components.

NOTE: _____

The following procedure applies only to top and bottom correction. For side correction problems, see Sec. 9-6.

Depending on the type of circuit, a general troubleshooting procedure is as follows:

1. Touch up any sweep related adjustments: height and width controls, linearity, raster centering, raster tilt, as needed. Correct any deficiencies.

2. Scope check the butterfly-wing pattern at the vertical output. If there is only a sawtooth wave, the pincushion circuit is not working. If there is a normal pattern, try the pincushion adjustments, as described later.

3. Scope check the horizontal sweep voltage applied to the pincushion circuit. Normally (if there is adequate raster width) this voltage will be good unless there is a coupling defect or loading by the pincushion circuit. Consult schematic for proper amplitude and waveform. If necessary, trace the sweep voltage to its source, which may be the horizontal output transistor or a tap on the flyback transformer.

4. Make an ohms-continuity check of all interconnected coils, the vertical yoke, pincushion, and convergence coils. Sometimes they are all in series and one reading (around 20 Ω) is all that is necessary. If it is a sweep, convergence, or pincushion problem, one approach is to place a jumper alternately across each coil while observing the results. With such a series hookup, *any* open coil causes a complete sweep loss, and this defect is normally found when troubleshooting for *that* condition.

5. If the circuit has a *pincushion amplifier,* make an input-output signal check followed by checking the transistor, its operating voltages, and all associated components as appropriate. If it is a circuit where the horizontal sweep voltage feeds a *vertical* driver (see Fig. 7-15a), check that circuit in the same manner.

6. If the circuit uses a *saturable reactor* (as in Fig. 7-15b), generally a continuity check of the windings is all that is required.

Pincushion Adjustments. The adjusters have different names, such as amplitude, phase, and tilt. Some are pots; others are slug-adjusted coils. A generalized adjusting procedure follows, but for specifics, check the service literature on the particular set.

1. Produce a bar or crosshatch pattern on the screen using a dot-and-bar generator.

2. While viewing the screen at normal eye level, alternately adjust all adjusters to make the center of the raster at top and bottom bow outward to obtain the straightest possible lines. Too much correction will produce the opposite effect, a barrel-shaped raster. If a generator is not available, reduce the height slightly and adjust for the straightest possible scanning lines. Where the phase coil has little or no effect, check for shorted windings or a broken or missing slug.

7-25 DYNAMIC CONVERGENCE PROBLEMS

For dynamic convegence of the three beams of a delta-type color tube, the three sets of coils on the convergence yoke (see Sec. 15-46) must be supplied with parabolic-shaped sweep voltages. The parabolic *vertical* waveform is produced by combining out-of-phase signals from two points at the vertical output stage. This waveform also ties in with the vertical yoke and the pincushion circuit (see Fig. 7-15). There is also the convergence *panel* containing all the necessary adjusters. Because of this interrelationship, an open convergence coil (or pincushion coil) will often result in loss of deflection, and the trouble will be found while investigating *that* condition.

Where good convergence cannot be obtained *for all three* primary colors (see Sec. 15-121), check the parabolic voltage at its source where it will probably be good, assuming there is good vertical deflection. Look for a loss of the voltage somewhere between that point and where it is split (in the convergence panel) prior to routing to the three vertical coils on the convergence yoke. A common problem is an extender cable that has simply become unplugged.

When there is a convergence problem with one color only, check for parabolic voltage across the corresponding coil on the convergence yoke. If it is good, check for an open coil, which is fairly common. If there is no voltage reaching the coil, scope trace back to its source.

If it is an in-line color tube, parabolic voltage may be fed to a special winding *inside* the *deflection yoke* or on the tube neck. Some current-model sets dispense with convergence circuitry entirely in favor of adjustable ring magnets (Fig. 15-15 and 15-27). Troubleshooting for *horizontal* misconvergence follows the same general procedure. Convergence adjustment procedures are covered in Sec. 15-123.

7-26 TROUBLESHOOTING FOR IC DEFECTS

Modern sets may use several ICs in the vertical section, for example, an IC containing the conventional oscillator and driver components and those used with the digital countdown system. Since ICs are not repairable, the internal circuitry is of no great concern to the technician. But he or she *must* understand its overall function and know what prior tests to make to determine its condition before considering replacement.

In general, when an IC is suspect, follow the procedures described in Secs. 5-4, 5-5, and 5-9. Specific tests in regards to the *countdown IC* are as follows:

1. Scope check the composite video or sync input signal while tuned to a station. If it is weak or missing, check back to the source, the video amplifier, or detector if necessary. *Reminder:* Either with or without direct coupling, a short-circuit upstream from the IC can be the cause. Try isolating the IC by disconnecting the coupling to its input.

2. Scope check the vertical output signal. If it is weak or missing, loading by the following stage can be the cause. Isolate, as explained previously, and recheck for a signal.

3. If there is still no output or it is degraded in some manner, check the B voltage feed to the IC. If it is good, remove the IC and make further tests from the socket pins. Check for the presence of dc voltage where it should not be. If it occurs, locate and correct the cause before replacing the IC. Test all external components, as appropriate, that connect to the IC. *Reminder:* Some defects, like an open capacitor, do not show up on a dc voltage check. With some circuits there is direct coupling from the vertical output stage to the IC, and a shorted transistor can destroy the new IC. Verify this before making a replacement. As a final precaution, make a resistance check from *all* IC pins to ground. If any reading is suspiciously low, check with the schematic before proceeding.

7-27 INTERMITTENTS AND OTHER PROBLEMS

For intermittent troubles that show up *abruptly* and are difficult to isolate because of their short duration, proceed as described in Sec. 5-13. In particular, watch for transistors with intermittent internal connections, intermittent shorts that blow fuses, controls that cut out, and a defective CRT setup switch. These slide-type switches give a lot of trouble with dirty or defective contacts. Try cleaning.

CAUTION:

Solvent can create leakage, so allow time for evaporation *before* power turn on.

Some sets feature automatic shutdown (screen goes dark) in the event of a vertical sweep failure. Check its operation by disabling the sweep (as by grounding out the sawteeth with a suitable capacitor).

A common cause of fuse blowing is a shorted output transistor. If it is intermittent, temporarily connect a lamp, as an indicator, in place of the fuse. Don't leave set unattended when the fuse is *defeated* in this manner.

7-28 AIR CHECKING THE REPAIRED SET

Don't neglect this important last step of all repair jobs. With the set operating for a minimum of 30 min. or so, make a *critical* evaluation. Is there adequate raster height and good linearity? Is the sync stable or just marginal? Check with the strongest and weakest channels. Watch for problems that may have previously gone unnoticed. Now is the time to make any last minute touch-up adjustments that may be required. For other considerations, see Sec. 5-15.

7-29 SUMMARY

1. The sawtooth-forming capacitor is charged from the B voltage through one or more resistors while the oscillator (or switching transistor) is turned off. It is discharged through the low-resistance conducting path of the transistor when the transistor is turned on.

2. The *R/C* time constant (TC) of the sawtooth-forming circuit establishes the sawtooth amplitude and waveshape. A short TC (small value capacitor and/or resistor) results in high amplitude but a degraded waveshape. If TC is too long (large R and C), the amplitude is reduced but the waveshape is improved. The discharge path (the C/E transistor junction) must be a low resistance to ensure maximum amplitude and a rapid retrace.

3. A blocking oscillator has but one transistor and is readily identified by its small *transformer*. A *basic* MV requires two transistors. Feedback may be from the output of one to the input of the other, or from the voltage developed across an unbypassed resistor common to both emitter circuits. Most MVs in current use have only one transistor with feedback from a subsequent stage. All stages and components within this *closed loop* are suspect if the oscillator stops working.

4. A *closed loop* in *any* circuit poses a special troubleshooting problem. When an oscillator depends on feedback from a driver or output amplifier, it is customary to *first* eliminate the amplifiers as the possible cause before checking out the oscillator stage.

5. The countdown sweep system has only one oscillator serving both the vertical and horizontal sweep circuits. The oscillator operates at 503.5 kHz. The required vertical and horizontal sweep signals are produced with *frequency dividers*. Computer-type logic circuits are used, which are contained in an IC. Hold controls are not required because of the high degree of frequency stability. The vertical output from the IC is typically about 2 to 4 V p-p.

6. The time allotted for vertical retrace is about 1.2 ms. Actual retrace time varies somewhat for different sets. Vertical foldover results if the allotted time is exceeded.

7. In most cases, abnormal dc voltage readings at an oscillator transistor when the stage is not working are the *result* rather than the *cause*. Such readings are useful, however, to make sure that voltage is applied to the stage.

8. To obtain vertical sync, two conditions must be satisfied: the oscillator free-running frequency must be correct (+ or - one cycle of 60 Hz), and the oscillator must be supplied with triggering pulses of the correct polarity and amplitude.

9. Where the vertical roll is *upward*, the oscillator is running too *slowly*; if the roll is *downward*, it is too *fast*. Knowing this can help in analyzing a problem. If the roll can be stopped momentarily, but the pix will not lock, the frequency is correct, but the pulses (if there are any) are not doing their job of triggering.

10. Many things influence the frequency of the oscillator, particularly components that determine the *RC* time constant of the circuit. To allow for slight variations, the pix should lock at the approximate mid-setting of the hold control.

11. The integrator is a low-pass *RC* network at the sync input to the vertical oscillator. Its function is to block the horizontal sync pulses (and noise), while building up a cumulative vertical triggering pulse. An integrator defect can cause partial or complete loss of sync and, in some cases, results in a dead oscillator and no vertical deflection.

12. If the pix barely locks at one extreme setting of the hold control, it indicates oscillator trouble. Unless corrected, frequency drift may result in loss of sync with no margin for adjustment.

13. Oscillator drift (that necessitates periodic adjustment of the hold control) is often caused by a bad transistor or a component changing value (see thermal intermittents, Sec. 5-13).

14. As a check on vertical sync, first adjust the hold control. If a single stationary pix cannot be obtained (even momentarily), the oscillator stage is at fault. When rolling can be stopped but the pix will not lock, check for weak or missing sync pulses. *Reminder:* It is impossible for even strong pulses to trigger an oscillator that is off frequency.

15. Height, linearity, and vertical peaking controls often interact. The height control normally has the greatest effect on the bottom of the screen; the linearity control, the top portion. They are usually adjusted together. A touch-up of these controls will often improve sync stability.

16. Bias problems on the output stage are usually caused by a bad output transistor. With direct coupling, however, a bad *driver* transistor can create the same problem.

17. To obtain good response and linearity, negative feedback is often used on driver stages. Sometimes the feedback loop may embrace all driver stages and the output amplifier. A trade-off (with negative feedback) is a reduction of stage gain.

18. Drivers and power output transistors often operate as emitter followers with the load supplied from the emitter circuit. This configuration has no voltage gain, but it does provide a *power* gain (i.e., output current is greater than the input current).

19. Shorted transistors are the most common trouble in an output stage. If overload protection is provided, it may trip a circuit breaker or blow a fuse; if not, other components may be destroyed, such as resistors, the output transformer, and even the yoke. When one of two transistors checks bad, they should both be replaced along with resistors that show indications of overload.

20. Thermistors and VDRs are often used to stabilize vertical deflection to offset the reduced sweep caused by yoke heating. Check both the hot and cold resistance of the units and compare with the schematic.

21. Crossover distortion occurs in a dual-output stage unless there is a smooth transition from one transistor to the other. It shows up as a notch on the sweep waveform and a horizontal bright line across the center of the screen. Check bias components, check and/or replace both transistors as a matched pair, and adjust the bias control if there is one.

22. Because of the *heavy current demands* of the vertical output stage, good regulation and filtering of the B supply are essential, especially if other stages are fed from the same source. Using a CRO or ac EVM, check for a high ac component on the dc bus. Defective zener diodes and filter capacitors are the most common causes.

23. Depending on whether the same or different transistor types are used, a complementary symmetry circuit may supply both transistors in series from a single dc source; or two PSs of opposite polarity may supply them individually. In the former case, the dc voltage measured at their junction should be half the dc source; with separate PSs, the voltage at their junction should be zero. Such readings will show whether or not the transistors are *balanced.*

24. DC current in a yoke creates a raster centering problem. The greater the current is, the bigger the problem. Whether the raster is too high or too low depends on the *direction* (polarity) of the current. If minor, the trouble may be corrected with centering adjustment. With BW sets, this is a *centering ring* on the CRT neck; for a color set, it is a centering potentiometer that introduces some compensating dc into the circuit. Common causes are a leaky electrolytic in series with the yoke or an unbalanced output stage.

25. A keystone-shaped raster occurs when the deflection produced by both halves of a yoke is not equal. Common causes are an open or shorted yoke coil, an open damping resistor across one coil, and in some cases an open pincushion circuit.

26. With vertical retrace blanking, a high amplitude vertical sweep pulse is used to overbias the CRT to cutoff during retrace periods. Troubles with this circuit can cause visible retrace lines when the brightness is turned up or a shadowy black area at the top of the screen. Sometimes a separate *blanker transistor* is used, sometimes in conjunction with *horizontal blanking*.

27. Sweep power may be transferred from the output transistor(s) to the yoke using *direct* coupling, a high value (around 2000 μF) coupling capacitor, or with a transformer. In a color set, the yoke, convergence, and pincushion circuits are all interconnected.

28. Power transistors *must* have an efficient heat sink to dissipate heat as fast as it is developed; otherwise, the transistor will be destroyed. When making a replacement, be sure to use liberal amounts of silicon grease (if called for) and tighten all attaching screws.

29. When there is a loss of vertical deflection, the first step is to turn down the brightness to prevent burning the screen phosphor(s). Some sets have an automatic disabling circuit that serves this purpose.

30. Thermal runaway is a common problem with vertical output transistors. The symptoms are a gradual reduction in raster height and overheating of the transistor. Typcial causes are transistor leakage, bias circuit defects, or a misadjusted bias control. When runaway is suspected, monitor the dc *IR* drop across emitter and collector circuit resistors.

31. Pincushioning is where the raster (on large-screen rectangular tubes) bows inward on all four sides. The bending of the horizontal lines is really a vertical sweep problem, and distortion on the sides is a horizontal problem. Correction is made by automatically changing the relative amounts of sweep for different portions of the screen.

REVIEW QUESTIONS

1. Sawteeth that are truncated or flattened at the top causes pix compression at the bottom of the screen. Explain.

2. What event terminates the charging of the sawtooth-forming capacitor?

3. Does increasing the forward bias on a PNP oscillator make it run faster or slower?

4. State possible causes of poor interlace.

5. An oscillator is free-running at the correct frequency but does not lock. What is wrong?

6. Pix should lock with the hold control near mid-setting. (a) Why? (b) What is wrong if it doesn't?

240 THE VERTICAL SWEEP SECTION

7. Is the oscillator fast or slow if the roll is upward?

8. Explain the merits of the countdown system for developing sweep voltage.

9. How does the sawtooth amplitude and waveshape relate to height and linearity of raster scanning?

10. What happens to the amplitude of the oscillator signal if values of the sawtooth-forming capacitor and resistor are increased? Explain why.

11. When is an oscillator considered free-running?

12. Explain how sync pulses trigger the vertical oscillator.

13. How can you judge linearity by observing the horizontal scanning lines?

14. Explain the purpose of the following controls: vertical hold, height, linearity, vertical peaking, bias.

15. Explain the function of the integrator network.

16. What are the main factors that determine the free-running frequency of the vertical oscillator?

17. A medium-*power* transistor is sometimes used in the last driver stage. Why?

18. (a) What is negative feedback and why is it used?
(b) How does it affect stage gain?

19. A given feedback loop uses a capacitive voltage divider consisting of two capacitors connected in series to ground. What changes in their respective values will increase or decrease the amount of feedback? Explain.

20. An emitter-follower driver has a resistor and a decoupling capacitor in its collector circuit. How would the signal output be affected if the capacitor opened up? Explain.

21. Explain the difference between regenerative and degenerative feedback and where each is used.

22. Why is a Darlington transistor sometimes used as a driver?

23. A MV oscillator is not working. It gets its feedback from the output stage. What circuit would you troubleshoot first? Explain why.

24. There are indications of thermal runaway in the output transistor. What tests would you make?

25. What is the result of too little or too much yoke damping?

26. Explain how a built-in thermistor in the yoke helps to stabilize the sweep.

27. How would you check a yoke for the following: (a) an open circuit? (b) shorted windings? (c) grounded windings? (d) leakage between vertical and horizontal coils?

28. Usually the vertical centering control connects to *two sources* of dc, + and − dc, or two slightly different voltages of the same polarity. Why?

29. For sweep problems, make a touch test of the output transistor(s) after they have been operating for a while. If cold, it means they are not conducting. If too hot, the current is excessive. What simple dc voltage test will determine the true condition?

30. Small voltage-amplifying transistors should never get hot. Give several possible causes when they do.

31. With a dual-output stage, one transistor is responsible for developing the top half of the raster, the other, the bottom half. This can be a big help in troubleshooting. How would you determine from the schematic which circuit is at fault?

32. Where a dual-output stage is supplied from one power source, the yoke must be driven via a series blocking capacitor. What is the effect on raster centering if the capacitor leaks or shorts out? Explain.

33. The junction of a dual-output stage is directly connected to the yoke, and there is a centering problem. What determines *which way* the raster is off center?

34. What is the result of too much top–bottom pincushion correction?

35. Can you expect to have good interlacing on an off-channel raster? Explain.

TROUBLESHOOTING QUESTIONNAIRE

1. A high-amplitude ac ripple or sweep voltage is measured on the dc bus. What is the probable cause?

2. What would you suspect when there is a higher than normal sweep voltage at the vertical output and there is no deflection?

3. State a logical procedure and possible causes when an output transistor overheats.

4. Overloaded components are discovered in the PS that feeds the vertical output stage. The circuit is not fused. What is a likely cause and how would you proceed?

5. A set has a ½-in. black border at the top of the screen. It could be a height or linearity problem, poor centering, shading due to retrace blanking, or unequal sweep from the two output transistors. How would you decide?

6. How much signal voltage increase would you expect between input and output of (a) a predriver, (b) a countdown system IC, (c) an emitter-follower driver, (d) a common-emitter output amplifier, (e) an emitter-follower complementary symmetry circuit?

7. A receiver suddenly develops symptoms of thermal runaway in the output stage. Would you try to compensate by adjusting the bias control? Explain.

8. An oscillator that obtains its feedback from the output stage is dead. How would you proceed?

9. What circuits would you check when there is loss of vertical sync and (a) it is only possible to roll the pix in one direction with the hold control? (b) pix will roll in both directions but will not lock? (c) there is also a loss of horizontal sync?

10. Give a step-by-step procedure for replacing an output transistor.

11. How would you proceed when a blown fuse is found in the output stage?

12. After replacing a yoke, what could be the cause of the following: (a) pix is upside down? (b) pix is transposed left to right? (c) both of the preceding?

13. Full B voltage is measured at the junction of two output transistors fed from a single supply. State the probable cause.

14. How would you check the frequency of an oscillator when there is no raster?

15. A CRO check shows no output from a countdown IC. Would you try a replacement and recheck, or remove the IC and make voltage and other tests *before* substitution? Explain why. *Reminder:* When replacing an IC, make sure the set is turned *off*. Better still, remove the power plug.

16. The pix locks OK but the vertical hold must be periodically adjusted. What is wrong?

17. When a defective driver transistor is the cause of reduced height, would you expect it to test (a) shorted? (b) open? (c) leaky? Explain your reasoning.

18. A pix occasionally flips frames, especially when changing channels, and the sync is not too good. The pix can be made to roll either way. State the probable cause.

19. A regulated PS feeding the output stage has poor regulation.
(a) Would you suspect the PS, the output stage, or both?
(b) How would you proceed?

20. No signal can be measured at either end of a feedback loop. Would you check (a) the stage where feedback originates? (b) the stage being supplied with the feedback voltage? (c) the loop circuit? (d) all of the preceding?

21. (a) What is meant by a *reluctant* oscillator?
(b) List several possible causes.

22. Give a step-by-step procedure for replacing a yoke on a BW set, including the precautions to observe.

23. A PNP output transistor runs cold but it checks OK. A voltage check shows a reverse bias (B to E) of +10 V. State the probable cause.

24. A pix will not quite lock with the hold control at its limit. Altering the value of a fixed series resistor restores operation. Is this advisable as a permanent fix? Explain your answer.

25. A set has a slight loss of height, and the dc bus to the vertical section checks a little low. How might you quickly determine if low B voltage is the *cause?*

26. A color set has a trapezoidal waveform at the sweep output but no evidence of the expected butterfly-wing pattern. What is wrong?

27. State four possible causes of a keystone-shaped raster and explain each.

28. The emitter resistor in an output stage is defective and a new one also burns up. What is wrong?

Troubleshooting Questionnaire 243

29. Full B voltage is measured across an emitter resistor. Give the reason for the high reading and the original cause of the condition.

30. A raster is off center and beyond the range of the centering control. Give several possible causes.

31. Where output transistors must be periodically replaced, yet the set appears to work well, what would you do?

32. A number of directly coupled transistors are found to be open or shorted. Where would you start looking for the cause? Explain your reasoning.

33. An NPN output transistor runs hot. A check on dc operating voltages shows + 20 V at the collector, + 15 V on the base, and 0 V at the emitter. State the probable cause.

34. Trouble has been localized to the vertical output section of a color set, but everything in the *sweep* circuit seems to be good. Give five *interfacing circuits* where the trouble might be found.

35. Name two kinds of component defects in an integrator network that can kill or weaken the sync pulse and the the test procedure for each possibility.

36. What was thought to be a CRO display of sync pulses at the oscillator input is found to change with the hold control. Explain. How would you proceed?

Chapter 8

HORIZONTAL SWEEP SECTION:
OSCILLATOR, AFC, AND DRIVER STAGES

8-1 INTRODUCTION

The horizontal sweep circuits, especially the output stage (of color sets in particular) are the most complex and hardest worked of any in the entire receiver. Small wonder then that this is where the vast majority of defects occur. The major stages of the horizontal sweep section of a set (including the HV and some interfacing circuits) are shown in the block diagram of Fig. 8-1. Except for the heavily outlined blocks (which are peculiar to color sets only), most of the stages are found in both BW and color receivers.

    This chapter is concerned mainly with development and synchronization of the sweep signal, and the stages involved are the oscillator (or countdown circuit), the AFC (automatic frequency control) circuits, and the driver amplifier(s). Although AFC performs a *sync* function, it is considered at this time because of its close relationship to the oscillator, which it controls. Other sync information is contained in Chapter 10. The output stage and related circuit functions are covered in Chapter 9. Horizontal deflection of the CRT beam is the *primary* purpose of the sweep section, but there are numerous "spinoffs," as indicated.

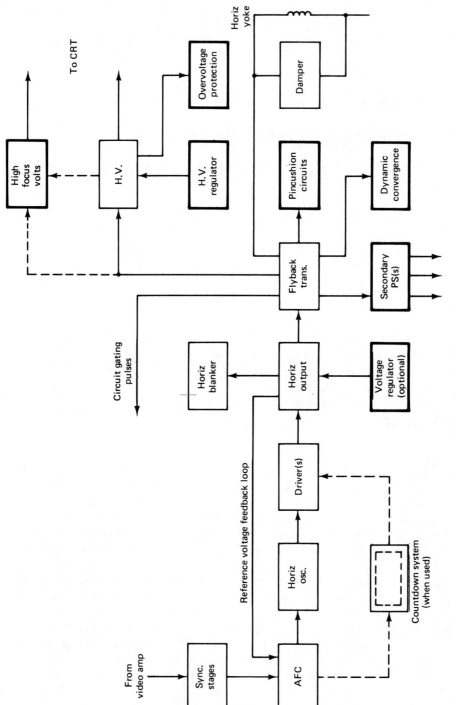

Fig. 8-1 Horizontal sweep section and related circuits. (Heavy outlined blocks are for color sets only.)

246

8-2 FUNCTIONAL DESCRIPTION OF HORIZONTAL STAGES

A brief overview description of all blocks is as follows. There are three separate categories: beam deflection circuits, synchronizing circuits, and sweep-dependent circuits.

- **Beam Deflection Stages**

This includes the stages that generate and amplify the sweep voltage to drive the yoke, as described next.

Horizontal Oscillator. This stage *generates* the sweep voltage. The frequency of oscillations (which must be very stable) can be manually adjusted with the horizontal hold and/or frequency controls. The sweep frequency for a BW set is 15,750 Hz, and slightly lower (15,734 Hz) for color. Instead of the conventional oscillator, some modern sets use the *countdown system,* as described in Sec. 7-9.

Horizontal Driver Amplifier(s). There may be one, two, or even three such stages. As with vertical drivers (Sec. 7-11), the purpose is twofold: to amplify the weak sweep voltage generated by the oscillator to adequately drive the output stage, and to act as a *buffer* to prevent loading of the critical oscillator circuit. Some drivers also act as *waveshapers* to develop the sawtooth waveform required for linear beam deflection, but most develop a *square wave* for switching the output stage. Instead of the conventional oscillator–amplifier arrangement, some sets use *SCR switching circuits,* as will be described later.

Output Stage. This is a *power* amplifier that drives the deflection yoke via the output (flyback) transformer to sweep the CRT beam the full width of the screen. The amount of deflection may be varied with a *width control.* The *damper* is a diode that prevents excessive ringing in the yoke. Other blocks (Fig. 8-1) associated with the output stage perform a variety of functions, as will be described later.

- **Beam Synchronization**

To synchronize the developing pix with the station, the oscillator is controlled by the incoming sync pulses via the sync stages and some form of AFC (automatic frequency control) circuit, as follows: The AFC circuit is responsible for keeping the horizontal oscillator locked in *frequency* and *phase* with its counterpart at the station. It does this by comparing the incoming sync pulses with a sampling of the generated sweep voltage.

- **Sweep-Dependent Circuits**

These sweep-related circuits are tied in with the output stage as shown in Fig. 8-1. Some are used in both BW and color sets; some are found in color sets only. They are briefly described as follows:

Horizontal Retrace Blanking. As with the vertical sweep (Sec. 7-14), the CRT beam must be blanked out during *horizontal* retrace intervals; and it is done in the same manner, by applying a high-amplitude horizontal sweep pulse to overbias the CRT at the proper times and for the duration of beam retrace. Early-model sets fed the blanking pulse to the CRT directly; late-model sets (with direct coupling) feed the pulse to the video amplifier.

Control Pulses. A number of circuit functions (especially with color sets) are *gated* or *keyed* by the horizontal sweep. These control pulses are generally supplied from tertiary windings on the flyback transformer. The pulses must be of specific amplitude and polarity according to the circuit being controlled. More on this later.

Dynamic Convergence (color sets only). As with the vertical (Sec. 7-14), a modified sweep waveform is applied to the horizontal coils on the convergence yoke of a triad-type CRT to maintain good convergence of the beam at the screen edges.

Horizontal Pincushioning. This circuit also has its counterpart in the *vertical* sweep section (Sec. 7-14). It is found on large-screen sets to prevent raster distortion where raster tends to bow inward or outward on the left and right sides. Correction is accomplished by feeding a portion of the *vertical* sweep voltage to the *horizontal* yoke.

Sweep-Derived Power Supplies. A number of dc supply voltages are derived from the horizontal sweep section. They are as follows:

1. *HV supply:* The HV required by the CRT is obtained by rectifying the stepped-up voltage from the flyback transformer. With a BW set (depending mostly on the screen size) the HV is usually about 15 kV or so. For a color set it is much higher, 25 to 30 kV or more. The current drain (beam current), however, is very low, seldom exceeding 200 uA for a BW set and 1 mA for a color CRT. With color sets, the HV is *regulated* and adjustable. There are also special *disabling* or *shutdown* circuits as protection against potentially harmful x-radiation should the HV become excessive.

2. *Focus voltage:* A color CRT requires 1000 V or so for the focus anode. It is obtained by rectifying the high pulse voltage of the horizontal output or from a separate winding on the flyback transformer. For beam focusing, the voltage is varied either with a potentiometer control or an adjustable coil shunted across a flyback transformer winding.

3. *B-Boost and other secondary power supplies:* More and more sets are utilizing the horizontal sweep section to supply the dc current requirements of various stages. Commonly known as *scan rectification*, different amounts of voltage are provided by individually rectifying the outputs of separate windings on the flyback transformer. Except where high *current* is involved, the use of scan rectification has several advantages over the conventional 60-Hz LVPS: greater efficiency (less heat and power consumption), and less filtering required because of the high ripple frequency.

8-3 HORIZONTAL SWEEP CIRCUIT FUNDAMENTALS

The horizontal sawtooth waveform is developed in the same manner as the vertical sawtooth (Sec. 7-6), that is, by alternately charging and discharging a capacitor in either the oscillator stage or one of the driver amplifiers. The charge cycle (while the transistor is nonconducting) corresponds with the active (left to right) trace of the CRT beam, and the discharge cycle (initiated when the transistor is turned on) relates to horizontal retrace. The amplitude of the waveform, and hence the raster width, is determined by the amount the capacitor becomes charged and discharged in the allotted time, as established by the time between horizontal sync pulses. Since the time intervals are much shorter than with the vertical sweep, the C and R values in the circuit are considerably smaller.

As with the vertical sweep, the *shape* of the waveform (applied to the yoke) determines the linearity of the beam trace and, in this case, the horizontal proportions of the pix. Figure 8-2 shows some examples where abnormal waveshapes stretch or compress the pix in different areas of the screen (also see Fig. 7-3). Most solid-state receivers do not have a horizontal linearity control.

For the AFC circuit to lock the pix in sync, the oscillator must be adjusted to free-run at the correct frequency. The *hold* and *frequency* controls are provided for this purpose.

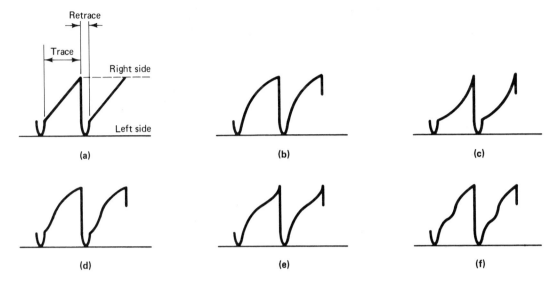

Fig. 8-2 Waveshapes and horizontal linearity conditions: (a) good linearity; (b) stretching on left side, crowding on right side; (c) crowding on left side, stretching on right side; (d) crowding on both sides, stretched in the middle; (e) stretched on both sides, crowding in the middle; (f) alternate stretching and crowding.

8-4 TYPICAL HORIZONTAL OSCILLATOR CIRCUITS

The multivibrator and blocking-type oscillator circuits used for vertical sweep may also be used to develop the horizontal sweep voltage. In addition, another circuit, known as the Hartley or *sine-wave oscillator* is widely used. Each type of oscillator is used with a particular type of AFC circuit. Because of the higher frequency, it is more difficult to maintain good frequency stability for a horizontal oscillator than for a vertical oscillator. Some oscillator circuits are more stable than others. As an aid to frequency stability, the horizontal oscillator is usually fed from a *regulated dc* supply. Thermistors and varistors are often used to stabilize bias voltage. And many oscillator circuits include a sine-wave frequency *stabilizer* or ringing coil.

The output of a typical horizontal oscillator is very low, seldom exceeding 1 or 2 V p-p. Unlike an amplifier, most oscillators operate class C (with reverse bias). Whereas a *vertical* oscillator often relies on positive feedback from a driver or output stage to sustain oscillations, a horizontal oscillator does not. Most circuits, however, do employ feedback, but for a different purpose—a sweep circuit *sampling voltage* for AFC (see Sec. 8-5 and Fig. 8-3). With many modern sets, the oscillator, AFC, and at least one driver stage are combined in a single IC.

250 HORIZONTAL SWEEP SECTION

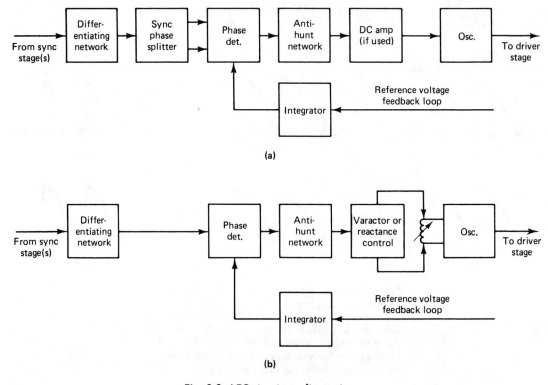

Fig. 8-3 AFC circuit configurations.

- **Sine-Wave Stabilization**

To improve frequency stability, many oscillator circuits incorporate an adjustable coil, called a *stabilizing* or *ringing* coil. Shunted by a small capacitor, it is made to resonate at (or close to) the horizontal sweep rate. When shock-excited by the oscillator switching action, damped sine-wave oscillations are produced in the coil, which tends to stabilize the oscillator frequency.

- **Blocking Oscillator**

This is perhaps the most widely used horizontal oscillator because of its good frequency stability, and because it is readily controlled by a simple AFC circuit. It can be recognized by its small transformer, which usually has an iron core but is *not* adjustable. Oscillations are produced by inductive feedback in the transformer. Its free-running frequency is determined mainly by the transistor operating voltages and the RC time constant of the base–emitter circuits. The transistor operates with a high reverse bias, which in many cases is adjustable with the hold control, which controls the blocking duration for each cycle.

The oscillator transformer may have anywhere from two to four separate windings, for feedback, coupling from the AFC, and coupling to the next stage, the first driver. Sometimes an autotransformer is used having a single tapped winding. The output of a blocking oscillator is more or less a square wave. The operation of a basic circuit is described in Sec. 7-8. A typical AFC control circuit is shown in Fig. 8-4.

8-4 Typical Horizontal Oscillator Circuits

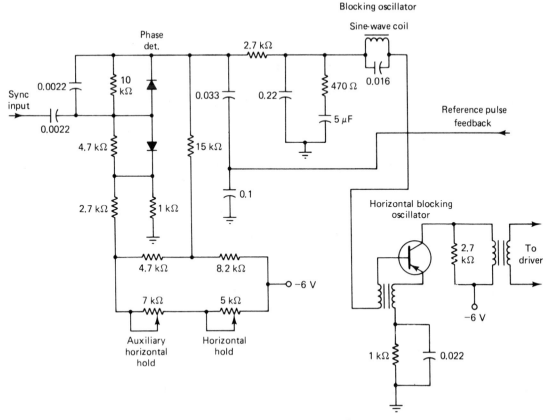

Fig. 8-4 AFC control of blocking oscillator.

- **Horizontal Multivibrator**

The MV is not used too often in modern sets. With most circuit designs the frequency stability is rather poor, and it requires two transistors. Oscillations may be produced either with feedback from one transistor to the other or by having a single unbypassed resistor common to both emitter circuits. As with the blocking oscillator, its frequency is determined by the dc operating voltages and base–emitter time constants, either of which may be adjustable with the hold control. With this circuit, a ringing coil (the only coil in the circuit) is a *must*. Oscillator output is a modified square wave. The operation of a basic MV is described in Sec. 7-8.

- **Sine-Wave Oscillator**

This popular circuit has extremely good frequency stability and does not require a ringing coil. Like the blocking oscillator, it too has a transformer for developing oscillations. Unlike the latter, however, the transformer is *tuned* (to establish the frequency). The adjustable core slug functions as a *coarse* frequency control. The horizontal hold serves as a *fine* or *vernier* control over the frequency.

There are two basic circuits, the Hartley and the Colpitts oscillator. The main difference between the two is the manner of obtaining feedback. The Hartley oscillator has a tapped coil with the tap connected to the transistor emitter.

252 HORIZONTAL SWEEP SECTION

Feedback is the result of electron coupling in the transistor and inductive coupling between one section of the coil and the other. The turns ratio establishes the amount of feedback.

In lieu of a coil tap, the Colpitts oscillator has two series-connected capacitors shunted across the coil with their junction connected to the emitter. It serves as a capacitive voltage divider and their ratios establish the amount of feedback. The output from both circuit variations is a sine wave. Because of its excellent frequency stability, sine-wave oscillators are not normally controlled *directly* from the dc output of a conventional AFC circuit. Instead, a reactance circuit is required (see Sec. 8-7 and Figure 8-5).

- **Countdown System**

As explained in Sec. 7-9, this system uses only *one* oscillator to develop both vertical and horizontal sweep voltages (see Fig. 7-6). The fundamental frequency of the oscillator (which is often crystal controlled) is 31,468 kHz (exactly twice the normal sweep frequency for a color receiver). This signal is subsequently reduced by a 2:1 frequency divider, as shown in Fig. 7-6. This arrangement provides excellent sync stability with pull-in and *locking* for a frequency drift equivalent to ±6 diagonal bars as observed on the screen. There is no need for the usual vertical and horizontal hold controls. Most of the components are contained in one or two ICs.

8-5 OSCILLATOR SYNCHRONIZATION

Because of its higher frequency (where frequency drift becomes a problem) and its susceptibility to random triggering by noise impulses, it is not practical to control the horizontal oscillator *directly* by sync pulses as is done with the vertical oscillator. Instead, it is done *indirectly* in a roundabout manner with some form of AFC circuit. In general, an AFC circuit compares the generated sweep voltage (in both frequency and phase) with the incoming sync pulses to develop a dc *error* or *correction voltage* if and when it is needed. This dc voltage either controls the oscillator *directly* or via a *reactance control* stage. The heart of the system is a *frequency-* and *phase-sensitive* circuit known alternately as a *phase detector, discriminator,* or *comparator.* There are two basic systems, as shown in the block diagrams of Fig. 8-3. With both systems, the phase detector is supplied with two kinds of signals as shown: sync pulses from the sync stages and a sampling of the horizontal sweep voltage that serves as a *reference signal.* The dc *error voltage* (if any) is the result of comparing these two signals.

Where the pix is locked in sync, no error voltage is developed (or required) and, therefore, no change to the oscillator is needed. If the oscillator tends to speed up, however, an error voltage of the proper amount and polarity is developed to force it to slow down. Conversely, if the oscillator is running slowly, a voltage of a different amount or polarity is produced to speed it up, until sync lock occurs. For such automatic control, the oscillator must be manually adjusted (with hold control) to free-run *at* or *close to* the proper frequency.

In Fig. 8-3a, the error voltage controls the oscillator with variations of transistor bias. In Fig. 8-3b, either a reactance stage or varactor control is used. Circuit functions are described more fully in the following sections.

- **Oscillator Frequency**

The oscillator may be off frequency or out of phase for any number of reasons. If it is a minor condition and the error is slight, it may produce such symptoms as *horizontal pulling.* If more severe, there may be pix *tearing,* where the pix is broken up by a number of sloping diagonal bars. The bars are produced by the horizontal

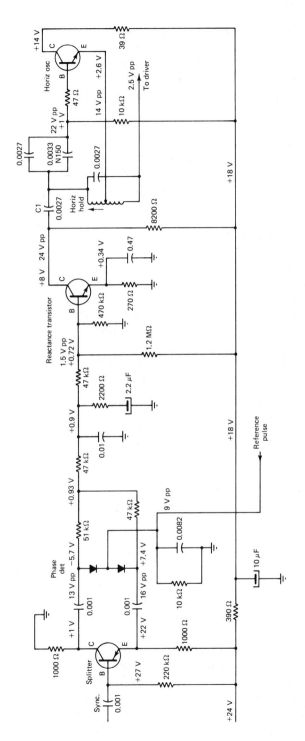

Fig. 8-5 Reactance control of oscillator.

blanking and sync pulses, and each bar is equivalent to one field, or in *time,* 1/60 sec. Thus the number of bars provides an indication of how far an oscillator is off frequency. For example, 6 bars means the oscillator is off by 6 × 60, or 360 Hz. A great many bars equates with a drastic frequency shift, and just a few bars means the oscillator is almost on frequency. A properly operating AFC circuit has a pull-in range up to 2 or 3 bars.

Which way the bars *slope* is an indication of *which way* an oscillator is off frequency. If the bars slope *upward* to the right, the oscillator is too *slow.* If the slope is *downward* to the right, the oscillator is running *fast* (frequency is high). When an oscillator is off by an even multiple of the normal sweep rate, it is possible to get a number of complete stationary pictures side by side, each separated by a blanking bar.

8-6 BASIC PHASE DETECTOR

Simplified versions of phase detectors are shown in Fig. 8-6. Some circuits use a pair of *matched* diodes; others, a transistor. A dc error voltage is developed across the load resistor(s) by the conductance of the diodes (or transistor). The amount of conductance (and error voltage) is determined by the *amplitude, polarity,* and relative *timing* between the sync pulses and the sweep reference voltage. Silicon diodes start to conduct at about +0.5 V.

• **Sync Input**

The block diagrams (Fig. 8-1 and 8-3) show how sync pulses reach the phase detector. Both vertical and horizontal sync pulses are present at the output of the sync separator, after they have been *stripped* from the composite video signal. At this point the pulses are separated *from each other.* The low-frequency vertical pulses are fed to the vertical oscillator via a *low-pass filter* (the integrator network). The higher-frequency horizontal pulses reach the phase detector via a *high-pass filter* called a *differentiating network.* This *RC* type of differentiator shapes the square-wave horizontal pulses into short-duration *spikes* as required for precise AFC control, at the same time rejecting vertical pulses and low-frequency noise bursts that might otherwise cause sync problems.

REMINDER: _____

Horizontal sync must be maintained *at all times,* even during vertical retrace, during which time the AFC is supplied with a variety of pulses for horizontal control: odd- and even-line equalizing pulses, vertical sync pulse serrations, and regular horizontal pulses (see Sec. 1-7 and Fig. 1-6).

Some phase detectors require *two out-of-phase* sync input signals. These two signals may be supplied from opposite ends of a coupling transformer or from a *sync inverter-phase splitter* stage (see Fig. 8-6). The pulses must be of equal amplitude and opposite polarity. For a single sync input, the polarity may be positive-or negative-going, depending on the circuit. The amplitude may be anywhere from 1 V to about 20 V p-p. There is no voltage gain because of the unbypassed emitter resistor.

• **Sweep Reference Voltage**

The reference voltage required by the phase detector may be obtained from various points in the sweep circuit, driver, or output amplifier, but most often from a winding (or tap) on the flyback transformer. The voltage must be shaped from a pulsed waveform into a sawtooth. This is done with either a simple integrator network or a waveshaping transistor circuit in the feedback loop. Most phase detectors use a single-ended reference voltage, but some require a dual input of opposite polarities. For the latter, a phase inverter circuit may be used. As with sync pulses, the polarity of the reference

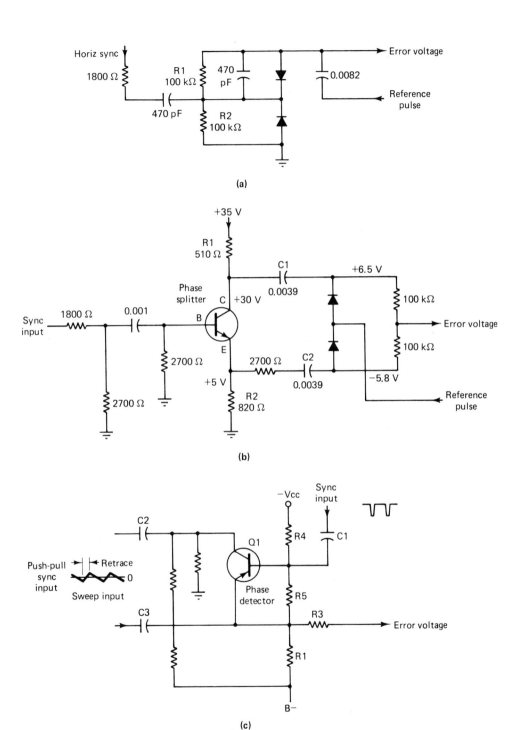

Fig. 8-6 Basic phase-detector circuits.

voltage may be either positive- or negative-going according to the type of phase detector. The amplitude of the feedback voltage is usually quite critical for any given receiver. It may be as much as several hundred volts p-p at the source, but is typically around 10 to 20 V p-p at the phase detector due to the insertion loss of the waveshaper in the servo feedback loop.

• **Phase-Detector Operation**

In the basic phase detector circuit of Fig. 8-6a, the two diodes are connected back to back and, acting as simple rectifiers, independently develop dc voltage drops across their respective load resistors *R1* and *R2*. The *net* voltage across the two resistors in series represents the *error-correcting* voltage used to control the oscillator. This circuit has single inputs for both the sync and the reference voltage. The two diodes conduct according to the instantaneous values of the combined signal that each receives at any given instant. Conditions for diode conduction and oscillator control are shown in Fig. 8-7. This shows the sync pulse superimposed on the reference voltage at each diode for different sync conditions. With the pulse at position 1, both diodes conduct by the same amount, developing equal and opposite voltages across the split load resistors, with a *net* voltage of *zero*. This is the condition when the oscillator is locked in sync and no corrective action is required.

If the oscillator is running *slowly,* the pulse is late and appears at position 2 for the two diodes. As can be seen, the positive-going pulse adds to the reference voltage and *D1* conducts heavily. By comparison, the positive pulse opposes the negative peak of the reference voltage applied to *D2* and the conductance of *D2* is reduced. The result is unequal voltage drops across the two load resistors with a net *negative* voltage in reference to ground. This forces the oscillator to speed up.

If the oscillator is running *fast,* the pulse is early and appears at position 3 for the two diodes. This is opposite to the preceding condition, and *D2* will conduct more heavily to develop a net voltage of a *positive* polarity to force the oscillator to slow down. The opposite is true if the diodes are reversed.

The circuit shown in Fig. 8-6b has the two diodes connected series-aiding, and there are *two* out-of-phase sync inputs, as supplied from a sync inverter (phase splitter) stage. These two equal and opposite sync signals are developed across the inverter split load resistors *R1* and *R2* and feed the phase detector via coupling capacitors *C1/C2*. The sweep reference voltage feeds the junction of the two diodes. As with the previous circuit, each diode conducts and contributes to the development of an error-correcting voltage in accordance with the amount of ac signal each receives. This amount in turn depends on the instantaneous values of sync and sweep voltages. One difference from the previous circuit is that the diodes function as *shunt* rectifiers *for the sync pulses,* and *series* rectifiers *for the sawtooth signal*. With the pix locked in sync, the diodes conduct equally, developing equal and opposite voltages across their two load resistors, with zero voltage at their junction. As with the previous circuit, this *error* voltage will vary in either a positive or negative direction should the oscillator change phase or its frequency drift higher or lower (by 100 cycles or so) from its normal rate.

The circuit in Fig. 8-6c uses a transistor instead of matched diodes. It is alternately switched on and off according to the phase relationship between the sync and reference pulses. With the oscillator locked in sync, the transistor is slightly conductive, producing a small *IR* drop across the emitter resistor. If the oscillator is running slowly, conduction increases and the increased negative error-correcting voltage causes it to speed up. Conversely, if the oscillator is fast, the *duration of conductance* is reduced and the lower error voltage causes the oscillator to slow down. The base input resistors establish the relative amounts of sync and sweep

8-7 Oscillator Control

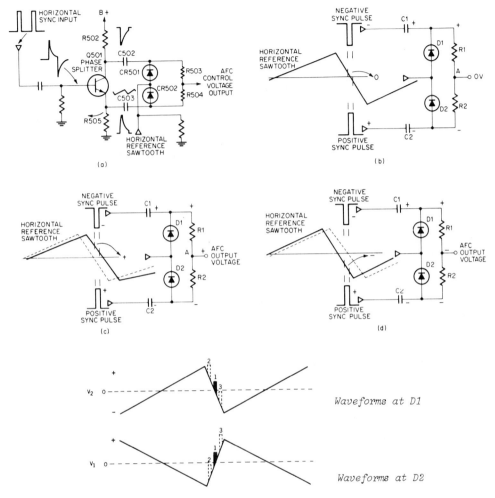

Fig. 8-7 Phase-detector operating conditions: (a) and (b) pix in sync; (c) oscillator frequency too low; (d) oscillator frequency too high.

voltage, which is quite critical. The *RC* components in the emitter circuit function as an *antihunt* network.

8-7 OSCILLATOR CONTROL

The error-correcting voltage can be used to control the oscillator in one of two ways, by *direct* variations of the oscillator bias, or with a *reactance control* circuit (Fig. 8-3). The latter may use either a transistor or a *varactor diode*.

The block diagrams (Fig. 8-5) show an *antihunt filter* ahead of the oscillator. This is simply an *RC* network in the dc error voltage circuit. It performs a dual function: to filter out the ripple component of the rectified horizontal sweep voltage and provide some immunity against noise impulses that might cause random triggering, and to prevent hunting, where the AFC is trying to regain control when there is loss of sync. The *RC* time constant establishes the AFC *reaction time*. If either too short or too long, there may be unstable sync and such symptoms as "piecrusting". Component values are critical. For better control, some sets use a dc amplifier to boost the dc error voltage.

- **Oscillator Bias Control**

This method is used mainly with MV and blocking oscillators, (most sine-wave oscillators have too much inherent stability to be controlled in this manner). Figure 8-4 shows a typical partial circuit using a blocking oscillator. Although the phase control may develop a zero dc voltage during sync lock, the error voltage measured at the oscillator base is usually slightly positive or negative by a small amount because of the applied bias. When sync is lost, the variations in error voltage either add to or subtract from the normal bias by small amounts, although the polarity remains constant.

REMINDER:

An increase in forward bias speeds up the oscillator, and vice versa; whether this requires a + or a − voltage depends on the type of transistor used, an NPN or a PNP.

- **Reactance Control**

Figure 8-5 shows a sine-wave oscillator controlled by this method. In effect, the control transistor is shunted across the oscillator tuned circuit. The dc error voltage applied to the transistor base controls the C/E conductance, which acts as a variable resistance series with the coupling capacitor (*C1*) to change the resonant frequency as required. An increase in forward bias means greater conductance, a reduced X_c, and a speedup of the oscillator, and vice versa.

- **Varactor Control of Oscillator**

Figure 8-8 shows a circuit where a varactor diode performs the control function. A varactor acts as a *simulated variable capacitor* (see Sec. 5-9), which is controlled by small changes in applied dc voltage, in this case the error voltage developed by the phase detector. The varactor (in series with the coupling capacitor *C1*) is shunted across the oscillator tuned circuit. When the oscillator is running *fast*, a dc error voltage is applied to the varactor to simulate an increase in capacity to force the oscillator to slow down. A voltage of opposite polarity does the reverse. The error voltage *reverse biases* the varactor, and the actual polarity depends on the phase detector and which way the varactor is connected in the circuit.

- **AFC Control of the Countdown System**

The method of control for this digital system is quite conventional, and most sets use the phase detector described in Sec. 8-6 that has single inputs of both sync and sweep reference pulses. The dc error voltage is applied directly to the oscillator. Although the oscillator operates at 31,468 kHz, the reference voltage is at the normal sweep rate since it is obtained from a later sweep stage. (It thus resembles a phase-locked-loop (PLL) system with oscillator and detector at different frequencies). The system has excellent pull-in, and sync lock-in can normally be obtained at up to six blanking bars (on the screen). Most of the AFC components are usually contained in the countdown IC.

8-8 BUFFER-DRIVER AMPLIFIERS

Many sets use only one horizontal driver; some have two or three. The first stage is often called a *predriver*. A typical two-stage driver is shown in

8-8 Buffer—Driver Amplifiers

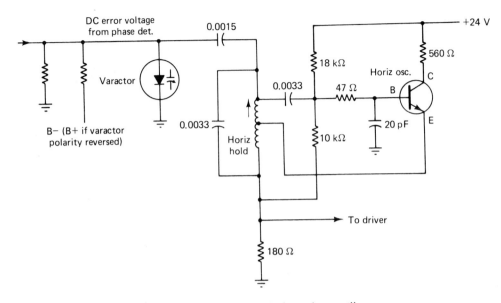

Fig. 8-8 Varactor control of Hartley oscillator.

Fig. 8-9. *Direct* interstage coupling is used (which is customary) except for the driver transformer, which feeds the output transistor. The main purpose of drivers is to amplify the sweep signal and provide a certain amount of drive power to the output stage. Note the p-p signal levels at various circuit points in the diagram. However, these levels vary greatly with different circuits. In some cases the gain of a predriver may be as high as × 150 or more and for following stages as low as × 2. Where a driver is an *emitter follower,* it has *no voltage gain* at all. Stage gains in Fig. 8-9 are × 5 and × 16, respectively. Both stages are conventional common-emitter amplifiers, which is typical.

Typically, a *medium-power* transistor (with or without a heat sink) is used in the last driver stage, since it is required to supply some *power* drive to the output transistor. Such a driver may draw about 80 mA at 60 V, representing a dissipation of nearly 5 W.

REMINDER: _____

A distinction must be made between *voltage* and *current gain.* If the last driver is an emitter follower, it provides a current (power) gain even though the voltage gain may be unity or less. A further increase in current drive to the output transistor is obtained by using a driver transformer with a step-down turns ratio. Typical ratios vary from about 4:1 to as much as 20:1 or greater.

The secondary winding of the transformer is a low impedance to match the low impedance of the output transistor base circuit. In Fig. 8-9 note the low resistance of the two windings. In some cases, the driver transformer resembles an interstage flyback transformer; in addition to coupling, it may have one or two extra windings to provide dc *scan-rectified* voltages for other stages of the set.

Precautions are usually taken to prevent the driver transformer from being shock-excited into oscillations by the sharply pulsed sweep voltage. Such *ringing* is overcome with either a diode or an *RC* network shunted across the transformer primary circuit. In addition to certain trouble symptoms, ringing is a common

260 HORIZONTAL SWEEP SECTION

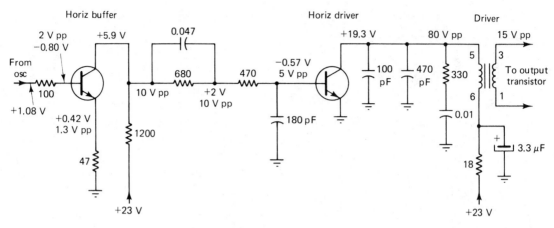

Fig. 8-9 Typical driver amplifiers.

cause of component failures. Note the antiringing *RC* components across the transformer in Fig. 8-9.

The dc operating voltages of driver transistors vary greatly; typically, collector voltages range from less than 10 V to as much as 100 V or more, with the greater voltages on the last driver. With direct coupling, bias is supplied from the previous stage. In many cases, very little, if any, forward bias is used and the driver transistor conducts only on the peaks of the applied sweep voltage pulses.

• **SCR Switching Circuit**

This is an unconventional sweep system used by some sets; SCRs are used as *switching devices* in place of the usual transistors in the drivers and output stage (see Sec. 9-3).

8-9 TYPICAL COUNTDOWN SWEEP CIRCUIT

Following is a brief functional description of the initial horizontal sweep stages of a late-model color set. Included are only those circuits up to the output stage. A description of the output stage and related circuits for this set is covered in Sec. 9-16.

As explained earlier (see Secs. 7-9 and Sec. 8-4), the horizontal sweep originates with a 31,468-kHz oscillator, and then is reduced with a 2:1 frequency divider (in IC 501 on the schematic). This 15,734-kHz signal (with an amplitude of 1.0 V p-p) leaves the IC at pin 8 to feed the single drive transistor Q 580. The collector of the driver is supplied from a +55-V source. Operation is as a conventional common-emitter amplifier. Resistor R 1552 and coil L 1701 in the transformer secondary circuit form an antiringing network.

Control of the oscillator is with a conventional phase detector. It is supplied with 6.5-V p-p sync pulses via IC pin 14 and a 1.25-V p-p reference signal from the flyback transformer via pin 1.

8-10 TROUBLESHOOTING THE HORIZONTAL OSCILLATOR, AFC, AND DRIVER STAGES

Troubles associated with the horizontal sweep section are many and varied, for example; *no raster* (due to loss of HV), reduced width, linear-

8-10 Troubleshooting the Horizontal Oscillator, AFC, and Driver Stages

ity and foldover problems, partial or complete loss of horizontal sync, and many more that are *indirectly* related to this section, such as those involving stages *keyed* with horizontal pulses and stages that obtain their B supply from *scan rectification*. This chapter deals only with troubles attributed to the oscillator, AFC, and driver stages. Troubles associated with the horizontal output circuits and HV are covered in Chapter 9.

After considering the trouble symptoms and identifying the horizontal circuits and components on the chassis, make such tests as appropriate (see Chapter 4) to *localize* the problem to this section of the set. Be guided by the symptoms listed in Table 8-1. Study the schematic and inspect the set for possible clues. If it is a condition hazardous to components, like arcing or overheating, turn the set off and proceed with caution. Try to localize the problem to a particular stage or section of the horizontal sweep, first, by analytical reasoning, then by performing the tests described in the following paragraphs. Once localized, the object is to *isolate* the fault *to the component level*.

A troubleshooting flow diagram is shown in Fig. 8-10. The tests are performed in a logical sequence based on specific symptoms; but the sequence may vary somewhat depending on circumstances and other considerations. For the most part, locating a *dead* stage is relatively simple, but troubleshooting for *marginal* conditions can be more difficult, for example, slight loss of width, oscillator drift, poor linearity, sync instability, and so on.

Table 8-1 Trouble Symptoms Chart

Trouble Condition	Probable Cause	Remarks and Procedures
No raster (or dim raster). Sound is OK unless stages passing sound signal are powered by *scan* rectifiers.	Loss of HV. Defective CRT. Improper CRT operating voltages.	Observe whether CRT heater lights up (it may not be supplied from the flyback transformer (HOT). Check for HV, but *not by arcing*. Loss of sweep voltage at any stage will kill the HV. Check CRT operating conditions.
Delayed or intermittent raster.	Poor soldering anywhere in sweep circuits. Defective CRT. HV cuts out. Intermittent oscillator.	Monitor dc at various circuit points. Monitor sweep signal at input and output of each stage. Oscillator may require a surge to start it; look for causes of oscillator loading, such as low B volts, leaky transistor, shorted coil windings, defects in adjacent stages, etc.

HORIZONTAL SWEEP SECTION

Table 8-1 Trouble Symptoms Chart *(cont.)*

Trouble Condition	Probable Cause	Remarks and Procedures
Loss of width (But no linearity or foldover problems).	Any condition in any stage that reduces sweep amplitude. Low B voltage. Oscillator far off frequency. Gradual loss of width may be caused by thermal runaway in horizontal output transistor. Low amplitude sweep voltage.	Check signal amplitude at each stage. Adjust width for some overscan to allow for line voltage variations.
Reduced width *and* height (small raster).	Low B voltage, excessive HV, or excessive loading.	Height is affected if vertical circuits are powered by scan rectifiers.
Poor horizontal linearity or foldover.	Improper waveshape developed by some sweep stage. Excessive horizontal retrace time.	Not too common with solid-state sets. Not to be confused with loss of width.
Trips CB or blows fuses	B voltage short in LVPS or voltage-distribution network. Overload condition in some stage, often a shorted or leaky transistor. Momentary power surge. Defective fuse or CB. Check output transistor for shorts or thermal runaway.	Look for overheating components. Make ohmmeter checks on suspected circuits. Measure and compute current drain. May be a dead oscillator stage. Arcing.
Raster wobble (undulating ripple on raster edges). May also be hum bars.	60-Hz modulation of horizontal sweep voltage, a filtering problem of the dc power.	Not to be confused with horizontal pulling (a sync problem), which is not observed off channel.
Drive bars (vertical white bar or bars near left side). Otherwise normal raster.	Excessive sweep drive to output stage. Oscillations in driver or output stage.	Not too common with solid-state sets. Check sweep amplitude and waveforms with CRO.

8-10 Troubleshooting the Horizontal Oscillator, AFC, and Driver Stages

Table 8-1 Trouble Symptoms Chart *(cont.)*

Trouble Condition	Probable Cause	Remarks and Procedures
Loss of horizontal sync (oscillator off frequency). Vertical sync OK. Pix tearing (sloping bars) or multiple pictures.	Defect in oscillator or AFC circuit. May be *any* component including shorted windings in oscillator coil.	Unable to obtain one pix with hold control or frequency adjustments. Check oscillator with AFC disabled.
Loss of horizontal sync. But oscillator *can* be made to free-run at the correct frequency. Symptoms may be horizontal pull, pix bending, or phasing problems like two half pixs with blanking and sync bar visible, horizontal roll, or *"piecrusting"* (jagged pix outlines).	Defect in AFC or related circuits.	Not an oscillator problem if it can be made to free-run at correct frequency where one pix can be obtained momentarily with hold control. For "piecrusting" check antihunt components.
Gradual or intermittent loss of sync. Vertical sync OK.	May be due to oscillator frequency drift often caused by an unstable oscillator transistor or other component in either the oscillator or AFC section.	Monitor operation of oscillator and AFC stages. Dress components away from any heat source.
Marginal sync stability. (oscillator frequency is stable). Pix contrast normal.	If vertical sync also marginal, trouble may be in sync stages, video section, or AGC. If vertical is OK, it is probably an AFC problem.	Verify proper strength of pulses feeding phase detector. Check dc error voltage, etc.
Raster or pix tearing with *very bright* streaking. (oscillator *squegging*, double-triggering, or *"Christmas-tree"* effect).	Defect in oscillator–AFC section. Misadjusted stabilizing coil.	Check oscillator with AFC disabled. Check for driver oscillations. Condition can destroy driver, output transistor, and other components.

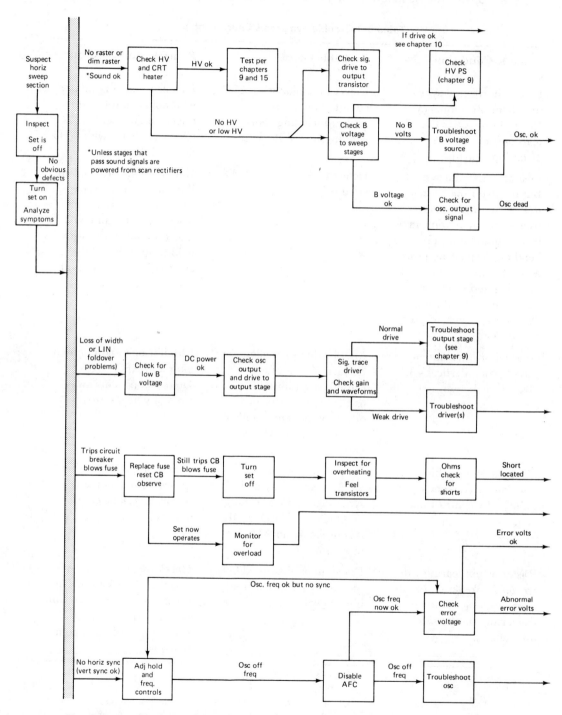

Fig. 8-10 Troubleshooting flow diagram.

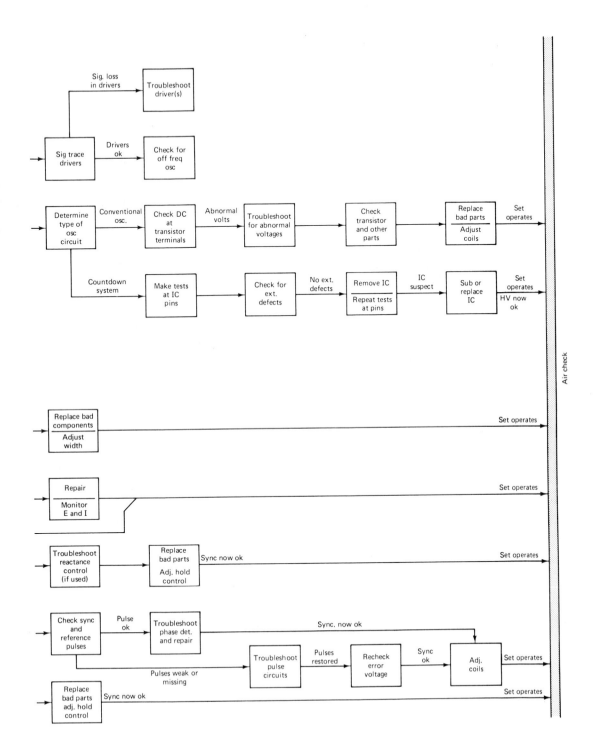

265

8-11 LOCALIZING THE PROBLEM

The tests to make and in what order are determined largely by the particular trouble *symptom* encountered. Usually, only two instruments are required: an EVM for checking dc and sweep voltages, and a CRO for signal tracing and waveform analysis. The general procedures are as follows.

• **No Raster or Dim Raster**

There are many possibilities involving different circuit functions, the condition of the CRT, and its operating voltages, which includes the *HV*. Here we are concerned only with the latter, since it in turn is dependent on the sweep section, in this case the horizontal oscillator, AFC, and driver stage(s) in particular. In general, a defect in these stages (resulting in *loss* of sweep voltage) means a complete loss of HV. Troubles that *weaken* the sweep voltage *reduce* the HV and result in a dim raster and reduced width. Two additional possibilities for loss of HV are automatic shutdown of the sweep section by the X-radiation protective circuit (when used) and a horizontal oscillator that is far off frequency.

No raster and no *sound* (a dead set) normally mean trouble with the LVPS. However, where stages passing the sound signal are powered by *scan* rectifiers, the horizontal sweep section becomes suspect.

A good starting point (see Fig. 8-10) is to check the HV. This should be done with a meter and *not* by arcing, which can destroy transistors. Also observe whether the CRT heater(s) are lit. Modern sets often supply the CRT heater from a winding on the flyback transformer, and where there is heater voltage it means the sweep circuit must be working. Where there is HV but no raster, perform the tests described in Chapters 9 and 15.

Once it is established that the trouble is in the sweep section, it is a good idea to check the dc power feeding all stages. If no dc (or low dc), check back to its source and look for such things as shorted capacitors, transistors, and zeners in the dc circuit, and the possibility of excessive loading by the sweep stage being supplied.

If dc power is good, check for sweep voltage at the base of the output transistor. For loss of drive, the trouble is probably in the oscillator or driver stages or loading by a defect in the output stage. If there is normal drive and no HV, check the HV PS (including the regulator and shutdown circuit).

CAUTION: _____

With some circuits, loss of drive can destroy the output transistor. Feel transistor. If overheating, turn set off *immediately*.

When there is no drive to the output stage, check for oscillator output. Normal output from the oscillator seldom exceeds about 1 V p-p and a reading or CRO display can be misleading, for example, hum pickup by the meter or CRO test leads. If in doubt, verify oscillator operation with dc voltage checks (see Secs. 4-9 and 5-4). If the oscillator is working, the driver stage(s) are suspect.

If there is no oscillator output, examine schematic to determine the type of oscillator and AFC circuit that is used. Verify that the oscillator is supplied with dc power of the proper amount. Check for possible loading of the oscillator by the first driver stage. If there is abnormal oscillator bias, determine if the phase detector or reactance control stage is the cause. Disable the AFC by grounding the dc error voltage *at the phase detector*. If the oscillator now operates, the AFC is at fault; otherwise, concentrate on the oscillator circuit.

REMINDER: _____

An oscillator that has output but is far off frequency can result in no HV. Since there is no raster, this possibility can only be checked by *measurement* using a CRO. If the CRO frequency calibration is reliable, obtain a single waveform display and take a reading from the sweep dial, which should indicate close to 15,750 Hz or 15,734 Hz as appropriate. Another method is to obtain a Lissajous figure with an accurately calibrated AF generator fed to the horizontal input of the CRO.

- **Loss of Width**

Assuming no linearity or foldover problems, signal trace for a *weak* sweep voltage between the oscillator and the output stage. *Carefully* measure p-p amplitudes at the input and output of each stage and compare with the schematic. Check applied dc power to sweep stages; if low, try increasing the voltage with a line booster to see if it is the cause of reduced width. Try the width adjustment.

- **Linearity or Foldover Problems**

Compare waveforms at each stage with the schematic. Check for excessive ripple voltage on dc power feeding the sweep stages. Try adjusting linearity control if there is one.

- **Circuit Breaker (CB) Trips or Blows Fuses**

Quickly observe any signs of smoke, arcing, or overheating components. Replace fuse and/or reset CB *once only* while noting anything abnormal when set is turned *on*. Unless raster appears, or for indications of overload, quickly turn set *off*. If receiver operation seems normal, it could be a defective fuse or CB or an intermittent problem that may subsequently reappear. While air-checking the set, check for overheating of the output transistor and monitor for possible overload condition. When troubleshooting for this problem, it is a good idea to bridge the open fuse or CB with a 100-W lamp as a visual indicator of the relative current and to protect against component damage until the cause is found and corrected.

Common causes where the CB trips or fuse blows are shorted filters, zeners, and output transistors.

- **Loss of Horizontal Sync (Vertical Sync OK)**

Analyze symptoms, noting whether it is a drastic or a minor sync condition. Try adjusting the hold and frequency controls. If unable to free-run the oscillator at the correct frequency (where one pix can be stopped momentarily within range of the controls), the trouble is with the oscillator *or* AFC circuits, and the first step is to determine *which*. This is done by disabling the AFC as described earlier. If the oscillator will now operate at the correct frequency, the trouble is in the AFC section. When the oscillator is still off frequency, concentrate on the oscillator stage.

- **Intermittent Problems**

Proceed as described in Sec. 5-13. Try to create the trouble condition. Analyze the symptoms. Monitor dc power and sweep voltages, noting anything unusual or abnormal that might *localize* the problem area.

8-12 ISOLATING THE DEFECT

As for other sections of the set, once the problem has been localized to a single stage, pinpointing the defect follows the same general procedures as described in Chapter 5: dc voltage checking at transistor terminals, resistance tests, checking components, and making adjustments.

REMINDER: ⎯⎯⎯⎯⎯⎯⎯⎯⎯⎯⎯⎯⎯⎯⎯⎯

Avoid surging the circuit. Even small voltage surges can damage or destroy components. Always turn set *off* when substituting or replacing components and when connecting or disconnecting certain test instruments. Use an isolation transformer if either the receiver chassis or the test instrument is hot (one side of the 120-V ac line grounded).

8-13 TROUBLESHOOTING THE OSCILLATOR

Troubles that can be caused by a malfunctioning oscillator stage are partial or complete loss of HV, reduced width or linearity problems, and loss of horizontal sync. What seems like an oscillator problem is often caused by a defect in the AFC section, which includes the phase detector and its interface with the last sync stage, the reference voltage circuit, and the dc error voltage amplifier and/or the reactance control circuit when used. Start by verifying whether it is an *oscillator* or *AFC* problem by disabling the AFC as described in Sec. 8-11. Where the particular phase detector has a normal output of zero dc, ground the circuit directly. In other cases a small error voltage serves as bias for the oscillator. In this case, *clamp* the dc circuit with a *bias box* that has been adjusted for the proper dc output and polarity. Testing of the oscillator is performed in accordance with the nature of the problem as described next.

• **Oscillator Has Little or No Output**

Measure and record the dc operating voltages at transistor terminals and compare with the values specified on the schematic. Since the voltages vary with the state of oscillations, some discrepancy can be expected. Also, abnormal readings may be either the *cause* of the problem or the *result* of some defect and will return to normal when such defect is corrected. If there is a bias problem originating from the phase detector or control circuit, try using a bias box (set for the specified voltage) and see if the oscillator starts operating or its output increases. Where direct coupling is used, the following stage (driver amplifier) could be the cause of an incorrect emitter or collector voltage.

NOTE:

Bias voltage is usually quite critical. Too much forward bias can damage or destroy the transistor and make the oscillator run fast. Too much reverse bias can make it inoperative, intermittent in starting, and cause it to run slowly.

Resistance checking from each terminal of the transistor (to ground) can often provide a clue to the defect. Make sure the set is turned *off* and the capacitors have lost their charge (see Sec. 5-5). Study the schematic, noting the conductive components between each terminal and ground. Compare anticipated measurement with the actual reading obtained. Try to account for any great differences. A little analytical reasoning here can often save much time in locating the fault.

REMINDER:

Resistance readings at diode or transistor terminals depend on meter polarity. Take each measurement twice, once with the test leads reversed. This will also show up a defective transistor. If suspect, remove or disconnect transistor and recheck.

Make resistance checks on all resistors and coils in the circuit, again making allowances for the shunting effect of diodes or transistor junctions. Watch for any high-value, low-wattage resistors that appear to be erratic and change value. A scorched appearance or changes in the coded colors is always a good clue. Make sure coils are tested on the lowest meter range. This is mainly for *continuity* only. Shorted coil windings are fairly common but seldom show up on an ohm's check. A few shorted turns raises the oscillator frequency and reduces its output. A more severe short will usually kill the oscillations entirely. The effects vary somewhat depending on whether it is a blocking oscillator, a sine-wave oscillator transformer, or a stabilizing (ringing) coil.

An ohm's check may show up a shorted or badly leaking capacitor; to make sure, disconnect one end and recheck as in Sec. 5-8. To check for a possible *open* capacitor, turn the set *off* and bridge with a suitable substitute, noting any improvement in oscillator operation.

For a *countdown system* (or whenever circuit functions are combined) using ICs, proceed as described in Sec. 5-9. Before replacing a suspected IC, make voltage and resistance checks at the connector pins, and check for defective components external to the IC.

REMINDER: _____

If sweep components are on a plug-in module, check for pin connectors not making good contact, a common problem. A typical IC (with pin voltages) is shown in Fig. 8-11.

When an oscillator intermittently cuts out or is slow in starting up, the most likely causes are low *B* voltage, an intermittent transistor (which may be difficult to verify except by substitution), a bad oscillator coil, a crack in the PC wiring, poor pin contacts if it is a replaceable PCB, poor soldering or a solder splash shorting the wiring, or incorrect oscillator bias. For intermittent operation, it is a good idea to constantly monitor the oscillator output.

• **Oscillator off Frequency**

Provided there is a raster, this condition is obvious by the sloping bars that tear up the pix. As explained earlier, the *slope* of the bars is an indication of whether the frequency is too high or too low, and the *number* of bars of how much the frequency is off. The first step is to adjust the hold control and, if necessary, the slug on the oscillator coil if there is one. If the slug must be adjusted to either extreme to obtain sync, it indicates circuit troubles, which should be investigated further. If it is possible to obtain a momentary pix and bars that slope in either direction, the oscillator stage is probably good. If the pix will not lock or remain locked, it is an AFC problem, and the checkout procedures are described in Sec. 8-14.

REMINDER: _____

Normal sync is where a pix pulls in at two or three bars and remains in sync at all times.

In general, the same kinds of defects that can kill an oscillator or reduce its output can also throw it off frequency. So the testing procedure is essentially the same. In particular, check for such things as abnormal transistor bias, a transistor having excessive leakage, open or leaky capacitors, resistors changed value, and shorted coil or transformer windings. For the latter, a good clue is when the oscillator is running too fast and, in the case of a sine-wave oscillator, when the coil slug must be turned *full in* in develop a near-normal frequency.

Another frequency problem is *drift,* where the hold control must be periodically readjusted. Check for variations in transistor operating voltages. When the *source* voltage changes, investigate the voltage regulator, if there is one. Also test for resistors that change value, components close to a heat source, or an unstable oscillator transistor. Dirty hold controls are quite common. If critical to adjust, try cleaning.

REMINDER: _____

An oscillator that is *far* off frequency can reduce or even kill the HV. When suspected, check the frequency as described in Sec. 8-11.

Under certain conditions the horizontal oscillator may produce *several* frequencies simultaneously. This is alternately known as *double-triggering, squegging,* or *Christmas-tree effect.* The latter name aptly describes the symptoms:

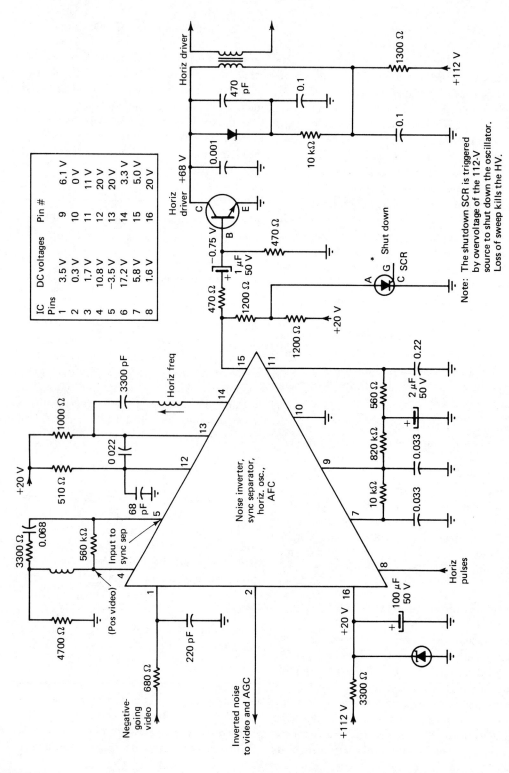

Fig. 8-11 Typical horizontal sweep IC and operating voltages.

bright horizontal streaking concentrated near the center of the screen. Sometimes it is accompanied by a high-pitched squeal emanating from the transformers. The most common causes are general oscillator instability, weak AFC control, or a misadjusted ringing coil.

When all oscillator components check out but it is still off frequency, it is probably caused by a malfunctioning AFC section, either the phase detector or reactance control stage, if there is one. The clue here is improper oscillator bias, and/or an unbalanced phase detector whose output is not close to 0 V dc.

8-14 TROUBLESHOOTING THE AFC

Even a *properly operating* AFC circuit cannot perform its function unless the oscillator is made to free-run at (or close to) the correct frequency. So once again make certain the oscillator is not responsible for the sync problem by attempting to get a single floating pix within range of the hold and/or frequency controls. If this is *not* possible, go a step further (as explained in Sec. 8-11) to disable the AFC and repeat the adjustments.

- **Checking the Phase Detector**

First determine if the detector is *balanced* (both diodes conducting equally *off channel*). Check the *error voltage*. Readings should be taken (and recorded) under three conditions: (1) off channel, (2) with hold control adjusted to obtain a single floating pix, and (3) with hold control misadjusted slightly in both directions. Off channel, or with a single pix displayed, the error voltage should be close to zero. With the oscillator made to run fast, then slowly, the voltage should change about ± 4 V either side of zero.

Switch to a blank channel. If phase detector output is greater than about 0.5 V, the circuit is unbalanced. Carefully check and compare the two diodes. They *must* be a matched pair. Test and compare the two load resistors. They too must be matched (unless the schematic shows otherwise). This is a *critical* circuit, so replace any and all components that are off tolerance. Another possible cause of an imbalance is leakage in one of the two coupling capacitors from a sync inverter stage, when used. Such leakage will cause one diode to conduct either more *or* less than the other depending on the *polarity* of the diodes and the leakage current. When replacing a bad diode, make sure its connections are not reversed, a common human error.

REMINDER: _____

Diodes may be series aiding or back to back depending on the circuit configuration.

To double-check for a balanced condition, switch to an active channel and adjust to obtain a floating pix. Again check the error voltage. It should be zero or near zero, the same as with a blank channel. If there is a difference between the readings, the phase detector is not operating properly in accordance with the applied pulses. The two kinds of pulses are checked separately as described next.

- **Sync Pulses**

While tuned to a strong channel, check for sync pulses (with an ac EVM or CRO) at the phase detector input(s). Make sure (by switching on and off channel) that it is the *sync* pulses being observed, and not ac line pickup or other interference. Amplitude of normal pulses is typically about 5 to 10 V p-p (check with the schematic for the precise amount). Where a sync inverter (phase splitter) is used, the pulses at *both* inputs should be of equal amplitude and opposite polarity. If *not* equal, check the inverter load resistors and coupling capacitors. When there are no pulses (or weak pulses) at the phase detector input(s), signal trace back through the differentiating network to the sync stages as explained in Chapter 10.

HORIZONTAL SWEEP SECTION

• Reference Pulses

Switch off channel and check for reference voltage at the phase detector input(s). Verify it is truly a reference voltage display by noting if it changes when the hold control is adjusted. Check amplitude and waveshape as compared with the schematic. Pulses should be sawtooth shaped and roughly 5 to 10 V p-p or slightly stronger than the sync pulses. If pulses are weak, missing, or distorted, signal trace back to the source, in most cases a winding on the flyback transformer. Pulse amplitude at the *source* is considerably stronger than at the phase detector, sometimes as much as several hundred volts. Common causes of weak or missing pulses are an open or shorted flyback transformer coil or a bad component in the integrating network of the feedback loop.

• DC Error Voltage

Provided the phase detector is *balanced* and supplied with the *proper* pulses, it *should* produce a dc error voltage that varies with changes in phase between the two kinds of pulses; in other words, if the oscillator drifts off frequency by a small amount in either direction. To verify, switch to a moderately strong channel and monitor the error voltage while adjusting the hold control. With the pix locked in sync (or a single floating pix), the voltage should be close to zero. Adjust the control to alternately speed up then slow down the oscillator, as indicated by the sloping diagonal bars as the pix loses sync. The same number of bars (about two or three) in both directions should produce equal amounts of dc of opposite polarities.

When voltage *changes* are small, the pull-in/lock-in range of the AFC will be limited and the hold control quite critical, with general sync instability. The causes for this may be weak pulses or excessive loading of the error voltage circuit between the phase detector and the oscillator or reactance stage, if there is one. In particular, look for leaky capacitors to ground, and leakage in a dc amplifier, reactance control transistor, or varactor diode when used.

For normal voltage changes at the phase detector output, check to see that the voltage is reaching the next stage. A moderate *insertion loss* can be expected by the resistors in the antihunt filter network, but it should not be excessive. When error voltage is lost or greatly reduced, check components in the antihunt circuit and the transistor or diode in the following stage. Certain defects in the antihunt network can cause *hunting* which sometimes produces the symptom called "piecrusting." The most frequent cause is an open or dried-up electrolytic.

Where a dc amplifier is used to boost the error voltage for better oscillator control, check and compare voltage readings at the input and output of the transistor. When there is little or no increase, check the transistor and other components in the circuit as appropriate.

8-15 REACTANCE CONTROL CIRCUIT

A sine-wave oscillator usually (but not always) requires a reactance control stage (Sec. 8-7). A transistor, when used in series with a capacitor, functions as a variable reactance shunted across the oscillator tuned circuit. Variations of the dc error voltage create corresponding changes in frequency. These same variations in error voltage create changes in *simulated capacity* when a *varactor* is used. When there is loss of sync and the error voltage from the phase detector checks normal and the oscillator is good, the reactance stage is probably at fault.

It is a simple circuit to troubleshoot. Start by checking (under different sync conditions) the normal bias and changes in bias created by the error voltage. While switched off channel, measure the normal bias from B to E of the transistor (or across the varactor). A transistor normally has a small amount of forward bias, and a varactor is *reverse biased*. Although the error voltage from the phase detector is constantly changing polarity, the *polarity* of the bias is not permitted to change, only the *amount*.

To check the operation of a transistor reactance stage, check for variations of the collector and/or emitter voltage as the bias is made to vary (indirectly) with the hold control. Test transistor and associated components if there is little or no change. Where a varactor is used, test it the same as conventional diodes.

8-16 OSCILLATOR AND AFC ADJUSTMENTS

To provide maximum range of the hold control, adjust the core slug on the oscillator coil or transformer for pix lock-in *at mid-setting* of the potentiometer control. Where there is a stabilization (ringing) coil, adjust as follows: Using two short jumpers, ground out the sync input to the phase detector and short out the terminals of ringing coil. Adjust the oscillator for a *floating pix*, (one that slowly drifts). Remove the short across the coil. Adjust the ringing coil for floating pix. Unground the sync input and the pix should lock in sync.

NOTE:

Unless there are other circuit defects, the oscillator should still function with the coil shorted out.

If the ringing coil has little or no effect, the cause may be shorted windings, an open shunting capacitor, or a broken or missing core slug.

Check the sync stability. The pix should pull-in (and lock abruptly) when the hold control is *misadjusted* (in either direction) and then adjusted to produce about 2 or 3 sloping bars. Try switching channels and switch the set off and on to make sure the pix remains locked at all times. If sync is marginal, there may be occasional symptoms of horizontal pulling, phase displacement, or horizontal roll. In general, the more critical the hold control is, the poorer the sync.

8-17 TROUBLESHOOTING THE DRIVER STAGE(S)

A dead driver stage results in loss of HV and no raster. A *weak* stage results in reduced HV and, usually, some loss of width. Complete loss of the horizontal sweep voltage at some point is easily determined when signal tracing with a CRO or ac EVM. A small reduction in stage gain, however, is not so easy to pinpoint, except by careful measurements of input-output signals, as compared to those shown on a schematic. The amplification to expect from typical drivers is shown in Fig. 8-9. Note the reduction in voltage drive to the output stage due to the step-down turns ratio of the driver transformer. The dc bias on a driver transistor provides a rough estimate of normal signal input requirements. The p-p signal level is usually about two to three times the bias value. (By comparison, the output stage requires a drive signal up to 50 times the bias value). The driver output must be a *square wave* for proper switching of the output stage.

REMINDER:

With some circuits, loss of drive turns *off* the output transistor; in other instances it results in loss of reverse bias, causes thermal runaway, and the eventual destruction of the transistor. When working on any sweep stage, periodically check the transistor (by touching) for signs of overheating. If this occurs, turn the set *off immediately* and disable the output stage until the problem is corrected. Disabling the *HV* is also a good idea (when a raster is not needed) to reduce shock hazard.

In general, troubleshooting the drivers follows the same procedures as for the *vertical* drivers (Sec. 7-20). Once the trouble is localized, check the applied B voltage and the transistor operating voltages. Compare with the schematic. Such voltages, especially the fractional bias voltage, are usually *critical*. As common-emitter amplifiers, the typical driver operates with a small amount of forward bias (as measured from base to emitter). With *direct coupling,* the bias depends on the operating conditions of the previous stage. Similarly, the collector or emitter voltage can be influenced by defects in the following stage. The most common cause is a shorted or leaky transistor.

To see if a transistor is conducting, and by how much, compute the current from the measured *IR* drop across a known value of emitter or collector resistor. Compare with the current computed from values shown on the schematic. Typical predriver current ranges between 10 and 50 mA. The final driver may draw up to 400 to 500 mA. Common causes of no current or low current are a defective transistor, incorrect operating voltages (loss of forward bias in particular), or a break in the PC wiring.

An ohm's check from the transistor terminals to ground will often show up a bad transistor or other component. Check readings against the schematic. Open coupling capacitors are common and an open emitter bypass will result in loss of stage gain. When suspect, bridge with a suitable substitute.

In addition to the above defects, a gain loss in the last driver stage can be caused by a leaky ringing diode across the driver transformer, shorted windings in the transformer, or excessive loading caused by some defect in the output stage. A shorted transformer is often difficult to prove. One useful test is to disconnect the transformer and feed its primary with about 6 V, 60 Hz ac. Measure the output voltage and compute the voltage ratio. Compare with the ratio using the input-output signal levels on the schematic.

The oscillator and driver(s) are often contained in an IC. If testing shows little or no drive to the output stage, the IC is suspect. Before replacing, make all preliminary tests described in Secs. 5-9 and 7-26.

8-18 INTERMITTENTS

For intermittent conditions associated with the horizontal sweep, try to create the trouble *symptoms* while monitoring the operation of each stage individually. When the trouble is localized, proceed as described in Chapter 5.

8-19 AIR CHECKING THE REPAIRED SET

During or after a minimum burn-in period of 30 min or so, critically evaluate set operations as it relates to the horizontal section. Is the raster full size and of normal brightness? Is the horizontal linearity OK? Pay particular attention to sync stability. Does the pix lock at mid-setting of the horizontal hold control? If not, make a touch-up adjustment of the oscillator coil. Is the sync stable even under adverse conditions, as on the weakest (or strongest) channels, during noise or other interference? For other considerations, see Sec. 5-15.

8-20 SUMMARY

1. Besides beam deflection, the horizontal sweep circuit performs a number of secondary functions: developing the HV and focus voltage for the CRT; producing B voltage (scan rectification) to supply various stages, control pulses for keying various circuits, and horizontal blanking pulses; and for color sets, pincushion correction and convergence.

2. The sweep voltage originates with the horizontal oscillator, is amplified by the driver(s) and the output stage, and drives the yoke to deflect the CRT beam. Anything that reduces the amplitude of the sweep voltage will reduce the deflection and the HV, affecting the raster accordingly.

3. Beam deflection must be synchronized with the station. To this end, the incoming horizontal sync pulses control the frequency and phase of the oscillator with some form of AFC (automatic frequency control) circuit.

4. For the AFC to function, the oscillator must be made to operate at (or close to) the correct frequency. The horizontal hold control is a user control for *small* frequency variations; the adjustable oscillator coil is a coarse frequency adjuster.

5. Loss of sync is caused by either an off-frequency oscillator *or* a malfunctioning AFC circuit. To perform its function, the AFC circuit must be supplied with two kinds of pulses: sync pulses (from the station) and a sampling of the generated sweep voltage.

6. The main part of an AFC circuit is the *phase detector,* which may use either a transistor or a pair of matched diodes. The circuit *compares* the sampling pulse (reference voltage) with the sync pulses. Any frequency or phase difference produces a dc error voltage, which, directly or indirectly, forces the oscillator back in sync.

7. The dc error voltage may control the oscillator in one of two ways: by direct control of its bias, or by a *reactance control* stage. For better control, sometimes a dc amplifier is used to boost the error voltage.

8. Reactance control is used *only* with a *sine-wave* (Hartley) type of oscillator circuit. It uses either a transistor or a varactor diode shunted across the oscillator-tuned circuit. Variations in error voltage applied to a transistor create changes in reactance, producing frequency or phase changes in the oscillator. The error voltage applied to a *varactor* causes it to behave as a variable capacitor, creating small changes in the oscillator resonant frequency.

9. Some phase detectors operate by comparing two out-of-phase sync pulse signals with the sweep reference voltage. Other circuits compare two out-of-phase reference voltages with a single sync input. The more popular circuits, however, compare one to one.

10. There are three kinds of oscillator circuits in common use: the *blocking* oscillator (identified by its *blocking oscillator transformer*), the *multivibrator* oscillator (which has no transformer), and the sine-wave (Hartley) oscillator (which has a tuned resonant circuit that can be recognized by its *tapped-coil*).

276 HORIZONTAL SWEEP SECTION

11. For optimum frequency stability, most MV and blocking oscillators have an adjustable resonant circuit. The coil is called a *stabilizer* or *ringing* coil. When shock-excited, the circuit rings, producing damped sine-wave oscillations that tend to stabilize the oscillator.

12. The dc error voltage developed by the phase detector is fed to the next stage by an *RC* filter network, commonly called the *antihunt* circuit. In addition to filtering out noise bursts and other interference, it serves to prevent *overreaction* and *hunting* by the AFC circuit.

13. The *countdown circuit* (which is rapidly replacing the conventional sweep oscillator) generates a frequency at twice the normal sweep rate, which is subsequently reduced by a 2:1 divider using computer-type logic circuits. The system is characterized by its high sync stability and requires no hold control. The circuit is contained in an IC and is not repairable, except for a few external components.

14. Pix *tearing* with bars sloping upward to the right means the oscillator is running *slowly*. If the slope is downward, the oscillator is running *fast*. The amount of frequency error is indicated by the number of bars; the fewer the number, the closer the oscillator is to being locked by the AFC. Normally, an oscillator should pull-in on at least two or three bars.

15. Usually, the sweep reference voltage is supplied from a winding on the flyback transformer and feeds the phase detector via an *integrator circuit*. The integrator (consisting of either a transistor or an *RC* network) shapes the sweep voltage into a sharply peaked *sawtooth* wave.

16. The diodes of a phase detector conduct according to the relative amount of voltage each receives from the combination of sync pulses and reference voltage. Conduction of the diodes produces dc voltage drops across two equal-value load resistors. With a pix locked in sync, there is *zero* dc at their junction and no change to the oscillator. With the circuit *unbalanced* (where the oscillator is off frequency or out of phase) a + or – error voltage is developed (depending which way it is off) to bring the oscillator back into sync.

17. An off-frequency oscillator may be caused by defects in the oscillator circuit itself or the AFC circuit. It is easy to determine which. Simply disable the AFC. If the oscillator can now be made to free-run at the correct frequency, the AFC is at fault. If still off frequency, the trouble is in the oscillator stage.

18. Most solid-state sets require one or more driver stages to properly drive the horizontal output transistor. Predrivers are essentially voltage amplifiers; the final driver stage must also produce a certain amount of drive *power*. The driver output is a *square wave*.

19. The last driver stage is transformer coupled to the output transistor. It is a step-down transformer with a low-impedance secondary to match the low impedance of the output transistor and to supply the necessary drive *current*.

20. The most common causes of a dead oscillator are a defective transistor, an open or shorted oscillator coil, or loading by a driver stage defect.

21. The most common AFC problem is *minor* sync instability (e.g., horizontal pull). Check for leaky diodes or coupling capacitors in the phase detector and the strength of the sync and reference pulses.

22. Small sync problems are usually due to *slight* circuit anomalies and may be difficult to isolate. More drastic symptoms like pix tearing and complete loss of sync are caused by more *definite* component failures and are easier to pinpoint.

23. When the oscillator is working, signal tracing the sweep section is best done with an ac EVM or, preferably, a CRO. A *signal generator* can be used to locate a *dead* stage, but accurate gain checks are not possible unless the proper waveforms are used.

REVIEW QUESTIONS

1. What is a ringing coil and why is it used?

2. The vertical oscillator is directly triggered by sync pulses; the horizontal oscillator requires an AFC stage. Why?

3. Explain *how* and *why* shorted windings in an oscillator coil raise its natural resonant frequency.

4. Explain how a phase detector is able to sense changes in frequency and phase.

5. Explain the function of the antihunt circuit.

6. *Raster* size is decreased when a pix is received. Explain why.

7. (a) What is the approximate frequency of the oscillator if the pix is torn out with *six* bars sloping upward to the right?
(b) If bars slope downward to the right, is the oscillator running too fast or too slow?

8. Under what conditions does a phase detector develop *zero* error voltage?

9. Explain how a transistor reactance stage controls a Hartley oscillator.

10. Some phase detectors are preceded by a sync phase splitter. Why?

11. What oscillator frequency produces two complete pictures separated by a blanking bar?

12. State several symptoms caused by poor horizontal sync.

13. Explain how a phase detector develops either a + or − error voltage.

14. An oscillator directly controlled from the phase detector has a normal bias of −1.5 V. If oscillator drift results in a variation of error voltage from +1.2 to −1.2 V, what are the changes in bias voltage measured at the transistor?

15. Is the AFC a likely suspect when there is loss of both vertical and horizontal sync? Explain your answer.

16. Identify all the pulses that control horizontal sync during vertical retrace.

17. Why is it important that the diodes (and usually the load resistors) of a phase detector be precisely *matched*?

278 HORIZONTAL SWEEP SECTION

18. State the purpose of the differentiating network at the sync input to the phase detector.

19. Is a blocking oscillator conducting or nonconducting during each active horizontal trace of the CRT beam?

20. What polarity of transistor bias causes a PNP oscillator to speed up? Explain.

21. What is the purpose of a diode shunted across a driver transformer feeding the output stage?

22. Why does a driver transformer have a step-down ratio?

23. Some driver transistors require a heat sink. Why?

24. What is the purpose of a varactor in an AFC circuit?

25. Explain three tests to determine whether an oscillator stage is oscillating.

26. It is often wise to disable the horizontal output and HV when troubleshooting oscillator and driver stages for loss of sweep. Explain.

27. While adjusting the hold control, which condition represents the greater sync stability: (a) when the pix locks when there are about six sloping bars? (b) when lock occurs only when adjusted to one bar? Explain your answer.

28. What do you observe on a BW set if the oscillator is running at (a) 3937.5 Hz? (b) 31,500 Hz? (c) 16,350 Hz? (d) 15,150 Hz? (e) if the frequency is correct (15,750 Hz), but the phase is off?

29. A vertical black bar may split or divide a pix due to a phasing problem or an off-frequency oscillator. Explain the difference in the two symptoms.

30. The CRT heater (when supplied from the flyback transformer) is lit. What does this tell you about the operation of the sweep section?

31. Horizontal pull and an off-center raster can be mistaken for one another. How would you determine which condition exists?

32. Why must the driver stage develop a *square wave?*

TROUBLESHOOTING QUESTIONNAIRE

1. When the oscillator cannot be made to free-run at the correct frequency, how do you determine which is responsible, the *oscillator* or the *AFC* stage?

2. How would you check the oscillator frequency if there is no raster?

3. What stage(s) would you logically check if there is (a) loss of both vertical and horizontal sync? (b) poor vertical sync only? (c) poor horizontal sync only?

4. What is the next logical step when an off-frequency oscillator can *almost* be locked at one extreme setting of the hold control?

5. How would you proceed when there is poor sync but a single floating pix can be obtained at mid-setting of the hold control?

6. By mistake, a phase detector diode is installed backward (reversed polarity). State the effect on set operation, and explain why.

7. Off channel, a considerable voltage is measured at the junction of the phase detector load resistors.
(a) What trouble is indicated?
(b) State three possible causes.

8. As a check on the phase detector, a bias box is connected to the source of error voltage and slowly adjusted. If the oscillator cannot be controlled, where would you look for the cause?

9. Trouble is localized to the phase detector. Off channel, a normal zero voltage is developed, but on channel the sync is very poor. What tests would you make?

10. State several possibilities when an oscillator is running too fast and the core slug either has very little effect or must be screwed fully into the coil to achieve sync.

11. What stage(s) would you check for (a) normal sync but reduced width? (b) loss of width and horizontal sync problems? Explain your answers.

12. There is an undulating waviness on the left and right sides of a pix. State the possible cause and the tests to make if (a) the off-channel raster is normal with straight edges, (b) the effect is also present on the raster, with indications of hum bars. Explain.

13. What components would you check for symptoms of "piecrusting"?

14. A set has marginal sync, a very limited lock-in range, and only small variations of error voltage when the hold control is adjusted. How would you proceed?

15. State the procedure for adjusting an oscillator coil, ringing coil, and hold control.

16. A phase detector has a push–pull (out-of-phase) sync pulse input, but one signal is much weaker than the other. State several possible causes of this unbalance.

17. How would you test a reactance control circuit that uses a (a) transistor? (b) varactor?

18. A CRO shows a strong 60-Hz sine wave superimposed on the horizontal sweep waveform. How would you proceed?

19. State the probable cause and what to do about it where there is normal sync but hold control is critical and the pix shimmers when the control is adjusted.

20. State several possible causes of a reluctant oscillator that does not always start up.

21. After replacing a flyback transformer, the AFC reference voltage is found to have the wrong polarity. State the probable cause and what can be done.

22. A black vertical bar on the left side of the screen could mean loss of width, horizontal phasing problems, poor linearity, or a centering problem. How would you decide which?

23. For such symptoms as reduced width or loss of HV, what single test will determine whether the cause is *ahead of* or in the output stage or beyond?

Chapter 9

HORIZONTAL OUTPUT AND HV SECTION

9-1 INTRODUCTION

As described in Sec. 8-2, the horizontal sweep voltage generated by the oscillator is amplified by the driver stage(s) to drive the horizontal output amplifier. The amplifier, in turn, drives the horizontal deflection yoke, either directly or via the flyback (horizontal output transformer, or H.O.T.). In addition to beam deflection, there are numerous sweep-dependent stages as shown in Fig. 8-1 and described in Sec. 8-2.

9-2 HORIZONTAL OUTPUT STAGE

Actually, *all* sweep-dependent circuits can be considered part of the output stage since the sum total of all power must be supplied by the output transistor. In addition to yoke current, this includes CRT beam current (and in some cases CRT heater power), the total current drain of all stages being supplied with dc scan power, and current for the pincushion and convergence circuits when used.

• Output Transistor Operating Conditions

As with vertical sweep, the horizontal output transistor may operate either as an emitter follower or in the common-emitter configuration in class *B*. Most sets use the latter hookup. The transistor is a high-power type, and because of high dissipation, a heat sink is required. It also has high amplification, developing pulse voltages up to 5kV or more. For this reason, dc measurements should not be taken at the collector unless the circuit is known to be inoperative.

282 HORIZONTAL OUTPUT AND HV SECTION

The dc power for the transistor is supplied from a high-current source in the LVPS. Usually, it is regulated and is called the *HV regulator*. The dc polarity may be either + or −, depending on the transistor type and the circuit configuration. The transistor operates with a small amount of *reverse bias,* typically about 1 V or less. Most circuits depend on *excitation bias* produced by rectification of the drive signal by the B/E junction. Unlike other transistor circuits, the *base* current is considerable (around 100 mA), which is one reason for the low resistance of the driver transformer coil and why the driver must furnish *some* sweep *power*.

Like the preceding driver stage, the output transistor is driven by a *square wave,* which switches the transistor alternately between its two states, on or off. The maximum amount of drive to the transistor input is not especially critical in most cases. However, a drastic loss of drive will either turn the transistor off (depending on the circuit) or result in reduced width, poor linearity with foldover, and sometimes, shortened life of the transistor. Excessive drive is normally taken care of by clipping action of the B/E junction. The drive signal must overcome the bias before the transistor will conduct. Conduction takes place for only the duration of each incoming pulse. With some circuits, complete loss of drive can cause thermal runaway and destruction of the transistor; in most cases, however, the transistor is turned off and there is no problem. The current drain of the transistor is typically about 600 mA for a BW set and 1.5 A for a color set.

• **Driving the Yoke**

With some sets, coupling from the transistor to the yoke is via a winding on the flyback transformer. Most sets, however, use *direct drive* for greater efficiency. Figure 9-1 represents a typical circuit. With this arrangement, the transistor collector current is split two ways between the flyback transformer primary and the yoke. The return circuit is via the pincushion–convergence circuitry.

The horizontal yoke (like the vertical yoke)

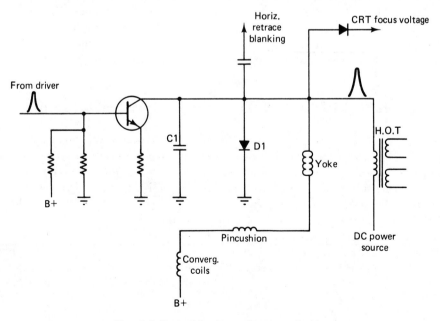

Fig. 9-1 Typical horizontal output circuit.

consists of two windings; depending on the set, they may be connected either in series or parallel. The horizontal yoke coils are the inner coils closest to the CRT neck and are positioned at right angles to the vertical windings (see Fig. 15-2). Yoke coils have a very low impedance to match the output transistor. This, in turn, means a high yoke *current,* which is typically about 4 A for a BW set and about twice that for a color receiver. Raster width is controlled by changing the amount of yoke current. The most common method is with an adjustable coil in series with the yoke. The *inductance* of the yoke transforms the *pulses* of sweep *voltage* into a *sawtooth current* for linear deflection of the beam.

• **The Damper and B-Boost**

Each horizontal pulse *shock-excites* the yoke into oscillations. *Controlled* ringing (for a half-cycle) is desirable since it ensures a rapid retrace of the beam. Sustained ringing, however, produces vertical drive bars on the left side of the screen. After one half-cycle of oscillation, the ringing is suppressed with a *damping diode* ($D\,1$ in Fig. 9-1) connected across the transistor output circuit. With the rapid collapse of flux around the yoke during each retrace period, a high, self-induced voltage is developed in the yoke windings. This stored energy produces the *first half of each horizontal trace* until the transistor starts conducting to produce the second half of the trace, as shown in Fig. 9-2. Thus the energy of the oscillations is absorbed or used up in a useful manner. In some sets, the high amplitude pulses rectified by the damper provides a source of increased voltage called *B boost.*

The abrupt switching of the transistor tends to produce another type of oscillation (in the transistor itself). Such oscillations may be *radiated* by the wiring, picked up by the tuner, and appear on the screen as one or more thin vertical bars (sometimes refered to as "spooks" or "snivets"). The problem is largely overcome with one or more *ferrite beads* on the B, E, or C leads to the transistor. A ferrite bead is a small sleeve of powdered iron. When slipped over a wire, it behaves as a small inductance. It is mostly effective at frequencies greater than 50 MHz (harmonics of the horizontal sweep). Unlike a coil-type choke it does not have any self-resonance, which would create other problems.

The capacitor (C1 of Fig. 9-1 and Fig. 9-5) is sometimes called a *safety* capacitor. It is part of the HV hold-down system that prevents excessively high HV and the dangers of x-radiation. Should this capacitor open up, the HV would rise by several thousand volts. To minimize the danger, sometimes several smaller capacitors are connected in parallel. If one opens up, there is still enough capacity to offer protection. Some sets use a special capacitor with a jumper wire inside. Should the capacitor become open, the sweep circuit is interrupted. To withstand the high pulse voltage, a replacement safety capacitor must have a rating at least as high as the original.

To develop the necessary sweep *power,* the output transistor, in addition to supplying high *current,* must also have a high voltage amplification. As stated earlier, stage gains vary greatly. Note the gain of the output stage in Fig. 9-1, according to the p-p levels of the input and output signals. In some cases, this pulse voltage at the output may run as high as 1000 V p-p, or greater, and special precautions must be observed when taking measurements.

• **H.O.T. or Flyback Transformer**

There is little resemblance between a flyback transformer and the more conventional transformers found in other circuits. Instead of a laminated core for example, an H.O.T. has a solid, powdered-iron or ceramic type of core to reduce losses due to the high frequency. To feed a variety of diversified circuits, the H.O.T. has a number of special-purpose windings. Unlike 60-Hz sine-wave transformers, voltage and turns ratios are not proportional because of the

HORIZONTAL OUTPUT AND HV SECTION

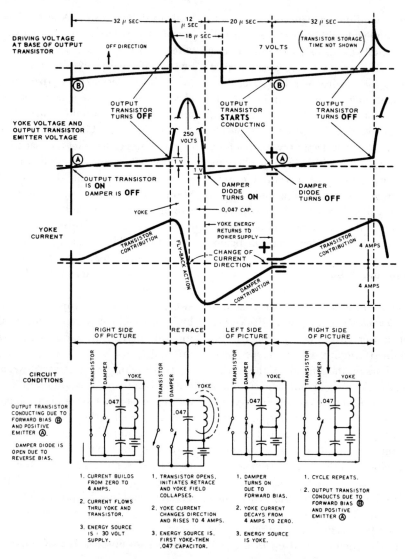

Fig. 9-2 Yoke current development. Relative time relationship of the horizontal-output operational cycle.

pulsed waveform. A typical transformer is shown in Fig. 9-3. It is generally enclosed in a shielded box (HV cage) along with other HV components as a precaution against x-radiation and interference that might be transmitted to other circuits. The transformer coils, whose ohmic resistances are shown on most schematics, are described as follows:

1. *Primary winding:* It is fed directly from the output transistor as shown in Fig. 9-1. It has a low impedance that closely matches the tran-

Fig. 9-3 Typical flyback transformer.

sistor output impedance for maximum power transfer.

2. *Yoke feed winding:* For transformer coupling of sweep voltage to the yoke when direct drive is not used. It has a low impedance to match the yoke.

3. *CRT heater winding:* When the CRT heater(s) are supplied from the sweep circuit instead of the LVPS.

4. *HV (tertiary) winding:* This is the outer coil of the transformer that has many thousands of turns to develop the high voltage (40 kV p-p or so) that is subsequently rectified to feed the CRT. This *high,* pulsed, kickback voltage is produced during horizontal retrace intervals by the rapidly collapsing flux. The coil is usually surrounded by an insulating tire or boot as protection against arcing and corona discharge.

5. *Keying-pulse windings:* The purpose of these coils is to provide keying or gating pulses for the control of certain circuits, such as AFC, keyed AGC, and color killer. Such voltages are critical and are one reason for using *exact* replacement transformers.

6. *Secondary PS windings:* There may be several such windings to furnish the dc *scan-rectifier* power required by some of the receiver stages. Some sets have as many as four or five of these power supplies of different voltages and polarities. Because of improved efficiency (less power loss, less heating) over conventional LVPS sources, they are used extensively in most modern receivers.

7. *Convergence and/or pincushion windings:* Some transformers have separate coils to provide these voltages.

For optimum efficiency in developing the sweep and HV, the flyback transformer, yoke, and all associated coils and capacitors lumped together are made to resonate at approximately 70 kHz or the third harmonic of the horizontal sweep rate. An adjustable coil is provided for tuning the circuit.

In some cases, the *driver* transformer has extra windings and takes over some of the functions of the flyback transformer, such as developing B *boost* and dc scan power.

9-3 SCR SWEEP SYSTEM

This system, which replaces the conventional drivers and output stage, was used by a number of sets in the recent past. Instead of transistors, it uses SCRs (silicon-controlled rectifiers) and diodes, acting as switches, to control trace and retrace of the CRT beam. The energy required for trace and retrace comes from the *charge* on capacitors and the energy stored in the yoke, which in turn is furnished by tuned circuits that are made to oscillate or *ring* momentarily by the abrupt switching of the SCRs and diodes. The system is activated by sweep signals generated by a conventional horizontal oscillator, and the output drives the deflection yoke in the usual manner.

Figure 9-4a shows a simplified version of the system. Understanding its operation requires a knowledge of SCRs and the fundamentals of ringing. Note that the system is divided into two parts: the components for developing beam *trace* and those for developing the *retrace,* which are independent functions. The voltage regulator (that supplies the B voltage) is also shown. See Sec. 9-12.

In the trace section, *SCR 1* and diode *D1* control the yoke current during the *trace* interval; and in the *retrace* section, *SCR2* and *D2* control yoke current for retrace. Each pair (SCR and its associated diode) working together controls the ringing of the tuned circuit by switching on and off at the correct times.

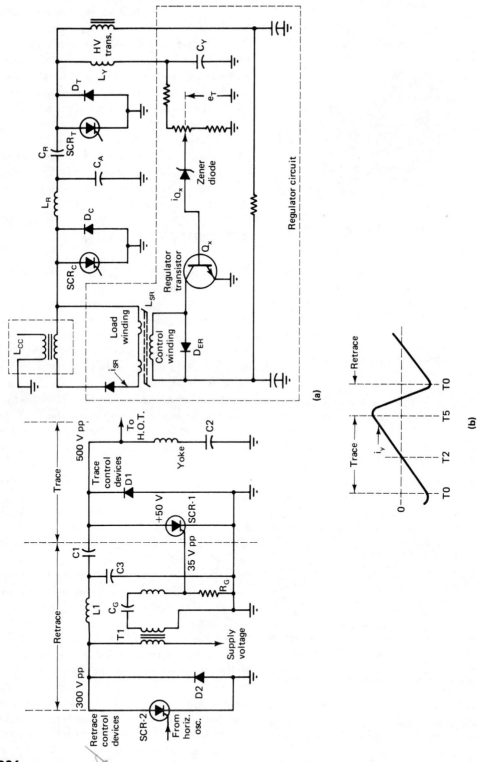

Fig. 9-4 SCR sweep system.

9-3 SCR Sweep System

Ringing occurs in the series resonant circuit consisting of the yoke and capacitors $C1$ and $C2$. Note the relative capacity values. When $SCR1$ is turned on (and acting as a closed switch), $C1$ is effectively shorted out, and the resonate frequency is relatively *low* (about 5 kHz) to develop the left-to-right *active trace* of the CRT beam. For beam *retrace*, $SCR1$ is switched off (an open circuit), and with $C1$ now in the circuit, the resonant frequency increases to about 35 kHz. Thus, during switching operations, the ringing frequency is constantly changing, and $C1$ and $C2$ become alternately charged and discharged. The voltage across $C1$, the smaller of the two, is fairly high (about 500 V p-p), and it is this voltage that drives both the yoke and the flyback transformer. The flyback transformer has no bearing on system operation and can be disregarded at this time.

A brief review of fundamentals might be in order at this time.

SCRs. As explained in Sec. 5-9, an SCR is one of several types of *thyristors*. It is a special type of diode that can be turned on or off under certain controlled conditions. The *control element* of an SCR is called the *gate*. Normally, an SCR operates with a fairly high + voltage on its anode (the outer *case* of the device). The SCR will not conduct, however, until a + voltage (at least 0.8 V) is applied to the gate. The actual amount of gate voltage required to switch the SCR on depends on the amount of voltage on the anode.

When gated into conduction by a positive-going pulse (as in the SCR sweep system), the SCR becomes essentially a *short circuit*. When switched off, it becomes an *open circuit*. Sometimes the on and off switching is called *latching* and *unlatching*. Once triggered into conduction, the SCR can only be switched off when the current drops below a certain minimum level. Voltage conditions for turn on are quite critical.

Ringing. Abrupt changes in switching can shock-excite a tuned circuit into a state of oscillation. The frequency is determined by the L and C values of the resonant circuit. The waveform is a sine wave. In a high impedance, high-Q circuit, the oscillations may be quite strong and continue for many cycles. Shunting the circuit with a low-impedance device like a diode, SCR, or transistor causes the oscillations to *rapidly decay*. This is called *damping*. In the SCR sweep system, ringing is permitted for less than one half cycle of 8 kHz and 35 kHz. Detailed operation of the system can best be explained in sequential steps, as follows:

1. *First half of the horizontal trace:* This is from the left side to the center of the screen, up to $T2$ as indicated in Fig. 9-4b. At time T_{-0}, the yoke *retrace current* is maximum (from retrace functions of the previous cycle, to be described later). The magnetic field about the yoke now starts to decay and the yoke current gradually decreases to zero (time T_{-2}). The yoke current charges capacitor $C2$ via the trace switch diode $D1$.

2. *Second half of trace interval:* The yoke flux is no longer a source of energy but the capacitor $C2$ is now fully charged, and *its current* flows back through the yoke (via $SCR1$) to provide the remainder of the trace up to time T_{-3}. Meanwhile, capacitor $C1$ has become charged from the B supply (via transformer $T1$, the commutating coil $L1$, and $SCR1$) and is ready to furnish power for *retrace*. A positive-going pulse is induced into the secondary of $T1$ to *unlatch* the gate of $SCR1$.

3. *Start of retrace interval:* The retrace switch ($SCR2$) is turned on by a positive-going pulse from the horizontal oscillator applied to its gate. The charge on $C1$ dissipates (via $SCR2$ and the commutating coil $L1$). Coil $L1$ is permitted to *ring* for one half-cycle. $SCR1$ becomes *unlatched* at time T_{-4} as indicated in Fig. 9-4b. Note the overlap from the trace cycle and that actual retrace *has yet to begin.*

4. *First half of retrace interval:* Energy stored in the yoke inductance provides current, but $SCR1$ is turned off and current flows in the commutating loop consisting of $L1$, $C1$, and $SCR2$. The time is still T_{-4}. The decreasing yoke current charges $C1$ (with reversed polarity), creates ringing (briefly) in the resonant circuit, and moves the CRT beam to the left. The time is now T_{-5}, the yoke current is zero, and $C1$ is fully charged.

5. *Second half of retrace:* Yoke current is provided by the charge on $C1$, via the retrace diode $D2$.

SCR2 is turned off. The time is T_{-6} and beam retrace is completed. The retrace interval is very *fast* because the ringing frequency is *high* (35 kHz). The yoke current now flows through trace diode *D1, the time is* T_{-0}, and we are ready to start another trace interval.

In the forgoing explanation, note the following: Retrace supplies part of the trace interval; the charge on *C1* is constantly being replenished from the B supply; and each pair of switches (SCRs and diodes connected back to back) conducts *alternately,* to initiate trace and retrace.

9-4 HORIZONTAL SWEEP-DEPENDENT CIRCUITS

As briefly described in Sec. 8-2, numerous receiver functions are dependent on and operate in conjunction with the horizontal sweep section. Some are shown in Fig. 8-1 and some have counterparts in the vertical sweep section (see Fig. 7-15). Each is described as follows in reference to the simplified schematic diagram of Fig. 9-5.

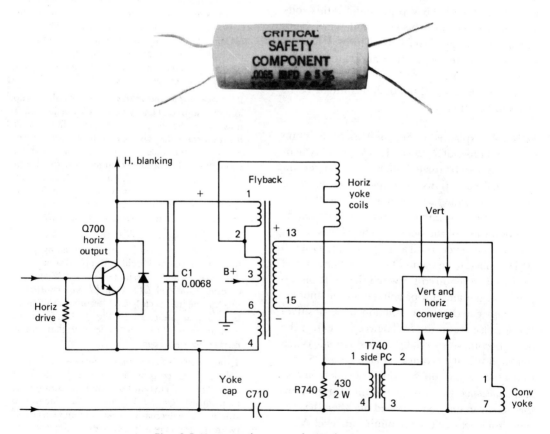

Fig. 9-5 Horizontal sweep-dependent circuits. Note the special type "safety capacitor" C1.

9-5 HORIZONTAL RETRACE BLANKING

The vertical and horizontal blanking pulses that are part of the composite video signal should be strong enough to darken the screen during beam retrace periods. However, if the signal is weak and/or the brightness control is set too high, the CRT may become unblanked. Vertical blanking symptoms are described in Sec. 7-14. Loss of *horizontal* blanking can cause foldover effects on the left side of the screen.

As with the vertical, retrace blanking is accomplished by feeding high-amplitude sweep pulses of the correct polarity to the CRT input circuit in one of two ways: direct to the CRT (grid or cathode depending on pulse polarity), or to a video amplifier stage (the *Y* channel in a color set). The actual p-p amplitude of the pulses varies with different sets according to the source of the pulses and to what point they are fed. In any case, the pulses must be strong enough to overcome the normal bias of the CRT (which varies with the setting of the brightness control) to produce beam current cutoff during retrace. In some sets a *blanker amplifier* is used to boost the pulse amplitude. In color sets, the same pulses may also be fed to the chroma section to disable the chroma amplifiers during beam retrace.

The blanking pulses may be supplied directly from the horizontal output transistor (as shown in Fig. 9-1) or from a winding on the flyback transformer.

9-6 PINCUSHION CORRECTION

As explained in Sec. 7-14, a problem peculiar to large-screen sets is where the raster tends to bow inward (or outward) at top and bottom or the sides. Bowing *outward* is caused by overcorrection by the antipincushion circuit and is called the *barrel effect*.

Horizontal pincushioning occurs when there is greater horizontal deflection at the top and bottom compared to the vertical center of the screen. Correction is made either by reducing the sweep at top and bottom or increasing it in the middle.

Generally, the problems for *side correction* are less formidable than for top–bottom correction. One method uses a modified sampling of the vertical sweep voltage (parabolic waveform) to regulate the HV applied to the CRT at different times of the vertical trace period. As determined by the waveform, *loading* of the horizontal sweep becomes greatest at the beginning and ending of each vertical trace to reduce the amount of horizontal sweep to match that at the center of the screen.

Another method (see Fig. 9-6a) uses the vertical sweep voltage to control the collector current of the horizontal output transistor. It operates as follows. Two driver amplifiers are used to control a HV-regulating transistor. Transistor *Q1* (a common-emitter amplifier) is driven by pulses from the emitter of the vertical output transistor. Pulse amplitude (and therefore the amount of correction) can be regulated by potentiometer control *R1*. With a negative-going input, the output of *Q1* drives *Q2* with a positive-going signal. The positive-going output of *Q2* (an emitter follower) increases the forward bias of *Q3* (the regulator), increasing its conductance. With the regulator connected in series with the B+ feed to the horizontal output transistor, an increase in conductance means greater sweep (raster width) for the center portion of the screen. Because of the parabolic shape of the vertical waveform, there is no increase in width at the top or bottom. This waveform has the characteristic butterfly-wing pattern. The bias control (*R2*) varies the gain of *Q1* and is used as an overall control of HV and raster width.

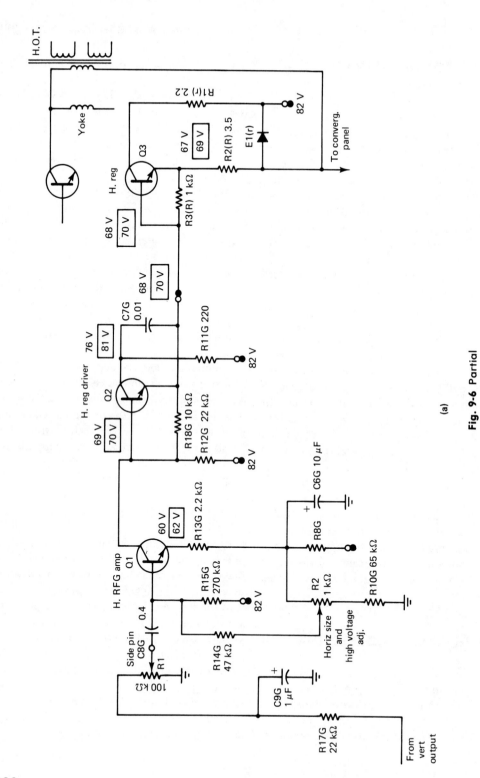

Fig. 9-6 Partial

9-7 Dynamic Convergence

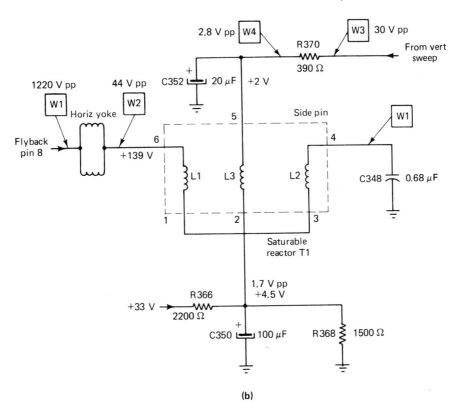

Fig. 9-6 Antipincushion circuits.

Figure 9-6b shows another type of antipincushioning circuit. This circuit uses a *saturable reactor* (as described in Sec. 7-14). Transformer windings $L1$ and $L2$ are connected in series with the horizontal yoke. The degree of core saturation (and hence the yoke current) is governed by the amount of dc current flowing through coil $L3$. A residual amount of magnetism is furnished by current from the +33-V source. Saturation comes from the applied vertical parabolic waveform, increasing the inductance when the beam trace is near the top and bottom, reducing the width for those areas of the screen. No pincushion adjusters are provided.

9-7 DYNAMIC CONVERGENCE

As explained in Sec. 7-14, the convergence assembly on the neck of a delta-type CRT has three pairs of coils that correspond with, and are mounted adjacent to, the three guns. One coil of each pair is energized from the vertical sweep section (see Fig. 7-16b). The other coil of each pair is energized from the horizontal sweep. Controls are provided to establish the amount of current through each coil and its influence on the CRT beams at the edges of the screen. Convergence for in-line CRTs is a good deal less complex. For a detailed description of convergence for both types of tubes, see Chapter 15.

9-8 HORIZONTAL OUTPUT CONTROL CIRCUITS

A number of sweep-related control functions are described next. Some sets provide one or more of these controls; with other sets, they may be dispensed with. Some representative partial diagrams are shown in Fig. 9-7 and 9-8.

Flyback and Yoke Tuning. Many sets resonate the horizontal output circuit to the third harmonic of the sweep frequency. This is done to improve the HV regulation and to reduce the chances of transistor breakdowns by limiting the amplitude of the flyback pulses. This is a design factor in choosing values of L and C for the overall output circuit. In some cases the circuit is fine-tuned by an adjustable coil. In some sets it is used as a *linearity* adjustment much like the efficiency coil found in many tube-type receivers. Because resonance of the output stage is fairly critical, repairs should be made using *exact* replacements of all components, particularly the yoke and flyback transformer.

Width Control. Some sets use an adjustable coil series with the horizontal yoke. Increasing the inductance lowers the yoke current and reduces the raster width. Another system of control is shown in Fig. 9-7. Switching is done with a short connector lead and pin jacks or lugs. Grounding out the pulses in the flyback primary (via $C1$) reduces the width. With a boost from the transformer secondary ($L1$), the width is increased.

Horizontal Centering. As with *vertical* centering (Sec 7-15), a small amount of dc current is made to pass through the yoke coils; the *direction*

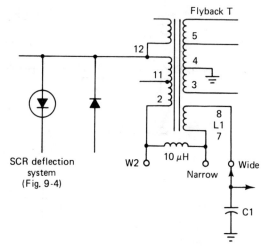

Fig. 9-7 Raster width control.

of the current determines which way the raster is moved, to the left or the right. The dc is developed by rectifying some of the horizontal sweep voltage. Typical circuit arrangements are shown in Fig. 9-8.

In part a, a *pulse transformer* is driven from the output transistor. Low-voltage ac from the secondary is rectified by a pair of diodes with reversed connections to provide both a + and a − dc polarity. A centering control pot feeds the yoke. Mid-setting of the control represents zero dc with maximum + or − at either extreme. Note the low value (10Ω) of the control.

The circuit in part (b) rectifies the pulse voltage developed across capacitor $C1$ in the yoke return circuit, and no transformer is required. A jumper connection and pin jacks take the place of a control pot to provide a +, −, or zero dc voltage.

9-9 SWEEP-DEPENDENT POWER SUPPLIES

The horizontal sweep section is the source for three kinds of dc power: (1) secondary power supplies (often referred to as *scan rectification*),

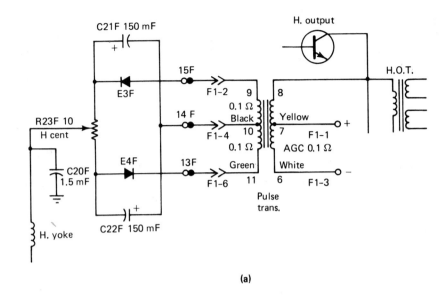

(a)

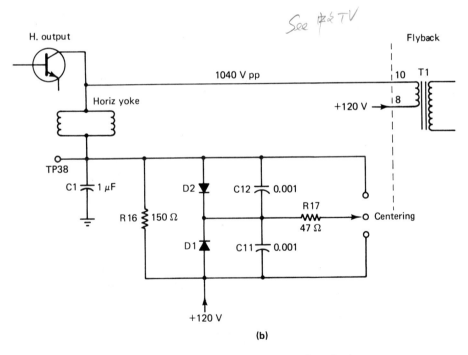

(b)

Fig. 9-8 Horizontal centering control methods.

294 HORIZONTAL OUTPUT AND HV SECTION

(2) HV dc required by the CRT (up to 30 kV or so for large-screen color sets), and (3) a medium focus voltage (around 5 kV for a color set).

• **Scan Rectification**

Practically all modern color sets use these sweep-derived power supplies to meet the dc requirements of certain stages. As a by-product of the horizontal sweep section, a scan-type PS has several advantages over the 60-Hz LVPS when the load current is not too great. For example, less filtering is required because of the high ripple frequency. No filter chokes are needed, and capacitor values can be relatively small. Also, scan rectification is more efficient, with reduced power consumption and less heat. The LVPS (as the *source* of *all* power) is still required, but with less dependency on it than with earlier sets. Naturally, any defects in the sweep circuit will have a detrimental affect on the output of the scan PSs. This in turn can create problems in diagnosing troubles by symptoms, depending on which stages are powered by the scan rectifiers.

Scan rectifiers are supplied with ac pulses from windings on the flyback transformer. Figure 9-9 shows several of such PSs to furnish different stages with different amounts of voltage with both + and − polarities. Each PS consists of a single diode (for half-wave rectification) and a filter capacitor. Capacity values range from a fraction of a microfarad to 20 μF or more. The latter type are electrolytics. The only difference between a + and a − supply is how the diode is connected and the polarity of electrolytic filters when used.

The diode rectifies the peak value of the positive- or negative-going pulse from the transformer coil and not the full p-p value as indicated on a CRO. This is because the pulse is actually ac, with some of the voltage (of opposite polarity) below the zero reference line. The amount of dc under load depends on the *width* or *duration* of the pulses and, as with any power source, the amount of load current. Instead of separate PSs, sometimes several different voltages are obtained from only *one* transformer winding (and one or two diodes) by using series-dropping resistors. Scan-derived voltages vary

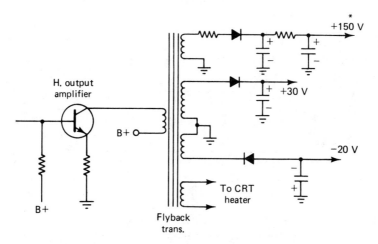

*Some outputs may be regulated.
Note polarity of diodes and filter capacitors.

Fig. 9-9 Secondary power supplies (scan rectification).

greatly among different sets, from as low as 10 V or so in some cases to as much as several hundred volts depending on the requirements of the circuit being supplied. Where a higher voltage is needed, a *boost voltage* (up to 1 kV or more) is obtained by rectifying the high-amplitude pulses direct from the output transistor (see Fig. 9-1). This voltage is used for *focus* on a BW CRT or for the screen grids of a color tube. For this purpose, very little if any filtering is required.

The output of some scan PSs is regulated, usually with a simple zener regulator as described in Sec. 6-14. Some sets go so far as to power the horizontal oscillator from scan rectifiers. This is made possible by an oscillator *start-up* circuit, where power from the 60-Hz LVPS is *momentarily* applied, then disconnected; once started, the oscillator runs from the scan dc source. More on this later.

9-10 HV AND FOCUS VOLTAGE POWER SUPPLIES

A CRT requires a very high *dc* voltage to attract electrons over the relatively great distance between the *source* (the CRT cathode) and the screen. With a BW set, good brightness is obtained with a HV between 10 and 20 kV and a beam current of a few hundred μA. By comparison, a color CRT requires between 25 and 35 kV for a beam current of 1000 μA or more for the same degree of brightness. These figures represent a *power* expenditure of 4 W or so for a BW tube and about 30 W for a color tube. This power must be supplied by the horizontal sweep circuit via the flyback transformer and the HV rectifier.

As shown in Fig. 8-1, the HV section for a BW set is represented by only *one* block. With a color set, there are a number of related functions, as indicated. The purpose of these subsections is briefly described next.

Focus Voltage. A typical BW CRT operates with a dc focus voltage of around 100 V, and sometimes none at all. For good beam focus, a *color* CRT requires a much higher focus voltage (usually about 20% of the amount of HV), between 5 and 7 kV in most cases, depending on the screen size, type of tube, and other factors.

HV Regulation. Unlike a BW set (where the HV is not too critical), the HV for a color set must be held constant regardless of varying load conditions. This requires some form of *regulating circuit*. Without such a regulator, the HV would rise to a very high value (as much as 50 kV in some cases) when there is little or no load, for example, during dark scenes or when the CRT screen goes completely dark. For maximum load (high beam current during bright scenes or when the brightness control is turned up), the voltage tends to drop (as with any power source). Such a fluctuating voltage creates problems: poor focus, poor color due to misconvergence, blooming, and variations in brightness and pix size.

Overvoltage Protection. Excessive HV due to failure of the regulating circuit can have serious consequences: arcing, damage to components, and the hazards of x-radiation. All modern large-screen color sets are required to have some form of protection against this possibility. In the event of excessive HV (regardless of the cause), the set is forced into a *shutdown mode* and becomes disabled until the fault is corrected. *Methods* of disabling vary with individual sets.

Automatic Brightness Limiting (ABL). Most color sets have some form of ABL circuit to limit the beam current, which, if excessive, can damage the CRT. The circuit senses any abnormal increases in the HV/CRT beam current to automatically bias the CRT (via the manual brightness control and the video amplifier section) to reduce the screen brightness to a safe level. For further details on ABL, see Sec. 14-15.

9-11 DEVELOPING THE HV

As with an automotive ignition system, *some* voltage is generated by the ignition coil (and induced into the HV secondary winding of a TV flyback transformer) during the relatively slow buildup of magnetic flux when current is applied. In both cases, however, the voltage is *much greater* during the *rapid collapse* of flux (as when the ignition points open up to fire the spark plugs) and to develop a high pulse voltage in the flyback transformer, which is subsequently rectified and fed to the second anode of the CRT. The rapidity of the flux collapse (and therefore the amount of pulse voltage is further increased by *kickback action* from the self-induced voltage from the yoke (ringing) during horizontal retrace. *Excessive* ringing is prevented by a *damper* diode. The p-p voltage from the HV winding of a flyback transformer is typically around 30 kV for a BW set; for a color set it might be double that amount or more depending on the set and the HV rectifying system.

HV Rectification. BW sets and some color sets use a single diode as a half-wave rectifier of the high-amplitude pulses from the flyback transformer. They are special diodes, able to withstand the high inverse voltage being applied. Essentially, a HV rectifier consists of numerous junctions *in series* with the voltage divided between them. The amount of *dc* available at the output of the rectifier and at the CRT is considerably lower than the applied pulse voltage and tends to vary with changes in the *load* (the CRT beam current). Modern sets have no HV filter capacitor as such; instead, they rely on the built-in capacity of the CRT (between the inner and outer aquadag coatings). For added protection against breakdown, some color sets use two or more rectifiers connected in series.

Voltage Multipliers. As in the LVPS (Sec. 6-4), many color sets use this method of increasing the voltage. Doublers, triplers, and even quadruplers are used extensively in present-day receivers. These rectifier circuits require two or more diodes and a number of HV capacitors. Sometimes the components are accessible, but more often they are contained in a sealed unit that is not repairable.

A typical HV tripler circuit is shown in Fig. 9-10. It uses 6 HV diodes and an equal number of capacitors. Its operation is explained as follows.

Assuming a 10-kV p-p input (5-kV peak value), on the positive half-cycle, $C1$ charges via $D1$ with the polarity as indicated. When the ac reverses, the input is series aiding with the charge on $C1$ to charge $C4$ (via $D2$) to twice the input (10 kV). On the next alternation, the voltage on $C4$ is additive to the input, and $C2$ charges to 15 kV via $D3$. When the ac reverses, this voltage adds to the input, charging $C5$ to 20 kV via $D4$. On the next reversal, $C3$ is charged to 25 kV via $D5$; then $C6$ is charged to 30 kV via $D6$. The voltage regulation of multipliers is not as good as with a single rectifier, but it permits using a transformer with a lower step-up, thus reducing the chances of breakdown. Transformer output is typically around 9 kV p-p.

9-12 HV REGULATOR CIRCUITS

Practically all color sets regulate the HV *indirectly* by regulating the B supply to the horizontal output stage. Where the regulation of the LVPS is extremely good, sometimes such a regulator can be dispensed with. Because of the great variations in the high current drain of the output transistor, a simple zener regulator (Fig. 6-14a) is inadequate. Transistor regulators (Fig. 6-15) are often used and sometimes more elaborate circuits that employ several transistors that can sense changes in the *HV* and can handle the high current. Some sets have a *HV adjust* control in the regulator circuit, but most

9-13 Overvoltage Protective Circuits 297

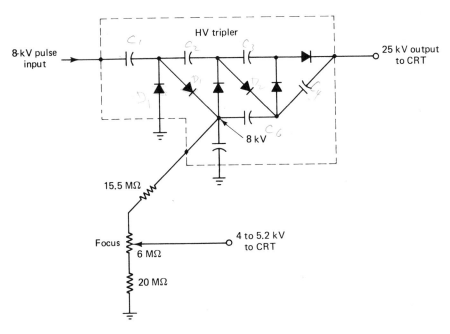

Fig. 9-10 HV tripler circuit.

modern receivers no longer provide such a control. Where there is a control, it *must* be adjusted for the correct amount of HV specified by the manufacturer.

The HV regulator circuit for the SCR system of deflection (Sec. 9-3) is included in Fig. 9-4. Its operation is as follows. Capacitor $C2$ is in the ground return of the yoke, and the voltage across it is proportional to the amount of HV being generated. A portion of this voltage is fed to the regulating transistor ($Q1$) via the HV adjust control. Thus the conductance of $Q1$ is made to vary in accordance with the amount of HV. The collector current of $Q1$ is used to vary the degree of saturation of the saturable reactor $T2$ and, therefore, the inductance of its secondary winding. This winding is in parallel with $T1$ that feeds control pulses to the gate of the trace switch (SCR1). The two inductances (along with capacitors $C1$ and $C3$) form a resonant circuit that is made to *ring*, producing a high-voltage charge on $C1$, as explained in Sec. 9-3. The amount of this charge determines the amount of sweep energy available for beam trace.

To summarize, an *increase* in HV causes an *increase* in the transistor current. This *increases* the saturation of the saturable reactor, *decreasing* the inductance of its secondary winding. This *raises* the frequency of the resonant circuit, *decreasing* the voltage across $C1$, and *lowers* the HV. Conversely, a *decrease* in the HV produces a reverse action to *raise* the HV. The setting of the HV adjuster determines the level of conductance for the regulating transistor $Q1$.

9-13 OVERVOLTAGE PROTECTIVE CIRCUITS

Besides arcing and damage to components, voltages in excess of 25 kV or so can produce harmful radiations. The radiation hazard is reduced or prevented by proper shielding and

298 HORIZONTAL OUTPUT AND HV SECTION

the use of overvoltage protective circuits, sometimes refered to as *fail-safe, shutdown,* or *hold-down circuits*. Their function is to automatically disable the receiver in some manner if the HV exceeds safe limits. Various methods are used to disable the set: forcing the horizontal oscillator off frequency so a pix cannot be viewed; blanking out the raster or the video; killing the HV; or upsetting the vertical sweep. Normally, disabling circuits are inoperative unless the HV becomes excessive. Although their functions are similar, protective circuits should not be confused with a HV regulator, which they resemble.

9-14 FOCUS VOLTAGE CIRCUITS

Modern BW CRTs use *electrostatic* fixed focus, and no focus control is required. The amount (and polarity) of the focus voltage is not particularly critical. Figures 9-10 and 9-11 shows different focus circuits in common use. In Fig. 9-11a, the CRT focus grid is supplied with B *boost* (anywhere from about 200 V to 1 kV) using a rectifier and *RC* filter direct from the horizontal output transistor. A lower voltage may be obtained from a tap on the flyback transformer or from the LVPS.

In Fig. 9-11b, either a switch or pin jacks are used to provide alternative connections to obtain best possible focus.

A color CRT requires several thousand volts, and a focus control is usually provided. The circuit in Fig. 9-11c is supplied from a winding on the flyback transformer using scan rectification. The focus control is part of a voltage-divider network to obtain a variation from about 4 to 6 kV. Because of the high ripple frequency, a filter capacitor is seldom used. A *spark gap* (SG) is connected to the focus grid as protection against arcing. More on this in Sec. 9-15.

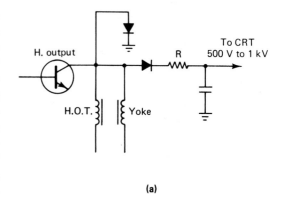

(a)

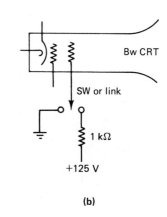

(b)

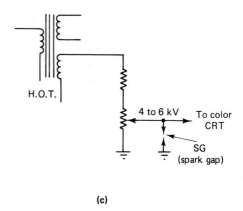

(c)

Fig. 9-11 Focus voltage circuits.

The circuit in Fig. 9-10 shows the focus voltage supplied from an 8-kV tap on the voltage multiplier that develops the HV for the CRT. Here again the focus control forms part of a voltage-dividing network. Note the very high value resistors that are used.

9-15 HV ARCING AND CORONA DISCHARGE

Special precautions must be taken in the design and servicing of TV receivers to prevent arcing and corona, particularly with large-screen color sets that require very high voltages for the CRT. A distinction must be made between the two conditions. *Arcing* occurs when there is a complete breakdown of insulation. It may be an insulation failure under normal voltage conditions or when the voltage becomes excessive. Other considerations may be involved, such as the *weather,* where high humidity is often a factor. Or there may be a leakage path provided by foreign particles or carbonized deposits. HV arcing is usually accompanied by loud crackling sounds.

Corona discharge occurs when air particles become *ionized* (forming a high-resistance conducting path) between HV points. It is characterized by a thin, bluish-colored discharge that streams out from any sharply pointed HV contact to any nearby object or the chassis. When there is a lengthy air gap, the discharge may be barely noticeable except in a darkened room; but its presence may be detected by a faint *hissing* or *sizzling* sound and the sweetish, pungent odor of *ozone.* A *stronger* discharge may be easily observed and in extreme cases develops into a *sustained arc.* Design criteria that prevent or reduce the chances of arcing include the use of plenty of high-quality insulation at critical points, adequate spacing between HV points and proper lead dress, the use of components conservatively rated to withstand the HV, avoiding sharp points at HV terminals and soldered connections, and the use of *spark-gaps* and *arc gates.* Further details on arc prevention are given next.

HV Coil of the Flyback Transformer. Generally, a boot or tire of a high-quality insulation encircles the winding to prevent corona discharge. When the transformer is (or has been) overloaded, there is often a telltale accumulation of wax drippings and the chances of breakdown are increased.

HV DC Circuit. This includes all components and wiring between the HV rectifier and the CRT. The rectifier is usually mounted on a well-insulated portion of the flyback assembly. HV multipliers are enclosed in a plastic box. The HV lead to the CRT is specially insulated and treated to withstand the HV, but sometimes becomes brittle and cracked. This represents a potential shock hazard and the risk of arcing. The *ultor* button contact on the CRT is located several inches from the aquadag coating, and the HV lead terminal is fitted with a large rubber suction cup.

To function as a HV filter capacitor, the outer aquadag coating of the CRT must be grounded. The coating is grounded with flat, spring strips, or most often a long spiral spring stretched across the bell of the CRT. As a capacitor, the CRT is very efficient and can retain a charge for a very long time.

Spark Gaps. An arc gap consists of two closely spaced contacts used as protection against momentary arcing that can damage or destroy components, particularly the CRT. Most color CRTs have *spark gaps* connected between the various grids and ground. They are often enclosed in the CRT socket.

300 HORIZONTAL OUTPUT AND HV SECTION

9-16 TYPICAL HORIZONTAL OUTPUT AND HV CIRCUIT

Following is a brief functional description of this section of Sample Circuit A at the back of this book. The oscillator, AFC, and driver stages were described in Chapter 8.

Sweep voltage from the driver stage feeds the base of the output transistor (Q 1701), which operates with a small amount (-0.4 V) of reverse bias. The transistor develops 800 V p-p of pulse voltage at its collector. Ferrite beads are used on both the collector and emitter leads to prevent excessive ringing (see Sec. 9-2). The horizontal yoke is connected in series with the pincushion transformer T1810 and is *directly* driven from the transistor. In the horizontal scanning process, the circuit is made to *ring* at two different frequencies (10 and 40 kHz) as energy is alternately exchanged between the yoke and capacitor C1707. Switching is performed by the output transistor and the damper diode Y 1701. A more detailed description follows, in reference to the simplified diagram shown in Fig. 9-12.

Trace and retrace of the CRT beam is accomplished in *four* operations. Using the center of the screen as a starting point, the output transistor is turned on by a sweep pulse at its input. While conducting, the transistor effectively shorts out the *timing capacitor* (C1703). Energy from C 1707 (which was previously charged) flows into the yoke to sweep the beam to the right side of the screen. Energy is now stored in the magnetic field about the yoke. At this point the transistor is turned off, and with C 1703 now in the circuit, the resonant frequency is raised from 10 kHz to about 40 kHz. As the yoke flux collapses, the current *rapidly* decays, sweeping the beam back to the center of the screen. Capacitor C 1707, which is now charged, discharges through the yoke to move the beam to the extreme left of the screen. Energy is once more stored in the yoke. This time the *damper* provides a short across C 1703 as it is *forward biased* by the voltage applied to its cathode. With C 1703 out of the circuit, the resonant frequency is reduced to 10 kHz, and energy stored in the yoke discharges into C 1707 to move the beam from the left to the center of the screen, thus completing one trace and one retrace period. With the arrival of another sweep pulse, the transistor is again turned on to initiate another sequence.

With the rapid collapse of flux during horizontal retrace, a very high pulse voltage is induced into the HV winding of the flyback transformer. The HV is rectified by *three* series-connected diodes located within the transformer assembly. The HV dc (30.5 kV) is fed directly to the CRT. The HV is not adjustable. Focus voltage for the CRT is supplied from the HV via a voltage-divider network and is adjustable with the focus control pot.

HV regulation is accomplished by regulating the B supply feeding the output stage, as explained in Sec. 6-10.

Scan Power. Sweep derived scan power is obtained from the horizontal output circuit, and most of the scan related components are contained on a separate *scan module*. An exception is the screen grid supply voltage for the CRT where the 800 V p-p pulse at the output of Q 1701 is rectified by diode Y 1703. This voltage can be varied with the master screen control R 1781.

The scan module contains two sweep derived PSs which are supplied from windings on the

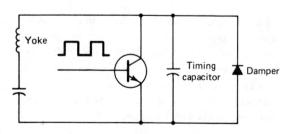

Fig. 9-12 Simplified horizontal output circuit.

flyback transformer. One PS delivers −25 V dc via rectifier diode Y 1804, and is reduced to −15 V by resistors R 1810 and R 1811. The 25 V is for vertical centering and the 15 V is for electronic tuning (channel selection) when provided. The other PS develops about +75 V dc which when added to the +125 V already in the circuit, furnishes +200 V for the video/chroma output transistors. The components Y 1800, R 1801, R 1802, and C 1800 are part of the horizontal centering circuit. The flyback transformer is protected against overload by resistors R 1875 and R 1879 which are designed to open up in the event of an overcurrent. The scan module also provides a horizontal sync pulse output for AFC and blanking.

With this set, no dynamic convergence circuitry is required, and a simplified type of antipincushioning circuit is used. Transformer T 1810 and the horizontal yoke are in series, providing a path for the output transistor current, which is supplied from the +115-V source via the horizontal centering control. Horizontal sweep voltage from the secondary of T 1810, in turn, is series connected to the *vertical* yoke to ensure uniform horizontal deflection over all portions of the screen. The centering control does its job by regulating the amount of *dc* flowing in the yoke coils.

9-17 TROUBLESHOOTING THE HORIZONTAL OUTPUT, HV, AND RELATED CIRCUITS

Because of the high voltages and high-amplitude pulses that are present, component breakdown and other problems are quite common to this section of the set. Table 9-1 contains a listing of trouble symptoms most frequently encountered and the nature of such problems. A troubleshooting flow diagram is shown in Fig. 9-13. The first step is to *localize* the trouble when possible, and then to *isolate* the defect. Most of the information in the following paragraphs applies to both BW and color sets, unless otherwise specified. Before proceeding, a review of Sec. 8-10 is suggested.

TABLE 9-1 Horizontal Output–HV Trouble Symptoms Chart

Trouble Condition	Probable Cause	Remarks and Procedures
No raster (or dim raster)	See Table 8-1.	
Delayed or intermittent raster	See Table 8-1.	Monitor HV and drive to output transistor. Monitor operation of output transistor, HV rectifier and regulator, and the CRT. For SCR system, monitor input and output of both SCRs.
Reduced width	See Table 8-1.	
Poor linearity or foldover	See Table 8-1.	
Horizontal pincushioning	Misadjusted pincushion magnets if BW set. Pincushioning circuit defects if color set.	Troubleshoot or adjust side pincushioning circuit. Adjust magnet positioning if BW set.

Table 9-1 Horizontal Output–HV Trouble Symptoms Chart (*cont.*)

Trouble Condition	Probable Cause	Remarks and Procedures
Keystone-shaped raster	Shorted horizontal yoke coil.	Verify by testing and replace yoke.
Raster off center horizontally	Misadjusted centering control. Circuit defect causing abnormal dc current in yoke. Misadjusted centering ring on BW CRT.	Adjust control. Check for dc in yoke circuit. Check for loss of width, poor linearity, or foldover.
Receiver inoperative	Set may be in shutdown mode because of excessive HV.	Disable or defeat shutdown circuit and check for excessive HV.
Poor focus	Defective CRT or incorrect CRT operating voltages. HV is too high or too low. Improper focus voltage. Misadjusted focus control. Blooming.	Check CRT and its voltages. See that focus voltage varies with control adjustment.
Blooming	Any condition that causes poor HV regulation. Excessive CRT beam current.	Note other symptoms caused by poor HV regulation. Check HV variations while adjusting brightness control. Check CRT operating voltages and beam current.

9-18 LOCALIZING THE PROBLEM

After signal tracing and performing other tests as described in Chapters 4 and 8, presumably the trouble has been localized (at least tentatively) to this section of the set. The procedures that follow are more specific. Although we are concerned at this time with the condition of the *raster*, it must be remembered that raster problems also show up on a *pix*.

CAUTION: ─────────────────

Be careful to avoid shocks when working around HV circuits. When probing or making voltage checks, keep body and one hand clear of the chassis. Don't trust the insulation on the HV lead. Avoid contact with the CRT ultor terminal. The CRT can hold a charge for a *very long time*. Ground the ultor before handling the CRT or before removing or reconnecting the HV lead.

Common trouble symptoms and *localizing procedures* are described next.

• **No Raster or Dim Raster**

If there is no sound from the speaker (i.e., the set is completely dead), it is probably one of three conditions: trouble in the LVPS that disables the B supply or trips the breaker or blows a fuse; horizontal sweep troubles that kill the scan PSs if one supply also feeds the audio section; or the set may be in the *shutdown mode* because of excessive HV. Check outputs of the LVPS. If the breaker is tripped or a fuse is blown, reset the breaker or replace the fuse (for *one time* only). If the trouble is still present, proceed as in Figs. 6-17, 8-10, and 9-13. Check the schematic to determine *how* the set is disabled by the overvoltage protective circuit.

If there is sound but no raster or a dim raster, the horizontal sweep, the HV section, and the CRT are all suspect. Start by checking for HV dc at the ultor contact on the CRT. Use a HV probe for this test.

CAUTION: _____

Never check the HV by arcing, as is commonly done with tube-type receivers. Voltage surges can damage or destroy transistors and ICs.

If a HV probe is not available, a sensitive indicator can be made up by attaching an NE-2 neon lamp to the end of a long insulating rod. The brightness of the lamp when held against or near a HV lead or terminal provides a relative indication of the amount of voltage. Although *arcing* is not advisable, another method is to create a faint corona discharge by bringing a well-insulated screwdriver (or grounded jumper wire) to within 1 or 2 in. of the HV point. Don't get close enough to create an *arc* and take care to avoid being shocked. Listen for the hissing sound. The intensity of the corona is a good indication of the amount of pulse voltage.

If HV checks OK, check the CRT, its operating voltages, and the brightness control circuit as described in Chapter 15.

If there is little or no HV dc, check for high-amplitude pulses at the collector of the output transistor and/or at the input to the HV rectifier(s). Never connect a CRO or EVM to these points when pulses are present. Use a neon lamp as described. If pulse amplitude appears OK and there is little or no HV dc, remove HV lead from the CRT and recheck HV at the *lead*. If the HV checks normal, the CRT may be gassy or shorted, reducing or killing the HV when the lead is attached. If there is still no HV on the *lead* (when disconnected from the CRT), the HV rectifier(s) are probably bad.

If there are strong pulses at the output transistor but pulses at rectifier input are weak or missing, the flyback transformer may be defective.

If there are little or no pulses from the output amplifier, check for drive signal at the base of the transistor. If drive is OK, check the transistor and its dc operating voltages. Check collector-emitter *current* (either by computation or direct measurement with a low-range ammeter).

REMINDER: _____

To be turned on (and draw current), the transistor requires a *drive signal* to its input.

CAUTION: _____

It is OK to check for dc at the transistor *collector* or *emitter*, provided no high-amplitude pulses are present. To avoid damage to an EVM, verify the absence of pulses *before* attempting to measure dc.

If there is no B voltage at the collector terminal, check back to the LVPS, the source of the dc power. There may be an *open* circuit, or the voltage may be *grounded out* by a shorted damper diode or its shunt capacitor, or even the transistor itself.

If the transistor and its dc operating conditions appear good, but the pulses at its output are weak or missing, the cause may be *loading* due to a shorted flyback or yoke, a shorted HV or focus voltage rectifier, or a leaky damper. For this possibility, try substituting the horizontal yoke, and/or the flyback primary with a suitable inductance while observing any increase in pulse amplitude from the transistor.

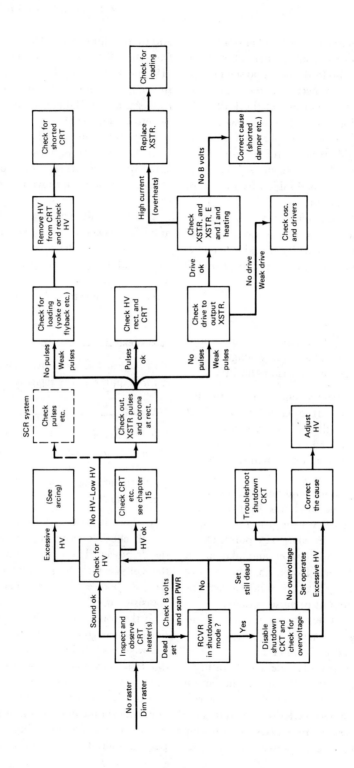

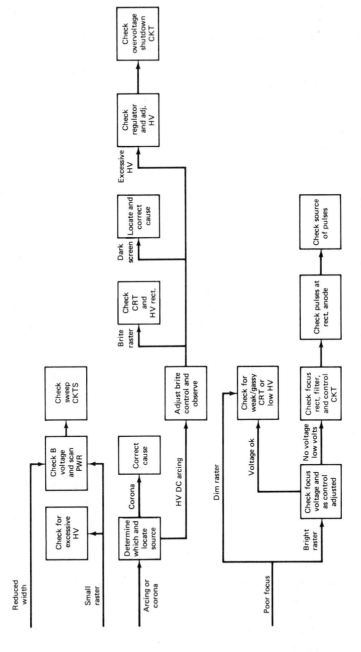

Fig. 9-13 Horizontal sweep and HV troubleshooting flow diagram.

Where CRO or ac EVM shows no drive to the transistor input, it could be due to a leaky or shorted output transistor or trouble in the oscillator-driver stages. Remove or disconnect the transistor and recheck for drive signal. For trouble in previous stages, see Chapter 8.

- **Small Raster, Reduced Width, or Linearity-Foldover Problems**

These conditions are often interrelated and the test procedures are essentially the same for each.

Small Raster (Reduced Height and Width). Usually, it is a LVPS problem such as low line voltage. Measure the line voltage and, if marginal, try a booster. Check the dc power common to both the vertical and horizontal sweep circuits. Another possibility is excessive HV, since the amount of HV has a direct bearing on the CRT deflection sensitivity. Check the HV, and if it is greater than specified in the service data, adjust; or troubleshoot the overvoltage protective circuit as appropriate.

HINT:

A small raster caused by low B voltage will also suffer loss of brilliance; if caused by excessive HV, it will be *bright*.

Reduced Width. This can also be caused by low B voltage, but only the power supplying the *horizontal* circuit. Measure the p-p drive signal at the base of the output transistor and compare with the specified value. For a weak drive signal, troubleshoot the oscillator and driver stages (see Fig. 8-19). *Complete* loss of sweep results in no HV and therefore no raster.

If normal drive, check the amplitude of the sweep pulses at the output of the output transistor. A CRO or EVM may be used provided the p-p voltage shown on the schematic does not exceed the input rating of the instrument.

Where a very high pulse voltage is specified, you can use a 2:1 *divider* consisting of two 1-MΩ resistors in series from transistor terminal to ground with the CRO connected to their junction. Low output from the transistor can be caused by incorrect operating voltages, a bad transistor or damper diode, a defective flyback, or anything causing abnormal loading of the transformer. In some cases, an open yoke or pincushion transformer can result in a *complete* loss of width, where the CRT produces only a thin, bright, vertical line. For minor width problems, adjust the width control.

Linearity-Foldover Problems. Make the same tests as for loss of width. This condition on the left side of the screen is usually caused by the damper or its associated components; if on the right side, it is probably the output transistor circuitry.

REMINDER:

Where there is black showing on only one side of the screen, there are several possibilities: a *centering* problem, loss of width, poor linearity, or a combination of all three.

- **Blooming**

This is a very common problem where the raster (and pix) expands as the brightness control (or color control) is turned up. As the raster grows larger, it becomes dimmer and the focus is impaired. In extreme cases the raster completely fades out and disappears. Blooming is caused by a reduction in HV, and the lower the HV is, the greater the amount of beam deflection. The reverse is also true. As the brightness is turned up, the increasing beam current represents a greater load on the HV power supply and its output tends to drop, increasing the size of the pix. The poorer the regulation of the HV supply, the greater the voltage variations are between no load and full load and the greater the effect. Hence anything contributing to poor regulation will aggravate the problem. The most common causes are a bad HV rectifier or troubles in the HV regulating circuit. Another possibility is excessive beam current such as caused by insufficient CRT bias.

• **Poor Focus**

When there is also some loss of brightness, check for a weak or gassy pix tube, excessive bias, or insufficient HV. If there is normal brightness, check for low focus voltage and for voltage variations as the focus control is adjusted. Check the focus voltage rectifier. Typical focus voltages for both BW and color sets are described in Sec. 9-14.

REMINDER:

Good focus is when raster scanning lines appear sharp and distinct. It should occur somewhere between the two extremes of the control setting.

• **HV Arcing or Corona Discharge**

Arcing and *corona* (as described in Sec. 9-15) are very common problems, and it is important to recognize the difference between the two. There is also a difference between HV *dc* arcing and the arcing of unrectified HV *pulses*. HV *dc* produces loud crackling or popping sounds, whereas the arcing of *ac* is "fat" and "flamelike" with very little sound. Arcing can be a very serious problem that demands immediate attention. It not only causes damage and destruction of components and insulation; it also represents a distinct *fire hazard*.

Corona. This is easily recognized by the hissing sound, the smell of ozone, and frequently by visible *streamers* emanating from some HV point. Often it is difficult to locate the source except in a darkened room. A stethoscope consisting of a length of rubber or plastic tubing can be useful for this. With one end held to an ear, listen for the loudest sound as the other end is moved about. Another trick is to blow through the tubing while listening for a *change* in the sound. A sheet of plastic can create the same effect by breaking up the discharge, but be careful to avoid getting shocked.

Corona discharge most often occurs between the HV coil of a flyback transformer and any nearby object, including the cage, or from *sharp* points in the HV *dc* circuit or a cracked HV lead.

Arcing. Usually the source can be located visually or by the sound. When the arcing is *invisible* (as *inside* a component such as the yoke or flyback), a stethoscope tube can be helpful. Start by adjusting the brightness control to both extremes while observing any change. If the arcing occurs *only* with a bright raster, the cause is related to the CRT *beam current* at maximum intensity. Check for such things as a bad CRT, HV rectifier, or HV filter resistor, or a break in the flyback HV winding. See if arcing stops when the HV lead is removed from the CRT.

If arcing persists with a *dark* screen (when HV is *maximum* because there is no load on the HV supply), there are many possibilities, including a shorted CRT. For this, simply remove the HV lead and see if the arcing stops. Or the arcing could be caused by *excessive HV*. Measure the HV (for conditions of raster brightness as specified) and compare with the normal value shown on the schematic. If excessive, try adjusting the HV control if provided, and see if arcing ceases. If normal voltage cannot be obtained, check the HV control circuit, the HV regulator, overvoltage protective circuits, and the critical safety capacitor shunting the damper. To check HV regulation, simply monitor the HV while adjusting the brightness control for the two extremes of a bright and a dark screen. Normally, the HV should not vary more than 1 or 2 kV. If it does, there will also be severe blooming.

308 HORIZONTAL OUTPUT AND HV SECTION

Check the overvoltage protective circuit by grounding or defeating its function in some other manner appropriate to the circuit. If the set continues to operate with arcing and excessive HV, this shutdown circuit is at fault.

Even with normal HV, there are many places where arcing may occur. Typical problem areas are described next.

The CRT. Sporadic arcing *for a short time* in a *new* CRT is quite common. It is usually caused by small particles that are quickly burned away. If the arcing persists, in lieu of replacement the tube may be flashed (see Sec. 15-112). Continued arcing in the CRT neck (along with a bluish glow) may indicate a loss of vacuum, as caused by a small crack in the glass. The clue here is when the neck gets warm and the CRT heater(s) is not visibly lit, yet shows continuity and there is no raster. Arcing can also take place around the HV ultor contact, between the outer aquadag coating and the grounding spring and in built-in spark gaps in the CRT socket.

Inside Components. Arcing is often accompanied by *smoke,* which often aids in pinpointing the defective part. Arcing inside a yoke or flyback is quite common. If necessary, remove a suspected yoke and, while still connected with the set turned on, observe it closely. Remember to turn down the brightness to prevent burning the screen. Turn the set off and check the coil or transformer (by feel) for signs of overheating. Actually, this is a condition of B+ or B boost arcing, and not the HV, but the effects are the same.

Some BW sets still use a HV filter capacitor, which can break down and arc internally. When suspect, simply disconnect to see if arcing stops. The same holds true for a focus voltage filter of a color set.

REMINDER: _____

Arcing can destroy components. Operate the set only long enough to locate the source.

- **HV Substitution**

When a set has been triggered into the automatic shutdown mode or the HV section is inoperative or suspect, a raster may be obtained (for test purposes) using the HV from another set, provided the voltage requirements are similar. The two chassis are connected together with the HV lead of the one connected to the CRT of the set under test.

- **Circuit Breaker Tripped or Fuse Blown**

One common cause is a drastic overload in the horizontal output section, such as a shorted transistor, damper diode, flyback, or yoke. Reset the CB or replace the fuse. If the set now operates, monitor key checkpoints for a possible reoccurence. Troubleshoot as described in Sec. 8-11 and Figs. 8-10 and 6-17. Remember, surges from LV or HV arcing can also cause this problem.

- **Receiver Shutdown**

It is not always easy to determine whether a set is inoperative because of automatic shutdown (due to excessive HV), because of a defect in the overvoltage protective circuit itself, or because of a defect in some other circuit directly related to the trouble symptoms. Generally, it is the *latter,* but if testing of suspected stage(s) proves fruitless, it is time to consider the other alternatives.

Testing procedures vary according to the type of shutdown circuit, the method used for *sensing* an overvoltage condition, which stage or circuit of the set is disabled, and the manner of disabling. To protect the user, various methods are used to disable a set as described in Sec. 9-13. When a shutdown condition is suspected, but there is a raster, a logical first step is to measure the HV. If excessive, try the adjustment, if provided. If unable to obtain the correct voltage, chances are the shutdown circuit has been enabled to kill the video or throw the horizontal out of sync, as the case may be. Possible causes of excessive HV are described in Sec. 9-19. Most modern receivers disable the *HV* in some manner, which kills the raster and in some cases the *sound*.

Next, examine the schematic to determine the *source* of the *sensing voltage* used to monitor the HV. The sensing voltage may come from a tap on the flyback transformer, a scan rectifier, B boost, or any source that changes along with the HV. Measure the sensing voltage and compare with the schematic.

The next step is to *defeat* the shutdown circuit in some manner to see if receiver operation can be restored. The method used depends on the particular circuit, and may involve grounding it out, or (as applicable to most circuits) opening the circuit between the input of the shutdown circuit and the sensing voltage source.

CAUTION: _____

Depending on the fault that created the overvoltage condition, restoring set operation in this manner can cause damage to components. Defeat the shutdown circuit *only long enough* to monitor the HV and to observe the operation of the stage being disabled and the set in general.

During this testing, the stage or circuit being disabled (horizontal oscillator, horizontal sweep, blanking circuit, video amplifier, etc.) should be monitored with a CRO or EVM as appropriate.

If, after defeating the shutdown circuit, normal set operation is restored and the HV is *not* excessive, it can be assumed that the defect is in the shutdown circuit (i.e., it triggered shutdown without good cause). If the HV is still excessive, the shutdown circuit is exonerated and the trouble will be found in the horizontal output-HV section.

- **Troubleshooting the SCR Sweep System**

Trouble symptoms are not too different from the conventional driver-output amplifier circuits using *transistors* (loss of HV, no raster, etc.) and, with a fair understanding of system operation, localizing a defect is not too difficult. As a start (depending on the symptoms), check the dc supply voltage (see Fig. 9-4). If the circuit breaker is tripped, reset it and repeat the test. Partial or complete loss of dc at this point can be due to troubles in the LVPS or in the regulator circuit feeding the sweep section (not shown in the diagram), or to an overload-type defect in the sweep circuit itself. Also check for B + at the anode (the case) of each of the two SCRs. Normal voltage on the *retrace* SCR is about equal to the supply voltage (+ 150 V), with + 50 V on the *trace* SCR. Check the SCRs and their associated diodes for heating. Overheating (thermal runaway) can cause CB tripping and component damage. An SCR that runs *cold* is *not conducting*. An SCR test circuit is shown in Fig. 5-9.

Check for oscillator drive to the gate of the retrace SCR. Lack of drive could mean the SCR is shorted (to ground) or the oscillator stage is dead. Normal drive signal is about 8 V p-p. Check the gate signal of the *trace* SCR, which is normally about 35 V p-p. Check for pulses at the anode of each SCR. The retrace and trace SCRs should develop about 300 and 500 V p-p, respectively. Waveform checking will usually localize a defect to either the trace or the retrace sections of the circuit. The *retrace* section will operate independently of the trace section, and further tests can be made by disabling the trace section by grounding out the anode of the trace SCR. With the circuit grounded, it should be possible to obtain output from the retrace SCR; if not, the fault is with the retrace circuit. In cases of overload where the CB trips, but it no longer trips when the trace circuit is disabled, then the defect is in the trace circuit. When the CB still trips, remove the ground jumper and *momentarily* disconnect the yoke. If the CB now does not trip, the yoke is shorted. If the CB still trips, reconnect the yoke and disconnect the flyback transformer primary. If the CB no longer trips, it is a bad flyback or an overload on its output circuits. If the CB still trips, reconnect the flyback and individually disconnect each output circuit, including scan PSs and the HV and focus voltage PSs, while monitoring for the overload condition.

CAUTION: ─────────────────────────

To avoid damage to components, don't operate set with the yoke or flyback disconnected for long periods. If the testing is prolonged, it is a good idea to drop the line voltage to about 100 V and bridge an open CB or fuse with a 100-W lamp (as an indicator) until the trouble is corrected.

Where overload still exists with all flyback load circuits disconnected, the flyback transformer is probably shorted, especially if it runs hot and there is an accumulation of wax drippings.

For obscure problems, it is possible to make a static check on the switching action of the SCRs and their diodes. Normally, the switching rate is too fast to be detected. Start by removing the gate drive from the retrace SCR by disabling or disconnecting the horizontal oscillator. Next, check the ON/OFF state of both SCRs with a dc voltage check from each anode to ground. With the SCRs turned off (and no gating pulse), normal B+ should be measured at the anodes. With the meter on the retrace SCR, momentarily apply a simulated gating voltage of about +10 V dc from an external PS. This will trigger a good SCR to the ON state, and its anode voltage should drop to near zero.

REMINDER: ─────────────────────────

Once triggered, the gate loses control and the SCR will continue to conduct until the anode voltage is removed or drops to a low value.

Check the triggering action of the trace SCR in the same manner. Since there is no dc path to the anode, B+ must be applied with a jumper wire. When gated with ± dc from an external PS, its anode voltage should drop as just explained. To reset the SCRs, simply turn the set off. When an SCR cannot be gated, it is probably open or shorted.

9-19 ISOLATING THE DEFECT

Once localized, it is relatively easy to pinpoint a defect in the output-HV section of any receiver. Isolation of a problem, however, can be complicated by the interdependence of one circuit function on another and by the effects of *loading*. For example, troubles in the output amplifier adversely affect the operation of all circuits beyond, such as scan power and HV. At the same time, many defects in these *downstream circuits* can either *kill* or reduce the amplifier output.

With many sets, most (or all) the components of the sweep circuits (as with other receiver sections) are contained in a replaceable module, and the temptation is to replace a module after a minimum of testing effort. In some cases (after weighing and comparing labor and module costs), this is an acceptable procedure, provided all preliminary testing has been done as described in Sec. 5-10. The procedures that follow, however, are for isolating a failure to the *component* level.

Arcing and the overheating of components are common problems with this section of a set, demanding instant action of the highest priority. Unless protected by a tripped CB or blown fuse, turn the set off immediately to avoid further damage. For such conditions, apply power only long enough to perform the necessary tests, and shut the set off *between tests*. Also, it is a good idea to either reduce the line voltage or bridge the fuse or CB with a 100-W lamp until the problem is resolved.

• **Dead Set**

Many modern sets feed the audio section, the CRT heater, and even the dial lights from secondary (scan rectifier) power, making it impossible to determine if a set is on without at least one voltage check. Check for B voltage in the sound section. If there is no dc power, determine from the schematic the *source* of the dc, the *LVPS* or a *scan PS,* and make further tests as appropriate. Chances are loss of sound *and* raster are caused by the same defect.

Automatic *shutdown* of the set is another possibility, as caused by an unsafe overvoltage condition. Try to restore set operation by defeating the shutdown circuit as explained earlier. Check the HV; if excessive, locate and correct the cause. If the shutdown circuit is at fault, check the sensing voltage input; make other voltage tests as appropriate, followed by the testing of individual components.

9-20 TROUBLESHOOTING THE OUTPUT AMPLIFIER

As stated previously, the operation of the amplifier is dependent (to a great extent) on its *load,* which is the combination of *all* sweep-related circuits. For the purpose of troubleshooting, each, including the output amplifier, is considered separately. The amplifier (so far as beam deflection is concerned) consists of the transistor, damper, flyback transformer, yoke, and all components directly associated with their operation. For the following, it is assumed that a trouble has been definitely localized to this section.

• **Little or No Output from the Amplifier**

This means no pulses or weak pulses at the transistor collector (or emitter in some cases) as indicated with an EVM or CRO. Check the drive voltage at the transistor base. If there is little or no drive, the fault could be in the oscillator-driver stages or a shorted or leaky output transistor. Disconnect or remove the transistor and repeat the test. At the same time, measure the no signal bias (between base and emitter). A small reverse bias is considered normal in most cases. A *forward bias* indicates the presence of a drive signal, and the greater the bias, the greater the drive. Other important considerations of the drive signal are its *frequency* and *waveform,* both of which should be verified at this time.

REMINDER: _____

Loss of drive or too much drive does not usually endanger components, because the transistor is cut off. However, *reduced drive* can damage or destroy the transistor, the flyback, and so on. The usual symptoms for this are loss of width and extreme foldover.

After verifying there is *no pulse voltage* at the transistor collector, check for B voltage at this point. If strong pulses *are present,* there is *no need* to check for dc. Also, such a check would risk damage to the test equipment. The amount of no-pulse dc collector voltage is usually equal to the maximum output of the LVPS.

If there is little or no collector voltage, check for dc at the cold end of the flyback where it is supplied from the LVPS via the HV regulator circuit. If it is considerably higher at this point, check for a leaky or shorted output transistor or damper or an open flyback primary. Because of the low resistance of the flyback coil, there should be very little *IR* drop in the collector circuit, unless the current is excessive.

For loss of B voltage, check the dc at the input and output of the HV regulator, and check its operation. One method is to monitor changes in regulator output voltage *and* the HV feeding the CRT as the brightness control is adjusted. The change in both should be minimal. Symptoms of poor HV regulation are blooming, variations of pix size and brightness with changes in video, poor convergence, and excessive HV with arcing. Troubleshoot the regulator circuit as described in Sec. 6-13.

When there is dc collector voltage (allowing for variations caused by abnormal operating conditions), check the collector–emitter current, which, for a color set, is normally around 1.5A.

REMINDER: ───────────────

The operating temperature of the output transistor should be periodically monitored by touching. A *cold* transistor is symptomatic of *low current* or *no current,* and an *overheating* transistor is indicative of *high current.* The latter usually leads to thermal runaway, tripping the circuit breaker, and fuse blowing. The most common cause of high *transistor* current is a bad transistor, which in turn may be caused by *some other defect. Before* replacing a transistor, checkout this possibility to avoid damage or destruction of the replacement. If a mica insulator is used, make sure it is coated with liberal amounts of silicon grease and that the transistor makes good contact with its heat sink.

Damage or destruction of the flyback is often caused by a current overload through its primary coil, and the most obvious indication is an accumulation of melted wax. But high *transistor* current is not the only possible cause. A shorted or leaky damper or safety capacitor results in a high current through the transformer to *ground.* The quickest check is to disconnect them to see if the overload is relieved. Breakdown of these two components is very common because of the high-amplitude sweep pulses at the transistor output.

One of the more difficult problems to resolve is when there is little or no sweep output from the amplifier, and the transistor, its operating voltages, and the drive signal all seem to check OK. Under these conditions, if the transistor current is *high,* the most likely and common cause is circuit *loading* by such things as shorted windings in the flyback or yoke. A shorted yoke produces a keystone-shaped raster, but if only a few turns are shorted out, the symptoms may go unnoticed. In some cases you can get *some* sweep and HV with the yoke disconnected, and *any* improvement in amplifier output when the yoke is disconnected indicates a bad yoke. Testing horizontal yoke coils is the same as for a vertical yoke (see Sec. 7-21). Arcing within the yoke or between yoke coils and the grounded CRT aquadag is quite common. Remove the yoke and test as in Sec. 7-21. A repair is seldom permanent and replacement is advisable.

• **Testing the Flyback Transformer**

Since this is one of the more costly items in a TV set, its replacement should not be considered until extensive tests have been made. Even then there is often some doubt that the flyback is really the cause of the problem, particularly since so many other defects can produce the same symptoms. These symptoms are many and varied: a completely dead set, reduced width, loss of brightness and focus, and so on. Another consideration is that a bad flyback is nearly always the *result* of some other defect, a problem that *must* be corrected *before* the flyback is replaced. The most common causes are a leaky or shorted transistor, damper, or safety capacitor, or arcing.

A seldom encountered problem is an *open* winding. When suspected, check the continuity of the coil(s) using the *lowest* ohmmeter range. Compare with resistance readings shown on the schematic. When in doubt, disconnect other circuits that may influence the measurement.

The most common defect is *shorted* windings, the aftermath of overheating. With the set off, examine for a scorched, charred appearance, and deposits of melted wax. An ohmmeter check is not practical where shorted turns are suspected.

Where amplifier loading is suspected, disconnect one end of the flyback primary and substitute with a similar flyback (primary coil only) or some suitable inductor of near the same impedance and ohmic resistance. In the absence of HV or a raster, results are judged by *any increase in pulse amplitude* at the amplifier output. If there is no increase, the flyback or its load circuits are *not* the problem. When there *is* an increase, the cause is either a bad flyback or circuit loading on one of its secondary windings. The test procedures as follows will decide *which,* after the flyback connections are restored.

While monitoring for pulse amplitude at the transistor output, disconnect the various loads, one at a time, from the flyback secondaries. This includes the scan PSs, the HV and focus PSs, the yoke (in some cases), and circuits feeding gating pulses to different stages. Check for any increase in amplifier output after *each* circuit is isolated. Start with any loads where the most obvious abnormal voltage readings were obtained, for example, a winding feeding a scan rectifier whose output is particularly low or absent. A quick check for HV loading is to remove the lead from the CRT.

Where abnormal loading is relieved by disconnecting a particular circuit (as indicated by an increase in amplifier output), troubleshoot *that circuit* for a short or other defect. If, after disconnecting *all* transformer loads, there is *no increase* in amplifier output and the transformer gets hot, the flyback is probably defective.

REMINDER: ⎯⎯⎯⎯⎯⎯⎯⎯⎯⎯⎯⎯⎯⎯⎯⎯⎯⎯

Unless previously tested, don't forget to check the damper diode and its associated safety capacitor, both of which can simulate a bad flyback.

There are different ways to test a suspected flyback transformer. One method is by ringing as described in Sec 7-21, using either a flyback-yoke tester or an AF generator and a CRO. For this test a flyback must be completely isolated from its loads. An alternative test consists of feeding the primary with about 6 V of 60-Hz ac and making relative comparisons of all secondary voltages. Make sure there is no overheating (which could indicate shorted coils) after operating for about 5 minutes without a load.

When replacing a defective flyback, always use an *exact* replacement. Any other may adversely affect the operation of various dependent circuits that it feeds, because of mismatching, ringing, and other problems.

• **Troubleshooting B Boost and Scan PSs**

Measure the dc output from each of these PSs and compare with the schematic. Anything that upsets the normal operation of the sweep circuit will also affect these voltages. Therefore, if *all* scan outputs are uniformly low, look for the cause elsewhere.

If only *one* PS is affected, perform tests as described in Chapter 6. If there is little or no dc output, check for a leaky or shorted rectifier or filter capacitor, an *open* filter, a weak rectifier (one with a high forward resistance), or excessive load current due to problems with the circuit being supplied with the dc power. Such a defect can load the flyback, reducing its output, and cause damage by overheating. Check the resistance to ground with the set turned off. If it is greater than a few hundred ohms, look for an open circuit between the rectifier and the load, possibly in the regulator circuit if there is one. An open circuit results in a *high* dc output. A low ohms reading indicates a short. Make sure of the polarity when replacing a rectifier or electrolytic.

• **Troubleshooting HV and Focus Voltage PSs**

When there is low HV or no HV at the CRT and little or no pulse voltage feeding the rectifier, disconnect the HV lead from the CRT and recheck. If the pulse voltage is still weak or missing, the trouble is a bad flyback or some other sweep circuit defect. When there is a strong pulse but little or no dc output, check for an open or shorted rectifier or filter (if there is one). To check a filter capacitor, simply cut it loose and see if the HV is restored. HV rectifiers are best tested by substitution or as described in Sec. 5-9. A weak or near-open rectifier is a common cause of *blooming*. A HV *tripler* uses *several* rectifiers; generally, it must be replaced as a unit.

For poor HV regulation (more than a 2-kV drop from a dark screen to full brightness), check for a weak HV rectifier, troubles in the HV regulator circuit, or excessive CRT beam current. For excessive HV, check the HV adjustment and operation of the HV regulator. Also check for an open capacitor in the damper circuit. The value of this safety capacitor is very critical; if open or too small a value, the HV will rise to dangerous levels. Because of the high-level pulses, capacitor breakdown is common. Use an *exact* replacement.

REMINDER:

When excessive HV is measured (and the set continues to operate), always check the operation of the shutdown circuit; if necessary, troubleshoot as described in Sec. 9-19. Adjust the HV (if a control is provided) to the amount specified in the service data for the set.

Focus Voltage. A weak or gassy CRT is a common cause of poor focus. For a BW set, try a different voltage selection, if a choice is provided. For a color set, check the focus voltage at the CRT for *normal raster brightness*. Loss of voltage can be caused by a bad rectifier, a defective focus control or its voltage-dividing network, a shorted or leaky filter, or a shorted CRT. For the latter, remove the CRT socket and recheck. Check the focus control for smooth operation. If it is erratic or cuts out, replace. When focus voltage is supplied from the HV and both check low, there is obviously a HV problem that must be corrected.

• **HV Arcing**

Follow the procedures described in Sec. 9-19 to locate the source. For corona discharge, round off any pointed contacts and apply a silicone-rubber sealant. If necessary, spray HV insulation on the flyback windings or other surfaces, or replace the insulating boot around the HV coil. Dress the HV lead clear of the CRT neck. If the HV lead is cracked, replace. When there is arcing around the CRT ultor contact, clean the glass. Even fingerprints here can cause trouble. If the rubber suction cup is decomposed and conductive, either replace or trim the edges so they do not contact the glass. For sparking around the aquadag coating, check for good contact from the grounding spring or relocate as appropriate. Check for carbonized tracks around insulation strips or boards, particularly in the flyback area. Clean thoroughly by scraping or replace. Replace any components where internal arcing is suspected. Spark gaps sometimes break down and must be replaced. When a set has been exposed to dampness, allow it to dry *before* applying power. Place a 100-W lamp near the HV section and clear of combustibles or components that might be damaged. Cover the set or chassis with a blanket or carton as appropriate. Don't overheat or leave unattended.

9-21 AIR CHECKING THE REPAIRED SET

This is the time when previously unnoticed symptoms often show up or when an intermittent condition makes itself known. If not done previously, verify the correct HV and adjust if needed. Make final adjustment of focus and other controls as necessary.

9-22 SUMMARY

1. Horizontal beam deflection (particularly for a large-screen set) requires a strong magnetic field from the yoke. This, in turn, requires considerable sweep *power,* which is supplied by the power transistor in the horizontal output stage.

2. Besides beam deflection, the output stage provides for other receiver functions: developing HV and focus voltage for the CRT, dc scan power for various stages, keying or gating pulses for a variety of circuit functions, and others.

3. In the absence of a drive signal, the horizontal output transistor is turned off by a near-zero bias. Forward bias is developed by B/E rectification of sweep pulses (from the oscillator or driver stage), causing the transistor to conduct heavily. Conduction lasts only for the brief duration of the input pulses.

4. Most sets feed the yoke *directly* from the output transistor in parallel with the primary of the flyback transformer. Other sets supply the yoke from a winding on the flyback.

5. Brief periods of ringing (oscillation) in the yoke are necessary to produce a high HV and rapid beam retrace. Excessive ringing is undesirable and is prevented by a *damper* diode shunted across the transistor output. The damper is responsible for the first part of each horizontal beam trace. The transistor supplies the sweep power for the remainder of the trace (on the right side of the screen) *and* energy (stored in the yoke) for the start of the next trace.

6. An important component in the output stage is the safety capacitor shunted across the damper. Its value is critical and has a great bearing on the amount of sweep and HV developed. If the capacitor is too small or becomes open, the HV rises to dangerously high levels. The capacitor must withstand the maximum p-p amplitude of the sweep pulses, which is usually quite high.

7. The flyback transformer (or H.O.T., horizontal output transformer) has numerous windings as the source of HV, dc scan power, and to supply other sweep-dependent circuits. The flyback is subject to damage by current overloads often caused by a shorted output transistor or damper. For efficient operation, the circuit comprising the flyback, yoke, and other components is made to resonate at about 70 kHz.

8. The SCR sweep system uses SCRs (silicon-controlled rectifiers) and diodes (instead of the usual driver and output transistor) to develop sweep power. The active units act in *pairs,* one SCR and its associated diode developing beam *trace,* and the other pair producing beam *retrace.* It is an on and off *switching action,* where energy is alternately transferred between the charge stored on capacitors and that stored in the magnetic field of the yoke. Switch *timing* is governed by oscillations in the trace circuit (about 5 kHz) and faster ringing (about 35 kHz) for beam retrace.

9. Without some form of pincushion correction, the raster on large-screen sets tends to bow inward on the left and right sides. The method of correction employed by most color sets uses a sampling of vertical sweep voltage to control the output of the horizontal sweep amplifier to increase the horizontal beam deflection for the vertical-center areas of the screen.

10. The modern color set supplies most of the stages with dc scan power obtained by half-wave rectification of sweep voltage from the flyback transformer. There may be one or more of such secondary PSs with different output voltages of either + or − polarity. Some sets even operate the horizontal oscillator (where horizontal sweep *originates*) from scan power after the oscillator is *started* with dc from the conventional LVPS.

11. The HV dc required by the CRT comes from rectifying high-amplitude pulses from a tertiary winding on the flyback transformer. For large-screen color sets, the HV may be as great as 30 kV or more. The *current* drain from the HV supply is represented by the beam current of the CRT (up to 1.5 mA for maximum brightness).

12. Besides creating problems of arcing, HV in excess of 20 kV or so poses an x-radiation *health hazard*. Modern sets are required to disable the set in some manner if the HV for any reason rises to an unsafe level.

13. Many sets use a *voltage multiplier* (usually a tripler) instead of a single HV rectifier diode. A typical tripler consists of a sealed unit containing a number of separate diodes. This permits using a flyback transformer with a lower HV pulse output.

14. For good convergence and to reduce blooming, the HV of a color set must be well regulated. Most sets control the HV by regulating the B supply feeding the horizontal output transistor. Failure of the regulator may result in excessive HV and receiver shutdown by the overvoltage protective circuit.

15. Most BW CRTs employ fixed focus. A color CRT requires from 5 kV to around 8 kV to obtain good beam focus. The focus voltage may be obtained in several ways: by rectifying the pulse voltage direct from the output transistor or from a tap on the flyback transformer, or by reducing the HV with a voltage-divider network of which the focus control is a part.

16. HV arcing or corona discharge occurs when insulation (air or some solid-dielectric material) breaks down, especially if the voltage becomes excessive or under high humidity conditions. It often takes place between sharply pointed contacts or terminals and some nearby conductor or chassis. Arcing creates interference, destroys components (including the CRT), and represents a serious fire hazard. Spark gaps are connected to CRT terminals to take care of momentary flashovers.

17. Vertical and horizontal centering on a BW set makes use of a magnetic centering ring on the neck of the CRT. With color sets, a small amount of controlled dc is sent through the yoke coils. The direction of current (polarity) determines the direction of beam movement, up and down or sideways.

18. Ferrite beads (small sleeves of powdered iron) are used over wiring to the output transistor to prevent radiation of high harmonics of the horizontal sweep voltage. Such interference may be picked up by the tuner and degrade the pix. The metal cage around a flyback and other HV components serves the same purpose, in addition to preventing x-radiation.

REVIEW QUESTIONS

1. (a) What is the primary function of the horizontal output stage?
(b) List a number of secondary functions.

2. (a) What constitutes the load on the HV power supply?
(b) Under what conditions is the load current greatest?

3. (a) How much sweep power is developed by the output stage of a typical large-screen color set?
(b) What is the efficiency of the circuit (in percent) if the output transistor draws 1 A at 20 V dc?

4. How is it possible to operate the horizontal oscillator from scan power when one depends on the other?

5. When there are symptoms of overload, why is it often desirable to bridge a CB or fuse with a lamp? Give two reasons.

6. A shorted output transistor may destroy the flyback transformer, and vice versa. Explain.

7. List several symptoms that might be caused by shorted coils in the flyback.

8. Explain the purpose of an overvoltage protective circuit.

9. Explain *blooming,* listing several possible causes.

10. Explain spark gaps and where and why they are used.

11. What is meant by B-boost and how is it developed?

12. Name three ways to obtain focus voltage for a color set.

13. A typical output transistor is supplied with reverse bias. What makes it conduct?

14. Briefly explain the operation of an SCR sweep system.

15. What is meant by yoke ringing and what purpose does it serve?

16. Identify the circuit configuration where the output transistor collector goes to ground via the H.O.T., and voltage is applied to the emitter.

17. Give three trouble symptoms resulting from poor HV regulation.

18. Explain how and why a horizontal sweep problem can sometimes affect other receiver functions such as sound, vertical sweep, and so on.

19. (a) Why does the HV vary as the brightness control is adjusted?
(b) Is the HV maximum with a bright screen or a dark screen?

318 HORIZONTAL OUTPUT AND HV SECTION

20. Testing HV by arcing is not advisable for solid-state receivers. Explain why.

21. A HV probe will measure the HV at the CRT but not at the rectifier input. Why?

22. Why are several high-value resistors sometimes used in series in the HV circuit instead of one resistor of the equivalent value?

23. Shorted windings in a yoke or flyback cause an increase in output transistor current. Explain why.

24. State several advantages of scan power over dc power obtained from the LVPS.

25. Sparking is observed in a newly installed CRT. Should it be replaced? Explain.

26. The output transistor conducts for only a part of the horizontal trace interval. Explain how the rest of the trace is developed.

27. (a) What is the function of a damper diode?
(b) Explain its operation.

TROUBLESHOOTING QUESTIONNAIRE

1. State the probable cause of arcing in the CRT only when (a) brightness is turned up full; (b) when the screen is dark.

2. When the CRT filament is supplied from a winding on the flyback, how would you proceed if (a) heater glow is visable? (b) there is no visible glow? Explain your reasoning.

3. Give four possible causes of poor focus.

4. What would you look for if there are flashes on the screen and a strong smell of ozone?

5. How would you proceed when the HV is excessive but the set appears to work OK?

6. A HV rectifier measures infinity in both directions on an ohmmeter. Is it defective?

7. Explain two methods for checking for shorted coils in a yoke or flyback.

8. Give three trouble symptoms that may be caused by leakage of a yoke coupling capacitor.

9. How would you determine if loss of raster is caused by a sweep or HV defect, or from activation of the shutdown circuit?

10. Symptoms: No raster, HV and all other CRT voltages check OK, the CRT checks bad and its heater does not glow but has continuity, and the neck of the tube is warm. State the possible cause. *Hint:* The HV lead is resting against the tube neck.

11. It is unwise to connect an EVM or CRO to the base or collector of a horizontal output transistor in the presence of high amplitude pulses. Explain why.

12. A HV reading of negative polarity is obtained after replacing a HV rectifier and there is no raster. State the probable cause.

13. What would you check if (a) there is normal pulse voltage from the output transistor but no HV dc? (b) the HV checks OK but there is no focus voltage at the CRT? (c) the HV drops to zero when the lead is connected to the CRT?

14. What would you do if good focus is obtained only when the HV is adjusted higher than normal?

15. The output transistor and flyback overheat. State several possible causes and how you would check for each.

16. No dc is measured at the collector of the output transistor.
(a) Give several possible causes.
(b) What if there is voltage on the emitter?

17. There is low HV and very little drive measured at the base of the output transistor. Give two possible causes and how you would proceed.

18. When there are severe HV fluctuations with changes in pix brightness, how would you determine if this is caused by a HV defect, the HV regulator, or excessive loading due to excessive CRT beam current?

19. A good "beam current" arc is obtained between the HV lead and the CRT. State the probable cause if (a) you are unable to darken screen with the brightness control; (b) there is no raster, and the arc does not change with the brightness control.

20. Explain why the CRT ultor contact should always be grounded *before* removing or connecting the HV lead or handling the tube.

21. In the SCR sweep system, the CB trips unless the anode of the trace SCR is grounded. What part of the circuit is responsible?

22. There is little or no pulse voltage at the collector of the output transistor. The flyback is cut loose from the transistor and a substitute inductor is used as a load. Where would you look for the trouble if (a) a good pulse is now obtained? (b) there is still no pulse at the transistor output? Explain your reasoning.

23. There is corona discharge between the flyback and the HV cage. What would you do?

24. Give four possible causes where there is a good pulse voltage at the rectifier anode but little or no HV dc at the CRT.

25. A set is troubled with sporadic arcing or corona but only in damp weather. What would you do about it?

26. What are the symptoms of excessive HV? State possible causes.

27. How would you check the operation of the HV regulator by monitoring both the B voltage feed to the output transistor and the HV at the CRT?

28. A set is troubled with poor HV regulation and blooming, and the HV coil of the flyback measures several thousand ohms. What is wrong?

29. An output transistor has B voltage at the emitter terminal and zero dc at the collector. State two possibilities.

Chapter 10

SYNC SECTION

10-1 INTRODUCTION

To reproduce a picture, beam deflection at the receiver must be *exactly* in step with raster scanning of the camera at the TV station. To this end, vertical and horizontal sync pulses are generated at the station and transmitted as part of the *composite video signal* (see Fig. 1-6). Without control by these pulses, the sweep oscillators would be free running, with a resulting loss of sync.

In the video amplifier section of the receiver, some of this signal that feeds the pix tube is extracted at the point of *sync takeoff,* which may be anywhere beween the video detector and the CRT input (see Fig. 10-1).

This composite signal (video and sync information) is then fed to a *sync separator* stage (also known as a *clipper* or *stripper*), whose purpose is to separate the pulses from the unwanted video, which, if permitted to reach the sweep oscillators, would result in random triggering and loss of sync.

Noise impulses can also upset the sync, and most sets provide some form of *noise immunity* circuit as some protection against this problem. As shown in Fig. 10-1, a noise-prevention stage (sometimes called a noise gate, noise inverter, or noise canceller) operates in conjunction with the sync separator.

Although most modern sets have only a sync separator and a noise circuit, other sync arrangements are sometimes found. Some sets follow the separator with an amplifier stage as shown in Fig. 10-1. In some cases, *two* separators are used for more complete stripping of the pulses from the video.

Coexisting at the output of the sync separator are both *vertical* and *horiztonal* pulses, which much be separated *from each other*. This is accomplished by simple *R/C* networks: the *integrator* (as described in Sec. 7-10) and the *differentiator* (Sec. 8-6). Sharply spiked vertical pulses at the integrator output trigger the vertical oscillator *directly.* Control of

321

the horizontal oscillator is *indirect* by the horizontal pulses from the differentiator and the AFC circuit (see Sec. 8-2). With some AFC circuits, a *phase splitter* or *inverter* is used.

Depending on circuit configurations, the *polarity* of the pulses feeding the sweep circuits may be either positive- or negative-going, and any change in normal polarity results in a complete sync loss. Other pulse considerations are amplitude (which is quite critical), waveshape (a sharply spiked pulse is essential for precise control), and freedom from noise or other interference.

Sync circuits are often combined with other stage functions in a single IC or module, especially *AGC,* although there is no direct relationship between the two, except they are both supplied with a sampling of the composite signal. Don't confuse *beam synchronization* with *color sync,* two completely unrelated functions.

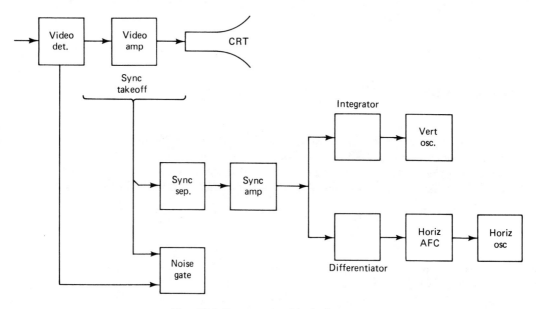

Fig. 10-1 Sync section block diagram.

10-2 SYNC SEPARATOR

The sync pulses represent about one-third of the total composite video signal, extending beyond the highest levels of video and the blanking pedestal (see Fig. 1-6). The separation of pulses from the video is made possible by this difference in amplitude. It is desirable that pulses be clipped at or near the blanking level. If only the sync *tips* are recovered, the result is unstable sync, especially on weak signals. If the clipping level is too *low,* the *video* peaks will also get through, triggering the sweep oscillators at random intervals.

A simple diode rectifier circuit can be used for sync clipping, as shown in Fig. 10-2a. The diode is reverse biased (by signal rectification), and conduction occurs only for the sync tips of the positive-going incoming signal. Signal polarity depends on which way the diode is connected. With a diode, there is no polarity reversal between input and output.

As a sync separator, a transistor is preferred over a diode since it also provides a signal in-

10-2 Sync Separator

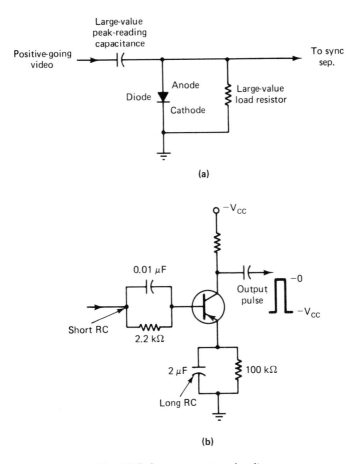

Fig. 10-2 Sync separator circuits.

crease. A typical basic circuit is shown in Fig. 10-2b. Pulse separation takes place in the B/E circuit, which is biased to cutoff or beyond (class B or C), and the pulses are amplified in the collector circuit.

Signal Input. Loading of the video amplifier by the first sync stage can adversely affect its frequency response and degrade the pix. This is prevented with some form of isolation between the sync input and the point of sync takeoff. If sync takeoff is a high-level point (such as the CRT input), a high-value isolating resistor may be used. Sometimes the resistor is part of a voltage-dividing network for biasing the sync transistor. Some sets use a sync amplifier immediately following the sync takeoff as a buffer stage. Loading effects are reduced if the sync takeoff is at a low-impedance point such as the detector, but the low signal level requires greater amplification in the sync stages that follow.

Signal *polarity* is another consideration, as determined by the type of transistor used in the first sync stage. An NPN requires a positive-going signal to turn it on. The reverse is true for a PNP. This assumes the signal is fed to the transistor *base*. The opposite condition prevails if fed to the emitter. Pulse polarity must also be considered in all subsequent sync stages to obtain the desired polarity at the input of the vertical oscillator and the horizontal AFC.

Sync Separator Biasing and Pulse Separation. It is desirable for the bias to change in accordance with signal strength to maintain the same relative clipping level for both weak and strong signals. This precludes the use of fixed bias, which would either clip much of the pulse amplitude or allow video to get through. Self-bias, with a resistor and capacitor in either the emitter or base circuit, is often used, but since a tube or transistor *cannot cut itself off,* most circuits also depend on *excitation bias* as developed by B/E rectification of the incoming signal. Bias developed in this manner automatically adjusts to the signal level.

Figure 10-2b represents a typical circuit. Most of the bias is established by the *IR* drop across the emitter resistor when the transistor is conducting. The emitter bias plus the excitation bias developed by rectification of the signal peaks maintains the transistor at cutoff in the absence of a signal and between pulses. This is a *reverse* bias. When it is overcome by the negative-going sync tips of the incoming signal, the transistor conducts (for the duration of the pulses), and the positive-going pulses appear at the collector output. Between pulses, the emitter bias is maintained relatively constant by the charge on the capacitor and the time constant of the circuit. In general, the greater the bias is, the stronger the signal that is required for conduction. The result is usually a weaker but *cleaner* pulse output. Also, how *much* of the sync tips that produce conduction is determined largely by the time constant of the B and E circuits. With a short TC, conduction may be excessive with video contaminating the output. With a long TC, conduction time may be short and the pulse output will be weak.

Saturation and Noise Clipping. Saturation is where the collector current has reached a *maximum, steady* value and further increases are not possible. Most sync separators operate with a fairly low collector voltage so the transistor will saturate at some predetermined signal level. The stage thus operates as a noise clipper or leveler where all signals (including noise and other interference) have the same amplitude. This can only occur, however, when the incoming signal is strong enough to drive the transistor into saturation. Noise pulses at the output are not *eliminated,* but being reduced to the same level as the sync, are less likely to create sync problems.

10-3 NOISE IMMUNITY

Interference (generally referred to as noise) not only causes a snowy pix and scratchy sound; it is also a cause of poor sync. The problem tends to be worse in noisy locations where the signal-to-noise ratio is low and the receiver is operating at maximum sensitivity due to AGC action. Whenever possible, serious noise problems should be overcome at the *source.* Where that is not possible, make changes to the antenna system, as appropriate.

In addition to *noise clipping,* various *noise immunity* circuits are found in modern sets. For strong impulse-type interference, the simple diode circuit shown in Fig. 10-3a can be quite effective. Noise pulses stronger than the video

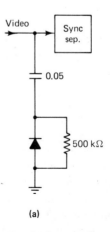

(a)

Fig. 10-3 Partial

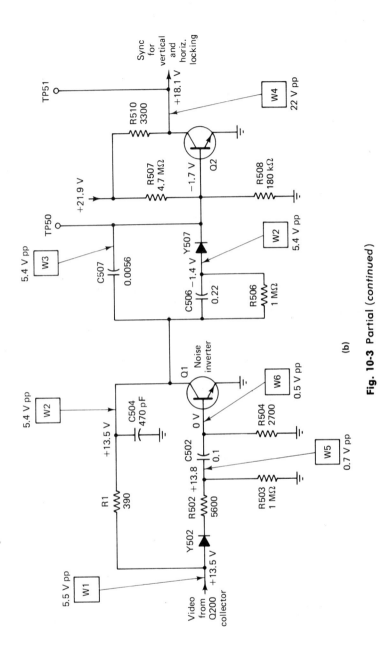

Fig. 10-3 Partial (continued)

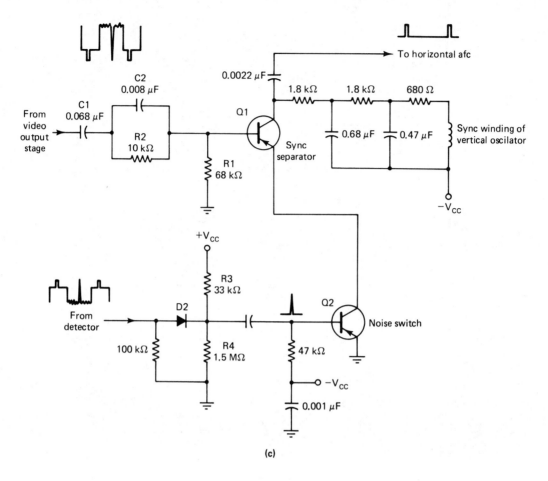

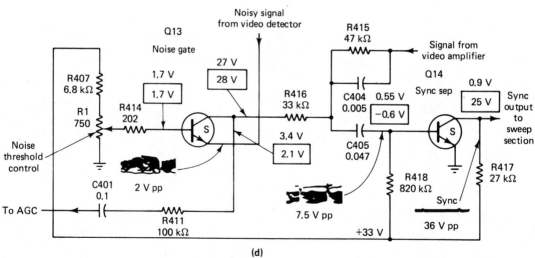

Fig. 10-3 Typical noise-immunity circuits.

are shorted to ground when the diode conducts. The circuit shown is for a negative-going signal. For a positive signal, the diode is simply reversed. In the absence of noise (diode not conducting) there is very little loading of the video. When warranted, the circuit can be easily added to most sets.

Noise Canceller-Inverter Circuits. Noise cancellation circuits are used in most sets. One version is shown in Fig. 10-3b. Here, the video input to the sync separator is via a noise inverter circuit. Normally (in the absence of noise pulses), *Q1* is biased to cutoff and the video gets to *Q2* via resistor *R1*. For low-intensity noise, *Q1* amplifies and *inverts the polarity* of the noise. At the collector of *Q1*, the inverted noise cancels out the noise in the video arriving via *R1*.

For *high-level* noise bursts, *Q1* is driven into saturation and noise *and video* are momentarily shorted to ground via the C/E junction. The normal stability and flywheel effect of the sweep oscillators in most cases is enough to carry them through these short periods without sync loss.

Some noise inverter circuits have *two* inputs. In Fig. 10-3c, high-amplitude video is fed to the sync separator *Q1* in the normal manner. The *noise switch* or gate (*Q2*) is supplied with low-level video of opposite polarity from the video detector. Note that the C/E junctions of both transistors are *in series*. Normally (in the absence of noise), *Q2* is forward biased into conduction, providing a path for the *Q1* emitter to ground, and *Q1* functions normally as a sync separator. The presence of strong noise bursts switches *Q2* off, opening the C/E circuits and disabling the sync separator momentarily. The diode *D2* aids in the operation by conducting only when the noise reaches a certain predetermined value.

Figure 10-3d shows another circuit having two inputs. A positive-going signal from the video amplifier is fed to the sync separator via an *RC* isolation network. This NPN transistor is self-biased (in a reverse direction) by excitation, and the transistor conducts only on the positive-going sync tips. Note the two bias voltages on the diagram: -0.6 V off channel (no pulses) and $+0.55$ V on channel (with sync pulses). The noise gate *Q13* is normally turned off and conducts only when a strong noise pulse is received. A negative-going noise signal is fed to the emitter *Q13*, which has the same effect as a positive signal fed to the base of a transistor. When *Q13* conducts (from noise pulses), the negative-going noise impulses at the collector output (no inversion with this configuration) *cancel out* noise in the positive-going signal feeding *Q14*. Note the noise *threshold control* (*R1*) as found on some circuits to establish the noise level at which *Q13* operates.

10-4 SYNC AMPLIFIER

For optimum sync stability, some sets use one or more sync amplifiers to increase the pulse level. Other considerations are the signal level at the point of sync takeoff, signal gain of the sync stages and the level required for triggering the sweep oscillators, and signal polarity at the various points between the sync takeoff and the oscillators. In some cases a sync amplifier is used *ahead of* the sync separator to isolate it from the video section (to prevent loading) and for signal gain when the point of takeoff is the low-level detector stage. The circuit is much like a *video* amplifier except for the absence of peaking coils. Although serving no purpose, the stage also passes video information.

In some sets, a sync amplifier comes *after* the separator, in which case it amplifies pulses *only*. Other sets may use selective amplifiers: a *vertical* sync amplifier with a *long* time constant for the benefit of the vertical pulses, a horizontal amplifier having a shorter TC, or *both*.

Whereas a sync separator operates at cutoff (with low collector voltage so it will saturate readily), a sync amplifier usually operates as a conventional class A amplifier with a small *for-*

328 SYNC SECTION

ward bias and a fairly high collector voltage. Also, with a separator the pulse input turns the transistor on, but with an amplifier the signal drives it toward cutoff.

Depending on its operating conditions, a sync amplifier may perform other functions according to its name, such as sync *leveling* or *limiting,* so that the amplitude of pulses will be relatively constant regardless of variations of the received signal, and *waveshaping,* where it is desired to square off the pulses.

Whether or not a sync stage inverts the signal between input and output depends on the circuit configuration (e.g., common emitter, common base, or common collector).

10-5 INTEGRATOR AND VERTICAL SYNCHRONIZATION

It is not enough that the vertical and horizontal sync pulses be separated from the video; they must also be separated from each other. This is accomplished by the integrator and differentiating networks. The integrator is essentially a low-pass filter that passes the low-frequency vertical pulses, while rejecting the higher-frequency horizontal pulses. The differentiator is described in Sec. 10-6.

A basic single-section integrator consists of nothing more than a resistor and capacitor, as shown in Fig. 10-4a. Although typical of that found in most modern sets, many older models use two or more such circuits in *cascade* for greater effectiveness. The integrator may consist of individual components, but more often it is a self-contained encapsulated unit or part of an IC.

As described in Sec. 7-10, part of the composite video signal consists of a complex series of pulses that occur after the completion of each scanning field immediately prior to (and during)

the relatively long vertical retrace interval. The vertical oscillator is triggered by a single, sharply-spiked pulse developed from the integration of serrations of the vertical sync pulse as shown in Fig. 10-4b. The vertical pulse, in effect, actually consists of a train of six individual pulses. Their purpose is twofold; individually, they function as *horizontal* sync pulses to control the *horizontal* oscillator during vertical retrace. Collectively, they develop an accumulating voltage for triggering the *vertical* oscillator.

The operation of a single-section integrator is explained as follows. *All* pulses (vertical, horizontal, and equalizing pulses) are present at the integrator input. With the arrival of the first (of the six) serrated pulses, the integrator capacitator *C1* becomes charged, as shown in Fig. 10-4b. Each successive pulse *adds to* the charge buildup. The circuit has a long time constant, and considerable voltage is developed by each of the wide vertical pulses, whose duration is about 27 μs. During the short interval (about 5 μs) between the serrations, very little of the voltage is lost. Normally, a triggering voltage for the vertical oscillator is developed after about two or three pulses. The arrival of any remaining pulses serves no purpose except to maintain control over the *horizontal* oscillator. Which of the six pulses actually triggers the oscillator is determined by signal strength. For a strong channel it may occur with the first pulse; for a weak station, it may require all six pulses to do the job. Where the signal is extremely weak, there may be no control at all and the oscillator will lose sync.

Because of the long TC of the integrator, the short duration of the narrow horizontal and equalizing pulses, and the long interval between these pulses, they do not contribute to the charge on the integrator capacitor(s) since each small charge quickly decays in the time between pulses.

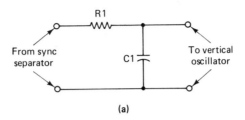

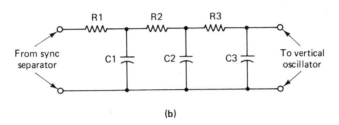

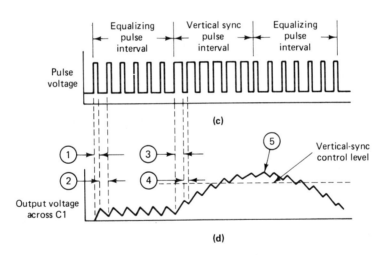

Fig. 10-4 Integrator network: (a) integrating circuit; (b) cascade integrating circuit; (c) vertical sync-pulse group; (d) output of integrating circuit.

From:
Television Service Manual, ©1977, by Robert G. Middleton, used with permission of the publisher, The Bobbs-Merrill Company, Inc.

329

10-6 DIFFERENTIATING NETWORK AND HORIZONTAL (AFC) CONTROL

As with the integrator, the differentiator consists of a resistor and capacitor. The only difference is their relative values and that the positions of the two components are interchanged: the capacitor is in series (like a regular coupling capacitor) and the resistor is shunted *across* the circuit, like a conventional load resistor. Basically, this *RC* combination represents a high-pass filter and waveshaper whose purpose is to produce sharply spiked pulses from *all* sync pulses, not only the regular horizontal sync pulses, but also *all* pulses superimposed on the vertical blanking pulse, that is, the equalizing pulses, vertical sync pulses (serrations), and regular horizontal pulses. The object is to control the horizontal oscillator (via AFC action), as described in Chapter 8.

Both vertical and horizontal pulses are present at the input of the differentiator. Since both have the same amplitude, separation of one from the other makes use of the difference in their *duration*.

The circuit action (see Fig. 10-5) is described as follows. Sync pulses are essentially square (or rectangular) waveforms, and the abrupt increase in voltage represented by the *leading edge* of a pulse *rapidly* charges the capacitor C1 to the peak value of the input. The relatively high charging current in turn develops a voltage across the resistor, but *only while the current is flowing*. This is the pulse voltage that is passed on to the next stage. When the capacitor becomes charged, current *ceases,* and the *IR* drop across the resistor *gradually* drops to zero, as shown in Fig. 10-5b. With small values of *R* and *C* (short TC), the rise time is almost instantaneous. The time required for drop-off corresponds with the duration of the input pulse (about 5 μs for horizontal pulses and 27 μs for vertical pulses).

This difference is of no consequence since it is the *leading edge of all pulses* that is used for controlling the horizontal oscillator.

In the time *between pulses,* the capacitor discharges through the resistor, developing a pulse of *reversed polarity*. These pulses (of opposite polarity) serve no purpose and have no detrimental effect on AFC control, since they occur at the wrong times, the scan periods when the oscillator is insensitive to change. In Fig. 10-5b and c, note how horizontal pulses are developed from the vertical pulses.

REMINDER: _____

Although every equalizing pulse and serrated vertical pulse results in a *spiked* differentiated pulse, for interlaced scanning only every *alternate* pulse is used for horizontal control while scanning the odd and even fields.

The differentiated pulses are fed directly to the horizontal AFC circuit or (depending on the type of AFC) to a sync inverter stage. These circuits and horizontal synchronization are described in Chapter 8.

10-7 TYPICAL SYNC CIRCUIT

The sync section of Sample Circuit B at the back of this book is mostly conventional. The composite video signal is applied to the base of the noise gate Q 505 (via the isolation components Y 502 and R 502, and coupling capacitor C 502), and also to the base of Q 510, the sync separator, at which point noise cancellation takes place. Note the amplitude of the output signal from the gate (5.5 V p-p) and the output of the sync separator (15 V p-p), which are *pulses only,* after being stripped from the video. From the collector of Q 510, the vertical and horizontal pulses are split to control their respective sweep circuits contained in the countdown IC (IC 501), as shown in Fig. 7-6.

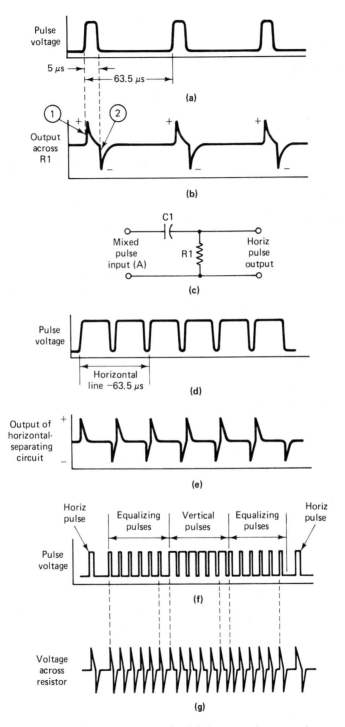

Fig. 10-5 Differentiator network: (a) horizontal sync pulses; (b) output of differentiating circuit; (c) differentiator circuit; (d) vertical-sync pulse with serrations; (e) output of horizontal-differentiating circuit; (f) vertical pulse group; (g) output of horizontal-differentiating circuit.

10-8 TROUBLESHOOTING SYNC STAGES

In general, there are three basic sync problems, those involving *both* vertical and horizontal sync and those affecting one *or* the other. Although the circuitry is relatively simple and uncomplicated, sync problems are not always easy to *accurately diagnose* or resolve. In some cases this is because the symptoms are misunderstood. For example, unstable vertical *or* horizontal sync, where one is more critical than the other, may actually be a problem affecting *both*, but to a different degree. Then there is the problem of deciding whether sync loss is because of an oscillator defect or a problem with the sync pulses. To further complicate matters, sync loss is often related to signal strength, sometimes showing up on *strong* channels and sometimes with the *weakest* stations. Another difficulty is that sync loss can often be caused by stages and conditions *other than* the sync circuits themselves, in particular, video amplifiers and AGC, and in some cases problems with hum voltage, arcing, and noise or other interference. Thus, it is apparent that a careful analysis of the symptoms, followed by a few preliminary tests, will go a long way in resolving sync problems and will reduce the chances of being sidetracked into a dead end.

10-9 LOCALIZING THE FAULT

Sync loss is often a matter of *degree*. There may be a *complete loss,* where the pix continuously rolls or is torn up by diagonal bars, or the trouble may be less severe, one of many possible sync-related symptoms which are often more difficult to correct. The troubleshooting approach in each case depends on the particular symptoms as described in the following paragraphs. See Fig. 10-6 and Table 10-1.

10-10 LOSS OF VERTICAL AND HORIZONTAL SYNC

Sync loss is often accompanied by *other* symptoms that should be carefully considered. Such conditions (which often provide a clue in troubleshooting for loss of both vertical *and* horizontal sync) are treated separately.

- **Sync Loss but Normal Pix and Sound**

Normal pix in this sense is in reference to contrast *only,* an indicator of signal strength, and even with an unrecognizable, out-of-sync pix, contrast can be judged by the darkness of the bars and portions of the pix. This depends on

Table 10-1 Sync Troubleshooting Chart

Trouble Condition	Probable Cause	Remarks and Procedures
Loss of vertical sync (horizontal sync OK)	Trouble with vertical oscillator or its input circuits. See Chapter 7.	May be complete loss of vertical sync, or sync instability. Symptoms may take many forms (e.g., vertical roll, vertical jitter, poor interlace, multiple pixs). See Table 7-1.

10-10 Loss of Vertical and Horizontal Sync

Table 10-1 Sync Troubleshooting Chart (*cont.*)

Trouble Condition	Probable Cause	Remarks and Procedures
Loss of horizontal sync (vertical sync OK)	Trouble with horizontal oscillator or AFC stages. See Chapter 8.	May be complete loss of horizontal sync, or sync instability. Symptoms may take many forms (e.g., multiple pixs, horizontal roll, tearing, floating pix, horizontal pull or bending, flagwaving, oscillator squegging). See Table 8-1.
Loss of vertical *and* horizontal sync and weak pix	Sync circuits are probably OK. If weak pix without snow, investigate video amplifier(s). For weak pix with snow, suspect tuner, antenna system, or reception problems.	Sync will probably clear up when cause of weak pix is corrected. Try adjusting AGC threshold control. May be complete loss of sync *or* sync instability.
Loss of vertical *and* horizontal sync. Pix *contrast* is normal	May be caused by pulse clipping in video amplifier, loss of pulses in sync stages, or interference.	Check pulses at sync takeoff and observe hammerhead display on the screen. Signal trace pulses through sync stages.
No vertical or horizontal sync with a negative pix	Troubles in video section causing reversed signal polarity. Misadjusted AGC.	Sync will be restored when cause of negative pix is corrected. Adjust AGC.
Unstable vertical and horizontal sync with excessive pix contrast and smearing.	Too much video gain because of AGC troubles or improper setting of threshold control. May also be poor low-frequency response of video amplifier.	Sync will be restored when other symptoms are cleared up.
Unstable vertical and horizontal sync with streaks or flashes on screen	May be LV or HV arcing, or noise or other interference.	Correct the problem and sync will clear up.
Unstable vertical and horizontal sync with hum bars on the screen	Poor filtering of dc power circuits.	Note whether hum bars are present at all times or only on a pix.
Intermittent sync	Many possibilities, particularly bad PC soldering.	Not to be confused with sync *instability*.

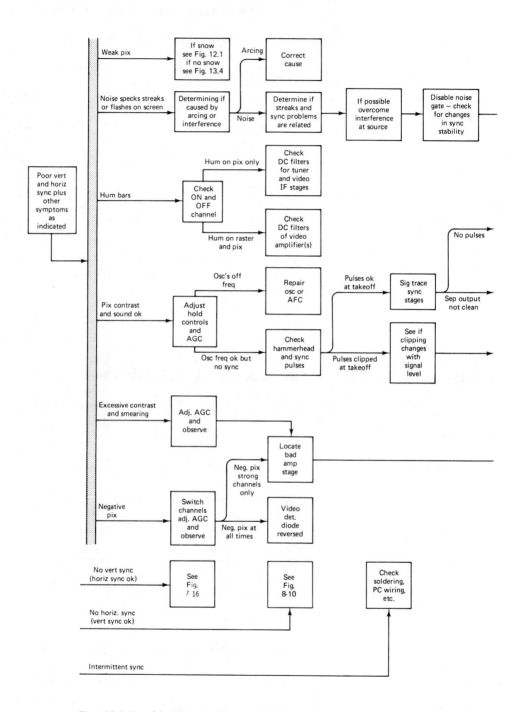

Fig. 10-6 Troubleshooting flow diagram.

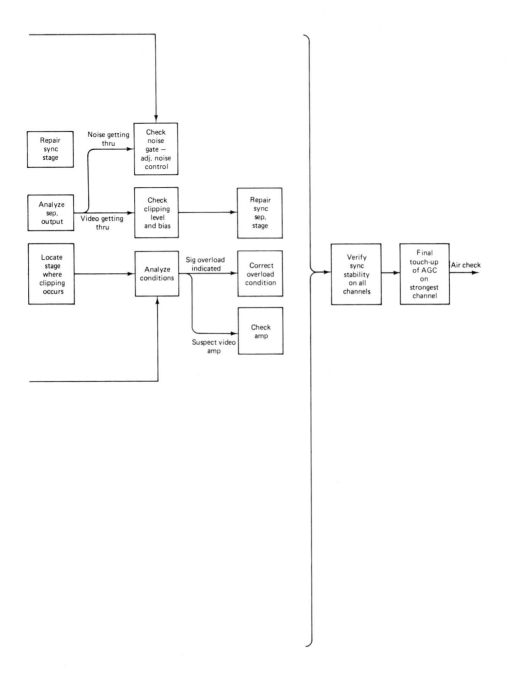

335

the relative settings of the brightness and contrast controls.

Start by adjusting the vertical and horizontal hold controls. *Both* may have been tampered with, or you may find it is a problem with only *one* circuit, the vertical *or* the horizontal, calling for a different approach.

REMINDER: ─────────────────────

For the vertical, it should be possible to make the pix roll *in both directions* and to stop a pix (momentarily) near mid-setting of the hold control. If it is not, try adjusting the height and linearity controls, which often have a bearing on frequency. If unsuccessful, the *oscillator* stage is at fault.

For the *horizontal*, it should be possible to get diagonal bars sloping in *both* directions and to stop a pix (momentarily) near mid-setting of the hold control and frequency adjustment if there is one. If unsuccessful, the horizontal oscillator or AFC is at fault. For oscillator troubles, see Chapter 7 or 8 as appropriate.

If both oscillators are vindicated, the problem involves the sync pulses, which have either been lost or attenuated at some point ahead of the sweep input circuits. A good starting point is to check for dc power supplying the sync stages. If OK, the next logical step is to check the signal at the sync takeoff point, which can be quickly determined from the schematic or by tracing backward from the first sync transistor to the video section.

Since the pulses are part of the composite video signal at the point of sync takeoff, it would seem that one could not be affected without the other, and considering in this case that the video must be normal since the pix is also. But this is not necessarily true, mainly because the pulses represent the *peak* of the signal where they are vulnerable to *clipping* or *compression* when a transistor approaches saturation or cutoff. This does not affect the *video*.

Figure 10-7a shows a normal signal in the video section. In Fig. 10-7b, the pulses are compressed (or weakened). Should this occur in a stage prior to sync takeoff, the result is unstable sync, especially on weak stations. If it occurs *downstream* (in the video amplifier) from the point of sync takeoff, sync stability is not affected.

In Fig. 10-7c, the pulses are completely *clipped,* and if this occurs ahead of the sync takeoff, the result is *complete loss* of both vertical and horizontal sync.

Pulse clipping in a video amplifier is usually caused by *video overload* (overdrive), and there are two possibilities: troubles in the video stage itself resulting in clipping or compression at even normal signal levels, or *excessive* signal drive (with which the amplifier cannot cope). This is due to excessive amplification by some previous stage(s), in most cases caused by problems with the AGC, which is responsible for regulating the overall sensitivity of the receiver. More on this in Chapter 11 and Sec. 14-37. There are two methods of checking for weak or missing pulses caused by troubles *prior to* the point of sync takeoff.

Method 1, with a CRO. Connect the scope to sync takeoff and adjust the receiver for a moderately strong station. The CRO display should appear as in Fig. 10-7a with the sync tips extending about 25% beyond the blanking level. A quick-look estimate can be made by observing the two bright horizontal retrace lines and their separation, which represents the blanking and sync tip levels. Low-amplitude pulses mean signal *compression* is taking place. Complete *loss* of pulses (nothing beyond the blanking level) is a sign of sync *clipping.* In both cases the sync is affected to some degree but with little or no effect on the *video signal* reaching the CRT. To help localize a faulty amplifier stage, connect the scope to various checkpoints, work-

10-10 Loss of Vertical and Horizontal Sync   337

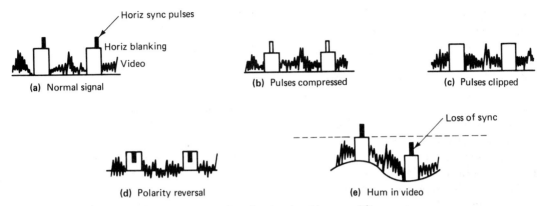

Fig. 10-7 Pulse clipping in video amplifiers.

ing back toward the video detector while looking for a normal display.

In most cases of sync clipping or compression, the relative amplitude of the pulses (as compared to the video) will change with variations of signal input. This can be determined by observing the display as the signal strength and/or the gain of the receiver is varied in some manner, such as by adjusting the contract or AGC threshold control, orienting a rabbit-ears type of antenna, or switching channels. The object here is to determine if there is clipping or compression and, if so, *where* it takes place and how it relates to signal strength.

Adjust the receiver for normal pix contrast. If there is clipping or compression under these normal signal conditions, chances are the amplifier stage is in trouble. If it occurs only with stronger signals, it is a case of signal overdrive (video overload). The usual cause of this is AGC problems. Note the setting of the threshold control for best sync stability. If the control must be turned well down (for lowest receiver sensitivity) to obtain good sync, it is a case of signal overload. If the control must be at maximum, then the pulses are simply weak and the pix contrast will be low. To verify suspected AGC problems, try clamping the AGC bus with a bias box, as explained in Chapter 11.

Since there is a limit to the amount of signal *any* amplifier can handle before overloading and clipping occur, the trick is in deciding normal and maximum permissible levels at the input of the stage where the overload is taking place. And, once again, if clipping occurs with a normal input signal, the amplifier stage is at fault. If there is trouble only with the strongest signals, it could be either the amplifier or it is being overdriven.

While monitoring the scope display and varying the signal input to the set or the receiver gain, as just explained, indications of pulse clipping or compression usually worsen with increases in the signal, going progressively through the different phases shown in Fig. 10-7a, b, and c. If carried to the extreme, the pulses are compressed to the blanking level, and then appear to be swallowed up. At this point the entire signal may become *inverted,* resulting in complete loss of sync and a *negative picture.*

Method 2, Observing Pulse and Blanking Bars on the Receiver Screen. Under normal conditions, blanking pulses drive the pix tube to cutoff, darkening the screen during vertical and horizontal retrace intervals. Sync pulses extending beyond the blanking level (in the "blacker-than-black" region) are permitted to reach the

CRT, where they serve no purpose, but neither do they affect the pix since they also occur during beam cutoff. Blanking, sync, and other pulses produce bars on the screen during retrace and can be used to confirm their presence or absence and to evaluate their relative strength. The procedure is as follows.

Misadjust the vertical hold control to make the pix roll slowly downward. Observe the blanking and sync bars (as shown in Fig. 10-8c) that separate the pix fields. The wide portion of the horizontal bar represents the blanking bar, and the long and darker, narrow bar, the vertical sync pulse(s). Leading and lagging *equalizing pulses* produces what is known as the *hammerhead,* as shown. The vertical bar seen in Fig. 10-8a can also be observed. It becomes visible by misadjusting the *horizontal* hold control.

Adjust brightness and contrast controls until the blanking bar(s) are the same shade of gray as the darkest portions of the pix. Under these conditions, strong sync and equalizing pulses show up *black* against the gray background, as shown in Fig. 10-8a. With a color transmission, the color burst also shows up as a gray vertical bar.

Complete loss of pulses (as caused by clipping in the video amplifier) is apparent when no sync bars are visible within the blanking bars. *Weak* sync pulses (as caused by sync compression) are indicated by a sync bar not much darker than the blanking bar (see Fig. 10-8b and c). Repeat the observations under different conditions by switching channels and adjusting the AGC threshold control.

In some cases, vertical retrace blanking pulses can make it difficult to evaluate the pulses. This can be overcome by disabling the blanking circuit with a large capacitor connected at the pulse input to the CRT to ground.

REMINDER: _____

This test does not isolate a defect, but it does provide a quick evaluation of pulses as they exist at the point of sync takeoff. When the sync bar is weak or missing from the display, the cause will usually be found in the video amplifier or AGC stages. When pulses appear normal, troubleshoot the sync stages *beyond* the sync takeoff point.

• **Signal Tracing the Sync Stages**

After confirming the presence of strong sync pulses at the sync takeoff, the next logical step is to signal trace the sync section (which is probably responsible when there is loss of both vertical *and* horizontal sync). This includes all sync separators, amplifiers, and noise circuits. The object is to localize the trouble to the stage where the pulses are lost, weakened, or degraded in some manner.

Start by examining the schematic to determine exactly what stages are involved and what kind of response should be expected at each checkpoint. Signal tracing can be performed with a CRO, an ac EVM, or some type of signal tracer, although the scope is the instrument of choice. Connect the CRO to the sync takeoff or the input of the first sync stage. Select a channel that gives a good pulse display. If sync is OK on some channels only, choose a channel where sync stability is marginal. If necessary, adjust the AGC threshold control, which is usually quite critical. Either vertical *or* horizontal pulses can be observed (by choosing the proper timing (sweep) frequency on the scope.

Alternately connect the scope to the input and output of each sync stage. Besides checking for loss or attenuation of the pulses, there are several other things to watch for. Make sure the output of sync separator(s) is clean, that is, free of video information and noise. Check the display on both weak and strong stations. Verify signal *polarity* at each checkpoint in accordance with the schematic. Make sure there is no hum voltage, as indicated by a thickening of the display (with the scope set for 7867 kHz) or wavy undulations of the base line. When sus-

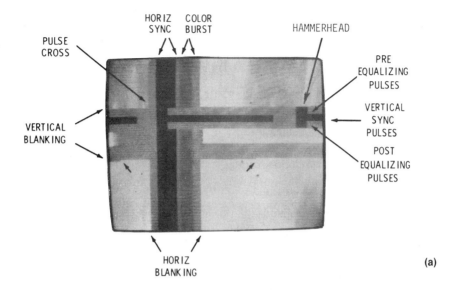

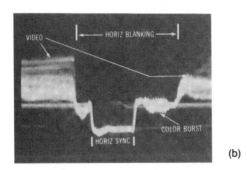

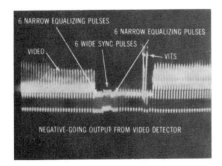 

Fig. 10-8 Hammerhead display. The composite video waveform [(a) and (b)] obtained from the video detector has been locked at 60 Hz and widened for clarity.

pected, check further for hum with the CRO set for 30 Hz. Check for ac ripple voltage on the dc supply feeding the sync stages.

Typical pulse amplitudes for various checkpoints are shown in Fig. 10-3b and d, and Sample Circuit B on page 823.

Actual values depend on the particular set and variations in signal strength and receiving conditions. Note the fairly high outputs from the sync separators. Some separator circuits using the emitter-follower configuration, however, contribute little or no gain.

Follow the signal up to the integrator and differentiator networks and, if necessary, the inputs of the vertical oscillator and AFC circuit. A leaky or shorted capacitor, diode, or transistor in this area will often upset both vertical and horizontal sync. Don't mistake the signal generated by the vertical oscillator for a vertical pulse. Try adjusting the vertical hold, which should *only* affect the oscillator display; also note that *sync* pulses exist only on an active channel.

To determine if a noise gate is upsetting the sync, try disabling it (by shorting the base to emitter) to see if the sync improves. Most sets will operate without the noise circuit operating.

An alternative approach is to signal trace in reverse, working from the oscillator inputs back toward the sync takeoff. When sync pulses check normal throughout the circuit, investigate the possibility of *multiple troubles* involving *both* the vertical and horizontal sweep circuits.

- **Poor Sync and Weak Pix**

Chances are there is only one underlying cause. Correct the loss of contrast and the sync will probably clear up. There are many possible causes, as described in the following chapters. Usually it is due to loss of signal strength caused by reception problems, the antenna system, trouble in some stage between the antenna input and the sync takeoff, or an AGC defect. Try adjusting the AGC threshold control. Signal trace

up to the sync takeoff. Troubles between the takeoff point and the CRT input will affect the *pix* but *not the sync*. If the pix is *snowy*, the cause is probably the antenna system or the tuner.

- **Unstable Vertical and Horizontal Sync, Pix Normal**

The sync disorder may manifest itself in any number of ways as described in Table 10-1. Sync is usually lost momentarily when the set is first turned on and when switching channels. Adjustment of the hold controls is usually critical. Signal trace sync stages as just described, checking for *weak* pulses rather than missing pulses. Also watch for excessive noise or hum voltages. Make sure the signal is normal at the sync takeoff with no indications of pulse compression. Try adjusting the threshold control and any noise immunity control that is provided.

- **Negative Pix and Loss of Sync**

A negative pix is when pix phase is reversed; that is, white objects appear black and black objects, including the blanking and sync bars, show up white. It may be caused by a polarity reversal in a video amplifier stage or a video detector diode connected in backward. The reversed signal also causes loss of vertical and horizontal sync if it takes place prior to the point of sync takeoff. Sometimes it shows up only on the strongest channels, and may change when the fine tuner, contrast, or AGC threshold controls are adjusted. Poor AGC action is usually the cause.

- **Poor Sync with Hum Symptoms**

The hum (ac ripple problem) may produce a wavy raster and/or hum bars on the screen. Check for ac ripple voltage on the dc power source feeding the video and sync stages.

• **Noise and Interference Problems**

This is a common cause of sync loss, especially in a weak signal area when the set is operating at maximum sensitivity. RF interference (often caused by amateur radio or CB stations) produces a bar pattern on the screen. A streaky pix indicates car ignition, electrical appliance interference, or arcing within the set.

10-11 POOR VERTICAL SYNC

Assuming the horizontal sync appears normal and the vertical pulses are good up to the integrator input, the vertical sync problem is localized to either the integrator or the oscillator. Check for pulses at the integrator output. If they are weak or missing, check the integrator and other components in the area.

REMINDER:

The oscillator signal may be mistaken for pulses. If pulses reaching the oscillator appear normal, try readjusting the height and linearity controls since their relative settings often have a bearing on sync stability. Normal sync is when the pix locks *abruptly* when made to roll slowly upward with the hold control. *Marginal* sync is when the roll is almost continuous with only a brief hesitation between frames. When adjusting the hold control, it should be possible to make the pix roll in *both directions* and to stop the roll (at least momentarily) near the mid-setting of the control. If this cannot be done, troubleshoot the oscillator as in Chapter 7.

10-12 POOR HORIZONTAL SYNC

As with vertical sync, there may be a *complete* loss of sync or some degree of sync *instability*. Normally, when adjusting the hold control, a pix should lock in when the oscillator is slightly off frequency as indicated by at least three or four sloping bars. With *marginal sync,* the hold control has a narrow lock-in range and is very critical; the pix may be unstable with a tendency to float or waver, and the pix does not lock (if at all) until the oscillator is almost on frequency with only one or two sloping bars.

Start with a critical adjustment of the horizontal hold. It should be possible to obtain bars sloping in either direction for the opposite extremes of the control and a single pix at mid-setting. If it is not possible, try adjusting the slug-tuned horizontal frequency coil if one is provided. If you are unable to obtain a single pix (at least momentarily) or if it occurs at one *extreme* of the control(s), the oscillator and/or the AFC circuit are suspect. Troubleshoot as described in Chapter 8.

Where the oscillator function seems normal, signal trace the horizontal pulses from the differentiator network to the AFC input. If the AFC is preceded by a phase inverter stage, check and compare its two outputs; they should be equal and of opposite polarity.

If the pulses check OK and the oscillator free runs at the correct frequency, but the pix will not lock or is unstable, the AFC stage is probably at fault. Troubleshoot per Sec. 8-14.

10-13 ISOLATING THE DEFECT

Assuming a sync problem has been localized to a single stage, further troubleshooting is normally confined to that stage only. The testing of individual stages is described next.

• **Troubleshooting the Sync Separator**

In general, there are two possible trouble conditions peculiar to the sync separator: a *dead* stage (there is *no* output and therefore a complete loss of both vertical and horizontal sync) or the separator output is not *clean,* that is, some video or noise is getting through or the signal is modulated by power supply ripple (hum) voltage. The separator output must be clean over a very wide range of signal strengths. Troubleshooting a *dead* stage is comparatively easy; sync *instability* problems can be more difficult to resolve.

For a dead stage with no output, the usual procedure is to check the dc operating voltages at the transistor, followed by testing the transistor and other components as appropriate. For unstable sync, critically examine a scope display of the output signal to determine if it is a noise problem, hum, or whether the pulses are clipped at the wrong level, resulting in either *weak* pulses or noise contamination.

Improper clipping is the most common problem. It may be caused by video overload (excessive signal drive), a bad transistor, or incorrect operating voltages. For the latter, refer to the schematic.

To ensure saturation at low signal levels, a fairly low collector voltage (between 10 and 20 V) is usual. If it checks high when receiving a station, the transistor is not conducting enough, if at all. The cause may be a bad transistor, too little signal input, or excessive reverse bias. Try a jumper between the base and emitter to reduce the bias to zero. If the collector voltage drops as it should, the transistor is probably OK and the cause is too much reverse bias. On a blank channel, verify that the transistor is cut off (collector voltage and the supply voltage are the same).

Because the bias voltage varies with signal strength, it should be checked both off channel and when tuned to a strong station. Three readings should be taken for each condition: B to ground, E to ground, and B to E, which is most important. Typically, a bias variation from near zero (no signal) to 3 or 4 V (on a strong station) can be expected.

Incorrect bias readings may be caused by too little or too much signal input, a defect in the previous stage where direct coupling is used, or a leaky coupling capacitor, a leaky transistor, or defective resistors or diodes in the base and emitter circuits. Try to relate the conditions to the signal level, bearing in mind that the stronger the signal is, the greater the bias that is developed by rectification by the base-to-emitter junction. Where the CRO shows weak pulses are getting through, the possibilities are insufficient drive or excessive reverse bias. Conversely, if the output pulses are strong, but with video and/or noise getting through, check for excessive drive or too little bias. Excessive drive also produces a contrasty pix and the AGC is often responsible. Critically adjust the threshold control on the strongest channel and just below the point where the sync becomes unstable.

Where a diode clipper is used, it may be open, shorted, or leaky. If suspect, clip one end loose before testing. Verify correct polarity when making a temporary substitution or a permanent replacement.

- **Troubleshooting the Noise Immunity Circuit**

An inoperative noise stage may be responsible for loss of sync during periods of noisy reception. A component breakdown here can also cause sync problems, even during normal reception since it interfaces so closely with the sync separator stage. For the former problem, check the functioning of the noise gate under simulated noise conditions. Noise may be generated locally by operating electrical devices such as razors, hairclippers, or certain appliances near the antenna input. Monitor the signal at the separator input or output. If the gate is not doing its job, verify the presence of pulses of the proper amplitude and polarity at the gate input(s). Check the transistor dc operating voltages, particularly the bias. Compare with

10-13 Isolating the Defect

values shown on the schematic. If there is too much reverse bias compared with the signal input level, the transistor will be cut off and inoperative at all times. Compare collector voltage with the dc source; it will be lower than the source if the transistor is conducting, but readings will be the *same* if the transistor is turned off.

If there is too *little* reverse bias, the transistor may be conducting even under no-noise conditions, and the video and pulses may be permanently grounded out. In this case the CRO will show no signal at the separator input. See if sync is restored when the gate is disabled, either by opening its circuit, shorting base to emitter, or by applying a high reverse bias with a bias box or by some other means.

REMINDER: ─────────────────────

Since a noise gate operates only in the presence of noise, the set will operate normally during quiet periods when the gate is disabled.

For improper bias, check all components in the gate input circuit and critically evaluate the signal input(s). As in testing the sync separator, check bias under both no-signal and maximum signal conditions, by switching on and off channel, and for different settings of the AGC threshold and noise control.

In severe cases where the noise circuit is unable to cope, try to identify the type of interference and, if possible, correct at the *source*. Try relocating the antenna and lead-in. Try a shielded lead-in and a power line filter.

- **Troubleshooting the Sync Amplifier**

Troubleshooting this stage, as with any amplifier stage, is mostly straightforward. Start by scoping the signal at the amplifier input and output as described in Sec. 10-10. Note the amplitude, polarity, and waveform as shown on the schematic. If the amplifier *precedes* the sync separator, both pulses *and video* are amplified. If it follows the separator, its output *must be clean*.

Pulse distortion or compression is often caused by incorrect operating voltages (particularly the bias) or signal overload. Try adjusting the AGC control.

Check the transistor operating voltages and compare with the schematic. Make resistance checks, followed by component testing as necessary.

REMINDER: ─────────────────────

Incorrect voltages can be caused by a bad transistor or other component in the circuit, by too little or too much signal drive, or by defects in a directly coupled stage preceding the amplifier or immediately following it.

- **Troubleshooting Integrator and Differentiator Networks**

Scope check for normal pulses at the input of each network. Verify that there is no significant amount of *vertical* pulse voltage at the output of the differentiator circuit (input of the AFC) and little or no horizontal pulses getting through the integrator to the vertical oscillator. Pulses must be sharply spiked and of the correct polarity. Check with the schematic. When pulses are lost or attenuated, see if they are restored when the network(s) are isolated from the sweep circuits. Check resistors and capacitors and other interconnected components in the area. In practice, these circuits give very little trouble, and component breakdown is rare.

- **Troubleshooting a Sync Inverter (Phase Splitter)**

Scope check horizontal pulses at the base input of the inverter transistor. Check for equal and opposite pulses at the inverter outputs.

REMINDER: ─────────────────────

No amplification can be expected from this stage.

344 SYNC SECTION

If there is little or no output, check dc operating voltages, the transistor, and other components. If the two outputs are not balanced, check the circuit where the reading is lowest. The most common defects are a load resistor that has changed value, or a leaky or shorted coupling capacitor or phase detector diode. See Sec. 8-14.

10-14 INTERMITTENT SYNC

Weak sync is often mistaken for an intermittent condition since sync loss (either vertical, horizontal, or both) is often sporadic. Evalute sync stability by switching channels, turning the set off and on, and adjusting the hold controls, which should not be too critical. For weak sync, proceed as described earlier. For a truly *intermittent* condition, follow the procedures described in Sec. 5-13. The most common problem is poor soldering, intermittent PC wiring in particular. Where sync loss is attributed to arcing or interference, investigate these possibilities as appropriate.

10-15 AIR CHECKING THE REPAIRED SET

After making repairs, verify both vertical and horizontal sync stability under all conditions, especially when switching channels. Where receiving conditions are poor (weak snowy pix), some sync instability can also be expected.

10-16 SUMMARY

1. A sampling of the composite video signal feeds the sync stages from the *sync takeoff* point in the video section. The vertical and horizontal sync pulses are separated from the video by a sync separator (clipper) stage. Vertical and horizontal pulses are separated *from each other* by *RC* networks (filters), an *integrator* circuit that passes only the vertical pulses to control the vertical oscillator, and a *differentiator* circuit that routes horizontal pulses to the AFC circuit to control the horizontal oscillator.

2. Most modern sets have only a sync separator stage and a noise gate. Others have additional stages, such as sync amplifiers and a sync inverter. Amplifiers are used to obtain sync stability under weak signal conditions. An inverter provides two out-of-phase signals required by one form of AFC circuit.

3. Strong noise impulses can function as sync pulses to cause sync instability. *Noise immunity circuits* (such as a noise gate) can help overcome this problem. Such circuits function by either grounding out the noise bursts or by using two out-of-phase noise bursts to cancel each other.

4. Depending on signal strength, the vertical oscillator may be triggered (to initiate vertical retrace) by any one of six *serrated* vertical sync pulses. To prevent loss of *horizontal* sync during the relatively long vertical retrace interval, the serrated equalizing pulses also function as *horizontal sync pulses*, alternately controlling the scanning of odd and even fields.

10-16 Summary

5. The setting of the AGC threshold control has a great bearing on sync stability. If set too high, insufficient AGC action results in signal overdrive and pulse clipping or compression in a video amplifier stage. Normal setting of the control is just below the point where the sync becomes unstable on the strongest channel.

6. The amplitude of sync pulses is roughly one-third of the total composite video signal. Ideally, the sync separator should clip the sync pulses at or slightly above the blanking pedestal. If clipped *below* this level, peaks of the video signal will get through and may cause random triggering of the sweep oscillators. If only the tips of the pulses are clipped, the result is unstable sync because of *weak* pulses.

7. *Fixed* bias is not used on a sync separator transistor. Maintaining the proper clipping level for both weak and strong signals requires a variable *automatic bias* that changes with the signal. This bias is obtained by rectification of the applied signal by the base–emitter junction of the transistor. A *reverse bias* is normally used to prevent conduction except when the signal level exceeds the bias. Thus only the pulses get through to the next stage.

8. The output of a sync separator is said to be clean when the pulses are free of video information, noise, hum, or other interference that may cause sync instability.

9. Troubles that occur at some point prior to sync takeoff naturally affect both the pix quality *and* the sync. A defect in the video section *after* sync takeoff affects only the pix. Poor sync is often caused by video overload. This in turn may be due to AGC problems.

10. Some sets have a *sync blanker* stage that grounds out the signal immediately after the vertical oscillator is triggered for each field. This is a noise immunity precaution to prevent double triggering of the oscillator. Troubles causing the blanker to conduct at all times result in complete loss of sync.

11. The sync circuits are often incorporated in modules or ICs along with the components of other stages. This may limit troubleshooting to checking input–output signals, checking for applied dc power, followed by substitution or replacement. Make sure the trouble is not caused by a defect *external* to the module or IC.

12. Localizing the trouble when there is loss of both vertical and horizontal sync usually calls for signal tracing (with a CRO) between the point of sync takeoff (in the video section) and the inputs to the vertical oscillator and the AFC circuit. Also observe the sync-blanking bars on the screen and try adjusting the AGC threshold.

13. If there is loss of vertical and horizontal sync, there is no point in checking the *sync stages* if there are indications of video overload (e.g., excessive pix contract, no sync bars or weak bars observed on the hammerhead display, or a *negative pix*).

14. Hum voltage getting into the sync circuits causes sync problems. While signal tracing, watch for the signs: a thickening of a horizontal pulse display or an undulating reference line on the CRO. Verify while observing a *vertical pulse* display. Poor filtering of the dc source for the sync or video stages is the usual cause.

15. The dc operating voltages of both a sync separator and a noise gate will change with the signal. Therefore, these voltages should be measured both off channel and when tuned to a strong station. Between these two extremes, a 3- or 4-V bias variation can be expected.

16. Like a sync separator, a noise gate is biased to cutoff or beyond, conducting only when noise is received. One test is to make sure the gate is not conducting when switched off channel. To turn the gate on, noise can be developed locally using a suitable noise generator. When the gate circuit is suspect, it can be disabled without affecting the normal operation of the receiver. The presence of noise is indicated by dots or thin horizontal streaks moving across the CRT screen.

17. Any amplifier can be overloaded by excessive signal drive, and any stage defect reduces the amount of signal that can be handled. In a video amplifier, pulse clipping normally occurs at full setting of the AGC threshold control. When clipping occurs even at *low* signal levels, look for a defect in the stage, especially conditions affecting bias, such as a leaky transistor.

REVIEW QUESTIONS

1. (a) Why is it necessary to separate (clip) the sync pulses from the video? (b) What stage performs this function?

2. What circuits are responsible for separating the vertical and horizontal pulses *from each other?* How is this accomplished?

3. Does a noise prevent strong noise impulses from reaching the (a) CRT? (b) sync circuits? or (c) both?

4. Why does one type of noise gate have two signal inputs from different points in the video section?

5. In the absence of test equipment, how can you determine whether pulses are weak or missing up to the point of sync takeoff?

6. Explain why a pix loses sync when the AGC threshold control is adjusted for the darkest pix.

7. What sweep (timing) frequency of a CRO produces a display of (a) one vertical pulse? (b) two vertical pulses? (c) two horizontal pulses?

8. Since a scope display may be either upright or inverted, how can you determine signal polarity for a given test point in a circuit?

9. Inadvertently connecting a video detector diode in backward results in a negative pix *and complete loss of sync.* Explain why.

10. Explain how a sync separator does its job.

11. (a) Explain why an AGC malfunction can be responsible for loss of sync. (b) How would this affect the action of the AGC threshold control?

12. Video overload in some stage prior to sync takeoff can cause loss of sync. Explain.

13. Explain the purpose and function of the *equalizing pulses* that precede and follow the six serrated vertical sync pulses on the composite video signal.

14. Unlike the vertical oscillator that is triggered *directly* by the sync pulses, the horizontal oscillator must be controlled *indirectly* with the AFC circuit. Explain why.

15. Explain the integrator network, how an oscillator triggering pulse is developed from the serrated pulses, how the horizontal pulses are rejected, and why its output is sharply spiked as required.

16. Explain the action of a differentiating network and how it produces sharply spiked horizontal pules. (b) Explain how and when the 6 vertical pulses also contribute to horizontal sync.

17. Besides a smeary pix, poor low-frequency response of a video amplifier can also cause sync instability. Explain why.

18. What changes are normally observed in the bias and collector voltage of a sync separator transistor while switching on and off channel? Explain.

19. An NPN sync separator feeds the base of a PNP vertical oscillator. What is the signal polarity at the sync takeoff?

20. (a) The AGC threshold control should be adjusted on the strongest channel and preferably at the location where the set is to be used. Explain why.
(b) What is the proper setting for this control?
(c) What if it is set too high? Too low?

21. What is the advantage of a transistor sync separator over a simple diode?

22. Sync pulses along with the blanking pulses and video are permitted to reach the CRT input. Do they (a) serve any purpose? (b) create any problems? Explain. (c) How can they be observed on the CRT screen?

23. Explain the causes and results of sync clipping and compression that may occur in a video amplifier.

24. A critical hold control with a limited lock-in range is one indication of unstable sync. Explain.

25. Momentary loss of sync when switching channels or during brief periods of interference is another indication of sync instability. Explain.

26. Visible indications of poor sync can be many and varied. Name five possible horizontal sync symptoms.

27. Name five possible symptoms of poor vertical sync.

28. Explain automatic bias as required for a sync separator.

29. A set has good pix contrast but poor sync. Does this necessarily mean the pulses are present at the point of sync takeoff? Explain.

30. What is meant by a *floating pix* and under what conditions does it occur?

31. Briefly explain the operation of a typical noise gate.

348 SYNC SECTION

TROUBLESHOOTING QUESTIONNAIRE

1. Sync clipping in a video stage is suspected. How would you proceed?

2. List several possible causes of video overload and sync clipping in a video stage.

3. What is the probable effect on the sync separator output if the separator transistor has (a) excessive reverse bias? (b) too little bias? (c) How would sync be affected in each instance?

4. A CRO shows considerable signal or hum on the dc bus feeding the sync section and the set has poor sync. State the probable cause.

5. Pulses are being clipped in a video stage. How would you determine if that particular stage is at fault or if the stage is overdriven because of troubles elsewhere?

6. Where sync is unstable on the strong channels but appears normal on weak channels or when the antenna is disconnected, what would you suspect and how would you proceed?

7. If the pix locks only at one extreme of the vertical hold control and periodically rolls, what stage is suspect?

8. While examining the hammerhead display on a CRT, what condition is evident if (a) the sync bar is considerably darker than the blanking bar? (b) only slightly darker? (c) no sync bar is visible? (d) What stages could be the cause of poor sync in each instance?

9. A floating pix can be obtained at mid-setting of the horizontal hold control but the pix will not lock. What stage is suspect?

10. Where would you check (relative to the point of sync takeoff) (a) for negative pix and loss of sync? (b) for negative pix but normal sync? (c) if pix is normal except for loss of sync?

11. What stage is suspect if (a) the pix rolls and one pix cannot be obtained within range of vertical hold control? (b) the vertical roll can be stopped momentarily but vertical hold control is critical and horizontal sync is also unstable?

12. Pulses observed at the input of the vertical oscillator may be either sync pulses or those generated by the oscillator. Give two ways to identify them.

13. Where would you check if symptoms are poor sync, hum in the audio, and an undulating raster with hum bars?

14. It is easy to mistake horizontal pulling (a sync problem) for a wavy raster (a filtering problem). How would you distinguish one from the other by switching on and off channel?

15. Would you start by checking the sync stages when the symptoms are poor sync and a weak snowy pix? Explain.

16. What stage(s) are suspect if trouble symptoms are (a) poor vertical and horizontal sync? (b) loss of vertical sync only? (c) loss of horizontal sync only?

17. A set develops a negative pix and loss of sync after replacing a video detector diode. State the probable cause.

18. For excessive pix contrast with loss of sync, start by adjusting the AGC threshold control. How would you proceed if (a) sync is restored but only when pix contrast is sacrificed? (b) no improvement in sync when control is turned down? (c) control has little or no effect? (d) adjustment results in normal contrast and sync but for some channels only?

19. Normal setting of the vertical size and linearity controls results in poor vertical sync. The pix can be made to lock OK by misadjusting the controls with a sacrifice of raster height. What stage is at fault?

20. For loss of horizontal sync, how would you determine if it is caused by a defect in the oscillator stage, the AFC circuit, or from loss of the horizontal sync pulses?

21. Pulses are being monitored for possible clipping or compression as the AGC control is adjusted. What would you decide if (a) clipping occurs only at maximum setting of the control? (b) clipping occurs even at a low setting where pix lacks contrast?

22. A set periodically loses sync when switching channels. Would you look for an intermittent trouble or some condition causing a loss of sync stability?

23. A set has multiple pictures. What stage is at fault when a single pix cannot be obtained within range of the hold control?

24. Explain a logical troubleshooting procedure when pulses appear clipped at the point of sync takeoff.

25. State the probable effect on set operation for a leaky capacitor in the integrator net-pedestal (see Fig. 1-7). The separation of pulses

26. There is evidence of sync instability when a set is tapped or jarred. State the possible causes and how you would proceed.

27. Signal tracing shows pulses are lost in the separator stage. What follow-up tests would you make to isolate the cause?

28. A pix appears off center horizontally. It is corrected by a slight adjustment of the horizontal hold control, but horizontal sync becomes unstable. How would you proceed?

29. Vertical sync can often be restored by a touch-up adjustment of height and linearity control if the hold control has a limited range. Explain.

Chapter 11

AUTOMATIC GAIN CONTROL

A TV receiver must be able to function over a wide range of signal levels at the antenna, from as low as 200 μV or so in a fringe area to as high as 200 to 300 MV in some urban areas. Its sensitivity must be high for the reception of the weak signals and at the same time be able to handle signals from strong local stations without overload. This requires some form of automatic gain control (AGC) that automatically regulates the gain or amplification of certain stages of the set in accordance with the strength of the received signal.

Ideally, an AGC system should maintain uniform pix contrast and sound levels for all stations, from the weakest to the strongest. Without AGC, it would be necessary to frequently adjust a number of controls, contrast, color intensity, volume, sync, and so on, when switching channels. There would be problems of poor pix contrast with snow on the weaker channels and video overload on the strong stations, where the symptoms are a harsh contrasty pix with unstable sync.

An AGC system develops a *dc control bias* (AGC voltage) to regulate the gain of transistors in the RF/IF stages of the set. The amount of the AGC voltage developed depends on the strength of the received station carrier. If the signal is weak, little or no AGC voltage is produced and the receiver operates at maximum sensitivity. With a strong carrier, more AGC is developed to reduce the gain of the controlled stages. In this way, the system continuously and automatically compensates for variations of the received signals.

Most sets use AGC control for the RF stages in the tuners and two of the video IF stages. The RF and IF stages are generally controlled separately, with most of the control done by the IFs for weak to moderately strong signals. RF gain is controlled only on the strong stations.

11-1 BASIC AGC CIRCUITS

The block diagram of Fig. 11-1 shows the elements comprising a basic AGC system. Here the video detector serves double duty by developing the AGC voltage. The AGC voltage is

352 AUTOMATIC GAIN CONTROL

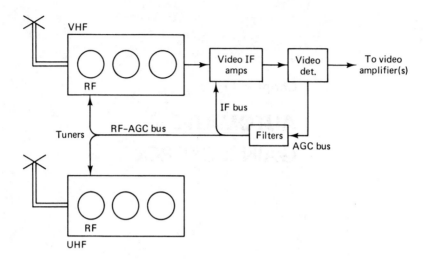

Fig. 11-1 Basic AGC block diagram.

fed, via an AGC *bus* to a network of *R/C* filters. At this point the AGC voltage is carried by an *IF bus* to the IF stages being controlled, and by an *RF bus* to the RF stages in the tuners. The entire system can be considered as a closed loop with each element dependent on the others.

The simplified circuit of Fig. 11-2a shows how the AGC is developed. The video detector, a simple diode, demodulates the IF carrier, producing an output across its load resistor *RL*. The video component is passed on to the video amplifiers and eventually to the pix tube. The *dc* component is the source of the AGC voltage.

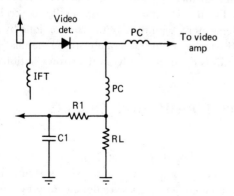

Fig. 11-2 Basic AGC circuit.

The amplitude of this dc varies with the average changes of the video. Its polarity is determined by which way the diode is connected. In this case the AGC voltage will be positive in reference to ground. The capacitor $C1$ and resistor $R1$ act together to filter out the video fluctuations that would otherwise create problems.

A basic circuit is seldom used because the AGC voltage is too weak for good control of strong signals; and because as an average-level rectifier, the AGC tends to fluctuate with the video variations and does not reflect the true strength of the received carrier; the result is a pix that changes with the average brightness level of the televised scene, and at such times there may be momentary loss of sync. Another problem is the long time constant of the AGC filters, which makes the system incapable of responding to rapid changes of the received signal.

There is another shortcoming of the circuit just described. Even with a weak signal, *some* AGC voltage is developed, which further reduces the receiver gain. Another problem is *noise*. Strong noise impulses can override the station signal, developing a high AGC voltage that reduces the gain. Practically all modern sets overcome these problems by using what is known as *keyed AGC*.

11-2 FEEDING THE CONTROLLED STAGES

The transistors in the controlled stages are supplied with two kinds of voltage, a dc *bias voltage* and the AGC voltage. Each transistor is provided with a small amount of *forward bias*. This voltage comes from the PS via one or more voltage-dividing networks, as shown in Fig. 11-3. Each transistor is biased for *maximum gain*. When a signal is received, the *AGC voltage changes the net bias* to reduce gain in accordance with the strength of the received signal.

The AGC is routed from its source to a main AGC bus from where it divides to control the RF and IF stages, as shown in the diagram. The main bus contains R/C filters whose purpose is to smooth out fluctuations due to video, pulses, and noise. Both solid dielectric and electrolytic capacitors are used.

Each branch line also contains R/C networks. Their function is to decouple or isolate the stages from each other. They also provide additional filtering. The "lumped" values of all R and C components represent the time constant (TC) of the circuit. The AGC time constant is fairly critical. It must be reasonably short (small values of C and R) for the system to follow rapid variations of signal strength. Too short a TC, however, reduces the effectiveness of the filters.

11-3 CONTROLLING TRANSISTOR GAIN

Transistors in the controlled stages are forward biased for maximum gain (highest receiver sensitivity) under no-signal conditions when no AGC is being developed. The gain is reduced when a signal is received. The stronger the signal is, the greater the amount of AGC produced, and the greater the reduction in gain. The no-signal bias is obtained from the PS via

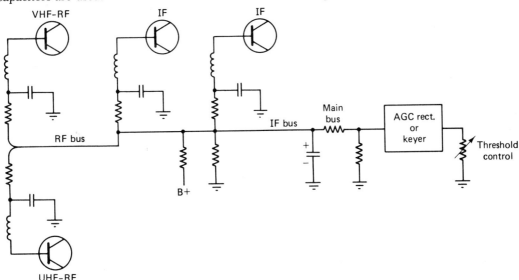

Fig. 11-3 Feeding the controlled stages.

voltage-divider networks. The polarity of the PS may be *either* + *or* −, regardless of whether the transistor is a PNP or an NPN.

There are several ways of obtaining this bias. If it is a PNP type, for example, apply −dc to the base, or apply +dc to the emitter, or use voltages of the same polarity applied to *both B and E*. For example, if a PNP transistor has +2.3 V on its base and +3.1 V at the emitter, the net forward bias (base relative to the emitter) is −0.8 V. The net polarity is determined by which of the two voltages is the greater. For an NPN transistor, the conditions are the same, except the polarities are reversed. The no-signal, maximum sensitivity bias for transistors in controlled stages is typically between 2 and 7 V. The polarity depends on the transistor *type*.

When a signal is received, this no-signal bias is either increased or decreased by the amount of the AGC for the purpose of *decreasing the gain*. With a *tube-type* receiver, increasing amounts of *negative* bias are used to reduce the gain. With solid state, gain may be reduced in one of two ways, depending on the receiver design: by *increasing* or *decreasing* the forward bias. Where gain is reduced by *increasing* the bias, it is called *forward AGC*. Where gain is reduced by *decreasing* the forward bias, it is called *reverse AGC*. Reverse AGC must not be confused with reverse *bias* since there is *no change in polarity*. Some sets use forward AGC exclusively, some use reverse AGC, and some use a combination of both, for example, forward AGC for the RF stage and reverse AGC for the IF stages.

- **Forward AGC**

Forward AGC makes use of the *saturation* characteristics of a transistor. Starting with a no-signal bias that corresponds with maximum gain, if the bias is increased in a *forward* direction, the increasing transistor current develops an increasing *IR* drop across a series load resistor, causing the collector *voltage* and the gain

to decrease. When fully saturated, the transistor gain would be zero. This condition is avoided, however, to prevent transistor destruction by the high current. The relative gain reduction obtained with forward AGC is shown by the curve of Fig. 11-4a. The *negative* voltages indicate that a PNP-type transistor is used.

With forward AGC, the no-signal bias and the AGC voltage are *additive* since they are the same polarity. The net bias obtained for different amounts of AGC voltage can be seen from the graph of Fig. 11-4b. The graph is for a PNP with an arbitrary no-signal bias of −4V. In the example, saturation occurs with a forward bias of −8 V.

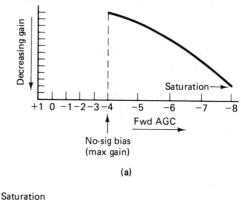

(a)

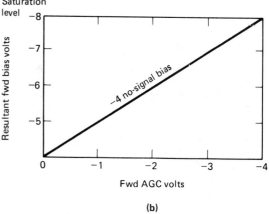

(b)

Fig. 11-4 Forward AGC operating conditions.

• **Reverse AGC**

Reverse AGC makes use of the *cutoff* characteristics of a transistor. Starting with a no-signal bias that corresponds with maximum gain, if the forward bias is *decreased*, the current drops toward *cutoff*, at which point the gain becomes zero. The gain reduction for reverse AGC is shown by the curve of Fig. 11-5a. As before, the transistor is a PNP type. Note the drop-off for reverse AGC is more abrupt than for forward AGC.

With reverse AGC, the no-signal bias and the AGC voltages are *subtractive* since they are of *opposite polarity*. Figure 11-5b shows the net bias obtained for different amounts of AGC voltage. As in the previous example, the no-signal bias is −4 V. Cutoff occurs at approximately 0 V.

To summarize, when tuned to a station, the true operating bias for an AGC-controlled transistor *is not* the no-signal bias on the bus *or* the amount of the AGC. With forward AGC, it is the *sum* of the two. With reverse AGC, is it the net difference between the two with one voltage opposing the other. Stated differently, with forward AGC, the AGC and the no-signal bias are of the same polarity. With reverse AGC, the polarities of the AGC and the no-signal bias are opposite.

AGC may be applied to either the base or emitter of a controlled transistor. A positive-going AGC voltage applied to the base of a transistor accomplishes the same thing as a negative-going AGC to the emitter. The reverse is also true. These things must be considered when determining the method of control used for a given set. Consult the schematic, noting the polarity of the AGC, the type of transistor being controlled, and whether it is fed to the B or the E. For example, if AGC is positive-going and it feeds the base of a PNP, then *reverse AGC* is being used.

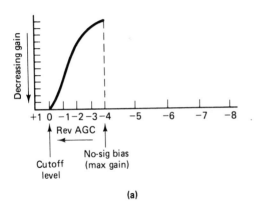

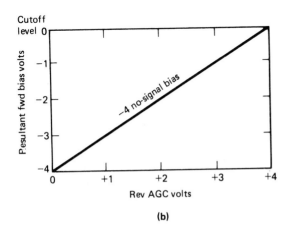

Fig. 11-5 Reverse AGC operating conditions.

11-4 CONTROLLING A FET

Dual-gate MOSFETs are often used in the tuner RF stage and sometimes in IF preamps of modern sets; they have a number of advantages over conventional bipolar transistors. Some of the desirable characteristics are a high input impedance, high gain, low noise, less crosstalk, and practically no loading of the signal input or AGC circuits, which means that no AGC *power* is required. Particularly at high frequencies, it combines the best features of both a transistor and a vacuum tube.

With a dual-gate FET, one gate receives the input signal and the other is supplied with AGC.

The AGC voltage controls the total current from the *source* to the *drain*. Depending on the type of FET, the no-signal bias for maximum gain is between 5 and 10 V, the polarity depending on the type. *Reverse AGC* is used for gain reduction, and cutoff occurs at about 5 V *of reversed polarity*. Depending on design, the AGC may be either positive- or negative-going. Figure 12-10 represents a typical AGC-controlled RF stage.

11-5 KEYED AGC

Some form of keyed or gated AGC is used in practically all modern sets because it overcomes most of the shortcomings of simple AGC systems previously described. It is relatively insensitive to *noise*. It develops the fairly high AGC voltage required for good control of strong signals. Because of a short TC, it is fast acting, all but eliminating the problems associated with fading and fluctuating signals, especially a condition called "airplane flutter." With keyed AGC, the AGC voltage *is not* developed continuously. It is produced at the rate of 15,750 times per second by sampling each horizontal sync pulse that accompanies the incoming signal.

The keyer or *gate* transistor as it is often called produces an output only when periodically keyed into conduction by gating pulses provided by the horizontal sweep circuit.

A block diagram of a keyed AGC system is shown in Fig. 11-6. Note the two signal inputs, a sampling of the composite video taken from the video amplifier section, and the gating pulses, which come from a winding on the flyback transformer. The *threshold control* is a service adjustment for setting the threshold level at which the keyer conducts and the required amount of AGC is produced.

A keyer is normally biased to cutoff and is made to conduct *only* when *both* signals are present, that is, for the brief periods when the incoming horizontal sync pulses and the gating pulses from the sweep circuit occur *simultaneously*. This puts a constraint on the sweep circuit, which *must be* operating at the correct frequency and phase relative to the incoming station signal.

When gated into conduction, the keyer produces an AGC output that varies with the strength of the sync pulses that correspond with the station carrier. AGC is developed *only* for the brief duration of the sync pulses during each horizontal retrace interval. This greatly reduces the chance of the AGC being influenced by noise and fading signals; and when it does occur, it is not observed because the screen is blanked out at such times.

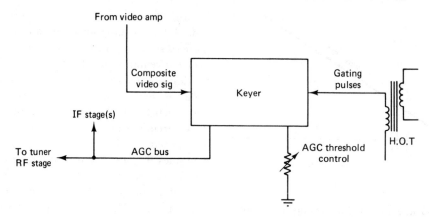

Fig. 11-6 Keyed AGC block diagram.

Although AGC is produced only at intervals and for short periods, it is *sustained* for the duration of each horizontal *trace* as the raster is scanned.

11-6 KEYER INPUT SIGNALS

A keyer will only conduct and produce an AGC output *under certain conditions:* when provided with the correct amount of *reverse bias,* and when supplied with the two signals described next. A simplified keyer circuit is shown in Fig. 11-7.

• Composite Video Signal

The video signal usually comes from a takeoff point at the first video amplifier. In most cases it is fed via an isolation resistor to the base of the keyer transistor as shown. Sometimes it is fed to the emitter. The resistor prevents loading of the high video frequencies being passed on to the CRT. With some sets, the video must go through a noise gate or inverter before reaching the keyer for additional noise immunity under severe noise conditions, although a keyer by itself is relatively noise immune.

The video signal is a *forward-biasing* signal whose polarity is determined by the type of transistor. It must be positive-going for an NPN and negative-going for a PNP when feeding the base, and the opposite polarity when feeding the emitter. The amplitude varies with the strength of the received carrier. Although both video and sync pulses exist together, the video itself serves no useful purpose since the keyer responds only to the *peaks* of the signal, that is, the *pulse tips.*

• Gating Pulses

The collector voltage for most transistors comes direct from the PS. With a keyer transistor, it comes from the horizontal sweep circuit in the form of pulses, specifically from a small winding on the flyback transformer. It feeds the collector via a diode called a *pulse gate* (see Fig. 11-7). The purpose of the pulses is to key or trigger the transistor into conduction at the horizontal sweep rate. Pulses may be supplied to the keyer in a variety of ways: directly through a coupling capacitor, or where the flyback coil is in *series* with the keyer collector circuit. Sometimes there is a capacitor to ground (as a capacitive voltage divider) to establish the correct amount of pulse amplitude, which is typically around 30 V p-p, depending on the set. In some cases the

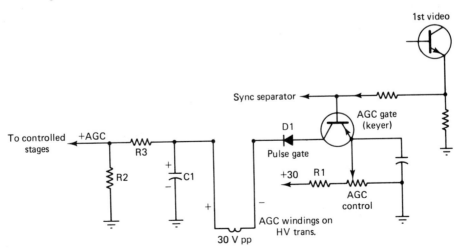

Fig. 11-7 Simplified keyer circuit.

358 AUTOMATIC GAIN CONTROL

amplitude is much higher at the source, then subsequently reduced to the desired value. Pulse amplitude is indicated on most schematics. As for the video signal, the polarity of the pulses is determined by the type of transistor. The polarity must be correct to *forward bias* the transistor into conduction. The *frequency* of the pulses must also be correct as determined by the horizontal sweep rate.

• **Functional Operation**

Off channel, and with no signal received, a keyer is biased to cutoff. Cutoff normally occurs with a reverse bias of about 1 V. This voltage, as measured between B and E is furnished by the PS feeding the B, the E, or both. The bias is usually adjustable by the threshold control connected in either the B or the E circuit. The control is adjusted to the point where only the *peaks* of incoming sync pulses cause the keyer to conduct and produce an AGC output. The adjustment is quite critical. At one extreme setting, the excessive reverse bias makes it impossible for the keyer to conduct, and the result is *no AGC*. At the other extreme, too little bias enables the keyer to conduct *at all times* and the result is *excessive AGC*. This is called an *overactive keyer*. The normal setting is somewhere between these two extremes (see Sec. 11-21).

As stated previously, a keyer is made to conduct when simultaneously supplied with video and gating pulses. Neither one nor the other is sufficient *by itself* to turn the transistor on. They are *both* required. And when these conditions are present, they work together to overcome the reverse bias.

The polarity of the video, the gating pulses, and the developed AGC depend on the type of transistor used as a keyer. A PNP requires negative-going pulses and produces a *positive going* AGC voltage. An NPN requires positive-going pulses and produces a *negative-going* AGC.

In reference to Fig. 11-7, the pulse gate diode $D1$ plays an important role in the operation. It serves as a rectifier of the gating pulses and also as a one-way blocking device that prevents the developed AGC from being shorted out by the B/E junction of the transistor. It also prevents the B/C junction from being *foward biased* in the intervals between pulses.

The keyer collector voltage as indicated on a schematic *for no-signal* conditions is produced by rectification of the pulses. When a signal is received, the polarity of this dc is *reversed*. Here is what takes place:

When an incoming signal causes the keyer to conduct, the energy supplied by the flyback transformer produces a current flow via the diode, the E/C junction, the threshold control, and $R1$ to charge the capacitor $C1$ to the peak value of the sync pulses. Note the *polarity* as determined by the direction of the current. The capacitor charge is the *source of AGC* voltage that supplies the bus. When the gating pulse is gone, the transistor stops conducting and the capacitor slowly discharges via $R2$ and $R3$ during the interval between pulses, the active scan time. The C/R time constant is such that little of the $C1$ charge is lost, and the AGC voltage on the bus remains relatively constant for the duration of the trace period. Stated differently, the charge on the capacitor maintains the AGC between pulses when the keyer is not conducting until the charge is replenished by the next gating pulse.

AGC voltages (as indicated on a schematic) typically range from about 1 to 5 V for weak and strong signals, respectively.

• **Noise Immunity and Airplane Flutter**

With a simple AGC system (Fig. 11-2), *noise* acts like a signal and the AGC voltage developed reduces the receiver sensitivity, creating changes in pix contrast and the like. *Keyed* AGC is largely unresponsive to noise because the AGC can only be influenced for very brief periods, the 5-μs duration of each horizontal sync pulse. Since this represents only about one-tenth of each trace–retrace period in the scanning of a raster, the chances are slight for noise occurring at these particular times. Noise received during the relatively long periods between pulses when the keyer is not conducting will have no effect on the AGC.

Airplane flutter is a problem created by aircraft, especially in the vicinity of an airport when a simple AGC system is used. Ordinarily, most TV signals reach the receiver via a direct line-of-sight path. When an aircraft passes overhead, some signal is reflected from the craft to arrive at the receiver by an indirect route. Since the craft is moving and the distance traveled by the indirect signal is constantly changing, there is a changing phase relationship between the two signals arriving at the receiver. When they are *in phase* they reinforce each other, creating an increase in pix contrast. When they are out of phase, they tend to cancel and the contrast is reduced. The effect created is a sort of rapid, in-and-out flutter to the pix. This is largely overcome with a *keyed* AGC system for two reasons.

1. Because of its short TC, the keyer circuit is able to respond to rapid signal fluctuations, maintaining a uniform pix regardless of such signal variations.
2. Because the AGC changes only during horizontal retrace when the screen is blanked out, the result is not visible to the eye.

11-7 KEYER CIRCUIT VARIATIONS

The circuit shown in Fig. 11-8a uses three transistors to develop an AGC voltage. Transistor $Q1$ is an AGC inverter, $Q2$ is a gated amplifier, and $Q3$ is the keyer. The video signal is amplified and inverted and then fed to the base of $Q2$, which is gated into conduction by $Q3$. The keyer, $Q3$, conducts only for the duration of the gating pulses applied to its base. The E/C junctions of $Q2$ and $Q3$ are in *series*, and both must conduct together or not at all. When $Q2$ conducts after being enabled by $Q3$, a negative-going AGC voltage is developed across $R1$ and then fed to the bus.

The circuit shown in Fig. 11-8b uses four diodes instead of a transistor. The diodes are connected in a *bridge configuration.* Gating pulses are fed to the bridge via a small transformer. Video is fed to the junction of two of the diodes with the AGC output taken from the opposite arms of the bridge. Off channel, the bridge is balanced and there is no output. When tuned to a station, a negative-going signal is applied to the left side of the bridge. However, this signal voltage is *blocked* by $D1$ and $D2$ and cannot reach the output until a gating pulse is present. When a gating pulse is present across the two vertical arms of the bridge, $D1$ and $D2$ are keyed into conduction, along with $D3$ and $D4$, which conduct in a forward direction. The dc output of this circuit is limited by the signal input (typically less than about 0.5 V), and additional AGC amplification is required for control of strong signals.

A circuit used extensively in early-model sets is shown in Fig. 11-8c. With this arrangement, a sampling of the video signal is taken from the last stage of the video IF strip to feed the keyer. The keyer is gated by pulses applied to its collector via $D1$, the pulse gate diode, and $T1$, a conventional, tuned IF transformer. As a departure from other circuits described, the IF signal itself is gated. The output of the keyer consists of a series of IF *pulse bursts* at horizontal retrace intervals, which are subsequently rectified by diode $D2$ and then applied to the bus. The AGC output of this circuit is low, so additional amplification is required.

11-8 AMPLIFIED AGC

Where the AGC output from a keyer is not adequate to prevent overload on strong signals, an *AGC amplifier* is often used between the keyer (or

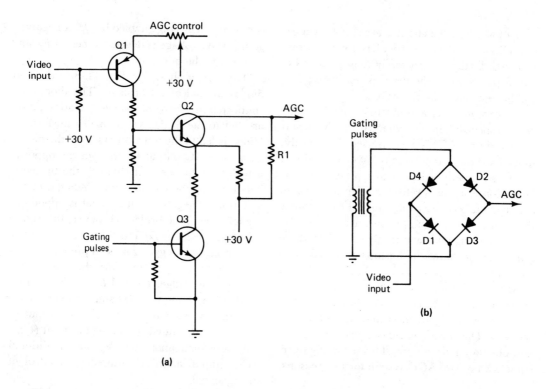

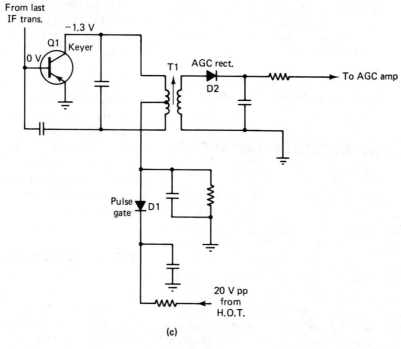

Fig. 11-8 Keyer circuit variations.

360

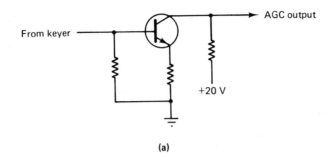

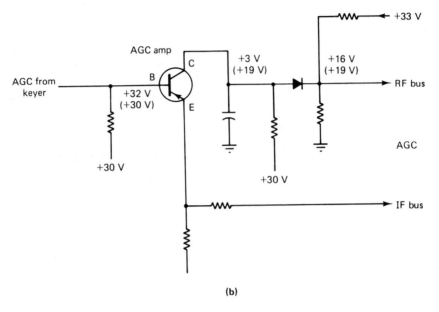

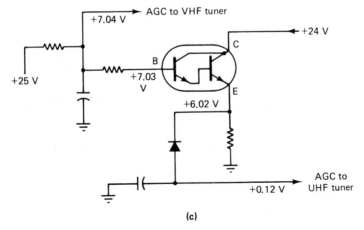

Fig. 11-9 Typical AGC amplifiers.

other AGC source) and the controlled stages. A typical amplifier is shown in Fig. 11-9a. This is a conventional linear dc amplifier with a voltage gain of about 3 times. Since dc voltage is being amplified, *direct coupling* must be used in both the input and output circuits. As with any amplifier of this configuration, a polarity reversal occurs between input and output. The input polarity must be compatible with that of the keyer output. The amplifier output polarity must meet the requirements of the controlled stages, which are dictated in turn by the types of transistors used and whether forward or reverse AGC is employed.

Figure 11-9b is an amplifier circuit with *dual* outputs having opposite polarities. As with Fig. 11-9a, an inversion takes place between the input and the collector output. The output from the *emitter* is the *same* polarity as at the input. This arrangement is used where one controlled stage requires forward AGC and the other, reverse AGC, or where different polarities are required when one controlled transistor is a PNP and the other an NPN. Figure 11-9c shows an amplifier that uses a Darlington-type transistor for additional gain.

An AGC amplifier is normally biased to cutoff with a B/E reverse bias of about 0.2 V, and requires a *forward-going* input voltage to turn it on.

A variety of circuit configurations are in use to satisfy the many variables and polarity requirements of different sets. Most are shown in Figure 11-10 to illustrate polarities at each stage.

In Figure 11-10a, no AGC amplifier is used and the negative-going output of the keyer provides reverse AGC to the controlled stage. In circuit (b), the positive-going output of the keyer is used as forward AGC for the controlled transistor. In part (c), a PNP transistor is supplied with positive-going reverse AGC. In part (d), a positive-going keyer output provides foward AGC to an IF amplifier, which also serves as a dc amplifier. The positive output from its emitter provides forward AGC to another IF stage. In part (e), a negative keyer output drives an AGC amplifier, whose negative output in turn is used to feed two controlled stages simultaneously with forward AGC. In part (f), a negative voltage is used as reverse AGC for two stages. In part (g), a positive voltage is used as reverse AGC for two stages. In part (h), the negative output of the keyer supplies reverse AGC to one stage, and the positive output from the amplifier supplies reverse AGC to the other stage. In part (i), forward AGC is used to control the two transistors. In part (j), the opposite polarity outputs of the amplifier provide forward and reverse AGC to two transistors. In part (k), a negative-going AGC is used to control a FET.

11-9 CONTROLLING THE IF STAGES

The AGC-controlled transistors are biased for maximum sensitivity under no-signal conditions. This is accomplished with a small amount (around 0.5 V) of forward bias applied either to the bus or directly to the B or E of the transistor(s), which amounts to the same thing. The voltage comes from the PS via voltage-dividing networks. When a signal is received, the bias is made to change by the amount of AGC superimposed on the bus. Although AGC voltage always *increases* with the signal, the bias itself may either increase or decrease depending on whether forward or reverse AGC is used.

Because of their different requirements, the RF and IF stages are usually controlled *separately*, with the IFs receiving the maximum AGC available on the bus.

As shown in Fig. 11-11, there are several alternate ways of feeding AGC to the IF stages. In Fig. 11-11a, the first and second IFs are fed in parallel through individual isolating

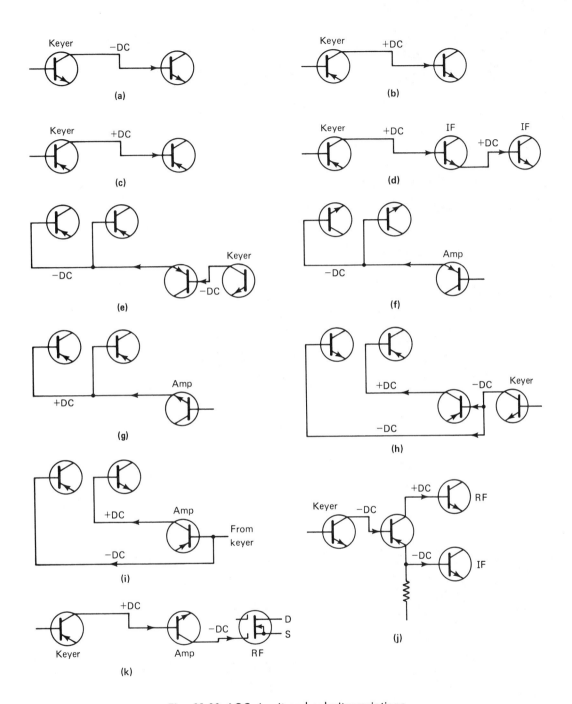

Fig. 11-10 AGC circuit and polarity variations.

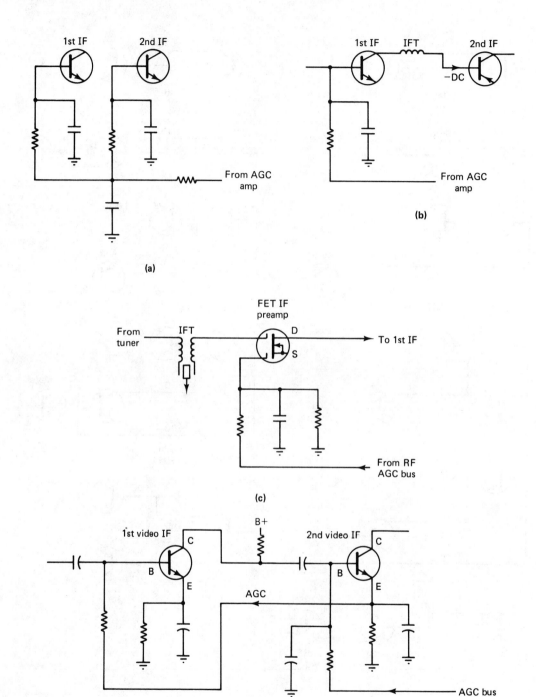

Fig. 11-11 Controlling IF stages.

364

(decoupling) networks from a common source. Since the AGC polarity is the same for both stages, either both transistors use forward AGC or two different transistor types are used, one receiving forward AGC, the other, reverse AGC. The AGC may be applied to either a base or an emitter, although the former is most common.

Figure 11-11b shows a *cascaded* arrangement, where one IF stage is directly coupled to the next. AGC is applied only to the first stage, which in turn controls the second stage. Sometimes a FET is used as an IF *preamp*, as shown in part (c). The *stacked* arrangement shown in part (d) is currently the most popular. The AGC is applied to the base of the second IF transistor only, and the dc component developed across its emitter resistor is used as AGC for the *first stage*.

11-10 CONTROLLING THE RF STAGE(S)

Just as the gain of the IFs must be controlled on strong signals to prevent overload of the downstream stages, so must the gain of the RF stage in the tuner(s) be controlled to prevent overload of the *mixer* stage. Control of an RF stage on *weak signals* is not only *unnecessary*, it is *undesirable*. But without some control of the *strong* signals there is a chance of mixer overload and problems of *cross modulation* (crosstalk) between adjacent stations, which shows up as a "windshield-wiper effect," (see Table 11-1).

As with the IF stages, an RF stage transistor is forward-biased for maximum sensitivity under no-signal conditions. The amount of the bias is typically about 1 V, about half the amount applied to the IF transistors. The polarity may be either + or − depending on the type of transistor being controlled. When a moderately strong signal is received, the AGC voltage applied to the RF bus is also less than for the IFs. For strong signals, however, the RF AGC may be greater than the IF AGC. Either forward or reverse AGC may be used. Where the UHF tuner has an RF stage, AGC is simultaneously applied to both tuners. Most modern sets use a FET in the RF stage.

- **Signal-to-Noise Ratio and the Snow Problem**

Picture-degrading interference, or *white noise* as it is sometimes called, may be received as background to a station signal, or it may be generated within the set itself. It shows up as *snow* or *confetti* (colored snow) on the screen of the pix tube, particularly on weak channels. Most of this noise is generated in the mixer stage. To keep the noise (and snow) to a minimum and obtain the highest possible signal-to-noise ratio, it is necessary to drive the mixer with the strongest possible signal. The factors that contribute to signal strength are the receiver location, the antenna installation, the tuner input circuits, and the *gain of the RF stage*.

RF gain is controlled by AGC. Although a weak signal develops only a small amount of AGC, this can seriously weaken an already weak signal. The result is a washed-out pix with excessive snow. This means that *no* AGC should be applied to the RF stage when receiving a weak station. This is accomplished with a *delay circuit* that delays the application of AGC to the tuner until the signal input is at least 500 μV or more. Note that this is a *voltage delay*, not a *time* delay.

- **Delayed AGC**

All but the most inexpensive sets currently use some form of AGC delay. There are three basic delay circuits, as shown in Fig. 11-12. The circuit in Fig. 11-12a uses a *clamping diode*. The diode, when tuned off channel or to a weak station, is forward biased into conduction by the +10 V from the PS. This *clamps* the RF bus to

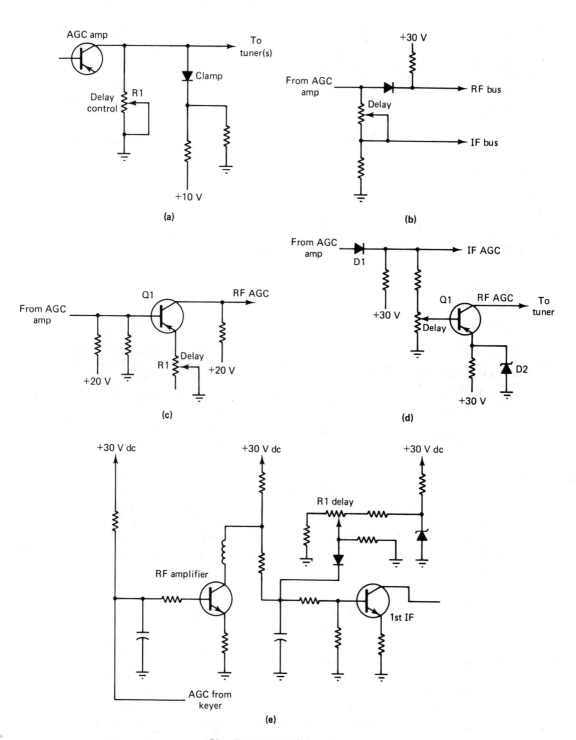

Fig. 11-12 AGC delay circuits.

11-10 Controlling the RF Stages

approximately *zero volts* until the received signal is strong enough to develop at least 4 or 5 V of AGC at the *IF* bus. When this occurs, the AGC voltage overcomes any clamping voltage, the diode is reverse biased into nonconduction, and the RF transistor is supplied with AGC. Thus, for the weak signals there is *no RF AGC* and the stage operates at maximum gain, with all the control provided by the IFs. For moderate-to-strong signals, the RF AGC takes over to provide most of the gain reduction. This is shown by the graph of Fig. 11-13.

Resistor *R1* of Fig. 11-12a is called the *AGC delay control*. It is factory preset to establish the delay threshold level, the point at which AGC is supplied to the tuner. It could be called a *snow control* because, in practice, it is adjusted for minimum snow on a weak channel. For precise adjustment procedures, see Sec. 11-21. With the control set to one extreme, the bus is grounded for maximum delay. At the other extreme, the delay is reduced and the RF stage becomes controlled for weaker signals.

The delay circuit of Fig. 11-12b uses a *blocking diode* (sometimes called a gate) connected *in series* between the AGC source and the controlled stage. On a weak signal the diode is reverse biased (nonconducting) by the applied B+ and no AGC reaches the RF stage. With a strong signal, the AGC overcomes the delay voltage and the diode conducts and feeds AGC to the RF transistor. Here, again, a delay control sets the threshold level.

The circuit of Fig. 11-12c uses a transistor for delay. With a weak signal, the delay transistor *Q1* is biased to *cutoff* by the voltage-divider network in the base circuit and the directly coupled voltage from the AGC amplifier. The turn-on bias is adjustable with the delay control *R1*. For a moderate signal, the AGC output from the amplifier to the IF bus is about 6 V, and the IF gain is minimum. For a stronger signal, the increasing AGC voltage offsets the delay transistor reverse bias, the transistor is turned on, and the increasing voltage at the collector drives the RF transistor toward *saturation*, decreasing its gain.

Another circuit variation is shown in Fig. 11-12d, which uses both a blocking diode for both RF and IF delay, plus an additional delay provided by a transistor. On weak signals, the delay transistor *Q1* is saturated and no AGC reaches the tuner. On strong signals, *Q1* begins to conduct to control the RF stage. The emitter voltage of *Q1* is stabilized with a zener diode, *D2*.

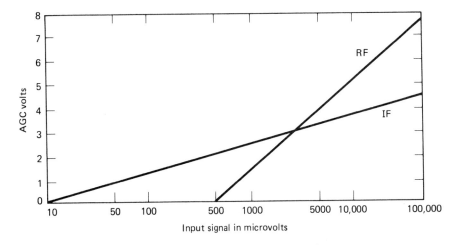

Fig. 11-13 RF versus IF gain control.

368 AUTOMATIC GAIN CONTROL

Figure 11-12e shows still another circuit variation. In this configuration, the full AGC output from the keyer is applied directly to the RF stage instead of the IFs. The RF transistor serves a dual purpose. In addition to its *normal* function, it also acts as a dc amplifier-inverter to supply AGC to the first IF stage. The positive-going AGC developed by the keyer is used as forward bias to reduce the gain of the RF stage. The inverted negative-going output of the RF transistor controls the IF transistor with *reverse* AGC. The bias control $R1$ establishes the threshold delay level for the RF stage.

- **AGC Voltage Stabilization**

AGC voltage is sometimes stabilized to limit the voltage variations that might otherwise destroy sensitive transistors and FETs in the controlled stages. A typical regulating circuit using two zener diodes is shown in Fig. 11-14. This circuit develops either a + or a − voltage according to signal strength, ranging from approximately +8 V for weak signals to about −6 V on strong signals. Off channel, *TP1* and the AGC bus are maintained at +5 V, which comes from a 150-V source via voltage-dividing resistors $R1/R2$. Also, 25 V of B+ is supplied to the junction of the back-to-back zener connections.

Diode $D1$ stabilizes the +5 V at *TP1*, and diode $D2$ stabilizes the +14 V at *TP2*. Each diode conducts only when the voltages rise above these levels. If the voltage at *TP1* rises above 7 V on a weak signal, $D1$ conducts, maintaining the voltage at +6 V. If on strong signals the bus voltage goes higher than 4.5 V in a *negative* direction, the bus is clamped at that level. Hence, when the keyer is cut off on weak signals, the bus voltage swings *positive* to *raise* the gain of the controlled stages. On strong signals when the keyer is turned on, it develops an increasingly *negative* output to oppose the positive voltage, and the bus goes *negative* to *reduce* the gain.

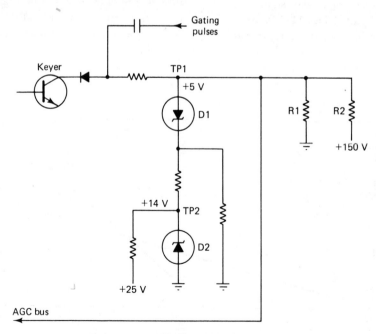

Fig. 11-14 AGC voltage regulating circuit.

11-11 INTEGRATED-CIRCUIT AGC

Modern sets incorporate most of the AGC circuitry in an *IC* along with other TV functions, such as sync and noise limiters. Since the internal circuitry of an IC is not shown on a schematic, our only concern, from a troubleshooting point of view, is with the terminal connections and any associated external components and circuitry. A typical IC is shown in Fig. 11-15. Note the pin connections associated with the AGC functions. Connections for other functions have been omitted for clarity.

11-12 TYPICAL AGC SYSTEM

Following is a brief functional description of the AGC section of a late-model color set. See Sample Circuit C at the back of this book. This AGC system does not use a keyer since there are no gating pulses applied to the IC (IC 101). Video is supplied internally to the AGC rectifier via a noise-limiter stage. The output of the *IF* AGC block controls the first and second IF stages. The output of the *RF* AGC block controls an *IF preamp* (using a FET), in addition to the RF stages in the tuner(s). Separate IF and RF AGC controls are provided. The RF control

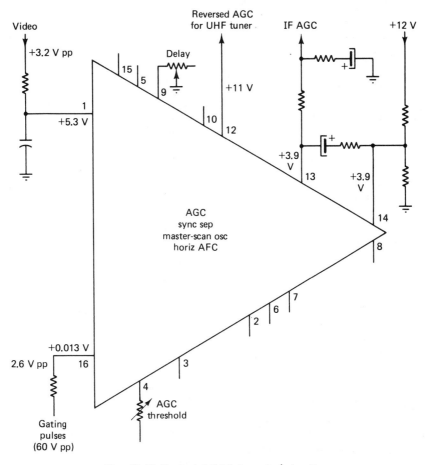

Fig. 11-15 Typical AGC integrated circuit.

370 AUTOMATIC GAIN CONTROL

establishes the dc threshold level (at pin 22) to about 8 V. The IF control sets the level (at pin 19) to about 2 V. The *polarity* of the AGC is indicated by the capacitor C148.

11-13 TROUBLESHOOTING AND REPAIR

Troubleshooting the AGC can be somewhat more difficult than for other sections of a set. There are a number of reasons for this. First there is the problem of accurately diagnosing the symptoms and deciding if it is an AGC problem or a defect elsewhere in the set. And it is not always easy to distinguish one from the other, at least not until a few preliminary tests have been made. Also involved is the question of cause and effect, since an AGC failure can create problems in other parts of the set, and other stages in turn can cause AGC problems. For example, excessive AGC to a controlled stage can weaken or kill the signal. On the other hand, a defective signal stage can weaken or kill the AGC action.

Another difficulty is the result of the AGC system being a closed loop embracing several stages and each interdependant on the other. This sometimes necessitates *breaking into the loop* and checking each section individually.

Another problem is the critical nature of the AGC system, where *fractional* voltage discrepancies can often spell the difference between a properly operating receiver or one with a *variety* of trouble symptoms. In addition, since all signal stages between the antenna input and the pix tube are influenced by the AGC action, an AGC malfunction can result in abnormal signal and voltage indications all down the line.

To complicate matters further, most AGC problems are related to signal strength, with some troubles showing up only on the weaker channels and others on the strong channels. Also, there is the problem of AGC polarity and which of two systems is used, forward or reverse

AGC, one aspect of solid state with no counterpart in a tube-type receiver.

Once an AGC trouble is localized, however, isolating and correcting the defect is no more difficult than for other sections of the set.

11-14 ANALYZING THE TROUBLE SYMPTOMS

Of the many trouble symptoms that may be caused by an AGC malfunction, most of them fall into one of two categories, those caused by *loss of AGC,* and those caused by *excessive AGC.* A listing of the most common symptoms and typical causes is shown in Tables 11-1 and 11-2. Most of the these symptoms, however, *can also be caused by problems in other sections of the receiver.*

Where AGC troubles are *suspected,* start by noting the presence or absence of *snow* on a blank channel. Then switch channels, noting the degree of pix contrast and the amount of *snow on the weak channels,* and any signs of overload on the stronger stations. Video overload usually shows up as partial or complete loss of sync, along with a dark contrasty pix and high color intensity.

Frequently, some of these problems can be cleared up with a slight adjustment of the AGC control(s). Before adjusting, however, mark the original settings so they can be restored if the troubles persist. A misadjusted control will *compound the problems.* Adjust for minimum snow on weak signals and no overloading on the strong channels.

To verify suspected overload, try disconnecting the antenna and substituting it with a short piece of wire. An overload is indicated where set operation improves.

REMINDER: ⎯⎯⎯⎯⎯⎯⎯⎯⎯⎯⎯⎯⎯⎯⎯

One important objective of the initial testing is to determine *whether or not* it is truly an AGC problem.

Table 11-1 Symtoms Chart for AGC Trouble Caused by Loss of AGC

Trouble Condition	Probable Cause	Remarks and Procedures
No pix, raster OK	Loss of AGC resulting in video overdrive, or trouble in one of the signal stages between the tuner and the CRT.	Try adjusting the AGC controls. Check for AGC voltage. Signal trace signal stages.
Excessive contrast, buzz in sound, and poor sync	Loss of AGC resulting in video overdrive.	Check for AGC voltage. Adjust AGC controls.
Loss of sync and excessive contrast	As above. If AGC OK, then trouble in sync stages.	As above.
Negative pix and sync loss	As above. Could also be reversed signal polarity in video detector or amplifier stages.	As above. Also check for reversed signal polarity.
Loss of raster on strong signals. Off-channel raster OK	As above. Could also be RF/IF oscillations.	Check for AGC voltage. Adjust AGC controls. With direct coupling, the CRT may be biased to cutoff on strong signals.
Windshield-wiper effect (adjacent channel crosstalk). Vertical and horizontal blanking bars superimposed on pix	Loss of RF AGC resulting in mixer overload. Could also be interference from strong local station or RF/IF misalignment.	Check for loss of RF AGC. Adjust delay control.
Fluctuating pix	Airplane flutter or a wobbly antenna.	Check operation of keyer stage. Check for open AGC filters. Secure antenna.
Loss of pix when switching channels	Possible tuner trouble or AGC *lockout*.	AGC lockout can occur momentarily as the RF delay threshold is reached when RF gain is at a maximum and IF gain is minimum.

372 AUTOMATIC GAIN CONTROL

Table 11-2 Symptoms Chart for AGC Trouble Caused by Excessive AGC

Trouble Condition	Probable Cause	Remarks and Procedures
No pix or weak pix; raster OK, with no snow; poor sync	Excessive IF AGC or trouble in a video stage.	Check for excessive AGC. Adjust threshold control. Signal trace video IF and amplifier stages.
No pix or weak pix and excessive snow	Excessive RF AGC or tuner trouble. May also be poor location or antenna troubles.	Check for excessive RF AGC. Adjust delay control. Check antenna and tuner.
Set works OK in shop but not in user's home	Intermittent condition or abnormal reception conditions.	Adjust AGC controls. Air check for intermittent problems.

11-15 LOCALIZING THE TROUBLE

The fundamentals for troubleshooting a typical AGC system can be learned from the simplified flow drawing of Fig. 11-16 and the block diagram of Fig. 11-17. The system uses keyed AGC, an AGC amplifier, and a delay circuit. The trouble symptoms for this example are no pix, no sound, and no snow. The first thing is to determine whether or not the AGC system is at fault. There are two ways of doing this. If the set uses plug-in transistors, simply remove the ones from the AGC amplifier and delay circuits. This removes any abnormal AGC voltage that is causing the problem and permits the RF/IF stages to operate at maximum gain. If the pix is restored, it proves the AGC is at fault. If there is no improvement, the trouble is elsewhere. If it is not convenient to remove the transistors, the alternative is to *clamp the bus* with a bias box, as explained in Sec. 11-18.

When AGC trouble is indicated, the next step is to switch to a blank channel and check for dc at the input to the AGC amplifier if there is one. Refer to the schematic of the set for the normal off-channel voltage. If voltage is *excessive*, the possibilities are a defective voltage-dividing resistor feeding B voltage to the bus or trouble with the keyer circuit, more specifically, an *overactive keyer* (one that is conducting at all times or excessively).

REMINDER: _____

The keyer *should not be conducting when no signal is being received.*

Typical causes are a leaky or shorted keyer transistor or loss of reverse bias. The threshold control is usually part of the biasing circuit.

Where normal dc exists at the keyer output and the input to the AGC amplifier, check the amplifier output where it feeds the bus. Refer to the schematic for the proper amount and polarity. Loss of voltage at this point or an insufficient increase indicates an amplifier defect. Start by checking the transistor.

Where the bus voltage is normal *with no signal* received, the next step is to check for AGC action when tuned to a station. Check for dc voltage variations on the RF bus and the IF bus when receiving weak and strong signals. Refer to the schematic for typical readings. Adjust the AGC controls if required.

REMINDER: _____

Proper adjustment is for minimum snow on weak signals with no overload on strong signals.

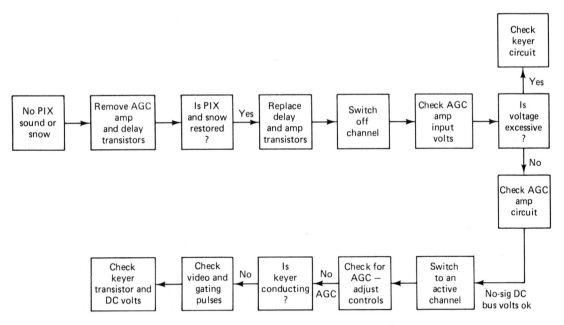

Fig. 11-16 Simplified troubleshooting flow diagram.

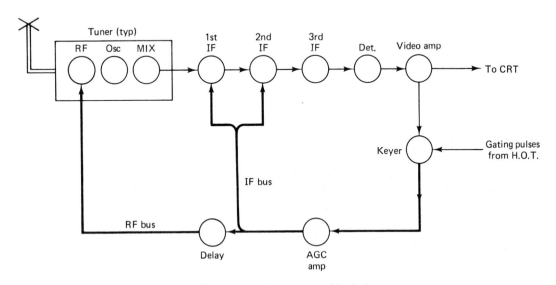

Fig. 11-17 Basic troubleshooting block diagram.

373

In lieu of a station signal, an RF signal generator may be used if desired. Ideally its output should be calibrated in microvolts.

NOTE: _____

The dc measured on a bus with no signal received *is not* an AGC voltage. When tuned to a station and the AGC circuit is operating normally, the developed AGC voltage should either *increase* or *decrease* the bus voltage, depending on whether *forward* or *reverse AGC* is used.

If there is no AGC action (no dc variations, either *increasing* or *decreasing*) with a signal, check for a keyer that is not conducting or not enough. Typical causes are a defective keyer transistor or gate diode, too much reverse bias, or loss of either video or gating pulses.

REMINDER: _____

Sync pulses and gating pulses must occur simultaneously, so make sure the horizontal sweep frequency is correct.

11-16 COMPLEXITIES OF AGC TROUBLESHOOTING

Figure 11-18 shows a more detailed flow diagram that can be used to troubleshoot most any type of AGC failure. Although a keyed AGC system is indicated, most of the procedures also apply to other types of AGC systems. The drawing is in two parts. Sheet A includes trouble symptoms and testing procedures primarily under no-signal conditions. Sheet B deals with abnormal conditions when tuned to a station. No-signal tests are normally performed first. On-channel testing assumes normal dc voltage on the bus when no signal is received. Depending on circumstances, testing may begin at any point in the flow.

IMPORTANT: _____

The following notes are intended as an aid when following procedures shown on the flow drawings. They should be read and considered before getting too involved with the testing.

1. A set may experience any *one or more* of the trouble symptoms shown on sheet A. When there are multiple symptoms, there may be more than one defect. Chances are there is but *one* underlying cause common to all.

2. All voltage readings, both on and off channel, should be recorded for later evaluation. For *precise* off-channel measurements, remove the antenna and short the terminals. Compare readings with those shown on the schematic. Most dc voltages indicated on a schematic are for off-channel conditions with no signal received and all controls at their normal setting. Also indicated is the average "range" of AGC voltages to expect when tuning between weak and strong stations.

3. There should be a dc voltage on the bus *at all times*. The amount is critical and the polarity is that required to *forward bias* the transistors in the controlled stages, both with and without a signal. With a signal, the dc reading may or may not represent a true AGC voltage. A true AGC voltage changes between off and on channel and with changes of signal strength. *Loss of AGC* can be interpreted to mean loss of AGC *action* and not the steady dc voltage normally present off channel. The term *excessive AGC* does *not* mean excessive dc that may be present on the bus *off-channel*.

4. *Excessive AGC* can mean either *too much* or *too little* voltage *of either polarity* depending on whether *forward* or *reverse* AGC is used. Excessive dc on the bus (either on or off channel) can reverse bias the controlled transistors to cutoff or forward bias them to saturation, depending on polarity, type of transistor, and whether forward or reverse AGC is used.

5. Don't attach too much significance to the many voltages that may disagree with those shown on the schematic. Abnormal AGC

11-16 Complexities of AGC Troubleshooting

action affects many voltages in a great many circuits. The trick is to distinguish between cause and effect.

6. The primary reason for clamping a bus is to determine whether or not the trouble symptoms are caused by an *AGC failure*. The amount and polarity of the clamping voltage required to restore receiver operation provides a clue to the nature of the problem. When clamping a bus, adjust the bias box to the proper voltage and polarity shown on the schematic *before* connecting to the set.

7. Loss of pix can be caused by *either* loss of AGC or excessive AGC. Loss of AGC causes overdriving and either cutoff or saturation of some video stage. Excessive AGC can bias a controlled stage to cutoff or saturation, depending on polarity. Loss of AGC can also result in no *raster*, where a directly coupled CRT is driven to beam current cutoff.

8. Snow off channel is a *normal* condition. No snow indicates trouble in the video IF, detector, or amplifier stages.

9. A polarity reversal should take place at the keyer collector between off channel and when a signal is received. A PNP keyer normally develops a +AGC voltage and an NPN keyer, a −AGC voltage. A nonconducting keyer can be made to conduct temporarily with a bias box connected B to E. Observe polarity.

10. An overactive keyer is one that is conducting excessively or at all times. This affects the bus voltage both on and off channel.

11. Exercise extreme caution when taking measurements or clamping a bus, especially when a FET is used. Sometimes the slightest surge, the wrong voltage or polarity, or an inadvertent short can destroy one or more transistors.

- **Troubleshooting Procedures**

With reference to the troubleshooting flow diagram (sheet A), the first thing, as with any TV problem, is to analyze the symptoms. Switch channels. Observe whether there is snow on the raster and on the weak channels. After marking the original setting, make tentative adjustment of the AGC control(s). It may be possible at this point to make a guess as to whether the AGC is at fault or if there is trouble elsewhere, and also, assuming there *are* AGC problems, if the symptoms are caused by *loss* of AGC or *excessive* AGC. Speculation, however, must be followed by actual *testing*.

Observing the presence or absence of *snow* can be very helpful in diagnosing AGC troubles. A moderate amount of snow off channel is normal. If there is no snow on the raster, it means the signal path has been interrupted somewhere in the video section. This, then, is the logical place to start.

Check for snow on a weak station. If it is excessive, the possibilities are antenna trouble or a poor location, a defect in the tuner, particularly the RF or mixer stages, or excessive AGC feeding the RF stage as from a misadjusted delay control. In this case, the front end of the set is the place to start. After marking original settings, try adjusting both the threshold and delay controls. The delay control has a great bearing on the snow level.

To localize trouble further to a weak or dead stage, signal trace through the *video stages* with either a CRO or by injection with a signal generator. Tracing the signal path in the *front end* can best be done with a generator unless the CRO has high sensitivity and a good demodulator probe is available.

The most common trouble is *loss of AGC*. This invariably causes overload problems on strong signals. This can be verified where set operation is restored when the input signal is weakened, as by replacing the antenna with a piece of wire.

The threshold control also provides a clue. If the set operates better with the control turned cw, the cause is probably *excessive* AGC. If operation improves with the control turned ccw, the problem is *loss of AGC*.

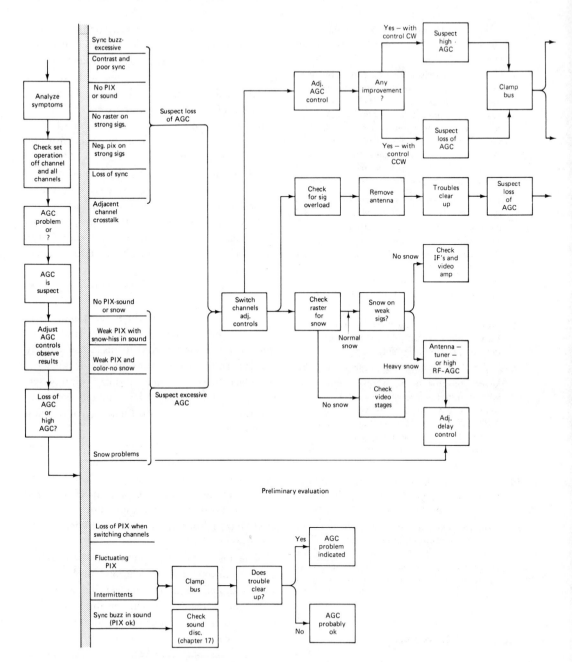

Fig. 11-18 AGC troubleshooting flow diagram (sheet A).

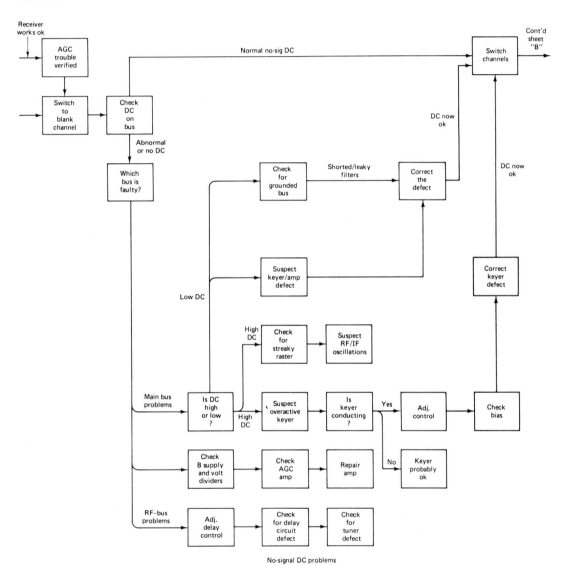

No-signal DC problems

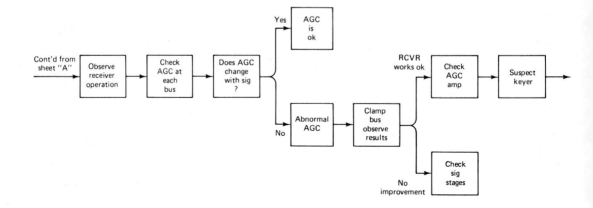
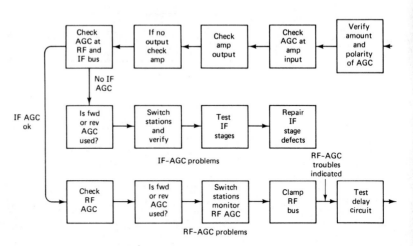

Fig. 11-18 AGC troubleshooting flow diagram (sheet B).

378

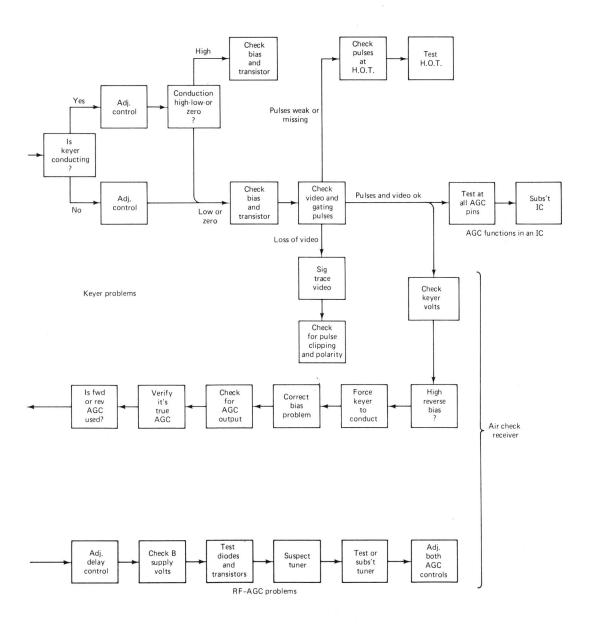

• Checking the Bus Voltage

At this point, switch to a blank channel and measure the dc on the AGC bus. Compare the amount and polarity with that shown on the schematic. The polarity is the same as its source, the PS. For maximum gain with the least amount of snow on weak channels, this voltage (the source of forward bias for the controlled transistors) can be quite critical. Make *accurate* measurements. Check and compare voltages on the RF and IF buses. Where no voltage or an abnormal voltage is measured, it must be corrected before proceeding. Typical causes are described in Sec. 11-23.

11-17 AGC TESTING WITH A SIGNAL RECEIVED

Refer to the troubleshooting flow diagram, sheet B for the following procedures.

REMINDER: ────────────────

At this point, it is assumed there is normal dc on the bus when switched off channel.

Start by monitoring (and recording) voltage variations (if any) on the main AGC bus when switching between weak and strong channels. Compare with the normal range of voltages shown on the schematic. By observing the schematic, determine whether the set uses forward AGC, reverse AGC, or both.

Measure voltage at a number of checkpoints: the keyer output, input and output of the AGC amplifier, the RF and IF bus. If voltage distribution to the controlled stages is interrupted, check for an amplifier defect or the AGC filters. If voltage is abnormal (too high, too low, wrong polarity, or it does not change with the signal) it is probably not a *true* AGC voltage. The next step is to clamp the bus with the *proper* voltage to see if normal set operation can be restored.

11-18 AGC CLAMPING WITH A BIAS BOX

Grounding the bus is a common test procedure with tube-type receivers. It is a fast way to obtain zero bias for the controlled tubes to see if the trouble symptoms clear up. This does *not* work, however, with solid-state receivers because a transistor, in order to amplify, must be provided with a small amount of forward bias. Grounding the bus would negate this bias, and the transistors would be driven to cutoff.

Clamping is a means of stabilizing the bus voltage at some fixed desired value to determine if the AGC is causing the trouble symptoms. Clamping the bus defeats the AGC action. When a set has multiple troubles, including the AGC, clamping is often done while other troubles are being investigated or corrected.

Almost any source of dc can be used for clamping: a battery (when shunted with a potentiometer) or an adjustable, well-filtered power supply. A *bias box* (see Sec. 3-20) is designed for the purpose. A good bias box has the following characteristics: It provides well-filtered dc of either polarity. It is adjustable. It has a low internal resistance for good voltage regulation and is able to supply a certain amount of *power*. It has an accurate built-in voltmeter so the output can be monitored at all times.

To use the bias box, first check to make sure the receiver bus is not grounded. With the set *off*, connect the dc leads to the bus and ground. Make *certain* of the polarity. Adjust the dc to the proper value specified on the schematic. Make sure there are no shorts at the connections, switch the set *on* and observe results. With most sets the voltage is quite critical and should be adjusted accordingly. In some cases, a deviation as low as 0.1% can either drive the controlled transistors to cutoff, or more disastrously, to saturation. Monitor the dc *at all times*.

Poor sync is a common by-product of AGC failure. And since a keyer can not function unless the horizontal sweep frequency is correct, adjust the horizontal hold before removing the bias box and before continuing with AGC testing.

11-19 KEYER PROBLEMS

When the bus voltage is abnormal and clamping restores set operation, the keyer circuit becomes a prime suspect. Determine whether the keyer is conducting or not conducting, and, if conducting, whether too little or too much. This can be determined by the *IR* drop across the keyer load or emitter resistors or by a polarity check at the collector. Try adjusting the threshold control.

Using a CRO or ac EVM, check for video at the keyer input. Critically evaluate the sync pulses for possible clipping or compression. For loss of video signal or pulse clipping, trace signal back to the video amplifier.

Using a CRO or ac EVM, check for *gating pulses* at the keyer input. Refer to the schematic for proper amplitude and polarity. For missing or abnormal pulses, trace back to the source, the flyback transformer.

When a keyer is not conducting or is conducting excessively, forcing it to conduct or not conduct can provide a useful clue. From the schematic, determine the desired bias conditions. Connect a bias box, adjusted for the proper voltage and polarity, between the keyer B and E terminals. Check the effect on keyer operation.

REMINDER: ────────────────

With a signal received, the keyer should be conducting. When tuned off channel, it should *not* be conducting.

11-20 CHECKING AGC DELAY ACTION

NOTE: ────────────────

Before checking for delay action, first verify normal AGC voltage on the *IF* bus.

When normal AGC is supplied to the IF stages, check for AGC voltage on the RF bus feeding the tuner. This is usually a white or yellow wire. There should be little or no AGC on a weak channel. If AGC is abnormal (too high, too low, wrong polarity, or no transistion when switching from a weak to a strong signal), switch to a blank channel and repeat the tests described in Sec. 11-16. Tentatively adjust the delay control.

If the RF bus voltage appears OK with no signal received, tune to an active channel. Connect voltmeters to both the RF and IF bus. To check the delay action, tune to a *weak* station. If necessary, replace the antenna with a short piece of wire. Under these conditions there should be a small AGC reading on the IF bus and *no* AGC on the *RF* bus.

NOTE: ────────────────

If desired, an RF signal generator can be used for this test in lieu of the station signal.

While monitoring the two meters, slowly increase the signal input to the tuner, using either the fine tuner or the RF attenuator if a generator is used. The IF AGC voltage should slowly increase. At some value of IF AGC (see the schematic), an *RF* AGC voltage should appear. With further increases in signal, both meter readings should increase proportionately.

REMINDER: ────────────────

Depending on whether forward or reverse AGC is used, a bus voltage may either increase *or decrease* with *increases* in signal strength.

382 AUTOMATIC GAIN CONTROL

Trouble in the delay circuit is indicated if there is no delay with a measurable amount of AGC on the RF bus on a weak station and excessive snow on the pix, if there is too much delay with no AGC to the tuner on a moderate-to-strong station, or if there is too much difference between RF and IF AGC readings on a strong signal. For any of these conditions, try adjusting the delay control.

If desired, the RF bus can be clamped to the voltage and polarity indicated on the schematic for a particular signal level while noting any improvement in set operation. If clamping does not help and symptoms persist, the tuner may be defective. Try a substitute tuner (see Sec. 12-22).

11-21 CONTROL ADJUSTMENT PROCEDURES

NOTE: _____

AGC controls should be adjusted at the location where the receiver is to be used. The alternative is to make adjustments using average signal levels for typically weak and strong stations. Final adjustments are made after all AGC problems have been corrected.

A suggested procedure is as follows:

1. Tune to a strong station and adjust the fine tuner.
2. Adjust the delay control to approximate mid-setting.
3. Slowly adjust the threshold control clockwise for increased pix contrast until some instability is noted.
4. If the receiver has a *noise control,* adjust for pix stability.
5. Repeat steps 3 and 4.
6. Adjust the delay control for minimum snow on a weak station.
7. If there are problems of adjacent channel interference (windshield-wiper effect), a compromise may be necessary with the delay control adjustment.
8. Check *all* channels for snow, pix contrast, and sync stability. Make final touch-up adjustments as required.

11-22 LOCALIZING AGC TROUBLES TO AN IC

Many sets use ICs for the AGC and other circuit functions, and it is not always easy to determine if a fault is with the IC or an associated external circuit. Never substitute or replace an IC without first making some preliminary tests. A defect *external* to the IC can destroy the new unit. Test procedures are as follows:

1. Measure and record all AGC-related dc voltages and signals entering and leaving the IC. Refer to the schematic for values and terminal identification. Particularly note the amount and polarity of video and gating pulses. Check for dc on each AGC bus and compare with the schematic. Check for AGC action on weak and strong signals.
2. If AGC troubles are indicated and the IC is suspect, remove the IC and recheck the signals and voltages at the terminals as before. If no external troubles are suspected, monitor the terminals where readings are abnormal and substitute with a good IC. If normal readings and set operation are not immediately restored, shut the set off and rethink the problem. Failure to do so can result in destruction of the new IC.
3. When repairs are completed, adjust the controls as in Sec. 11-21, and check set operation on all channels.

11-23 ISOLATING THE DEFECT

Because the AGC interfaces with so many other circuits, and especially where direct coupling is used, it is no easy matter to narrow the problem down to one functional area. But once localized, pinpointing the cause is no more difficult than for other sections of the set. Follow the usual procedures of scoping signals, making voltage and resistance checks at the trouble spot to locate a bad diode, transistor, capacitor, or resistor until the problem is resolved.

Procedures for isolating faults within each functional AGC subsection are described in the following paragraphs, starting at the *source* of the AGC and working toward the controlled stages.

- **Simple AGC Circuit**

This combined video detector–AGC circuit (Fig. 11-2) seldom gives any trouble. For loss of pix, verify that video from the last IF stage is reaching the detector. Check for demodulated video at the detector output. If there is no output or weak output, check the detector diode (with one lead disconnected). If in doubt, try a substitute, making sure of the correct polarity. If the diode is enclosed in a can, make all tests from the external connections.

When tuned to a station, check for dc developed across the load resistor, and at various points on the bus, up to each controlled stage. Monitor voltage changes when switching channels. Check for normal dc on the bus when switched to a blank channel. If necessary, check peaking coils and all resistors in the circuit. If there is any *ac* on the bus, try bridging the filters and decoupling capacitors. Make sure polarity of the electrolytics is consistent with the polarity of the dc on the bus.

REMINDER: ─────────────────────

A shorted or leaky transistor in a controlled stage will reduce or even kill the bus voltage.

- **Keyer Circuit**

The majority of AGC problems originate here because of the critical nature of the circuit and its dependence on the video and gating pulse inputs. The first check should be for dc at the keyer output when switched off channel. The amount (as specified on the schematic) is that furnished from the PS via the voltage dropping network. If abnormal in any way, it must be corrected before proceeding with AGC voltage testing per se. If voltage is excessive, the most common causes are an overactive keyer (conducting *all the time*) or a shorted or leaky keyer transistor, both of which will feed an additional, undesired dc to the bus. Check the transistor. When removed or disconnected, see if the bus voltage returns to normal.

Improper bias is the usual cause of an overactive keyer. Measure bias between B and E.

REMINDER: ─────────────────────

Operating bias is the *difference* between B and E voltages as shown on the schematic.

Depending on the set, the typical off-channel bias for a keyer is close to zero for the normal setting of the threshold control. See if normal bias can be obtained within range of the control. If not, the cause is either too little or too much voltage on either the B or the E with respect to ground. Determine which condition exists. The most likely problem is with the *B* voltage, which usually orginates from the directly coupled video amplifier. If so, the obvious next step is to troubleshoot that stage.

Assuming normal dc on the bus and a keyer that is not conducting (as it should not be) when switched off channel, the next step is to see if the keyer *is* conducting when a station is received. Determine this by the *IR* drop across any resistor in the collector or emitter circuits. The keyer may be nonconducting or conducting too little or excessively.

- **Overactive Keyer**

Sometimes the keyer conduction is not sufficient to upset the off-channel dc on the bus and shows up only when a signal is received.

384 AUTOMATIC GAIN CONTROL

REMINDER:

When tuned to a station, the base voltage changes with the incoming video signal, and the C voltage changes with the applied gating pulses. If either is too strong, the keyer will conduct heavily. Measure both signals with a CRO or EVM and compare with schematic.

Adjust the threshold control cw for *more reverse bias* or *less forward bias* to reduce the conductance and the developed AGC. Note the polarity reversal at the collector, which normally occurs when a signal is received.

• **Insufficient or No Conduction**

This is the most common problem, and the typical causes are a defective keyer transistor or pulse gate diode, too little forward bias or too much reverse bias, or the applied video and gating pulses are either missing or too weak. Each possibility must be checked individually.

Check the bias between B and E while adjusting the control ccw. Is there a bias problem that could prevent normal conduction? If so, make additional tests as described earlier. Double-check the transistor and pulse gate diode. These items are under considerable stress and leakage is a common problem. Replace if *any leakage* is indicated. A clue here is when a threshold control has little if any effect. Don't blame the control; it is the keyer that is not functioning.

• **Video Signal**

If there is little or no conduction, check for weak or missing video at the keyer base terminal. Compare p-p amplitude and polarity with that shown on the schematic.

REMINDER:

It is the *sync pulses,* not the video, that is important. For weak or missing pulses, scope trace the signal to the takeoff point at the video amplifier and, if necessary, beyond.

• **Gating Pulses**

Check for weak or missing gating pulses at the collector terminal. Compare p-p amplitude and polarity with that shown on the schematic. Trace the pulses to their source, the flyback transformer. Check for an open pulse gate or coupling capacitor. Make sure the gate diode is not connected backward. A common problem is shorted windings in the flyback coil. An ohmmeter check here is seldom conclusive. Use a flyback-yoke tester or *ring* the winding (see Sec. 9-21.) When testing or replacing the flyback, verify the connections. A keyer cannot conduct with pulses of the wrong polarity. Where a new transformer has been installed and the polarity is wrong, simply transpose the connections.

11-24 AGC AMPLIFIER

When there is normal AGC at the keyer output and the input to the AGC amplifier, measure the dc at its output. The increase between input and output is typically about two to three times. Check the schematic for exact amounts. Depending on the set, the output may be taken from the E, the C, or both. With the former, input and output polarities should be the same. For the latter, a reversal should occur.

The most common trouble is a bad transistor. Where dc readings at the input and output are the same, the transistor is probably shorted. It will show up with a simple ohmmeter check. As with the keyer, no leakage can be tolerated. Replace if the slightest leakage is indicated.

The amplifier is normally biased to cutoff in the absence of a signal, that is, no dc at its input. If the transistor is conducting off channel or, when tuned to a station, conducting excessively or not at all, and the transistor is OK, it is probably a bias problem. Check between B and E and compare with the schematic. If abnormal, locate and correct the cause.

11-25 AGC BUS PROBLEMS

Most bus problems show up when making off-channel dc measurements. For loss of voltage or abnormal readings, assuming the amplifier output is OK, check for troubles at the inputs to the controlled stages (particularly a leaky or shorted transistor). Check for a bad resistor in the PS voltage divider. Check for a leaky or shorted filter or decoupling capacitor. When there is an ac voltage on the bus, no matter how small, suspect an open or dried-up electrolytic. Check by substitution while observing correct polarity. When the faulty unit is bridged, the ac reading should drop to zero. Measure and compare AGC readings at the base inputs of each controlled IF stage. In a cascade arrangement where one stage controls the other (see Fig. 11-11d) and there is loss of dc bias for the *first* IF stage, suspect the *second* stage, in particular, the transistor or the emitter load resistor. Another possibility, when a *bifilar*-type IFT is used between the stages, is that primary-to-secondary leakage can create bias problems for the second stage. The best check for this is a dc voltage test at the base terminal *with the load removed* (i.e., with the transistor removed or disconnected from the IFT).

REMINDER:

An ohmmeter check from bus to ground can be misleading because of the shunting effect of transistor junctions. Take two measurements while reversing the meter leads. The higher reading is the correct one.

11-26 TROUBLESHOOTING THE DELAY CIRCUIT

Assuming that trouble has been localized to the delay circuit and the integrity of the tuner has been verified, proceed to *isolate* the defect. Switch to a blank channel and check for normal bias on the RF bus feeding the tuner terminal, which is usually accessible. If there is no voltage or an abnormal voltage, turn the set off and check the resistance to ground for possible leakage or shorts. Try reversing meter leads since the RF transistor junctions are essentially in parallel with the bus circuit. If necessary, disconnect the AGC bus from the tuner and repeat the resistance tests from the bus to ground and from the tuner terminal to ground. Try clamping the terminal with the proper amount and polarity of fixed bias.

CAUTION:

Excessive voltage can destroy the RF transistor, especially if it is a FET.

When trouble is isolated to the *bus* and not the tuner, check the delay diode or transistor, the associated decoupling and filter capacitors, and the voltage-dividing network from the PS source. Try adjusting the delay control. When tests are inconclusive, replace all components that are suspect.

REMINDER:

The no-signal bias on the RF bus is very critical, and accurate measurements are essential for the highest possible signal-to-noise ratio. Unfortunately, the *true B to E* bias on the RF transistor cannot be checked except from *inside* the tuner in most cases.

Where a series blocking diode or a transistor is used for delay, check to see that they are not conducting off channel or when a weak signal is received. If they *are* conducting, check for loss of reverse bias or the presence of forward bias on the diode or transistor. Where a *clamping diode* is used, check to see that it *is conducting* off channel and on a weak signal. Try adjusting the delay control.

Switch to a strong station to see if the delay diode or transistor is switched on or off, as the case may be, to feed an AGC voltage to the tuner. If this does not occur, either the diode or transistor has too much reverse bias or the AGC from the bus is not strong enough to do its job.

REMINDER: _____

For *forward* AGC, the polarity of the AGC is the same as for the no-signal dc bias on the bus. For *reverse* AGC, the polarity of the AGC and the dc bias are *opposite*. Verify this condition by measurement and checking with the schematic. With most sets, the normal voltage variation on the RF bus is from about 3 V with little or no signal to about 5 V on a strong channel that activates the delay circuit.

After correcting a defect, make final touch-up adjustments to both the AGC threshold and delay controls. Remember, the delay control is adjusted for the least amount of snow on a weak station. If the control is set too high (no delay), the RF stage is supplied with AGC at the same time as the IFs, and the result is a weak pix with snow. If set too low (too much delay), you get a good pix without snow, but the possibility of mixer overload on strong signals. In an extreme *fringe area*, see if clamping the RF bus to some value other than specified reduces the snow. If so, consider changing the RF bias using different-value resistors.

11-27 TROUBLESHOOTING AGC VOLTAGE REGULATORS

Where a regulator circuit is used to stabilize the bus voltage (as in Fig. 11-14) and voltage levels are not maintained within specified limits, perform the tests described in Sec. 6-13. Check for open, shorted, or leaky zener diodes, the most common cause of failure. If defective, check further for the possible *cause* of the failure, such as a leaky or shorted capacitor, or a *result* of the zener breakdown, such as a damaged resistor or a destroyed FET in the tuner.

REMINDER: _____

Special techniques are required for testing a FET (see Sec. 5-9).

11-28 AGC COMPONENT REPLACEMENT

AGC circuits are critical. When replacing transistors in the AGC or the controlled stages, use only transistors having the same characteristics as the original. Resistor and capacitor values should be within tolerance. Replace any diode whose leakage checks excessive. *Double-check* for correct polarity.

Electrolytics often become dehydrated and open up. Intermittently open capacitors are also common. A quick check for this condition is bridging with a suitable substitute. For a suspected leaky or shorted capacitor, one end must be cut loose prior to testing.

A replacement flyback transformer sometimes has a gating pulse winding with reversed polarity. In this case, simply reverse the coil leads. Where the gating pulse amplitude is excessive compared with the schematic, it can be reduced by adding a capacitor, such as a capacitive voltage divider, from the keyer collector to ground. The value should be determined by trial and error.

When repairs are completed, make a final touch-up adjustment of the AGC controls.

11-29 INTERMITTENT AGC TROUBLES

Intermittent problems are not uncommon, but don't confuse this with a set that has *real* but *marginal* AGC control resulting in *occasional* and intermittent trouble *symptoms*. For suspected intermittent AGC troubles, monitor the AGC voltage during receiver warm-up and longer if necessary. If the problem area has been localized, monitor that section for any abrupt voltage changes. For additional procedures, see Sec. 5-13.

REMINDER:

Abnormal changes of voltages in the AGC section can be either the *cause* of the problem or the *result* of troubles in some other circuit.

One solution to a difficult intermittent problem is to use the "shotgun approach," replacing all suspected components without bothering to test them.

11-30 MULTIPLE TROUBLES

Unless serviced when troubles occur, a set may develop a variety of symptoms over the years. Among these are AGC problems. When evaluating the receiver performance and diagnosing troubles, it is often helpful to stabilize the AGC by clamping the bus while considering the other receiver functions.

11-31 AIR CHECKING THE REPAIRED SET

Check receiver operation on *all* channels. Watch for excessive snow on the weak stations and signs of overload, especially sync instability on the stronger stations. If necessary, make final touch-up adjustments of the AGC controls. Some compromise may be necessary in areas of high interference or where there are both weak and strong signals.

11-32 SUMMARY

1. With AGC, the gain or amplification of a receiver is made to change in accordance with the strength of the received signal. The gain of the transistors in the controlled stages (the tuner and IF strip) is regulated by changes of bias. When a strong signal is received, a relatively high AGC voltage is developed, which when applied to the controlled transistors reduces their gain. For a weak signal, little if any AGC is produced and the amplification of the stages is increased. In this way, the signal level in the video amplifiers is maintained relatively constant.

2. With a *simple* AGC system, an AGC voltage is developed across the video detector load resistor, filtered, and applied as a variable bias to the controlled transistors. There are two serious problems with this basic system: (a) Noise impulses, acting like a signal, develop an AGC voltage, reducing the receiver sensitivity and gain. (b) The AGC produced tends to fluctuate with variations of *video* information, which it not a function of *carrier* strength. A good AGC system uses the *sync* pulses whose level remains constant for a given station signal.

3. AGC voltage is routed from its *source* to a *main AGC bus,* and then splits into an *RF bus* and an *IF bus* to feed their respective stages. Each bus contains one or more *RC* filtering networks to smooth out any fluctuations as caused by hum voltage, pulses, video, and the like. The networks also provide for *decoupling* the controlled stages from each other.

4. A *time constant* for the AGC system is established by the combination of all resistor and capacitor values. The TC should be reasonably short (using small values of R and C) so the system can respond to rapid variations of signal strength. At the same time, it must be long enough to ensure good filtering of low-frequency impulses.

5. Simple AGC systems have a number of shortcomings. They have no built-in noise immunity, and noise behaves like a signal, developing AGC voltage, which reduces the receiver sensitivity, a serious problem with weak reception. Also, such circuits by themselves do not develop enough AGC voltage for good control of strong signals.

6. The purpose of any AGC system is to control the gain of the RF/IF stages to prevent overload on strong signals. For maximum sensitivity on weak stations, there should be little or no control.

7. The amount of AGC voltage developed by any AGC system varies in accordance with the strength of the received signal. Off channel and when receiving weak stations, little if any AGC is developed and the sensitivity of the set is greatest. With strong signals, increasingly greater amounts of AGC reduce the gain of the controlled stages.

8. There are two systems of AGC, forward AGC and reverse AGC. With forward AGC, a forward-biasing voltage is developed, which reduces gain by biasing the controlled transistor toward *saturation.* With reverse AGC, a reverse biasing voltage is developed, which opposes the normal forward bias on a transistor, reverse biasing it toward *cutoff.* A PNP in a controlled stage always requires a negative-going AGC voltage, and an NPN, a positive-going voltage, regardless of whether foward or reverse AGC is used.

9. The highest gain is obtained from a transistor when supplied with a small amount of forward bias. The gain can be reduced by either increasing *or* decreasing the forward bias from this amount. Zero gain is obtained at either cutoff or saturation.

10. An AGC system should respond only to the peaks of the composite video signal, the tips of the sync pulses that have a constant amplitude, changing only with variations of the received *carrier.* Basic AGC circuits that rectify the average level of signal produce a fluctuating AGC voltage, and the receiver gain will vary with brightness changes of the televised scene.

11. Maximum gain from a transistor occurs at a certain level of conduction between the extremes of E/C current *cutoff* and *saturation.* This operating point is established by the B/E *forward* bias. The normal no-signal bias for transistors in AGC controlled stages is provided by the PS using voltage-dividing resistors. When tuned to a station, the bias is made to increase or decrease with the signal level.

12. Unlike a vacuum tube, the gain of a transistor can be reduced by either increasing *or decreasing* its conductance. The conductance is varied (by AGC action) to either increase or decrease the *forward bias*. Both systems are used by modern sets. When gain is reduced by *increasing* the conduction (toward saturation), it is called *forward AGC*. When gain is reduced by *reducing* the conduction (toward cutoff), it is known as *reverse AGC*.

13. AGC variations between minimum and maximum signal levels are fractional, seldom exceeding 2 or 3 V. Off-channel voltages and the range of variations with a signal are shown on most schematics. Whether forward or reverse AGC is used can be determined by the polarity and the type of transistors used in the controlled stages.

14. Most sets use keyed AGC. With this system, the AGC is developed by a *keyer stage* that is gated into conduction only during horizontal retrace periods. Because of this, a set is almost immune to the problems of *noise* and *airplane flutter*. The system also develops a fairly high AGC for good control.

15. The keyer stage is normally biased to cutoff when no signal is received. The cutoff level and the amount of AGC are determined by the setting of the AGC *threshold control*. To conduct (and develop an AGC voltage), the keyer must be simultaneously supplied with two kinds of signals: a sampling of the composite video signal and *gating pulses* supplied by the horizontal sweep section. The amount of AGC produced varies with the amplitude of the incoming sync pulses.

16. Keyer gating pulses are obtained from a winding on the flyback transformer. Pulse amplitude is typically around 25 V p-p. Pulse amplitude and polarity are determined by the number of windings on the transformer and which way it is connected. A PNP keyer transistor requires a negative pulse, and an NPN type, a positive pulse. With insufficient amplitude or the wrong polarity, the keyer cannot conduct.

17. Many sets use a dc amplifier to boost the AGC for better control over strong signals and/or to obtain a polarity reversal if needed. Like the keyer, an AGC amplifier is usually turned off until a signal is received. The base or emitter is driven with a foward-biasing voltage from the keyer. The amplified output may be taken from either the emitter, the collector, or both, depending on the requirements of the controlled stages.

18. *Both* IF stages may be controlled individually from the bus. A more common arrangement is to feed AGC only to the second IF transistor and use *its* output to control the first IF stage. In effect, the second IF stage is operating as an AGC *amplifier*.

19. Either forward or reverse AGC may be used to control the RF/IF stages. Some sets use a combination of *both* systems working together to reduce the gain.

20. Most sets use *delayed AGC* to delay control over the RF stage until a reasonably strong signal is received. Without such a delay, even a small amount of AGC voltage will reduce the RF gain somewhat, resulting in a weak pix with *snow*. AGC control is necessary on *strong* signals, however, to prevent mixer overload.

21. With most modern sets, the AGC circuit and most of the components are included in an *IC* along with other circuits. As with any IC, troubleshooting is limited to checking the external signals, voltages, and components, followed by replacement if necessary.

22. Because one tends to emulate the other, it is sometimes difficult to determine if trouble symptoms are caused by an AGC defect *or* by some other section of the receiver.

Thoughtful analysis followed by a few systematic, preliminary tests, including clamping of the bus, will usually determine the cause.

23. In general, AGC troubles fall into one of two categories: those that result in *loss of AGC,* and those that produce *excessive AGC.* An AGC voltage check and analyzing the *symptoms* will quickly determine which condition exists.

24. There are two main steps in troubleshooting any AGC system: (a) checking for normal fixed bias voltage on the bus when switched to a blank channel, and (b) checking for *changes* in the dc created by AGC action while switching from weak to strong stations. Important considerations are the *amount* and *polarity* of each reading.

25. A bias box can be very useful for AGC troubleshooting. If clamping the bus with dc of the proper amount and polarity improves receiver operation, an AGC defect is indicated. If there is no improvement, the AGC is absolved as the cause.

26. The trickiest circuit to troubleshoot is the keyer. In general, there are three kinds of keyer problems: (a) a keyer conducts *off channel* when no signal is received, (b) a keyer that does *not* conduct, and (c) an *overactive keyer,* where conduction is excessive. If conducting off channel, check for loss of reverse bias or a misadjusted threshold control. If there is little or no conduction, there may be too much reverse bias, loss of video or gating pulses, or the transistor may be bad.

27. The AGC threshold control is usually quite critical and should be adjusted, if possible, at the location where the set is to be used. If advanced too far (CW), there may be video overload and sync problems on strong signals. If set too low, there will be loss of pix contrast and color intensity.

28. The AGC *delay* control is also critical. At one extreme there will be no delay, resulting in a washed-out snowy pix on weak channels. At the other extreme setting, there may be problems of mixer overload. Remember, for weak-to-moderate signals, only the IF stages are controlled. The RF stage is controlled *only* on the strongest signals.

29. A fair amount of snow on an off-channel raster is considered normal. Lack of snow indicates trouble in some video stage following the tuner. Excessive snow indicates a noisy location. A washed-out pix with snow usually means a weak signal input to the tuner, an RF or mixer stage defect, excessive RF AGC, or a misadjusted delay control.

30. DC on an AGC bus is not necessarily an *AGC* voltage. An AGC voltage always varies with the strength of the received signal. The no-signal dc on the bus is provided by the PS to bias the controlled transistors for maximum gain. Too much or too little voltage can mean either a *loss of AGC* or *excessive* AGC, depending on whether forward or reverse AGC is employed.

31. The AGC output from a keyer always increases with the signal, and when forward AGC is used, so does the bus voltage. With *reverse* AGC, the forward biasing voltage on the bus is reduced and may even go through a polarity reversal. The reduction occurs because the AGC voltage polarity is opposite to that of the no-signal dc on the bus.

32. The terms *loss of AGC* and loss of *off-channel bus dc* are not synonymous, and each can develop problems. Certain troubles can cause the off-channel dc to be zero, too high, too low, or the wrong polarity. If this voltage is abnormal, there can be little or no AGC control.

33. Loss of AGC can cause video overdriving and clipping of sync pulses. Pulse clipping in turn can prevent the keyer from working, resulting in no AGC. This poses the question of which is the cause and which the result. To decide, try clamping the bus.

34. One way to determine whether there is loss of AGC or excessive AGC is to adjust the threshold control. A cw rotation usually decreases the AGC, and if conditions improve, excessive AGC is indicated. The opposite holds true for a ccw rotation.

35. It is customary first to correct any off-channel dc problems before being concerned with AGC. Typical causes of abnormal dc or loss of dc on the bus are problems with the PS source or voltage dividers feeding the bus, leaky or shorted AGC filters, leaky or shorted transistors such as the keyer, AGC amplifier, delay transistor, or in one of the controlled stages, an overactive keyer or AGC amplifier that is conducting at all times, or voltage reaching the bus in some roundabout way.

36. Excessive IF AGC reduces the gain of a controlled stage and in extreme cases makes it completely inoperative. The result is a weak pix or no pix, without snow. Too much RF AGC results in no pix or a weak pix with heavy snow.

37. Loss of IF AGC causes overdriving of the video amplifiers, resulting in excessive pix contrast and color intensity, sync pulse clipping with loss of sync, and sometimes a negative pix. In extreme cases the *raster* may be blanked out on strong signals. Loss of RF AGC may not be apparent on weak signals. On strong signals there could be mixer overdriving, resulting in adjacent channel interference or crosstalk.

38. A keyed AGC system is almost immune to noise because the AGC is developed for very brief periods and then only when the screen is blanked out. To further reduce the noise problem, some sets feed the video to the keyer via a noise inverter stage.

39. Fading or fluctuating signals caused by aircraft, a wobbly antenna, and so on, have minimal effect on the pix when a keyed AGC system is used. Keyed AGC is a fast-acting system because of its short TC and is able to follow rapid variations of signal. With the AGC stabilized, the pix contrast remains relatively constant.

40. Loss of pix can result from either too much or too little AGC. Excessive AGC will weaken or kill a controlled stage. Loss of AGC results in video overload on strong signals to weaken or kill the pix, and sometimes the raster, when direct coupling is used.

41. The keyer collector voltage shown on a schematic is under no-signal conditions with the keyer not conducting. When keyed to conducting when a signal is received, the AGC developed causes a reversal or polarity at this point. Such a polarity check is useful in determining whether or not a keyer is conducting.

42. For a keyer to conduct and develop an AGC voltage, the timing of the gating pulses must coincide with that of the incoming sync pulses. This means the horizontal sweep frequency must be correct when troubleshooting the AGC. This can be a problem is there is no pix on the screen.

REVIEW QUESTIONS

1. In what ways does AGC contribute to the performance of a TV receiver?
2. Briefly explain the operation of a basic AGC system.
3. What purpose is served by each of the following: (a) keyer stage, (b) gating pulses, (c) AGC filters, (d) decoupling networks, (e) AGC amplifier, (f) delay circuit?
4. What is the function of (a) AGC threshold control? (b) delay control?
5. What are the two important characteristics of an AGC voltage?
6. What is meant by *forward AGC* and *reverse AGC*?
7. What stages of a set are usually controlled by AGC? State the reasons.
8. In the absence of a signal (no AGC developed), the transistors in the controlled stages are biased for maximum gain.
(a) Where does this fixed bias come from?
(b) What changes this voltage when a signal is received?
(c) Does it *increase* or *decrease*?
9. Loss of signal in the tuner results in reduced contrast or no pix *with snow*. For loss of signal in an *IF* stage, there is *no snow*. Explain why.
10. Fill in the missing words: With forward AGC, the developed AGC _____ the normal *forward* bias of a controlled transistor. With reverse AGC, the *forward* bias is made to _____.
11. Explain how the gain of a transistor can be *reduced* with either an increase or decrease of forward bias.
12. (a) How can the gain of a first IF stage be controlled when the bus feeds only the second IF stage?
(b) Explain how it works.
(c) What is the advantage of this method?
13. (a) Explain how sync pulse clipping in a video amplifier stage can cause loss of AGC.
(b) How does loss of AGC in turn result in pulse clipping?
14. State reasons for the use of an AGC amplifier-inverter.
15. Is there a need for AGC when tuned to (a) a weak station? (b) a strong station? Explain.
16. (a) Explain why loss of AGC causes video overload and sync clipping on strong signals.
(b) What are the symptoms for this condition?

17. Besides a weak or missing pix, tuner trouble or excessive AGC to the RF stage usually produces *snow*. Explain why.

18. What are the advantages of keyed AGC over a simple AGC system?

19. It is important that *no AGC* be supplied to the RF stage when receiving weak signals.
(a) Explain why.
(b) What circuit normally prevents this from happening?

20. Wrong component values can change the time constant of any AGC system. What troubles may result if the TC is (a) too long? (b) too short?

21. What is a clamping diode? Explain its function. Some sets use a transistor for this function. Why?

22. How does AGC control of a FET differ from that of conventional transistors?

23. How can *both* + and − AGC voltages be obtained from an AGC amplifier?

24. AGC voltage may be fed to either the B or E of an AGC amplifier, and the output may be taken from either the E or the C. Explain the polarity–phase relationships for each of these configurations.

25. Under no-signal conditions, is a transistor in a controlled stage biased to cutoff, saturation, or for maximum gain?

26. (a) Does a PNP transistor in a controlled stage require a negative-going AGC voltage *or* a positive-going voltage?
(b) Which would be used, *forward or reverse* AGC?

27. Some sets use forward AGC for controlling one stage and reverse AGC for another. How is this possible when both are fed from a common bus of one polarity?

28. Explain how keyed AGC helps to prevent airplane flutter.

29. Explain the operation of an AGC keyer circuit.

30. Explain why the horizontal sweep frequency must be correct in order for the keyer to do its job.

31. (a) What is an *overactive keyer?* (b) What are the causes and results?

32. Why does a keyer not conduct when no signal is received?

33. (a) What polarity of video and gating pulses is required to activate a PNP keyer transistor?
(b) What is the result if one or the other has the wrong polarity?

34. (a) In what way does the threshold control control the keyer?
(b) How is keyer conductance affected if the control is turned fully cw? fully ccw?

35. AGC control of the RF stage is *delayed* until moderate-to-strong signals are received. Why is this desirable and how is it accomplished?

36. A schematic shows a negative (off-channel) AGC voltage on the bus feeding an NPN IF transistor. Does the set use *forward* or *reverse* AGC?

37. What polarity of AGC is required to feed the emitter of a PNP controlled transistor using forward AGC?

38. What is the source of the gating pulses. Give three characteristics of the pulses.

39. Explain the need for the pulse gate diode in a solid-state keyer circuit and how it contributes to the development of an AGC voltage.

40. Explain why a keyed AGC system has a high degree of noise immunity.

41. What polarity of AGC is developed by a (a) PNP keyer? (b) an NPN?

42. What is the purpose of AGC delay for the tuner?

43. What determines the amount of AGC developed by a keyer, the amplitude of the *sync* pulses or the strength of the *gating* pulses?

44. Name the three essential keyer operating conditions that must be satisfied for a keyer to conduct and produce an AGC output.

45. What are the expected results for the following keyer conditions: (a) gating pulses too strong? (b) too weak? (c) no pulses? (d) strong pulses but reversed polarity?

46. Why are the RF and IF stages controlled *separately*?

47. The polarity of the dc at a keyer collector changes when the tuner is switched from off channel to a station. Explain the reversal. Is this dc an *applied* or a *developed* voltage?

48. Explain how either *loss* of AGC or *excessive* AGC can cause loss of the pix.

49. Explain why self-oscillations in a tuner or IF stage cause excessive AGC that weakens or kills the pix, and sometimes the raster.

50. With no signal received, the voltages measured at the transistor in an AGC controlled stage are as follows: $+9.8$ V at the base and $+10$ V at the emitter.
(a) What is the actual operating bias (between B and E) when tuned to a station that produces 2 V of forward AGC.
(b) Is the transistor a PNP or an NPN?

51. A PNP controlled IF transistor is forward biased with a small negative voltage from the PS.
(a) Where *reverse* AGC is used, what is the polarity of the AGC feeding the bus?
(b) Does the bus voltage *increase* or *decrease* with the signal?

52. How can increase in AGC voltage *decrease* the bias on a controlled transistor? Illustrate with an example.

53. An AGC bus is supplied with B+. How is it possible to simultaneously forward bias both an NPN and a PNP transistor in the controlled stages?

54. How would you determine from the schematic whether a set uses forward or reverse AGC?

55. Exactly why does snow show up on a weak signal?

56. (a) What is the source of the no-signal dc on an AGC bus? (b) What purpose does it serve?

57. Explain *how* the AGC threshold and delay controls perform their respective functions.

58. Is the keyer transistor enabled by voltage supplied to the collector from the PS or from the flyback transformer?

59. Fill in the missing words: A keyer conducts only when supplied with two kinds of signals, a sampling of the _____, and _____. The two signals must occur _____. This means the horizontal sweep _____ must be correct. A keyer conducts and produces AGC only for the duration of the _____. The AGC voltage is sustained between pulse intervals by the _____.

60. Since a keyer is turned off in the absence of a signal, how do you account for the dc collector voltage as indicated on a schematic?

61. When is the sensitivity of a receiver the greatest, on channel or off channel? Explain why, relating it to the amount of snow observed on the screen.

62. Fill in the missing words: No delay to the RF AGC results in a _____, pix with _____ on a _____ channel. No RF AGC on strong signals results in _____. No AGC on the IF stages causes _____, especially on _____ signals.

63. Explain why loss of AGC causes poor sync, especially on strong channels.

64. Fill in the missing words: With forward AGC control of a PNP, an increase of + AGC _____ the gain. With reverse AGC, the gain of a PNP is _____ by a negative-going AGC.

65. A short or leakage in a controlled transistor will upset the no-signal dc on the bus. Explain why. Will the bus voltage increase or decrease? Explain.

66. Why is there more snow and noise off channel than when a signal is received?

67. Explain why the no-signal polarity of the dc at a keyer collector is opposite in polarity to the AGC developed at the collector.

68. A particular set uses NPN transistors in the controlled stages. All voltages shown on the schematic have a positive polarity. Is forward or reverse AGC being used? Explain your reasoning.

TROUBLESHOOTING QUESTIONNAIRE

1. Why is it customary to begin AGC troubleshooting *off channel* when no signal is received?

396 AUTOMATIC GAIN CONTROL

2. A schematic shows +2.5 V at the base of a PNP IF amplifier and +3.2 V at the emitter. When a weak signal is received, approximately how much dc is required from a bias box when clamping the bus to obtain maximum gain? What polarity?

3. State the probable cause if the dc at the collector of an AGC amplifier measures the same as at the base; (b) if the reading is zero volts?

4. What is the effect, if any, on the off-channel bus voltage if the keyer is conducting at all times?

5. A set has multiple troubles, including problems with the AGC. How would you proceed?

6. State the possible causes when gating pulses at the keyer collector are (a) weak, (b) missing, (c) fluctuating, (d) wrong polarity.

7. A scope shows good video at the base of the keyer transistor but the sync pulses are weak or missing. State the probable cause.

8. When clamping a bus, state the probable result if the bias box (a) has reversed polarity; (b) polarity is correct but output is excessive when the set uses *forward AGC;* (c) as in (b) when *reverse AGC* is used?

9. As a threshold control is turned cw, pix contrast increases as it should. At the extreme setting, pix loses sync as expected but the contrast is poor. What would you suspect?

10. How would you check the operation of the delay circuit?

11. State several possible causes where the *off channel* voltage on the bus checks (a) zero, (b) too high, (c) too low, (d) wrong polarity.

12. State a logical troubleshooting procedure when there is a normal raster but no picture or sound, and it is not yet known if the fault is with the AGC or elsewhere.

13. Does a set use *forward* or *reverse* AGC if the AGC from the keyer is positive going and the fixed bias on the bus is *negative*? Explain your reasoning.

14. List several possible causes where a keyer is (a) conducting off-channel; (b) not conducting when a signal is received.

15. After replacing a flyback transformer, it is discovered that the AGC is not working. Assuming an association, what could be the cause?

16. What AGC problems will cause (a) loss of pix? (b) loss of raster? (c) excessive contrast and poor sync? (d) excessive snow on weak channels?

17. What initial tests would you make when the symptoms are (a) no snow on the off-channel raster? (b) excessive snow off channel? (c) weak pix with snow? (d) weak pix with no snow? Explain your reasoning in each case.

18. Fill in the missing words (a) No snow off channel usually means the AGC is _____ and trouble is in the _____ section. Normal snow off channel

and loss of pix on strong signals means it is probably an _____ problem.
(b) Two symptoms of _____ AGC are loss of signal and _____ snow on an unused channel.
(c) An indication of loss of gain in the IF section is _____ snow on a blank channel. For normal gain, there should be _____ snow off channel because receiver sensitivity is at a _____.

19. What precautions must be observed when clamping a bus with a bias box?

20. State several possibilities when the threshold control has little or no effect on set operation.

21. The AGC functions are incorporated in an IC. Voltages at all pins check normal, except there is no AGC output. How would you proceed?

22. Give the proper procedure for adjusting the (a) AGC threshold control, (b) delay control. (c) What troubles might be expected if either is misadjusted?

23. When the symptoms are loss of pix, no AGC, and all voltages on the keyer check normal, yet the keyer is not conducting, what could be wrong?

24. How would you determine if a keyer is (a) conducting? (b) not conducting? (c) conducting but excessively? (d) not enough? How would the dc on the off-channel bus and the AGC on channel be affected in each case?

25. Off channel a bus voltage checks normal, and the complaint is no pix or sound. Clamping the bus restores set operation. What is the problem?

26. When clamping an AGC bus, there is a drastic reduction in output from the bias box. State the possible causes.

27. (a) When there is no video at the keyer input and no pix on the screen, what stages are suspect?
(b) What would you check for loss of pix if there is video at the keyer?

28. The symptoms indicate possible AGC trouble, but clamping the bus shows no improvement. What would you do next?

29. What circuits would you suspect for the following trouble symptoms: (a) poor contrast and little or no AGC? (b) poor contrast and normal AGC? (c) poor contrast and excessive AGC?

30. (a) How would you check for suspected shorted windings in the pulse pickup coil of a flyback transformer?
(b) Why is an ohm's check seldom conclusive?

31. For consistently weak fringe-area reception, a receiver operation may sometimes be improved by disconnecting the AGC from the tuner. Explain.

32. A set has a washed-out pix, poor sync, and excessive snow. What would you suspect if the AGC checks (a) low? (b) high?

398 AUTOMATIC GAIN CONTROL

33. A video detector inadvertently connected backward results in a negative pix and loss of AGC. Explain why for each condition.

34. The suspected cause of excessive AGC is oscillations in the tuner or IF section.
(a) How would you verify?
(b) Why do oscillations in a signal stage result in a high AGC?

35. When is AGC voltage the greatest, with the threshold control fully cw, or ccw? Explain.

36. A set has a weak pix with snow. Clamping the RF bus restores normal operation. What would you check for? Explain your reasoning.

37. (a) What is the effect on set operation if a bus voltage-dividing resistor to ground opens up or increases in value? Explain why.
(b) What if it is the dropping resistor from the PS?

38. A raster and/or pix is lost when tuned to a strong station. What trouble is indicated? Explain why this occurs.

39. What would you suspect when sync problems clear up when the antenna is disconnected or replaced with a short piece of wire?

40. State the probable cause when there is considerable AGC on the RF bus on a weak station.

41. (a) Where the complaint is no pix or a weak pix and there is no measurable AGC on a weak channel, would you suspect AGC trouble or trouble in one of the signal stages?
(b) What if there is a *high* AGC voltage? Explain your reasoning.

42. How can you distinguish between dc furnished by the power supply and the dc that is truly an *AGC* voltage?

43. Where the complaint is a weak pix with snow and clamping the bus gives no improvement, what is wrong?

44. A set in a fringe area has the classic symptoms of weak pix with snow. No AGC can be measured at the bus.
(a) Is the AGC circuit suspect?
(b) What would you do?

45. (a) When AGC troubles persist after replacing a *defective* IC, how would you proceed?
(b) Name two trouble conditions that may still exist.
(c) What precautionary tests should be made *before* trying a new IC?

46. State the possible causes when there is a reduction in AGC voltage between the keyer and the controlled stages.

47. (a) Explain the procedure for checking AGC action using an RF signal generator instead of the station signals. (b) What precautions must be observed?

48. What tests would you make if a keyer is not conducting and both video and gating pulses at the keyer check normal?

49. (a) When the AGC checks low and there is loss of pix contrast and no snow, what would you suspect?
(b) What if there is a weak pix and no snow, but the AGC is *normal*?
(c) What if the AGC is low and the pix is weak with excessive snow?

50. What are the symptoms of an open or shorted gating pulse coil on the flyback?

51. Give several possible causes of an overactive keyer.

52. (a) State the possible causes when no dc can be measured off channel on the bus.
(b) What would you check first?

53. The AGC is fed to a second IF stage, which in turn controls the first stage. If there is loss of pix but no snow, and the AGC is normal on the IF bus, but the first stage is not being controlled, where would you look for the cause?

54. What simple test will establish whether the delay circuit is operating or not?

55. Explain clamping of the AGC and why it is sometimes done when troubleshooting.

56. State possible causes of low output from the AGC winding of a flyback transformer.

57. The result of a reversed video detector diode is a negative pix and loss of sync. Explain why for each condition.

Chapter 12

CHANNEL SELECTORS TUNING SYSTEMS, AND REMOTE CONTROL

12-1 INTRODUCTION

Whereas the foregoing chapters dealt with signals generated *within* the set, the tuner or channel selector (sometimes called the front end) is the first of several stages to process the *station signals* picked up by the antenna. Each of the two tuners has two primary functions: to select and amplify a desired station signal (to the exclusion of all others), and to convert it to a lower carrier frequency for additional and more efficient amplification by the tuned video IF stages that follow. Each station signal actually comprises *two* signals, the picture and sound RF carriers with their modulation sidebands.

By way of review, each TV station has an assigned channel number in either the VHF or the UHF band. For VHF, the channels are numbered 2 through 13, and for UHF, 14 through 83, as shown in Table 1-1. Each station occupies 6 MHz of the available RF spectrum.

12-2 GENERAL PHYSICAL CONSIDERATIONS

In the past few years there have been many improvements and innovations in tuner design: the all-electronic *varactor* tuner that dispenses with troublesome switches; push-button channel selection from remote *keyboards,* elimination of the fine tuning control in some cases, and improved methods for displaying channel numbers, to name a few. Despite such changes, all tuners still perform their same basic function as just described. A modern tuner assembly is shown in Fig. 12-1.

401

402 CHANNEL SELECTORS AND TUNING SYSTEMS

Fig. 12-1 Typical VHF-UHF varactor tuners.

Except for varactor tuners (which may be located anywhere on the main chassis) most tuners consist of a separate subassembly mounted behind the front panel and connected to the chassis with extended wires and cables. The VHF and UHF tuners are located side by side with either combined or separate tuning controls. With some sets, both the VHF and UHF (varactor) tuners are combined in the same shielded box.

Panel Controls. A typical VHF tuner has two control knobs, one for channel selection and the other for fine tuning. The selector has a mechanical *detent* that stops the selector at each channel position coinciding with the switching contacts. A UHF tuner usually has a two-speed drive that dispenses with a separate fine tuner.

A varactor tuner uses touch-tune push buttons for channel selection and may or may not require a fine tuner. Instead of, or in addition to, a fine tuner, some sets have an optional feature, automatic frequency control (AFT) that automatically controls the tuning. An AFT *defeat switch* is located on the panel for switching the circuit on or off as desired.

Channel Identification. Older sets used etched dials or translucent windows with channel numbers lighted from behind. Modern sets use *LED* readouts or display the numbers on the pix tube screen.

Antenna Input. Most sets have four antenna terminals on the back cover panel, two for VHF and two for UHF. Small table model sets and portables usually have some form of built-in antenna, a single rod (monopole) or dual rabbit ears. The rods are oriented toward the station(s) for best reception. Some sets have a single loop for UHF reception.

Tuner Input-Output Connections. Terminals on the tuners provide for interconnection between the VHF and UHF units and with the main chassis. Voltages such as AGC and the dc power connect to *feed-through capacitors* that serve a dual purpose, wire termination and bypassing or decoupling. A typical feed-through capacitor as shown in Fig. 12-1 consists of a small piece of ceramic tubing (the capacitor dielectric) enclosing an inner connecting wire with a terminal at each end. They are used at various places in the tuner.

Short sections of ribbon line feed the station signals from the antenna terminals to the tuner inputs. The signal output from the UHF tuner connects to the VHF tuner via phono-type pin jacks and a short length of coaxial cable. The signal output from the VHF tuner goes to the IF amplifiers on the main chassis, also via coax.

Varactor tuners have additional connections for bandswitching and tuning voltages. Modern sophisticated tuning systems require still another connection for sampling the signal generated by the tuner oscillator.

12-3 BASIC TUNER FUNCTIONS

A basic block diagram of a TV front end is shown in Fig. 12-2. The VHF tuner has three stages, an *RF* stage, a *mixer* (frequency converter), and an *oscillator*. Each stage contains a transistor and a number of tuned circuits. The main part of the tuner is a *channel selector,* which operates in conjunction with the tuned circuits to respond to the desired station signal.

A TV receiver requires many amplifier stages to boost the weak signals intercepted by the antenna to the level needed for good picture and sound. Unfortunately, there is a practical limit to how much amplification can be obtained at high (VHF) frequencies, and even more so with UHF. This problem has been overcome in

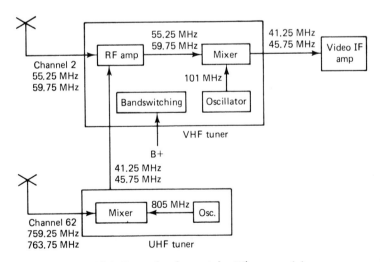

Fig. 12-2 Tuner fundamentals. When receiving UHF, the VHF RF and mixer stages function as IF amplifiers and the VHF oscillator is disabled.

TV and radio receivers by converting the selected station signals to a *lower frequency* called the *intermediate* (IF) *frequency*. The advantages of amplifying at IF frequencies (compared to RF) are greater *efficiency* and *selectivity,* the ability to accept only one signal while discriminating against all others. Tuned circuits in the IF amplifier section are *fixed-tuned* to *one* frequency, the IF, which does not change regardless of which station is received.

As with *radio* receivers, the front end of a TV set works on the *heterodyne principle,* where an unmodulated *continuous-wave* (CW) signal generated by a local oscillator combines (in the mixer stage) with the received station signals to produce signals at the IF frequency. Except for the lower carrier frequencies, in the conversion process, the picture and sound information remains unchanged.

A TV station sends out two signals, a pix RF carrier and a sound RF carrier. Modulation sidebands of the pix carrier contain information for producing a pix in BW and color, plus synchronizing information. Sound signals exist as sidebands to the sound RF carrier. The two RF carriers are separated by 4.5 MHz.

With *radio,* having only one station carrier, only one IF signal is produced in the conversion process. With TV, having *two* carriers, *two* IF signals are developed, a pix IF carrier and a sound IF carrier, with the same 4.5-MHz separation as before and containing the same pix and sound information.

The oscillator frequency is always *higher* than the RF station carriers by the amount of the IFs. During conversion, the single oscillator signal *beats* or *heterodynes* with the station carriers, producing the IF signals, which are the *numerical difference* between the oscillator frequency and the pix and sound RF carriers.

Early-model sets used IF frequencies in the 20-MHz range. With modern sets, the pix IF is standardized at 45.75 MHz and the sound at 41.25 MHz. As an example, suppose we are tuned to channel 5, which has a pix carrier of 77.25 MHz and a sound carrier of 81.75 MHz (see Table 1-1). To produce a pix IF of 45.75 MHz, the oscillator frequency must be 77.25 + 45.75, or 123 MHz. Similarly, the *same* oscillator frequency simultaneously produces a *sound* IF carrier, that is, 81.75 + 41.25, which is 123 MHz.

Since the pix and sound signals travel together through the video IF amplifiers, the IF section must have a *bandpass* of at least 4.5 MHz, with a response centered at approximately 43 MHz. At the output of the mixer, on the tuner, is a tuned circuit also resonant to this frequency.

Besides a numerical difference signal, other signals are produced in the mixer, such as the *sum* of the oscillator signal and the station carriers, and from beats with the oscillator *harmonics.* These serve no purpose and are rejected by the tuned circuits at the mixer output and IF amplifier. Note that the pix and sound signals have become transposed in the mixing process. Prior to conversion, the pix carrier was *lower* than the sound carrier. After conversion, a reversal has taken place, with the sound on the *low* side. If television used the sum instead of the difference signal for the IF, there would be no reversal.

To produce good pix and sound, the IF frequencies produced in the tuner must coincide exactly with the tuning of the IF amplifier section. To develop the IF signals, considering that the IF section is fixed-tuned to one band of frequencies, a different oscillator frequency must be developed for each station received. And the frequency must be *exact*. If the oscillator is *far off* frequency (by several megahertz, for example), the IF signals produced will be off by the same amount and will not get through the IF strip. The result is complete loss of pix and sound. Where the oscillator frequency error is small, the result is a degraded pix, color, or sound, depending on whether the frequency is too high or too low. The same problems can be expected if the *IF amplifiers* are mistuned, even though the oscillator frequency is correct and the proper IF frequencies are being developed.

• **UHF Reception**

Except for the absence of an RF stage and the difference in frequencies, operation of the UHF tuner shown in Fig. 12-2 is essentially the same as for VHF. When a UHF channel is selected, the CW signal generated by the local UHF oscillator beats with the two station carriers to produce the normal pix and sound IF carriers of 45.75 and 41.25 MHz, respectively. Instead of feeding the IF strip direct, however, the signals are routed through the VHF tuner RF and mixer stages, which now operate as broadband IF amplifiers for additional gain. With the VHF tuner selector at the UHF position, the VHF oscillator is disabled, since no VHF conversion is required. From the VHF mixer, the UHF signals go through the 43-MHz tuned transformer, then via coaxial cable to the IF amplifier in the normal manner.

If the UHF tuner is set for channel 62 for example, the UHF oscillator frequency would be 805 MHz, which combined with the pix and sound RF carriers of 759.25 and 763.75 MHz produces the required IF carriers of 45.75 MHz and 41.25 MHz. For UHF reception, a changeover switch on the VHF tuner disables the oscillator, disconnects the RF stage from the VHF antenna circuit input, and feeds dc power to the UHF tuner. With varactor tuners, these operations are performed by *switching diodes*.

12-4 TUNED CIRCUIT FUNDAMENTALS

The difference in assigned carrier frequencies makes it possible to select one particular station signal from all the others present at the tuner inputs. This is accomplished with *tuned circuits* that are *resonant* to the selected station frequency. Besides the tuner, tuned circuits play an important role in other sections of a TV receiver: the video and chroma stages, the sound section, and others. Hence a good understanding of the fundamentals is *essential*, particularly when alignment of these tuned circuits becomes necessary. Following is a brief review of these fundamentals.

• **Inductance**

Inductance is the property of a coil (inductor) and is designated as L, that opposes the flow of an ac current. The basic unit is the *henry* (H). The very small coils found in TV tuners are rated in microhenries (μH). Factors that determine the inductance of a coil are the number of windings, physical dimensions, and the type of core if there is one. A large coil of many turns with an iron core has considerable inductance. Conversely, a small coil of a few turns with no core has very little inductance. When used in a tuned circuit, the greater the inductance is, the lower the frequency, and vice versa.

• **Capacitance and Capacitors**

Capacitance (C) is the property of a *capacitor* that enables it to store an electrical charge. There is also the capacitance that exists between any two objects, wires, or components in close proximity, which must be taken into account in high-frequency circuits as in tuners. Most tuned circuits consist of a coil and a capacitor. Sometimes, however, the only capacitance is that which exists between coil windings and the *distributed capacity* of the circuit.

The basic unit of capacitance is the farad (F). Small capacitors found in tuners are measured in picofarads (pF). A basic capacitor may consist of two or more metallic plates separated by a thin insulator called the dielectric. With solid-dielectric capacitors the insulation may be mica, ceramic, or other material with good insulating qualities. Factors that determine the capacitance of a capacitor are the number of plates and their surface area, the separation of the plates (thickness of the dielectric), and the type of dielectric used. For a given dielectric, a capacitor consisting of many large closely spaced plates would have a relatively high capacitance. A small *variable* capacitor on the other hand with well-separated plates and air as the dielectric would have a low capacitance. There are many types of capacitors used in television, some with a fixed value, others adjustable. In a tuned circuit, the greater the capacitance, the lower the resonant frequency, and vice versa. Where a very small capacitance is needed in a tuner, sometimes a *gimmick* capacitor is used. A gimmick (Fig. 12-8) consists of two short pieces of insulated wire twisted together. The more twists, the greater the capacitance is.

• **Reactance**

Reactance (X) is the opposition a coil or capacitor offers to the flow of alternating current. A coil in a dc circuit limits the current by the amount of its *ohmic resistance*. That same coil in an ac circuit has a greater opposition to current. It is called *inductive reactance,* designated X_L. The reactance varies with the frequency of the ac, as well as with inductance. The higher the frequency and/or inductance, the greater the reactance and the lower the current. Reactance is the result of opposing magnetic fields set up in the coil windings.

Because of its dielectric insulation a *capacitor* acts as an open circuit to dc so there can be no current flow. In an ac circuit, a capacitor in effect *passes* an ac current in accordance with the amount of its *capacitive reactance,* designated X_C. Capacitive reactance varies inversely with the frequency of the ac and the value of the capacitor. Besides their use in tuned circuits, capacitors are often used where both ac and dc exist together in a circuit to pass the one while blocking the other.

• **Resonance**

A combination of a coil and a capacitor represents a tuned circuit that is resonant to a specific frequency. The resonant frequency depends on the values of L and C in an inverse relationship where an increase in either one or the other or both represents a *lower* frequency, and vice versa.

At resonance, voltages developed across the coil and capacitor are equal and opposite and cancel out. For this condition the circuit is purely *resistive*. At frequencies *below* resonance, the inductive reactance predominates and the circuit is said to be *inductive*. The opposite is true for conditions above *resonance,* where the circuit becomes *capacitive*. Some circuits use fixed values of L and C and are resonant to but one fixed frequency. In tuned circuits, the resonant frequency may be changed by varying the values of L or C or both. In high-frequency tuner circuits, L and C are very small. Sometimes there is no capacitor as such, the only capacity being the *distributed capacity* of the coil windings and the capacities represented by associated components and wiring.

There are two kinds of resonant circuits, *parallel resonant* (where L and C are connected in parallel) and *series resonant* (where L and C are connected in *series* with each other). The impedance (*total* opposition to current) of a parallel circuit *at resonance* is *maximum,* developing maximum voltage for ac signals at the resonant frequency. At the same time, the impedance to signals off frequency (in either direction), and therefore the voltage developed, becomes progressively less the further their frequency is from resonance. This is the basis of a tuned circuit's ability to select a desired signal (according to frequency) while rejecting all others.

At resonance, the impedance of a *series-resonant* circuit (and therefore the voltage developed across the combination of *L* and *C*) is *minimum,* with increasing impedance to signals *off resonance.* Thus this type of tuned circuit *accepts* signals of resonant frequency while rejecting all others.

Both types of circuits are found in tuners and other TV stages. Figure 12-3a shows a parallel-resonant circuit shunted *across* a tuner stage for channel selection. In Fig. 12-3b, a series-resonant circuit accomplishes the same results in a different way. The same circuits can be used as *wave traps* to trap out undesired signals. They are connected *opposite* to parts (a) and (b). In Fig. 12-3c, the parallel-resonant trap rejects the undesired signal at resonant frequency while passing desired signals either side of resonance. The series-resonant trap shunts the undesired signal at resonant frequency to ground with negligible effect on signals off resonance.

• **Tuning Methods**

The resonant frequency of a tuned circuit may be varied in a number of ways. A common method is by changing the inductance of the coil with an adjustable core, a threaded *slug* of either powdered iron or some nonmagnetic material such as brass. An iron core *increases* the magnetic flux around the coil, and therefore its inductance, resulting in a *lower* frequency. The opposite is true for materials like brass, where induced eddy currents develop opposing magnetic fields, *reducing* the inductance, and *raising* the frequency. A single loop of wire inductively coupled to the coil and simulating one or more shorted windings accomplishes the same effect, a fairly common defect with coils of all types.

The frequency of a tuned circuit is varied as the core is moved in and out of the coil by varying amounts. Screwing an iron core into a coil *lowers* the frequency; a brass core *raises* the fre-

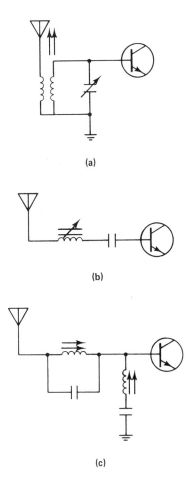

Fig. 12-3 Series- and parallel-resonant circuit applications: (a) high impedance at resonance and maximum signal feeding the transistor; (b) low impedance to resonant frequency and maximum rejection of signals off resonance; (c) series- and parallel-resonant circuits used as wave traps.

quency. During tuner manufacture, coils are sometimes resonated by either crowding (compressing) or spreading the end windings of a coil. The more crowded the windings, the lower the frequency is.

Variable capacitors and "trimmers" are also used in tuned circuits, for example, in the tuning capacitor found in most AM/FM radios and

UHF tuners. With *varactor*-type tuners, the varactor diodes simulate capacitors whose value is changed by the amount of applied dc.

• **Selectivity and Bandpass Considerations**

When only a high-frequency *carrier* and no sidebands are involved, a tuned circuit can be made very *selective* and efficient. Sharply tuned, highly selective circuits, however, cannot be used in the RF–mixer stages of a tuner because of the wide band of frequencies that must be covered. When tuned to channel 3, for example, the circuits are broadly resonant to approximately 63 MHz and are required to pass a band of frequencies extending from 60 to 66 MHz. This can be seen by the broadly tuned response shown in Fig. 12-4a. Here the optimum bandpass taken at the 50% level is 6 MHz, with the pix and sound RF carriers, which are 4.5 MHz apart, at the two peaks where they receive maximum amplification. By comparison, note the narrow, sharply peaked curve, which would provide higher selectivity and gain, but would

be unable to pass the pix and sound sidebands. Hence a trade-off must be made, sacrificing gain and selectivity in order to pass the all-important sidebands.

On the other hand, if the bandpass is *too wide*, gain and selectivity will be unnecessarily reduced, creating problems of adjacent channel interference. In Fig. 12-4a, the relative locations of the adjacent channel frequencies that could be troublesome are shown at the base line and beyond the response where they receive no amplification, the sound carrier for channel 2 on the left and the pix carrier for channel 4 on the right.

To further ensure against adjacent channel interference, the video *IF* stages contain traps tuned to these frequencies after conversion from RF to IF. A wide bandpass is obtained by shunting a tuned circuit with resistors or relying on the low shunting impedances of transistor junctions. With tuned *transformers,* the coils are closely spaced (*overcoupled*). Loose coupling and high-efficiency coils with *no* shunting are required for trap circuits.

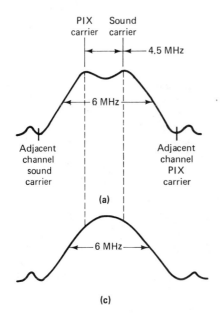

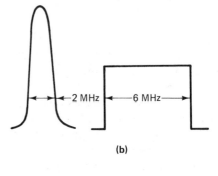

Fig. 12-4 Front-end response curves: (a) double-humped response; (b) ideal response; (c) haystack curve. A wide bandpass is required to pass all the sound and pix information of a TV signal.

12-6 RF-Mixer Tuned Circuits of a UHF Tuner

• **Response Curves**

A response curve, as shown in Fig. 12-4, is a convenient way of showing bandwidth and relative gain for frequencies on and off resonance. The ideal (but impossible-to-obtain) response for the RF-mixer stages would be a square- or rectangular-shaped response, flat on top to indicate equal amounts of gain for all frequencies within the bandpass (see Fig. 12-4b). The closest approach is the double-humped curve of Fig. 12-4a, or more often a single-humped curve commonly called a *haystack* response (Fig. 12-4c). Points on the curve representing different frequencies are called *markers*.

Let us now apply the foregoing fundamentals to the tuned circuits in TV tuners.

12-5 RF-MIXER TUNED CIRCUITS OF A VHF TUNER
(RF + MIXER)

Because of the extreme frequency range between channels 2 and 13, it is necessary to cover the VHF channels as two bands of frequencies, a *low band,* channels 2 through 6, and a *high band,* channels 7 through 13. The channels are thus known as either low-band or high-band channels. A separate set of coils is used for each channel for both the RF and mixer stages. Each coil is part of a tuned circuit that is resonant to the center of the response for whatever channel is selected (e.g., 57 MHz for channel 2, 63 MHz for channel 3). Each tuned circuit must have a bandpass of 6 MHz in order to pass all the pix and sound information being transmitted by the station.

A tuner may have either two or three channel-selecting tuned circuits at the RF stage input, its output, and sometimes at the mixer input. Channel selection is accomplished by either switching coils or, in the case of varactor tuners, with dc-controlled varactor diodes. Tuned circuits are provided for all 12 channels, although some are unused, depending on geographic location and the stations within range.

Besides the channel-selecting tuned circuits, one or more tuned traps are usually found at the tuner input between the antenna terminals and the RF stage. Their function is to prevent interference from FM stations, signals at the receiver IF frequency, or *image* frequency signals. At the mixer output is a tuned coil or transformer resonant to the approximate center of the IF (about 43 MHz).

All the preceding coils are factory tuned and unless damaged or tampered with seldom if ever require readjustment. Generally, the RF-mixer tuned circuits are not considered too critical because of their wide bandpass.

12-6 RF-MIXER TUNED CIRCUITS OF A UHF TUNER

Conventional tuned circuits cannot be used in UHF tuners because of the extremely high frequencies to be covered, 470 to 890 MHz for channels 14 through 83. At such frequencies a coil of only one turn may be one too many, and even a short length of wire may have considerable inductance, not counting distributed capacities, which can be significant.

Except for varactor tuners, tuning is accomplished with a continuously variable tuning capacitor in conjunction with tuned lines instead of coils. Some UHF tuners have an RF stage; some do not. FETs are often used instead of bipolar transistors because of their high gain and low noise at these frequencies.

As with the VHF tuner, the RF-mixer tuned circuits resonate to the desired channel frequency. Tuning is somewhat more critical than for VHF.

12-7 VHF AND UHF OSCILLATOR TUNED CIRCUITS

In both VHF and UHF tuners, the frequency generated by the local oscillator is higher than the received station carriers by the amount of the IFs. Thus, to produce the proper IF frequencies, the VHF oscillator must cover a range from 101 to 257 MHz, and the UHF tuner from 517 to 931 MHz (see Table 1-1).

A VHF oscillator coil has fewer windings than for the RF-mixer coils because of the higher frequencies. Unlike the RF-mixer circuits, however, there is no necessity for a wide bandpass since the oscillator only produces an unmodulated CW signal. It is the oscillator frequency *that determines which channel is received,* and *not* the frequency represented by the RF-mixer tuned circuits. For example, it is common to receive the wrong station on the wrong setting of the channel selector simply because the oscillator frequency is off.

Because of the *fixed-tuned* video IF stages, a different oscillator frequency must be generated for each channel received. When switching channels, the RF-mixer-oscillator tuned circuits are all changed at the same time. The three circuits are said to be *ganged,* with the oscillator always tracking with the signal stages by the correct amount.

The oscillator frequency, especially for the high channels and even more so for UHF, is extremely critical. Unless the frequency is *exact,* the pix, color, and sound will be adversely affected. Since oscillator *drift* can be a problem, especially at high frequencies, some means must be provided to vary and control the frequency. This is done with either a front panel fine tuning control or with AFT (automatic fine tuning). The fine tuner on a VHF tuner usually consists of a cam-operated slug that moves in and out of the oscillator coil, changing its frequency by a small amount.

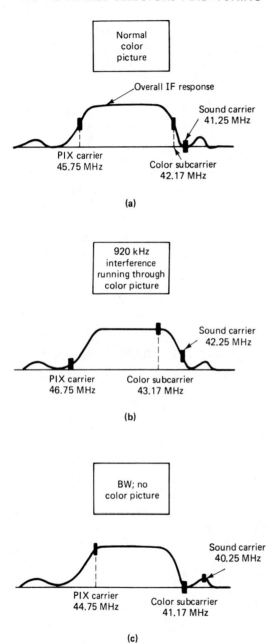

Fig. 12-5 Oscillator frequency variations and their effects: (a) normal osc. frequency, channel 3, 107 MHz; (b) osc. frequency high, channel 3, 108 MHz; (c) osc. frequency low, channel 3, 106 MHz.

The fine tuner for a varactor tuner is a potentiometer which varies the dc applied to the varactor diodes. Continuous UHF tuners do not require a fine tuner. Instead, a two-speed channel selector, having a very high (20:1) ratio to obtain the desired vernier control of frequency, is used.

Figure 12-5 shows how an incorrect oscillator frequency affects reception and the reproduction of pix, color, and sound. Figure 12-5a represents normal conditions when the oscillator frequency is correct and the pix IF carrier, the color subcarrier, and the sound IF carrier are all correctly located on the *IF response curve*.

In Figure 12-5b, the oscillator is 1 MHz too high, which places the pix marker too low on the curve, where it receives little or no amplification. The color subcarrier has moved to the top of the curve, where it receives too much gain relative to the weakened BW signal. The sound marker also is higher on the curve where the increased amplification can produce problems of *sync buzz* and a 920-kHz beat pattern on the screen due to the heterodyning of the sound carrier and the 3.58-MHz color subcarrier.

In Fig. 12-5c, the oscillator frequency is 1 MHz too low, which shifts the markers in the opposite direction. The effects produced are excessive pix contrast with smearing of large objects and complete loss of both color and sound.

The proper adjustment of the fine tuner is near mid-setting, where pix, color, and sound (the color tones in particular) are correct.

12-8 FUNCTIONAL OPERATION OF TUNER STAGES

A detailed block diagram of a typical front end is shown in Fig. 12-6. A description of each functional block follows.

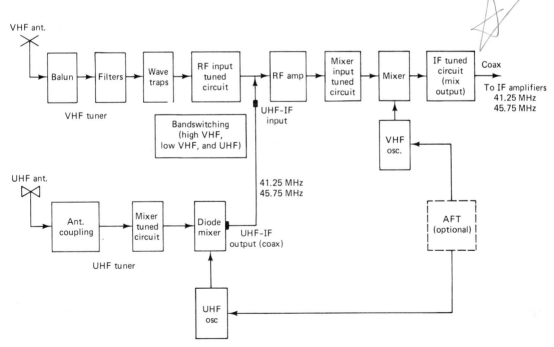

Fig. 12-6 VHF/UHF tuner functions.

• VHF Tuner Input Circuit

From the antenna terminals, the incoming signal goes through an impedance matching device known alternately as a *balun* or *elevator transformer*. Everything possible must be done to obtain the highest possible signal-to-noise ratio up to and including the RF stage. Thus, for maximum signal transfer, it becomes important to have a 1:1 impedance match between the antenna lead in and the tuner. A mismatch here creates a high standing wave ratio (SWR) because some of the received signal is reflected back up to the antenna and then returns, each *reflection* creating ghostlike images on the pix.

The typical input impedance is 300 Ω for the flat-ribbon-type lead in and 75 Ω for shielded coax. Some sets provide for both. A balun (balanced-input, unbalanced-output) is a versatile matching transformer that can accommodate either impedance according to how it is connected. Actually, the device is not really a transformer in the true sense; its coils, which are cut to a precise length, behave as two sets of inductively coupled parallel-tuned lines, much like the tuned circuits in a UHF tuner.

The tuner input is said to be *unbalanced* because one side connects to ground. The same is true for a coax cable, often used as a *transmission line* (antenna lead in) to reduce noise pickup. A 300-Ω ribbon line, on the other hand, is *balanced* because each of the two wires has the same distributed capacity to ground, and any signal pickup by the wires is cancelled out. Figure 12-7 shows two impedance matching hookups. In Fig. 12-7a, a simple balun is used to match a monopole antenna or a 75-Ω unbalanced antenna line (coax) to a 300-Ω balanced input. Note the grounded center tap on the coil. In part (b), the balun is reversed to go from a balanced antenna to the unbalanced tuner input. The most common configuration uses a four-coil balun to match a 300-Ω line to the tuner.

Certain safety devices are often used in the input circuit, for example, *spark gaps* connected

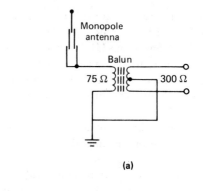

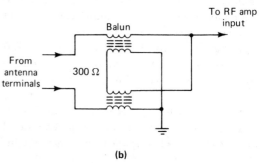

Fig. 12-7 Antenna-tuner impedance matching. (A balun is located on the outside of a tuner, see figures 12-1 and 12-8.)

from each side of the circuit to ground to protect easily damaged FETs and other components from static buildup on the antenna or lightning discharges.

Another precaution protects against shock hazard where the chassis is hot; that is, one side of the power line is connected to the chassis. Low-value capacitors are inserted in each input line to pass the signal, with high-value shunting resistors to drain away any accumulated charge.

• Wave Traps and Interference Filters

Both series- and parallel-resonant traps (Fig. 12-3) are used in the input circuit to reduce or eliminate certain kinds of interference. Parallel traps are in the signal path where they have high opposition to unwanted signals at their resonant frequency. Series traps shunt such signals to ground. These traps are used against interference from FM stations whose operating frequencies (88 to 108 MHz) fall within the VHF spectrum, between channels 6 and 7, and signals close to the IF frequency of 45.75 MHz. The traps are either fixed-tuned or adjustable. Their adjustment is critical. If tuned too high, they will attenuate the signals of channel 7. If tuned too low, there will be problems receiving channel 6. Such interference *must be* trapped out in the front end. If permitted to reach the IF section and beyond, it results in an interference pattern that seriously degrades the pix.

The input circuit of most sets also contains one or more *high-pass filters*. Their job is to pass the high-frequency TV signals while attenuating all frequencies lower than channel 2, which includes signals at IF frequencies. Some interfering frequencies *cannot* be eliminated with filters or traps in the tuner (because the TV station signals themselves would be rejected). This includes *image* frequencies and adjacent channels. An image signal is one whose frequency is higher than the tuner oscillator by the amount of the IF. For example, when receiving channel 2 the oscillator frequency is 101 MHz and the image frequency is 101 + 45.75, or 146.75 MHz. If a strong signal of this frequency reaches the mixer, it will beat with the oscillator to produce the 45.75-MHz IF and travel along with the desired station signal as a source of interference. Image, however, is seldom a problem because the high IF frequencies used puts it beyond the frequency range of most TV channels.

Interference from adjacent channels depends on the selectivity of the RF-mixer tuned circuits and traps in the IF amplifier section.

12-9 STAGE FUNCTIONS

A brief functional description for each of the three tuner stages is as follows. Circuit details for typical tuners are described in Sec. 12-10.

• **RF Amplifier Stage**

Most VHF tuners (and some UHF tuners) have an RF stage and for good reason. Besides selecting and amplifying the weak signals before they can be lost or overridden with background noise, it discriminates against *image* and other kinds of interference; it also helps to prevent the local oscillator signal from backing up to the antenna, where it would be radiated and interfere with the reception of nearby receivers.

The main items in the RF stage are the channel selector with its tuned circuits and the transistor, both of which contribute to signal amplification. Typically, the stage has two sets of tuned circuits for each channel, one at the transistor input and one at its output. One coil of each circuit is selected, along with those in the mixer-oscillator stages when the tuner is switched to any given channel.

A tuned circuit contributes to signal amplification by the amount the signal is boosted at resonance. This also represents an increase in *selectivity* because of the attenuation of undesired off-resonant signals. Up to a point, the greater the number of tuned circuits, the better the selectivity is.

In a tuner, the resonant signal voltage developed across the first tuned circuit drives the transistor for further amplification. Maximum amplification of the desired signal is obtained because the load impedance is greatest at resonance. From the RF stage, the signal drives the mixer, using either direct, capacitive, or inductive coupling.

Most of the background noise that produces snow on a weak signal is generated in the mixer stage. Therefore, to override the noise and obtain the highest possible signal-to-noise ratio, the mixer must be driven with a *strong* signal, and anything that reduces the drive, such as antenna problems or troubles in the tuner input circuits of the RF stage, will produce a weak pix with snow.

On the other hand, *excessive* drive creates overload problems on *strong* signals. As explained in Chapter 11, the AGC is responsible for controlling the RF amplification in accordance with the strength of the received signal. Off channel, and for weak signals, the RF transistor is forward biased for maximum amplification. Gain is reduced on strong signals by either increasing or decreasing the bias, depending on whether a *forward* or *reverse* AGC system is used. Some modern tuners have an AGC *delay control* to establish the level that gives the least amount of snow on weak channels.

Any *positive* (regenerative) *feedback* from output to input of the RF stage results in a narrowing of the response and general instability. Too much feedback results in *oscillations* and more severe problems. This is prevented by introducing a small amount of *negative feedback* to cancel out the positive feedback, using a *neutralizing capacitor*. The capacity of this unit is very small, typically about 1 pF). It may be a fixed value, a trimmer capacitor, or in some cases a *gimmick capacitor*. The amount of feedback, determined by the capacitor, is very critical. Too much *negative* feedback causes loss of gain, and *too little* feedback does not solve the problem. To prevent feedback, RF and mixer transistors are sometimes connected as common base amplifiers.

Some modern tuners use a FET in the RF stage because of its high sensitivity and signal-to-noise ratio.

• **Mixer Stage**

The mixer, like the RF stage, also contributes some RF gain and selectivity, but its main function is to develop the pix and sound IFs by combining the received station signals with a CW signal generated by the local oscillator. The transistor can be considered as a diode mixer followed by an amplifier. Mixing action makes use of the nonlinear characteristics of the B/E junction when supplied with a small amount of forward bias. The signal from the RF stage may be applied to either the *B* or the *E*. Oscillator injection may be to any of the three transistor elements. For good *conversion gain,* the oscillator signal should be considerably stronger than the station signal, typically 0.1 to 0.3 V rms at the injection point, the amount needed to overcome the mixer bias.

The RF tuned circuit at the mixer input is the same as those found in the RF stage. Some tuners have only one set of coils to simultaneously tune the RF stage output and the mixer input; others use separate tuned coils for each circuit. At the mixer output is a tuned circuit broadly resonant to about 43 MHz, the approximate center of the IF bandpass. Numerous *heterodyne beats* are produced in the mixing process, but only one is accepted by the output tuned circuit, the numerical difference signal comprising the pix and sound IFs. From here, the IF signal is fed to the IF amplifiers via coax cable.

Some tuners use a FET in the mixer stage, and in some cases an additional IF *preamp* stage for extra gain. Generally, there is no need to neutralize the mixer stage since its input and output circuits are at different frequencies. However, some feedback is possible under certain conditions. To prevent this and obtain increased stability with greater conversion efficiency, some tuners use a *cascade* type mixer, with two transistors in series. One transistor operates in the common base configuration with its base grounded for RF to prevent feedback. The other transistor functions as a pre-IF amplifier.

• **Local Oscillator**

Unlike the RF–mixer stages, the oscillator generates its own signal, an unmodulated (CW) signal that combines with the station signals in the mixer to produce the IF signals as previously described.

A tuner oscillator must have exceptionally high frequency stability and NTC (negative temperature capacitors) are used to reduce drift. The frequency is established by the tuned circuits, one for each channel, when the selector is turned. Since the RF, mixer, and oscillator tuned circuits all change at the same time, each oscillator coil must be adjusted to track with the station signals at all times to develop the proper IFs. The oscillator coil (or a portion of it) is usually made adjustable by in-and-out movements of an iron slug operating from the fine tuner control. With a *varactor* tuner, a different system of fine tuning is employed.

REMINDER:

Although it is the RF–mixer tuned circuits that resonate with the station signals, it is the *oscillator* frequency that determines which station is received.

12-10 VHF TUNER TYPES

There have been many changes in tuner design over the years to obtain the highest possible signal-to-noise ratio, increased reliability, and long life. Current trends seem directed toward user convenience. There are four basic tuner types: the *wafer-switch* type, the *drum* or *turret* type, the *continuous tuner* now used mostly for UHF, and the *varactor tuner*. The first two use mechanical switching arrangements for channel selection, switching different coils in and out of the circuit. The continuous tuner uses a variable capacitor, as with most radio receivers. The varactor tuner, currently the most popular, uses varactor diodes for tuning.

• **The Wafer–Switch Tuner**

This tuner, once used by many sets, is rapidly becoming obsolete. Channel switching is performed by waferlike switches that turn with the channel selector control. Most such tuners have four ganged wafers, two for the RF stage input and output, one for the mixer, and one for the oscillator. Each wafer has 13 switch positions for channels 2 through 13 and for UHF. Mounted directly on each wafer and connected to the switch terminals are the channel-selecting coils. In most cases, the coils are connected in series with taps from each coil junction connected to the switch contacts. When channel 2 is selected, the required inductance for each stage is represented by the *total* inductance of all the interconnected coils. When switching to higher channels, the inductance becomes progressively less as more and more coils are *shorted out*. For channel 13, only a small two or three-turn coil is used on each wafer. For UHF, the switch position is between channels 2 and 13.

At the UHF position, the VHF oscillator is disabled, the UHF-IF signal (from the UHF tuner) is switched to the RF stage input, with the RF–mixer stages operating as IF amplifiers, and the UHF tuner is supplied with B voltage.

The oscillator coils are individually adjusted with small brass screws accessible from the front. With such a series-connected coil arrangement, the adjustments are interdependent with the adjustment for any given channel affected by the tuning of all *lower* channels. Fine tuning is accomplished with a small variable capacitor device.

• **Detent Systems and Drive Mechanisms**

A variety of systems have been used for turning the channel selector. Some sets use separate knobs for the VHF and UHF tuners; others combine them for single-dial control using slipping clutches, planetary drives, or gear trains. In some cases, the fine tuner is part of the assembly. Channel indications may be on the selector dial or, more often, on a translucent drum, illuminated from behind.

A tuner requires some means of holding the selector at the desired channel position relative to the positioning of the channel-switching contacts. This is called a *detent*. Most detents consist of a spring-loaded steel ball or roller that slips into an indented notch that coincides with each channel position. Sometimes it is adjustable, and, if misadjusted, stations will be received at some point *between* the detents, at the wrong dial setting.

Drum or Turret-Type Tuner. A diagram of a typical drum tuner with its associated schematic is shown in Fig. 12-8. The tuner box that encloses the drum is divided into two compartments, with the front section housing the oscillator components and the RF-mixer circuits at the rear. The drum contains *removable* coil strips, oscillator coils at the front, and RF-mixer coils at the rear. The windings of each coil terminate at silver-plated contact points, which make contact with corresponding stationary contact strips on the tuner assembly. Sometimes blade-and-wiper contacts are used. As the drum is rotated to each channel position, a new set of coils is switched into the circuit. Each oscillator coil can be individually adjusted with threaded brass slugs accessible from the front. With some tuners these slugs have sprocket heads driven by nylon gears that engage when the knob is pushed in, thus serving as a fine-tuning control. Another type of fine tuner has a cam-driven iron slug that moves in and out of the oscillator coil. At the top of the tuner is a trimmer capacitor that acts as a *master* oscillator adjustment for *all* channels. Circuit operation is as follows:

The incoming signal is routed through the balun transformer T1, the high-pass filter network, feedthrough capacitor C 23T, coupling capacitor C2T, to the antenna coil L4T. From the antenna coil the signal is inductively coupled to the RF coil L5T, which feeds the RF transistor. The amplified signal is inductively coupled to the mixer coil L6T to feed the base of the mixer transistor. The two coils are broadly tuned to the desired channel frequency with small trimmers.

The oscillator is a Collpits type with feedback through the B/E junction of the transistor and the small capacitor C 10T. About 1V of forward bias is provided by resistors R10T and R11T to ensure start-up. After oscillator start-up, all voltages change somewhat according to the strength of oscillations. The oscillator coil L7T is slug tuned to bring in the desired channel. Capacitor C12T serves as a front panel fine tuner. Note its small value.

Oscillator injection is by inductive coupling between the oscillator and mixer coils. The adjustable coil T2T at the mixer output is resonant to about 43 MHz to accept the IF signal developed in the mixer. From a phono-type jack, the signal is routed to the video IF strip via coaxial cable.

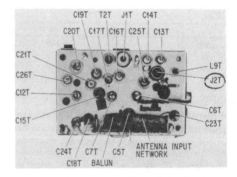

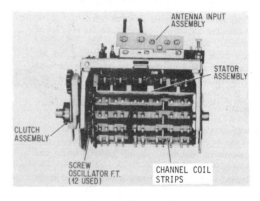

Fig. 12-8 Partial

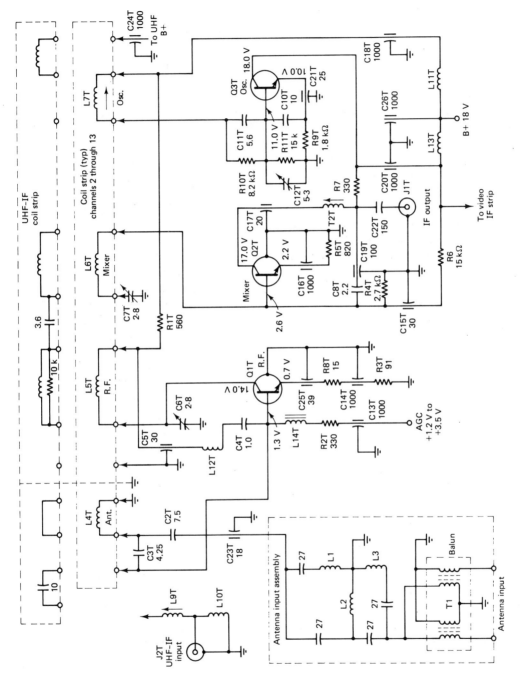

Fig. 12-8 Drum tuner.

When the UHF coil strip is in position, note how the oscillator is disabled, how switching of the antenna inputs occur, and how the UHF RF and mixer coils take the place of the VHF coils.

Varactor Tuner. Mechanical-type channel selectors sooner or later develop trouble from dirty or worn switch contacts. This is no longer a problem since the advent of the *varactor*-type tuner now used in practically all modern sets. Whereas other tuners require separate coils for each channel, this tuner has only *one* tuned circuit for each of the three stages. Channels are selected from a keyboard arrangement of push buttons on the front panel. The channels are selected *electronically*, which allows the tuner to be located at any convenient spot on the main chassis.

Channel selection (tuning) is accomplished with varactor diodes connected to the tuned circuits of the RF, mixer, and oscillator stages.

As described in Sec. 5-9, a varactor is a diode that simulates a capacitor. Its capacity can be varied by applying different amounts of dc voltage, as a *reverse bias*. The capacity varies inversely with the applied voltage. An increase in voltage, for example, reduces the capacity, raising the resonant frequency of the tuned circuit to which the varactor is connected. Conversely, reducing the voltage lowers the frequency. The dc is known as the *tuning voltage*.

Since a varactor has a limited tuning range, the VHF channels must be covered in two steps, the *low band* (channels 2 through 6) and the *high band* (channels 7 through 13). Voltage variations to cover all channels of each band ranges from about 1 to 30 V. Band changing is accomplished with *switching diodes* controlled by three different voltages that correspond with the three bands, low VHF, high VHF, and UHF. The UHF voltage activates a varactor-type UHF tuner.

A simplified schematic of a varactor tuner with its associated channel selector buttons is shown in Fig. 12-9. There are four varactors, one for each tuned circuit. Each is supplied with a dc tuning voltage from a common, regulated source. The correct amount of dc simultaneously applied to all four varactors tunes their respective tuned circuits to whatever channel is selected. A different and *precise* amount of voltage is required for each channel. These voltages are supplied from special preset potentiometers, as shown. There are normally 12 such pots (one for each channel) as a subassembly. They are adjusted by the user for the channels he wishes to receive. Fine tuning is accomplished with a front panel potentiometer control that varies the tuning voltage by a small amount.

When channels 2 through 6 buttons are pressed, extra contacts on the buttons feed a specific amount of voltage to the tuner to select the low-band channels. When channels 7 through 13 buttons are pressed, a different voltage switches the tuner to the high band. Similarly, pressing a UHF button, activates the UHF tuner.

Figure 12-9b shows one method of bandswitching. For the low VHF band, the bandswitching diodes *D1/D2* are reverse biased into cutoff, and the circuit inductance is represented by *L1, L2,* and *L3* in series. For the high band, a dc of opposite polarity forward biases the diodes into conduction, shorting out coils *L2* and *L3,* leaving only *L1* in the circuit. The same circuit configuration is used in each of the three tuner stages.

For channel indication, most modern sets use LED displays that are activated by the bandswitching-tuning voltages. A detailed schematic of a VHF varactor tuner is shown in Fig. 12-10. It uses a FET in the RF stage.

12-11 UHF TUNERS

A pictorial diagram of a *basic* UHF tuner is shown in Figure 12-11. A schematic is shown in Fig. 12-12. The side covers have been removed

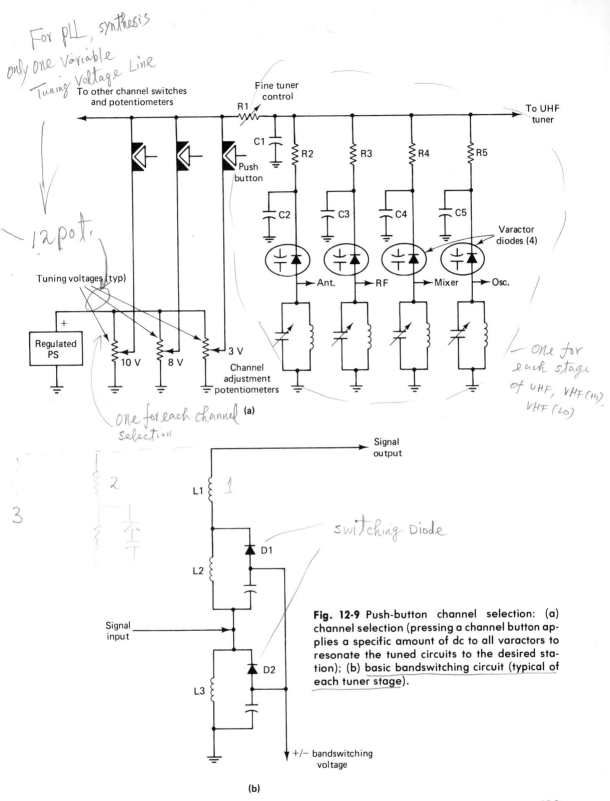

Fig. 12-9 Push-button channel selection: (a) channel selection (pressing a channel button applies a specific amount of dc to all varactors to resonate the tuned circuits to the desired station); (b) basic bandswitching circuit (typical of each tuner stage).

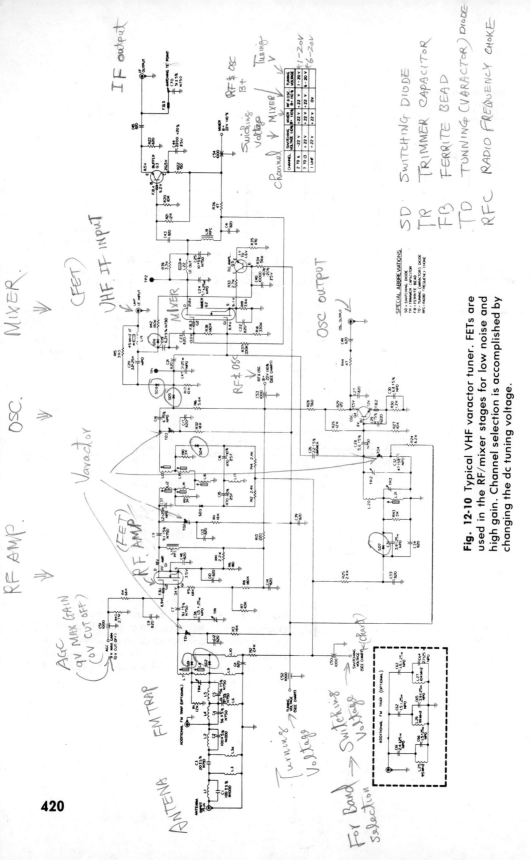

Fig. 12-10 Typical VHF varactor tuner. FETs are used in the RF/mixer stages for low noise and high gain. Channel selection is accomplished by changing the dc tuning voltage.

12-11 UHF Tuners 421

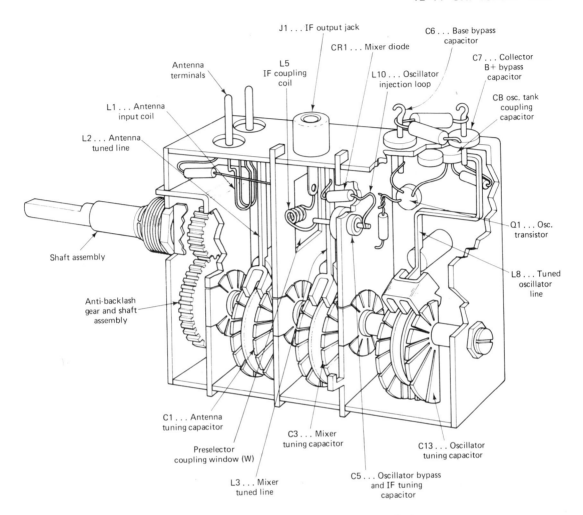

Fig. 12-11 Basic UHF tuner. Component and wiring placement on a UHF tuner is extremely critical.

to show the internal structures and for component identification. This is a *continuous* tuner, which uses a three-ganged variable capacitor for channel selection. There is no detent. Tuning is extremely critical, and for fine tuning, a two-speed, high-ratio gear drive is used.

The tuner is divided into three compartments (cavities) to isolate the antenna, mixer, and oscillator circuits. There is no RF stage and only one transistor, the oscillator. A high-frequency *diode* is used as a mixer. Instead of conventional coils, inductance is provided by three tuned lines ($L2$, $L3$, $L8$), as indicated on the schematic. In Fig. 12-11 the tuned lines can be seen as extensions of the stator plates of tuning capacitors $C1$, $C3$, and $C13$.

The signal from the UHF antenna is coupled to the tuned line ($L2$) via a small hairpin coil, where it is tuned by $C1$. The signal reaches the mixer through a window in the baffle plate separating the two stages. The mixer tuned circuit consists of the tuned line $L3$ and capacitor

422 CHANNEL SELECTORS AND TUNING SYSTEMS

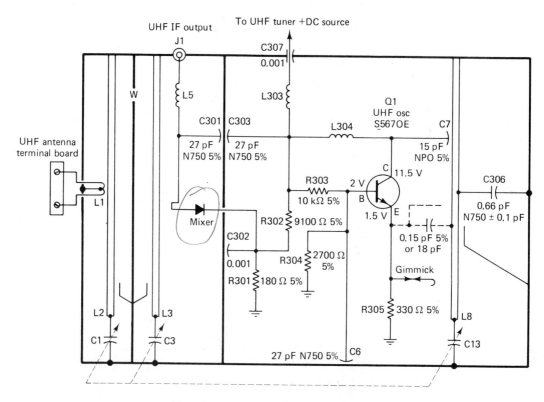

Fig. 12-12 Basic UHF tuner schematic.

C3. The mixer diode is critically located in an opening between the mixer and oscillator compartments, where, because of its proximity to both stages, it picks up the received signal and the oscillator signal. The oscillator is a Colpitts type using a silicon-planar-type transistor, and its tuned circuit consists of the tuned line *L8* and capacitor *C13*.

The IF signal developed in the mixer is coupled via the IF coupling coil *L5* (which is broadly resonant to 43 MHz), and then leaves the tuner from a phono-type jack and a short length of coax. With this tuner there is no amplification, hence the use of the RF–mixer stages in the VHF tuner.

- **A Modern UHF Tuner**

Figure 12-13 shows a typical varactor-type UHF tuner now used in most receivers. Gain is controlled by AGC. The ganged tuning capacitor is replaced by varactor diodes that tie in with the four tuned lines. When the tuner is enabled with the proper bandswitching voltage, the varactors are supplied with tuning voltages ranging from about 1 to 30 V, the same as for VHF. This UHF tuner has an RF stage.

12–12 AUTOMATIC FINE TUNING (AFT)

As an optional feature on many TV receivers, AFT is used to maintain a constant oscillator frequency with no need for periodic (manual) adjustment of the fine-tuning control. Oscillator frequency tends to *drift*, especially on the high

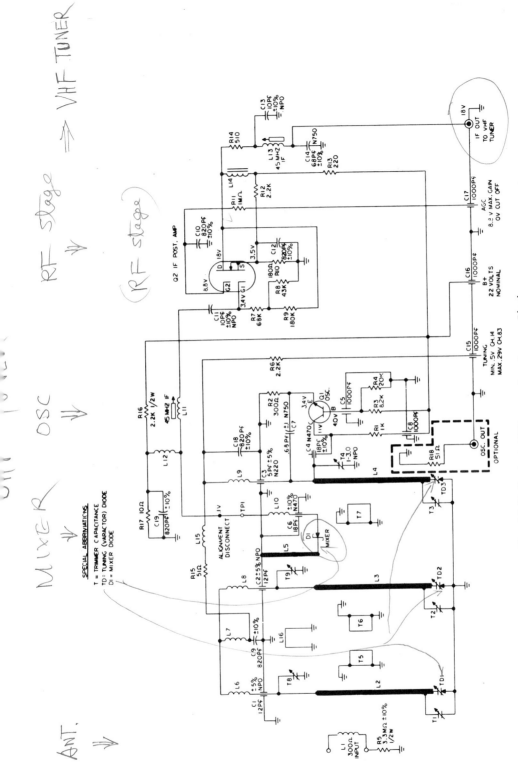

Fig. 12-13 Typical UHF varactor tuner. Black sections are the four tuned lines.

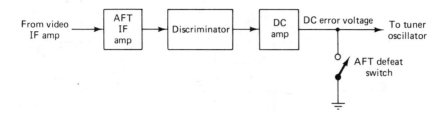

Fig. 12-14 Tuner AFT system.

VHF channels, and even more so with UHF. This results in poor reception of pix, sound, and *color*, in particular, which is *most* affected by a frequency error. This is because of the location of the color sidebands on the IF response, as shown in Fig. 12-5.

A set equipped for AFT has an OFF-ON *defeat* (disabling) *switch* on the front panel. When switched off, the oscillator is manually controlled by the fine tuner in the normal manner. The AFT components are located on the chassis, not the tuner. With modern sets, most of the AFT circuitry is contained in an *IC*.

A block diagram of a basic AFT system is shown in Fig. 12-14. Its operation is not too different from the *AFC* system used to control the horizontal sweep oscillator (Sec. 8-5) with a dc error (correction) voltage.

The heart of any AFT is a *discriminator circuit* whose function is to compare the resonant frequency of its discriminator-tuned circuit with the video IF frequency developed at the tuner. If the oscillator frequency is correct, the IF will also be correct and no AFT action is required. If the oscillator drifts off frequency in either direction, so will the IF. The discriminator circuit senses any frequency deviation and develops a dc correction voltage, which, applied to the tuner oscillator, forces it back on frequency.

A schematic of a typical AFT system is shown in Fig. 12-15. This circuit has a dc amplifier to boost the error voltage for better oscillator control. Some circuits also have an IF amplifier between the takeoff point in the third video IF stage and the discriminator.

The discriminator circuit as shown uses a pair of *matched* diodes to produce the dc error voltage by rectification of the incoming IF signal. The signal is fed to the primary coil of the discriminator transformer ($T1$) and reaches the diodes in *two ways:* by *induction* and *directly* from a tap on the secondary winding. At any given instant, the *induced* voltages at opposite ends of the secondary are 180° *out of phase* with a + voltage applied to one diode and a − voltage to the other. At the same time, an *in-phase* voltage is simultaneously applied to the cathodes of both diodes from the transformer tap via capacitor $C1$.

The diodes conduct according to the relative amount of signal voltage each receives at any given instant, which is determined by phase differences. For example, if the *induced* voltage and the *direct* voltage feeding $D1$ of Fig. 12-15 are both positive going while an out-of-phase condition feeds a predominately negative voltage to $D2$, then the conduction of $D1$ will be greater than for $D2$, developing a *positive* dc error voltage across the split load resistors $R1/R2$. For the opposite condition where $D2$ conducts more than $D1$, a *negative* error voltage is produced. Where the two diodes conduct equally, the net voltage will be zero. This is the *ideal* condition that corresponds with best pix, color, and sound.

Phase relationships, and therefore the amplitude and polarity of the dc error voltage, change in accordance with the resonant frequency to which the transformer is tuned (45.75 MHz) and the frequency of the IF signal

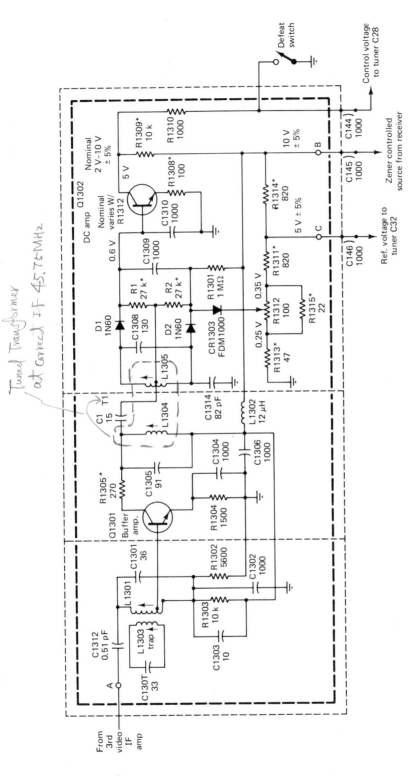

Fig. 12-15 Typical AFT circuit. AFT automatically compensates for oscillator frequency drift and eliminates the need for a fine tuner.

developed at the tuner, which varies with the oscillator frequency. If the oscillator frequency (either VHF or UHF) is too *high* (so the developed IF does not match the tuning of the IF strip), the phase difference between the *induced* and the *direct* voltage feeding *D1* becomes *less* than 90°, while the phase difference for *D2* is *more* than 90° to produce a positive error voltage that causes the oscillator to slow down by the correct amount, that is, to develop the correct IF frequency as recognized by the tuning of *T1*. If the oscillator is running slow (frequency too low), the negative error voltage produced causes the oscillator to speed up. When the oscillator frequency (and therefore the IF frequency) is *correct,* the signals feeding the two diodes have the same 90° phase relationship; they conduct equally with a net dc output of zero. This creates no change in oscillator frequency since none is required.

In effect, the discriminator functions as a *converter,* changing *frequency* variations into *amplitude variations.* The *further* the oscillator is off frequency, the greater the amplitude of the dc error voltage. This same circuit is used in the sound section of some receivers to convert frequency variations of the FM signal into amplitude variations at audio frequencies.

Some circuits have a *balance control* to balance the outputs of the two discriminator diodes. Its adjustment has a critical bearing on the AFT pull-in range, which is normally about ± 1 MHz. In some cases, transistors are used instead of diodes. The dc error voltage from the discriminator (or dc amplifier when used) controls the oscillator frequency with a varactor or, in some instances, a transistor, using only the B/C terminals with the emitter left floating. With a varactor tuner, the AFT error voltage varies the amount of *tuning voltage* as supplied from the channel select potentiometers.

The receiver fine tuning control (if there is one) is not used when the AFT is operating.

AFT setup procedures are described in Sec. 12-37.

12-13 PHASE-LOCKED-LOOP (PLL) TUNING SYSTEMS

Channel selection (tuning) for the more expensive of modern receivers is accomplished by sophisticated tuning systems using computer technology and a varactor-type tuner. In addition to channel selection and the precision control of the tuner oscillator (so a fine tuning control is no longer required) are a number of innovations made possible by such systems, for example, instant channel selection from a telephone-type keyboard, either from the receiver panel or from a remote-control unit; a system where a number of selected channels can be rapidly scanned and momentarily reviewed in rapid succession; new methods of displaying channel numbers; and remote-control systems that do not require motors, latching relays, and so on.

Such tuning systems are known by various names: digital touch tuning, frequency synthesizing, star, to name a few. The basis of most of these systems is the *phase locked loop* operating in conjunction with a *microprocessor,* which acts as a sort of command center. Almost all of the operations are performed by ICs. Learning how these systems work requires a good understanding of PLL fundamentals.

12-14 PLL FUNDAMENTALS

With a varactor-type tuner, channels are selected by applying dc voltages to the tuner varactors. Ordinarily, these voltages (a different amount for each channel as established by the

12-14 PLL Fundamentals

adjustment of preset potentiometers) are selected by channel selector push buttons. Additional control of the tuner oscillator may be provided by AFT. A *PLL system* accomplishes the same thing but in a different way.

Operation of a PLL system may be likened to the control of the horizontal sweep oscillator (Sec. 8-5). However, the horizontal AFC has only *one* frequency to control, whereas the PLL system is faced with the added task of developing and controlling *many* frequencies, a different frequency for each channel received.

A block diagram of a basic PLL system is shown in Fig. 12-16. Note the name changes and other differences compared with horizontal AFC. With PLL, the oscillator being controlled, the tuner oscillator, is called a *VCO* (voltage-controlled oscillator). In place of a horizontal phase detector we have a *comparator*. Where horizontal AFC uses the station sync pulses as a frequency reference, PLL uses a crystal-controlled *reference oscillator*. In addition, a PLL system has two or more *frequency dividers*.

• **VCO (Tuner Oscillator)**

The PLL system develops a dc *tuning voltage* that is applied to the varactors in each stage of both tuners. Although it is the *RF-mixer* circuits in each tuner that are resonant to the station signals, it is the *oscillator* frequency that determines which station is received. A sampling of each tuner oscillator is fed to the comparator via two *feedback loops,* where they are continuously *monitored*. Although the VCO is the source of the feedback signal, the *frequency* is reduced (by dividers) before it reaches the comparator.

• **Reference Oscillator**

This is a highly accurate crystal oscillator that generates a *basic* reference signal, which is fed to the comparator via a *reference loop*. The oscillator frequency is usually around 4 MHz, depending on receiver design. Like the feedback loop, this loop also contains one or more dividers, which reduces the frequency for comparison with that developed by the VCO.

• **Frequency Dividers**

A frequency divider (sometimes called a *counter*) uses computer-type logic circuits to reduce the frequency of an applied signal by a given amount, commonly referred to as the *divide-by* ($\div$) *ratio*. Some dividers have a *fixed* ratio;

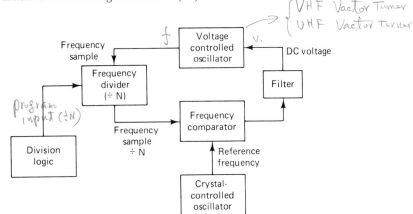

Fig. 12-16 Basic PLL system.

others are programmable (by a microprocessor) for different ratios. Dividers may be used to reduce the frequency of a reference oscillator, a VCO, or *both* depending on how they are connected in the circuit.

The purpose of the dividers is to establish a 1:1 frequency relationship between the two signals supplied to the comparator (via the *reference* and *feedback* loops), *when the VCO frequency is correct.* To this end, dividers are used in the reference loop to produce a VCO frequency *lower* than the reference signal, and in the feedback loop to produce a VCO frequency *higher* than the reference signal. The use of dividers makes it possible to obtain practically any multiple of the reference oscillator frequency as required (by the VCO) for the reception of all VHF and UHF stations. As each channel is selected, the appropriate ÷ factor for that channel is programmed by logic circuits in the microprocessor.

- **Frequency Comparator**

This stage has two inputs: the reference signal (after going through one or more dividers), and also via dividers, a sampling of the signal generated by the VCO tuner oscillator. Although it operates differently, the comparator serves the same purpose as a *discriminator circuit:* the comparison of two signals and the development of a dc error voltage when any difference in frequency or phase is detected. Unlike most discriminator circuits, however, these two signals do not represent the desired VCO frequency; their frequency is arbitrarily chosen for comparison purposes only.

The dc error voltage (commonly called the *tuning voltage*) is the means of controlling the VCO of both the VHF and UHF tuners. When the VCO frequency is incorrect for the reception of a desired station (as indicated by a *difference* between the two signals at the comparator input), the comparator develops a dc error or tuning voltage of the proper polarity to force the VCO to either slow down or speed up as required until the two inputs to the comparator are alike. The *amount* of the error voltage is not too important. Controlling an oscillator in this manner is known as *frequency synthesis,* and, once *locked,* the accuracy and stability of the VCO is as good as the *reference oscillator.*

- **Filter**

The low-pass *R/C* filter integrates the pulsed output of the comparator and filters out any fluctuations to produce a pure dc tuning voltage. Typical tuning voltages are from near zero to about 30 V, the low voltages for the low channels and higher voltages for the high channels. Whereas a varactor tuner not controlled with a PLL system requires *precision* tuning voltages for each selected channel, the *amount* of tuning voltage with a PLL system is not critical since the tuner responds to *changes* in voltage and polarity, not to actual *amounts.*

- **Functional Operation of a PLL System**

Let's summarize the operation of a typical PLL system with some examples. Where the reference oscillator frequency is 4 MHz and a ÷ 200 divider is used, the reference frequency at the comparator input will be 0.02 MHz (20 kHz). The comparator will sense any difference between this frequency and that of the VCO feedback signal. If there is a difference, a dc error voltage is developed to force the VCO to change frequency until a *matching* 20-kHz *feedback* signal exists at the comparator input. When the two comparator inputs are identical in frequency and phase, the system becomes *stabilized,* with the VCO operating at the correct frequency for the selected channel.

In the foregoing example, where a divider with a *fixed* dividing factor is used, a VCO can only be locked to *one* frequency. Since a different tuner oscillator frequency is required for each channel being received, the divide-by factor must be changed when switching channels. This is taken care of by the *programmable divider* whose ÷ factor can be changed when instructed by a *microprocessor*. Some sets have an additional programmable divider called a *prescaler*. Its job is to relate the selected channel with one of the three bands (low or high VHF and UHF) and to initiate the appropriate bandswitching voltage. Each band, as with channel selection, requires a different divider ratio, which is usually indicated on a schematic. ⟫

Divide-by ratios for the reference and feedback loops are chosen to obtain matched frequencies at the comparator input when the tuner oscillator frequency is correct. It is important to note that this frequency relationship *does not change* for different channels, although different frequencies are used depending on receiver design.

As a general rule concerning dividers, the higher the ÷ ratio in the *feedback loop*, the higher the frequency of the VCO when the system becomes stabilized, and vice versa. Conversely, the higher the ratio in the *reference loop*, the *lower* the frequency of the VCO is.

12-15 TYPICAL PLL TUNING SYSTEM

The block diagram for a typical late-model set is shown in Fig. 12-17. As is customary, both fixed and programmable dividers are used in both loops. The reference oscillator signal goes through two dividers to reach the comparator, a ÷ 1 divider, followed by a ÷ 3667 divider. The resultant frequency is shown at the comparator input.

REMINDER: ⎯⎯⎯⎯⎯⎯⎯⎯⎯⎯⎯⎯⎯⎯⎯⎯⎯

This frequency is peculiar to a *particular* receiver and it remains the same, regardless of which channel is selected. To obtain frequency lock of the VCO, this frequency must be matched by an identical frequency from the feedback loop.

In the *feedback* loop, the VCO goes through *three* dividers to reach the comparator; a ÷ 256 prescaler, a ÷ 4 unit, and a ÷ 101 divider (when programmed for the low VHF band). Multiplied together, this represents a total division factor of 103,424, and the 101-MHz VCO frequency divided by this figure produces the proper comparison frequency at the comparator input. Should the VCO change frequency for any reason, a corresponding frequency change takes place in the feedback signal at the comparator input. As these two differing signals are compared, an increase or decrease in the tuning voltage forces the VCO back on frequency, at which time the two comparator inputs will again agree. The range of tuning voltages is from near zero to approximately 25 V.

When a different channel is selected (one of the UHF channels for example), the ÷ 1 divider in the reference loop is enabled along with the ÷ 3667 unit to produce the frequency indicated at the comparator input. At the same time, the ÷ 931 unit in the feedback loop is enabled to produce the same frequency, provided the VCO frequency is correct.

• **Keyboard Control**

All VHF and UHF channels are selected from a *keyboard*, an arrangement of push-button switches on the front panel. Some sets have a button identified with the channel number; others use a telephone-type keyboard with buttons numbered 0 through 9. For single-digit channel numbers you press the required number. For double-digit numbers (channel 23 for example), you press 2, then 3. The keyboard switching matrix produces a binary (BCD) coded output, which is recognized by the logic circuits in the microprocessor, which selects the proper ÷ factor from a programmable divider. Besides channel selection, this coded signal also selects the proper *band* for the selected channel and commands a readout of the channel number. In these operations, diode and tran-

430 CHANNEL SELECTORS AND TUNING SYSTEMS

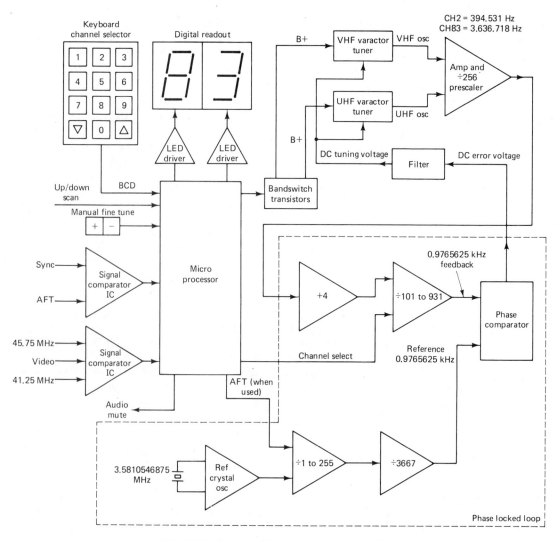

Fig. 12-17 Typical PLL tuning system. All PLL functions are performed by one or more ICs.

sistor switching is so fast that results are almost instantaneous.

• **Bandswitching**

This is an automatic function when any channel button is pressed. The binary code from the keyboard tells the microprocessor which of the three bands to select, and an appropriate bandswitching voltage is developed to control the switching diodes in the tuner.

• **Channel Number Readout Display**

In place of indicator lamps, most modern sets use a two-digit *LED* display. The binary output from the keyboard feeds a decoder whose discrete outputs energize the appropriate segments of the numeric display.

• **Up-Down Channel Scanning**

As an alternative to channel selection from the keyboard, this feature permits the user to preview a number of preselected channels in rapid succession. The user programs the tuning voltages for his preferred channels into a memory bank, where they can be recalled on demand, when an UP/DOWN button is pressed. The UP button searches from the low to the high channels; the opposite is true for the DOWN button. Scanning can be in any desired sequence. A *muting system* kills the sound while skipping the unprogrammed channels. With some sets, the memory is *volatile* and the pre-programming information is lost when the set is unplugged.

• **Other Features**

A PLL system also provides for other user conveniences, remote control, for example, where all or most of the front panel control functions can be duplicated by a remote transmitter unit. With PLL, the fine-tuning control can be dispensed with, except in some cases, to provide for cable reception, video games, and other features.

12-16 STAR SYSTEM

The STAR (silent tuning at random) system is a forerunner of the PLL system of tuner control. It also makes use of computer-type logic circuits contained in ICs. The basic difference between the two systems is the methods used for frequency synthesis. Whereas the PLL system compares a reference frequency with a sampling of the tuner oscillator signal, STAR makes use of the *harmonics* of a 6-MHz reference oscillator.

The system uses a 24/6 MHz *harmonic comb generator* to develop all harmonics, 6 MHz apart, that fall within the VHF/UHF TV bands. The tuner oscillator is *swept* through the harmonic spectrum, in both directions, until a harmonic closest to the required oscillator frequency is recognized, for example, the seventeenth harmonic (102 MHz) for channel 2.

The output of the comb generator along with a sampling of the VCO signal is fed to a *mixer,* which identifies the desired harmonic, and as the VCO sweeps within ± 1 MHz of each harmonic, 1-MHz beat notes, called *birdies* are produced. These birdies are continuously counted by a *counter* until the harmonic corresponding with the desired channel is identified, at which point the sweep is halted. The mixer feeds a *ramp generator,* which develops the proper tuning voltage for the VCO.

12-17 REMOTE-CONTROL SYSTEMS

This optional feature permits the user to control the receiver from a remote, hand-held *transmitter* unit. A *basic* remote-control system is limited to essential functions only, such as channel selection, ON/OFF, and volume. The more elaborate systems can control other functions, such as brightness, contrast, and color. Of the many systems in use, all make use of a remote transmitter unit and a remote receiver unit contained in the TV set. the main difference between the various systems is the *means* of conveying commands from the transmitter to the receiver. Mediums of transmission in common use are *acoustical,* a *light beam, infrared light beam,* and *RF.* For receiver *turn on* from a remote unit, the *remote receiver* in most sets must be operating at all times, even when the set is turned off.

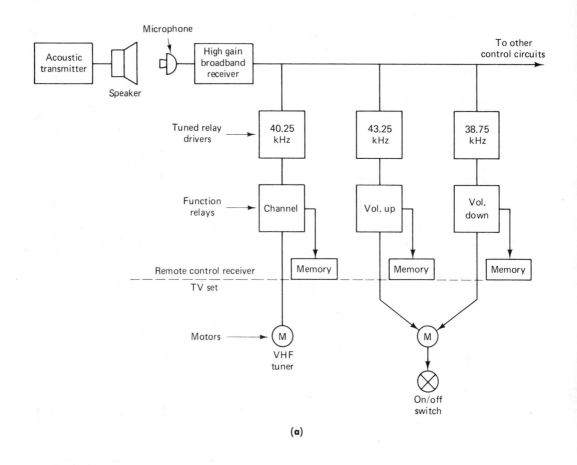

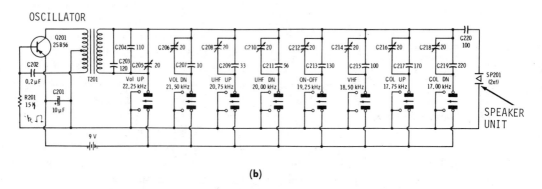

Fig. 12-18 Acoustical remote control system.

12-17 Remote-Control Systems

• Acoustical Remote-Control System

This system (as shown in the block diagram of Fig. 12-18a) has been in use for many years. Inaudible *ultrasonic sound waves* is the medium of transmission. Two methods are commonly used for generating the sound waves at the transmitter, mechanical, and electronic.

A mechanical-type transmitter contains a number of tuned metal rods. Each rod produces a different ultrasonic tone when struck by a hammer when a spring-actuated button is pressed. The tones, one for each function, are in the 30- to 50-kHz range.

The electronic transmitter (see Fig. 12-18b) uses an oscillator circuit to develop the supersonic frequencies, which are emitted by a special high-frequency speaker called a *sonic transducer*. The transmitter unit has a built-in battery.

At the receiver, a piezoelectric-type microphone picks up the sound waves, converting them into electrical impulses that are fed to a high-gain broadband amplifier, sometimes contained in an IC. From the amplifier, the signals go to a number of *tuned filters,* each resonant to a frequency generated by the transmitter. Discrete outputs from the filters go to *driver amplifiers* to actuate relays and reversible motors that perform the various functions. The remote-control circuits are all contained in a separate subassembly.

The motor-driven ON/OFF and volume functions are usually combined. When the PWR ON/volume UP button on the transmitter is pressed, a *power ON relay* in the receiver unit supplies the set with ac power. When pressed again, the volume slowly starts to increase. When a volume DOWN button is pressed, the volume is gradually reduced until the set goes off. Other functions are controlled in the same manner: up-down (increase-decrease) of brightness, contrast, color, and so on. Older sets used a motor to turn the channel selector shaft with an indexed switching arrangement to stop the motor at each active channel. With varactor tuners, channel selection is made with dc tuning voltages and no motor is required.

• Memory

It is desirable to have some means (either mechanical or electrical) to remember the last setting of the channel selector and other controls when the set is turned off; otherwise, it becomes necessary to readjust the controls when the set is switched on. Each function requires a separate *memory.*

One type of *electronic memory* (see Fig. 12-18a) employs low-leakage memory capacitors that retain a charge long after a set has been turned off. With the volume function, for example, the capacitor becomes charged to a voltage and polarity corresponding with the last setting of the volume level. When a volume DOWN function is initiated, the capacitor becomes charged with the reverse polarity. When the set is turned off, the charge remains, ready to restore the original setting according to the amount and polarity of the voltage. Other functions are remembered in the same way.

• Modulated RF Carrier System of Remote Control

This system has a transmitter that radiates an RF carrier in the 300- to 400-kHz range. The carrier is modulated by frequencies around 40 kHz, each frequency relating to a particular function. At the receiver, the signal is picked up by a ferrite rod antenna, detected, amplified, then selected by tuned filters to drive amplifiers that perform the desired functions.

• Infrared Carrier System

This system is used by modern sets in conjunction with the PLL tuning systems described earlier and also in Sec. 12-18, to follow. A block

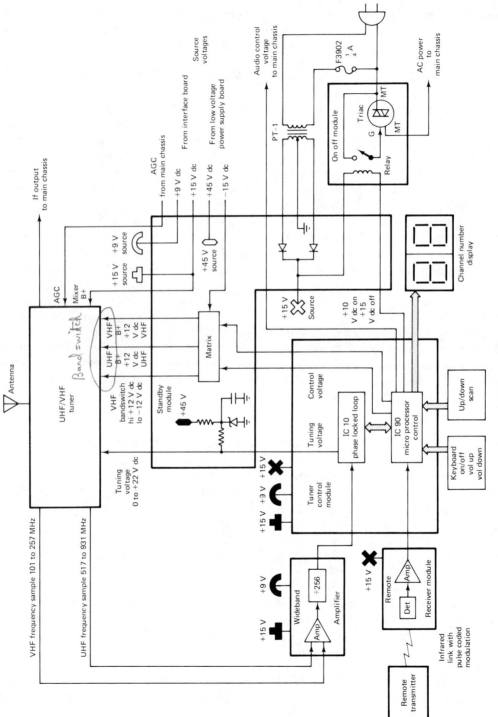

Fig. 12-19 PLL tuning system with infrared remote control.

12-18 Another Front-End PLL Tuning System   435

diagram is shown in Fig. 12-19. All functions normally controlled from the receiver panel can be duplicated from the remote transmitter unit. The transmitter is battery operated and contains binary logic circuits as found in the receiver. In addition, there is a crystal-controlled oscillator operating at approximately 33 kHz. Each function initiated from the keyboard channel selector or other controls is encoded into a shift register (temporary memory) to produce a train of BCD pulses that are then emitted by a LED as invisible infrared light-beam pulses.

At the receiver, the infrared carrier is detected by a light-sensitive diode and *visible* light is removed by a special filter. The signal is then demodulated to recover the coded signals representing the various functions. No motors or relays are used with this system, and all functions are performed by dc voltages. For the UP/DOWN control of volume, color, and so on, variations are produced with changes in transistor biasing, similar to AGC action. Channel selection by the varactor tuner uses tuning voltages developed by the PLL system. Channel numbers are displayed in the normal manner.

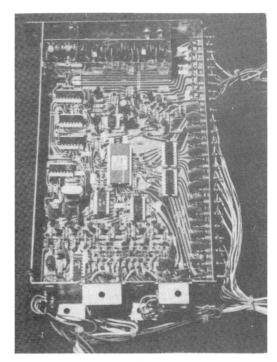

12-18 ANOTHER FRONT-END PLL TUNING SYSTEM

The tuners used in Sample Circuit C at the back of this book are both varactor types and use the PLL tuning system described in Sec. 12-13. A brief block diagram description of the tuning system is as follows (see Fig. 12-19).

Most of the functional blocks can be recognized from the basic PLL system described earlier. Note the various inputs and outputs of the tuner control module (the control center), which contains the microprocessor and PLL ICs (see Fig. 12-20). Also note the inputs and out-

Fig. 12-20 Typical tuner control module and remote transmitter. All PLL logic circuits are contained in 10 ICs.

puts of the combined VHF/UHF tuners represented by a single block on the diagram.

When a channel is selected from the keyboard, its binary-coded output instructs the microprocessor to initiate the following actions:

1. *Channel number readout:* The signal is *decoded* into BCD (binary-coded decimal); discrete outputs feed driver amplifiers to energize the appropriate segments of the LED readout to correspond with the channel selected.

2. *Power ON/OFF:* An ON/OFF keyboard button sends a signal via the microprocessor to the ON/OFF module, which turns on a triac to supply ac power to the main chassis.

3. *VHF/UHF switching:* Two outputs from the microprocessor feed a matrix in the *standby module*. In accordance with the channel selected, the matrix feeds B+ to either the VHF or the UHF tuner as appropriate. Another output from the matrix is for *bandswitching* high VHF, low VHF, or UHF, again depending on which channel is selected. Note the amount and polarity of the bandswitching voltages.

4. *Audio muting:* An output from the microprocessor goes to the main chassis to reduce sound level during channel switching.

5. *Channel selection:* A tuning voltage for the varactor tuners is developed by the PLL (*IC10*). Included in this IC are the crystal-controlled reference oscillator, dividers, and the comparator. The oscillator frequency is 4 MHz, which is subsequently reduced to 3906.25 Hz, the comparison frequency at the comparator input. A sampling of both the VHF and UHF tuner oscillator signals feeds a wideband amplifier as part of the *feedback loop*. The selected frequency is then reduced (by a ÷ 256 divider) to feed the comparator in *IC10*. As a result of the frequency comparison, a dc tuning voltage is developed, which is routed to the tuners via the standby module.

The remote-control system uses an infrared light beam, pulse-coded for the various functions. The signal is picked up and processed by a *remote receiver module* at the TV receiver. From this module the signals feed the microprocessor to command the same functions as controlled from the main receiver panel.

12-19 TROUBLESHOOTING CHANNEL SELECTORS AND TUNING SYSTEMS

Tuners are often mistakenly blamed for troubles that exist elsewhere in a set, and when this results in sending it out for repair and, later, after reinstalling, discovering the troubles are still present, it can be very costly. Such mistakes can be avoided by careful and accurate analysis of the problem and performing a number of tests before jumping to conclusions.

12-20 CONSIDER THE TROUBLE SYMPTOMS

Devote as much time as necessary in considering *all* the symptoms, bearing in mind the tuner is not the only possible cause; the video section, for example, and the AGC, AFT, and PLL circuits are all possible suspects.

Typical trouble symptoms are complete loss of pix and sound on some or all channels, weak reception with or without *snow,* fading, and intermittent problems. Some troubles are obvious, like those caused by dirty or worn channel selector contacts or mechanical defects.

Check for snow on the off-channel raster and on a weak station. *No snow* and a weak pix or no pix usually mean trouble in the IF section or video amplifiers. *Excessive* snow on a weak pix *could* be a bad tuner. It can *also* be an AGC problem.

12-21 WHICH TUNER IS SUSPECT?

Check reception on *all* VHF and UHF channels. Do the symptoms relate to one tuner or the other, or both? Remember UHF is dependent on the proper operation of the VHF tuner, so for loss of VHF *and* UHF reception, only the VHF tuner is suspect. For normal VHF reception, but no UHF, the trouble is with the *UHF* tuner or its interfacing circuits.

12-22 FAULTY TUNER OR CHASSIS PROBLEMS?

Except for the more obvious defects, certain tests must be made to determine if one of the tuners is the cause or if some other circuit or section of the set is at fault, and, if there is a bad tuner, to decide which one. Such tests are described in the following paragraphs. For these tests and as an aid in troubleshooting, refer to the troubleshooting flow diagrams (Fig. 12-21) and the trouble symptoms chart (Table 12-1). These and other tests can usually be made from the tuner terminals and without removing it from the cabinet. Typical tuner inputs and outputs are shown in Fig. 12-22.

• **DC Power Test**

When there is loss of pix and sound, check for B+ at the suspected tuner. In some cases there should be B voltage at two or more terminals. Consult the schematic. Check the B voltage

(cont'd page 453)

Table 12-1 Front End and Remote Control Trouble Symptoms Chart

Trouble Condition	Probable Cause	Remarks and Procedures
No reception (VHF and UHF)	Dead video IF, amplifier, or VHF tuner. Missing or wrong voltages applied to tuner (B+, AGC, or bandswitching/tuning[a] voltages. AGC control misadjusted.	Localize trouble to video stages, tuner, or PLL system. If bad tuner suspected, verify by substitution. Localize trouble to RF, mixer, or oscillator stage. For PLL system, check bandswitching and tuning voltages and replace any suspect modules or ICs. Check for leaky or shorted switching diodes or varactors in tuner.
Weak pix with snow (VHF)	Weak video IF amplifier. Weak mixer stage. Weak or dead RF amplifier. Incorrect AGC action or misadjusted delay control. Antenna system defect.	Localize the trouble. Adjust AGC controls. Substitute antenna. Substitute tuner. Check RF-mixer alignment.
Weak pix (no snow)	Weak video IF or amplifier stage.	

Table 12-1 Front End and Remote Control Trouble Symptoms Chart *(cont.)*

Trouble Condition	Probable Cause	Remarks and Procedures
Loss of some channels only (VHF)	Misadjusted oscillator. Dirty or damaged channel select contacts. Defective oscillator or mixer coil(s). Missing or incorrect bandswitching/tuning voltages. Misadjusted channel select potentiometers.	Adjust oscillator slugs. Inspect and clean switch contacts. Correct cause of missing or incorrect bandswitching/tuning voltages.[a]
Some VHF stations received at wrong channel positions	Dial slippage. Misadjusted oscillator slugs. Incorrect tuning voltages.[a]	
Reception erratic or noisy when switching channels; snow	Dirty, worn, or damaged switching contacts.	Clean and lubricate contacts. Inspect for bent, damaged, or improperly aligned contacts. Check by moving and rotating tuner shaft.
Loss of low VHF channels only	Misadjusted oscillator slugs. Bandswitching defect. Missing or incorrect bandswitching/tuning[a] voltages.	
Loss of high VHF channels only	As above.	
Mechanical defects		Check for damaged or missing detent parts. Inspect for worn drive gears or sprockets.
Need to frequently adjust fine tuner, especially on the high channels	Oscillator drift. Defective fine tuner. AFT troubles.	Adjust all oscillator slugs so stations are received at mid-setting of fine tuner control. If necessary, adjust master oscillator slug or compress/stretch coil turns. Replace oscillator transistor. Inspect for fine tuner defect such as a loose or missing slug. Check regulation of tuning voltage.[a] Monitor tuning voltage during oscillator drift. Check AFT action. Check for loose oscillator coil slugs.
Fine tuner has little or no effect	Defective fine tuner mechanism.	Inspect for worn or broken parts, loose or missing coil slug, damaged oscillator coil.

438

Table 12-1 Front End and Remote Control Trouble Symptoms Chart (*cont.*)

Trouble Condition	Probable Cause	Remarks and Procedures
Pix or sound mistracking (best pix and/or sound obtained for different settings of fine tuner)	Tuner or IF misalignment.	Make sure stations are received at mid-setting of control. Adjust oscillator slugs with fine tuner at mid-setting. Check the RF/mixer overall response. Check for misaligned IF strip.
Weak pix with snow, pix streaking, and herringbone pattern on screen, mostly on the high channels	Oscillations in RF amplifier stage or IF strip.	Neutralize. If necessary, dress RF circuit wiring. Make sure baffle plates and cover shield are securely in place. Check for open or intermittent ground connections, especially coax cable feeding the IF strip.
Interference from adjacent channels, FM, amateur, CB, or other strong local stations	Defective or improperly adjusted antenna input circuit traps. Antenna problems.	Adjust filters and traps at tuner input circuit and IF strip. Relocate antenna and use shielded lead in. Add a filter or trap at receiver input. If windshield-wiper effect, check for loss of AGC to RF stage. Try adjusting delay control.
Ineffective AFT control action	Oscillator misadjustment. Defective AFT circuit.	Verify all stations received with fine tuner at mid-setting and AFT switched off. Check for missing or incorrect dc error voltage at tuner, with AFT switched on; if not, troubleshoot the circuit as in Fig. 12-21f. If IC is used, substitute.
Intermittent reception	Dirty or worn channel select contacts. Poor soldering. Loose connections. Shorts caused by solder splashes.	Inspect and clean switch contacts. Inspect for poor soldering, loose connections, especially cable connectors and module pin contacts. Inspect for solder splashes in wiring. If necessary, monitor operation of all tuner stages and tuner control circuits, particularly bandswitching/ tuning voltages.[a]

Table 12-1 Front End and Remote Control Trouble Symptoms Chart (*cont.*)

Trouble Condition	Probable Cause	Remarks and Procedures
No UHF reception (VHF OK)	Mixer or oscillator stage defect. Defective diode or transistor.	Verify an IF signal gets through from the UHF-IF coax with VHF tuner set for UHF. Verify correct B+, AGC, and bandswitching/tuning voltages[a] at UHF tuner. Try substitute tuner. If tuner is bad, localize the defect as in Fig. 12-21c. Replace diodes, varactors, and transistors as appropriate.
Weak UHF reception (VHF OK)	Poor antenna or location. Weak RF or mixer stages.	Try substitute antenna. Replace diodes, varactors, and transistors in RF and mixer stages. Check AGC voltage.
UHF stations received at wrong dial positions	Dial slippage. Oscillator drift or misadjustment. Tuning voltage error.[a]	Adjust oscillator. If necessary, replace oscillator transistor and/or varactor.[a] Correct cause of tuning voltage error. "Knife" oscillator tuning capacitor plates if required.
Missing or incorrect bandswitching voltage[a]	Defect at the *source* or internal short or leakage in tuner.[a] Suspect bandswitching diodes.	Loss of or incorrect voltage can be caused by tuner loading. Check for leaky or shorted bandswitching diodes. For PLL system[b] replace suspected modules or ICs as in Fig. 12-21d.
Missing or incorrect tuning voltages[a]	Defect at the source or loading by the tuner.[a] Suspect leaky or shorted varactor diodes.	Check tuning voltage(s) at source.[a] Check voltages at channel select buttons and individual pot controls. Adjust pots. For PLL system[b] replace suspected modules or ICs as in Fig. 12-21d. If loading by the tuner, check for a leaky or shorted varactor, or its associated transistor.
Channel readout display problems[a,b]	Defective LED display. Loss of energizing voltage(s). Defective keyboard switches of PLL system module or IC.	Check for defective LED readout. Check energizing voltages for each selected channel. Check for keyboard outputs. Replace suspected

440

Table 12-1 Front End and Remote Control Trouble Symptoms Chart (*cont.*)

Trouble Condition	Probable Cause	Remarks and Procedures
Channel scan or search problems[b]	Defective scan memory or microprocessor ICs. Loss of volatile memory when set unplugged.	modules and ICs as in Fig. 12-21d.[b] Replace suspected modules or ICs. Reprogram memory with favorite channels.

Remote Control Systems (all receiver functions are normal using front panel controls)

Trouble Condition	Probable Cause	Remarks and Procedures
No power turn-on by remote control	Remote receiver not turned on. No turn-on command generated by transmitter. Defective turn-on relay or triac in receiver.	Check transmitter. Replace battery if weak. Check for closure of turn-on relay or a bad triac.
Loss of sensitivity (transmitter must be held close to receiver)	Weak battery. Low output from transmitter oscillator. Low transmitter output due to defective ICs if PLL tuning system.[b] Weak remote preamp in receiver. Defective pickup sensor in receiver.	Check and replace battery. Substitute, repair, or replace transmitter unit. Make stage-by-stage gain check of receiver preamp. If it is an IC, replace. Test or substitute pickup sensor.
Loss of all functions	Weak or dead transmitter. Defective pickup sensor or receiver preamp.	Substitute transmitter. Check receiver sensor and preamp.
Loss of some functions only	Defective transmitter. Defective or misadjusted tuned filters. Defective driver amplifiers. Defective relays or motors.	Substitute transmitter. Check for defective filters. Tune filters if required. Check driver amplifiers, relays, and motors that correspond with lost function. If PLL system, verify troublesome functions are OK using *receiver* controls. Replace PLL system modules or ICs as appropriate.
Intermittent remote control	Suspect transmitter unit. Could be a defect in any stage or unit of the system.	Substitute transmitter. Monitor signals for all functions at various stages in receiver. Verify integrity of all interconnecting cables and connectors in receiver.

[a] Varactor-type tuner.
[b] PLL tuning system.

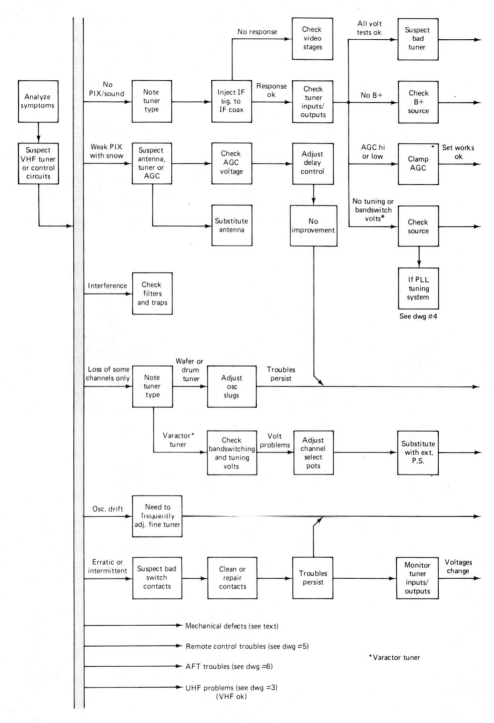

Fig. 12-21a Localizing front-end and tuning system malfunctions.

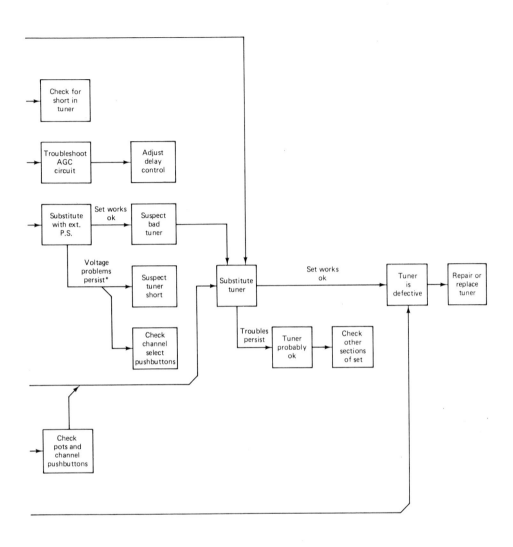

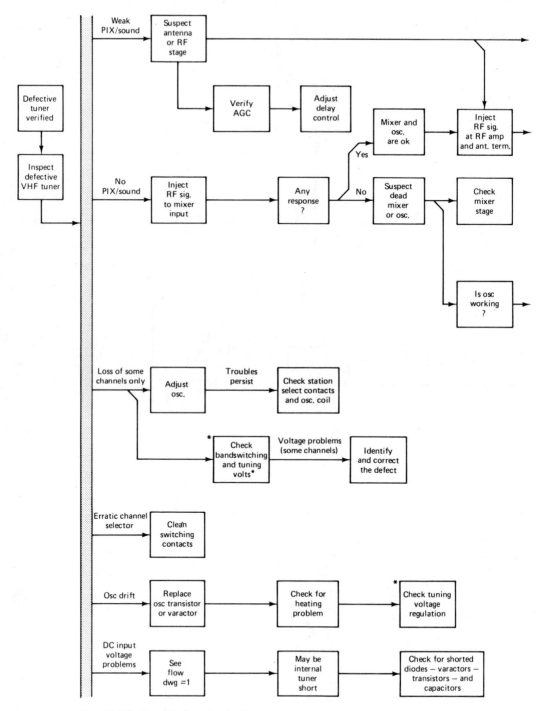

Fig. 12-21b Troubleshooting VHF tuners.

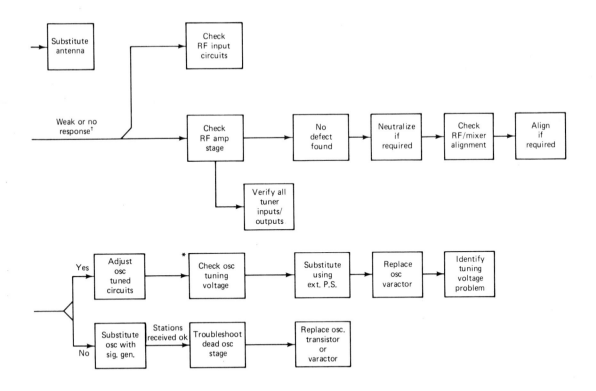

*Varactor tuner

†Note: A dead RF amplifier may or may not pass a signal depending on nature of the defect.

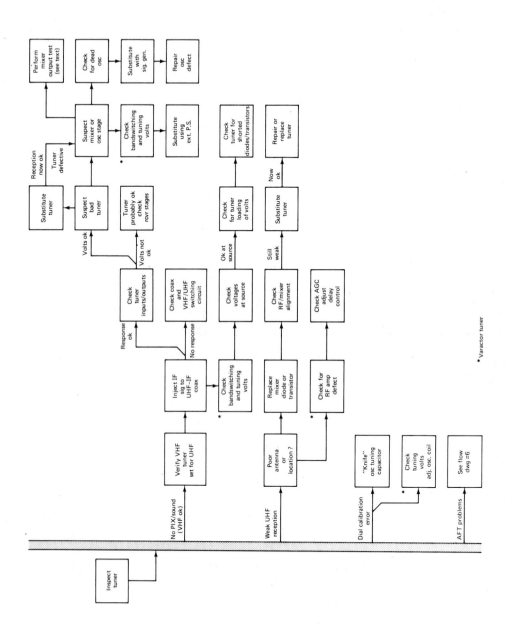

Fig. 12-21c Troubleshooting UHF tuners.

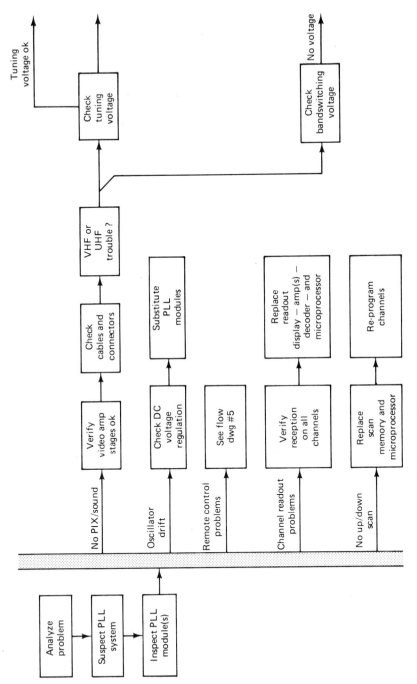

Fig. 12-21d Troubleshooting PLL tuning systems.

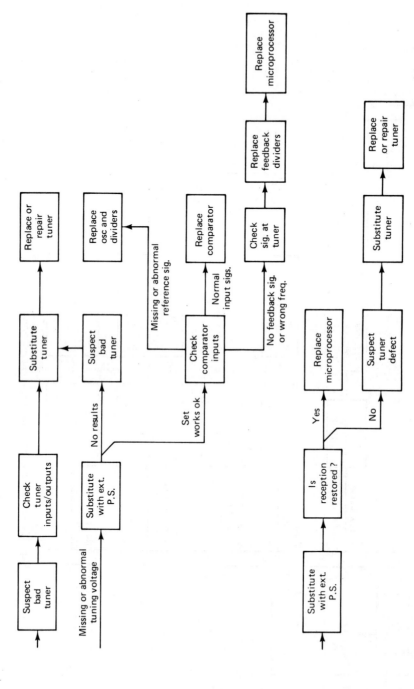

Fig. 12-21d (cont.)

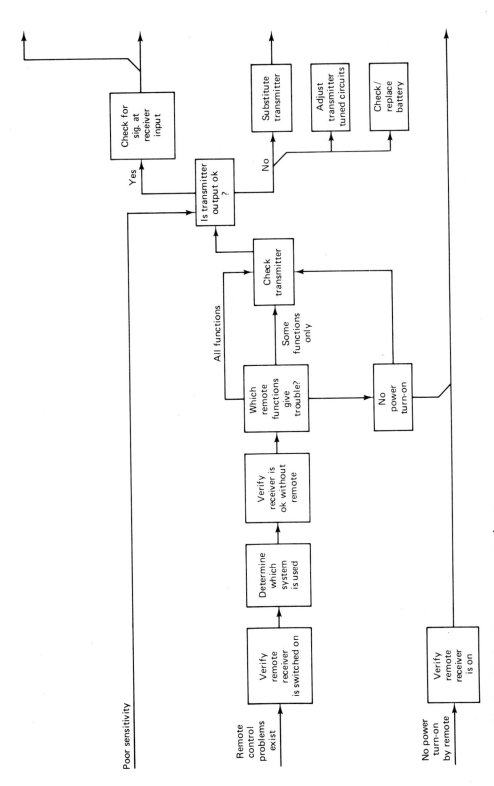

Fig. 12-21e Troubleshooting remote-control systems.

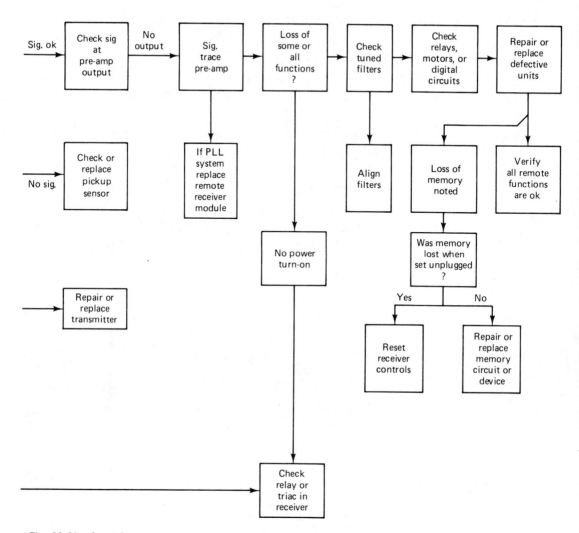

Fig. 12-21e (cont.)

450

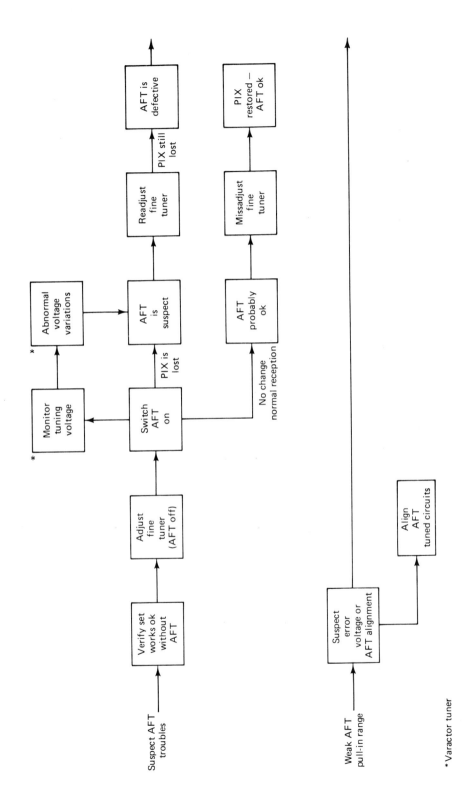

*Varactor tuner

Fig. 12-21f Troubleshooting AFT systems.

451

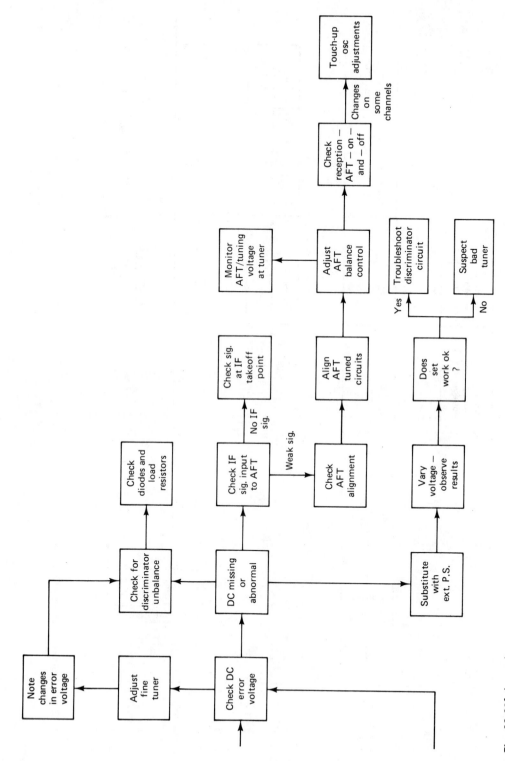

Fig. 12-21f (cont.)

12-22 Faulty Tuner or Chassis Problems? 453

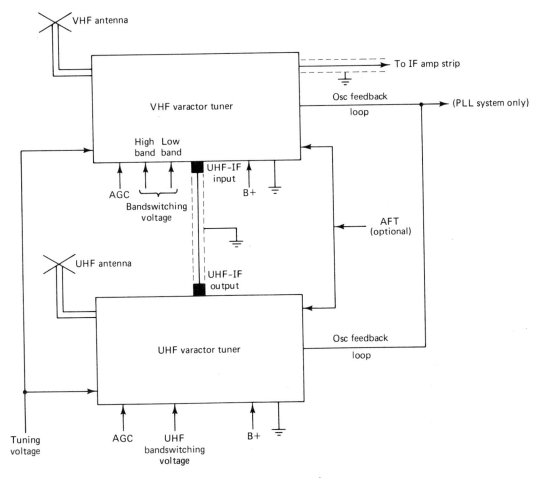

Fig. 12-22 Tuner inputs and outputs.

switching to the two tuners. Don't expect voltage at the UHF tuner unless the VHF tuner is at the UHF position. Low voltage or loss of voltage can be caused by trouble at the *source,* a defective series-dropping resistor, or loading by the tuner due to an internal short or excessive current drain. For such troubles, try disconnecting the B supply lead to see if the loading is reduced, and make resistance checks from the lead and/or the tuner terminal to ground.

Where a digital tuning system is used, check the ICs and circuits that route the B voltage to the individual tuners.

- **AGC Voltage Test**

When symptoms are loss of pix and sound, weak reception with snow, or video overload, an AGC defect is *often* the cause. Measure the AGC voltage at the suspect tuner, off channel, and for both weak and strong signals. Check against the schematic for polarity, normal reading off channel, and to determine whether *forward* or *reverse* AGC is used. Either too little or too much AGC will weaken the pix and increase the snow. When switching between weak and strong stations, typical AGC variations are 1 to 4 V for bipolar transistors and approximately double these amounts where a FET is used.

REMINDER: _____

AGC is supplied to a *UHF* tuner only if it has an RF stage. When it does, both tuners may be supplied with the same voltage or different amounts or of opposite polarities; check with the schematic.

When the AGC is in doubt, try clamping with a bias box. If set operation is restored, the AGC is at fault and not the tuner.

- **Bandswitching and Tuning Voltages (Varactor Tuner Only)**

Check the three bandswitching voltages when switching between the low and high VHF channels and UHF. Compare amounts and polarity with those indicated on the schematic. For loss of these voltages the trouble may be at the source or a shorted bandswitching diode in the tuner.

Check tuning voltages at the suspect tuner while switching channels. Normal voltages range from about 1 or 2 V for the lowest channel (low VHF band, high VHF band, or UHF) to about 30 V for the highest channel. Compare with the schematic.

If there is trouble with *all* channels, check for tuning voltage at its source. It could also be a shorted or leaky varactor diode in the tuner. Check with ohmmeter to ground. Try substitution using a bias box or an adjustable dc PS, set for voltages indicated on the receiver schematic.

For no or incorrect voltages for some channels only, check voltages at the setup control pots and push-button channel selectors corresponding with the troublesome channels.

Where a PLL tuning system is used to develop the bandswitching and tuning voltages, almost all ICs in the system become suspect.

- **Signal Injection**

Verify the IF strip and all downstream stages are working by feeding a 45.75-MHz signal from an RF signal generator to the IF input and the tuner end of the coax. Note the strength of the bar pattern on the pix tube screen. When there is a UHF problem, try feeding the signal to the UHF-IF terminal on the VHF tuner with the VHF tuner set for UHF reception.

When the VHF tuner is suspect, feed an RF signal corresponding with the frequency of a selected channel to the antenna terminals. Repeat for all channels as appropriate. Check the operation of the fine tuner. When a dead tuner oscillator is suspected, try using the signal generator as a substitute (see Sec. 12-28).

- **Antenna Substitution**

Troubles with the antenna system can cause symptoms of weak or no reception on some or all channels, snow, weak or missing color, and so on. Try a substitute antenna, such as rabbit ears, while noting *any* improvement. For poor UHF reception, try a jumper from the VHF to the UHF terminals.

- **Tuner Substitution**

This is a time-honored way to determine if a tuner is defective. Use either a *subber* (see Sec. 3-20), another tuner known to be trouble-free, or connect the set back to back with another receiver provided it is not a hot chassis. Disconnect the IF coax from the suspect tuner and connect it to the substitute tuner. When a separate tuner is used, it must be supplied with B voltage. Clamp the AGC with a bias box. For back-to-back testing with another set, the chassis must be connected together and both sets plugged in.

When only the UHF tuner is suspect, connect a substitute tuner to the UHF-IF input of the VHF tuner of the troublesome receiver.

REMINDER: _____

Substitution test results may not be valid if the substitute unit has its own PS and the cause of the problem is loss of B voltage, AGC, and so on.

12-23 TWO OPTIONS IF TUNER IS DEFECTIVE

Unless the trouble is obvious and minimal, such as dirty contacts or oscillator touch-up adjustments, it may be too costly in time to attempt repairs. The option is to send it to a tuner repair station for repair or replacement on an exchange basis. Many defects, however, can be corrected using ordinary troubleshooting techniques, and time permitting, it may prove worthwhile to attempt your own repairs.

12-24 TOOL REQUIREMENTS

Tuner repair requires considerable patience, expertise, and manual dexterity, working in confined areas with small and easily damaged components. Proper tools are a *must*. Some of the more useful items are a flashlight, magnifier lens or jeweller's loupe, a sharp pick, small but strong nonmetallic prods, small diagonal pliers, needle-nosed pliers, tweezers, a small desoldering sucker, and a low-wattage soldering iron with a small tip (see Sec. 5-9).

12-25 INSPECTION OF FAULTY TUNER

With the set off, make a visual inspection for such things as physical damage, stripped channel selector drive gears, broken switch wafers, fine tuner defects, cracked feed-through capacitors, scorched resistors or balun coil, and solder splashes.

12-26 TROUBLESHOOTING AND REPAIR OF VHF TUNERS

The most common trouble with wafer- and turret-type tuners is bad switching contacts. The symptoms are easily recognized: noisy erratic reception when switching channels, intermittent pix and sound, snow, microphonics when tapped, the inability to receive some channels unless the tuner knob is jiggled or turned slightly off the detent position. For such problems, it is a good idea to clear them up *before* getting further involved.

12-27 CLEANING AND REPAIR OF CHANNEL SELECTOR CONTACTS

Sooner or later, wafer and drum tuners develop problems where the silver-plated contacts become dirty, tarnished, or oxidized and make poor electrical contact. Repair usually consists of a good cleaning job, using a suitable solvent or degreaser followed by lubrication to prevent wear. An aerosol spray may be used, or apply directly using a hypo syringe or a small brush. Use sparingly and wipe up any drips. Avoid spraying coils, trimmers, neutralizing capacitors, transistors, and varactors. Excess solvent seeps into components and leaves a residue that detunes the circuits, especially on the high channels. Some solvents like alcohol can dissolve plastic parts, the enamel on coil windings, and cause deterioration and eventual destruction of components. Spraying or squirting solvents indiscriminately into tuner openings may be expedient, but inadvisable. Remove the shield cover. If it is a drum tuner, remove several coil strips for access to the stationary contacts.

After applying the solvent or degreaser and before turning the set on, operate the channel selector for at least 1 min. or longer and the contacts will usually clean themselves.

Don't be alarmed about oscillator drift where some of the high channels can no longer be received, and don't be hasty in making oscillator adjustments or in neutralizing the RF stage. Allow plenty of time for the solvent to evaporate and chances are no adjusting will be needed. Evaporation can be speeded up with a hair dryer or a lamp. For stubborn contact problems, persevere; the cleaning procedure may have to be repeated.

Where repeated cleaning does not get results, brush the contacts with a stiff-bristled brush dipped in solvent or as a last resort, use an abrasive such as silicone carbide paper or a pencil eraser. In rubbing and burnishing the contact surfaces, exercise extreme care so as not to bend, distort, or damage the contacts. One misaligned contact can result in serious damage when the selector is rotated. Remove any accumulated brush hairs or abrasive with solvent. Check results on all channels and by rocking the channel selector knob up and down and sideways while observing the pix.

Where troubles still persist, make a *close* inspection of the contacts. When set for the most troublesome channel, try to localize the defect by gently probing contacts with an insulated probe. Avoid any unnecessary pressure since the tension and alignment of the contacts is very critical.

When a bad contact is located, clean, straighten, or align the surfaces as appropriate. *Slowly* rotate the selector while observing the opening and closing of the contacts. Check results on all channels.

Sometimes the stationary contacts on a drum tuner become broken or lose their tension. Remove as many coil strips as necessary for easy access. Carefully bend the spring loops to make good contact with the corresponding contacts on the coil strips. Slowly rotate the drum to check the wiping action. Reinstall all coil strips and check operation on all channels.

Most tuner solvents contain a lubricant. If not, follow the cleaning process with a spot of vaseline or light oil applied to each contact. Reinstall the tuner shield cover prior to alignment.

12-28 LOCALIZING VHF TUNER TROUBLES

The first step in tuner troubleshooting is to localize the trouble to one of the three stages. Start by removing the cover shield for access to the interior. When trouble symptoms are weak pix with snow, the defect is probably in the tuner input circuits or the RF stage. Using a modulated RF signal generator, inject a signal at various points between the antenna terminals and the mixer input. Use a frequency corresponding with the selected channel. Tune the generator either side of the channel center frequency while observing the response on the pix tube. There should be strong sound bars even with an attenuated generator output. If there is a good response with signal fed to the mixer but a weak response from the RF stage input, the RF stage is at fault. If signal is lost between the antenna input and the RF input, the balun or filters are suspect.

NOTE:

A dead *RF* stage may or may not pass a signal depending on the nature of the defect (e.g., a *shorted* transistor versus an *open* circuit).

When the symptoms are *no* pix or sound, the mixer and oscillator stages are suspect. Inject a 45.75-MHz signal to the input of the IF strip, the tuner end of the coax, and to the collector of the mixer transistor. Use insulated miniconnectors to avoid shorts, and make and break connections only with the set turned *off*. A good response should be obtained at each of the checkpoints mentioned. When the IF response is OK, but an *RF* signal will not go through from the mixer input, the mixer or oscillator is at fault.

A quick check for a possible dead oscillator stage is to substitute using an unmodulated RF generator signal. Loosely couple the generator to the mixer circuit or the regular *injection point*, as indicated on the schematic. Tune the generator 45.75 MHz higher than the selected channel video carrier frequency (or 41.25 MHz higher than the sound carrier). If pix and sound for that channel are obtained, then the oscillator is either far off frequency or not working. Repeat the test for one or more channels of both the low and high bands. If it is a varactor tuner, verify the proper tuning voltages as described earlier.

If the set has AFT, check its ability to control the oscillator, assuming a pix can be received. Check the AFT pull-in and lock-in range by operating the defeat switch as the fine-tuning control is misadjusted by different amounts.

12-29 ISOLATING VHF TUNER DEFECTS

Except for the difficulty of working in a confined area, fault finding in a tuner is no different than for other circuits. The following procedures for pinpointing defects assume the trouble has been previously *localized* to one of the three stages. Alignment procedures for the tuned circuits are also included.

- **Tuner Input Circuit**

Component breakdowns in this area are rare, but there are two things to watch for: a balun burned open from a lightning discharge, and broken connections. An ohmmeter check will show up most defects.

- **RF Stage**

Is there a problem with only the low channels, the high channels, or both? Check the transistor, preferably without removing it.

REMINDER: ⎯⎯⎯⎯⎯⎯⎯⎯⎯⎯⎯⎯⎯⎯⎯⎯⎯

Tuner transistors are special high-frequency types and are not interchangeable with others in the set. Most have a fourth terminal lead that must be grounded.

Where a FET is used, leakage or shorts are very common. Make an ohms check from each gate to ground.

REMINDER: ⎯⎯⎯⎯⎯⎯⎯⎯⎯⎯⎯⎯⎯⎯⎯⎯⎯

It is very easy to damage or destroy a FET. Take extraordinary precautions when testing or making a replacement.

Check dc operating voltages at transistor terminals and compare with the no-signal voltages indicated on the schematic.

REMINDER: ⎯⎯⎯⎯⎯⎯⎯⎯⎯⎯⎯⎯⎯⎯⎯⎯⎯

For maximum sensitivity with minimun snow, the *RF bias* is extremely critical. Make precise measurements. Typically, an RF transistor operates with a small forward bias (as measured between B and E) under no-signal conditions. Abnormal voltage readings can be expected if the transistor is open, leaky, or shorted.

Measure the bias variations when receiving both weak and strong signals. On weak stations (where delayed AGC is used) the reading should be the same as off channel. On strong signals it should either increase to as much as 5 or 6 V maximum or decrease to zero, depending on whether *forward* or *reverse* AGC is used. This can be determined from the schematic. The polarity depends on the type of transistor, NPN or PNP.

REMINDER: ─────────────────────────

A bias deviation of as little as 1 V or less can often make a difference between normal reception, a weak snowy pix, or complete loss of pix and sound. When the symptoms are weak pix with snow, the cause may be excessive AGC. For mixer overload, the cause may be too little AGC on the strong stations. For weak pix with snow, try adjusting the AGC delay control (if provided) as described in Sec. 11-21.

Where the tuner is a varactor type and reception is weak, check bandswitching voltages and the tuning voltage at the RF stage varactor. Compare with the tuning voltage applied to the other stages and as specified by the schematic for the various channels. If only the RF tuning voltage is abnormal or missing, the varactor is probably defective.

When a bad transistor, varactor, or other diode is discovered, make additional tests to determine the cause. Also check for other components such as resistors that may have been damaged by a shorted diode or transistor. More often than not, because of cause and effect, two or more parts will be found defective.

- **Oscillation and Neutralizing**

Regeneration (positive feedback) in the RF stage causes instability; carried further into oscillations, it results in streaks and herringbone patterns on the screen. A clue to this problem is when symptoms change drastically when a finger is brought close to the RF components or wiring. Feedback can be caused by improper lead dressing (after replacing a transistor or other component) or excessive stage gain due to insufficient AGC. The feedback may be neutralized using one of the following procedures:

Method 1. Locate the neutralizing trimmer capacitor and adjust until the trouble symptoms disappear. Where a gimmick capacitor (see Sec. 12-4) is used, add more or fewer twists while observing the results. In extreme cases where there are no provisions for adjustment, disconnect the original fixed capacitor and install a low-value trimmer or gimmick capacitor.

Method 2. Disable the RF stage by disconnecting the dc power feeding the transistor. Do not remove the transistor. If neutralization is required, some pix and sound will be observed on the strongest stations. Adjust the neutralizing capacitor for *minimum* pix contrast and *maximum* snow. Reconnect the B supply.

- **RF Stage Alignment**

Some misalignment of the RF tuned circuits can be tolerated because of the wide bandpass, and alignment is *seldom required*. Serious misalignment, however, will reduce the signal-to-noise ratio and the selectivity of the stage. When it becomes necessary, the RF and mixer circuits are usually aligned at the same time.

- **Mixer Stage**

With a dead mixer stage you lose pix and sound. A weak mixer results in a weak pix with some snow. Verify it is not a weak IF section by feeding a 45.75-MHz generator signal to the tuner end of the coax.

Other tests are the same as for the RF stage (i.e., checking the transistor and its operating voltages). The mixer is not supplied with AGC. The typical forward bias on a mixer is about 1 V.

Check for oscillator injection voltage at the B or E depending on the set. It is normally around 2 V. Loss of this voltage usually indicates a leaky or shorted mixer transistor or a dead oscillator stage.

For a varactor tuner, check the bandswitching and tuning voltage applied to the mixer stage, and compare with other stages and the schematic. For an abnormal reading when tuning and bandswitching voltages on other stages are OK, check for a leaky or shorted varactor or bandswitching diode.

12-30 ALIGNMENT OF THE RF-MIXER STAGES

Realignment of these two signal stages is seldom required. Even when misalignment is suspected, first *verify* the need before making any adjustments. A review of the fundamentals (Sec. 12-4) is suggested before proceeding. The *desired* response of the RF-mixer stages is indicated by the double-humped response curve shown in Fig. 12-4, with the pix and sound RF carrier frequencies at the two peaks where they receive maximum amplification. The bandpass at the 50% level should be 6 MHz. The center of the curve should correspond with the center frequency of whichever channel is selected, for example, 57 MHz for channel 2.

If the tuned circuits are misaligned so the bandpass is either too wide or too narrow, the pix and/or sound will be adversely affected. Other problems occur if the circuits are tuned either too high or too low in frequency and beyond the limits of the selected channel. If tuned too *low*, for example, the center frequency of the channel will not correspond with the center of the curve, and the sound carrier will be at or near the bottom of the curve, resulting in weak sound. If tuned too *high*, the pix carrier will be down on the opposite slope, resulting in a weak pix. Note that the two carriers move *together* relative to the curve since their 4.5-MHz separation is established by the *station* and cannot be altered by conditions in the receiver. The two abnormal conditions described here can be simulated during reception by misadjusting the fine-tuner control in both directions from its normal setting.

12-31 CHECKING THE RF-MIXER RESPONSE

The accuracy of alignment can be no better than the calibration accuracy of the signal generator used. *Some* calibration error (especially at VHF/UHF frquencies) can be expected from all but the more expensive generators, so don't be surprised if *apparent* misalignment of a receiver turns out otherwise. Where misalignment is suspected, but not yet proved, there are three possibilities: the generator previously used by others to align the set was in error, the generator *being used* for checking is not 100% accurate, or the alignment adjustments have been tampered with.

There are two methods of *checking* and *performing* alignment: using an accurate, AM-type RF signal generator, and *sweep alignment* using a sweep generator, a marker generator, and a CRO.

Method 1: Using an RF Signal Generator. The following procedures are used only for *evaluating* alignment conditions and no adjustments are made at this time.

1. To defeat the AGC action, clamp the main AGC RF/IF bus (with a bias box) to the no-signal level and polarity indicated on the schematic.

2. Connect the signal generator, using a low-impedance termination pad, to the VHF antenna terminals.

3. Connect an EVM with a demodulator probe to the mixer output or the coax feeding the IF strip. Select a low dc voltage range.

460 CHANNEL SELECTORS AND TUNING SYSTEMS

4. Disable the tuner oscillator. One way is with a jumper from B to E.

5. Tune the generator to the approximate center frequency of the selected channel while observing the meter. If the meter pegs, reduce the generator output rather than select a higher range. Always use the lowest possible signal level to prevent overload.

6. Slowly tune the generator about 2 MHz lower in frequency to get a peak reading that corresponds with the pix carrier frequency for that particular channel. Does the meter peak at *that* frequency? It should.

7. Tune the generator 2 MHz higher than center frequency to see if a peak reading is obtained for the sound carrier frequency of the selected channel.

8. Check the bandwidth by tuning the generator both higher and lower than center frequency to get two meter indications 50% lower than the peak readings. The bandpass is the difference between the two generator frequencies. If necessary, check the response for two or more channels.

Method 2: Using Sweep Equipment. This is the preferred method. It requires three instruments, a sweep generator, a marker generator, and a CRO. The instruments may be *separate* or *combined* (see Sec. 3-12). The equipment setup using separate instruments is shown in Fig. 12-23. With this method, the sweep generator supplies an FM signal that *sweeps* over the frequency range of the RF-mixer bandpass to produce a response curve on the CRO. The marker generator identifies frequencies represented by the curve.

NOTE:

Results depend on the calibration accuracy of the *marker* generator. Extreme accuracy of the *sweep* generator is not essential.

1. To avoid interference with the response, disable the tuner oscillator as in method 1. Clamp the AGC as described earlier. With the sweep generator tuned to the approximate center frequency of a selected channel, a display should be obtained on the CRO.

2. Center the display using the generator tuning dial. With the CRO adjusted for high vertical gain, slowly attenuate the generator output until the response almost fills the CRT vertically. Use the lowest possible generator signal (and highest CRO gain) to avoid overload problems. However, too much vertical gain on the scope results in instability and hum pickup.

3. Set the sweep (frequency deviation) on the generator to about 12 MHz or until the response almost fills the CRT horizontally. Too much sweep compresses the curve and sweeps a wider band of frequencies than necessary. Too little sweep sweeps only a *portion* of the bandpass.

4. Tune the marker generator and observe the marker pip that moves over the response curve. The strength of the marker signal is often critical. If too weak, it will not be observed, especially on the slopes of the curve. If too strong, it will tend to flatten and distort the curve, giving misleading indications.

5. Identify the marker locations on the curve. For marker signals representing the pix and sound carrier frequencies of the selected channel, the markers should appear at the two peaks as indicated in Fig. 12-4, with a marker for center frequency midway between the peaks.

6. Check the bandwidth by locating two markers at the 50% level on both sides of the curve. For further details on sweep alignment, see Sec. 13-41 and Fig. 13-17.

12-32 RF-MIXER ALIGNMENT PROCEDURES

When the previous evaluation tests indicate a need for alignment, proceed as follows, using either of the two methods described.

1. Decide what adjustments are required to obtain a normal response. For example, if the pix and sound markers are on the low-frequency side of the curve, the curve must be moved over until the markers appear on the two peaks. This is done by tuning the RF-mixer coils *lower* in fre-

12-32 RF–Mixer Alignment Procedures

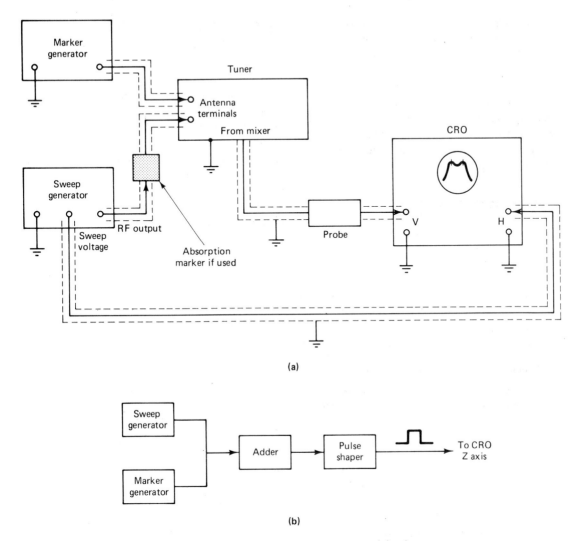

Fig. 12-23 Sweep alignment setup: (a) the marker and sweep generators are usually combined in one instrument; (b) post-injection (intensity) marker system.

quency. Conversely, if the markers are located on the high-frequency side of the curve, the coils must be tuned *higher* in frequency.

2. A *tuning wand* can be a useful aid in checking and performing alignment. This device consists of a short length of nonmetallic rod or tubing with a powdered iron slug at one end and a brass slug at the other. As explained in Sec. 12-4, an iron core lowers the resonant frequency of a coil and a nonmagnetic core raises the frequency. To use the wand, hold the iron slug near the coil under test or insert it if possible. If the response improves, the coil needs to be tuned lower in frequency. Repeat the test with the brass end of the wand. Where response improves, the coil needs to be tuned *higher*. Where both slugs produce opposite effects but with no improvement in results,

the coil is tuned correctly and no adjustments are required. Repeat this check for all channels, as appropriate.

Where no coil adjustments are provided, they may be tuned by either *stretching* or *compressing* two or three turns at the end of the coil(s). Compressing the windings lowers the frequency, and spreading the turns apart raises the frequency.

- **Adjusting the Tuner IF Coil**

   1. Connect an RF signal generator through a small blocking capacitor to the mixer input. Tune the generator to 43 MHz or the frequency specified on schematic.

   2. Connect an ac EVM to some convenient test point at the video detector or beyond.

   3. Tune the generator to see that the meter peaks at or near the frequency of the IF transformer. Select a low-meter range and attenuate the generator for a low output.

   4. With the generator set for the proper frequency, peak the transformer for maximum meter reading.

If the coil will not resonate properly, suspect a broken or missing slug, shorted coil windings, or loading caused by some defect in an associated circuit.

- **Adjusting Input Circuit Traps**

Identify and locate any adjustable traps in the tuner. Connect an RF signal generator to the antenna terminals. Connect an EVM with a demodulator probe to the mixer output. For an IF trap, tune the generator to 45.75 MHz. For an FM trap, tune the generator to approximately 98 MHz (or the frequency of any interfering FM station). Adjust traps for *minimum* indication on the meter.

12-33 TROUBLESHOOTING THE OSCILLATOR STAGE

A dead oscillator (no oscillations) results in loss of pix and sound. This can also occur if the oscillator is far off frequency. A *slight* frequency error usually causes loss of one or several channels and imperfect reception on the others. Frequency *drift* requires periodic adjustment of the fine-tuning control.

The following test procedures assume the trouble has been tentatively localized to the oscillator, the most critical stage in the tuner.

- **Dead Oscillator**

To determine if a *vacuum-tube* oscillator circuit is working, you simply check for a negative dc voltage at its control grid. With a transistor oscillator, especially a high-frequency tuner oscillator, test results are sometimes inconclusive. One way to check for oscillations is to use a *counter*. A sensitive ac EVM will also give an indication, depending on the meter. Since the dc operating voltages of an oscillator transistor change slightly between an oscillating and a nonoscillating state, look for such changes as the circuit is loaded as by touching a finger to one or more transistor terminals. No change in voltages or the absence of the oscillator dc injection voltage at the mixer means the circuit is not oscillating. This can be verified by substitution, as explained in Sec. 12-28. When an unmodulated signal generator substituted for the tuner oscillator (and tuned to the correct frequency) restores reception, the tuner oscillator is at fault. Make sure the VHF tuner is not set for UHF, at which point the VHF oscillator is normally disabled.

To pinpoint the cause when the circuit is not oscillating, start by checking the transistor, preferably in circuit so as not to unnecessarily disturb the circuit and risk damaging the transistor or other components. when replacement becomes necessary, avoid altering the lead dress and placement of components, which will necessitate later readjustment of the tuned circuits. For a bad transistor, use only an exact replacement.

Check all dc operating voltages at the transistor and compare with the schematic. For abnormal voltages, verify the condition of the transistor and other associated components.

If the tuner is a varactor type, check for bandswitching and tuning voltages, comparing with the voltages and polarity indicated on the schematic. If the voltage at the oscillator varactor is lower than at other tuner varactors, chances are the oscillator varactor is leaky or shorted. An *open* varactor will kill the stage without changing the tuning voltage. For loss of tuning or bandswitching voltages, check back to the channel selector pots, the channel selector, and if necessary, the source. Look for such things as a shorted bandswitching diode, a bad selector pot, or a defective zener or transistor in the voltage-regulating circuit.

• **Incorrect Oscillator Frequency**

Pix and sound reception is only possible (for any selected channel) when the oscillator frequency is correct. For loss of *all* channels, the frequency error may be extreme. A *slight* frequency error is more common.

Start by adjusting the fine tuner. Is it possible to receive some channels, but only at one extreme setting of the fine tuner? The cause could be normal aging of components (which calls for a touch-up adjustment of the master oscillator adjustment or the individual channel slugs), a defect associated with the fine tuner, a less than perfect oscillator transistor, a resistor or capacitor that has changed value, or, for a varactor tuner, incorrect tuning voltages.

Inspect the fine tuner assembly. Typical defects include a damaged coil or a broken or missing tuning slug. For such troubles, replacement rather than repair is advisable.

For a varactor tuner, measure tuning voltages for all channels and compare with the schematic. Check the source voltage and at several points between the source and the tuner.

REMINDER:

Low tuning voltages can be caused by a *leaky* varactor. When only some channels are affected, check the associated channel selector pots.

When the cause of the frequency error is obscure, perform all the tests described for a *dead* oscillator, as appropriate. For a PLL system, see Sec. 12-38.

• **Oscillator Drift**

Frequency drift is most noticeable on the high channels and as a set warms up. A slight drift, although annoying, can be compensated for with the fine tuner or the AFT. For severe frequency drift, the adverse affect on pix, color, and sound progressively worsens, and one by one the high channels may be lost. In checking for oscillator drift on a set provided with AFT, switch the AFT switch to *off.*

The cause of drift is often difficult to pinpoint. Start by replacing the oscillator transistor even though it checks OK. Is there a heat source near the oscillator components? A resistor that normally runs warm should be relocated, if possible. Check the security of the oscillator coils and their adjusting slugs. Where there are stripped threads, a slug can move slightly as it heats up. Check the fine tuner for an unstable coil or loose slug.

For a varactor tuner, monitor the tuning voltage on the highest active channel. No variations should be tolerated. If voltage does change, check the regulating circuit. Where a PLL tuning system is used, refer to Sec. 12-38. If warranted, try an external dc supply as a temporary source of tuning voltages.

For stubborn cases of drift, replace all capacitors in the oscillator circuit, the NTC types in particular.

• **Oscillator Alignment**

Which stations are received, if any, is determined by the *oscillator* frequency and not by the tuning of the RF-mixer stages, as might be expected. Before making any oscillator adjustments (which are extremely critical), clear up any problems with noisy or erratic channel selector switches as described in Sec. 12-27. Allow plenty of time after a cleaning job before making adjustments. Make sure all tuner repairs are completed and the cover shield is in place before adjusting. Where possible, use a nonmetallic tool for the slug adjustments to avoid damage and detuning effects.

When all channels are received at the wrong channel number readout positions, chances are the number readout has slipped. When one or more stations are received at inactive channel positions, make the necessary adjustments so these positions are blank, as intended.

Most tuners have a *master* oscillator adjustment that affects *all* channels. It should be used only as necessary when *all* channels are off tune by some amount. When only some channels require tuning, the individual channel adjustments are used. Specific alignment procedures vary somewhat according to the type of tuner, as follows.

Drum Tuner Oscillator Alignment. The master oscillator adjustment is usually at the top of the tuner. Individual channel slugs are adjusted through holes at the front. Channels can be adjusted in any desired sequence. Adjust fine tuner control to mid-setting and don't touch until alignment is completed.

Using the proper alignment tool, carefully adjust each channel slug, as required, for best pix, color, and sound.

REMINDER: _____

Screwing a brass slug into a coil raises its resonant frequency, and vice versa.

Stripped threads on an adjusting screw are quite common and must be corrected. A quick fix is to remove the slug, insert a small strip of photographic film negative, nylon thread, or the equivalent, and reinstall the slug. For a missing slug or a damaged coil, use the coil strip from an inactive adjacent channel. When a channel cannot be received within the range of its slug adjustment, try stretching or compressing coil turns as described in Sec. 12-32.

Check reception on all channels. With accurate adjustment of the channel slugs, it should be possible to change channels and get the best reception *without using the fine tuner*.

Wafer-Switch Tuner Oscillator Alignment. Unlike the drum tuner, the coils are interdependent and must be adjusted in a definite sequence, starting with channel 13 and working progressively to channel 2. With the fine tuner at mid-setting, adjust the coil slug for channel 13 (or the highest active channel for the location) for the best pix and sound. Select the next (lower) active channel (channel 11, for example) and adjust its coil in the same manner. If it cannot be received, go back and adjust channel 12 (in this example) in conjunction with the channel 11 slug. If channel 11 still does not come in, make a *slight* touch-up adjustment of the *master* oscillator adjustment if there is one.

Proceed with channel 9 adjustment. If necessary, also adjust the inactive channel 10 slug. Adjust all active channels (in conjunction with the adjacent higher channel), finishing up with channel 2. If possible, avoid use of the master adjustment since its use necessitates readjusting all *individual* channel slugs. Another option here is to stretch or compress coil turns as described earlier.

Verify reception on all active channels, making touch-up adjustments as necessary. Touching up a low channel, however, may require readjusting all higher channels.

Varactor Tuner Oscillator Adjustments. If AFT is provided, all adjustments are made with the AFT switched *off*. Adjustments can be made in any desired sequence. With the fine tuner at mid-setting, adjust all channel selector pots to obtain the correct tuning voltages for all active channels as indicated on the receiver schematic.

REMINDER:

Low tuning voltages equate with the low channels and higher voltages with the high channels. Check reception on all channels.

If all channels are off by a *small* amount, make a touch-up adjustment of the oscillator coil. If there is a large error where all channels cannot be received with the specified tuning voltages, suspect a bad oscillator varactor. Replace it even though it checks OK. When there is a problem with some channels only, select the most troublesome channel, adjust its associated pot for the correct tuning voltage, and make a slight compensating adjustment of the oscillator coil. Check reception on all channels, making further adjustment of the pots as necessary.

12-34 TROUBLESHOOTING AND REPAIR OF UHF TUNERS

The usual trouble symptoms are no UHF reception or weak reception with snow. Before condemning and removing the UHF tuner, however, verify whether *it* is the cause. It is assumed that VHF reception is normal; otherwise, that problem should be considered first. If there is no UHF reception, verify there is B voltage at the tuner terminals when the VHF tuner is set for UHF. Check the integrity of the UHF-IF coax connected to both tuners. Try feeding a 43-MHz generator signal to the cable. Try substituting a generator signal for the UHF oscillator, as described for VHF tuners in Sec. 12-28. The frequency used must be 45.75 MHz higher than the video carrier of the selected station. Finally, try a substitute UHF tuner.

• **Troubleshooting the Basic UHF Tuner**

The simple two-stage tuner described in Sec. 12-11 gives very little trouble. The usual troubles are a bad oscillator transistor or mixer diode. One good test is to disconnect the UHF-IF cable from the VHF tuner and connect a low-range dc milliammeter from the UHF tuner IF output to ground. If the mixer–oscillator stages are OK, you should get a reading of 0.5 to 5 mA over the entire UHF tuning range. Either a high or a low reading can indicate trouble. If results are uncertain, try killing the oscillator (by touching the tuning capacitor or transistor terminals). If the oscillator is working, the reading should drop to zero. For a dead oscillator, replace both the transistor and the mixer diode without bothering to check them.

Check the oscillator operating voltages at the transistor terminals. For loss of B voltage check back to the source, which is typically between 10 and 40 V. Make sure the VHF tuner is set for UHF. Check operation of the VHF/UHF changeover switch on the VHF tuner.

Some basic UHF tuners are the varactor type. Check the bandswitching and tuning voltages and compare with the schematic. For these problems, check the varactor and bandswitching diode.

CHANNEL SELECTORS AND TUNING SYSTEMS

Weak UHF reception can be caused by a poor location, antenna or lead-in problems, or a faulty mixer diode. Try a substitute antenna or a jumper between the VHF and UHF antenna terminals. Replace the mixer diode using one of the *same type,* and observe the correct polarity.

REMINDER:

The placement of wires and components is extremely critical and they must not be disturbed when making tests and repairs. Even with normal precautions, some detuning can be expected after replacing a transistor or other components.

- **Alignment of Basic UHF Tuner**

Unlike a detent-type tuner, the tracking of the oscillator and the mixer tuned circuits is no great problem. Some dial calibration error (as much as one or two channels) is considered acceptable. Make sure there is no slippage of the channel indicator dial. Assuming no slippage and no disturbance of the mixer stage tuning, any severe calibration error can be corrected as follows.

Determine whether stations are received too high *or* too low on the dial. If channel 40, for example, comes in at 30 on the dial, the oscillator tuning capacitor must be *increased.* Conversely, if it comes in at 50 on the dial, the capacity must be *reduced.* To verify, try slipping a small sheet of plastic between the stator and rotor plates of the oscillator tuning capacitor to see which way the calibration changes. The plastic increases the capacity, shifting the calibration toward the high end of the dial.

The outer plates of each tuning capacitor are *slotted.* The capacity can be varied by bending the plates inward or outward in a procedure called *knifing.* Determined by which portion of the dial is off the most, bend the appropriate plate(s) by a *small* amount in the required direction and check the results. Allow for some variation when the tuner cover plate is reinstalled. Make sure no stator or rotor plates are touching over its tuning range.

When the calibration error has been corrected, check the resonance of the mixer tuned circuit. Tune to a weak station and insert the plastic tuning aid between the plates of the mixer capacitor. The pix contrast should decrease. If there is an increase in contrast, knife the rotor plates as described previously.

For weak UHF where no defect can be found, try improving the RF/IF conversion efficiency by changing the amount of oscillator injection. Vary the position of the injection *hairpin loop* relative to the oscillator tuned line. The adjustment is critical, so check results frequently on the weakest UHF channel.

For a dial calibration error on a *varactor* tuner, or when there is no reception for the highest or lowest channels, measure the tuning voltage and compare with the schematic. Check the operation of the fine-tuner control pot. Try replacing the varactor. Try using an external PS as a temporary source of tuning voltage.

- **Troubleshooting a Three-Stage UHF Varactor Tuner**

Compared to the basic UHF tuner, better results can be expected, mainly because of the AGC-controlled RF stage and the transistor mixer. For weak or no reception, troubleshooting procedures are essentially the same as for its VHF counterpart.

Make sure the VHF tuner is operating properly and that an IF signal from a generator gets through to the IF strip when the VHF tuner is set for UHF. While making this test, try peaking the mixer output transformer at 43 MHz or as specified on the schematic.

Check all applied voltages at the UHF tuner input, the B voltage, AGC, and tuning voltages for all channels. Compare with the schematic. For missing or abnormal voltages, check the possibilities previously described. As a check on the oscillator, try substituting with an RF signal generator. Check the signal path between the antenna terminals and the mixer input by injecting a UHF signal from a generator (tuned to some UHF channel) to various key checkpoints.

Test all transistors and varactors as appropriate, preferably in circuit (to avoid detuning and possible component damage). Defective transistors and varactors and incorrect tuning voltages are common problems. Check the operation of the fine tuner.

FETs are often used in the RF-mixer stages. Use extreme care when testing, making voltage checks, and making replacements (see Sec. 5–9).

Misalignment of the tuned circuits can cause weak reception or loss of certain channels. Most tuners of this type are provided with small trimmer capacitors for resonating the tuned lines. Alignment procedures are as follows.

Select a station from the push-button channel selector. Measure the tuning voltage for that channel and compare with the schematic. If the voltage is incorrect, adjust the appropriate channel selector setup pot to obtain the *correct* voltage. When this results in loss of the station, adjust the oscillator trimmer until the station comes in. Check the resonance of the RF-mixer tuned circuits using the sheet plastic tuning aid described earlier. If necessary, adjust the RF-mixer trimmers for maximum pix contrast with minimum snow. Check reception on all active channels. If necessary, make compromise touch-up adjustments to favor the station(s) of choice.

12-35 MECHANICAL PROBLEMS WITH VHF/UHF TUNERS

Due to wear and tear on moving parts, both drum and wafer-switch tuners sometimes break down. Some defects are repairable, some are not. Where damage is extensive and repairs time consuming, or where the permanence of repair is in doubt, consider sending the tuner out for repair or replacement on an exchange basis. Typical defects are as follows, with repair procedures for the do-it-yourselfer.

Dial Knobs. The built-in spring strips frequently become lost or lose their tension; the knob becomes loose and may not turn the channel selector shaft. Bend or replace the spring.

Broken Tuning Shaft. Tuner replacement is recommended.

Stripped Gears. Nylon gears, especially those used in fine tuning assemblies, frequently become stripped or worn along with other nylon parts. Replace as a unit. When dismantling tuner mechanisms, and no exploded drawings or photos are available, make your own sketch *as parts are removed.* Don't trust to memory.

Detents. A missing steel ball or broken spring is common. Replace.

Slipping Clutch. These are used with two-speed UHF tuners. Clean but do not lubricate. If necessary, replace.

Core Slugs Missing, Stripped, or Broken. With drum tuners, the fine tuner sliding iron core often falls out. Unless missing, replace using a *tiny* drop of cement. Don't enter the core into the coil until the cement is dry. Sprocket-driven fine tuning slugs of individual oscillator coils sometimes become stripped. Borrow slug from an unused adjacent channel coil or shim with a small piece of film. Wax having a high melting point can also be used. Do not use cement.

Switch Wafer Not Turning. The rotatable center part of a wafer may develop an enlarged hole and does not rotate with the shaft. When suspected, closely examine each wafer as the shaft is turned in both directions. When a slipping wafer is found, position it correctly relative to other wafers for normal reception of some active channel. Secure the moving element of the wafer to the shaft using liberal amounts of cement. Don't allow cement to run onto switch contacts. Allow cement to dry before changing channels.

Switching Contact Problems. For cleaning and repair procedures, see Sec. 12-27.

Loose Coil Strips on Drum Tuners. The retaining spring strips may be broken or lose their tension, allowing coil strip to hang free or drop out when the drum rotates. Secure with fast-drying cement. Don't get cement on the contacts.

Stiff or Noisy Drive Mechanisms. The intricate drive mechanisms of some tuners can become hard to turn and in some cases seize up when switching channels. For smoother operation and to prevent wear, apply light oil or vaseline to all moving parts except clutches. Wipe up any excess.

12-36 TUNER REPLACEMENT

Tuner repair stations often receive tuners that are *not* defective. This can be prevented by performing all tests as described in Sec. 12-22 before considering replacement.

To avoid costly mistakes and to save time, when replacement is necessary, make a sketch of all external connections (Fig. 12-22) *prior to removal*. Indicate all wire terminations and other terminals (whether used or not) and feedthrough capacitors. Don't rely too much on the color coding of wires, which is often nonstandard. Label the wires as appropriate. Even with normal handling, wires sometimes break loose and may become connected to the wrong terminal points.

When wires or terminals of a replacement tuner are different from the original, it may become necessary to identify each wire or cable connecting to the chassis. Make continuity tests with the aid of the schematic, using the *lowest* ohmmeter range. (When tracing wires using a *high* range, results can be misleading.) Resistance tests to ground can also aid in identification. With a voltmeter, identify the wires that carry voltage to the tuner. Later, after reconnecting wires to the replacement tuner, verify the accuracy of the connections *before applying power*.

When trouble symptoms persist after a tuner replacement, there are three possibilities: (1) Besides a bad tuner, the set had additional troubles that were overlooked. (2) The tuner was not defective and replacement was unnecessary. (3) An error was made in reconnecting the replaced tuner.

12-37 AFT TROUBLESHOOTING AND ALIGNMENT

It is normally a simple matter to determine if the AFT circuit is doing its job. With the AFT switched off, verify that the tuner is operating properly. If the oscillator adjustments for the individual channels need adjusting, this should be done before proceeding.

With the fine tuner set for best pix and sound on some high channel, switch the AFT on. There should be no change in pix or sound. If there is, the AFT is suspect. With AFT on, misadjust the fine tuner; there should be no effect on pix or sound until beyond the AFT pull-in limits, at which point pix and sound will *abruptly* be lost. Switch AFT off and misadjust fine tuner slightly from its normal setting. When the AFT is switched on, the oscillator should lock in, restoring normal pix and sound.

Whether the AFT is completely inoperative or has a limited lock-in range, the troubleshooting procedure is the same. With the AFT

12-37 AFT Troubleshooting and Alignment

switched on, check for a dc error voltage at the tuner or the AFT output. A normal reading is about 5 V, which should drop to zero when the AFT is switched off. Monitor the voltage as the fine tuner is adjusted. Check with the schematic for the proper voltage and polarity. If there is no error voltage, an abnormal reading, or the wrong polarity, the cause may be an AFT circuit defect or misalignment of the AFT tuned circuits. Try disconnecting the AFT from the tuner, using a bias box as a source of error voltage. See if the tuner oscillator can be controlled as the amount and polarity of the substitute voltage are changed. If not, suspect a tuner defect even though it operates normally with the AFT switched off.

Where a dc amplifier is used to boost the error voltage, and the voltage is OK at its input but not at the output, the amplifier circuit is at fault. Check the dc operating voltages, the transistor, and all associated components.

Check for the dc error voltage at the output of the discriminator circuit. If there is no voltage, an abnormal reading, or the wrong polarity, a discriminator circuit defect may be indicated. Defective diodes are the usual cause. For a balanced condition, the relative conductance of the diodes is very critical. When defective, they should be replaced as a *matched pair*.

For normal operation, equal and opposite voltages should be developed across the split load resistors. Where this is not the case, either the two diodes are not matched or one of the resistors has changed value.

Where a transistor instead of diodes is used and a *balance control* is provided, adjust for a zero error voltage when the AFT is switched on and the tuner oscillator is locked.

For loss of error voltage when the discriminator circuit is OK, check for the video IF signal (using an ac EVM or CRO) at the discriminator input. If the discriminator is preceded by an AFT/IF amplifier stage, check for signal at its input and output. For loss of signal, check back to the takeoff point at the video IF amplifier. If there is a weak signal, suspect misalignment of the AFT tuned circuits.

When an IC is used for the AFT function, troubleshooting is limited to checking input-output voltages, the external AFT components, and substitution with another IC.

• **AFT Alignment**

When no circuit defects can be found, check the AFT alignment using one of three methods:

Method 1.

1. Connect an RF signal generator to the input of the last video IF or to some convenient point upstream. Connect an EVM on a low dc range to the junction of the two discriminator load resistors.

2. Tune the generator to the video IF frequency (45.75 MHz).

NOTE:

For optimum AFT control it is essential that the AFT tuned circuits be aligned to the pix IF carrier frequency established by the tuner oscillator and alignment of the IF strip. Ideally, this is 45.75 MHz, but alignment errors are not uncommon. To compensate for possible frequency discrepancies, adjust the AFT using a frequency that precisely matches the IF strip.

3. Obtain a pix on the receiver screen. Using a low output, tune the generator to blank out the pix. Regardless of the generator dial calibration, this is the frequency to use for AFT alignment.

4. Peak all AFT tuned circuits (except the discriminator transformer secondary) for maximum reading on the meter.

5. Connect the meter *across* the discriminator split-load resistors. Adjust the secondary of the discriminator transformer for *minimum* reading on the meter.

6. Attenuate the generator to the lowest possible level and repeat steps 4 and 5.

When properly aligned, equal and opposite dc readings should be obtained when the generator is detuned by the same amount above and below the AFT frequency. When a coil cannot be made to resonate, suspect shorted windings, a broken or missing core slug, or loading due to a defective circuit component.

Method 2 (Sweep Alignment)

1. Connect a sweep generator to the video section as with method 1.

2. Connect a CRO to the AFT output.

3. Tune the generator to 45.75 MHz (or the *actual* IF frequency as in method 1). Adjust the generator for a sweep of approximately 5 MHz. An S curve (see Fig. 17-5) should appear on the CRO.

4. Loosely couple a marker generator to the video IF strip. Tune the generator to obtain a marker pip on the curve at 45.75 MHz (or the corrected IF frequency).

5. Alternately adjust all AFT tuned circuits to obtain two symmetrical peaks of maximum amplitude with a crossover at the IF frequency. Repeat all adjustments as often as necessary.

Method 3. In the absence of a signal generator, use a *station signal.* Simply peak the transformer primary and adjust the secondary for zero dc across the load resistors. Check AFT operation, as described earlier, for all channels. If there are any changes in pix, sound, or color when the AFT is alternately switched on and off, touch up the tuner oscillator adjustments as required.

12-38 TROUBLESHOOTING PLL TUNING SYSTEMS

Troubleshooting a PLL system is not too difficult considering its complexity. Troubleshooting and repair are on a *systems level,* and since most of the functions are performed by ICs (which cannot be repaired), it is simply a matter of localizing a problem to a particular module or IC with no need for checking and replacing individual components. The only essential test instrument required is a meter, although a counter can also be useful.

• **General**

There are four main steps in troubleshooting a PLL system: (1) Determine the nature of the problem and decide which function or functions are involved. (2) Identify the module or IC most likely at fault. (3) Check input and output signals and voltages at the suspected IC and all related external components. (4) Replace the suspected module or IC.

Where more than one function is affected, suspect modules or ICs that are common to both functions. Cables and their connectors and the connections to modules and ICs are another source of trouble. With some sets a great many interconnecting cables are used between the various functional areas and subsystems. In servicing as a *system,* it is often necessary to use extender cables.

CAUTION: _____

Some ICs contain FETs, which are easily damaged. Use extreme care in handling. Such ICs (as with individual FETs) are generally packaged in conductive foam and should not be handled except immediately prior to use. Before handling, touch some grounded object in case your body has accumulated a static charge. The soldering iron tip should be grounded. And never remove or replace an IC without first removing the ac power. It is not enough to switch the set off; *pull the plug.*

Typical trouble symptoms and procedures are described in the following paragraphs. Start by analyzing the problem(s) in reference to the block diagrams of Figs. 12-17 and 12-19.

REMINDER: _____

Don't condemn and replace a module or IC without first making the tests as indicated. This

12-38 Troubleshooting PLL Tuning Systems

includes the checking of input and output signals and voltages at the terminals of any suspected unit. Also check the integrity of interconnecting cables and connectors by gently moving them while observing any effect on set operation. Depending on the set, some of the tests described cannot be made if an IC contains more than one function.

• **No Pix or Sound**

There are many possible causes: a defective tuner, loss of tuner dc power or AGC, loss of or incorrect bandswitching or tuning voltages developed by the PLL system. It is assumed that the trouble has been previously localized to the PLL system.

Improper Bandswitching Voltages. Is there a problem with *one* or *all three* of the voltages for the three bands, low VHF, high VHF, and UHF? Check for reception on all three bands. Measure voltages and compare with those specified on the schematic. If one or more voltages are missing or abnormal, suspect the channel selector keyboard encoding matrix, the microprocessor, bandswitching amplifier where used, and the bandswitching diodes in the tuner.

• **No Reception on Some or All Channels**

For this problem, *all* functional areas of the PLL are suspect. Is there trouble with *some* or *all* channels? See if the problem relates to one specific digit of the keyboard; for example, a faulty channel 8 push-button switch would show up when selecting channels 8, 18, 28, 80, and so on. Check and compare the coded output from all buttons and compare with the schematic. Where a button is erratic, try cleaning with solvent.

Measure the tuning voltage (if any) at either the VHF tuner or the UHF tuner as appropriate. The reading should be anywhere from zero to about 25 V depending on whether the VCO is running fast or slow. With the oscillator locked at the correct frequency for any given channel, the tuning voltage is normally about 10 V.

If there is no tuning voltage or a grossly abnormal reading, disconnect the tuning voltage line from the tuner. As a check for possible loading caused by a defect (such as a leaky or shorted varactor) in the tuner, make an ohmmeter test from the tuner terminal to ground. To see if the VCO *can* be controlled, and to check the overall operation of the PLL, substitute the PLL-derived tuning voltage with dc from a variable external PS (bias box). Vary the voltage from zero to about 30 V. The oscillator should lock and restore set operation at some voltage within these limits.

REMINDER: _____

The actual amount of tuning voltage developed by a PLL system is not critical, since the system operates by comparing *frequencies,* and the dc voltage is only the *means* of forcing the oscillator to speed up or slow down as required. When an oscillator is running slow, an increased voltage is needed to speed it up, and if fast, a decrease to slow it down.

Missing or incorrect tuning voltages can be caused by a defect *anywhere* in the PLL system. And the tuning voltage problem can be *either* the cause or the result. Loss of tuning voltage, for example, results in the wrong VCO frequency at all points in the feedback loop. Similarly, with a defect in the PLL loops, the proper tuning voltage cannot be obtained. As with all closed loops, sometimes it is necessary to break the loop to perform the required tests.

With the tuner controlled from an external PS, and where the two inputs to the comparator are accessible and a counter is available, check and compare the frequencies of the two input signals. With a pix locked (tuner oscillator frequency correct), the frequencies should be the same. The comparison frequency is usually shown on a schematic of the set. Where the two inputs are the same and there are problems with the tuning voltage, suspect the comparator.

When the *feedback* signal at the comparator is missing or abnormal, check inputs and outputs of all dividers in the feedback loop. Replace any that are suspect. When a *programmable* divider has an incorrect output, both the divider and the *microprocessor* are suspect.

If the *reference* signal is abnormal or missing, check inputs and outputs of dividers in the reference loop, replacing any that are suspect, including the microprocessor. Check for a dead reference oscillator.

If necessary, repeat the preceding tests for all channels.

- **Channel Number Readout Problems**

Assuming there is no problem selecting channels, but the channel display is not operating properly, the most likely suspects are the microprocessor, LED driver amplifiers, decoders, or the LED readout itself. There may be one or more digits not operating or segments missing from the display. Localize the trouble by taking measurements at the microprocessor output, input and output of the amplifiers, and at LED terminals corresponding with the troublesome digits or segments.

- **Up-Down Channel Search (Scan) Problems**

Possible trouble symptoms are inability to select channels from the UP/DOWN panel buttons, unprogrammed channels received and programmed channels not received, and loss of memory that requires periodic reprogramming. Check for outputs from the UP/DOWN panel buttons and at the microprocessor input. If there is no output from the microprocessor, replace the IC. Replace the memory IC if there is loss of memory for one or more programmed channels. To program favorite channels into the memory, refer to the receiver operating instructions, according to the wishes of the user.

- **Loss of Volume Control Function**

If volume is controlled from UP/DOWN panel switches, check for switch outputs, and replace the remote decoder and microprocessor ICs as appropriate.

- **Loss of Remote-Control Functions**

For remote-control problems, it is assumed that all receiver functions initiated from the *receiver* control panel are normal. Note which functions do not work properly when the remote (transmitter unit) is used. Check the transmitter as described in Sec. 12-40. When trouble is at the *receiver,* probable suspects are the signal pickup device, the remote amplifiers, and the remote decoder. Also check for poor or intermittent pin contacts at modules and ICs and the cables and their connectors. Check with gentle pressure and by moving the suspected items while observing results.

12-39 TROUBLESHOOTING REMOTE-CONTROL SYSTEMS

Typical remote-control problems are complete loss of control for one or more functions, intermittent control, or loss of sensitivity where the transmitter unit must be operated close to the receiver. Before considering remote-control problems, first verify normal receiver operation when using the *receiver* controls.

REMINDER:

The receiver must be plugged in and the remote unit in the receiver switched on and operating for receiver *turn-on* by remote control.

12-40 Transmitter Testing 473

After determining which system of communication is used between the remote sending unit (transmitter) and the receiver, the first step is to localize the problem to either the transmitter or the receiver.

12-40 TRANSMITTER TESTING

Specific test procedures depend on which system is used. When a transmitter of the proper type is available, check by substitution; otherwise, test as follows:

- **Mechanical Acoustic Transmitter (See Sec. 12-17)**

Inspect for damage, especially if it has been dropped or abused in any way. Check the operation of all triggers and hammers and for the proper impact on each tone rod. Inspect the rod support gaskets for deterioration; make sure the rods are not touching any part of the mechanism or case. Check for missing or broken springs.

When the transmitter appears OK, check to see if the sound signals are being generated and picked up by the receiver, as each function button on the transmitter is pressed. As an indicator, use an ac EVM or a CRO connected to various test points, at the pickup device (microphone) and the high-gain preamp. Make sure the receiver remote unit is switched on.

REMINDER: ────────────────

Other transmitter types produce a *continuous* signal when the buttons are held down; the tone produced by a mechanical transmitter is only *momentary,* each time a button is triggered. In lieu of a transmitter, when in doubt, try jangling a set of keys in front of the receiver.

When the transmitter is known to be good but no signal voltage can be measured at the receiver, probable causes are a defective microphone or trouble with the amplifier. Try a substitute microphone if available or interconnect with another receiver of the same type.

Make a stage-by-stage check of the amplifier using an AF signal generator in the 40-kHz range with the lowest possible output to prevent overload and possible damage to the transistors. If a weak or dead stage is located, check the transistor, dc operating voltages, and the like, as appropriate.

When some but not all functions are operating, identify the tuned filter(s) corresponding with the troublesome function(s). Check for signal voltage at the input and output of the filter(s) when the proper buttons on the transmitter are pressed.

- **Electronic-Type Acoustic Transmitter (See Sec. 12-17)**

The most common trouble (especially for loss of sensitivity) is a weak battery. Measure the voltage *under load* while one of the function buttons is depressed. When in doubt, install a fresh battery.

Determine if the circuit is oscillating, as indicated by a CRO display or an ac voltage of 10 V or so at the transistor. The indication should be fairly constant as each button is pressed. Where readings are erratic for one or more functions, check for dirty or worn switch contacts.

When the oscillator is not working, check the transistor operating voltages against the schematic. Check the transistor and replace if questionable. Check for an open or shorted oscillator coil.

When the oscillator *is* working, check for output from the *sonic transducer,* which is normally around 10 to 20 V ac. If there is no output, the transducer is probably defective or there is a broken connecting lead. Check the frequency generated for each function using a counter or an accurate AF generator and CRO. When the frequencies are in error, adjust the oscillator tuning coil slug and/or trimmer capacitors if provided. When *all* frequencies are too *high,* and the adjustments are not effective, suspect shorted windings in the oscillator coil or a broken or missing core slug.

In lieu of a suspected transmitter, the receiver functions can be controlled using a variable-frequency AF generator feeding a tweeter-type speaker.

• **RF Transmitters**

Except for the difference in *carrier* frequencies, testing is the same as for the electronic acoustic transmitter. Use a low-range ac EVM or a high-gain CRO in conjunction with a demodulator probe to see if a signal is being generated. Place the indicator close to, but *not connected* to, the transmitter. Check for AF variations as the carrier is modulated when each function button is pressed.

• **Infrared Transmitter (see Sec. 12-17)**

This is a much more sophisticated transmitter than those described previously. Since it utilizes the same type of logic circuits found in the PLL tuning system of the receiver, troubleshooting procedures are essentially the same. The first step is to check the condition of the battery and to replace it if outdated or if the voltage is below 8.5 V *under load* (with a function button pressed).

If the transmitter is completely inoperative, check for dc pulses at the input of the infrared LED terminals, when one or all functions are initiated from the control buttons. If the pulses are OK, the LED may be defective or the trouble is at the receiver where tests for infrared *emissions* must be made. Make a back-to-front resistance check and replace the LED if questionable.

If no pulses are indicated at the LED input, most of the logic circuits are suspect. Check for output from the keyboard or other controls, depending on which function or functions are at

fault. Where the BCD voltages appear normal, replace the ICs. If trouble persists, check all circuit voltages, the diodes and transistors, and other components as appropriate.

12-41 TESTING REMOTE RECEIVERS

Before troubleshooting the remote-control circuits in the *receiver,* verify that the transmitter unit is operating properly, the remote receiver is turned *on,* and all interconnecting cables and wires are intact. In general, tests at the receiver include checking for signal output from the sensor or other pickup device at the receiver input; checking the progress of the signal through the amplifiers and filters; and, finally, if the signals are getting through, the testing of downstream circuits and devices directly related to the troublesome function(s).

Where a TV receiver cannot be turned on by remote control, and the remote unit is *switched on,* check for a bad *triac* if used. Check for closure of the ac power *turn-on relay.* Try closing the relay *manually;* then check further for the cause. Troubleshooting procedures for two typical remote systems are as follows.

• **Testing the Acoustic System Receiver (see Fig. 12-17)**

The microphone preamp seldom gives trouble, but a bad microphone is not uncommon. Check for a signal at its output as one of the buttons on the transmitter is pressed. When the microphone is suspect, a small speaker can be used as a temporary substitute, although its output will be low. Another alternative is to use a variable AF signal generator tuned to specific frequencies in the 40-kHz range. Feed the generator to the amplifier input and signal trace downstream stages using a CRO, or inject a signal at key checkpoints in the amplifier and filters, up to the relays or drivers that correspond with the troublesome function. If the signal gets through, check for voltage feeding the appropriate function control motor channel selecting device as applicable.

If there is trouble with one particular function, the corresponding filter may need adjusting. Feed an accurately calibrated AF generator to some point ahead of the filter to be tuned. Connect an ac meter, as an indicator, to a convenient downstream test point. With the generator tuned to the frequency specified for that particular function, peak the filter coil for maximum indication.

A more accurate method, especially if the accuracy of the generator is in doubt, is to first calibrate the generator using the frequencies developed by the transmitter as a reference. Connect a CRO to the microphone preamp. If it is a mechanical transmitter, repeatedly press an appropriate button while adjusting the CRO to obtain one or two sine waves. Connect the generator to the CRO and tune it to obtain the same number of cycles. Use *this* frequency in resonating the filter. Repeat for other functions as necessary. A *counter,* if available, can also be used.

Check the activation of relays corresponding with each troublesome function. If a relay does not close, try a jumper from B to E of its driver transistor.

NOTE: _____

The driver transistors are medium-power types that are normally biased to cutoff, conducting only when a signal is applied. If the relay will not close, its coil may be open or the transistor is defective. Try closing the relay manually to see if that function can be carried out.

As a transmitter button is pressed, check for energizing voltage at the appropriate motor terminals. If voltage is present but the motor does not revolve, it may be open, hung up, or seized.

If necessary, lubricate the bearings if accessible. Disconnect one lead and check the windings for continuity.

When one or more functions must be reset each time the set is turned on, check memory circuits as applicable. When troubles persist, check cable connectors and module pins for open or intermittent contacts.

• **Testing the Infrared System Receiver**

Remotely controlled functions make use of the same logic circuits as functions initiated from the *receiver* control panel. Therefore, if the transmitter checks OK and the TV receiver operation is normal, any remote-control problems make the infrared sensor and its associated preamp prime suspects.

After pressing a function button on the transmitter, check for a signal voltage at the amplifier input. If there is none, the sensor diode is probably bad. Check the diode for an open, short, or leakage. If tests are inconclusive, replace it.

Check for signal at the amplifier output. For loss of signal or a weak response, check stage by stage using signal injection or a CRO.

For troubles involving the logic circuits of the PLL tuning system, see Sec. 12-38.

12-42 INTERMITTENT TUNER OR RELATED SYSTEMS

For intermittent reception where the tuner or its associated control circuits are suspect, try to create the conditions long enough to make the necessary tests. Try applying heat or cold to suspected modules or components, while monitoring the circuit operation at key checkpoints. Try increasing or decreasing the line voltage. For additional procedures, see Sec. 5-13.

Intermittent tuner troubles are often caused by bad contacts, solder splashes, poor soldering, and cracked feed-through capacitors. Troubles external to the tuner include intermittent contacts at module and IC pins, loose cable connectors, cracked wiring, and solder splashes on PC boards.

12-43 AIR CHECKING THE REPAIRED SET

After all repairs and adjustments have been made, go through the following checklist:

Are all VHF and UHF channels received with good pix, color, and sound?

Is the amount of snow on weak channels consistent with receiving conditions for the location?

Are all channels received without using the fine tuner?

Is the channel selector noisy and erratic on one or more channels, indicating a need for further cleaning of the contacts?

Where AFT is provided, is it doing a good job with adequate pull-in and lock-in capabilities?

Are all channels adjusted so reception is the same with the AFT switched on or off?

Where a PLL-type tuning and readout system is used, is there instant and positive channel selection from the keyboard?

Do all other controls operate properly?

Has the UP/DOWN scan been programmed for the desired channels?

Are the channel number readouts correct and legible as each channel is selected?

When there is a push-button-controlled varactor tuner, are all channel preset controls properly adjusted?

When remote control is provided, is operation normal and reliable for all functions?

Before replacing the back panel on the receiver, verify that the short leads between the tuners and their respective terminals are *not transposed*.

12-44 SUMMARY

1. To obtain optimum signal-to-noise ratio, the RF amplifier is biased for maximum sensitivity when receiving weak stations. When a set uses *delayed* AGC, this condition is maintained until stronger signals that require AGC action are received. This *threshold* level is established with a *delay* control. The adjustment of this control is critical and has a great bearing on the amount of *snow* on a weak pix.

2. All modern sets have *two* tuners for the reception of VHF and UHF channels. The VHF channels (2 through 13) and the UHF channels (14 through 83) are listed in Table 1-1 along with their pix and sound RF carrier frequencies.

3. The purpose of a TV tuner or channel selector is to select and amplify the combined pix and sound signal of any given channel and to convert the two high-frequency carriers into two carriers having *lower* (IF) carrier frequencies. The pix IF carrier frequency of all sets is 45.75 MHz, and the sound IF carrier frequency is standardized at 41.25 MHz. These frequencies remain the same for each channel selected.

4. A typical tuner has three tuned stages: an RF amplifier, a mixer stage, and an oscillator stage. The tuned circuits in the RF–mixer stages are tuned to the center frequency of the desired station, with a bandpass of 6 MHz to pass all pix and sound infor-

mation. The oscillator does not handle the *station signal;* its purpose is to generate an unmodulated (CW) signal, which, combined with the pix and sound RF carriers in the mixer, produces the desired pix and sound IF carriers. The IF signals are sometimes called *heterodyne* or *beat* frequencies.

5. The frequency generated by the tuner oscillator is *higher* than the pix and sound RF carriers by the amount of the IF carriers for any given channel. Take channel 5, for example, where the oscillator frequency must be precisely 123 MHz. This frequency when combined with the 77.25-MHz pix RF carrier produces the required 45.75-MHz pix IF carrier. Simultaneously, the same 123-MHz signal beats with the 81.75-MHz sound RF carrier to produce the required 41.25-MHz sound IF carrier. To maintain this relationship when switching channels, the oscillator tuning must change along with the RF-mixer tuning.

6. Pix, color, and sound are greatly affected by the slightest variation in oscillator frequency. To compensate for *frequency drift,* most sets have a fine-tuning control to change the oscillator frequency by a small amount when required. Some sets use *AFT* for *automatic* control of the frequency. Many modern sets use a *PLL* tuning system; with this system there is no need for AFT or a fine tuner.

7. Most modern sets use *varactor* tuners, where each stage is tuned by a varactor diode. This device functions as a simulated variable capacitor when supplied with a dc voltage. Channels are selected from a keyboard arrangement of push buttons, each button supplying the tuner with a specific amount of *tuning voltage,* which corresponds with a selected channel.

8. A UHF tuner converts the station carriers into IF carriers in the same manner as a VHF tuner. The IF output of the UHF tuner feeds the VHF tuner, where its RF-mixer stages function as IF amplifiers to boost the signals before passing them on to the video IF stages. When receiving UHF, the VHF oscillator is disabled.

9. Too little signal drive to the mixer results in a weak pix with snow. *Excessive* drive can overload the mixer and/or one or more downstream stages, creating a variety of trouble symptoms. Loss of AGC or a misadjusted delay control is the usual cause.

10. Early-type UHF tuners had no RF stage and used a diode as a mixer, hence, no amplification to the weak UHF signals. Modern UHF tuners are usually varactor types with an AGC-controlled RF amplifier stage and a transistor mixer. FETs are often used to obtain the highest possible signal-to-noise ratio.

11. Snow on weak stations is caused by background noise, most of which is generated in the mixer stage. This is no problem on *strong* stations where the signal overrides the noise. To obtain a high signal-to-noise ratio (minimum snow on *weak* stations), the mixer must be driven with a strong signal. *Anything* that reduces the signal level prior to reaching the mixer weakens the signal and increases the snow.

12. Traps and filters are often used in the signal input circuit of a tuner to reduce or eliminate certain kinds of interference. FM broadcast stations represent one source of interference since the FM band is located within the VHF TV band between channels 6 and 7. Some filters are resonant to the receiver IF frequency. If permitted to reach the

mixer, the carrier (or harmonics) of an interfering station may beat with the oscillator to produce an IF signal that may get through the following stages to degrade the pix, color, or sound.

13. Other items found in the input circuit of a typical tuner include (a) a *balun* or elevator transformer, used to match the antenna impedance to that of the receiver (a *mismatch* tends to reduce the signal and produces *ghost images* on the pix); and (b) *isolation capacitors* where a chassis is hot, (i.e., the chassis is common with one side of the ac power line).

14. Regenerative feedback can be a problem with high-frequency amplifiers. To prevent oscillations, an RF stage may be *neutralized* by using out-of-phase negative feedback.

15. When any given channel is selected by the tuner control, the RF and mixer tuned circuits are in tune with that station. It is the *oscillator* frequency, however, that determines *which* channel is received. For example, the selector may be set for channel 4, but if the oscillator is off frequency, you may receive some other station, most likely 3 or 5. Similarly, channel 4 may come in at some other channel setting even though the RF-mixer circuits are not resonant to channel 4.

16. An AFT (automatic fine tuner) is used on some sets to maintain a constant oscillator frequency regardless of which channel is selected. There is an AFT *defeat switch* on the front panel so the AFT can be switched on or off as desired. In the off position the AFT is not operating and the tuner oscillator frequency is manually controlled by the fine tuner in the normal manner. When switched on, the AFT circuit takes over and the fine tuner is not used.

17. An AFT system uses a *discriminator circuit* that senses changes in the IF signal developed in the tuner if the oscillator drifts off frequency. From a takeoff point in the video IF strip, the IF signal is fed to a *discriminator transformer,* which is tuned to the *proper* IF frequency of 45.75 MHz. If the oscillator is *off frequency* (in either direction), the discriminator produces a net + or − error voltage, which forces the oscillator to speed up or slow down until the IF signal produced matches the resonance of the discriminator transformer. For proper AFT action, the transformer must be precisely tuned.

18. A PLL tuning system uses computer-type logic circuits to control the tuner oscillator for all channels. In a process called *frequency synthesis,* a comparison is made between a sampling of the oscillator frequency (either VHF or UHF) and the frequency developed by an accurate *reference oscillator* to produce a dc *tuning voltage* to control the (varactor) tuner oscillator. There is no need for a fine tuner.

19. Other features of a PLL tuning system include instant channel selection from a telephone-type keyboard, a scan-search system to provide the user with a momentary review of favorite channel programs, and a remote-control system with the same control capabilities as the receiver controls, and without the need for electromechanical devices to perform the functions.

20. Instead of a push-button channel selector where each button is identified with a specific channel number, the *keyboard* used with a PLL system develops a binary (BCD) code. Each button represents one digit of the channel number, making it possible to select *all* VHF and UHF channels *with only 10 buttons.* Channel numbers are usually displayed on a two-digit, seven-segment LCD readout.

21. A PLL tuning system consists of two *loops,* a *reference frequency loop* that feeds the signal generated by the reference oscillator to a *comparator,* and a *feedback loop* that feeds a sampling of the tuner oscillator (VHF or UHF) frequency to the comparator. The comparator compares the frequencies of these two signals to develop a dc *error* or *tuning voltage,* which forces the oscillator to speed up or slow down as required until the comparison frequencies are identical in frequency and phase. Each of the two loops contains one or more *frequency dividers* to obtain a frequency (lower than the fundamental reference frequency) that is the *same* for all channels. Logic circuits in a *microprocessor* change the ÷ factor of *programmable* dividers in accordance with the particular channel selected.

22. A basic remote-control system consists of a small hand-held *transmitter* unit and, at the *receiver,* a sensor or pickup device (which receives the commands from the transmitter). A basic system may initiate only esential functions like power turn on, channel selection, and volume control; more elaborate systems do more. The means of communication between the transmitter and the receiver may be ultra-sonic sound waves, RF, or an infrared light beam.

23. The first step in troubleshooting for suspected front end troubles is to determine whether it is tuner *trouble* or if the fault is elsewhere. This may require extensive testing. When in doubt, a substitute tuner should be tried.

REVIEW QUESTIONS

1. Explain the relationship between gain, bandpass, and selectivity of a tuned RF amplifier stage.

2. What results can be expected from an RF–mixer bandpass that is (a) too wide? (b) too narrow? Explain your answers.

3. Explain the purpose of each of the three tuner stages.

4. With a numerical example, briefly explain RF-to-IF conversion using the heterodyne principle.

5. (a) What comparison frequency is used in a set having a 2-MHz reference oscillator with a ÷ 20 divider in the *reference* loop?
(b) What is the comparison frequency of another set having a ÷ 200 divider in the *feedback* loop?
(c) Does the comparison frequency change as different channels are selected?

6. Briefly describe the operation of each of the three remote-control systems in current use.

7. Front end problems often result in a weak pix with *snow;* video amplifier troubles may cause a weak pix but *no snow.* Explain why in each case.

8. (a) Explain the need for *bandswitching* in a VHF tuner.
(b) How is this accomplished with a varactor-type tuner?

9. (a) Why do older sets have a fine tuner control?
(b) State the effect on the pix, color, and sound if the oscillator frequency is (1) slightly high? (2) slightly low? (3) off frequency by a large amount? Explain *why* in each instance.

10. What is meant by oscillator drift and how is it prevented?

11. Why is tuning of the oscillator circuit much more critical than for the RF-mixer stages?

12. Why can't conventional coils be used in the tuned circuits of a UHF tuner?

13. Identify the potential sources of interference that are normally trapped out in the (a) tuner? (b) IF strip) (c) Why can't all such signals be trapped in the tuner or IF section?

14. Under certain conditions it is possible to have interference from adjacent channels. Identify the interference (*pix* or *sound*) from the next (a) lower channel, (b) higher channel. (c) How is such interference normally prevented in the (1) tuner and (2) IF section?

15. (a) What causes regeneration or oscillations in an RF amplifier stage and how is it normally prevented?
(b) Why is this not a problem in the mixer stage?

16. (a) Which tuner stage generates the most background noise that shows up as snow on a weak pix? (b) What factors must be considered in obtaining the highest possible signal-to-noise ratio to override the noise?

17. What are the advantages of using a FET compared to a bipolar transistor in the RF-mixer stages?

18. With a numerical example, show how the *oscillator* frequency determines which station is received, rather than the RF-mixer circuits that are resonant to the selected channel.

19. (a) Explain how a parallel resonant circuit can be used to either accept or reject a particular frequency.
(b) Answer part (a) for a series resonant circuit.
(c) Explain *why* in each instance.

20. What are the advantages of the PLL system over AFT control of a varactor tuner?

21. What is meant by frequency synthesis?

22. Many signals of different frequencies are intercepted by the antenna. Explain how the receiver is able to select a desired station while rejecting all others.

23. (a) Radiations from the tuner oscillator must be prevented. Explain why.
(b) What design considerations reduce or prevent such radiation?

24. Although both sections of the set must pass the same pix and sound information, the bandpass of the IF strip is only about 4 MHz compared to the 6 MHz of the RF-mixer stages. Explain.

25. What change in frequency (if any) can be expected of a tuned circuit where (a) L is increased? (b) C is decreased? (c) both L and C are increased?

26. As a result of beating with the oscillator signal and its harmonics, numerous interference signals are developed in the mixer. What prevents them from getting through the IF strip?

27. Why is the frequency of a tuned circuit reduced when an iron core is inserted into the coil? Why is the frequency *increased* with a nonmagnetic core.

28. With a numerical example, explain *image* interference and how it is prevented in the receiver design.

29. (a) If a tuner is set for channel 3 but the oscillator frequency is 113 MHz, what channel will be received? Explain why.
(b) The picture will be weak with snow. Explain why.

30. (a) Why is it important to have *short* connecting wires in the tuned circuits, particularly the UHF tuner?
(b) Why do the tuned circuits have such small values of L and C?

31. (a) What is the effect on the frequency of a tuned circuit by stretching or compressing coil windings?
(b) What change in frequency occurs by tightening a trimmer capacitor?
(c) By inserting plastic (or other insulating material) between the plates of a tuning capacitor?
(d) Are these effects greatest at low or high frequencies?

32. Why is low-impedance coax used in feeding the signal from the tuner to the IF strip?

33. What is the approximate center frequency of the RF-mixer response when tuned to channel 5? (b) What is the approximate RF/mixer bandpass?

34. Why must the oscillator frequency be changed when switching channels?

35. Can the 4.5-MHz separation of the pix and sound carriers be charged by adjusting the tuner or IF circuits? Explain.

36. While tuned to channel 2, interference is picked up from a powerful local AM station operating on 246 MHz. Explain.

37. What is meant by oscillator *tracking*?

38. Explain the use of feed-through capacitors.

39. Explain the purpose and operation of a typical AFT circuit.

40. How does a varactor tuner differ from the other types of tuners? Name several advantages. (b) Explain how tuning is accomplished.

41. There is no need to locate a varactor tuner behind the front panel of the receiver. Why?

42. With varactor tuning, what is the purpose of (a) the panel push buttons? (b) the preset channel setup controls?

43. What is the range of *tuning voltages* for channel selection with a varactor tuner? Does an increase in tuning voltage tune the tuner stages to a *lower* or a *higher* channel?

44. How is bandswitching accomplished with a varactor tuner?

45. Why is the IF output of a UHF tuner fed to the VHF tuner rather than directly to the IF strip?

46. (a) There should be no AGC applied to the RF stage while receiving weak stations. Why?
(b) An increasing AGC voltage *is* required for the stronger stations. Why?
(c) What type of AGC circuit can satisfy both conditions?
(d) What AGC control is often provided for the critical adjustment of the AGC for the reception of weak signals?

47. Briefly explain the operation of a basic PLL tuning system.

48. How is it possible to select all VHF and UHF channels from a keyboard having only 10 buttons?

49. How does the comparator recognize when the tuner oscillator frequency is correct for any given channel?

50. Why are one or more *programmable* dividers required in a PLL system?

51. (a) Explain the operation of a two-digit LED channel readout display used with PLL tuning systems.
(b) How is the selected channel number recognized?

52. What is the scan-search mode? And what is meant by UP/DOWN scanning?

53. (a) How is the tuning voltage developed by a PLL system different from that used by a varactor tuner controlled directly from the panel push buttons?
(b) With the latter system, voltages are extremely critical; with the PLL system the tuning voltage is *not* critical. Why?

TROUBLESHOOTING QUESTIONNAIRE

1. A set has a weak pix with heavy snow. The AGC voltage at the tuner is low, but other voltages appear normal. What would you suspect and how would you proceed?

2. Where VHF reception is normal, but there is no UHF and UHF tuner tests OK, what would you check?

3. Name three tests to determine if the tuner oscillator is operating.

4. Several stations are received at the wrong positions on the channel selector. How would you proceed?

5. Name at least six precautions to observe when working on a tuner.

6. A set has normal response when an IF signal from a generator is injected at the tuner end of the IF input coax, but no response when an RF station frequency is fed to the mixer input. Symptoms are loss of pix and sound. State the probable causes.

7. A set has weak pix but no snow, and AGC at the tuner checks normal. What section of the receiver is suspect? Explain your reasoning.

8. The B voltage at tuner terminals is low or zero. When the dc line is disconnected from the tuner, the voltage on the line checks normal. What would you suspect and how would you proceed?

9. How would you substitute an RF signal generator for a suspected tuner oscillator?

10. (a) State the procedure for adjusting the AGC delay control.
(b) State some other possible causes for a weak pix with snow.

11. A tuner is returned from a repair station with a note stating it was not defective.
(a) How would you proceed?
(b) What have you learned from this experience?

12. How would you determine if the AFT circuit is operating properly? If it is not, what tests will localize the trouble?

13. How would you adjust the AFT discriminator transformer if an accurate signal generator is not available?

14. A receiver operates normally, but the varactor tuning voltages all measure high or low compared to the service literature. How do you account for the inconsistency and what would you do about it?

15. What would you suspect if a counter indicates loss of signal or the wrong frequency at each of the following points in a PLL system: (a) the reference oscillator? (b) the reference signal input to the comparator? (c) the VCO output at the tuner? (d) the VCO feedback signal at the comparator input?

16. A tuner shield cover is inadvertently left off while making oscillator adjustments. How will reception be affected when the cover is replaced? Why?

17. Considering that no instruments are required, how do you know when the tuner oscillator frequency is correct?

18. What is a *subber*? Explain how to use it or its equivalent in evaluating the condition of a tuner.

19. Explain the procedure for making oscillator adjustments on the following tuners: (a) wafer-switch type, (b) drum type, (c) varactor tuner. (d) Why should the fine tuner control be set at mid-position and not touched during the procedures?

20. What is the effect on pix, color, and sound if the tuner oscillator frequency is (a) slightly high? (b) slightly low? (c) far off frequency?

484 CHANNEL SELECTORS AND TUNING SYSTEMS

21. Sweep alignment setup shows misalignment of the RF-mixer tuner circuits with markers representing the pix and sound carriers on the low-frequency slope of the response curve. To locate the markers at the peaks, would you screw the *iron* slugs *in* or *out* of the RF-mixer coils? Explain your answer.

22. State two possibilities when some or all channels are received at the wrong positions of the channel selector.

23. Shifting tuner components from their original positions or careless probing, especially in the oscillator circuit, should be avoided. Explain why.

24. Compared to a good outdoor antenna, what are the expected results (on pix, color, sync, and sound) when an indoor antenna is used?

25. (a) What is a *signal splitter* and how is it used for receiving VHF and UHF stations from a VHF antenna?
(b) What are the alternatives?

26. (a) What are the symptoms of dirty or worn tuner contacts? (b) Name two problems often enountered in the use of *excessive* contact cleaner or *improper* solvents.

27. What trouble symptoms or test results warrant the substitution or replacement of ICs that perform each of the following PLL tuning system functions: (a) frequency dividers, (b) comparator, (c) microprocessor, (d) keyboard encoder, (e) channel readout decoder, (f) remote control ICs, (g) UP/DOWN scan memory.

28. There is weak reception with snow and the measured (no-signal) bias voltage on the RF transistor differs from that specified on the schematic. State several possible causes.

29. (a) Why is an ordinary CRO not suitable for signal tracing the RF-mixer stages of the tuner?
(b) Where a CRO *is* used, a demodulator probe is required. Why?

30. (a) Reception, although weak, is still possible with a *dead* RF stage in the tuner. Explain. (b) This is *not* true for the mixer and oscillator stages. Why?

31. What would you do if the fine tuner has to be adjusted as each channel is selected?

32. How is reception affected if the tuner oscillator (a) is dead, (b) is weak, (c) has frequency drift, (d) has the wrong frequency?

33. There is intermittent reception when the tuner is tapped, but cleaning contacts does not help. What would you look for?

34. State two possible causes of adjacent channel interference (crosstalk).

35. (a) There is no reception on channel 4. Adjusting the brass oscillator slug brings in channel 5. Is the oscillator frequency now too high or too low?
(b) To bring in channel 4, would you screw the slug in or out of the coil? Explain why.
(c) What would you do if the adjustment will not quite make it?

36. Under what conditions might you suspect RF-mixer misalignment?

37. There is no B voltage present at a UHF tuner, but VHF reception is normal. State the possible causes.

38. There is no reception with a PLL system, and the tuning voltage checks high for all channels.
(a) Does this make the tuner oscillator run fast or slow?
(b) Describe the procedure for substituting a bias box for the tuning voltage.
(c) If the substitution in part (b) restores set operation, is the tuner at fault or the PLL system?
(d) What if abnormally high or low tuning voltages are required to receive stations?

39. (a) For what trouble symptoms might you try clamping the AGC at the tuner?
(b) If reception improves, is the tuner at fault or is it an AGC problem? Explain your reasoning.

40. Briefly explain how to check and align the RF-mixer tuned circuits using (a) an unmodulated RF signal generator, and (b) sweep alignment equipment. (c) Which is the preferred method? Why? (d) With sweep alignment, which of the two generators *must be* accurately calibrated? Explain why.

41. (a) No tuning voltage is developed by a PLL system and a counter indicates different frequencies for the two signals at the comparator input. How would you proceed?
(b) What if the condition is true for *some* channels only?

42. State the possible causes for the following remote-control problems: (a) no signal can be measured at the output of the pickup device on the receiver; (b) trouble with one function only; (c) no power turn on by remote control; (d) need to reset volume each time set is turned on; (e) all functions working except transmitter must be held close to the receiver.

43. There are symptoms of video overload. Reception improves when the antenna is replaced with a short pickup wire. What do you suspect and how would you proceed?

44. What would you suspect when the tuning voltage at tuner terminals checks (a) low for all channels? (b) low for one channel only? (c) zero for all channels?

45. State the probable cause when interchanging the VHF and UHF antenna connections to the back of the set improves both VHF and UHF reception.

Chapter 13

VIDEO IF AND DETECTOR STAGES

13-1 INTRODUCTION

The function of the IF "strip" is to amplify the station signals processed by the tuner and to reject any undesired signals that manage to get through. After conversion from RF to IF in the mixer stage, these signals exist as a band of frequencies, each with a different function. They consist of the pix, color, and sound carriers and their modulation sidebands. The pix IF carrier frequency for all receivers is established at 45.75 MHz. The sound IF carrier is 41.25 MHz. With an intercarrier type receiver, the pix and sound signals travel together to the point of sound takeoff where they separate. This compares with early-type dual-channel (split-carrier) sets, now obsolete, where sound and pix were separated soon after leaving the tuner.

The *video detector* is a stage by itself whose input can be considered part of the IF strip and whose output relates to the video amplifier section that follows. The detector's function is to demodulate the video signal, extracting the modulation frequencies from their carrier. *Color signals* are part of these sidebands and are not demodulated at this time.

With a BW receiver, the detector also functions as a *frequency converter* to further reduce the FM sound signal to a new, and lower, sound IF carrier. A block diagram of an IF strip and detector is shown in Fig. 13-1.

13-2 VIDEO FREQUENCIES AND BANDWIDTH REQUIREMENTS

The video signals that make up a pix are contained in the sidebands of the pix IF carrier. There are two sidebands, the *upper* sideband (on the high-frequency side of the carrier) and the *lower* sideband. These sidebands extend outward from the carrier in both directions, with the low-modulation frequencies close to the carrier and the high frequencies further removed.

488 VIDEO IF AND DETECTOR STAGES

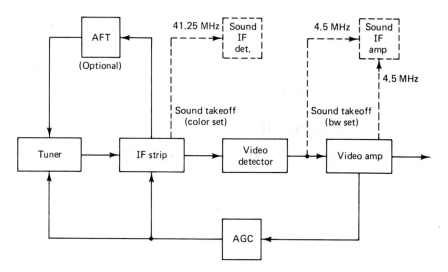

Fig. 13-1 Basic block diagram.

The higher the modulation frequency, the further it is from the carrier, and the greater the bandwidth requirements to pass such frequencies. Low video frequencies produce large pix objects and the high frequencies, the fine pix detail. Video frequencies range from about 30 Hz to slightly over 4,000,000 Hz (4MHz).

The bandpass requirements of the IF strip are determined partly by screen size. A large screen (a 25-in. pix tube, for example) requires a wide bandpass of at least 4 MHz to obtain maximum resolution of the fine pix detail, especially for close-up viewing. With a *small screen*, such detail is wasted, and the highest frequency needed for good resolution is only about 3 MHz or less. Thus a bandwidth of 3 MHz is more than adequate for such a receiver.

Where the IF bandpass of a receiver is less than required, the high video frequencies will be lost and, along with them, the fine pix detail. The bandwidth of the IF strip depends on the proper alignment of its *tuned circuits*.

13-3 IF AMPLIFICATION

Most of the signal amplification in a TV set is provided by the IF strip, and because of its wide bandpass, several stages are required. A typical large-screen set has from three to four stages; a smaller set with a narrow bandpass has fewer stages, in some cases only two.

REMINDER: _____

The narrower the bandpass is, the higher the gain per stage, and vice versa.

The average gain of a typical IF stage is between 15 and 20. Thus, for a three-stage IF strip, the overall gain may be 20 × 20 × 20, an overall amplification of 8000 times. Using these figures, where signal level at the tuner output is 1 mV, the voltage at the detector output will be 8000 mV or 8V. Stage gains for BW and color sets, both solid state and tube types, are about the same.

13-4 SOUND TAKEOFF

The sound signal undergoes two distinct changes, first, in the mixer stage, where it is converted from an RF to an IF frequency (41.25 MHz), and once again where the carrier is further reduced to 4.5 MHz. The 4.5-MHz signal is the result of mixing the 41.25-MHz sound IF carrier with the 45.75-MHz pix IF carrier, and represents the difference between these two frequencies.

The *sound takeoff* is the point where the sound signal separates from the video signal. For a BW set, where the 4.5-MHz signal is produced in the video detector, the sound takeoff may be at the detector output or at some point in the video amplifier stages that follow. With inexpensive small-screen receivers, the takeoff is often close to the CRT input for maximum amplification of the sound signal by the video stages. Large-screen quality receivers usually have the sound takeoff at the detector to reduce the possibility of *intercarrier buzz* when high-level video tends to cross-modulate and override the FM sound signal.

The 4.5-MHz sound signal (which conveys the same information as prior to conversion) is separated from the video with a tuned circuit adjusted to 4.5 MHz. This takeoff coil serves a dual purpose, as a resonant *tank* circuit for the sound IF strip, and as a *trap* to prevent the 4.5-MHz beat signal from reaching the pix tube. A small-value coupling capacitor (about 1 pF) is often used at the sound takeoff to help prevent the video signals from getting into the sound section.

The sound takeoff for a *color* set is *ahead* of the video detector, usually at the last IF stage. This is an added precaution against 4.5-HMz beat interference, which is particularly objectionable on a color pix. It also helps prevent 920-kHz beat inferference, which occurs when the 4.5-MHz sound IF and the 3.58-MHz color subcarriers are combined.

The sound is separated from the video with a tuned circuit adjusted to 41.25 MHz. This tuned circuit also serves two purposes, as a tank for the sound section and as a trap to attenuate the 41.25-MHz carrier before it reaches the video detector.

From the takeoff point, the sound signal goes to a *sound IF detector*. This stage functions *not as a demodulator*, but as a *frequency converter*, changing the sound IF from 41.25 to 4.5 MHz, as with the video detector of a BW set. Don't confuse the sound IF detector with the detector that demodulates the FM sound signal later in the sound section.

13-5 IF RESPONSE AND BANDPASS CONSIDERATIONS

A *response curve*, as described in Sec. 12-4, offers a convenient way for indicating the relative amplification of all video frequencies within the bandpass of the IF strip. For the typical response curve shown in Fig. 13-2a, the bandwidth is the frequency difference between two points, at the 50% level, on opposite sides of the curve. Whereas the RF stage in the tuner has a bandwidth of 6MHz to pass the pix and sound signals at RF frequencies, at IF frequencies, the ideal bandwidth for the IF strip is typically about 4 MHz in order to pass video frequencies up to that amount.

Bandwidth is determined by the degree of *coupling* between the IF stages and conditions of alignment. *Tight* coupling results in a wide bandpass, which is desirable. A wide bandpass, however, equates with poor *selectivity*, the ability to discriminate against interfering signals. *Loose coupling* provides increased selectivity and higher gain but at the expense of passing the

490 VIDEO IF AND DETECTOR STAGES

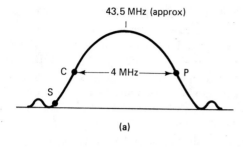

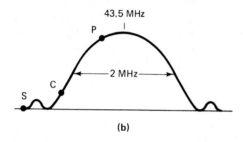

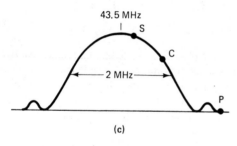

P, 45.75-MHz pix carrier marker
S, 41.25-MHz sound carrier marker
C, 42.17-MHz color subcarrier marker

Fig. 13-2 Pix and sound mistracking: (a) normal pix, color, and sound; (b) pix but no sound; (c) sound but no pix.

necessary band of frequencies, those in the 3 to 4-MHz range in particular.

• **Markers**

Markers are used to indicate points on a response curve that correspond with specific frequencies. Note the location of the markers in Fig. 13-2a. The 45.75-MHz pix IF carrier is at the 50% level on the high-frequency side of the response. The 41.25-MHz sound carrier is normally at or near the base line on the low-frequency side of the curve. The 42.17-MHz marker, representing the color subcarrier, is at or near the 50% level on the low-frequency side of the response. These are *carrier* frequencies, each accompanied by the modulation frequencies contained in their sidebands, as spread over various portions of the curve, where they receive varying amounts of amplification.

The frequency of these three markers is established by the tuner oscillator, and under normal conditions for optimum pix, color, and sound, their relative locations on the curve are as indicated. As the fine-tuner control is varied, the changing oscillator frequency changes the IF carrier frequencies that are developed and their locations on the curve. With the fine tuner turned in one direction, the markers will move to the low-frequency side of the response; turned in the opposite direction, they move toward the high side of the curve. Such misadjustments have an adverse affect on pix, color, and sound.

• **Pix and Sound Mistracking**

This is a condition where best pix and best sound are not received for the same setting of the fine-tuner control. It occurs when the IF bandwidth is too narrow. Figure 12-5 shows how changes in oscillator frequency affect reception under normal conditions of bandwidth. Figure 13-2 shows what happens with a narrowed IF bandpass. In Fig. 13-2b, one setting of the fine tuner places the pix carrier at the correct position (at the 50% level) for best pix, but the sound marker is *beyond the curve* where there is no amplification, and the result is loss of sound. A different adjustment of the fine tuner, as in Fig. 13-2c, moves the sound marker to the correct position for normal sound, but the pix carrier is now too low on the response where its sideband signals receive little or no gain, resulting in a degraded pix or, in extreme cases, no pix at all.

Because of the 4.5-MHz separation of the pix and sound carriers as established by the station, which cannot be altered, all markers, including the color subcarriers, *move together*. With a narrowed bandpass, the conditions for best pix, as just described, move the color carrier and its sidebands toward the base line, resulting in loss of color or weak color of poor quality. When adjusted for normal sound, the color signals will be too high on the curve where they receive *too much* amplification.

13-6 NORMAL AND ABNORMAL MARKER LOCATIONS

Problems resulting from misplaced markers are described next.

- **Sound IF Marker**

When two signals are combined in a mixing process, the resultant beat signal retains the characteristics of the *weaker* of the two signals. The mixing of the 41.25- and 45.75-MHz signals to produce a 4.5-MHz beat that is *frequency modulated* by the sound signal (and not amplitude modulated by the video) requires that the sound carrier be greatly attenuated. This is accomplished in two ways: with the use of traps, and by locating the 41.25-MHz marker low on the IF response where it receives very little amplification. With a BW set, the 41.25-MHz marker is usually located about 5% to 10% up on the low-frequency side of the response (see Fig. 13-3a). For a color set, it is positioned closer to the base line, as shown in Fig. 13-3b. Too much amplification of the 41.25-MHz signal produces a buzz in the sound and 4.5-MHz and 920-kHz beat interference on the pix. Some sets have as many as three 41.25-MHz traps in the IF strip.

- **Pix IF Marker**

This marker is located at the 50% level (on the high-frequency side of the curve) to compensate for the unequal sidebands produced by vestigial transmission. The lower sidebands receive high amplification where they approach the top of the curve compared to the upper sideband where it approaches the zero-level base line. Note that the lower sideband extends over most of the top portion of the curve.

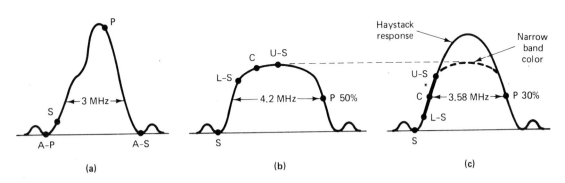

Fig. 13-3 Narrow- and broadband color systems: (a) response curve for typical small-screen BW set; (b) response and marker locations for wide band color set; (c) narrow band color and haystack response curves (heavy portion of curve *must be* linear for good color).

P, 45.75-MHz pix carrier marker
S, 41.25-MHz sound carrier marker
C, 42.17-MHz color subcarrier marker
U-S, upper color sideband marker (42.67 MHz)
L-S, 41.67-MHz lower color sideband marker
A-P, adjacent channel pix carrier marker
A-S, adjacent channel sound carrier marker

Vestigial Sideband Compensation

The useful pix information and most of the power is contained in the video sidebands. Both upper and lower sidebands contain identical information and *one* sideband is all that is necessary to produce a pix. To conserve space in the crowded TV spectrum, it would be desirable to transmit only one sideband, but this cannot be done. Instead, all the upper sideband is transmitted, but only a "vestige" of the lower sideband, ranging from 0.75 to 4.2 MHz. This means that the upper sideband is fully modulated from 0 MHz (the carrier frequency) to 4.2 MHz above the carrier frequency; but the lower sidebands only extend to 1.25 MHz (see Fig. 1-3). This overemphasis of low video frequencies compared to high frequencies would produce a pix of high contrast for large objects that would obscure the fine detail. Compensation is made at the receiver by locating the pix carrier at the 50% level on the curve.

42.17-MHz Color Subcarrier Marker

This marker, which is 3.58 MHz from the pix carrier, is located *within* the upper sideband limits of the video signal, along with the color sideband markers of 40.67 and 41.67 MHz.

Adjacent Channel Markers

Markers representing the pix and sound IF carriers of channels immediately above and below the selected channel are indicated at the base of the response, beyond the bandpass limits where they receive no amplification. The 39.75-MHz marker corresponds with the adjacent channel video carrier and the 47.25-MHz marker with the adjacent channel sound carrier. Traps are provided in the IF strip to eliminate these interfering signals.

13-7 COLOR SET RESPONSE CURVES

The 3.58-MHz color subcarrier at the station is modulated by signals whose lower sidebands (designated I) extend from 0 MHz (the carrier frequency) to 1.5 MHz. The upper (Q sideband) extends only to 0.5 MHz. Although the *full I* sideband is transmitted, most receivers (for reasons of economy) only use the I signals up to 0.5 MHz. Thus there are two systems in current use, commonly referred to as *broadband* and *narrowband* color. The bandwidth and the location of color markers on the IF response curve are different for each system, as shown in Fig. 13-3. Note that the color frequencies are identified in two ways: the color modulation frequencies centered at 3.58 MHz (the color subcarrier), and corresponding frequencies of the IF bandpass with the color subcarrier positioned at 42.17 MHz.

REMINDER: _____

The 3.58- and 42.17-MHz signals *are not transmitted*. These markers are for reference purposes only.

Broadband Color

The ideal response curve for a receiver designed for broad-band color is shown in Fig. 13-3b. With the color subcarrier marker at the top of the curve, both sidebands receive equal and maximum amplification. Note that the color sidebands are contained within the same bandpass as the BW video signals. Compare with the curve of a typical small-screen BW set (Fig. 13-3a).

Narrowband Color

The overall IF response for narrowband color is

shown in Fig. 13-3c. Here the color subcarrier marker is at the 50% level on the low-frequency side of the curve. This brings the color frequencies of the upper sideband, the frequencies around 0.5 MHz in particular, toward the base line where they receive little amplification. Loss of these frequencies, however, has little significance since color cannot be observed on the finest pix detail. The attenuation of the high color frequencies is partly compensated for in a *bandpass amplifier* later in the signal path.

With a narrower bandpass (3.58 MHz compared to about 4.2 MHz for the broadband system), fewer components are required in the IF strip. One shortcoming of the system is reduced amplification of the low-frequency sidebands (close to the carrier at the 50% level), which under weak signal conditions can result in excessive snow and unstable color sync.

- **Haystack Response**

Instead of a flat-topped response, many sets are aligned for a haystack-shaped curve, as shown in Fig. 13-3. For this type of response the pix IF carrier and the color subcarrier are still located at opposite points on the two slopes, but well down on the curve at about the 30% level. A receiver designed for this curve should not be aligned as in Fig. 13-3b, and vice versa.

13-8 INTERSTAGE COUPLING METHODS

Tuned circuits are used in coupling the signal between stages in the IF strip. Some sets use single tank coils; others use transformers. Each tank coil is tuned to some frequency within the IF bandpass. Sometimes the coils are shunted with small capacitors, but usually resonance is obtained using only the distributed capacity between coil windings, and the associated circuits. Wave traps of various types are used in conjunction with the tank circuits to eliminate or attenuate certain signals. The three coupling methods shown in Fig. 13-4 are described next.

- **Impedance Coupling (Fig. 13-4b)**

This system makes use of single, tuned, tank coils between each stage. Each coil represents the load across which its associated transistor develops maximum signal voltage for a particular resonant frequency. The signal is either directly coupled to the following stage or via a small coupling capacitor. Each coil is stagger-tuned to a different frequency about 1 MHz apart in the 40-MHz range. Although the bandwidth of each tuned circuit is relatively narrow (about 2 MHz), with staggered-tuning the overlapping curves provide an overall bandwidth of the correct amount (see Fig. 13-4a). With each circuit resonant to a different frequency, the chance of regenerative feedback and oscillations is greatly reduced. The frequencies of the various tank coils are usually shown on a schematic. For maximum amplification, high Q coils are usually employed with *swamping* resistors and the low-impedance transistor junctions to further broaden the bandpass. The *traps* are also instrumental in determining bandwidth.

The tuned circuit(s) at the input to the IF strip deserves special consideration. The tank coil (sometimes called a *link* coil) is located *on the tuner*, where it is sometimes overlooked. It represents the load for the mixer and is the first tuned circuit to accept the signals developed in the mixing process. The coil along with a small capacitor constitutes a low-impedance, series resonant circuit to feed the IF strip. The signal is routed from the tuner to the main chassis via coaxial cable to reduce circuit losses and to prevent the pickup of undesired signals and other kinds of interference.

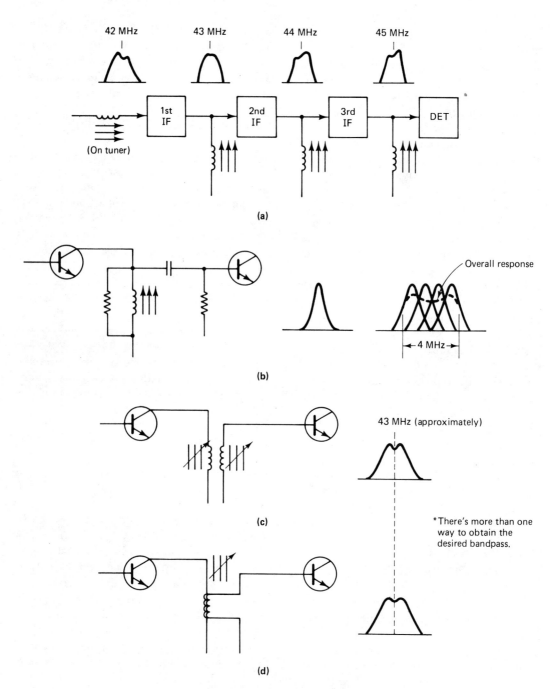

Fig. 13-4 Interstage coupling bandpass response: (a) stagger-tuning; (b) impedance coupling; (c) overcoupled transformer; (d) overcoupled bifilar transformer.

The signal is coupled to the IF input in one of two ways: with a tuned impedance-matching transformer, or using link coupling where the signal connects to a low-impedance coil tap, two or three turns from the ground end.

- **Overcoupled Transformers**

A circuit using overcoupled transformers is shown in Fig. 13-4c. The bandpass and the amount of gain obtained with this method are determined by the amount of coupling between windings of the double-tuned transformers. To obtain the necessary bandwidth, the primary and secondary coils are closely spaced (overcoupled). The tighter the coupling is, the wider the bandpass. A wide bandpass is synonymous with low gain, hence the need for several stages. With some sets the coils are all tuned to the same frequency (about 43 MHz), the center of the IF bandpass. At other times the coils are stagger-tuned for additional bandwidth.

- **Bifilar Transformers (see Fig. 13-4d)**

This special type of overcoupled transformer is often used. For maximum coupling, the coils are wound together (in parallel) on the same form. With such tight coupling, one slug simultaneously tunes both windings. The transformers are usually stagger-tuned to different frequencies.

With this type of transformer, phasing between the primary and secondary windings must be considered. If the connections to either coil are reversed, one winding bucks the other, decreasing the stage gain.

- **Tuned Circuit Loading**

Anything that reduces the Q of a tuned circuit will widen the bandpass. For this reason, high-Q tank coils are often shunted with swamping resistors. Values range from about 2 to 20 kΩ. The lower the value is, the greater the loading and the wider the bandpass, at the unavoidable expense of stage gain.

Traps are high-Q and, to be effective, are seldom loaded.

13-9 WAVE TRAPS

Traps are used extensively in the IF strip to shape the response and to reject certain frequencies that may interfere with the pix and sound. Such interference may be received along with the desired station or generated within the set. After the point of sound takeoff, for example, the sound signal must be kept out of the video section and the video signal from the sound stages. Sound IF (41.25 MHz) traps are used ahead of the sound takeoff to minimize buzz. A sound signal produces sound bars on the pix; video in the sound produces an annoying buzz. Different trap configurations are described in Sec. 13-10. The interference signals and their corresponding traps are identified as follows.

- **Adjacent Channel Traps**

To minimize the possibility of interference between adjacent channels, TV channels are allocated geographically so that stations received in any given area are separated by one or more channel numbers. There are times, however, when over-the-horizon reception does take place, creating problems. Sometimes the sound of the interfering station is heard, but more often sync and blanking bars can be observed as background to the pix being viewed. The sideways movement of vertical bars is called the *windshield-wiper effect*. Adjacent channel interference can be caused by either the pix carrier of the next *higher* channel or the sound carrier of the next *lower* channel. (See Fig. 1-3 and 13-3). Such interference can often be reduced or eliminated with a highly directional antenna or with proper orientation.

Because of the wide bandpass, adjacent channel signals have no trouble getting past the tuner, where traps for these frequencies cannot be used. The solution is to reject them in the *IF* strip, where traps tuned to 39.75 and 47.25 MHz reject adjacent channel signals regardless of which channel is selected for viewing. Usually the traps are located between the tuner and the IF input (Fig. 13-5); sometimes they are lumped with other tuned circuits in a module.

The adjustment of these traps has a considerable bearing on the IF bandwidth, as can be seen from their marker locations relative to the overall response shown in Fig. 13-3a. Located on the base line and beyond the limits of the IF bandpass, signals at these frequencies receive no amplification. Their adjustment is *critical*, and even in areas where interference is no problem they must be properly tuned to their respective frequencies. For example, if the 39.75-MHz adjacent channel video trap is tuned too *high* in frequency, besides not doing its job as a trap, it will cut into the response curve, altering its shape and adversely affecting the sound. Similarly, if the 47.25-MHz adjacent channel sound trap is tuned too *low*, the high-frequency side of the curve will be affected, degrading the pix. Where both traps are tuned in the opposite directions, the IF bandpass will be widened, the gain reduced, and all markers will be incorrectly located high on the curve.

• 41.25-MHz Sound IF Carrier Traps

Some 41.25-MHz sound IF carrier signal must get through the IF strip to combine with the 45.75-MHz pix carrier to produce the 4.5-MHz sound carrier. If too *strong*, however, it creates serious problems, such as buzz in the sound, 4.5-MHz interference on the pix, and, for a color set, 920-kHz beat interference on a color pix. A weak sound carrier is required for another reason: to ensure that the 4.5-MHz sound IF carrier is modulated by the FM sound signal and not by the AM video signal.

REMINDER:

In the conversion from a 41.25-MHz sound IF to the 4.5-MHz sound IF, the 4.5-MHz signal retains the characteristics of the *weaker* of the two signals being combined in the mixing process.

Normal location for the 41.25-MHz sound marker on the overall response of a BW set, as shown in Fig. 13-3a, is between the 5% and 10% level just within the IF bandpass where the signal receives very little amplification. For a color set (Fig. 13-3b), it is further down on the skirt, sometimes at the base line. As with the adjacent channel video trap, the adjustment of the 41.25-MHz traps aids in shaping and determining the bandwidth of the IF response.

The 41.25-MHz trap in the last IF stage of a *color set* serves a dual purpose, as a trap, and as the sound takeoff tank circuit at the input of the *sound IF detector*. Some sets have as many as two or three sound traps, including the sound takeoff if it is a color set. In some cases a *sound rejection control* is provided that works in conjunction with the traps for a more precise adjustment of the sound signal. Sometimes a variable resistor in a bridged-T type of trap serves this function.

• 4.5-MHz Sound Traps

With a BW receiver, the 4.5-MHz sound IF is developed in the video detector. With a color set, it is developed in the *sound IF detector*. The 4.5-MHz beat signal must be prevented from reaching the pix tube where it can cause a fine-grained interference pattern on the screen. This is normally prevented by the use of 4.5-MHz traps, one at the output of the video detector and in some cases additional traps in the following video amplifier stages. With a BW set, the first such trap is the sound takeoff. Although the sound takeoff for a color set is ahead of the video detector, it too requires these traps because of the beat produced in the *video* detector. Because the 4.5-MHz frequency difference between the 41.25- and 45.75-MHz carriers is *constant*, the traps, to be effective, must be *accurately* tuned to that frequency. With small-screen sets, the narrow 2 to 3-MHz response is enough to attenuate the 4.5-MHz beat signal, and traps are not required in the video amplifier.

- **920-kHz Beat Interference**

In a color receiver the 4.5-MHz signal can beat with the 3.58-MHz color subcarrier producing a difference frequency of 920 kHz. Such a signal produces a wormy or herringbone interference pattern on a pix much like that obtained when a fine tuner control is misadjusted. The beat signal is developed in the video detector and is normally prevented by the extensive use of 41.25-MHz traps in the IF strip to weaken the sound signal. Traps tuned to 920 kHz cannot be used since this frequency falls within the 0 to 4-MHz range of the video signal.

13-10 TRAP CIRCUITS

There are four basic types of traps in several typical circuits to be described.

1. **Series resonant traps** (*L203/C204* in Fig. 13-5): The series connected coil and capacitor are connected *across* the stage where its low impedance shunts the undesired signal to ground.
2. **Parallel resonant traps** (*L1* in Fig. 13-7): Here the parallel connected components are *in the signal path* from one stage to the next where its high impedance attenuates the undesired signal.
3. **Absorption traps** (*T1-9/10* in Fig. 13-6): This is a parallel resonant circuit inductively coupled to a tank from which it absorbs energy at its resonant frequency.
4. **Bridged-T trap** (sometimes called a *notch filter*) (*T2* in Fig. 13-6): Because of their sharp response, they are found in practically all modern sets. There are several variations of this trap, whose effectiveness is determined by the Q of the coil and the value of the notch resistors. The value of the resistors (which are sometimes variable) establishes the *depth* of the notch in the response.

13-11 PREAMPLIFICATION TUNING

Where ICs are used in the IF strip, most of the tuning is performed *ahead of* the high-gain IC amplifiers. Most of the tanks and traps are lumped together between the tuner and the IF input (see Fig. 13-5). In at least one late-model set, the tank and trap functions are performed by a plug-in tuning module with only one or two conventional tuned coils further downstream.

13-12 FEEDBACK, OSCILLATIONS, AND NEUTRALIZING

Inductive or capacitive feedback from output to input of an IF amplifier can seriously affect set operation. A *small amount* of in-phase feedback where the output signal reinforces the input signal makes the circuit *regenerative*. This results in a narrowing of the IF bandpass, video overload, poor pix quality, and general instability. Where the feedback is sufficiently strong, the circuit will start to *oscillate*. A circuit that is oscillating *generates its own signal* and can no longer function as an amplifier. The generated signal is usually many times stronger than the station signal and shows up as a high (10 V or more) signal voltage at the detector output. This in turn creates a high AGC voltage, which cuts off the IF stages or reduces their gain, resulting in poor contrast and color or complete loss of pix and sound. Other symptoms include a streaky pix of poor quality and squeals or "birdies" in the sound. In some cases, harmonics of the

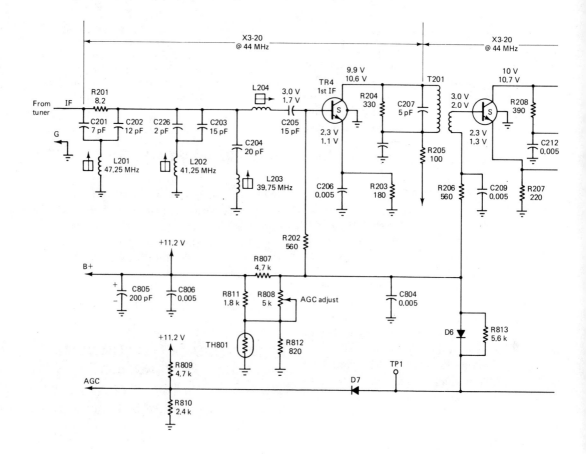

Fig. 13-5 IF and detector circuits (typical BW receiver).

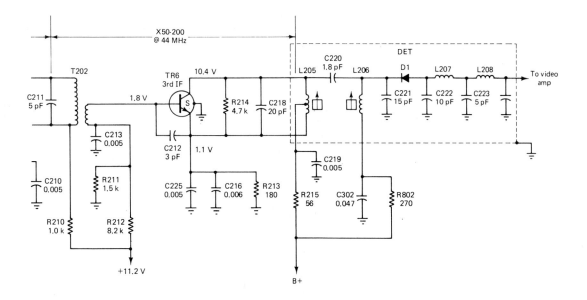

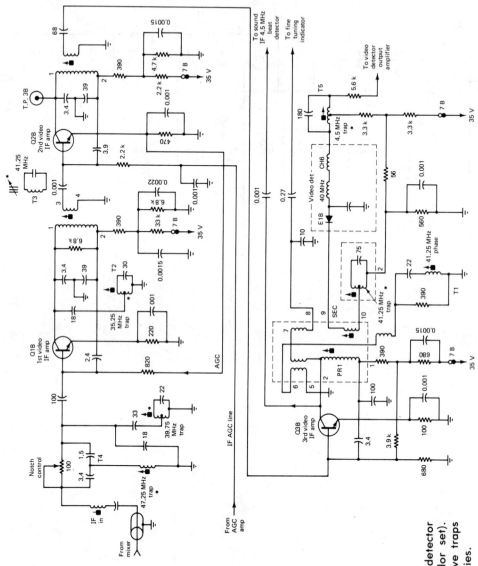

Fig. 13-6 IF and detector circuit (typical color set). Note the five wave traps and their frequencies.

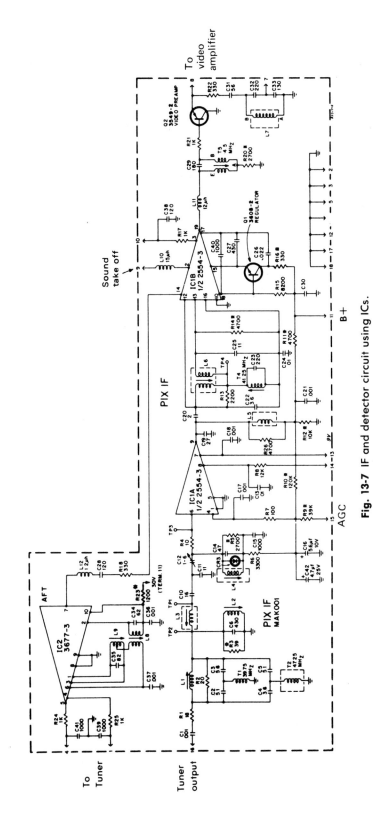

Fig. 13-7 IF and detector circuit using ICs.

generated signal are picked up by the tuner, then amplified, to produce an interference pattern on the pix. Strong oscillations can destroy sensitive transistors and other components.

Certain precautions are taken in receiver design to prevent feedback. These must be considered when servicing a set. They are (1) adequate spacing of input-output wiring and components, and (2) the use of separate *ground returns* for each stage rather than a "common ground" such as the chassis.

REMINDER: ────────────────────────

Chassis resistance is negligible for dc but may have high impedance at high frequencies across which considerable signal voltage may be developed. Such voltages may create either regenerative or degenerative effects.

Feedback is also prevented by (3) the use of shielding and baffle plates (sometimes the entire IF strip is enclosed in a shield), and (4) *neutralizing* the stage.

• **Neutralization**

This involves introducing some negative or *degenerative* feedback to cancel any positive feedback. Usually only the last IF stage is neutralized. Feedback occurs at high signal levels, and since the gain of the first and second stages is AGC controlled, neutralizing is unecessary.

A typical neutralizing circuit is shown in Fig. 13-8. In this common-emitter circuit, signals at the base and collector are 180° out of phase. Some of the output voltage developed across $R1$ is fed back to the base input via a *neutralizing capacitor*, $C1$. Note its small value. The amount of feedback is determined by the relative values of $C1$, $C2$, and $R1$. In some cases $C1$ is adjustable. The amount of feedback is critical. Too little feedback may not prevent oscillations; too much feedback causes a degenerative loss of stage gain. Some circuits obtain the feedback voltage from a tap on the tank coil. Neutralizing procedures are described in Sec. 13-25.

13-13 TYPICAL IF CIRCUITS

The IF and detector circuit of a typical BW set is shown in Fig. 13-5. The circuit uses overcoupled *untuned* transformers in the first and second stages. Note the sound IF trap and the adjacent channel traps at the IF input. The tank coil L204 is the mixer link coil. The two tank coils L205 and L206 between the third IF and the detector are adjustable. Only the third IF stage is neutralized, and C217 is the neutralizing capacitor. Note the gain figures for the three stages. The third stage runs "wide open" and has the highest gain because it is not controlled by AGC. The first two stages are controlled in parallel from the AGC bus. Note the variations of their dc operating voltages between off channel and with a signal received. Forward bias for no-signal operation is 0.7 V as measured between B and E. With a signal, the bias on the first transistor drops to 0.6 V and to zero for the second stage. This indicates *reverse* AGC is used.

The video detector is diode *D1*. Coils L207 and L208 are low-pass filters to prevent tweet interference on the pix. Coils L301 and L302 are peaking coils. The load resistor is R302. Note the two 4.5-MHz traps in the video ampli-

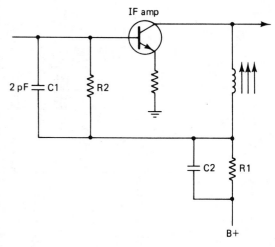

Fig. 13-8 Neutralization.

13-14 Basic Video Detector

fier stage. Trap L304 also serves as the sound takeoff.

The circuit for a typical color set is shown in Fig. 13-6. As with most receivers the transistors operate in the common-emitter configuration. With this set, AGC for the first IF stage is obtained from the emitter circuit of the second stage. All three stages are neutralized using feedback voltages developed across resistors in their collector circuits. Overcoupled tuned transformers are used between stages. Extra windings on the double-tuned IFT in the last stage furnishes a signal voltage in conjunction with a 41.25-MHz tuned circuit to operate a tuning indicator. Note the two adjacent channel traps in the IF input circuit. The 39.75-MHz trap is tapped to match the low-impedance coax from the mixer. The 47.25-MHz bridged-T trap has a variable resistor that functions as a notch control. The sound IF signal is attenuated by a 41.25-MHz trap in the detector circuit. The 41.25-MHz coil at the output of the third IF functions as a combination trap and a sound takeoff tank coil. A 4.5-MHz bridged-T trap is located at the detector output.

The color receiver circuit shown in Fig. 13-7 uses three ICs, two in the IF strip and one for AFT. All circuits shown are located on a single plug-in module. All large components such as tuned circuits are external to the ICs. Transistor $Q1$ is used to stabilize the dc power for the IF transistors. Video takeoff for the AFT circuit and the sound takeoff are from IC1B, which contains the last IF stage and the diode detector. The operation of AFT circuits is described in Sec. 12-12. Note the two parallel-connected, adjacent channel bridged-T traps at the IF input. Other trap and tank circuits are tuned to frequencies as indicated.

13-14 BASIC VIDEO DETECTOR

The primary function of any detector is to *demodulate* a high-frequency carrier to recover the modulation frequencies contained in its sidebands. For a video detector this is the composite video signal that makes up a pix and contains color and synchronizing information. Basically, the detector is nothing more than a half-wave rectifier that converts the ac carrier into unidirectional pulses that change in accordance with the frequency and amplitude variations of the received signal. Of the two signals fed to the detector (the 41.25-MHz sound IF carrier and the 45.75-MHz pix IF carrier), only the *amplitude-modulated* video signal is detected.

For a *BW set*, the detector has an additional function, developing a 4.5-MHz sound IF by the mixing of the sound and pix carriers. Although converted to a lower carrier frequency, the 4.5-MHz signal still retains the same FM sidebands as transmitted by the station. For a *color set*, the 4.5-MHz signal is developed by a *sound IF detector* from the sound takeoff in the last IF stage.

The circuit of a basic video detector is shown in Fig. 13-9. It contains relatively few components, and its operation is simple and straightforward. The diode $D1$ is the detector. The IF signal from the IF strip is fed to the diode via the tuned transformer $T1$. Connected as shown, the diode conducts during the negative half-cycles of the input signal to develop a negative-going pdc (pulsating dc) voltage across the load resistor $R1$. If the diode connections are reversed, the positive half-cycles of the signal would be rectified, producing a positive-going output.

The rectified signal has two components, ac and dc, and as such can be measured with either an ac or a dc meter at the output test point (TP). Note the components in the signal path, the IFT, diode, coils $L1/L2$, and the load resistor $R1$. The small capacitor $C1$ provides a low-reactance path for the IF signal. Early model sets used the dc as a source of AGC voltage. No signal voltage is developed across R2 and R3 because of the shunting effect of C5. Coil L1 is a three-section low-pass filter. Its purpose, along with the *ferrite beads*, is to prevent oscillations

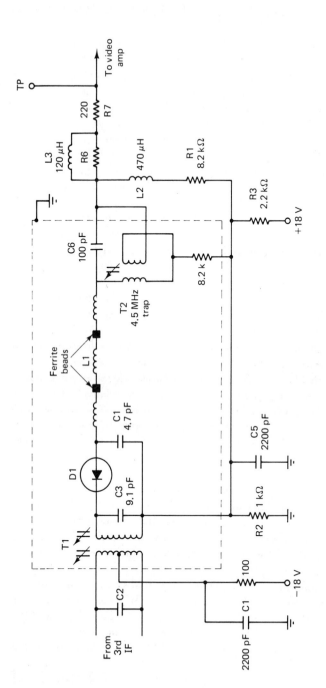

Fig. 13-9 Typical basic detector. Most detector components are enclosed in a shield can to prevent tweet interference.

and radiation of the high harmonics of the 45.75-MHz video signal. Unless suppressed, these signals will be picked up by the tuner and show up on the pix as tweet (tunable beat) interference. For this reason the detector is enclosed in a shield can.

Coils *L2* and *L3* are called peaking coils (PC); their purpose will be described later. Note the 4.5-MHz trap at the detector output, whose purpose is to prevent the 4.5-MHz signal from reaching the pix tube. In this circuit the trap also serves as the sound takeoff, which identifies the set as a *BW* receiver.

A diode does not amplify, and the output signal is divided between the diode resistance and the load resistor *R1*. In general, the higher the load resistance is, the more signal voltage that is available to pass on to the next stage. Typical values range from about 1 to 10 kΩ. To prevent attenuation of the low video frequencies, most sets use direct coupling from the detector on.

13-15 COMPOSITE VIDEO SIGNAL

Except for its low amplitude, the video signal as it exists at the detector output has all the characteristics necessary to produce a pix. Assuming normal bandpass of the IF strip, the signal at this point is made up of all frequencies ranging from near 0 Hz to about 4 MHz as required for good pix quality. Thus it is essential that all stages between the detector and the CRT input have a relatively flat frequency response over this frequency range, and any frequencies that are lost or attenuated cannot be regained. The biggest problem is preserving the *high* video frequencies, those responsible for the fine pix detail. These frequencies, in the 3 to 4-MHz range tend to be lost by the shunting effect of distributed circuit capacities. To compensate for such loss, they are given an extra boost starting with the detector.

• **High-Frequency Compensation**

Attenuation of the high video frequencies is minimized in the detector and subsequent video amplifier stages by using low-value load resistors and peaking coils. All things considered, the higher the impedance of a circuit is, the greater the bypassing of high frequencies to ground. Thus the choice of load resistor values becomes a trade-off, sacrificing signal voltage for what is more important, the preservation of high frequencies. For this reason, large-screen quality receivers tend to use resistors seldom exceeding 2 kΩ. Small-screen sets use larger values because of their less stringent frequency requirements and for increased amplification.

A *peaking coil* (PC) is a small inductance that is self-resonant to frequencies in the high (3- to 4-MHz) range. The detector circuit of Fig. 13-9 has two such coils, one in series with the load resistor, which is called *shunt peaking*, and another, called *series peaking*, in the signal path between the detector and the next stage. A PC is often shunted with a resistor to broaden its response. Sometimes the coil is wound directly on the resistor.

• **Video Peaking**

The higher the load resistance (or impedance) is, the greater the output from the detector. A resistor has the same opposition to all frequencies; a coil has the property of *impedance*, which *increases with frequency*. The detector load in Fig. 13-9 consists of the PC *L2* in series with the resistor *R1*. The ohmic resistance of the coil is negligible, as is the impedance of the coil at *low frequencies*. Thus the detector output at low and medium frequencies is determined only by the value of *R1*. At *high* video frequencies, the coil has considerable reactance, which added to the resistor value represents an increased load impedance and therefore greater output. At coil resonance, the impedance and output are maximum.

Series peaking also contributes to the high-frequency output by reducing the effect of stray capacities in the circuit. Without peaking coils in the detector and video amplifier stages, there is a serious roll-off for frequencies around 3 MHz and above. The boost obtained with PCs is shown by the frequency response curve in Fig. 13-9.

Increasing the level of the high video frequencies even beyond the low frequencies improves the contrast for fine pix detail and often results in a *sharper* or *crisper* pix. Because of this, many receivers provide a *peaking control* for use by the viewer to enhance the pix, especially old films. Some sets are designed for *overpeaking*, with a potentiometer control shunted across or in series with a PC in the detector or amplifier stage. More elaborate control circuits are described in Sec. 14-3. In some cases, the IF strip is aligned for a peaked response (see Fig. 13-4a).

• **Video Ringing**

This is a condition caused by *excessive* peaking, where abrupt amplitude variations of high-video frequencies shock-excite one or more PCs into oscillation. These are damped oscillations sustained only by the energy stored in the coil(s), which is quickly dissipated (see Fig. 13-9).

Video ringing produces ghostlike outlines around fine pix detail. They are sometimes referred to as *tunable* ghosts since they change with the fine-tuning control. As the oscillations are damped out, the repeat images trail off to the right and are lost. Sometimes they are mistaken for ghosts caused by multipath reception or an antenna-to-receiver mismatch. The usual causes of ringing are excessive IF amplification or when the IF strip is aligned for a peaked response. Resistors across PCs help to prevent this problem.

13-16 SYNCHRONOUS DETECTOR

This type of detector is now used in practically all late-model receivers because of improved performance over the conventional diode detector, which is rapidly becoming obsolete. A basic diode detector has a number of serious shortcomings: its operation is nonlinear at low signal levels; it has no gain; it produces beat frequency interference; and it accepts *any* signal (including adjacent channel interference) that is fed to it.

The synchronous detector overcomes these shortcomings to produce a better quality pix free of interference. It amplifies. It detects the weakest signals, which a diode overlooks. Its operation is *linear*, even on weak signals. And it responds *only* to signals synchronized to the *desired* 45.75-MHz carrier, which means fewer traps are needed in the IF strip.

All the detector functions are performed by ICs. Although the internal circuits are not accessible for testing, some understanding of its operation can be helpful, especially in troubleshooting.

Figure 13-10 shows a simplified schematic of this somewhat complex system of demodulation. There are three main items: a *reference amplifier* (sometimes called a carrier oscillator or generator), a *differential amplifier* and a *synchronous demodulator*.

The reference amplifier generates a signal that is synchronized with the 45.75-MHz carrier of the *desired* station signal. It has two 45.75-MHz square-wave outputs of opposite polarities, arbitrarily identified in Fig. 13-10 as outputs A and B. A high-Q tuned circuit ($L1/C1$) across the two outputs establishes the correct phasing and ensures that the outputs are not influenced by sideband modulation frequencies. Diodes $D1$ and $D2$ limit the amplitudes of the outputs.

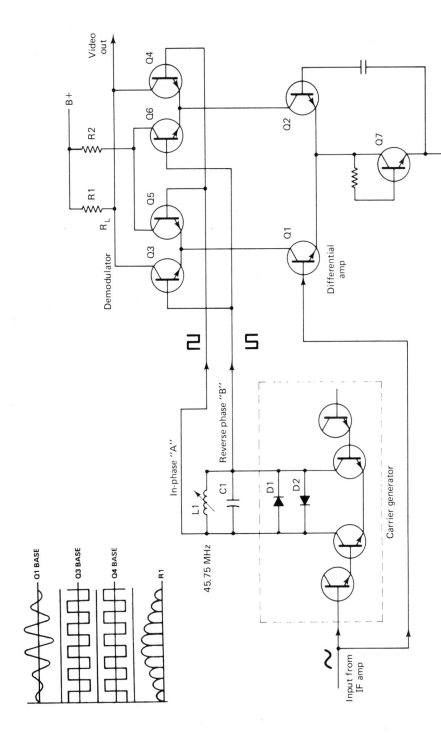

Fig. 13-10 Synchronous detector. A synchronous detector has many advantages over a basic diode detector.

The differential amplifier consists of two transistors ($Q1$ and $Q2$) that amplify the signal coming from the IF strip. It also has two outputs, 180° apart in phase.

NOTE:

A differential amplifier is a two-transistor amplifier having two balanced outputs of opposite polarities that are unaffected by variations of circuit parameters.

The demodulator consists of four switching transistors ($Q3$ through $Q6$) connected in a bridge circuit. The transistors are controlled by the square-wave outputs of the reference amplifier to alternately select one of two outputs of the differential amplifier. Its operation is similar to the bridge rectifier used in some LV power supplies. The demodulated output (the composite video signal) is developed across the load resistor $R1$. Transistor $Q7$ is called a *constant-current source*. Its job is to maintain a constant current through the differential amplifiers $Q1/Q2$.

The polarity of the reference amplifier outputs is constantly changing, along with the high-frequency ac video signal coming from the IF strip; therefore, the outputs are alternately in phase and out of phase with the station signal. Note that the two outputs (A and B) are fed to opposite arms of the demodulator bridge circuit, output A to $Q4$ and $Q5$, and output B to $Q3$ and $Q6$. The signal from the IF strip simultaneously feeds the differential amplifier $Q1$ and the reference amplifiers, where it serves as a synchronizing signal.

To summarize the operation, transistors $Q4/Q5$ and $Q3/Q6$ are alternately turned on and off by the A and B outputs of the reference amplifier. The differential amplifiers $Q1/Q2$ are similarly and alternately turned on and off by the AM video carrier of the incoming station signal.

On positive half-cycles of the station signal, $Q1$ and $Q3$ are turned on (by the positive reference signal from the A output), increasing the current through the load resistor $R1$, which is in series with the B supply.

On negative half-cycles of the station signal, the polarities of the A and B outputs are reversed, $Q1$ and $Q3$ are switched off (by the positive output from B), and $Q2$ and $Q4$ are turned on, increasing the current through $R1$ in the same direction as before. The circuit thus operates as a *full-wave* rectifier, developing an output for each half-cycle of the 45.75-MHz video IF signal whose amplitude is changing in accordance with the modulation frequencies. This is the composite video signal that is passed on to the video amplifier(s).

Note that on the positive half-cycles of the IF signal, the signal path is via $Q1$ and $Q3$ to the output. On negative half-cycles, the path is via $Q2$ and $Q4$ to the output.

A synchronous detector produces an output *only* for signals *that are synchronized* with the IF carrier, thus preventing detection of signals *other than* the desired station signal. Adjacent channel signals, for example, are *not* in step with the carrier and therefore do not appear at the output to be passed on to the next stage. In the demodulation process, both the 45.75-MHz pix carrier and the 41.25-MHz sound carrier are eliminated, and the only signal appearing at the output is the composite video signal and the 4.5-MHz beat signal. Since the circuit operation is *linear*, there are no problems with detector radiation and tweet interference on the pix. Proper operation of this type of detector depends on the precise adjustment of the 45.75-MHz tuned circuit at the output of the reference amplifier, which is *extremely critical*. Alignment procedures are described in Sec. 13-44.

13-17 TYPICAL IF STRIP USING ICs

A brief description of the IF and detector section of Sample Circuit C at the back of this book is as follows. Most of the functions are performed by

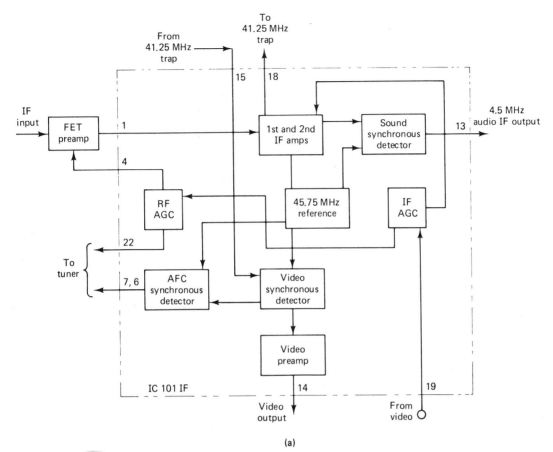

(a)

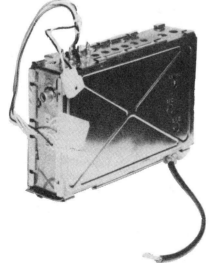

(b)

Fig. 13-11 Modern multistage IF and detector: (a) all circuit functions are performed by a single IC; (b) all components including tuned circuits are contained in one module.

a single IC (IC101) as depicted by the block diagram of Fig. 13-11. The IC and all associated external components are contained in the module shown in the photo.

The signal from the tuner feeds a FET-IF preamp (Q101) via the input circuit link coupling network. A FET is used to obtain a high signal-to-noise ratio. Most of the shaping of the IF response is accomplished *prior to amplification* by the lumped tuned circuits consisting of the two tanks (L101 and L102) and the three trap circuits. The input and output of the FET is tuned by tanks L103 and L107. From the FET, the signal goes to pin 1 of the IC.

The IC contains two stages of IF amplification and uses only one tuned tank (L108) and a 41.25-MHz sound trap (L109). Both IF stages are controlled by AGC. Note the two preset RF and IF AGC controls. The RF control adjusts the AGC threshold for the RF stage *and* the FET preamp. On strong signals the gain of the FET is reduced prior to that of the RF stage for lowest possible noise level.

The IC contains three synchronous detectors, for video, sound, and AFC, each requiring a synchronizing input from the reference amplifier, as shown. From the video detector the signal is boosted by a video preamp before leaving the IC. Video peaking is provided by a sharpness control farther downstream in a video amplifier stage.

13-18 TROUBLESHOOTING THE IF STRIP AND DETECTOR

Since *all* signals must pass through this section of the receiver, defects here can produce numerous trouble symptoms, such as loss of pix and/or sound, weak pix and sound, poor pix or sound quality, loss of color or weak color with poor quality, and partial or complete loss of sync, to name a few.

The *same* symptoms, however, *can* be caused by troubles *elsewhere* in the set. Logically, after considering the symptoms, the first step in troubleshooting is to determine whether or not the IF and detector section of the set is responsible for the problems. If it is responsible, tests are made to localize the defect to one particular stage and then to pinpoint the *cause*.

13-19 SYMPTOMS ANALYSIS

Examine the schematic. Note such details as the number of IF stages and tuned circuits, transistor identification, the point of sound takeoff, and the detector output polarity. Some of these things can have special significance when analyzing symptoms and deciding where to begin. Consider the following:

1. *No pix, no sound:* the signal is probably lost at some stage *ahead* of the sound takeoff, an IF stage or the tuner. The actual *cause* may be such things as loss of B voltage, an AGC problem, or an IF or tuner stage that is defective.

2. *No pix, sound OK:* the defect will normally be found in some video stage after the sound takeoff. There are exceptions, however, and the trouble may be *ahead* of the takeoff point because a faulty IF stage will sometimes permit some sound to get through.

3. *Pix OK, no sound:* the trouble probably is in the sound section after the sound takeoff.

4. *No snow, weak pix:* suspect video stages between the second IF and the pix tube.

5. *Weak pix, excessive snow, hiss in sound:* suspect the tuner, first IF stage, and AGC.

6. *Loss of AGC:* probably an interruption of the signal ahead of the AGC takeoff or an AGC circuit problem.

Suggested troubleshooting procedures are presented in the flow drawing of Fig. 13-12 in conjunction with the trouble symptoms listed in Table 13-1.

Table 13-1 IF and Detector Trouble Symptoms Chart

Trouble Condition	Probable Cause	Remarks and Procedures
No pix, no sound, or weak pix and sound	Defective stage ahead of sound takeoff, or excessive AGC. Misadjusted AGC threshold control.	Localize trouble by signal tracing station signal or by signal injection. Check AGC on and off channel. If excessive snow, suspect tuner, first IF, and AGC. If no snow, suspect stages between first IF and the sound takeoff. Check for signal at detector TP.
No pix, sound OK	Defective stage between the sound takeoff and the CRT. If sound slightly weak, could also be trouble *ahead* of sound takeoff. Excessive AGC.	Localize trouble by signal tracing or injection. Check AGC. Check for signal at detector TP. If synchronous detector is suspect, try replacing IC and/or check for misaligned reference coil.
No sound, pix OK	Defective sound stage between sound takeoff and the speaker. Also suspect IF alignment.	Verify signal at sound takeoff and signal trace sound stages.
No color or weak color, BW pix OK	IF misalignment. Misadjusted color killer control. Trouble in the chroma stages.	Check killer control adjustment. Check video display at detector TP for compression of pulses and color burst. Check location of color markers on overall response curve.
Unstable sync, pix contrast OK	Video overload caused by loss of AGC or peaked IFs. Trouble in sync stages.	Check hammerhead for loss of pulses. Scope check signal at detector TP and further downstream for pulse clipping or compression. If pulse clipping, check for loss of AGC or misaligned IF strip.

VIDEO IF AND DETECTOR STAGES

Table 13-1 (cont'd)

Trouble Condition	Probable Cause	Remarks and Procedures
Pix smearing, loss of resolution, buzz in sound, possibly negative pix; may also be loss of sync	Suspect video overload if trouble on strong stations only.	As above for pulse clipping.
Negative pix (black objects appear white and white objects are black).	If only on strong stations, suspect video overload. If on all stations and loss of sync, check for signal polarity reversal.	Check causes of overload as above. If reversed polarity at detector TP, suspect reversed diode connections. Adjust AGC.
Intercarrier buzz, pix and sync OK	Trouble in sound section.	Check sound discriminator alignment.
Loss of fine pix detail	IF misalignment. Poor focus. Video ringing. Poor high-frequency response of detector or video amplifiers.	Check overall IF response for narrowed bandpass or misplaced markers. Scope check for ringing. Adjust for *perfect* focus off channel. Check frequency response of video amplifiers.
Video ringing (visual echos on fine pix detail)	Excessive video peaking or IF misalignment.	Check video peaking circuit. Check IF alignment and possible causes of video overload.
IF oscillations (streaky pix and hash or birdies in the sound)	Excessive IF signal gain. Feedback coupling in IF strip. Poor grounds, including coax from tuner.	Verify by checking for excessive voltage at detector TP both on and off channel. If excessive IF gain, check for misalignment or partial loss of AGC. If warranted, change lead dress and parts placement in IF strip. Neutralize the troublesome stage(s).
Streaky pix	Arcing, loose connections, intermittent shorts, contamination in the CRT. IF oscillations.	Analyze nature of problem and localize the cause.

Table 13-1 (cont'd)

Trouble Condition	Probable Cause	Remarks and Procedures
Pix and sound mistracking (best pix and best sound occur at different settings of fine tuner)	IF misalignment.	Check overall IF response for narrowed bandpass.
Sound bars on pix (venetian blind effect) (horizontal black bars)	AF signal getting into the video section. IF misalignment. Poor LVPS regulation. Microphonics.	If bars fluctuate with station sound, check for IF misalignment, particularly the 41.25-MHz traps. Scope check for audio on B+ circuits. See if bars are affected by tapping and jarring.
Adjacent channel interference	Misadjusted traps.	Adjust appropriate traps; then recheck overall IF response.
Hum bars on pix (wide, black and white horizontal bar(s)	Hum modulation of video signal.	Check for ripple on PS. If bars on raster *and* pix, check ahead of detector. If only on off-channel raster, check after detector up to CRT input.
Tweet interference on pix (wavy lines in pix)	IF and detector radiations or coupling with tuner or antenna input.	Verify detector shield can in place and properly grounded. Possible defect with *L/C* filters in detector circuit. Lead dress including antenna lead-in.
Beat interference pattern on pix (wormy or grainy effect)	Misadjusted 41.25-MHz or 4.5-MHz sound traps.	Adjust traps. Touch-up IF alignment if required.
Intermittent problems	May be anywhere in the set.	Monitor AGC and voltage at detector TP, and possible changes in trouble symptoms.

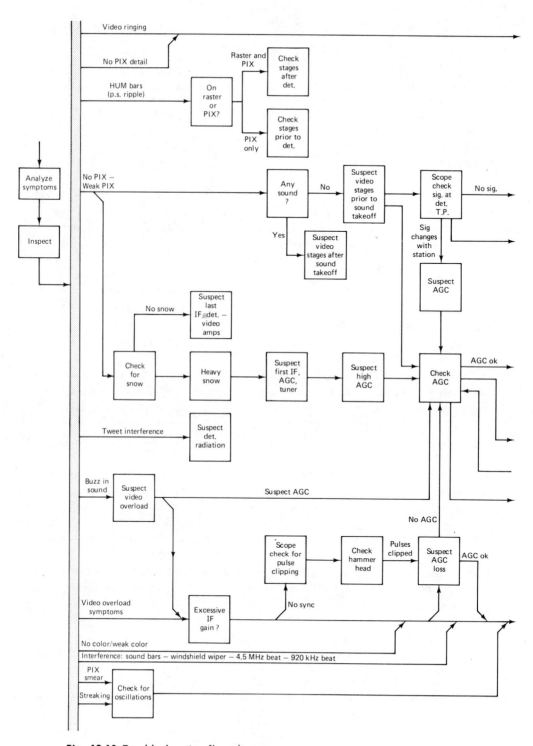

Fig. 13-12 Troubleshooting flow diagram.

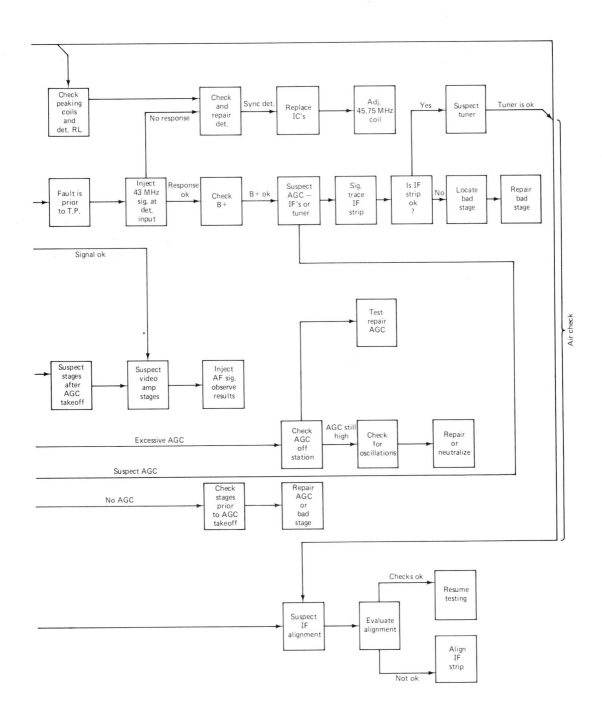

13-20 INSPECTION

Before testing begins, an inspection should be made in the suspected trouble area regardless of the nature of the problem. With the set off, look for obvious things like broken wires, and loose or damaged components. With the set on, look for overheating components and arcing, etc.

13-21 LOCALIZING THE FAULT

Where a *dead* stage is suspected, the following tests are normally made and in the sequence indicated: (1) check for a signal at the detector output, (2) check for B voltage supply to the IF strip, (3) check for AGC voltage on the AGC-IF bus and at the controlled IF stages, and (4) test to localize the defect to a particular stage.

• Detector Output Signal

Considering its location approximately midway in the signal path between the antenna input and the CRT and speaker, the video detector is a good place to start checking for loss of signal and other abnormal conditions. If the signal checks normal at the detector, the trouble is between that point and the CRT. If the signal is abnormal, the trouble is upstream: the IF strip, the tuner, or the AGC.

Although an EVM or other instrument can be used to check the signal developed across the detector load resistor, a CRO is preferred because it provides a visual indication of such things as interference, oscillations, and pulse clipping.

With no signal input (as when switched to a blank channel), there should be little or no indication except that caused by background noise, hash, or residual hum pickup. When tuned to a station of average strength, look for a peak-to-peak voltage between 2 and 5 V, either ac or dc.

Readings of 10 V or more usually indicate RF/IF oscillations or strong interference. The voltage should be fairly constant for both weak and strong signals; if not, the AGC is suspect.

Using a CRO, the composite video signal should appear as shown in Fig. 13-9. Depending on instrument design and the detector output polarity, the display may be either upright or inverted. Check for possible clipping or compression of the sync pulses. When pulses cannot be viewed distinctly, be guided by the two horizontal lines created by the CRO beam retrace.

• Checking for B Voltage

Verify B voltage is supplied to the IF strip. Refer to the receiver schematic for the proper amount and polarity. For loss of voltage, check for an open dropping resistor or a shorted capacitor or transistor. A *cold* or *overheating* resistor often provides a good clue.

• AGC Testing

Either too much or too little AGC voltage has an adverse effect on the operation of controlled stages. Loss of AGC causes video overload and poor sync on strong stations. Excessive AGC results in weak reception and in some cases complete loss of pix and sound. Check for AGC at the IF bus and at each of the controlled stages. Refer to the schematic for typical voltage and polarity both off channel and with a signal received. For abnormal AGC conditions, clamp the bus, noting any improvement in receiver performance. If AGC is normal, *all stages* ahead of the AGC takeoff in the video amplifier must be OK. For loss of AGC, suspect the AGC circuit or the signal stages ahead of the AGC takeoff.

• Locating the Faulty Stage

Localizing trouble to a single stage is usually

done in one of two ways, by *signal tracing* (following the progress of the *station signal* stage by stage using a CRO, EVM, or some form of tracer), or by *signal injection* with an RF/AF signal generator. Either or both methods may be used depending on circumstances and the equipment available.

• Signal Tracing Procedures (see Fig. 4-5)

With this method a suitable indicator is used to check the presence or absence of a *station* signal and its relative amplitude at the input and output of each stage. When a CRO is used as the indicator, a composite video signal should be observed at each checkpoint. An EVM (on a low ac range) may also be used as an indicator. When an aural-type tracer is used, the signal is heard as a characteristic buzz. When signal tracing *beyond* the detector, the indicator may be connected directly or using a low-capacity probe. When tracing *ahead* of the detector, a *demodulator probe* must be used with the indicator.

Signal tracing may be done in either direction, working from the tuner toward the detector (where the signal should get increasingly stronger), or working upstream from the detector back toward the tuner where the signal becomes progressively weaker as fewer stages contribute to the amplification.

To locate a weak or dead stage, tune to a reasonably strong channel and connect the indicator to the detector output test point (TP). If there is no signal or a weak indication, check at the detector input using a demodulator probe. If a normal signal is indicated, the detector stage is at fault. If there is no signal at the detector input, check for signal at the input and output of each stage working toward the tuner. Where there is a signal at the input of a stage but not at its output, that stage can be considered defective.

NOTE: _____

A weak third IF stage reduces the AGC, increasing the gain of the previous controlled stages. Where a weak stage is suspected, compare its gain with that of other stages. Typical causes of low gain are excessive AGC, misalignment, a weak (leaky) transistor, or incorrect transistor operating voltages.

When a CRO is used as the indicator and a signal is displayed, examine the display for abnormal conditions as follows:

PS Ripple Voltage. Inadequate filtering of the LVPS can cause hum modulation of the video signal. It shows up on the CRO as an *undulating time base*. Determine where the ripple first appears by scope checking at the detector output and the IF stages.

NOTE: _____

Hum voltage introduced at the detector or beyond shows up as hum bars on the pix tube on the off-channel raster and on a pix. Hum voltage introduced at some stage *ahead* of the detector requires a carrier to get through and only shows up on the screen *when a signal is received*. Other symptoms of hum modulation are unstable sync and hum in the sound.

Oscillation and Video Ringing. These conditions give an unnatural appearance to the composite signal being viewed on the CRO. If sporadic, they show up as short bursts of high-frequency interference where the individual cycles are too fast to be resolved. If strong enough, they may all but obscure the display. In some cases they may be observed when switched to a blank channel. With oscillations, a higher-than-normal voltage is measured at the detector output, with streaking on the raster and pix.

Pulse Clipping or Compression. Normally, the pulses should represent about one-third of the total amplitude of the composite video signal (see Fig. 10-7). For this condition, the pulses may be barely discernible or completely clipped at the blanking level, usually without affecting the video. Weakened pulses result in sync instability; *loss* of pulses causes complete loss of vertical and horizontal sync.

13-22 VIDEO OVERLOAD

Video overload (sometimes called signal overdrive) is a very common problem where the signal supplied to one or more stages is more than they can handle. There are a number of possible causes and more than a few trouble symptoms resulting from this condition. Understandably, most of the symptoms usually show up when strong stations are received. The more common symptoms are described next.

- **Intercarrier Buzz**

This is heard from the speaker as a rough raspy sound, particularly during commercial breaks in programming when lettering appears on the screen. Sometimes referred to as *sync buzz,* it is caused by the amplitude-modulated video signal getting into the sound stages. It may be due to such things as a misadjusted fine tuner, IF misalignment, loss of AGC, or a misadjusted or faulty sound discriminator stage.

- **Poor Pix Quality**

Signal overdrive causes excessive pix contrast and high color intensity that obscures the fine detail and in some cases results in a *negative pix*.

- **Loss of Sync**

When a stage is overdriven (toward cutoff or saturation), the vertical and horizontal sync pulses (which represent the peaks of the signal and therefore are the most vulnerable) are the first things to be affected. As shown in Fig. 10-7, the condition is one of degree, where a *small amount* of overload compresses the pulses only *slightly,* which may or may not result in sync instability. The greater the overload, the more the pulses are compressed and the poorer the sync. Where pulses are fully compressed (or clipped), there is complete loss of vertical and horizontal sync. The most extreme condition (Fig. 10-7d) is where the pulses become "swallowed up" and inverted and you get a negative pix. In the case of a color transmission, the *color burst* may also be weakened, causing loss of *color sync.*

These conditions are readily observed on a CRO, hence its value in signal tracing. When the CRO shows pulse clipping *at the detector,* try to localize the problem by scope checking (with a demodulator probe) at each IF stage. Once localized, check the stage for a defective transistor or incorrect operating voltages, the B/E bias in particular. For abnormal bias under *no-signal* conditions, suspect the voltage-dividing network, the transistor, or both. For abnormal bias *only* when a signal is received, suspect the AGC.

Although the CRO may show *no clipping at the detector,* the problems described can cause overdriving and pulse clipping at some point *in the video amplifier section,* which is *usually* the case. This will show up with a scope check at the pix tube input or by inspecting the hammerhead as described in Sec. 10-10 (see Fig. 10-8).

13-23 LOCALIZING TROUBLE BY SIGNAL INJECTION

This method of locating a defective stage (see Fig. 4-3) makes use of the signal supplied from an RF/AF signal generator. As an *indicator,* a CRO, EVM, or signal tracer may be connected to the detector output or some other convenient point downstream. Where only relative indications are needed (as when checking for a *dead* stage), a bar pattern produced on the pix tube by the generator modulation serves the purpose.

13-23 Localizing Trouble by Signal Injection

• **Preliminary Considerations**

For stage checking, the accuracy of generator calibration is not too important. Use a blocking capacitor in series with the generator cable. Start with a *low* output from the generator and avoid using full output at any time. Excessive signal can damage or destroy transistors and give false indications. To avoid interference, tune the receiver to a blank channel and if necessary disconnect the antenna. For precise gain measurements (when a *weak* stage is suspected), clamp the AGC.

• **Locating the Bad Stage**

1. Verify the video amplifier(s) is working by injecting an audio signal at the detector test point. Use either an AF signal generator or the 400-Hz output from an RF generator. Adjust the generator for an output of about 2 V. Where the pix tube is the indicator, you should obtain a strong (dark) bar pattern. No indication or a weak indication means trouble in the amplifier section. The *number* of bars (as determined by the AF frequency used) has no significance in this procedure.

2. Connect the RF cable of the generator to the input of the last IF stage. Switch the generator modulation on. With the generator tuned to approximately 43 MHz, you should get a bar pattern on the CRT, an indication on a meter if used, or a 400-Hz sine-wave display on a CRO. No indication means the detector stage is at fault.

3. Reduce generator output to get a small indication on the EVM or CRO. Using the bar pattern on the pix tube, adjust the receiver brightness and contrast controls until the bars appear *gray* (see Sec. 4-10).

4. Without changing the generator output, inject signal to the input of the next (upstream) IF stage working toward the tuner. The bars should change from gray to black. Tune the generator as necessary for maximum indication. If there is no indication, the IF stage is dead. If there is little or no change in the bars, the stage is weak.

5. Check the next upstream stage as in step 4. Start by adjusting for gray bars, which should become black if the stage is OK. In checking the first IF stage, inject signal at the mixer test point or the UHF jack on the VHF tuner. *Reminder*: Signal should be gradually attenuated as you approach the tuner, and the strongest indication should be with the signal fed to the first IF, with a *weak* output from the generator. If desired, stage checking can begin with the signal injected to the first IF and working toward the detector. Using this method, the indication will become *less* as you approach the detector because of the fewer stages involved.

• **Tracing a Generator Signal**

The IF strip can also be checked by signal tracing a generator signal instead of the station signal, in one of two ways:

Method 1. Inject signal at the mixer input and progressively follow its progress stage by stage using a CRO or other appropriate indicator.

Method 2. Inject signal to each stage individually with the indicator (using a demodulator probe) at the stage output. Repeat for each stage, comparing their relative gains.

• **Stage Jumpering**

Another way to locate a bad stage is to temporarily eliminate each stage in the IF strip, one at a time, while looking for any improvement in receiver operation. The simplest way to circumvent a stage in most cases is to bridge a low-value capacitor (about 10 pF) from input to output of a transistor. Where reception is restored or improved, the stage that is jumpered is probably defective in some way. A *normal* pix cannot be expected with one of the stages disabled.

• **Locating a Bad IC**

When a set uses one or more ICs and a module containing most of the IF tuned circuits, localize the defect using one of the procedures described previously. Before condemning an IC and making a replacement, verify all input-output signals and voltages as indicated on the schematic. Check all associated components not included in the ICs. When suspected, check the alignment of all tuned circuits external to the ICs.

13-24 INTERFERENCE PROBLEMS

Interference that disrupts normal receiver performance may come from outside to be picked up by the antenna or be generated within the set. Each interference problem is identified according to its cause and the symptoms it produces. Sometimes there is more than one symptom resulting from the same condition. Some problems affect the sound; most often the pix, sync, or color is affected. Common interference problems associated with the IF strip and the detector are described in the following paragraphs.

• Adjacent Channel Interference

Such interference may occur when conditions for beyond-the-horizon reception are favorable. Interference from the *video* of an adjacent higher channel shows up as the windshield-wiper effect, where sync and blanking bars of the interfering station can be observed moving slowly through the pix of the selected station. In some cases pix detail may be seen accompanied by the sound. *Sound* interference from an adjacent lower channel shows up as sound bars that fluctuate with the sound signal.

For such problems, check for excessive RF/IF bandwidth and adjust the adjacent channel traps as described in Sec. 13-39. Receivers using a synchronous detector are generally immune from this problem.

• Intercarrier Beat Interference

Any 4.5-MHz beat signal reaching the CRT gives the pix a wormy or grainy appearance or a weaving herringbone pattern of many finely spaced lines. The problem is normally overcome with one or more 4.5-MHz traps in the video amplifier section, and by attenuating the sound IF carrier with 41.25-MHz traps in the IF strip. For this problem, check the adjustment of the 4.5- and 41.25-MHz traps while aligning the IF strip as described in Sec. 13-39.

• 920-kHz Beat Interference

This problem also causes a fine-grained wormy effect on the pix. It is the result of the 4.5-MHz signal heterodyning with the 3.58-MHz color subcarrier in the video detector of a color set. The cause may be misadjustment of the fine tuner control or insufficient rejection of the 41.25-MHz sound IF carrier in the IF strip. For this problem, check the alignment, making sure the 41.25-MHz traps are properly adjusted where their marker appears close to the base line of the overall IF response.

• Tweet Interference

This shows up on the screen as one or more thin, dark vertical lines mostly on high channels and weak stations where receiver sensitivity is greatest. The interference changes with the channels selected and the adjustment of the fine tuner. It occurs when high-frequency harmonics of the beat signals are radiated from the detector to be picked up by the antenna leads or the tuner to be amplified by the following stages. Remedies include proper shielding of the video detector stage and the use of low-pass filters at the detector output. Make sure detector cover shield and baffle plates are in place and properly grounded and that the antenna lead is clear of the IF and detector section.

• Pix Streaking

There are many possible causes: car ignition interference, loose connections, arcing, internal shorts in the pix tube, or *IF oscillations*. Oscillations can also produce other symptoms: a background hissing noise in the sound, a weak pix, poor resolution, poor color, loss of sync, and in extreme cases loss of the pix and even the raster. For this problem, check lead dressing in the IF strip, misplaced components, excessive IF gain due to loss of AGC or a narrowed IF bandwidth, IF misalignment, or a neutralizing problem. Neutralizing procedures are described in Sec. 13-25.

• **Intercarrier (Sync) Buzz**

This interference is not to be confused with the buzz produced by vertical sweep getting into the sound section or hum caused by poor PS filtering.

Intercarrier buzz can be caused by video overload, IF misalignment, or troubles in the sound stages. In the IF section it occurs if the 4.5-MHz FM sound carrier is modulated by strong *AM* video signals or vertical sync pulses. Check the alignment, particularly the location of the 41.25-MHz marker on the overall response. Check for loss of AGC. Adjust the sound rejection control if one is provided. Where video overload is suspected, other symptoms include loss of sync and excessive pix contrast.

• **Sound Bars**

This horizontal bar pattern is usually caused by the sound signal getting into the video stages. For this condition the bars fluctuate in number and intensity along with the sound coming from the speaker. The bars are not to be confused with the horizontal bars on a color set resulting from loss of *color sync*.

Check the IF alignment, particularly the adjustment of the 41.25-MHz traps, and 4.5-MHz traps in the video amplifier. Where bars fluctuate with the sound of an adjacent channel, adjust the adjacent channel traps. Poor regulation of the LVPS can also result in sound bars. For this condition, audio signals will show up on the B supply feeding the IF strip or video amplifier stages. A less frequent problem is oscillations in the audio amplifier, which produce *stationary* sound bars.

Where sound bars occur only when the volume control is turned up, check for a microphonic condition aggravated by speaker vibrations.

• **Hum Bars**

These are wide, dark, horizontal strips across the pix, the raster, or both. They occur when PS ripple voltage gets to the pix tube via the video stages. For half-wave rectification in the PS, the ripple frequency is 60 Hz, which produces one bar for each vertical scan. For full-wave rectification, where the ripple frequency is 120 Hz, two bars appear on the screen. The bars are usually stationary, but sometimes they move slowly up through the pix. In extreme cases the bars may be prominent, at other times, barely discernible, and only show up at low settings of the brightness control or where the pix is made to roll vertically. Hum bars are usually accompanied by other symptoms, such as poor sync, poor pix and color quality, hum in the sound, and variations of vertical linearity of the scanning.

Hum bars may also be caused by cathode-heater leakage in the pix tube and, with tube-type sets, leakage in any tube between the antenna input and the CRT.

Check for hum bars on and off channel. Where bars are observed on the raster *and* a pix, check for hum voltage on the B supply feeding the video amplifier section. Where bars are observed only when a signal is received, check for PS ripple on the dc supply for the tuner or IF strip.

NOTE:

Hum modulation of the video signal prior to the video detector requires a *carrier* to get through the IF strip.

13-25 REGENERATION AND IF OSCILLATIONS

Regeneration is a condition where some of the signal at the output of an amplifier feeds back, in phase, to reinforce the signal at the amplifier input. In the IF strip it results in a narrowed bandpass and general instability. When the feedback is strong enough to overcome circuit losses, the stage begins to oscillate and generate its own signal. This usually occurs when receiving a strong signal or when the gain of the IF strip is excessive. Such *regenerative feedback* may be the result of poor lead dress or capacitive or inductive coupling between circuits. Such feedback is normally prevented by proper lead dress and shielding of components or wiring where necessary. However, some feedback usually takes place, and this is taken care of by the process called *neutralizing*. Some of the amplifier output signal is fed back (via a neutralizing capacitor) as *negative feedback* to the amplifier input to cancel out the in-phase feedback.

The symptoms of IF oscillations include a streaky pix with poor resolution, hash or birdies in the sound, loss of sync, no color or poor color, loss of contrast or sometimes loss of pix or raster, a critical fine tuner, and in some cases a negative pix. With oscillations, a reading of 10 V or more is obtained at the detector and, in some cases, off channel. An oscillating stage can be located with a CRO or by noting changes when the hand is brought close to the troublesome circuit.

Remedies for oscillation include the following:

1. Neutralizing: Where neutralizing is adjustable, follow the same procedures described in Sec. 12-29 for neutralizing the RF stage. If no trimmer is provided, try increasing the value of the neutralizing capacitor up to about 5 pF or substitute with a gimmick capacitor. *Reminder*: Too much negative feedback reduces stage gain.

2. Check for loss of AGC.

3. Try reducing the value of loading resistors across tank coils in critical stages.

4. Verify that all shields and baffles are in place and properly grounded. Check the grounding of all coax cables and chassis grounding lugs. *Reminder*: A good dc ground is not necessarily a good ground at RF/IF frequencies.

5. Check lead dress and component placement, especially after making replacements.

6. Introduce some degenerative feedback by disconnecting or reducing the value of the emitter bypass capacitor in one or more critical stages.

7. Touch up the IF alignment, particularly the mixer link coil on the tuner. *Reminder*: Feedback may occur if staggered-tuned circuits are adjusted too close to the same frequency.

13-26 ISOLATING TROUBLE IN THE IF STRIP

Assuming the trouble has been *localized* to a single stage, pinpoint the defect using standard troubleshooting procedures. For a *dead* stage, check for loss of B voltage, excessive AGC, a defective transistor, or incorrect operating voltages. Except for loss of B voltage, the same possibilities apply where the stage checks weak. A gain loss may also be due to misalignment of the tuned circuits in the stage.

Start by checking the transistor and its dc operating voltages. Switch off channel and compare the no-signal voltages with those shown on the schematic. Remember an abnormal voltage may be either the *cause* of the problem or the result of a defective transistor. Check the transistor bias between B and E. Typical no-signal bias is usually about 0.75 V, the amount required for maximum gain. Polarity depends on the transistor type. Either too much or too little bias can weaken or kill the signal, and with some sets, a voltage discrepancy as little as 0.1 V can be significant. When the bias reading is questionable, compare with voltages of an IF stage

13-26 Isolating Trouble in the IF Strip

that is operating normally and under similar conditions. If necessary, use a bias box, noting any improvement in receiver performance.

For an abnormal bias reading, verify the conditions of the transistor and make additional voltage checks from B to ground and E to ground. When the transistor is OK, check for defective resistors and bypass capacitors in the biasing network.

Check the AGC action by switching to an active channel. The bias should either increase or decrease depending on whether forward or reverse AGC is used.

REMINDER: _____

Loss of AGC causes video overload problems on strong stations; excessive AGC will weaken or kill the pix. Check for AGC variations when switching channels. Check signal voltage at the detector output. It should remain relatively constant.

Where AGC for the first IF stage is obtained from the emitter circuit of the second stage, the transistors and operating conditions for both stages should be checked when there is an AGC problem.

- **Tuned Circuit Problems**

Tank and trap circuits sometimes (but rarely) give trouble. They are suspect when adjustments have no effect or they cannot be tuned to the proper frequency. When suspect, check coil continuity on the lowest ohmmeter range. Normal reading is approximately 0 Ω. A higher reading indicates the coil is open. Sometimes a coil becomes intermittently open and the ohmmeter reading may be high and fluctuating. When suspect, disconnect coil from the circuit and *momentarily* apply about 10 V from an external source. Such an overload will not harm a good coil and will usually show up a defect.

A common defect with bifilar transformers is leakage between windings. This results in abnormal operating voltages for the transistor in the preceding stage and the one following. When suspect, disconnect the coil and check for leakage using the highest (most sensitive) ohmmeter range. If the test is inconclusive, repeat the check using a voltmeter and an external supply of 50 V or so.

When a coil will not tune properly or wants to resonate at a higher than normal frequency, suspect a broken or missing slug or shorted windings. When a slug will not adjust because of stripped threads, remove the slug, insert a piece of nylon thread or the equivalent, and replace the slug.

To check for shorted turns, remove the coil and test using one of the methods shown in Fig. 13-13. For the setup in Fig. 13-13a use the lowest dc range on the EVM. If the coil is OK, a sharp indication will be obtained as the generator is slowly tuned to the coil's resonant frequency. Try adjusting the slug. If the coil will not resonate or tunes too high, replace it.

Using a GDO (grid-dip oscillator) as shown in Fig. 13-13b, hold the instrument coil close to the coil under test and tune for an indication. Use the minimum coupling that gives a reading.

13-27 TROUBLESHOOTING THE BASIC DETECTOR

The components in the detector stage are under very little stress and seldom break down, and when there is trouble, the stage is often overlooked. Trouble symptoms depend on the set. For a BW set, a weak or dead detector affects both pix and sound. With a color set, only the pix is affected.

When the defect has been localized to the detector stage, start by checking the diode, the

524 VIDEO IF AND DETECTOR STAGES

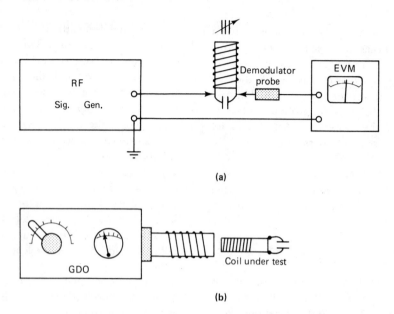

Fig. 13-13 Testing tank and trap coils: (a) checking coil resonance with signal generator; (b) checking coil resonance with a grid-dip oscillator.

most likely source of trouble. Unfortunately, the detector components are usually enclosed in a shield can and are not readily accessible for servicing. Usually, however, tests can be made from external terminals by referring to the schematic. A good diode should have a high back-to-front resistance ratio, measuring at least 1 MΩ in one direction and less than 500 Ω with meter leads reversed. When in doubt, replace the diode, being sure to observe the correct polarity. Almost any type of diode can be used as a replacement.

IF there is loss of fine detail, excessive contrast, or pix smearing, check the load resistor for an increased value. An *open* resistor is a rarity but is still possible. Check for an open peaking coil when there is loss of pix. An open coil is fairly common, usually from abuse while probing. Normal resistance of a PC is usually around 5 Ω. If it measures 1 kΩ or more, the coil is open and a shunting resistor accounts for the reading. The quickest check for an open coil is to short it out.

When receiver operation is restored, reinstall and ground the shield can.

• **Video Ringing**

Video ringing can be caused by misalignment and video overload. Where a crispening control is provided, adjust for minimum peaking. Check for open resistors across the peaking coils in the detector and video amplifier stages. Try lower-value resistors. Check alignment of the IF strip making sure the 45.75-MHz marker is not too high on the response.

13-28 TROUBLESHOOTING THE SYNCHRONOUS DETECTOR

The detector is suspect when signal tracing or injection shows little or no output. To verify, try substitution using a simple diode arrangement or a demodulator probe connected between input and output, noting any improvement in set operation.

Since all functions of this type of circuit are performed by ICs, testing is limited to checking the applied dc voltage and the input–output signals. Verify the 45.75-MHz signal from the last IF stage is applied to the IC. Before considering replacement of the IC, check all external components that tie in with the detector function. Refer to the schematic. The tuned circuit at the reference amplifier (oscillator) output deserves special attention. Its adjustment is extremely *important* and *critical* and should never be changed without good reason, and then only with the proper equipment and a full understanding of the procedure.

REMINDER:

This coil establishes the frequency and phase of the reference signal, which synchronizes and triggers the conductance of the demodulator transistors. If the signal is lost or of the wrong frequency and phase, the result is loss of pix or a pix of poor quality and contrast.

The coil is normally tuned to exactly 45.75 MHz, the pix IF frequency, assuming precise and accurate alignment of the IF strip. If the IF strip was aligned with a generator of questionable accuracy, the set must be realigned or the resonance of the reference coil must be changed to correspond. Generally, however, alignment is not considered until all testing has been exhausted and no defect can be found.

To check for a misadjusted reference coil, try adjusting the fine tuner control while checking for a restored pix or any improvement in pix quality.

EXPLANATION:

Changes in tuner oscillator frequency result in corresponding changes in the pix IF supplied to the IF strip *and the reference amplifier*, and a normal pix is possible *only* when the IF and reference frequencies are in agreement. Alignment procedures are described in Sec. 13-44.

13-29 IF ALIGNMENT

There are *many* trouble symptoms, too numerous to be described at this time, that *may be* attributed to misalignment of the IF strip. These same symptoms, however, *can be* caused by troubles other than alignment. Because of this, alignment should never be considered until all other possibilities have been exhausted and all known defects have been cleared up. Even when misalignment is suspected, don't be hasty in making adjustments until the need has been verified and the alignment conditions evaluated.

The fact is that the IF strip, like the tuner, *seldom if ever* requires realignment except where adjustments have been tampered with, the wiring disturbed, or components close to the tuned circuits replaced.

Alignment, when required, calls for considerable skill and know-how in addition to special equipment. It should never be undertaken lightly except after much thought. The surest way to compound a problem is to make adjustments when not needed or by turning a few screws in a hit-or-miss manner. In most cases only a touch-up alignment is needed, but unless performed correctly and accurately, what should be a simple job can quickly turn into a major project, with the possibility that the receiver performance will become worse instead of improved.

13-30 ALIGNMENT EVALUATION

There are two methods of alignment; each method has its variations and calls for different test equipment. The same equipment and setup is used for both *evaluation* and actual alignment when required. One method uses an accurate RF signal generator and an EVM. Another method is *sweep alignment* that requires a sweep generator, a marker generator, and a CRO.

526 VIDEO IF AND DETECTOR STAGES

Another method makes use of bar-sweep patterns (see Fig. 3-9). The variations of each method are (1) stage-by-stage alignment with the generator signal fed to the stage input with the indicator (EVM or CRO) at the stage output; (2) with the indicator at detector output and injecting the signal stage by stage, working progressively toward the tuner; (3) *overall* IF alignment with the indicator at the detector and the signal injected at the mixer input; and (4) overall *RF/IF* alignment with the signal fed to the antenna input terminals.

There are two important considerations to the proper alignment of a set: the accuracy of the signal generator and the skill of the technician. Unfortunately, many service-type generators are not 100% accurate and their calibration is subject to change over a period of time or to drift during warm-up. Such differences between one generator and another explain why a receiver aligned in one shop may appear to be misaligned (when it may not be) when checked by someone else using different equipment.

Fortunately, a receiver may still perform well despite a small frequency error, provided the alignment is done properly. In other words, there is no hard and fast rule that says the sound and pix IF carrier frequencies must be 41.25 MHz and 45.75 MHz as prescribed. In fact, sometimes it is expedient to deliberately choose frequencies as much as 1 MHz off to overcome stubborn interference that cannot be eliminated in any other way. In such cases the "error" must be *consistent* in the adjustment of *all* tanks and traps. An example showing how normal reception is possible with incorrect marker frequencies is shown in Fig. 13-14.

The tuned circuits involved in the alignment procedures to be described include all tank circuits, including the mixer output link coil on the tuner, the UHF-IF output coil on the UHF tuner, and AFT coils when AFT is provided; all traps in the IF strip, including the sound takeoff coil in a color set; all 4.5-MHz traps, including the sound takeoff in a BW set; and coils associated with a synchronous detector when used.

13-31 PRELIMINARY REMARKS AND PROCEDURES

1. The first step in alignment evaluation is to examine the schematic and the chassis to identify

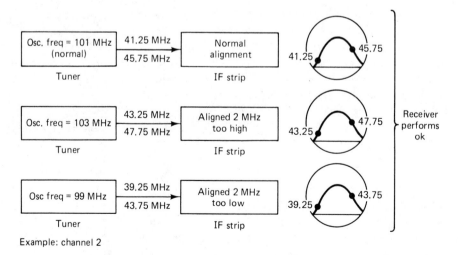

Fig. 13-14 Alignment with the wrong marker frequencies.

13-32 Alignment Evaluation with Signal Generator and EVM

and locate each tuned circuit and note its assigned frequency. To avoid a mixup, it is a good idea to make a sketch or mark them in some manner. Note whether the set employs staggered tuning or overcoupled transformers tuned to one frequency.

2. To prevent interference that may affect an EVM indication or CRO display, disconnect the antenna and short out the terminals; switch to a blank channel and/or disable the tuner oscillator; disable the horizontal output and HV; make sure all cables, instruments, and the chassis are well grounded (a metal-topped bench sometimes helps); turn off any fluorescent lights in the area.

3. One weak stage can seriously affect alignment making it impossible to obtain a normal overall response. Eliminate this possibility before proceeding.

4. If it is a hot-chassis receiver, use an isolation transformer.

5. Use a small blocking capacitor in series with the generator lead to avoid dc shorts and possible burnout of the generator attenuator control.

6. Clamp the AGC-IF bus to the no-signal voltage and polarity shown on the receiver schematic. This is necessary to ensure that indications are dependent *only* on alignment conditions and not influenced by AGC action. The clamp voltage is *critical* and has a bearing on the overall response for different signal levels.

7. To prevent signal overdrive and misleading indications, use the lowest possible generator output, and the lowest practical meter range or highest level of CRO gain.

8. *Regeneration and oscillation:* Misalignment can be either the *cause* or the *result* of oscillations. Try to correct the problem prior to attempting alignment. Oscillation often occurs when the bandpass is narrowed, causing excessive amplification, and when stagger-tuned coils are adjusted too close to the same frequency.

13-32 ALIGNMENT EVALUATION WITH SIGNAL GENERATOR AND EVM

NOTE: _____

This method is not recommended for *color sets*

where the *shape* and *bandwidth* of the response and the *marker locations* are of major importance.

• **Method 1: Stage-by-Stage Checking**

1. Identify all tuned circuits and their assigned frequencies as described earlier.

2. Connect the generator and meter to each of the IF stages individually, the same as for stage checking by injection (see Fig. 4-[a]). To prevent detuning effects introduced by the generator cable, the signal should be injected at least one stage ahead of the circuit under test. For the same reason, the EVM should be connected to the stage following the circuit under test. A *demodulator probe* (see Fig. 4-3) must be used with the indicator for all stages ahead of the video detector. With the meter at the detector output or beyond, use a direct connection or a low-capacity probe with the meter set for the lowest practical ac or dc voltage range.

3. Slowly tune the generator through frequencies assigned to the tank and any traps in the stage being checked. The meter should *peak* at the tank frequency and *dip sharply* at the trap frequency. If necessary, switch to a lower meter range and reduce generator output to obtain a *distinct* indication. Compare generator frequency with the specified coil frequency. If there is a discrepancy, either the generator calibration is in error or the coils need adjusting. Do *not* make adjustments at this time.

4. Check the resonance of each tank and trap of each stage in the same manner. To check the mixer *link* coil on the VHF tuner, inject signal at the mixer input test point or the UHF-IF input jack on the VHF tuner with tuner set for UHF reception. To check the UHF-IF coil on the UHF tuner, feed signal to the UHF antenna terminals with the meter connected to the output cable of the VHF tuner. To check the resonance of the 4.5-MHz traps, inject signal at the detector input with the meter at some convenient point in the video amplifier.

• **Method 2: Using a Grid-Dip Oscillator**

This method is not too accurate but can be useful as a backup for preliminary checking when a set is badly misaligned. It can only be used where the coils are exposed. With the receiver turned off, hold the instrument with its coil close to the coil under test. Tune the GDO for a dip (null) on its indicator. For best accuracy, repeat, using only sufficient coupling to obtain an indication. Compare the GDO dial reading with the frequency assigned to the coil under test.

- **Method 3: Stage-by-Stage with Meter at Detector Output**

   1. Using a direct connection or a low-capacity probe, connect the meter to detector output. To check the resonance of the detector input coil, inject generator signal to the input of the last IF stage. Tune the generator for a peak or dip on the meter as with the previous methods. Compare frequencies.

   2. Check the resonance of coil(s) in the last IF stage by injecting signal at the input of the upstream stage. Repeat for each stage, working toward the tuner.

- **Method 4: Checking the Overall IF Response**

With meter at the detector output, inject signal at the mixer input. *Slowly* tune the generator over the IF passband, noting each peak and dip indication, while relating them to the various tank and trap circuits shown on the schematic.

Plotting a Response Curve. If desired, a curve can be plotted while performing the overall alignment check. Using squared graph paper or the equivalent, plot amplitude against frequency with dots for each peak and dip indication for different frequencies taken from the generator tuning dial. Join the dots to form a curve, as in Fig. 13-15. Note the marker locations of the important frequencies: pix IF carrier, sound IF carrier, color subcarrier and its sideband limits, and the adjacent channel sound and pix IF carriers. The precise location of all markers is of great importance.

Checking Resonance of the 4.5-MHz Traps. Resonances of the 4.5-MHz traps in the detector and/or video amplifier section are checked individually as in method 1. See Fig. 13-16. Connect the meter downstream from each stage containing a trap, and inject signal at the preceding stage upstream. Tune the generator to obtain the expected dip at 4.5 MHz.

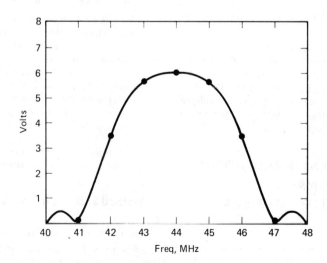

Fig. 13-15 Plotting a response curve.

13-34 Alignment Equipment Particulars **529**

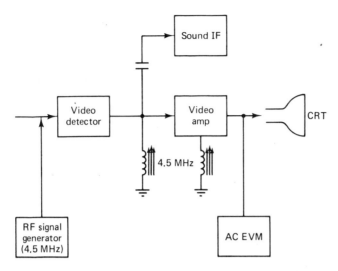

Fig. 13-16 A 4.5-MHz trap alignment setup.

REMINDER:

The 4.5-MHz frequency must be 100% accurate.

13-33 ALIGNMENT EVALUATION USING SWEEP EQUIPMENT

This is the preferred method for both BW and color sets because it provides a *visual* indication of alignment conditions, a response curve that shows bandwidth, the *shape* of the curve, and the relative location of all marker frequencies. The alignment equipment is connected as in Fig. 12-23 with the CRO as the indicator and the signal injected at some point in the IF strip, depending on the procedure used.

13-34 ALIGNMENT EQUIPMENT PARTICULARS

Test equipment required for the sweep method of alignment is described next.

• **Sweep Generator**

Unlike a conventional RF signal generator, which produces either a CW signal or an amplitude-modulated signal, a sweep generator develops an RF signal that is *frequency modulated*. With this method of alignment, an *FM* signal is required to sweep the *band of frequencies* represented by the bandpass of the IF strip.

The RF output frequency of a sweep generator varies continuously above and below its carrier or center resting frequency. The sweep rate is 60 Hz. The amount of frequency deviation is established by a *sweep control*, which is roughly calibrated from zero to about 20 MHz. As shown in Fig. 12-23, a sampling of the 60-Hz sweep voltage is fed to the CRO for horizontal deflection of its beam trace. The generator output *must be linear* with a constant output within the limits of its sweep.

Most generators have a *phasing control* for synchronizing the generator sweep with the horizontal retrace of the CRO. Some generators also have a *blanking control*. When switched on, it provides a zero reference base line for the IF response curve. Quality generators have built-in crystal markers for accurately identifying frequency points on a response curve. Alignment depends on the accuracy of *marker signals*; therefore, the calibration accuracy of the sweep generator is not too important.

• Marker Signals

The response curve developed by the sweep generator provides no information on alignment conditions unless the frequencies represented by certain points on the curve are accurately identified with markers. There are different ways of producing such markers.

Marker Generator. Any RF generator can be used for this purpose provided it is accurately calibrated. It is used with the modulation switched off. Connected as shown in Fig. 12-23, its CW signal beats with the signal from the sweep generator to produce a small *blip* on the response curve. The blip can be moved to any desired point on the curve by changing the generator dial setting. Some generators have built-in crystals for high accuracy at the more commonly required marker frequencies. The crystals may be the source of marker signals or a means of calibrating the tuning dial. For convenience, sweep and marker generators are often combined in one instrument.

Inexpensive generators are subject to frequency drift, especially during warm-up, and their calibration should be checked periodically by comparison with a generator of known accuracy or a crystal-controlled *multivibrator*.

Post-injection or Intensity Marker. This system uses a *marker adder*, which is supplied with signals from both the sweep generator and the marker generator. The marker signal from the adder connects to the Z axis of the CRO as shown in Fig. 12-23b. The marker shows up on the curve as either a bright *dot* or a short *gap* in the curve. Note that the marker signal does not pass through the IF strip, eliminating the possibility of signal overload.

Absorption Marker. This method uses a variable, accurately calibrated absorption-type trap connected between the sweep generator and the IF strip. It is essentially a passive system that produces a small marker *dip* on the response curve for whatever frequency it is tuned to.

Multiple Markers. Any of the preceding methods can be used to simultaneously produce two or more markers for monitoring changes in a response curve during alignment. Some generators provide separate crystals for this purpose.

• Oscilloscope

Almost any CRO can be used for alignment provided it has good low-frequency response. A *blanking control*, when provided, can be used to blank out the horizontal retrace when desired. When used for sweep alignment, the internal deflection oscillator is switched off, and voltage for the horizontal trace is obtained from the sweep generator.

13-35 SWEEP FUNDAMENTALS

The equipment is connected to the IF strip as for tuner alignment (Fig. 12-23). With the 60-Hz sweep voltage from the sweep generator producing horizontal deflection of the CRO, the timing of the beam trace coincides with frequency modulation of the generator. The width of the trace is adjusted with the CRO width control. The center of the trace corresponds with the carrier frequency being developed by the generator. The setting of the sweep control determines the frequencies represented by the left and right extremes of the trace. If the generator is tuned to 40 MHz, for example, and the sweep is set for 10 MHz, the trace limits correspond to 35 and 45 MHz. If the sweep is changed to 20 MHz, the trace represents a variation between 30 and 50 MHz.

When checking alignment, the amount of sweep used depends on the bandwidth of the circuit under test. The greater the bandwidth, the more sweep that is required. If the bandwidth is 2 MHz, for example, then a sweep of that amount would reproduce a curve, but only the portion above the 50% level. To view the entire curve where the response extends to 5 MHz, for example, at the base line requires a sweep of 5 MHz. Too much sweep covers *more* than the frequencies represented by the bandwidth, giving the curve a narrow, compressed appearance. In practice, a sweep of slightly more than the IF bandwidth is used.

The band of frequencies developed by the sweep generator is amplified in the IF strip in accordance with the resonant frequencies of the various tank and trap circuits. With the output of the IF strip feeding the vertical input of the CRO, the beam is made to deflect vertically, tracing a pattern whose amplitude varies with the voltages represented by the various tuned circuits.

13-36 OBTAINING A RESPONSE CURVE

Obtaining a curve of the proper proportions and other considerations is an important *first step* in evaluating conditions of alignment. This requires the proper adjustment of various controls on the different instruments. Misadjustment of the controls can result in inaccurate and misleading indications due to signal overload and other factors.

Do not connect the marker generator at this time. With the equipment set up as in Fig. 12-23, proceed as follows:

1. Disable the internal sweep of the CRO. Adjust the horizontal gain to obtain a line that barely extends across the tube face. Center the line as necessary.

2. Set the sweep control on the sweep generator to approximately 8 MHz. Adjust the generator attenuator for a moderate output level. Tune the generator to approximately 43 MHz, the center of the IF bandpass, or until a response is observed on the scope. Tune the generator to center the response on the tube face. Adjust the generator output and the vertical gain of the CRO until *all* the response can be viewed within the vertical limits of the tube. *Note*: Too much vertical scope gain results in an unstable response with the possibility of hum pickup. Too much generator signal will overload the IF stages and distort the curve. Where distortion is noted, attenuate the generator signal until it clears up. Increase the scope gain as necessary.

Figure 13-17 shows a number of abnormal displays that may be obtained. Figure 13-17a and b represent curves of normal proportions. Depending on signal polarity and the CRO, the curve may appear either upright or inverted. Either display may be used. Depending on the set, the curve may be flat-topped or have a "haystack" appearance as in Fig. 13-17c.

In Fig. 13-17d the curve is off center to the left and only part of the response is visible. Recenter the display with the sweep generator tuning control. In part (e), the curve is off center in the opposite direction. In part (f), there is either too much generator signal or the CRO vertical gain is set too high. In part (g), there is the opposite condition of either too little signal or not enough scope gain. It may also indicate a weak IF stage. In part (h) the curve is flattened, probably from signal overdrive. Try reducing the signal level. In part (i) there are two identical curves, one produced during the CRO retrace interval. Either eliminate the retrace curve with the CRO blanking control or adjust the phasing control on the sweep generator until the two curves overlap. Figures 13-17j and k represent the same condition but by different amounts of displacement. In part (l) there is not enough sweep for the IF bandpass and only the top por-

532 VIDEO IF AND DETECTOR STAGES

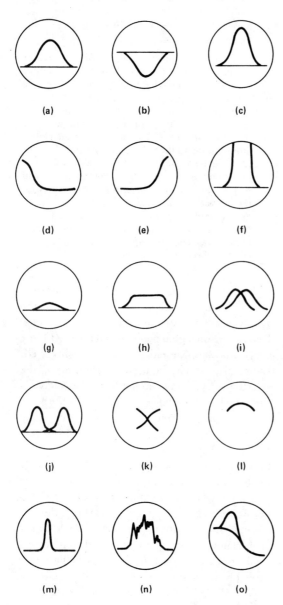

Fig. 13-17 Abnormal scope displays.

tion of the curve is being swept and displayed. It probably indicates excessive IF bandwidth. Increase the amount of generator sweep. Part (m) shows the opposite condition, either too much generator sweep is used or there is insufficient bandwidth of the IF strip. Reduce the sweep.

Figure 13-17n indicates interference or possibly oscillations. Figure 13-17o shows hum modulation of the display. Make sure the test equipment and cables are grounded and *bonded* if necessary. Try reducing the scope gain and compensating with a stronger signal.

NOTE: _____

The *appearance* of the curve is not indicative of bandwidth, since it can be altered by the relative adjustment of the CRO vertical and horizontal gain controls. A narrow-looking curve, for example, can be obtained with a wide bandpass, and what appears as a wide curve may actually represent a narrow bandpass. Also, the curve as it now stands gives no indication of alignment conditions. That can only be determined with a marker signal.

13-37 ALIGNMENT CHECKING WITH MARKER SIGNALS

To evaluate the overall IF response, loosely couple the marker signal to some convenient point at the tuner or IF strip input. Obtaining a good marker is sometimes a problem, being either too weak to be seen (especially on the slopes of the curve) or so strong it distorts the curve. The object is to obtain a sharply defined marker that does not affect the shape of the curve. Experiment with different signal levels and methods of injection. Try clipping the hot lead of the marker generator cable to the insulation of the sweep generator lead. Another method that often works well is to connect both the hot lead and the ground lead of the marker generator to two chassis points at opposite ends of the IF strip. This is called *ground loop injection*. With this method, a marker signal voltage is developed across the chassis resistance (which is high for RF) and simultaneously introduced into all IF stages. If the marker is too broad, try connecting a 0.01-mF capacitor across the vertical input of the CRO.

13-37 Alignment Checking with Marker Signals

When a good marker is obtained, tune generator to approximately 43.5 MHz. The marker should appear at the top center of the curve. Disregard any calibration difference between the sweep and marker generators. Slowly tune the marker generator, noting which way the marker moves on the curve. Determine which are the low- and high-frequency sides of the response.

• **Spurious Markers**

Spurious markers sometimes appear on the curve as a result of interference from a number of possible sources and are sometimes mistaken for the desired markers. Such markers, however, remain *stationary*. Try switching to another blank channel. Disable the horizontal output, HV, and tuner oscillator. Turn off any fluorescent lights in the area. If marker does not affect the shape of the response, it can usually be ignored.

• **Marker Locations**

Alternately tune the marker generator to each of the following frequencies, noting their marker positions on the response curve: 39.75, 41.25, 41.67, 42.17, 42.67, 45.75, and 47.25 MHz. Normal marker locations for receivers having different IF characteristics are shown in Fig. 13-3. The precise location of *each* marker is crucial to proper receiver performance.

Figure 13-3a represents the response for a typical small-screen receiver. Since the finest pix detail (corresponding with high video frequencies; 3 to 4 MHz) cannot be seen on a small screen, a wide bandpass is unnecessary. The narrowed response gives high amplification with fewer stages. Note the location of the pix carrier for maximum gain with a crisp pix.

Figure 13-3b shows the wideband, flat-topped response for a typical large-screen BW set and color sets using the wideband system. Note the location of the 41.25-MHz sound carrier, lower on the response than in Fig. 13-3a. Note the locations of the pix IF carrier and the color subcarrier at opposite sides of the curve at the 50% level.

Figure 13-3c shows a haystack response used for most modern color sets. This is the narrowband color system as indicated between opposite sides of the curve at the 50% level. With the pix carrier and color subcarrier located at this level, however, set performance would be poor. Note the *proper* location of these two markers, farther down on the curve, corresponding with the 50% level in Fig. 13-3b.

• **Wide- or Narrow-band Color?**

Before conditions can be evaluated or alignment performed, it is necessary to determine which of the two systems is used. Consult the receiver service notes which sometimes show the expected response. In any case it will soon become apparent since it is almost impossible to align a narrowband receiver for a wide bandpass while retaining the proper marker locations, or to align a wideband receiver for a narrowed bandpass. With *both* systems the pix carrier and the color subcarrier are *exactly opposite* on both sides of the curve *regardless of what level*. If, when evaluating the response, the curve appears flat-topped with both markers at the 50% level, chances are the wideband system is used. If the curve is more rounded with both markers exactly opposite but *below* the 50% level, the set uses the narrowband system.

• **Additional Marker Considerations**

Each marker must be treated individually, although the relative positions of several markers may be altered at the same time by changes in bandwidth and shape of the response. Problems resulting from mislocated markers are described next.

**39.75-MHz Adjacent Channel Pix IF Carrier. With the 39.75-MHz traps properly adjusted, this marker should appear at the base line beyond the response limits where station signals of this frequency receive zero gain. Adjustment of the traps influences the bandwidth and drop-off of the low-frequency skirt. If the trap(s) is tuned too *low*, there may be interference from the adjacent higher channel. If tuned too *high*, the bandwidth will be reduced.

47.25-MHz Adjacent Channel Sound IF Carrier. The marker for this frequency should appear at the base line beyond the high-frequency limits of the curve where there is zero amplification. The adjustment of the 47.25-MHz trap(s) influences the bandwidth and skirt drop-off on the high side of the response and changes the location of the pix IF marker. If the trap(s) is tuned too *high*, there may be interference from the adjacent lower channel. If tuned too *low*, the bandwidth will be reduced and the pix marker will drop toward the base line. The result is reduced pix contrast with smearing.

45.75-MHz Pix IF Carrier. The marker for this frequency should be on a linear portion of the curve at the 50% level to compensate for vestigial sideband modulation. Its relative position is influenced mainly by tanks and traps resonant in the 47.25- to 43-MHz range. If the marker location is too *high* on the curve, it can cause excessive pix contrast, video overload, and video ringing. If too *low* on the curve, the result is poor pix contrast, loss of fine pix detail, and poor sync. An exception is where the marker may be high on the curve in the case of small-screen receivers.

41.25-MHz Sound IF Carrier. For a BW receiver, this marker is normally located at the approximate 5% level on the low-frequency side of the response. Its relative location is determined by tanks and traps resonant in the 39.75-to 42-MHz range, the adjustment of 41.25-MHz traps in particular. For a color set, the 41.25-MHz marker *must be* at the base line for zero gain. If the marker is too *high* on the curve, it causes buzz in the sound, a 920-kHz beat pattern on the pix of a color set, and sound bars on the screen.

42.17-MHz Color Subcarrier. The location of this marker, at the 50% level on the low-frequency side of the curve, is *critical* for good color reception. It must be midway on a *linear* portion of the curve. Its location is determined by the adjustment of the same tanks and traps used for positioning the 41.25-MHz sound marker.

REMINDER: _____

When a color set is aligned with a haystack response, the pix IF carrier and the color subcarrier are exactly opposite but at a lower level than sets using broadband color.

41.67- and 42.67-MHz Color Sideband Markers. For uniform amplification of both upper and lower color sidebands, these markers must be located on a linear portion of the response. If the color subcarrier marker is too *high* on the curve, the upper sideband will receive too much amplification, resulting in poor color with smearing. If the subcarrier marker is too *low* on the curve, the lower sideband will receive little or no amplification, causing weak color of poor quality.

To summarize, markers correspond with specific carrier and sideband frequencies. The marker locations are influenced by the shape and bandwidth of the overall IF response. Shape and bandwidth in turn are determined by the markers. The slope of the skirt and the location of markers on the high-frequency side of the response determine pix contrast and the relative gain of low and high video frequencies and sync. The slope of the skirt and location of markers on the low frequency side of the response affect color intensity and quality and the sound.

• **Sweep Checking of Individual IF Stages**

The service notes for some receivers include response curves for each stage with markers to identify specific points on the curves. Inject the sweep signal ahead of each stage under test. Connect the CRO using a demodulator probe to the stage immediately downstream from the stage under test. Check marker locations on the curves obtained for each stage. Typical stage responses and how they combine to obtain the overall IF responses are shown in Fig. 13-4. Checking the overall response is a final step with this procedure.

• **Sweep Checking the Overall RF/IF Response**

With the CRO at the detector output and the sweep generator connected to the antenna terminals, tune the sweep generator to the center frequency of some selected channel. Set the sweep control for a sweep of about 10 MHz. The curve obtained should be similar to the overall IF response described earlier. Inject the marker signal to the IF strip input. Check for normal IF marker locations and the IF bandpass as for the overall IF response.

13-38 ALIGNMENT PRELIMINARIES

Once the conditions of alignment have been correctly evaluated, the actual alignment of a set is not too difficult. But first, consider the following:

1. Check the equipment setup for the method to be used. Verify adequate grounding of all items. Tune the receiver to a blank channel. Where interference is present, take necessary precautions as described earlier. With no signal applied, there should be little or no indication on the EVM or CRO.

2. Where only a touch-up alignment of certain coils is anticipated, keep track of the number of twists made on each slug and the direction they are turned so they can be quickly restored to their original settings if need be.

3. Use the proper alignment tool, usually a nylon hex "diddle stick," *never* an Allen wrench or a metal screwdriver that upsets the tuning and may damage the slugs.

4. A tuning wand as described in Sec. 12-32 can be useful where the coils are exposed.

5. Make sure all tanks and traps have been correctly identified along with their assigned frequencies. There is no room for error. Attempting to adjust a coil to the wrong frequency will seriously degrade receiver performance.

6. Common causes when a coil will not resonate to its assigned frequency are that the coil has been wrongly identified; the slug is broken, missing, or the threads are stripped; there are shorted coil windings or its shunt capacitor; there is loading caused by a leaky capacitor or transistor in an adjoining stage.

7. Tanks are adjusted for a peak indication; traps are dipped. When a meter is the indicator, the direction of pointer deflection may be either forward or reverse depending on the polarity of connections and the detector output polarity. For a backward indication, the tanks are dipped and the traps are peaked.

8. The indicator being used (whether an EVM or CRO) must be sensitive to the slightest change in coil adjustment. Use a low meter range and the CRO set for high gain. Be precise with each adjustment. Turning a slug as little as one half-turn in either direction may seriously affect the pix, sound, or color.

9. Use low output from a generator when adjusting tank circuits and *high* output for critically adjusting the traps.

10. Traps are high Q and their adjustment should be critical. Tanks are less critical to adjust. A critical *tank* often indicates regeneration or oscillations. Another indication is when the meter suddenly pegs while making adjustments. Adjust the slug until the circuit becomes stabilized; then readjust later as alignment progresses.

11. *Reminder:* Where single tuned coils are used

536 VIDEO IF AND DETECTOR STAGES

between IF stages (impedance coupling), the frequencies are *staggered*. With *overcoupled transformers*, they may be stagger-tuned or all tuned to the same frequency.

12. The IF tuned circuits are normally adjusted in sequence as follows: first the adjacent channel traps and the 41.25-MHz sound traps, then the tank circuits starting with the mixer output link coil, then the UHF-IF coil and AFT tuned circuits if necessary. The 4.5-MHz traps are not part of the IF strip and can be adjusted at any time.

13. When circuits are adjusted stage by stage using any method, it is advisable to make a follow-up check on the overall response using sweep equipment. Refer to response specified in receiver service notes if available.

14. When evaluating alignment, if all tuned circuits are far off frequency and adjustments have been tampered with, a complete realignment may be necessary. When symptoms are comparatively minor, a simple touch-up of certain coils is probably all that is required. When all tuned circuits are found to be off frequency by the same amount and in the same direction, don't be hasty in making adjustments. The calibration of the generator being used probably differs somewhat from the one used previously to align the set. If that is the case, realignment may not improve receiver performance significantly.

15. When using the sweep alignment method, it is desirable to have two or more marker generators for simultaneously monitoring changes at different points of the response curve. When calibrations of the marker sources are in disagreement, use the marker of known accuracy as a standard. If there is any frequency error, it must at least be *consistent* for each marker.

16. With sweep alignment, periodic readjustment of the sweep control may be necessary as alignment progresses. *Reminder*: With too much sweep the CRO display goes beyond the IF bandpass, giving the display a compressed appearance, If there is too little sweep, compared to the bandpass, the frequency deviation each side of the carrier will be small, limiting the display to the upper portion of the response, as in Fig. 13-17 l.

17. The overall IF bandpass is represented by the combined resonances of *all* tuned circuits in the IF strip. Because of this, the adjustment of *each* coil has a bearing on the overall response.

Therefore, unless each stage contributes its normal share of amplification, it will be impossible to obtain a proper response. Generally, *all* coils must be alternately adjusted at least *two or three times*.

13-39 ALIGNMENT PROCEDURES

When it has been determined that realignment is truly necessary, perform alignment using one or more of the procedures described in the following paragraphs. The procedures are of a *general* nature. For specifics on a given set, refer to the receiver service notes.

13-40 ALIGNMENT WITH GENERATOR AND EVM

• **Method 1: Stage-By-Stage**

1. Connect the generator and meter as described for alignment evaluation.

2. Verify changes in meter reading as generator output and frequency are varied.

3. Start with stages having traps. With the generator set for a fairly high output and tuned to the proper trap frequency, adjust the trap for lowest "dip" (null) on the meter.

4. Reduce generator output. Tune the generator to the specified frequency of the tank in the stage being aligned. Peak the coil for maximum indication on the meter.

5. Perform a repeat touch-up adjustment of both the tank and trap at least one more time. This is necessary because of interaction between the coils.

6. Change generator and meter connections to another stage. Adjust tanks and traps using the same procedures as above.

• **Method 2: Rough Alignment with a GDO**

With set off, position the GDO coil close to the

coil under test. Tune the GDO to the specified coil frequency. Adjust the coil slug for lowest dip on the GDO indicator.

REMINDER:

Greatest accuracy is with minimum coupling between the coils.

- **Method 3: Stage by Stage with Meter at Detector Output**

As the signal is injected into each stage, starting with the last stage and working toward the tuner, adjust each tank and trap as in method 1.

- **Method 4: Overall IF Alignment**

1. Connect the generator and meter as for overall alignment evaluation.

2. Verify that the meter changes as generator output and frequency are varied.

3. Tune the generator slowly over the IF range. At each specified trap frequency, adjust appropriate trap(s) for dip on the meter.

4. Once again tune the generator over the IF bandpass. At each specified tank frequency, adjust for maximum meter indication starting with the mixer output link coil.

5. Repeat steps 3 and 4 at least twice.

6. To align the UHF-IF output coil on the UHF tuner, inject signal to the antenna terminals. Adjust the coil for maximum indication at the specified coil frequency.

- **Adjusting 4.5-MHz Traps**

Connect the generator and meter as for alignment evaluation. Tune the generator to *exactly* 4.5 MHz, and adjust for a fairly strong output. *Critically* adjust each 4.5-MHz trap for lowest dip on the meter. This includes the sound takeoff coil if it is a BW set and any additional traps in the video amplifier.

REMINDER:

Some frequency error can be tolerated for the IF tuned circuits provided the error is consistent in the adjustment of all coils; the source of the 4.5-MHz signal, however, must be close to 100% accurate. If an accurate generator is not available, adjust the traps by ear, the takeoff trap for loudest sound, and additional traps until the beat pattern disappears.

13-41 OVERALL ALIGNMENT USING SWEEP EQUIPMENT

1. Connect equipment as for overall alignment evaluation described earlier. Verify that the response obtained represents the actual conditions of alignment with no distortion caused by signal overdrive. Check the markers and see that they change position with changes in marker generator tuning. Adjust for a suitable marker that is visible on the slopes of the curve but not strong enough to cause distortion of the response. With the markers, determine the high- and low-frequency sides of the response.

2. Carefully consider all marker locations and compare with their normal positions as shown in Fig. 13-18a. *Reminder:* Marker positions depend on receiver requirements as determined by screen size, and if a color set, if it is designed for narrowband or wideband color. Does the center of the response coincide with the center of the desired bandpass? If not, there are two possibilities. Either all tuned circuits are tuned too *high* or too *low* (depending on which way the curve is off center), or it is a problem of bandwidth, where the drop-off of one side or the other is either too steep or too shallow. For the latter, a simple touch-up of one or two coil slugs may be all that is necessary.

3. If the set is drastically misaligned and *all* markers are too far on the low-frequency side of the response (Fig. 13-18c), adjust all coils (by turning their slugs inward) until all markers are at their approximate locations with the response centered at approximately 43.5 MHz. *Note*: Since marker frequencies are *fixed*, the object is not to move the markers, but to *move the curve* until the markers assume their proper positions. As the curve moves to one side or the other, it can be recentered on the CRO screen using the sweep generator tuning dial.

538 VIDEO IF AND DETECTOR STAGES

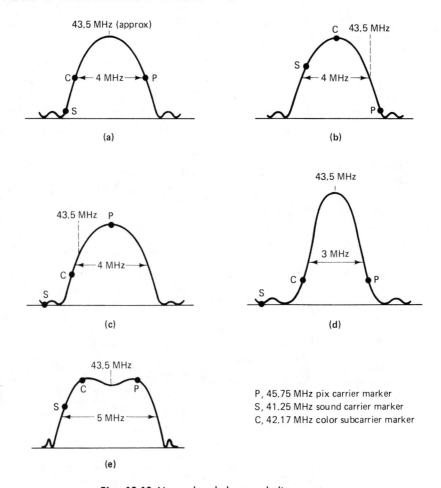

Fig. 13-18 Normal and abnormal alignment conditions: (a) normal response; (b) IF strip tuned too low (poor pix contrast, color smear, and sync buzz); (c) if strip tuned too high (excessive contrast, loss of fine detail, weak color and sound); (d) narrowed bandwidth (peaked response) (weak BW and color pix, no sound); (e) excessive bandwidth (weak pix with buzz).

4. If all markers are too far on the high-frequency side of the response (Fig. 13-18b), adjust all coils (by screwing their slugs out) until all markers are properly located with the response centered at 43.5 MHz.

5. Noting the exact location of all markers, the general shape of the curve, and the drop-off of the two skirts, arbitrarily choose which side of the curve to work on and the coils to start with. For the low-frequency side, coils having assigned frequencies in the 39- to 42-MHz range will have the greatest effect; for the high-frequency side, coils in the 43- to 47-MHz range will have the greatest effect in shaping the response.

6. For the low side of the curve, use markers of 39.75 MHz, 41.25 MHz, and if it is a color set, markers for the color subcarrier and its sideband limits. Using a trial-and-error approach, tentatively and alternately adjust the lowest-frequency coils by small amounts, noting the effect on the response. Keep track of the direction (cw or ccw) and the amount each slug is turned so they can be quickly restored to their original settings if there is no effect or if the desired results are not obtained. *Reminder*: The 39.75- and 41.25-MHz traps help shape the low side of the curve. As explained earlier, the precise locations of the sound IF marker and the color markers are especially important. If they are too high on the curve, they will drop down if traps are tuned to a higher frequency (slugs turned ccw). If too low on the curve, they will move up when the coils are tuned lower (slugs turned cw). Compare results with appropriate response shown in Fig. 13-18. *Reminder*: If the sound marker is too high on the curve, it produces buzz in the sound and a 920-kHz beat pattern on the pix. If the color subcarrier is too high, the 42.67-MHz sideband marker will be at the peak with too much amplification for the high-frequency sidebands. If the subcarrier is too low, the 41.67-MHz marker will be near the base line for loss of gain for those sidebands.

7. For the high side of the curve, use markers of 45.75 and 47.25 MHz. Alternately adjust the coils associated with the high frequencies, noting any changes in the response. *Reminder*: The 47.25-MHz trap helps to shape the curve on the high side of the response. The location of the 45.75-MHz marker is critical. Its proper location depends on receiver design (see Fig. 13-3). If too high on the curve, it will drop down when the coils are tuned lower (slugs screwed in). If too low on the curve, the marker will move up when coils are tuned higher (slugs screwed out).

8. Reexamine the low-frequency side of the response. Conditions have probably changed. Repeat steps 5 and 6. Reexamine the high-frequency side of the response where conditions may also have changed. Repeat step 7.

9. Check the bandwidth, the shape of the response, and the location of *all* markers and compare with the typical curves of Fig. 13-3.

10. Check the overall IF response with the VHF tuner switched to UHF and the signal injected at the UHF antenna terminals. If necessary, touch up the UHF-IF output coil.

11. Verify a normal overall RF/IF response as described earlier. Where final touch-up adjustments seem necessary, "tweak" the slugs by very *small amounts*; otherwise, conditions may worsen instead of improve.

12. Restore set operation, connect the antenna, and check performance on all channels.

13-42 ALIGNMENT OF LUMPED TUNED CIRCUITS

Where ICs are used in the IF strip, most of the tanks and traps are grouped together between the tuner and the IF input. The coils can be adjusted using either the generator–EVM method or with sweep equipment. Inject signal to the mixer input with the indicator (with demodulator probe) connected to the IC output. Correctly identify each coil and evaluate alignment conditions before making any adjustments.

Because of the interaction between the various coils, it is necessary to repeat each adjustment *several times* until further adjustments prove unnecessary. Final touch-up adjustments are critical and should be limited to fine tuning the coil slugs.

To adjust additional tuned circuits external to the ICs in later stages, move the EVM or CRO to the detector output. Where untuned plug-in filter units are used and there are problems obtaining the correct response, replace the unit.

13-43 BAR SWEEP ALIGNMENT

This method of alignment requires specialized equipment (see Fig. 3-9), and procedures are beyond the scope of this text. Refer to the test instrument instruction manual.

540 VIDEO IF AND DETECTOR STAGES

13-44 SYNCHRONOUS DETECTOR ALIGNMENT

There is usually only one tuned circuit (the 45.75-MHz reference coil) to adjust in a synchronous detector, but its adjustment is extremely critical. Since the operating conditions of the detector affect the overall IF response obtained with sweep alignment, the coil should be adjusted *before* aligning the IF strip, should that be necessary.

The symptoms of a misadjusted coil are loss of pix (and sound if a BW set) or a weak pix of poor quality. As with the IF strip, alignment is seldom if ever required unless tampered with. And since these same problems can be caused by defects elsewhere, don't be hasty in making adjustments without good reason.

Alignment of the demodulator with conventional sweep equipment is not too practical because of the need for an *accurate* and continuous 45.75-MHz signal, which a conventional sweep generator does not provide. At least one instrument is currently available, an all-purpose video analyzer (Fig. 3-9) that does have these capabilities.

In lieu of such special equipment, other methods can be used with varying degrees of accuracy. One method uses an accurate marker generator as the source of the required CW signal. It is used in conjunction with a regular sweep generator to obtain the overall response. With this procedure, the signal levels from both generators are quite critical and must be adjusted individually to obtain the desired results.

Another method calls for shunting the demodulator tuned circuit with a low-value resistor to disable the reference amplifier during alignment. The results are not too satisfactory, however, since two arms of the demodulator bridge are conducting at all times.

Another approach that does not require a generator makes use of the station signal and observation of the pix. For loss of pix, adjust the coil until a pix is obtained. A *rough* setting of the coil slug is midway between the two settings where the pix is lost. This normally represents about two full turns of the slug. To refine the adjustment, slowly turn the slug by small amounts in opposite directions while observing the pix. Adjust for maximum contrast and best pix quality. Check the results by switching channels and adjusting the fine tuner. If the pix drops out or the quality is impaired, make touch-up adjustments as required.

13-45 FRINGE-AREA PROBLEMS

Normally, good TV performance is only possible with a reasonably high signal-to-noise ratio as determined by such things as geographic location, the antenna system, local interference, and proper operation of the tuner. However, where reception is weak *on all channels* and snow on the pix is not excessive (and cable facilities are not available), certain modifications to the receiver can often prove beneficial. This includes aligning the IF strip for a *narrowed* bandpass.

Narrowing the bandwidth of the IF strip materially increases the gain and, in some cases, the signal-to-noise ratio. The trade-off is a reduction of fine pix detail and less-than-perfect color. However, where there is no alternative, a pix of impaired quality is better than no pix at all.

The IF strip is aligned for an overall response as shown in Fig. 13-3a, with the pix carrier located at or near the peak of the curve. Unavoidably, this places the color subcarrier and its sideband well down on the low-frequency side, reducing color intensity and, in extreme cases, eliminating color reception entirely.

Other things that can be tried to improve fringe-area reception include increasing the value of load resistors in the detector and amplifier stages, using a high-gain antenna with booster amplifiers, critically adjusting the AGC delay and threshold controls, disconnecting the AGC from the tuner, and altering the bias of IF transistors, as follows: Using a bias box, slightly increase or decrease the bias of each transistor to increase the amplification. Where any improvement is noted, change the value of voltage-dividing resistors as appropriate. Increasing the gain of the second and third IF stages may improve pix contrast, but, unfortunately, not the signal-to-noise ratio.

13-46 INTERMITTENTS

For intermittent problems in the IF-detector stages, monitor the AGC voltage and the signal at the detector output. To make the set act up, try tapping, applying heat, and so on, as described in Sec. 5-13. Check for poor soldering or cracked PC wiring by flexing the PC board and by probing. In hybrid sets, intermittent tubes and poor base-to-socket contacts are a common problem.

13-47 AIR CHECKING THE REPAIRED SET

After making repairs and adjustments, operate the set for an hour or so to make sure no problems have been overlooked and that none develop. Check reception on all channels. Critically evaluate such things as pix quality, color, sync stability, and sound. Check for trouble symptoms as listed in Table 13-1.

13-48 SUMMARY

1. The tuner converts the pix and sound RF carriers to pix and sound IF carriers, and passes them on to the IF strip where they are amplified along with their sideband frequencies. The IF strip provides most of the receiver gain, and its tuned circuits provide the necessary wide bandpass for amplifying all video frequencies up to about 4 MHz.

2. The amplitude-modulated pix IF carrier and the frequency-modulated sound IF carrier coexist in the IF strip up to the sound *takeoff*, the point of separation. The sound takeoff for a color set is *ahead* of the detector, usually the last IF stage. For a BW set, the sound takeoff is *after* the detector, at the detector output or beyond. Sound takeoff carrier frequencies are 41.25 MHz for a color set and 4.5 MHz for a BW receiver.

3. The IF carrier frequencies are standardized at 41.25 MHz for the sound IF carrier and 45.75 MHz for the pix IF carrier. The carriers retain their same 4.5-MHz separation as when transmitted by the station. The color IF subcarrier frequency is 42.17 MHz, which is 3.58 MHz lower than the 45.75-MHz pix IF carrier. The color subcarrier is *not transmitted*.

4. Both the 45.75-MHz BW carrier and its sidebands and the color sideband frequencies exist together within the 4-MHz bandpass of the IF strip. Both carriers have *upper*

and *lower* sidebands. The upper sideband of the pix carrier extends 4.2 MHz above the carrier. The lower sideband extends 0.75 MHz below the carrier. The upper color sideband extends up to 0.5 MHz above the subcarrier frequency; the lower sideband extends 1.5 MHz below the carrier.

5. Large-screen receivers require an IF bandpass of at least 4 MHz. Any reduction in bandwidth means the high video frequencies in the 3- to 4-MHz range will be lost and with them the fine pix details. For a color set, the loss of certain color frequencies results in loss of color or poor color reproduction. A small-screen receiver can get by with a narrower bandpass because loss of the highest video frequencies is not noticeable on the pix.

6. The sound signal goes through two frequency conversions, from RF to the 41.25-MHz sound carrier, and once again to a lower IF frequency of 4.5 MHz. The 4.5-MHz signal represents the beat produced by mixing the 41.25-MHz sound IF with the 45.75-MHz pix IF. With a BW set, the conversion takes place in the video detector. With a color set, it takes place in a *sound IF detector* stage.

7. With frequency conversion, the resultant beat signal retains the characteristics of the weaker of the two signals that are combined. Thus the 4.5-MHz signal is a frequency-modulated sound signal with little or no amplitude variations of the video signal.

8. Sound signals must be prevented from reaching the pix tube where they can produce sound bars and other interference on the pix. Pix signals must be kept out of the sound section where they can cause buzz from the speaker. This separation of signals is accomplished with tuned trap circuits.

9. An IF strip contains numerous tuned circuits consisting of *tanks* and *traps*, both operating together to determine the IF bandpass. The tanks resonate to frequencies within the IF passband; the traps, to frequencies beyond the bandpass. Traps tuned to 41.25 MHz limit the level of the sound IF signal to prevent buzz in the sound and beat interference on the pix. Other traps reject adjacent channel signals.

10. The mixing of the 4.5- and 3.58-MHz signals in a color set produces a 920-kHz beat signal that interferes with the pix. Since this frequency cannot be rejected with traps that would also reject video signals of that frequency, interference is minimized by drastically attenuating the 41.25- and 4.5-MHz frequencies with traps and proper alignment of the IF strip.

11. A response curve is a convenient way of representing the frequency response of the IF strip. Such a curve provides information on bandwidth; the locations of markers corresponding with the pix, sound, and color carriers and their sidebands; and the relative amplification of these frequencies.

12. The tuned circuits of the IF strip are aligned to obtain a particular overall response consistent with receiver design. In most cases, markers representing the pix carrier and the color subcarrier are located at the 50% level on opposite sides of the curve where their sidebands are amplified by varying amounts. The 41.25-MHz sound IF marker is located at or near the base line, where it receives very little amplification. Adjacent channel markers are located just beyond the response limits, where they receive no amplification.

13. A factor in determining bandwidth is the degree of coupling between the various IF stages. Some sets use *impedance coupling*, where single coils are stagger-tuned to obtain the desired overall bandpass. Other sets use *overcoupled* tuned transformers for a broadband response. The transformers may be stagger-tuned or all tuned to the same frequency, about 43 MHz, which corresponds with the center of the IF bandpass.

14. Since the IF strip is fixed tuned to accept a certain band of frequencies, it is up to the tuner to develop such frequencies. This requires a precisely accurate oscillator frequency as determined by the setting of the fine-tuner control. Assuming the IF strip is correctly aligned, the pix, sound, and color signals are all amplified by the correct amounts and the receiver performs as intended. As the fine tuner is varied in one direction or the other, *different* IF frequencies are developed which are *not compatible with* the IF alignment, adversely affecting the pix, color, and sound. With a color set, optimum setting of the fine tuner is for best color reception.

15. Vestigial sideband transmission by the station is compensated for at the receiver by locating the pix carrier at the 50% level of the overall response, where its sidebands are amplified by different amounts. The lower sideband, which was attenuated at the transmitter, receives maximum amplification where it extends to the upper portion of the curve, and the upper sideband receives less amplification where it drops toward the zero-level base line of the response.

16. The finest pix detail cannot be seen in color; therefore, some color sets limit the highest frequencies of the color sidebands by locating the color subcarrier marker at a lower level on the response. This is called narrowband color as opposed to broadband color, where all color frequencies receive greater amplification. Shortcomings of the narrowband system when reception is weak are reduced color intensity, unstable color sync, and in some instances loss of color.

17. In-phase feedback from output to input of a high-frequency amplifier causes regeneration, instability, and a narrowing of the IF bandpass. With strong feedback the circuit will oscillate. Oscillations can destroy transistors and produce many undesirable trouble symptoms. This is normally prevented by *neutralizing*, that is, feeding back some out-of-phase signal to counteract the regenerative feedback.

18. The purpose of the video detector is to demodulate the video IF carrier. This means extracting the sideband frequencies from the carrier. A basic detector uses a diode. With a BW set, the 4.5-MHz sound IF carrier is developed in the detector by the mixing of the 41.25-MHz and 45.75-MHz signals. With a color set, this occurs in the sound IF detector stage. The sound signal is *not demodulated* by either of these stages. Modern sets use a *synchronous detector*, which has many advantages over the simple diode circuit.

19. The *polarity* of the composite video signal is established by the detector. The detector output polarity may be either positive- or negative-going according to which way the diode is connected. Signal polarity must be considered up to the input of the pix tube. A reversed polarity produces a *negative* pix and loss of sync.

20. Loss of the high video frequencies in the 3- to 4-MHz range results in poor resolution of the fine pix detail. In the detector and subsequent amplifier stages, the high frequencies are given a boost by using low-value load resistors and the use of peaking coils that are self-resonant to about 3 MHz. Some sets provide a video peaking control that can be used to "crisp" a pix. Excessive peaking, however, results in *video ringing*, which produces ghostlike outlines on the fine detail of a pix.

21. A synchronous detector uses transistors that are triggered into conduction by a 45.75-MHz synchronizing signal. This assures the demodulation of only one signal, the 45.75-MHz pix IF carrier of the *desired* station. There are three functional sections to the detector, all contained within a single IC: the 45.75-MHz carrier generator (sometimes called a reference amplifier), a differential amplifier, and a four-transistor demodulator connected in a bridge circuit. Besides detecting the signal, it also amplifies. The most critical component is a tuned circuit that establishes the frequency and phase of the generated carrier signal(s).

22. Except for alignment, troubleshooting the IF section is no different than for other functional areas of the receiver. After observing the trouble symptoms, it is customary to check for voltage at the detector output and on the AGC bus. The presence or absence of these voltages provide good clues for localizing the trouble. Other things to check for are *snow* and whether the problem involves loss of pix, sound, or both.

23. Alignment of the IF strip is seldom required and should not be considered until all other possibilites have been checked out. Never adjust the coils indiscriminately in hopes of improving performance. Chances are you will end up with a major alignment job on your hands. Proper alignment requires the proper equipment and a thorough understanding of the procedures.

24. By careful observation and evaluation of the overall response curve and considering the symptoms, it is often possible to know which adjustments to make to correct an alignment problem. When a complete alignment is required, the normal sequence is (a) adjust all traps in the IF strip, (b) adjust all IF tank coils starting with the mixer link coil on the tuner and the IF input coil if there is one, (c) touch up the AFT tuned circuits if required, (d) peak the UHF-IF output coil on the UHF tuner if it is adjustable, and (e) adjust all 4.5-MHz traps.

25. The first tuned coil of the IF strip is *located on the tuner*. Its adjustment is considered part of the overall IF alignment.

REVIEW QUESTIONS

1. (a) Either too much or too little bias on an IF transistor will weaken or kill the signal. Explain why.

2. A receiver using a synchronous detector requires fewer traps in the IF strip and sometimes fewer stages than with a basic diode detector. Explain why.

3. Fill in the missing words: Under no-signal conditions the transistors in the IF strip are biased for maximum _____ and operate approximately midway between _____ and _____ .

4. (a) Why is it possible to measure the detector output signal with either an ac or a dc voltmeter?
(b) What reading would you expect (1) when switched to a blank channel? (2) when receiving a moderately strong signal? (3) if an IF stage is oscillating?

5. Where each of the three stages of an IF amplifier has a gain factor of 20 and the signal level at the mixer output is 1 mV, how much signal voltage would be measured at the detector output?

6. Why do most sets have one or more 4.5-MHz traps in the video amplifier section *beyond* the sound takeoff?

7. The value of coupling and by-pass capacitors in the IF strip is generally larger than in the tuner, but smaller than in the detector or video amplifier stages. Why?

8. Does IF alignment or the setting of the fine tuner affect the frequency of the 4.5-MHz sound IF carrier? Explain.

9. Each IF stage reverses the signal polarity; does this have any bearing on signal polarity at the CRT input? Explain.

10. (a) Explain why the quality of a color pix is affected by the location of the color subcarrier on the overall IF response.
(b) Why is color degraded if the bandwidth is too narrow?

11. Explain how vestigial sideband transmission is compensated for in the IF strip.

12. What changes take place in the video signal between the tuner, the IF strip, and the pix tube?

13. What changes take place in the sound signal between the tuner, the IF strip, and the speaker of a color set?

14. What is meant by pix and sound mistracking and why does it occur if the IF bandwidth is too narrow?

15. How is the gain of an IF stage affected if the transistor bias is (a) too high? (b) too low? Explain your answers.

16. What produces the best pix detail and color, a wide or narrow IF bandpass? Why?

17. What is the difference between a tunable ghost and a true ghost caused by multipath reception?

18. Why is the 41.25-MHz sound marker located at or near the base line of the overall IF response where it receives little or no amplification?

19. Explain how peaking coils and the use of a low-value load resistor in the detector help to preserve the high video frequencies.

20. Explain the difference between a low- or high-pass filter and a tuned trap circuit.

21. What is the tuner oscillator frequency for channel 5 when the IF strip is aligned with pix and sound IF carriers of 45 and 41 MHz, respectively? Would the set operate normally?

22. How is a pix affected by a reversed detector diode? Explain why.

23. (a) Name three factors that determine the polarity of the video signal as it reaches the pix tube.
(b) What is the polarity if signal is fed to the CRT cathode?

24. What is the frequency of the sound takeoff trap for (a) a BW set? (b) a color set?

25. (a) When the negative-going output of a video detector is direct coupled to the next stage, does that stage use a PNP or an NPN transistor?
(b) How would the collector current of the stage be affected if the detector diode were reversed? Explain.

26. Why is a BW set not troubled with 920-kHz beat interference?

27. Under what conditions is the IF strip operating at maximum sensitivity? When is the sensitivity lowest?

28. State two functions of a sound takeoff coil in a color receiver.

29. State three ways of obtaining a wide bandpass in the IF strip.

30. Tank coils in the IF strip sometimes have a tap connection. Give two possible reasons.

31. Why isn't a 920-kHz trap provided to prevent beat interference?

32. What are the two main differences between the IF strip of a BW set and a color set?

33. Explain how staggered tuning increases the bandwidth of the overall IF response.

34. The adjacent channel traps serve two purposes. (a) What are they? (b) Why aren't these frequencies trapped out in the tuner?

35. (a) Small screen sets do not require a wide IF bandpass. Explain why.
(b) Why does this reduce the number of IF stages that are required?

36. What are (a) the advantages and (b) the disadvantages of a narrow IF bandpass?

37. What determines the output polarity of a diode detector?

38. Why is the sound takeoff in a color set ahead of the detector?

39. Color sets usually have more 4.5-MHz traps than BW sets. Why?

40. Why is the video detector enclosed in a shield can?

41. (a) What is meant by video peaking?
(b) How is pix quality affected by excessive peaking?

42. Name two ways for increasing the "crispness" of a pix.

43. Are both the video and sound signals demodulated in the video detector?

44. What stage in a color set produces the 4.5-MHz sound signal?

45. What prevents (a) video getting into the sound section? (b) sound getting through the video amplifier stages?

46. (a) What is the difference between video peaking and video ringing?
(b) How does ringing show up (1) on the pix? (2) on a CRO?

47. (a) Why do most modern color sets use narrowband color compared to wideband color?

(b) How is this accomplished in the IF strip?

48. Adjacent channel traps are not always required when a receiver has a synchronous detector. Explain why.

49. What is a demodulator probe and why is it required when signal tracing stages ahead of the video detector?

50. Explain the purpose of the sound rejection control used in some receivers.

51. (a) How is the 42.17-MHz color subcarrier developed?
(b) What is the significance of the 41.67- and 42.67-MHz markers as observed on an overall IF response?

52. (a) Explain the operation of a synchronous detector.
(b) What are its advantages over a simple diode detector?

53. In the tuner the video carrier frequency is lower than the sound carrier; in the IF strip they are reversed. How did this occur?

54. (a) What is an overall IF response curve and how is it obtained?
(b) What purpose does it serve?

55. (a) Is 920-kHz beat interference peculiar to BW sets, color sets, or both?
(b) Explain how this beat frequency is produced and how the interference is prevented.

56. Compare the relative merits of wide- and narrowband color.

57. (a) It is common practice to feed AGC only to the second IF stage. With this arrangement, how is the first stage controlled?
(b) What is the purpose of doing it this way?

58. Why must the 41.25-MHz sound carrier be attenuated to a low level in the IF strip?

59. Proper adjustment of adjacent channel traps is necessary even for locations where there are no problems with interference. Explain why.

TROUBLESHOOTING QUESTIONNAIRE

1. Where would you check for loss of sync when a scope indicates pulses are (a) normal at detector output? (b) clipped at detector output? Explain your reasoning.

2. How is the AGC voltage affected by a (a) dead IF stage? (b) weak IF stage?

3. How does an inoperative video detector affect the *sound* of (a) a BW receiver? (b) a color set? Explain.

4. How does a weak third IF stage affect the gain of previous AGC controlled stages? Explain.

5. How can you check a detector diode without removing the shield can?

6. (a) State two common causes of pulse clipping.
(b) Clipping may be caused by troubles in the IF strip, but it usually *shows up* at some video amplifier stage beyond the detector. Explain.

548 VIDEO IF AND DETECTOR STAGES

7. A receiver has poor resolution of the fine pix detail and the off-channel raster appears blurred with no visible scanning lines. Would you check for loss of high video frequencies? Explain.

8. What would you suspect when a set has excessive pix contrast and poor sync, and the voltage at the detector output is about 8 V p-p on a moderate signal?

9. State three possible causes of video overload.

10. What is the effect on a color pix if the 42.17 MHz color subcarrier marker is: (a) Too low on the overall response curve? (b) Too high? (c) If that portion of the curve is not linear?

11. An oscillating IF stage can often be located by simply touching various circuit connections or components. Explain.

12. Explain why one weak IF stage can seriously affect the overall IF response.

13. (a) In testing the IF strip, when might you suspect leakage in a bifilar transformer? (b) How would you check for this condition?

14. (a) For sweep alignment, why is the calibration accuracy of a sweep generator not too important?
(b) The sweep, however, must be *linear*. Why?

15. During sweep alignment, what might you observe on the CRO if the generator sweep control is set (a) too high? (b) too low? Explain why in each case.

16. Why is a conventional sweep generator not suitable for aligning a synchronous detector?

17. Describe a procedure for adjusting the 45.75-MHz coil of a synchronous detector using only the station signal.

18. To observe the *entire* overall response curve when sweep checking alignment, it becomes necessary to use about 15 MHz of sweep.
(a) Is the IF bandwidth too wide or too narrow?
(b) How would you verify?
(c) As alignment progresses what change would you make in the amount of sweep? Explain.

19. Why is it necessary to clamp the AGC during IF alignment and when signal tracing by injection?

20. A receiver loses sync on strong stations and video overload is suspected. A scope display of the signal at the detector appears normal, but pulses are being clipped in a later stage. State three possible causes and how you would proceed.

21. How would you determine the amount and polarity of dc to use when clamping the AGC?

22. What is the expected result if the sound IF carrier marker is positioned (a) too high on the overall response? (b) too low on the curve? (c) as above for the video carrier marker? (d) for the color subcarrier marker?

23. (a) Name some symptoms caused by IF oscillations.

(b) How would you verify the condition?
(c) State possible remedies.

24. State three possibilities when the gain of *all* IF stages is below par.

25. State two undesirable results when stagger-tuned coils are all tuned too close to the same frequency.

26. When aligning a small-screen set would you strive for a 4-MHz bandwidth? Explain.

27. What would you suspect when trouble symptoms are (a) weak pix, snow, and background hiss in the sound? (b) weak pix without snow? Explain your reasoning.

28. A set has loss of pix and sound, which are restored when a jumper is connected from B to E of an IF transistor State the probable cause.

29. A receiver has loss of sync and a negative pix on strong channels only. Would you suspect the AGC or a reversed detector diode?

30. Give two reasons for using a weak generator signal during alignment and stage-by-stage checking.

31. Why should a generator be fed to some point at least one stage removed from a tuned circuit being aligned? *Hint:* The generator cable has considerable capacity and low impedance.

32. (a) Which way do markers appear to move on a response curve when all tank coils are tuned to a lower frequency? Explain your answer.
(b) Do the *markers* move or does the curve shift position?

33. What is the effect on IF bandwidth (a) if the 39.75-MHz trap is tuned too high in frequency? (b) if the 47.25-MHz trap is tuned too low? Explain why in each instance.

34. The AGC and signal at the detector are being monitored for intermittent loss of pix and sound. What stages are suspect (a) when both indications change with the intermittent condition? (b) when both indications remain normal and constant?

35. Which of the following front-to-back resistance checks indicates a defective detector diode? (a) 500 Ω/2 kΩ; (b) 200 Ω/2 MΩ; (c) 1 MΩ/infinity.

36. (a) Why is a demodulator probe required when signal tracing stages ahead of the detector?
(b) When can an amplifier-type probe be used to advantage?

37. The gain of the IF strip is affected by AGC defects. Troubles in the IF strip in turn affect the developed AGC voltage. How would you determine which condition exists when the symptoms are (a) video overload? (b) weak pix or no pix?

38. Either loss of AGC or excessive AGC can result in a weak pix or loss of pix. Explain.

39. A receiver has weak pix and sound and a reverse bias is measured on one of the IF transistors. State four possible causes.

40. Why is it so important to correctly identify all tanks and traps along with their assigned frequencies before aligning the IF strip?

(b) When might you suspect you have made a mistake?

41. Is the IF strip suspect when (a) there is loss of AGC voltage? (b) loss of pix but the AGC checks normal?

42. (a) When a set has loss of pix, at what point in the troubleshooting procedure would you suspect the detector stage?
(b) What verification checks would you make?

43. When a set has normal sound but no pix, what stages are suspect if (a) it is a BW set? (b) a color receiver? Explain your reasoning.

44. Why is it important to make all necessary repairs *before* aligning a set?

45. Why is misalignment one of the last things to suspect when a set develops troubles?

46. Alignment conditions should be evaluated before making *any* adjustments. Explain why.

47. Name three kinds of marker signals and how each shows up on a response curve.

48. Explain the problem and what might be done about it where a marker on a response curve does not change position with marker generator tuning, but may disappear when switching channels.

49. What alignment difficulties are experienced when a marker is (a) too strong? (b) too weak?

50. (a) With sweep alignment, explain why the CRO gets its horizontal deflection from the sweep generator.
(b) What is the function of the phasing control on the generator?
(c) What is the sweep *rate*, or does it change with the generator controls?

51. (a) With sweep alignment, what determines the width of the CRO trace?
(b) What band of frequencies is represented by the trace if the sweep generator is tuned to 50 MHz with the sweep control set for 5 MHz?
(c) What if the control is set for a 20-MHz sweep? (d) zero sweep?

52. State the probable cause when considerable voltage is measured at the detector output when tuned to a blank channel. How would you proceed?

53. A receiver has loss of pix and no AGC can be measured on the bus. How would you determine if the AGC circuit is the *cause* of the problem or if loss of AGC is caused by troubles in a signal stage?

54. What kind of pix images may be enhanced by video peaking?

55. (a) Explain why multiple ghost outlines appear around fine pix detail when there is video ringing.
(b) How would you pinpoint the cause and overcome this problem?

56. Aligning a set with the overall response peaked on the high-frequency slope is one way to "crispen" a pix.
(a) Explain why.
(b) What undesirable effects may result?

57. A pix carrier positioned too low on the IF response can result in poor sync and loss of fine pix detail. Explain why.

58. Usually there is no resistor or capacitor shunted across a bifilar type IF transformer. Explain why.

59. How is receiver performance affected if the IF bandwidth is (a) too wide? (b) too narrow? Explain why in each instance.

60. A weak or dead IF stage ahead of the sound takeoff can result in loss of pix but near normal sound. Explain this apparent paradox.

61. (a) When stage checking, a modulated signal is injected at the output of an IF stage and brightness and contrast controls are adjusted to get *gray* bars on the CRT. What change should occur in the bar pattern when signal is injected to the input of the stage? (b) What is the indication if the stage is weak? (c) dead?

62. How is sound and pix affected by a misadjusted reference amplifier coil in a synchronous detector circuit? Explain why.

63. Which results in the most snow on a weak pix, a weak RF stage or a weak IF stage? Why?

64. What would you suspect where there are symptoms of video overload on strong signals, which clear up when the antenna is disconnected?

65. Give five trouble symptoms caused by video overload.

66. State three trouble conditions in the IF-detector section that can cause video overload.

67. A receiver has symptoms of video overload. How would you proceed if (a) clamping the AGC bus restores normal operation? (b) clamping results in no improvement?

68. State five troubles that will result in poor resolution of fine pix detail.

69. How is pix quality affected by (a) loss of low video frequencies? (b) loss of the high frequencies? (c) What troubles in the IF-detector section will reduce the low frequencies? (d) the high frequencies?

70. (a) What are the symptoms of adjacent channel interference?
(b) Give the frequencies of the adjacent channel traps.
(c) How is it possible for two traps to reject adjacent channel signals for *all* channels?

71. Does video ringing result in ghost outlines around all pix objects, only large objects, or only the fine detail? Why?

72. (a) Explain the procedure for locating a troublesome IF stage using the signal injection method, and using bars on the pix tube as the indicator.
(b) Should the bars get darker or lighter when progressing from the tuner to the detector? Why?

73. One of the IF stages appears to be weak. How would you make a gain check using other stages for comparison?

552 VIDEO IF AND DETECTOR STAGES

74. What symptoms may be observed (a) if the resistor shunting a peaking coil opens up? (b) if the coil itself becomes open? (c) An open resistor here is unlikely. Explain why?

75. Intercarrier buzz indicates trouble with the AGC, the IF strip, or the sound section. How would you determine which is at fault?

76. Explain a quick fix for the following: stripped threads in an IF coil form; a frozen slug; a cracked or broken slug.

77. Why is visual sweep alignment preferred over the generator–EVM method?

78. (a) Why is accurate alignment of the reference amplifier coil of a synchronous detector *essential*? (b) What is the result if it is mistuned? (c) Explain why a conventional sweep generator cannot be used.

79. State two problems resulting from a misadjusted 41.25-MHz sound takeoff coil? (b) a 4.5-MHz sound takeoff coil? (c) other 4.5-MHz traps?

80. Obtaining a meaningful overall IF response curve requires a good balance between the signal output of the sweep generator and the vertical gain of the CRO. What problems may be encountered if (a) there is too much generator signal? (b) too little signal? (c) if CRO gain is set too high? (d) too low?

81. When there are hum bars on the screen, how can you determine if hum modulation of the signal occurs before or after the detector by switching from a blank to an active channel? What trouble would you suspect?

82. (a) State three possibilities when the adjustment of a tank or trap has no effect. (b) State three possible causes when the coil resonates at the wrong frequency or at some higher-than-normal frequency.

83. How much signal voltage would you expect to measure at the detector output under the following conditions? (a) off channel? (b) when tuned to a weak signal? (c) for a strong signal? (d) if an IF stage is oscillating? (e) Indications for (b) and (c) should be approximately the same. Explain.

84. Under what conditions might you suspect that the coax cable between the tuner and the IF strip is (a) open? (b) shorted? (c) not properly grounded?

85. (a) When might you suspect shorted windings in an IF tank or trap coil?
(b) Why is this difficult to verify with an ohmmeter check?
(c) How would you check for this condition?

86. Name three conditions in the IF–detector section that can cause loss of high video frequencies.

87. Considerable dc (but no ac) is measured at the detector output when switched to a blank channel. State two possible causes. *Hint*: A bifilar transformer is used in the last IF stage.

88. A mica or ceramic capacitor in an RF or IF circuit should never be replaced with a rolled-paper type. Why?

89. Give two good reasons for the use of a series blocking capacitor between a signal generator and the receiver.

Chapter 14

THE VIDEO AMPLIFIER

To obtain a pix, you must first have a *raster*. There are three main requirements to obtaining a raster: a functioning pix tube, beam deflection power, and the proper CRT operating voltages. The most critical of these voltages is the *bias*, which is adjustable with the brightness control. Adjusting the control fully cw reduces the bias to near zero, resulting in high beam current and maximum brightness. With the control fully ccw, the negative grid-to-cathode bias is increased to produce beam current cutoff and a blank screen, the equivalent to black in a pix. The amount of bias to obtain full brightness and cutoff depends on the type of CRT and to some extent on other applied voltages. A typical BW tube may cut off at approximately -30 V; a color CRT, between 100 and 150 V. The grid should never be permitted to go positive.

When tuned to a station, amplitude variations of the received signal cause the grid-to-cathode voltage to fluctuate in accordance with the bright and dark areas of the pix or scene being televised. The peaks of the signal (at the blanking pulse level) correspond to black objects and the lowest levels of signal to bright objects. See Fig. 2-3.

A pix of good contrast has detail ranging from black to the whitest whites, with full saturation of color. To produce such a pix, the signal must be strong enough to drive the CRT between its two extremes, cutoff, and near saturation. For a BW set, depending on the tube type and size, a typical drive signal is around 40 V p-p. For a color tube it is considerably higher, in the 100- to 150-V p-p range. It is impossible to obtain good contrast if the signal is weak at the CRT input. Under such conditions, white objects may appear normal by turning up the brightness control, but black objects will appear gray. If the control is turned down, black objects may appear black, but the white objects will be gray. The bias on the CRT for the normal setting of the brightness control is typically about -20 V.

The signal amplitude at the CRT input and, therefore, the pix contrast are adjustable with the contrast control. With some sets, the brightness and contrast controls interact and they must be adjusted together to obtain a normal pix.

Another consideration is the *polarity* of the signal at the CRT input. Polarity is established by the detector, and there may be several reversals before the signal reaches the pix tube. The signal may drive either the grid or the cathode, although the latter is most common. The signal must be negative going when applied to the grid or positive going when fed to the cathode. A reversed signal polarity produces a pix of reversed phase, a *negative pix*.

The *frequency* of the signal at any given instant corresponds with the size (width) of the object being reproduced. Low video frequencies produce large objects and high frequencies, small objects, the fine pix detail. To produce a pix with good contrast of both large and small objects requires a frequency response that is flat from about 30 Hz to 4 MHz.

Developing a *color pix* requires two kinds of signals, the BW video signal, often called the Y or luminance signal, and the chroma signals. The chroma signal (contained within the same bandpass as the video signal) is separated from the video at some point in the video amplifier. Chroma stages demodulate and process the signal into three discrete signals that correspond with the three primary colors, red, green, and blue. These signals (called the R-Y, G-Y, and B-Y signals) are then recombined in their proper proportions with the video signal to produce a normal color pix. The minus sign used with each of the three letters indicates color signals *without the Y component*.

14-1 VIDEO AMPLIFIER

NOTE: ─────────────────────────

Descriptive information contained in the paragraphs immediately following is of a *general* nature and unless otherwise stated applies to both BW and color sets. Information peculiar to color sets only is covered later, starting with Sec. 14-12.

The purpose of the video amplifier(s) is to boost the weak signal that exists at the detector output (about 2 to 3 V for normal reception) to the level required for good pix contrast. A typical BW receiver has two stages of video amplification; a color set has anywhere from five to ten stages, most of them contained in ICs. A typical amplifier used in a large-screen color set may increase the signal level as much as 80 times for an overall gain of approximately 30 dB.

A block diagram of a BW set is shown in Fig. 14-1. Note the alternate points of sound takeoff, which may be anywhere between the detector and the CRT input. Sound takeoff at

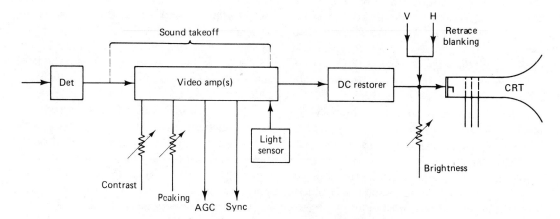

Fig. 14-1 BW set block diagram.

the CRT makes full use of the video amplifiers for sound amplification but has the potential problem of buzz in the sound and sound bars on the pix. The take off for sync and keyed AGC is at appropriate points as indicated.

- **Stage Coupling**

Coupling the signal from stage to stage may be *direct* or with coupling capacitors. With direct (dc) coupling, both the ac and dc components of the signal are transferred. With capacitors, only the ac signal gets through. Some sets use a combination of ac and dc coupling. Direct coupling is preferable since there is no attenuation of the low video frequencies, which may occur with capacitors. To obtain the lowest practical reactance, coupling capacitors when used are electrolytic types and have a fairly high capacity, usually between 4 and 10 μF.

14-2 DC RESTORATION

The video information developed by the cameras at the TV station is referenced to a dc voltage that corresponds with the average level of the signal and the level of background illumination of the scene being televised. Unless the same dc level is maintained at the receiver, the overall pix brightness will fluctuate, where dark scenes will appear brighter than normal (no black objects) and light scenes will be dark, with no white highlights. During dark scenes, retrace lines often become visible when the brightness control is turned up.

Figure 14-2 shows the effect produced with and without the dc component. In Fig. 14-2a, the dc reference level is close to zero during a bright scene; in Fig. 14-2b, this reference has moved closer to the blanking level during a dark scene. These are normal conditions where the dc component is present and the blanking level is *fixed* regardless of variations in scene brightness.

Figure 14-2c shows the effect when the dc component of the signal has been lost. Here the blanking and sync pulses are no longer lined up and are constantly changing with light and dark variations of the scene. For dark scenes, the average CRT bias is increased, reducing the beam current and the brightness on the screen. For light scenes, the average bias is decreased, increasing the screen brightness and producing abnormal shading of the pix.

With direct (dc) coupling between the detector and the CRT, the normal dc level is maintained and there is no problem. Where a coupling capacitor is used, the dc component is lost and must be restored. This is accomplished by rectifying the peaks of the sync tips at the CRT input using a simple diode circuit, as shown in Fig. 14-2d. The diode *clamps* the CRT bias to the blanking level of the signal. When the signal is positive going, as in the example, the inserted dc has a negative polarity. The reverse is true when the signal is negative going, as when driving the grid of a CRT. The time constant for the coupling capacitor and its associated resistors is fairly long, so the brightness level remains reasonably constant for the duration of each frame being scanned.

Some BW receivers using capacity coupling do *not* have a dc restorer. They depend on the small amount of dc developed by grid rectification by the pix tube. With such sets, visible retrace lines can be a problem when the brightness control is turned up. The dc reference level must also be maintained between the chroma stages and the CRT of a color set. If coupling capacitors are used, dc restorer diodes are needed at the inputs of each of the three guns.

14-3 FREQUENCY RESPONSE

For good pix quality, the video amplifier must have a reasonably flat frequency response, from 30 Hz to about 4 MHz depending on the size of

556 THE VIDEO AMPLIFIER

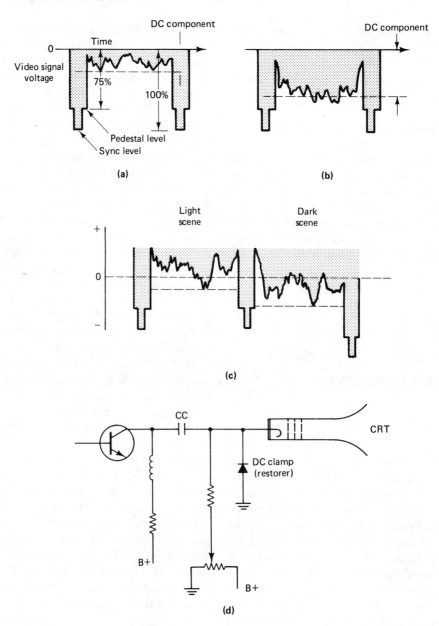

Fig. 14-2 Dc restoration is required when the transmitted dc level is blocked with a capacitor.

the pix tube. The ideal response curve for a typical large-screen set is shown in Fig. 14-3a. Unless compensated for with proper circuit design, the response tends to drop off at the low- and high-frequency extremes. Loss of low video frequencies reduces the contrast for large pix objects and causes smearing of the fine detail and poor sync stability (sync pulses are relatively low frequencies). Loss of high frequencies results in poor resolution and, in extreme cases, complete loss of the fine pix detail.

- **Low-Frequency Compensation**

The reactance of coupling and bypass capacitors increases inversely with frequency, causing a drop-off of low frequencies as shown in Fig. 14-3b. With direct coupling, this is largely overcome. Coupling capacitors, when used, should have as high a value as practical, and for this reason electrolytics are normally used. For optimum low-frequency response, the time constant represented by a coupling capacitor and the input resistance of the following stage should be high; the greater the values of C and R, the higher is the amplification at low frequencies.

To obtain a more uniform response to *all* frequencies, sometimes a resistor in an emitter circuit is left unbypassed, with a gain trade-off due to degeneration. Where a capacitor is used, it must be an extremely high value to be effective.

Some circuits obtain a low-frequency boost with a decoupling network in the output circuit of a transistor as shown in Fig. 14-3d. At high frequencies, the capacitor at the junction of the load resistor $R1$ and the decoupling resistor $R2$ grounds the signal at that point. At low frequencies, the reactance of the capacitor is high and $R2$ becomes part of the load. The lower the frequency is, the higher the load ($R1$ and $R2$ in series), increasing the gain to these frequencies.

- **High-Frequency Compensation**

The shunting effect of transistors and the distributed capacity of wiring and components tends to reduce the gain of high video frequencies in the 2- to 4-MHz range and, unless compensated for, results in a response curve roll-off as shown in Fig. 14-3c. As with the *video detector* (see Sec. 13-15), loss of high video frequencies in video amplifier stages can be prevented as follows:

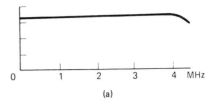

(a)

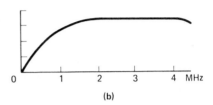

(b)

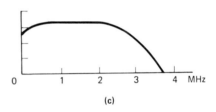

(c)

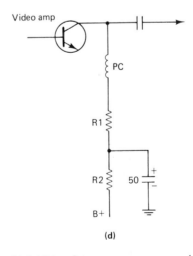

(d)

Fig. 14-3 Video frequency responses: (a) ideal frequency response; (b) low-frequency drop-off; (c) high-frequency drop-off; (d) low- and high-frequency compensation.

558 THE VIDEO AMPLIFIER

1. Using relatively low-value load resistors in the transistor output circuits, typically between 1 and 5 kΩ.

2. Extensive use of peaking coils, both shunt and series peaking, as described in Sec. 13-15. Such coils are self-resonant to about 3 MHz and are often shunted with damping resistors to prevent *ringing* and to broaden their response.

3. Using frequency selective degenerative feedback, sacrificing stage gain for all frequencies except the highs. This method is used instead of peaking coils in most IC chips, hence the need for more stages to obtain the necessary amplification.

4. Providing a video peaking (crispening) control.

• Video Peaking

Many receivers, color sets in particular, have a peaking control for use by the viewer to add crispness when a pix is lacking in fine detail. This is accomplished by accentuating the high frequencies in one of several ways. The simplest method is by making the detector load adjustable, as shown in Fig. 14-4a. The circuit shown in Fig. 14-4b uses a high-Q peaking coil (for overpeaking) shunted with a variable resistor. At maximum resistance the loading is reduced for an increase in peaking. With the circuit shown in part (c), the control is part of the emitter circuit of one of the video amplifier stages. An unbypassed resistor (as with the control arm farthest from ground) introduces a degenerative loss of gain for *all* frequencies. The small capacitor $C1$ has high reactance to all but the highest video frequencies, providing a variable amount of bypassing as the control is adjusted. With the control arm at ground, the degeneration is reduced for maximum amplification of the high frequencies.

The circuit shown in Fig. 14-4d is normally overpeaked with variable bypassing of the high frequencies to ground. A system using negative feedback is shown in part (e). The two resistors function as a voltage divider with the adjustable

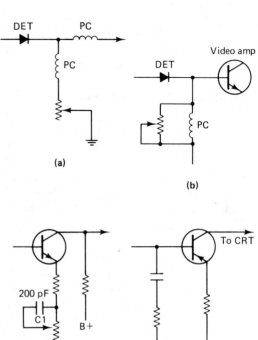
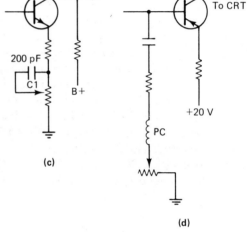
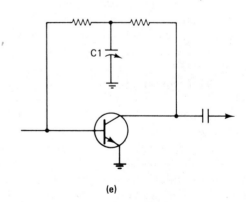

Fig. 14-4 Video peaking control circuits. Accentuating the high frequencies adds sharpness to the pix.

capacitor *C1* controlling the amount of feedback. At maximum capacity, degeneration for the high frequencies is reduced for maximum amplification.

Excessive peaking can create problems, such as *ringing* in the peaking coils, which produces ghostlike images around the fine pix detail (see Fig. 13-19). Overpeaking also extends the frequency response beyond 4.2 MHz, making it more difficult to keep the 4.5-MHz sound IF beat from reaching the CRT.

14-4 CONTRAST CONTROL

This is a *gain control* used to vary the amplification of the video signal being supplied to the CRT. So as not to affect the AGC and sync, the control is normally found in some stage after the AGC and sync takeoffs. In a color set it is downstream from the chroma takeoff so that its adjustment will not affect the chroma signal.

With direct coupling between stages, any change in transistor operating voltages will affect the CRT bias and therefore the raster and pix brightness. Hence, the contrast control should affect only the ac signal component. Typical contrast control circuits are shown in Fig. 14-5.

In Fig. 14-5a, the first video amplifier stage is an emitter follower with the load (the contrast control) in the emitter circuit, where it functions as a variable voltage divider. Because of the coupling capacitor (*C1*), the control has no effect on dc voltages and pix brightness. Maximum contrast is with the control arm farthest from ground.

In Fig. 14-5b, another ac signal voltage divider, the control is shunted across the load ($PC + R_L$) of the output transistor. Maximum signal transfer, and therefore pix contrast, is with the control arm at the top.

The circuit shown in part (c) makes use of degeneration produced by resistance in the transistor emitter circuit. The large value of capacitor *C1* has low reactance to all video frequencies and shorts out the signal developed across different portions of the control resistance. With the control arm at ground, there is maximum degeneration, low stage gain, and a loss of pix contrast. With the arm at the top, the control is shorted out by the capacitor to provide maximum gain and pix contrast. The trap circuit serves only to prevent 4.5-MHz beat interference on the pix.

The video gain of a video IC is usually controlled by varying the *dc* applied to the amplifiers as shown in Fig. 14-5d.

14-5 BRIGHTNESS CONTROL

The intensity of the CRT beam current and the average raster-pix brightness are controlled by variations of the grid-to-cathode bias, the function of the brightness control. Turning the control cw reduces the bias, which brightens the pix; with the control fully ccw, the CRT is biased to cutoff, resulting in a dark screen. Adjustments are normally made in conjunction with the contrast control for reproduction of both black and white pix objects. If the control is set too high, black objects will appear gray; if turned down, the intensity of white objects will be reduced. For a typical BW CRT, a bias of about -30 V is required for beam cutoff. For a color set is is much higher, 100 V or more in some cases.

With most BW sets, CRT bias is controlled *directly* with the control in either the grid or cathode circuit, as shown in Fig. 14-6.

In Fig. 14-6a (grid control), a variable amount of negative voltage is applied to the grid. In this case, bias variations range from 0 to -100 V when the control arm moves to the right. In most cases, the control is *part* of a voltage divider also comprised of fixed *resistors* to limit the range of the control.

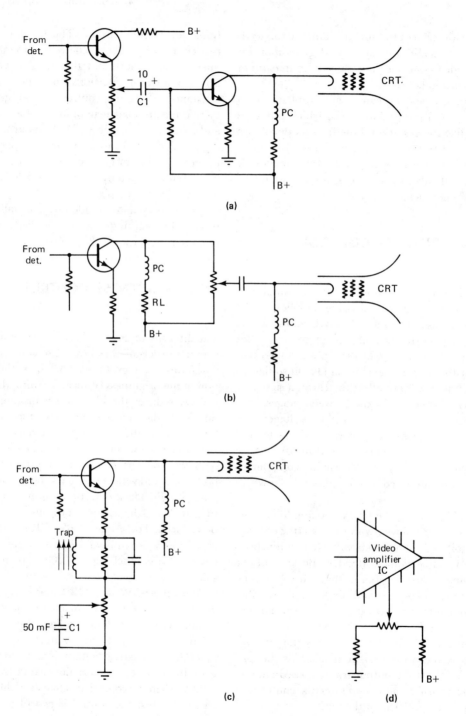

Fig. 14-5 Contrast control circuits.

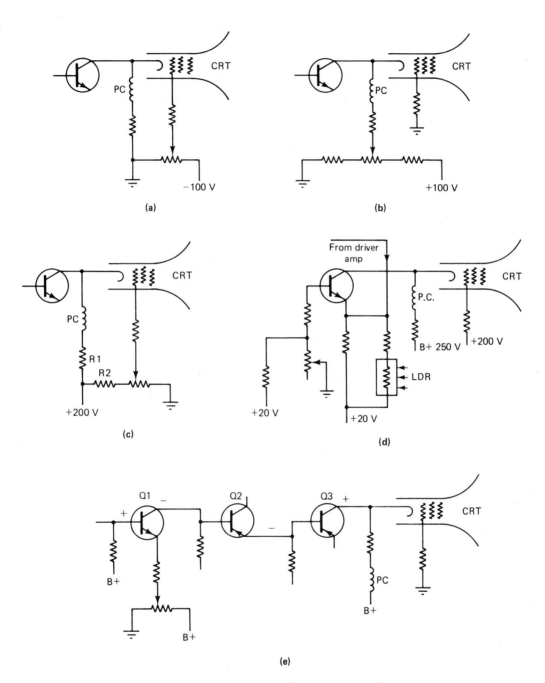

Fig. 14-6 Brightness control circuits. Bias is the potential difference between the grid and cathode voltages. The higher reading of the two establishes the polarity.

562 THE VIDEO AMPLIFIER

In Fig. 14-6b, a variable positive voltage is applied to the CRT cathode. Maximum bias for beam current cutoff is with the control moved to the right.

In Fig. 14-6c, a positive voltage is applied to *both* grid and cathode. Resistors *R1* and *R2* ensure that the cathode is *always* more positive than the grid. Minimum bias (for maximum beam current and brightness) is with the control arm to the left.

14-6 DC CONTROL STAGE

Where video amplifiers are directly coupled to the CRT, color sets (and some BW sets) have the control in one of the amplifier stages as shown in Fig. 14-6d. Since this is the stage where brightness control *originates,* it is sometimes called the *dc control stage.* Transistor bias variations produce voltage changes, which are passed on, (sometimes via several stages) as bias to the CRT. Increasing the control resistance in this case increases the forward bias, which increases the collector current, reducing the collector voltage and the positive dc on the CRT cathode. This reduction in CRT bias increases the CRT beam current and the brightness of the raster and pix. Turning the control ccw reduces the transistor bias, reduces the current, and increases the collector voltage and the positive dc at the CRT to decrease the screen brightness. In Fig. 14-6e the control is three stages ahead of the CRT. Note signal (and dc) polarities at the input and output of each stage.

With some sets using direct coupling, the control is two or more stages ahead of the CRT. With different circuit variations the factors to be considered are whether the grid or cathode of the CRT is being controlled, the number of stages between the control and the CRT, and what transistor configurations are used.

REMINDER: ─────────────────────

Between the input and output of a grounded collector (emitter follower) there is no polarity reversal and no gain.

14-7 AUTOMATIC BRIGHTNESS CONTROL

This feature (which is not to be confused with ABL, automatic brightness limiter circuits described in Sec. 14-15) is found in many modern receivers, particularly color sets. With the manual brightness control adjusted to its normal setting, this circuit automatically controls raster-pix brightness in accordance with ambient room lighting.

A photosensitive, light-sensing device (LDR, light-dependent resistor) is mounted on the front of the set where it monitors room lighting conditions. Figure 14-6d shows a typical control circuit, with the LDR in the emitter circuit of the dc control transistor.

The resistance of the LDR varies inversely with the amount of light it senses. In a dark room the LDR resistance is high, decreasing the forward bias of the transistor, lowering the collector current, and increasing the collector voltage (and the positive dc on the CRT cathode), reducing the light output of the CRT. For an increase in room lighting the LDR resistance is lowered and the reverse occurs, reducing the CRT bias and increasing the brightness. The reduced LDR resistance also reduces the amount of degeneration developed by the unbypassed emitter circuit, increasing the stage gain to increase the pix contrast.

14-8 SPOT KILLER

The glass envelope of a pix tube with the inner and outer graphite coatings (aquadag) constitutes a very efficient *capacitor* that can retain a HV charge for long periods of time. For some time after a set is turned off, when the CRT cathode has yet to cool down and the LVPS filters still hold a charge, beam current may continue to flow, producing a bright spot at the center of the screen. With the deflection circuits disabled there is no sweep. One way to quickly discharge the CRT and extinguish the glow is to increase the beam current by turning up the brightness control prior to switching off the set. Various methods have been worked out to accomplish this automatically.

Figure 14-7a shows a simple circuit using a neon glow lamp in the brightness control circuit. With the receiver operating, the gas in the bulb is ionized by the voltage across its terminals, providing a low-resistance path between B+ and the control. When the set is turned off, and the B voltage drops, the bulb ceases to conduct, reducing the CRT bias for a momentary increase in beam current, which rapidly discharges the CRT.

Another system in common use employs an extra switch ganged to the power ON/OFF switch. A circuit is shown in Fig. 14-7b. The spot killer switch $SW1$ is normally closed with the set operating to supply B voltage to bias the CRT. When the set is turned off, the switch is opened, removing the CRT bias and increasing the beam current to dissipate the HV. In some circuits, a positive voltage is momentarily applied to the CRT grid.

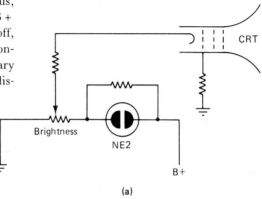

(a)

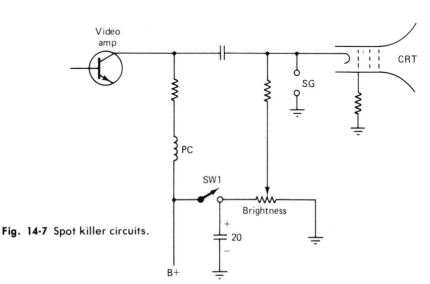

Fig. 14-7 Spot killer circuits.

(b)

14-9 SPARK GAPS (SG) AND HV ARCING

Sporadic arcing in a pix tube is quite common, especially in large-screen color CRTs where the HV is 20 kV or more. It occurs most often during the first few hours of operation when the tube is new. Arcing can also occur *external* to the CRT under conditions of high humidity or where there is a breakdown of insulation. Generally, HV arcing is only momentary and most often takes place when the HV is greatest, as when viewing dark scenes or when the brightness is turned down reducing the CRT beam current, the HV load. Such arcing is seldom harmful to the CRT itself, but the HV transients produced can damage or destroy components, particularly video and chroma output transistors

To prevent such damage, *spark gaps* (arc suppressors) are usually connected at critical points at the CRT inputs (see Fig. 14-15). Essentially, a spark gap is nothing more than two closely spaced wires or contacts. Frequently, they are molded into the CRT socket. A spark gap protects components in the same way a lightning arrestor protects a building by providing an alternate path to ground when voltages exceed a certain level. The gap is fairly critical. If too narrow, it can short out or cause periodic interruptions in receiver operation. If too wide, component breakdowns may occur. Spark gaps are rated between 2 and 5 kV.

Some sets provide additional protection using diodes in the video–chroma output stages, or transistors, called *arc gates,* in the horizontal sweep circuit to momentarily kill the HV when arcing occurs.

14-10 RETRACE BLANKING

Early-model receivers depended on the strength of the signal, specifically the blanking pulses, to darken the screen during vertical and horizontal retrace intervals. Unfortunately, this may not occur under weak signal conditions, especially when the brightness control is turned up. If the CRT is not biased to cutoff during beam retrace, the retrace becomes visible. Loss of vertical blanking shows up as bright, widely spaced lines sloping diagonally across the screen. The lines are superimposed on the pix and change with the settings of the brightness, contrast, and vertical hold controls. Loss of blanking during horizontal retrace (which is much faster) shows up as a haze over the pix and, for a color set, color contamination on the left side of the screen, which changes with the horizontal hold control.

To ensure adequate retrace blanking, all modern sets employ *blanking circuits.* Some BW sets use only vertical blanking. Color sets provide for both vertical and horizontal blanking. Retrace blanking is accomplished by supplying the CRT with high-amplitude pulses generated by the vertical and horizontal sweep circuits during retrace intervals. The pulses, which momentarily and periodically add or subtract from the video signal, serve to drive the tube to cutoff or beyond. Pulse *width* corresponds with the *duration* of retrace. If the vertical pulse is too narrow, the CRT may become prematurely unblanked, with retrace lines appearing at the top of the screen. If the pulse is too wide, blanking may continue *beyond retrace,* darkening the upper portion of the pix.

Similarly, a narrow *horizontal* blanking pulse does an incomplete job, causing a fuzziness on the left side of the screen; a wide pulse produces shading on the left side of the pix. The pulses must have the correct polarity to obtain beam current cutoff: positive going if fed to the CRT cathode or negative going if applied to the grid(s).

The pulses may be applied *directly* to the CRT (high-level blanking) or introduced to one of the video amplifier stages (low-level blanking). With high-level blanking (only used on BW sets), pulse amplitudes range from as low as

20 V p-p to as much as several hundred volts depending on the set and which element of the CRT is being driven. The control grid that has the greatest control over the electron stream requires the least amount of pulse voltage; the screen grid needs considerably more.

Blanking pulses (like the video signal) must have a positive-going polarity when fed to the CRT cathode, and negative-going when fed to the grid(s). Typical circuit configurations are shown in Fig. 14-8.

In Fig. 14-8a, both vertical and horizontal pulses are injected together at the CRT control grid. In part (b) they are supplied to the cathode. Vertical pulses may be obtained from any of the vertical sweep stages, but most often from the vertical output transistor. Horizontal pulses come from a tap or a separate winding on the flyback transformer.

In Fig. 14-8c, vertical pulses feed the control grid and horizontal pulses drive the screen grid (first anode). Note the difference in amplitude of the two kinds of pulses. The *RC* networks in both circuits (*pulse shapers*) provide isolation between the sweep circuits and the CRT and function as voltage dividers to establish the desired pulse amplitude at each point of injection.

The time constant (*TC*) of each *RC* network is quite critical, as is the pulse amplitude. A narrow pulse of low amplitude may be incapable of driving the CRT to cutoff during retrace intervals; a pulse that is too wide or too strong cuts off the beam beyond the retrace period, darkening a portion of the screen.

14-11 TYPICAL BW VIDEO AMPLIFIER CIRCUITS

The circuit shown in Fig. 14-9a is typical for most BW receivers. All the amplification is furnished by the output stage, which has a gain factor of approximately ×50 to develop a signal of about 100 V p-p at the CRT input. The first stage is an emitter follower with a gain of zero.

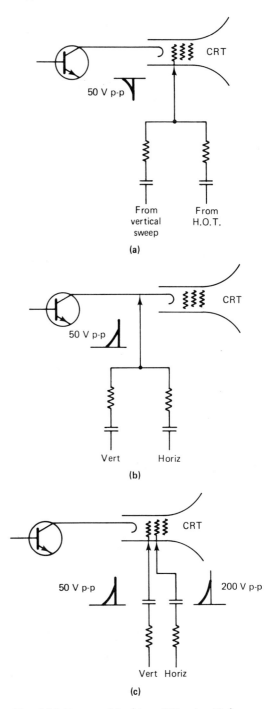

Fig. 14-8 Retrace blanking, BW sets. High-amplitude sweep pulses are required to darken the screen during retrace intervals.

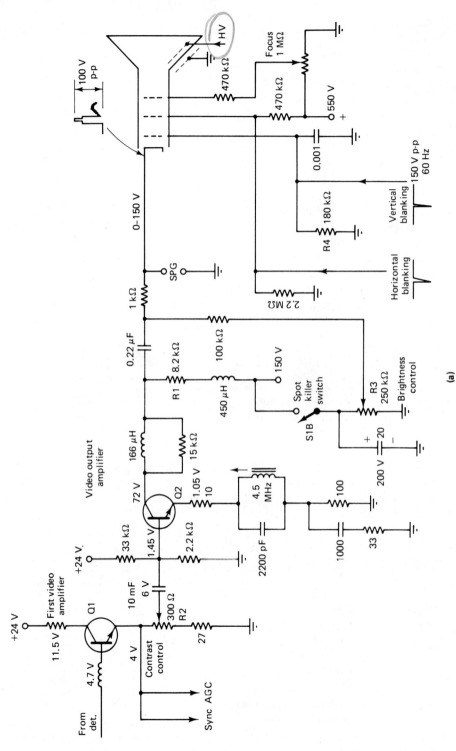

Fig. 14-9 Partial

14-11 Typical BW Video Amplifier Circuits

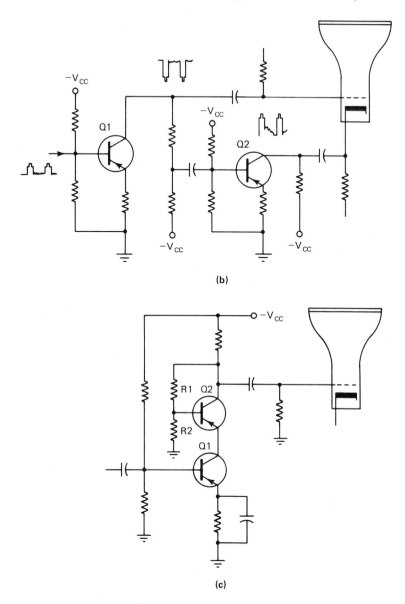

Fig. 14-9 (a) Video amplifier circuits, BW sets; (b) push-pull type of video amplifier; (c) series stacked type of video amplifier.

The contrast control in its emitter circuit functions as a simple voltage divider to vary the signal drive to the output stage. There is no phase reversal in the first stage, and the negative-going signal at the detector output becomes positive-going to drive the CRT cathode. The dc component of the signal is lost because of the two coupling capacitors, and dc

568 THE VIDEO AMPLIFIER

restoration is dependent on grid rectification by the CRT. Good high-frequency response is obtained by using series and shunt peaking, video peaking in the amplifier emitter circuit, low-value load resistors, and the low-impedance characteristics of the first stage. Sound takeoff is from the detector, and the 4.5-MHz trap shown provides additional protection against beat interference.

Both transistors operate with a small amount of forward bias as indicated. To obtain the necessary amplification, the output transistor, which draws approximately 10 mA, is supplied from a relatively high (150-V) source.

The brightness control varies the amount of positive dc applied to the CRT cathode for a bias variation from 0 to 150 V. Maximum brightness is when the control arm is at ground. Arc protection is provided by the spark gap (SPG). The spot killer switch S1B operates in conjunction with the power ON/OFF switch. With the contacts open as shown, the CRT is biased to cutoff until the charge on the PS filters is dissipated. High-amplitude, negative-going vertical and horizontal blanking pulses are applied to the CRT grids as indicated.

Two variations of output amplifier circuits are shown in Fig. 14-9b and c. To reduce the possibility of transistor breakdown from high-level signals, the circuit shown in part (b) uses two transistors operating in *push–pull*, one driving the CRT cathode, the other driving the grid. Assuming a 50-V p-p output from each amplifier, the grid-to-cathode drive to the CRT is 100 V p-p, the sum of the two outputs. Note the opposite polarity of the two signals.

The circuit shown in part (c) uses two direct-coupled transistors in the output stage operating in *cascade*. The transistors are stacked (i.e., connected in series across the B supply with each getting its share of voltage). As with the previous circuit, both transistors contribute to the CRT drive.

14-12 COLOR SETS

A color set must be capable of producing a pix from both BW and color transmissions. When receiving BW, the function of the video amplifier is essentially the same as the one in a BW receiver: to amplify the full range of video frequencies that make up the pix. When receiving *color,* the BW (monochrome) signals from the video amplifier serve two purposes: to provide *luminance* information to establish the brightness of objects reproduced in color, and to provide the high video frequencies that create the fine pix detail that is *not* reproduced in color.

The video amplifier of a color set (which is also known as the Y or *luminance* amplifier) is considerably more complex than the amplifier of a BW receiver. Functionally, there are some similarities and some differences. For example, a color CRT, which is less efficient than a BW tube, requires greater signal drive (up to 150 V p-p or so), which calls for more amplifier stages, sometimes as many as ten or more, including those of related functions. Most of the functions described for a BW set also apply to a color set; additional features, peculiar to color sets only, are described next.

Figure 14-10 shows a basic block diagram of a typical color set (the video output and chroma stages, which are not shown here, are described in Sec. 14-17). Color CRT functions are described in Chapter 15.

The composite BW signal at the detector output consists of video information and vertical and horizontal synchronizing pulses. When receiving color transmissions, there is also chroma information and the 3.58-MHz color burst that resides on the back porch of each horizontal sync pulse. The chroma signals, ranging from 0 Hz to 1.5 MHz, are the sidebands of a 3.58-MHz subcarrier. The subcarrier itself is not transmitted. The function of

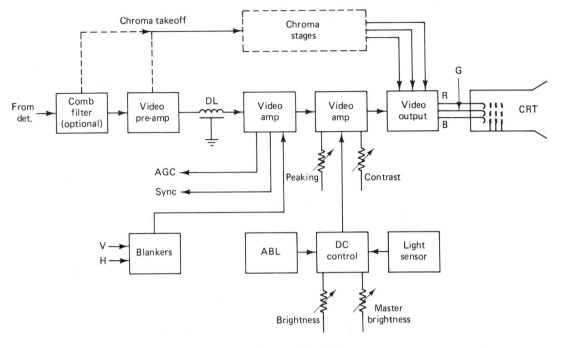

Fig. 14-10 Color set block diagram.

the chroma signals is to produce all but the finest pix detail in color. The burst signal has two functions, to turn on the chroma stages when receiving color and off for BW, and for color synchronization. At some point in the video amplifier section (usually the first stage) the BW and chroma signals become separated, each going its own way, ending up at the CRT.

Separation is accomplished in one of two ways: with a tuned circuit (filter) at the point of chroma takeoff, or with a *comb filter*. The coil is broadly resonant to approximately 3.58 MHz, the frequency of the color subcarrier and the burst signal. Its function is to reject all frequencies lower than about 3 MHz, allowing only the burst and chroma signals to enter the chroma stages. It also functions as a trap, preventing the burst and chroma signals from reaching the CRT via the video amplifier (luminance) stages. Most sets have additional 3.58-MHz traps farther downstream in the video section. The chroma takeoff coil is sometimes adjustable, sometimes not.

The first video stage, sometimes called a preamp, is often part of an IC. Any stages ahead of the chroma takeoff are usually referred to as video amplifiers, and stages downstream from the takeoff as Y or luminance stages. As with a BW set, sync and AGC takeoffs are from one of the amplifier stages. Also, there is one or more 4.5-MHz traps, although not for sound takeoff since sound is separated ahead of the detector. Two features unique to color sets only are the *comb filter* and the *delay line*.

14-13 COMB FILTER

Pix information contained in the modulation sidebands of the video carrier does not fill the

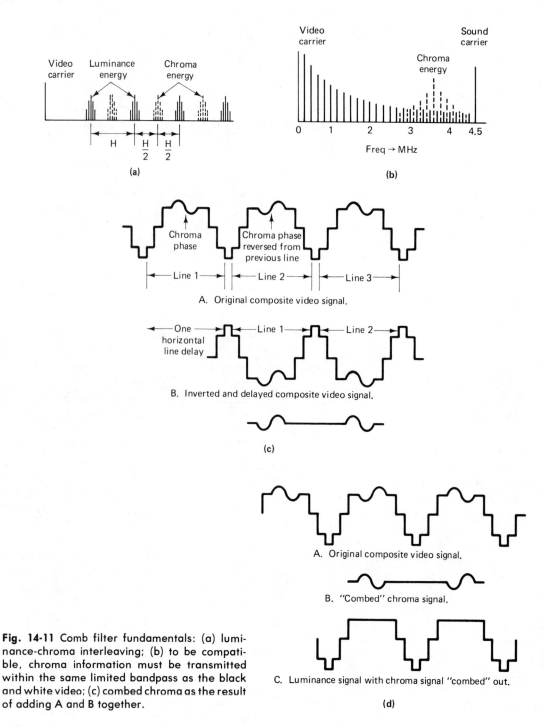

Fig. 14-11 Comb filter fundamentals: (a) luminance-chroma interleaving; (b) to be compatible, chroma information must be transmitted within the same limited bandpass as the black and white video; (c) combed chroma as the result of adding A and B together.

14-13 Comb Filter

entire spectrum of the video bandpass; it actually occupies less than 50% of the available frequencies from 0 to 4 MHz. The information exists as clusters of energy each separated by a frequency gap, as shown in Fig. 14-11a. When receiving color, the chroma sidebands, also in clusters, are sandwiched between the BW clusters in a process called *interleaving*. Since chroma frequencies only extend from about 3 to 4 MHz, the interleaving only occurs at the high-frequency end of the spectrum as shown in Figure 14-11b (note the resemblance to the teeth of a comb). Combining the BW and color signals in the *same* bandpass, however, creates a problem where the sidebands overlap at times, creating *beats* that show up on the pix as fine-grained interference or "rainbow whirls" on fine pix detail. A *comb filter* (as used in most modern sets) overcomes this and other problems with more *complete separation* of the BW and chroma signals in the video amplifier stages.

Another problem prior to the use of comb filters is chroma getting into the video stages and BW signals entering the chroma stages. Chroma reaching the CRT via the luminance stages causes the beat interference described. Until the advent of comb filters, it could only be prevented by limiting the high-frequency response of the video amplifiers. With a comb filter, this frequency response can be extended to better enhance the pix, specifically, to improve the sharpness of the fine pix detail.

Luminance signals reaching the CRT via the chroma stages can act as color signals, producing the color swirls described and wrong colors on fine pix detail. With better separation of the BW and chroma signals, this is prevented, at the same time permitting a wider bandpass of the chroma stages so smaller pix details can be reproduced in color.

- **Comb Filter Fundamentals**

A comb filter is *not a filter* in the true sense. Separation of video and chroma is accomplished in two ways: by *phase cancellation,* where signals of opposite phase and equal amplitude cancel each other out, and by *phase reinforcement,* where signals of the *same* polarity become additive.

The process of separating video from chroma, (or vice versa) is called *combing*. It is done in two steps, first to obtain a chroma signal without video contamination, and next to obtain a chroma-free video signal. This can be shown with the simplified block diagram of Fig. 14-12.

- **Developing the Chroma-Only Signal**

The composite video signal from the detector splits and is fed to an *adder stage* in two ways. One path routes it *directly* to the adder. Taking the other path, it goes through two blocks, a *delay device* and an *inverter* stage.

The delay device (DD) is a unit (see photo, Fig. 14-12) that *delays the signal* by 63.5 μs, the equivalent of *one horizontal scanning line* that makes up the pix. Basically the DD is a glass *acoustical chamber* where the delay is incurred by bouncing the signal around similar to the action in a microwave oven. The DD must not be confused with the *delay line* (Sec. 14-14).

The purpose of the inverter stage is to reverse the phase (polarity) of the signal reaching the adder via the DD. In the adder, the luminance information in one scanning line opposes the identical information in another line and they *cancel* out, leaving only the chroma information, which is then fed to the chroma stages.

This separation of the luminance–chroma signals is made possible by certain characteristics of the signal in the NTSC system of color transmission. For one, the chroma information *in each successive line being scanned is reversed in phase by 180° from the chroma in the previous line.* No such reversal occurs for the *luminance* signal, which remains *in phase* line by line.

Figure 14-11C shows how cancellation of the luminance signal takes place in the adder. The drawing shows three successive lines of the

A comb filter ensures good separation of chroma and video.

(a) DELAY DEVICE

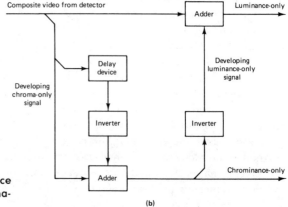

(b)

Fig. 14-12 Typical comb filter: (a) delay device and comb filter circuitry; (b) developing chroma-only signal.

composite video signal. For simplicity, the luminance information is represented by single sine waves. Comparison is made between line 1 (which was delayed) and line 2 (which was not delayed). Similarly, the delayed line 2 is compared with the nondelayed line 3. Because of the polarity reversal in the inverter stage, the luminance signals which are normally in-phase (line for line) now cancel each other. The *chroma* signals, which are now *in-phase,* do *not* cancel; in fact, they *reinforce* each other to increase the chroma-to-luminance ratio.

• **Developing the Luminance-Only Signal**

The combed chroma is now used to remove chroma information from the composite video signal. As before, a delayed signal and a nondelayed signal are recombined in an adder stage (see Fig. 14-12). The chroma-only signal, which is now in phase with the chroma in the composite video signal, is inverted and applied to the adder along with the nondelayed composite signal. In the adder, the out-of-phase chroma signals cancel each other line by line, to produce a chroma-free luminance signal, which is then fed to the video amplifier stages.

• **Another System**

Some receivers use another method of creating the one-line delay. This system uses an IC chip instead of the DD. The delay IC is synchronized to the 3.58-MHz color subcarrier using a *clock generator* operating at 10.7 MHz, three times the subcarrier frequency. Unlike the previous comb filter described, where the range of chroma frequencies is from about 3 to 4.2 MHz, this system delays the *entire* video spectrum from 0 to 4.2 MHz. For a more detailed description of a typical comb filter, see Sec. 14-20.

14-14 DELAY LINE

To produce a pix that is in perfect register (precise overlap of BW and color images), the BW and chroma signals must arrive at the CRT at precisely the same time. Unless a correction is made, the chroma signal would reach the CRT somewhat later than the BW signal because of the greater number of stages in the chroma section and its narrower bandpass. The result would be a double image or ghostlike effect where pix objects produced in color would be displaced to the right of the BW images, as much as $1/2$ in. or so in the case of a 25-in. CRT.

To ensure that the two signals coincide, the Y signal is delayed by the proper amount (about 1 μs); a *delay line* (not to be confused with the comb filter DD) is inserted in the signal path at some point after the chroma takeoff, usually between the first and second stages, as shown in Fig. 14-10. To achieve the necessary time delay, early-model sets used a length of coax whose conductor was spiral wound (like a coil), giving it the properties of a long transmission line.

A typical delay line (DL) used in modern sets consists of a $1/2$-in. coil, 3 to 6 in. long, with an outer wrapping of metal foil. The foil, which is grounded, serves as a ground plane to ensure that capacity to ground is uniformly distributed throughout the length of the coil. The delay is produced by the charging of the capacitor and the inductive reactance of the coil. The delay line introduces some unavoidable insertion loss, but is is essentially the same for all video frequencies.

To prevent coil ringing and line reflections (standing waves that create ghosts), the delay line is terminated at each end with loading resistors that match the impedance of the input–output circuits. The characteristics of a delay line are matched to the receiver and generally they are not interchangeable.

14-15 COLOR RECEIVER BRIGHTNESS CONTROL

In a color set, the CRT bias, and therefore the pix brightness, is controlled in various ways. The different control functions as shown in Fig. 14-10 and described in the following paragraphs are the front panel brightness control, master brightness control, automatic control with an LDR, ABL, and retrace blanking. The first two are manual controls; the others are automatic functions. These functions all work together to vary the bias of one or more video amplifiers (and, indirectly, the CRT bias) when direct (dc) coupling is used between these stages and the CRT. Voltage variations produced by changing the bias of the dc control transistor are passed on (sometimes via *several* direct-coupled stages) to the CRT to control its bias.

- **Front-Panel Brightness Control**

This is the user control, which manually controls the bias of one of the video amplifiers, (the dc control stage). With color sets, this is an indirect method of controlling the CRT bias (and pix brightness), as compared to most BW sets, where the control is at the CRT input.

- **Master Brightness Control**

This is a service type of adjustment found on some but not all color sets. It is known by various names: CRT bias, brightness range, for instance. Its function is to set the range limits of the front panel control. In some cases it is a pot; most often it is a two- or three-position slide switch to establish different levels of CRT bias.

574 THE VIDEO AMPLIFIER

Loss of emission as a CRT ages can be restored for a time by reducing the CRT bias with this control.

With some sets the control operates in conjunction with the front panel control circuit; sometimes it is at the CRT input (as with a BW set), where it directly and simultaneously controls the bias of all three guns. Other circuits that operate via the dc control stage to control the CRT brightness are shown in Fig. 14-10.

• Automatic (LDR) Control of Brightness

This optional feature found in many color sets to control pix brightness in accordance with ambient room lighting was described earlier for BW sets (see Sec. 14-7 and Fig. 14-6). It too operates by controlling the bias of the transistor in the dc control stage that is dc-coupled to the CRT.

• ABL (Automatic Brightness Limiter)

With the brightness control turned full on, the relatively high beam current of a color CRT (sometimes as high as 1 A) can cause blooming, defocusing, damage to the CRT and other components, and under certain conditions will overload and even kill the HV. An ABL circuit, as used in most color sets, monitors and automatically reduces the CRT beam current (and brightness) if it becomes excessive. This system (as with other circuits previously described) also controls the CRT bias from dc-coupled video stages, but it is not to be confused with automatic brightness control using an LDR sensor. Figure 14-13 shows simplified diagrams of several ABL systems in common use.

In Fig. 14-13a, the beam current circuit returns to ground via the flyback transformer, the resistor $R1$, and the LVPS. If the beam current is excessive, the negative voltage developed across $R1$ overcomes the B+, forward biasing the diode $D1$ into conduction. The negative dc forward biases the ABL transistor ($Q1$) into conduction, and its negative-going output alters the bias of the dc control transistor ($Q2$), which in turn increases the bias on the CRT to reduce its beam current and brightness.

The circuit shown in Fig. 14-13b also makes use of the *IR* drop across a resistor in the flyback return circuit to sense abnormally high beam current. Note the path of the beam current in this and other circuits: from ground, through the H.O.T., the HV rectifier, the CRT, the three video output matrixing transistors, and a video driver transistor ($Q4$), back to ground. The operation is essentially the same as in the circuit in part (a). The negative dc developed across $R1$ cancels out the B+, turning on the ABL transistor $Q1$, which is normally biased to cutoff and serves no purpose unless the beam current becomes excessive. If $Q1$ is turned on, its positive-going output is passed on to the dc control transistor $Q2$ to increase the CRT bias as previously described. The front panel brightness control is also connected at this point. Maximum CRT brightness occurs when the control resistance is reduced.

The circuit shown in Fig. 14-13c uses variations of the CRT *focus voltage* as an indication of beam current. Excessive beam current causes an increase in focus voltage, which applied to the ABL transistor $Q1$ increases its conductance and increases its positive output. The positive output of $Q1$ forward biases $Q2$, whose increased output reverse biases the three video output transistors, reducing their current and the beam current of the CRT. The output of $Q1$ is determined by the amount of resistance in its emitter circuit, which comprises the two controls as shown. Resistor $R1$ is a master control that establishes the range of the brightness control and the point at which the ABL system operates. The front panel brightness control is a fine adjustment for the CRT bias and its beam current.

The ripple component from a PS tends to increase with load current. In Fig. 14-13d, excessive beam current (the load on the HV

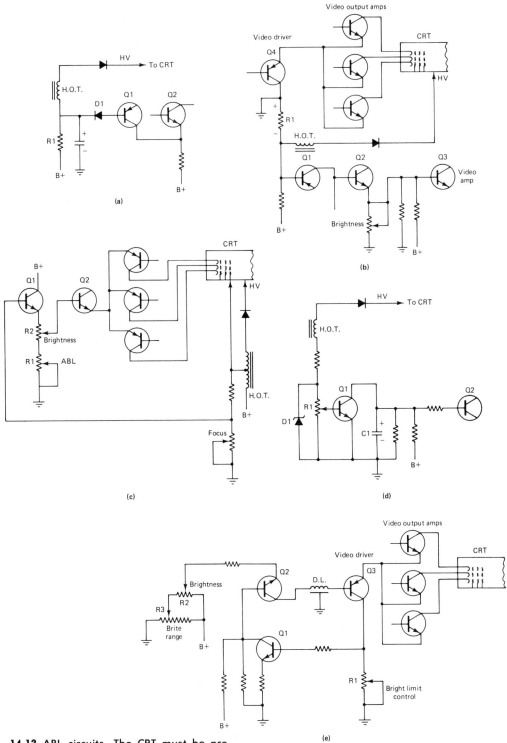

Fig. 14-13 ABL circuits. The CRT must be protected against excessive beam current.

supply) produces a positive-going ripple voltage across resistor *R1*, the ABL threshold control. The ABL transistor *Q1*, which is normally biased to cutoff by the negative dc developed across *R1*, is turned on by the ripple voltage when beam current becomes excessive. The reduced positive voltage at the *Q1* collector changes the bias on the dc control transistor (*Q2*) to increase the bias of the CRT. The filter capacitor *C1* removes the ripple component at the output of *Q1*. The zener diode *D1* prevents excessive voltage being applied to *Q1* in the event of arcing in the CRT.

In Fig. 14-13e, excessive beam current is sensed by the *IR* drop across a resistor (*R1*, the brightness limit control) in the emitter-collector circuit of video amplifier *Q3*, which is series connected with the three output transistors in the path of the CRT beam current. The ABL transistor *Q1*, which is normally turned off, is made to conduct by the positive voltage developed across *R1* under conditions of excessive beam current. The negative-going output of *Q1* changes the bias of *Q2*, reducing its output, which in turn reverse biases *Q3*, reducing its conductance and limiting the CRT beam current. Note the connections of the front panel brightness control and the master brightness or range controls used to manually vary the bias of *Q2*. The range control establishes the maximum brightness level; brightness adjustments within this range are made with the user control *R2*. The control *R1* establishes the ABL operating threshold.

• **Retrace Blanking Circuits**

Black and white sets use high-level blanking, with blanking pulses fed directly to the CRT. Color sets use low-level blanking, where the pulses are injected at one of the video amplifiers, usually via one or more *blanker* stages. Depending on the set, the function of a *blanker* is to provide isolation, amplification, or phase inversion of the pulses. Some sets have but one blanker to handle both vertical and horizontal pulses; most sets have separate stages, a *vertical blanker* and a *horizontal blanker,* each supplied with pulses from its corresponding sweep circuit.

Retrace blanking is associated with the various circuits described previously that control the CRT bias, and therefore the beam current and pix brightness. In most cases the blanker output(s) is tied in directly to the brightness control circuit to produce dc voltage variations that are passed on via dc-coupled stages to the CRT.

Some sets dispense with the blanker(s) by feeding low-level pulses directly to the input of an amplifier stage, where they combine with the composite video signal to be passed on to the CRT. Pulse levels at the point of injection are normally quite low, on the order of 2 or 3 V p-p, consistent with that of the video signal. Signal and pulse polarities are the same at the point of injection and when they reach the CRT. Factors considered in the design of a set are the polarity of the pulses supplied by the sweep circuits, the number of stages that invert the pulses prior to reaching the CRT, and which elements of the CRT are being driven. Different configurations are shown in Fig. 14-14.

No blankers are used in Fig. 14-14a, where both vertical and horizontal pulses are superimposed on the video signal at the output of the first video amplifier. Note the low amplitude of the pulses and their polarities. Since the first and second stages are capacitively coupled, the blanking inputs do not influence the CRT bias directly as when dc coupling is employed.

The circuit shown in Fig. 14-14b uses a blanker transistor with both vertical and horizontal pulses converging at its input. Note the difference in their amplitudes. The negative-going pulses are inverted in the blanker prior to combining with the video signal feeding the next amplifier stage.

The arrangement shown in Fig. 14-14c uses two blankers. Here the vertical pulses control the CRT cathode(s) via the video amplifier

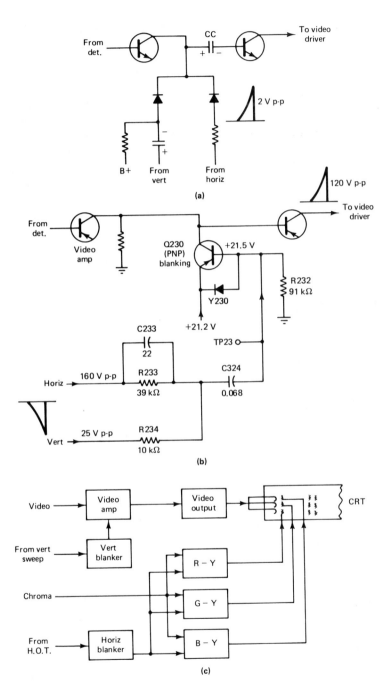

Fig. 14-14 Retrace blanking, color sets. With color sets, blanking pulses are injected at some stage ahead of the CRT input.

stages. The horizontal pulses control the *grids* via the three chroma output stages.

REMINDER: ⎯⎯⎯⎯⎯⎯⎯⎯⎯⎯⎯⎯⎯⎯⎯⎯⎯

These chroma stages are operating at all times, even when receiving BW.

Some color sets incorporate an additional feature called *raster blanking*. This system, operating in conjunction with the blanker stages, darkens the screen while switching channels and, in the event of a vertical sweep failure, to prevent burning the screen. Sensing signals are obtained from the tuner control circuitry and/or the vertical deflection yoke, as appropriate.

14-16 CHROMA CONSIDERATIONS

From the point of chroma takeoff (which is ahead of the delay line; see Fig. 14-10), the composite video signal goes to a comb filter or directly to the first of several chroma stages. Here the chroma and video become separated from each other, each taking a different route to the CRT. When receiving a BW transmission, the first chroma stage is disabled to prevent white noise from reaching the CRT via the chroma route. When receiving color, the chroma circuits are enabled and at some point downstream, (or at the CRT itself) the Y and chroma signals are recombined to produce a color pix. Since this mixing of signals sometimes occurs in the video output stage, the process is briefly described at this time. Further details are covered in chapter 16.

14-17 MATRIXING

Matrixing is the mixing of the chroma and luminance (Y) signals in the proper proportions to restore the original brightness levels of all colors.

Currently, there are two systems of matrixing in common use. They are briefly described next.

• **CRT Matrixing**

A block diagram of this system, which is found mainly in older-model receivers, is shown in Fig. 14-15. The video (Y) signal from the video driver (see Fig. 14-10) feeds an output stage whose function is to supply enough signal power to drive the three CRT guns between cutoff and near saturation. A medium-power transistor is used. Either direct or capacitive coupling is used depending on the set. The signal is fed to the three CRT *cathodes* via two or three *drive controls*. Some early-model sets have a master brightness control also tied in with the cathodes.

The chroma signals are processed into three discrete color signals $(R-Y, G-Y,$ and $B-Y)$ by three *color difference amplifiers* to drive the three *grids* of the CRT as indicated. The luminance (Y) signal and the color signals are combined (matrixed) *in the CRT* since both signals simultaneously control the beam current. Note the protective *spark gaps* connected at each of the three grids and cathodes.

The inner surface of the CRT screen is coated with three color phosphors in the form of *triads* or *strips,* depending on the type of tube. Because of their small size, each of the three color groups is viewed as one. When impacted by the electron stream, we see each group as a single color or a mixture of all three. To produce *white,* each of the three colors in a group must be energized in their proper proportions. To compensate for differences in the light-producing efficiencies of the three phosphors, these proportions are 30% red, 59% green, and 11% blue. An off-channel white *raster* is obtained by individually adjusting the beam currents of the three guns using R, G, and B *screen grid controls*.

To produce white objects in a *pix,* the three guns must be driven by Y signals having the same 30%-59%-11% relative amplitudes. This is the function of the drive controls mentioned

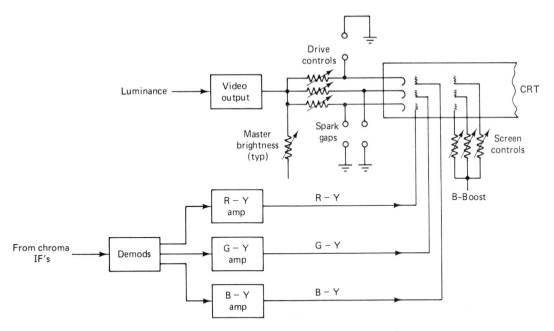

Fig. 14-15 CRT matrixing system. Variations of beam current are created by the positive-going Y signal and the negative-going chroma signals.

earlier. Note the signal levels at all six of the CRT inputs.

• **Drive Controls**

Some sets have *three* drive controls (also known as background controls) between the video output stage and the CRT inputs (Fig. 14-15). A *red drive* control varies the signal drive to the red gun, a *green drive* control varies the drive to the green gun; and a *blue drive* control varies the drive to the blue gun. These controls are preset service adjustments. They control the Y signal *only*, and are not to be confused with color adjusting controls. In practice, the drive controls are alternately adjusted to obtain the *whitest whites* in a pix. The adjustment procedure is known as *gray scale tracking*. When misadjusted, white objects appear off-white or with a color tint.

Most sets have only two controls, a green drive and a blue drive. They are adjusted for the proper balance relative to the signal feeding the red gun, which is a fixed level.

Instead of drive controls, some sets have a voltage-divider arrangement where signal levels are adjusted with pigtail connectors. With some late-model sets, drive adjustments are dispensed with entirely.

• **Developing a BW and Color Pix**

A white *raster* is a prerequisite to obtaining a *BW* pix; and a BW pix is prerequisite to a normal color pix. When receiving a BW transmission (assuming a normal *white* raster and proper adjustment of the drive controls), each of the three guns receives a proportionate amount of Y signal and they *operate together* to produce a pix in BW. Since the chroma stages are disabled, no signals are supplied to the CRT *grids* at this time.

When receiving *color* (assuming a normal raster and BW pix), the chroma circuits are

580 THE VIDEO AMPLIFIER

enabled, and the color signals at the three grids aid the Y signals on the cathodes in controlling the electron streams of the three guns, in accordance with the relative voltages on the grids at any given instant. As each particle of the three screen phosphors is impacted in the right proportions, pix objects are viewed as red, green, blue, or their complementary colors.

- **Pre-CRT Matrixing**

This system, used in most modern sets, is shown in Fig. 14-16a. In this system, the luminance and color signals are matrixed in three video output stages (matrix amplifiers) *ahead of* the CRT. The identity of these stages corresponds with the color signals they produce, red, green, and blue video output. Besides *mixing,* they provide signal power to drive the CRT. One advantage of this system is that transistor requirements have been reduced from four to three.

In this system, each of the three output amplifiers is supplied with both luminance and chroma signals. The R - Y, G - Y, and B - Y chroma signals from the demodulators are routed via three discrete lines to the base inputs of the output transistors. The Y signal from a video driver stage is fed, in parallel, to the emitters of the three transistors. The red, green, and blue matrixed outputs of the three power transistors (containing both chroma and luminance components) individually drive the three cathodes of the CRT. Unlike the older system, no signals are applied to the CRT grids.

Figure 14-16a shows the drive controls between the amplifiers and the CRT inputs. In some cases they are at the Y signal *inputs* of the three amplifiers. Their function is essentially the same for both configurations, that is, to set the relative drive to the three guns for producing the whitest whites in both a BW and color pix and the proper mix for good color rendition. Note the signal levels, typically about 5 V p-p,

at the amplifier inputs, with outputs of over 100 V p-p to drive the CRT.

A variation of this system is shown in Fig. 14-16b. Here the matrixing is performed at *low* signal levels in three small matrixing transistors *ahead of* the power output stages. As with the previous system, chroma is fed to the base inputs of the matrixing transistors and the Y signal to the emitters. Note the low signal levels (around 2 V p-p) up to the inputs of the CRT driver transistors.

14-18 SERVICE SETUP SWITCH

This is a two-position slide switch located on the rear apron of most color sets. It is used as an aid in making CRT setup adjustments to obtain a white raster (with the color screen controls) and for gray-scale tracking (with the drive controls). These adjustments can be best made when there are no signals to influence the beam currents, and with a thin horizontal trace rather than a full raster.

In a typical receiver, four things happen when the switch is moved to the service position: the video and chroma signals are disconnected from the CRT inputs and replaced with a dummy load resistor(s), the vertical sweep is disabled, and the brightness is reduced to prevent burning the screen, with the CRT bias clamped to a fixed level. Under these conditions the raster collapses into three thin horizontal lines (red, green, and blue) across the center of the screen. The lines are displaced by the amount of separation between the three color phosphors that make up each triad or color stripe. When the setup controls are misadjusted, the three lines can be observed, each of a different color. When properly adjusted, the colors blend into a single white line.

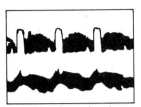

Matrixed waveforms of a colorcast. Top waveform is of the combined video and chroma, and the bottom waveform is the incoming chroma alone.

(a)

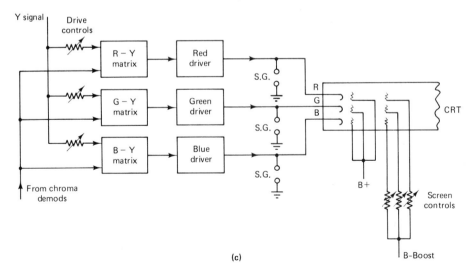

(b)

(c)

Fig. 14-16 Pre-CRT matrixing.

14-19 VIDEO AMPLIFIER OF A TYPICAL MODERN COLOR SET

The following is a brief description of the video amplifier section of Sample Circuit D at the back of this book. The chroma stages not covered at this time are described in Chapter 16. This receiver has 11 video stages. Two of the stages are located in an IF amplifier module; all other stages, including chroma circuits are in a signal module.

From the synchronous detector in IC 101 of the IF module, the composite video signal goes to a 4.5-MHz trap (T 175) and then to video preamp Q 102. From Q 102 the signal leaves the IF module to feed Q 220, the first stage on the signal board. The signal level at this point (TP 23) is 1.5 V p-p. This is the point of chroma takeoff (and sync takeoff via Q 200). Note the 3.58-MHz filter trap at the input to the bandpass amplifier (pin 15 of IC 300).

From Q 220 the path of the signal is as follows: the differential amplifiers Q 201/Q 203, Q 205, Q 207, the delay line, Q 209 (note the 3.58-MHz trap at its emitter), Q 211, to TP 25, at which point the signal amplitude is 5.5 V p-p. From Q 213 the signal goes to a video follower, Q 407, which feeds the *emitters* of the three power output amplifiers Q 401, Q 403, and Q 405 via the red, green, and blue drive controls.

The R - Y, G - Y, and B - Y demodulated chroma signals are fed to the *base* inputs of each of the three output transistors. The matrixed red, green, and blue outputs of the amplifiers drive the three cathodes of the CRT. Chroma signal levels at the amplifier inputs range from 1 to 3 V p-p. The Y signal inputs are about 1 V p-p. The level of matrixed output driving the red and green guns is 160 V p-p and 70 V p-p for the blue gun. Note the spark gaps built into the CRT socket.

• **Controls and Control Functions**

1. *Contrast (picture control):* The control is connected to the base circuit of Q 203 along with the LDR sensing circuitry, when used.

2. *Video peaking:* The *sharpness* control connects to the emitter output of Q 205.

3. *ABL:* Overcurrent sensing voltage from the ground return circuit of the flyback transformer is fed to beam-limiting transistors Q 210 and Q 212 to reduce the output of the differential transistor Q 201 to overbias the CRT.

4. *Brightness controls:* The front panel brightness control and the brightness range (master brightness) control both tie in with Q 262, the dc restorer. Voltage changes of Q 262 (which functions as a dc control transistor) varies the bias of Q 213 and the CRT via the directly coupled output stages. Three dc restorer diodes are used because of the coupling capacitor C 233.

5. *Retrace blanking:* Vertical pulses from the output of the vertical sweep amplifier Q 603 are routed via connector PG-7 to the vertical blanking gate Q 280 to combine with the video at the output of Q 213. Horizontal pulses from a tap on the flyback transformer are applied to the output of Q 213 via connectors PG-35 and PG-7.

6. *Drive controls:* The three drive controls are in the emitter circuits of the red, green, and blue video output stages. Operating as bias controls, they vary the gain of the amplifiers and therefore the amount of matrixed signal driving the CRT cathodes.

7. *CRT cutoff controls:* These three controls in the emitter circuits of the video output amplifiers operate in conjunction with the drive controls for gray-scale tracking.

8. *Master screen control:* This control is supplied from a + 125 V source. Only one screen control is required since the three screen grids are connected together. Beam control of the *individual* guns is accomplished with the drive and cutoff controls.

14-20 ANOTHER RECEIVER CIRCUIT

The circuit shown in Fig. 14-17 is typical of many late-model color sets. This set has a five-transistor comb filter (including two luminance amplifier stages; see photo, Fig. 14-12) and a video amplifier IC that drives the matrix output amplifiers.

• **Comb Filter Operation**

The composite video signal from the detector feeds the delay driver Q 404, which has two outputs. The composite video signal developed across its emitter resistor feeds the emitter of a luminance *adder* (Q 401) via a high-pass *RC* filter network whose function is to attenuate the luminance signal. The composite signal also feeds the base of the chroma buffer transistor (Q 402) via luminance amplitude control R 426.

The composite video signal from the collector of Q 404 feeds the delay device D 401 (see photo, Fig. 14-12). The *delayed* and *inverted* output of the DD combines with the nondelayed signal at the R 426 control, where the out-of-phase luminance signals in the two lines of video on being compared cancel each other. The *chroma-only* signal is then amplified by Q 402 and passed on to the chroma stages.

A sampling of the chroma-only signal is fed to the base of the Q 401 adder via a chroma amplitude control (R 401) and another *RC* filter network. The chroma feeding the base of Q 401 is out of phase with the chroma in the composite signal at the emitter, and the two cancel, leaving a chroma-free luminance signal at the transistor collector. From Q 401, the luminance-only signal goes to two emitter followers, Q 405 and Q 403, through the delay line (L 202) to the video amplifier IC. Note the optional 3.58-MHz series-resonant trap, which is *not required* in receivers having a comb filter.

The comb filter in this circuit has four adjustments, two control pots and two slug-adjusted coils. These coils (phase adjustments) are not required with comb filters having a clocked delay, as described earlier.

14-21 TROUBLESHOOTING THE VIDEO AMPLIFIER SECTION

Troubleshooting the video amplifier of a BW set is not too difficult; for a color set it can sometimes be a real challenge. It is especially true of late-model sets with their numerous dc-coupled stages, recent innovations, and associated fuctions, and the great variety of trouble symptoms that occur. As with other sections of the set, however, a good understanding of circuit functions and troubleshooting techniques goes a long way in correcting the faults that occur.

14-22 ANALYZING THE SYMPTOMS

As might be expected, most of the trouble symptoms associated with the video amplifier section can also be caused by defects in other areas of the receiver. And more often than not there are *multiple symptoms* caused by a single defect. Table 14-1 provides a listing of such symptoms with appropriate troubleshooting procedures for each. For multiple symptoms, the trick is in deciding which is most significant and the tests that should be made.

Start with a careful analysis of all symptoms, beginning with the raster if there is one. Remember, a normal raster is the *basis* of a good pix. Check for brightness, focus, snow, and the absence of hum bars. If it is a color set, is the

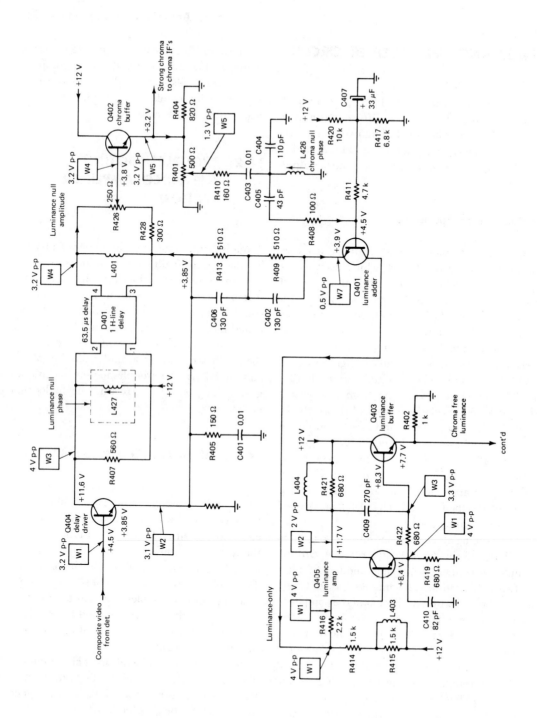

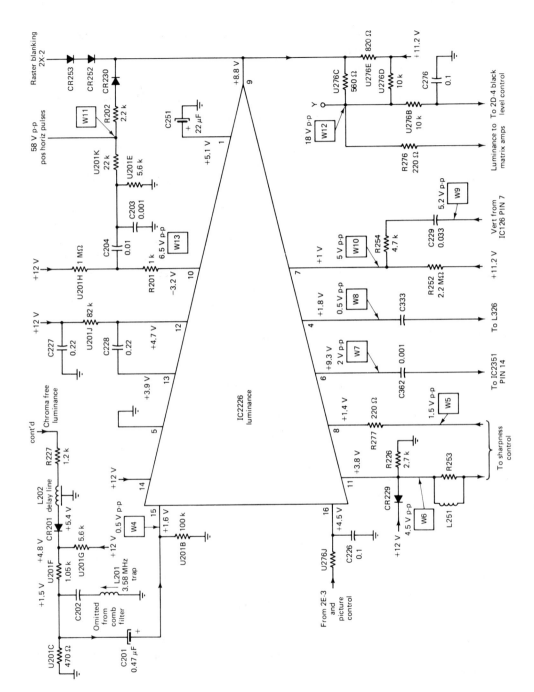

Fig. 14-17 Typical Y amplifier with comb filter.

raster *white* or is there color tinting or contamination? Check the action of the brightness control. Is there evidence of blooming or defocusing? Can the screen be darkened? If there is no raster, visually check for a CRT *heater* glow (*three* heaters if it is a color CRT).

Tune in a station. If there is a pix, is the contrast adequate but not excessive, causing smearing and loss of detail? Is the quality of a BW or color pix acceptable? Any problems with the sound (such as intercarrier buzz), visible retrace lines, overpeaking, sync instability? Check the action of other controls related to the problem: contrast, AGC, killer threshold, and so on. With color sets in particular, many problems are caused by improper *adjustments* rather than circuit defects.

14-23 INSPECTION

After observing and analyzing the symptoms and deciding the trouble is probably with the video amplifier or associated functions, examine the receiver schematic, noting such things as the AGC, sync, and chroma takeoff points; at what stage blanking pulses are injected; which stage or stages are controlled by the brightness control and ABL; whether the set has a comb filter; and which system of matrixing is used, among other things.

With set turned off, inspect the video amplifier modules, PCBs, and signal inputs to the CRT for the obvious: burned resistors, poor soldering, loose connections, loose cable connectors, damaged or intermittent PCB edge connectors, cracked or broken boards.

With set turned on, check for overheating resistors or transistors, arcing, and the like. If there is no raster, verify that the CRT heater(s) is glowing.

14-24 LOCALIZING THE FAULT

Signal tracing or dc voltage checking will usually localize the defect to the module level or a single stage or IC within the module. The initial troubleshooting procedures depend on the symptoms (see Tables 14-1 and 14-2). Refer to the block diagrams (Figs. 14-1 and 14-10) and the troubleshooting flow diagrams (Figs. 14-18 through 14-21).

Most TV troubles fall into one of the following categories: those affecting (1) the raster, (2) the BW pix, and (3) the color pix. A *raster* is prerequisite to obtaining a pix. Likewise, a normal *BW pix* on a color set is prerequisite to obtaining a good *color pix*. Testing, considering these priorities, follows the same sequence.

• **Raster**

The first step is to obtain a raster and to correct any abnormal raster conditions, such as reduced or excessive brightness, poor focus, and raster tinting. For each of these conditions, and more, there are many possible causes. Most raster problems can be solved by dc voltage tests alone.

• **BW Pix**

Obtaining a normal BW pix is the next step in troubleshooting both BW *and* color sets. Troubles causing loss of pix or an abnormal pix can usually be localized by signal tracing the stages between the detector and the CRT.

• **Color Pix**

Assuming a *normal raster* and *BW pix*, loss of color or an abnormal color pix indicates trouble in the chroma stages, which can generally be localized by signal tracing.

Table 14-1 Video Amplifier Trouble Symptoms Chart, BW Sets

Trouble Condition	Probable Cause	Remarks and Procedures

RASTER PROBLEMS[a]

No raster or dim raster; sound OK	Defective CRT. Loss of HV, focus, or screen voltage. Poor HV regulation causing blooming and possible loss of raster when brightness control is turned up. Excessive CRT bias, as caused by defects in brightness control circuit, or *any* circuit tied in with the CRT grid or cathode. This includes the video output stage and the dc restorer.	Check CRT and all applied voltages. Try a jumper from control grid to cathode. If raster appears or becomes brighter, it is a bias problem. If no raster, suspect the CRT.
Excessive brightness	If unable to darken screen, suspect loss of CRT bias caused by leakage or short in the CRT, troubles in brightness control circuit, or any circuit that connects to the grid or cathode.	A CRT short causes a bright raster with little or no pix visible. Condition is usually intermittent and shows up by tapping the gun. Check bias (between grid and cathode) with socket removed. Check both grid and cathode to ground to determine which circuit is at fault.
Blooming (raster and pix expand with loss of brightness and focus)	Most common cause is low HV or poor HV regulation where HV drops excessively under load (as brightness control is turned up). Can also be caused by too little CRT bias.	For HV problems, see Chapter 9.
No automatic brightness control with variations of room lighting	If brightness can be controlled normally with front panel manual control, suspect a bad LDR sensor or its associated sensitivity control circuit. If manual control is ineffective, suspect that circuit or dc-coupled video amplifier stages.	Check the LDR and/or try a substitute. Check voltage conditions between the LDR and the CRT.
Hum bar(s) on raster and pix; May also be waviness on left and right	Poor filtering of the B voltage supplying the video amplifier stages. For bars on *pix only,* check for ripple on dc feed to tuner and IF strip.	Sometimes the bars are very faint and show up only when the vertical hold is misadjusted. Check for ac ripple on

Table 14-1 (cont'd)

Trouble Condition	Probable Cause	Remarks and Procedures
sides (wide bright horizontal bar)		the B voltage line and try bridging filters.
Shading at top of screen	May be mistaken for a hum bar. Usual cause is excessive vertical retrace blanking.	Check for excessive blanking pulses at CRT input.
Clear raster, no snow	Usually indicates a weak or dead video amplifier stage. Normal snow or other interference indicates video amplifiers are working.	Localize defective stage by signal injection with an AF generator or by signal tracing with CRO etc.

PIX PROBLEMS[b]

Trouble Condition	Probable Cause	Remarks and Procedures
No pix or weak pix; raster OK	May be a weak or dead video amplifier stage or loss of signal at some prior stage, often caused by excessive AGC. If no pix but normal sound, trouble is probably after the point of sound take-off. If no pix and normal AGC, suspect stages after point of AGC takeoff. If no pix or sound, but some snow, suspect stages ahead of the detector. If no snow, suspect video amplifiers; if some snow, trouble is prior to the detector.	Try adjusting AGC threshold control if there is one. Localize faulty stage by signal tracing or injection with generator between detector and the CRT input.
Video overload symptoms including excessive contrast, buzz in the sound, loss of sync, pix smearing and loss of fine detail, and ghosting effects	Loss of AGC. Misadjusted AGC threshold control. IF misalignment (narrowed response). Ghosting (multiple images) may be caused by overpeaking (ringing) in video amplifier, or is due to mismatched antenna system or multi-path reception. These conditions also cause loss of fine detail.	Adjust AGC control below the point where sync becomes unstable. Check operation of AGC circuit. Check for IF misalignment. For pix smearing, suspect poor low-frequency response of video amplifier (often caused by near-open coupling capacitors). For loss of fine detail, check focus, possible impedance mismatching in antenna system, overpeaking, or a narrowed IF bandpass. For loss of sync,

		check for pulse clipping with CRO or by observing the hammerhead display.
Visible retrace lines (widely-spaced, bright diagonal lines)	If screen cannot be darkened with brightness control, suspect loss of CRT bias. Condition worsens as the CRT ages and brightness control must be turned up higher. The most common cause is weak or missing vertical blanking pulses.	For excessive brightness, check brightness control circuit and the CRT. Check amplitude and polarity of blanking pulses applied to the CRT grid or cathode.
4.5-MHz beat interference on pix (grainy or wormy appearance)	Misaligned IF strip or 4.5-MHz traps in detector or video amplifier stages.	Check proper adjustment of fine tuner. Check IF alignment and the 4.5-MHz trap adjustments.
Faint haze on pix and/or "windblown effect" on left side	Caused by insufficient blanking of horizontal retrace. As for vertical retrace problems, check for excessive brightness where screen cannot be darkened with control. The usual cause is weak or missing horizontal blanking pulses.	Check amplitude and polarity of horizontal blanking pulses at CRT input.

[a] Raster problems also show up on a pix.
[b] For problems with both raster and pix, consider the raster problems *first*.

Table 14-2 Video Amplifier Trouble Symptoms Chart, Color Sets

Trouble Condition	Probable Cause	Remarks and Procedures
RASTER PROBLEMS[a]		
No raster or dim raster; sound OK	In general, the same causes described for BW sets. For excessive bias (all three guns) the cause is usually a defect in a dc-coupled video amplifier stage or a chroma output stage or a defective dc restorer diode. Another possibility is the master brightness and/or the three screen controls are backed off too far. If controls are set too high or there is an ABL problem, excessive blooming can also kill the raster.	See Table 14-1. Check for heater glow on all three guns. Check for a defective CRT. Perform setup adjustments described in Chapter 15.
Excessive brightness	Same as for BW set (Table 14-1). Other possibilities are excessive HV,	Try backing off on master brightness and

	poor HV regulation, and ABL problems. A common cause is loss of CRT bias due to defects in dc-coupled video or chroma stages.	screen controls. Check for poor HV regulation. Check and measure HV and adjust to *value specified.* Check operation of ABL circuit. For loss of CRT bias, proceed as described for BW sets (Table 14-1).
Blooming	See Table 14-1. Also, suspect ABL problems, troubles with HV regulation, or the master brightness and/or screen controls set too high.	As for BW sets and excessive brightness, described above.
Raster streaking, HV arcing	Usually occurs when brightness control is turned down and often takes place in the CRT or components associated with the HV supply, such as the rectifier(s), regulator, and flyback transformer. HV may be excessive. Minor problems of corona arcing may take place only during humid weather.	Localize site of arcing by inspection in darkened room. May also be odor of ozone and hissing sounds. Adjust HV to value specified. See procedures described in Chapter 9. Arcing in a new CRT is usually temporary and no cause for alarm.
Colored raster	Proper setup adjustments will often cure this problem and produce a normal *white* raster. A magnetized CRT and the need for degaussing are a common cause. May also be due to one or more colors being weak or missing as caused by defective CRT or voltage problems associated with one or more guns.	Try the setup adjustments (see Chapter 15). Try degaussing the CRT. Determine if weak or missing colors. Check all three guns of the CRT and compare. Check for abnormal voltages on suspected gun(s). Check for bias variations as brightness control is adjusted, and screen voltage variations by turning screen controls.

PIX PROBLEMS[b]

No (or weak) BW or color pix (raster ok)	As described for BW sets (Table 14-1). Suspect a video stage between the detector and the chroma takeoff; also the comb filter when provided.	Evaluate other conditions such as sound, sync, AGC, and snow. Adjust AGC threshold.

		Localize faulty stage as described for BW sets.
No (or weak) color pix; BW pix OK	Loss of chroma signal in the chroma stages (see Chapters 15 and 16). Color intensity or color killer control turned off. IF misalignment.	Signal trace chroma stages. Turn up color and color killer controls. Check operation of comb filter if provided. Check IF alignment.
No BW pix; poor-quality color pix	Loss of Y signal in comb filter or video amplifiers.	Signal trace video stages between the chroma takeoff and the CRT. Check operation of comb filter if provided.
Poor-quality color pix; BW pix OK	Loss of one or more colors. Misadjusted drive controls. Defect in chroma stages (see Chapters 15 and 16).	Adjust drive controls per Chapter 15. For chroma stage defects, see Chapters 15 and 16.
Vertical retrace lines visible	As described for BW sets. Another cause is a defective blanker.	As for BW sets. Also check for loss of pulses in blanker stage.
920-kHz beat interference on pix (grainy or wormy appearance)	Misadjusted 4.5-MHz and/or 3.58-MHz traps. IF misalignment. Comb filter defect or alignment.	Align IFs and traps as needed.
Double images (BW and color pix do not overlap) or video ringing (ghosting)	Defective delay line or associated resistors. Defect in video peaking circuit.	Substitute delay line. Check peaking circuits.
Intermittent problems	Almost any circuit or component defect. See Chapter 4.	After creating the trouble symptom, proceed as for stable symptoms described above. A dual-trace CRO can be useful for simultaneously monitoring signals at different stages.

[a] A *color-free* raster is prerequisite to a good Bw or color pix.
[b] A good color-free BW pix is prerequisite to a normal color pix.

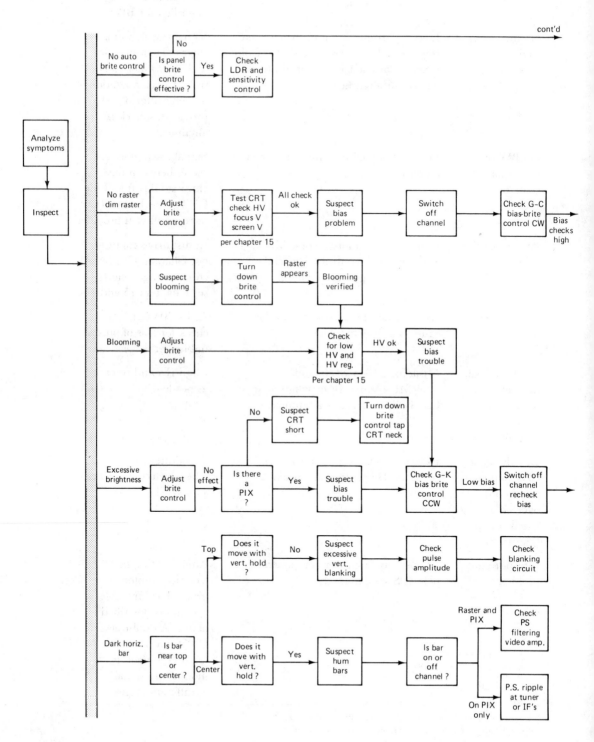

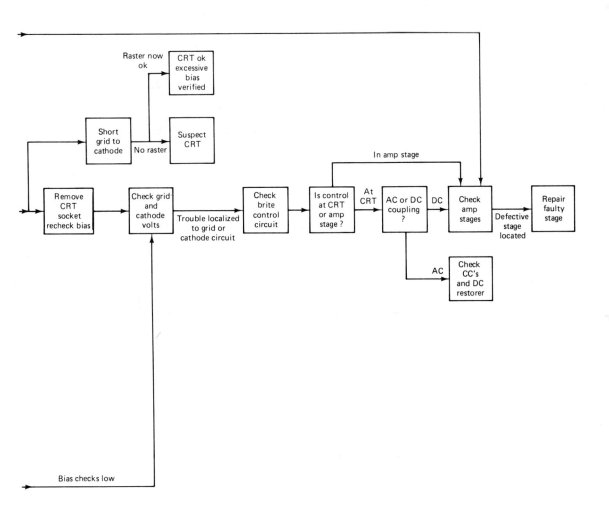

Fig. 14-18 Troubleshooting flow drawing, raster problems, BW sets.

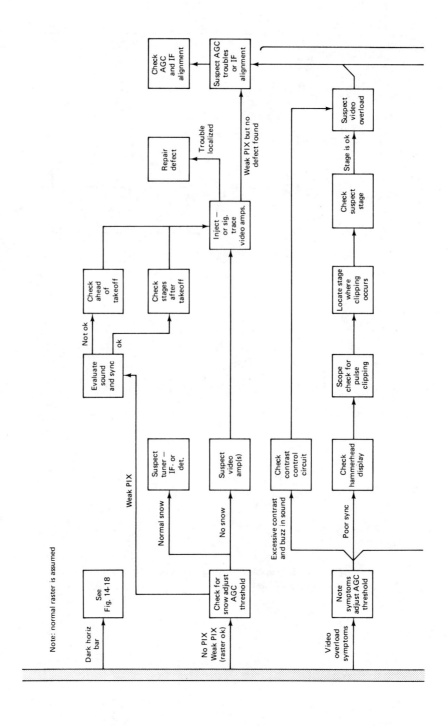

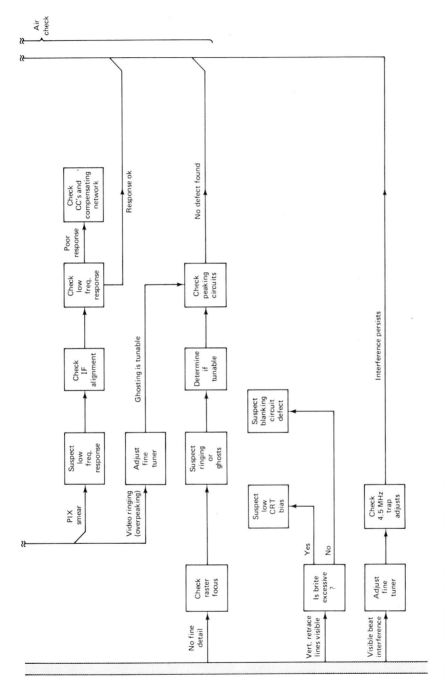

Fig. 14-19 Troubleshooting flow drawing, pix problems, BW sets.

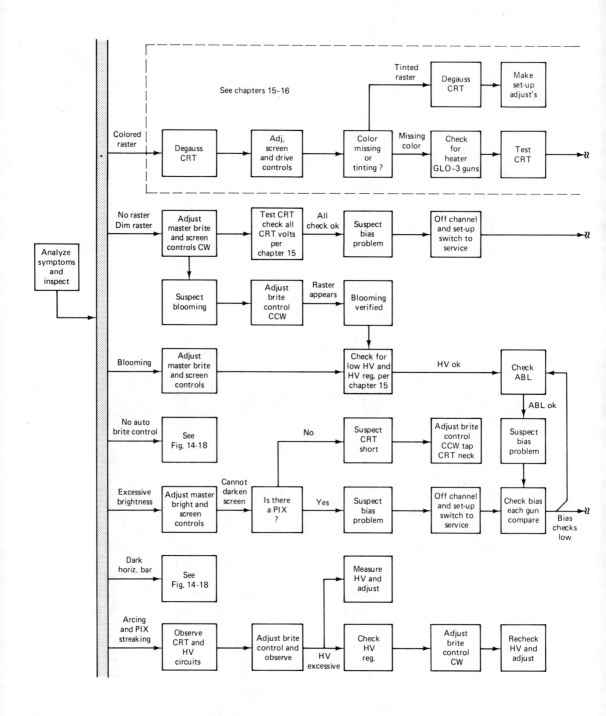

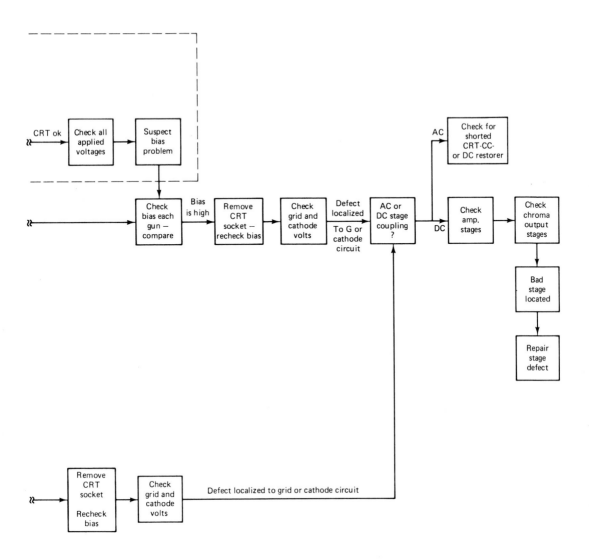

Fig. 14-20 Troubleshooting flow drawing, raster problems, color sets.

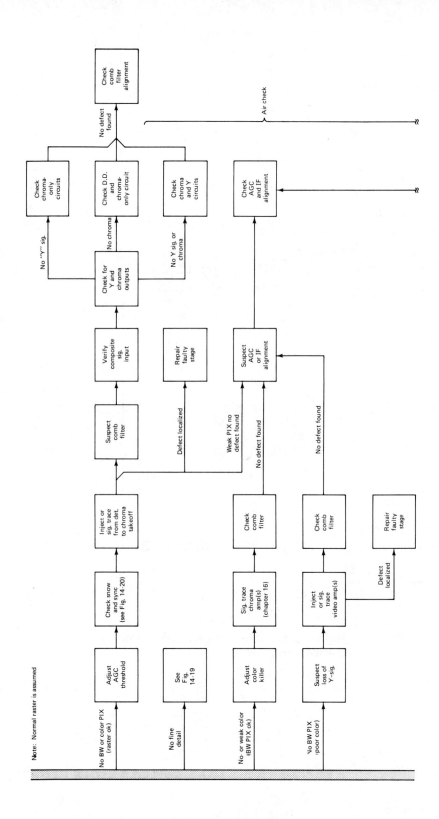

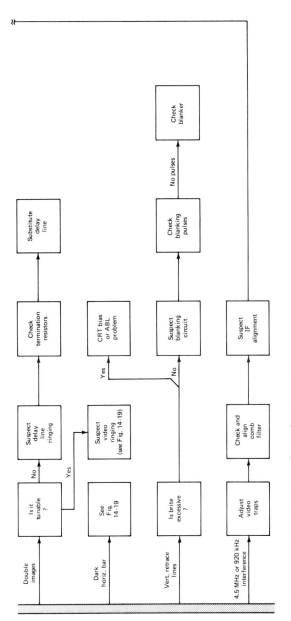

Fig. 14-21 Troubleshooting flow drawing, pix problems, color sets.

599

14-25 ISOLATING THE DEFECT

Once a trouble has been localized, the *cause* can usually be found by making dc voltage checks and by testing the transistor and associated components in the troublesome stage or module. When probing or attaching instruments, take *extra care* to avoid shorts, particularly at the closely spaced connections on PCBs and IC pins. In direct-coupled circuits, a momentary slip-up *can destroy, in domino-fashion, an entire chain of transistors.*

In the troubleshooting procedures that follow, BW and color sets are considered separately, although many troubles apply to both. Symptoms are categorized as either raster or pix problems.

14-26 TROUBLESHOOTING BW SETS

Most of the troubles afflicting a BW set (and the BW functions of a color set) can be considered as either *raster* or *pix* problems. Raster problems (which also show up on a pix) must be considered *first*. Most of the trouble conditions described are applicable to both BW *and* color sets.

14-27 RASTER TROUBLES

Of the many possible defects that can kill the raster or produce an abnormal raster, most involve the CRT, its operating voltages, or setup adjustments, or circuits that are dc-coupled to the CRT, specifically the video amplifier and chroma output stages. Raster problems are listed in Table 14-1; general test procedures are shown in the troubleshooting flow drawings.

- **No Raster or Dim Raster**

For loss of raster, it is customary to start by turning up the brightness control(s) (and the R-G-B screen controls if a color set) and checking for HV and the condition of the CRT, followed by other CRT voltage checks in that order (see Chapter 15). Of particular importance is the CRT bias, in this case, too much bias. A slight increase in normal bias will reduce the beam current and the raster brilliance. Excessive bias can result in beam current cutoff and a dark screen.

Bias is measured between the grid and cathode terminals of the CRT socket while adjusting the brightness control to both extremes. Normal reading should be within about 10 V of zero with the control turned full on. For this problem, look for a bias reading of 20 V or more.

- **Excessive Brightness**

With this condition, the screen cannot be darkened when the brightness control is turned down. There are two possible causes: a grid-to-cathode short in the CRT or loss of bias. As before, measure the bias between grid and cathode while adjusting the brightness control. Normal reading should be at least 30 V or more (depending on the tube) with the control turned ccw. For this problem, look for a reading of 10 V or less.

For a bright raster with little or no pix, and the control has little or no effect, suspect gas or a short in the CRT. Gas shows up as a bluish glow in the neck of the tube. A grid-to-cathode short is often intermittent. Turn down the brightness control and tap the neck until the screen goes dark. If the problem persists, try "flashing the tube," as described in Sec. 15-112.

For a suspected tube short, recheck the bias with the socket removed from the CRT, while adjusting the control. For a color set, where the

14-28 TROUBLESHOOTING CRT BIAS PROBLEMS

control has *some* effect but the raster cannot be totally blanked out, try turning down the master brightness and R-G-B screen controls.

For an abnormal bias condition (either too much or too little bias), start by making two follow-up voltage tests: from the CRT grid(s) to ground, and from the cathode(s) to ground, to determine which circuit is at fault. Compare readings with those shown on the schematic for both exteme settings of the brightness control. Measurements should be taken on a blank channel to prevent bias fluctuations due to the station signal. Subsequent testing depends on the type of control circuit, which is generally different for BW and color sets.

• **Brightness Control at CRT Input (BW sets)**

Most BW sets, as described in Sec. 14-5, have the brightness control in either the grid or cathode circuit of the CRT. Different circuit variations are shown in Fig. 14-6. Almost any defect in these circuits will create a change in bias conditions. This includes troubles in video amplifier stages that are dc-coupled to the CRT. Where ac (capacitive) coupling is used, a leaky or shorted coupling capacitor will cause the same problems. Other circuits that interface with the CRT, such as retrace blanking, are also suspect.

When the problem is no raster or a weak raster, try connecting a jumper between the grid and cathode of the CRT. This results in a zero bias condition. If the raster is restored, an excessive bias problem is indicated. If there is no improvement, the CRT or its operating conditions are suspect. With the jumper in place, no *pix* can be expected. This test is *not* advisable for *color sets.*

After determining which CRT element (grid or cathode) has an incorrect voltage, check for defective components in that circuit. In Fig. 14-6a, for example, if there is an abnormal voltage at the cathode, check for an *open* or *shorted* transistor *or any defect* affecting the transistor conductance or in dc-coupled stages upstream. An open resistor $R1$ is also a possibility. When the grid voltage is incorrect, the control pot, $R2$, and the applied B voltage are suspect.

In Fig. 14-6b, the control is part of a voltage-dividing network where the voltage at the control is largely determined by the relative values of the resistors $R1/R2$. If $R1$ opens up or increases in value, the result is a higher positive voltage on the cathode, which will weaken or kill the raster. A decrease in the value of $R1$ will produce the opposite effect to brighten the raster.

If $R2$ becomes open or its value increases, it results in less positive voltage or no voltage on the cathode, brightening the raster and, in some cases, making the control ineffective. If the value of $R2$ decreases, the cathode becomes more positive to darken the raster.

The same considerations apply to Fig. 14-6c. For example, an open $R1$ will remove the positive from the cathode and cause excessive brightness. Conversely, an open $R2$ removes the grid voltage, resulting in beam current cutoff. A leaky or shorted capacitor feeding blanking pulses to the CRT is quite common.

Control of brightness from a video amplifier stage (Fig. 14-6d and e) is used mainly with color sets. Troubleshooting procedures are as follows.

• **Brightness Control in Video Amplifier**

Any abnormal voltage condition in an amplifier that is dc-coupled to the CRT will have an adverse affect on bias. In Fig. 14-6d, for example, too much positive voltage on the CRT cathode probably means the transistor is not conducting, or not enough. Suspect a defective transistor, insufficient forward bias, or that the transistor is reverse biased. Verify with a voltage check from base to emitter. Check all circuit components and for loss of B voltage. If the CRT cathode voltage is too low, the transistor current is probably too high from too much forward bias, or the transistor may be shorted. For loss of B voltage on the CRT grid, suspect the series-dropping resistor.

In Fig. 14-6e, where the control is in a circuit three stages ahead of the CRT, the defect may be in *any* stage between that point and the CRT input. Start at the CRT and work upstream, taking dc voltage measurements at the input and output of each stage, comparing voltages with those shown on the schematic. Take special note of the required *polarity* at each point tested. In the example, there is a phase reversal in the first and third stages but not in the second stage, which is an emitter follower. With the control arm moved toward ground, the increase in bias should result in an increase in *Q1* current. The negative-going output of *Q1* (the dc control transistor) forward biases *Q2*, increasing its conduction. The negative-going output of *Q2* forward biases *Q3*, and its positive-going output supplied to the CRT cathode results in a decrease in brightness. The opposite conditions should occur when the brightness control is turned cw. When working backward from the CRT, checking voltages and polarities at each stage, the last stage encountered where readings are abnormal *is the one to suspect*. Check the transistor, dc operating voltages, and all components, as appropriate. One common complaint is when the control has a limited range. For this condition, which can be more difficult to troubleshoot, look for *small* deviations from normal voltages at each stage.

14-29 LDR BRIGHTNESS CONTROL PROBLEMS

As with the manual brightness control, the LDR automatic function also varies the CRT bias via a dc control stage in the video amplifier section. Don't confuse this with the ABL circuit. Troubles in the LDR control circuit can sometimes cause problems with the interconnected brightness control circuit, but most often do not.

To check the LDR function, look for variations in raster brightness as room lighting is changed or by covering up the LDR sensor, which should reduce the brightness. When troubles are suspected, first verify normal action of the front panel brightness control, which eliminates the video stages as possibilities. Where there are such problems, perform tests as previously described.

To check the operation of the LDR Fig. 14-6d), test for variations of the transistor emitter voltage with changes in ambient lighting. When the LDR itself is suspect, remove from the circuit and test with an ohmmeter. When in doubt, try a substitute.

14-30 BLOOMING

This is a condition where the raster grows larger and the brightness diminishes with loss of focus as the brightness control(s) and/or the color controls are turned up. In extreme cases the raster blooms into "nothingness," sometimes accompanied by loss of HV. Basically, it is caused by a reduction of HV, which *increases* the deflection sensitivity of the CRT. The most common causes are low HV, poor HV regulation, and anything that drastically increases the CRT beam current, such as a positive bias or, in the case of a color set, failure of the ABL circuit. Frequently, all that is needed is to back off the master brightness and/or the R-G-B screen controls if it is a color set.

14-31 ARCING

Arcing in and around the CRT is quite common, particularly in color sets where the HV may be as high as 30 kV or more. Sometimes the arcing occurs within the CRT, especially when newly installed. Usually the arcing is infrequent and eventually clears up. If it does not, try flashing the tube (Sec. 15-112). Where arcing is intense in the neck of the tube with a strong blue or reddish glow (gas ionization), the tube has lost its vacuum and must be replaced. Components interfacing with the CRT are normally protected by spark gaps at the CRT terminals. But full protection (as with lightning arrestors) is not always 100%.

Arcing outside the CRT represents a fire hazard and must be corrected. Usually it occurs when the brightness is turned down (condition for maximum HV) and when the weather is humid and conducive to arc-over. Procedures for correcting such problems are described in Sec. 9-20.

14-32 HUM BARS

Hum bars, as described in Sec. 13-24, occur when 60-Hz or 120-Hz PS ripple voltage is present at the CRT input(s). It is caused by poor filtering of the LVPS. With tube-type receivers, another cause is heater-cathode leakage in a tube in the signal path. Poor filtering of the B supply feeding the video and chroma amplifiers causes hum bars on the raster *and* pix. P.S. ripple in the tuned stages ahead of the detector produces hum bars *only on a pix*. Check for excessive ac ripple component (with CRO or ac EVM) at appropriate circuit points. For ac readings of 1 V or more, try bridging with a large-value filter capacitor while observing symptoms.

14-33 RASTER AND PIX SHADING

For this condition, the upper portion of the screen is darkened or completely blanked out. It can be mistaken for a hum bar except the edge of the dark area is more sharply defined. It is caused by excessive or prolonged vertical blanking. Check for excessive amplitude of the vertical blanking pulses applied to the CRT or one of the video amplifier stages. Look for a defective component anywhere between the vertical sweep circuit (the point of origin), blanker stages if used, and the point of injection. Compare with pulse levels shown on schematic.

• **Other Retrace Blanking Problems**

Vertical retrace lines that appear when the brightness control is turned up are a common problem that occurs when the CRT is not driven to cutoff during vertical retrace intervals. The lines are produced by *horizontal* beam tracing during this period. Their relatively few number and wide spacing are due to the short retrace time compared to the downward movement of the beam when the pix is being produc   Normally, as the beam is deflected upwar maintained by the vertical bl of the composite signal) and provided by the vertical sv amplitude and *width* must be great enough to sustain beam cutoff for the duration of retrace, *but no longer*. With a weak pulse, retrace lines become visible over most of the screen surface. With a moderately strong pulse the screen may become unblanked too soon, with lines appearing near the top. Advancing the brightness control tends to oppose the blanking by reducing the CRT bias. With a weak CRT, compensation is made by turning the control up full, which worsens the problem.

As for raster shading, scope trace vertical blanking pulses, checking their amplitude against the schematic. If pulses are obscured by video, switch off channel. Typical causes of weak or missing pulses are defective blanker, a leaky or shorted coupling capacitor (especially with high-level injection at the CRT), and *RC* component breakdowns in pulse shaping and attenuating networks.

Similar problems with *horizontal retrace blanking* results in beam current before the completion of each horizontal retrace. On a BW set it shows up as a streaky "windblown" effect on the left side with a faint haze covering the screen. With a color set there may be greenish-yellow smearing on the left side of the screen.

14-34 PIX TROUBLES (BW SETS)

Assuming a normal raster, most pix problems are *signal related*. The usual procedure is to *localize* the troublesome stage by signal tracing, followed by voltage and component testing to pinpoint the defect. Pix trouble symptoms are listed in Table 14-1. Troubleshooting procedures are described next.

14-35 NO PIX OR WEAK PIX

is indicates the signal has been lost or ned before reaching the CRT. Consider other symptoms and conditions that might prove helpful. For example, heavy snow on a raster eliminates the video amplifier(s) as a possibility. If there is no snow, the trouble is between the tuner and the CRT. For loss of pix with normal sound, the stages between the sound takeoff and the CRT are suspect. If there is no pix or sound, check the stages ahead of the sound takeoff. For a weak pix with good sync stability, suspect the stages downstream from the sync takoff. If sync is poor, the trouble may be ahead of the takeoff point.

With the contrast control fully cw, slowly adjust the AGC threshold control, if there is one. If sync becomes unstable before good contrast is obtained, restore the control to its original setting. To locate the stage where the signal is lost or attenuated, use one of the following procedures:

1. Signal trace all stages between the video detector and the CRT using a CRO or other suitable tracer. A demodulator probe is not required. Signal trace either from the detector, stage by stage to the CRT, or working backward from the CRT to the detector, as appropriate. Refer to the schematic for normal signal level and polarity at each checkpoint. Don't expect any amplification or phase reversal from an emitter follower. If there is sync instability, check for clipping or compression of the sync pulses. *Reminder:* When scope tracing between the stage supplied with blanking pulses and the CRT, the display consists of both video and blanking pulses.

2. *Signal injection:* Inject a signal from an AF signal generator working from the detector to the CRT, or in reverse as with signal tracing. Use a blocking capacitor in series with the generator cable. The indicator may be the bar pattern on the screen or a CRO or meter at the CRT input. With a bar pattern, the signal level is judged by the darkness of the bars. Avoid excessive signal that can damage components. Signal levels should be consistent with *station* signals as indicated on the schematic.

A quick way to localize trouble is to bridge a capacitor between the input and output of each stage individually until a pix is obtained or the contrast is improved. Normal results cannot be expected and, depending on the circuit, loss of sync or a negative pix may occur.

When the troublesome stage is located, isolate the defect by voltage and component testing. After repairs are completed, readjust the AGC threshold control.

14-36 EXCESSIVE CONTRAST

Excessive pix contrast is accompanied by smearing that obscures the fine detail. If the control has limited effect, check it and its associated components. Try adjusting the AGC threshold. The most common cause of too much signal is video overload as a result of AGC problems. Another symptom of overload is poor sync due to pulse clipping in a video amplifier stage. Other possibilities are a load resistor that has increased in value or a narrowed IF bandpass due to misalignment.

14-37 VIDEO OVERLOAD

This is a very common problem that has several possible causes. The usual symptoms are excessive contrast, pix smearing with loss of sync, and, in extreme cases, a negative pix. With a color set, color sync may be affected. The symptoms frequently clear up when the signal is reduced by disconnecting the antenna or replacing it with a short piece of wire. Try adjusting the threshold control.

For loss of sync, examine the hammerhead display for indications of pulse clipping (see Fig. 10-8). Pulse clipping usually occurs at one of the amplifier stages and becomes apparent when scope tracing the signal. The stage *where it occurs* may or may not be at fault. Either the stage has a defect resulting in clipping *at normal signal levels*, or it is being overdriven with too much signal. Check the signal level at the stage input and compare with the schematic.

REMINDER: _____

Pulse clipping beyond the sync takeoff has no effect on sync stability.

As described earlier, for signal overdrive, check for loss of AGC and the possibility of IF misalignment.

14-38 VIDEO RINGING

Sometimes mistaken for ghosting, the visual effects of ringing (high frequency oscillations around 3 MHz) are outlines around the fine pix detail. The repeat images are tunable (i.e., they change with the fine tuner). They can be caused by IF misalignment, overpeaking of the high video frequencies, or an open resistor across one of the peaking coils in the detector or video amplifier stages. To localize the problem, scope check the amplifier stages. Damped oscillations will be observed on the scope display when the pix on the screen has considerable fine detail. Check the shunt resistors across peaking coils in the troublesome stage. Where warranted, try lower-value resistors or replacing the coil(s). Check the peaking (crispening) control if there is one and its associated components.

14-39 HUM BARS ON PIX ONLY

When hum bars (see Sec. 14-32) are observed on a pix but not on an off-channel raster, check for poor filtering of the dc power supplied to the tuner and IF stages. Poor filtering of the B voltage feeding the video amplifier stages results in hum bars that *also appear on the raster.* The low-frequency ripple component cannot get through the high-frequency tuned stages except as modulation of the station signal that serves as a carrier, but 60 and 120 Hz is well within the bandpass of the video amplifiers.

14-40 PIX SMEARING

This can be caused by video overload (excessive contrast), along with symptoms of loss of sync, depending on the point of sync takeoff. See if the smearing and sync problems clear up when pix contrast is reduced. If there is no improvement, there are three possibilities: a minor ghosting problem caused by impedance mismatching in the antenna system, IF misalignment, or poor low-frequency response of the video amplifier(s). IF misalignment (where the pix IF carrier is located too low on the response) reduces the gain of the low video frequencies, which are close to the carrier. Sync pulses are low frequencies also, so they too are affected, along with the contrast of large objects.

When the low-frequency response of the video amplifier is suspect, check the relative gain of each stage by injecting a low-frequency signal (around 50 to 100 Hz) using an AF generator. When the trouble is localized, check for open or dried-up capacitors in the stage, particularly bypassing and coupling capacitors (electrolytics). Try bridging each suspect with a substitute capacitor of the correct value. Avoid surge damage to transistors by removing power while making and breaking connections. If warranted, perform an overall frequency response check (Sec. 14-59).

14-41 LOSS OF FINE DETAIL

Perfect focus is *essential*. Make sure the scanning lines are in sharp focus. Poor high-frequency response of the video amplifier results in poor resolution of the fine pix detail, but there are other possibilities: IF misalignment (narrow bandpass), an impedance mismatch in the antenna system, and video overpeaking or ringing.

When poor high-frequency response is suspected, check the relative gain of each amplifier stage by injecting a 3-MHz signal from an RF generator. A sensitive indicator is required because the output of most RF generators is quite low. When the stage with low amplification to high frequencies is located, check the transistor and, if questionable, replace it. Check for an increased value of load resistance.

14-42 INTERFERENCE

A 4.5-MHz beat signal is produced by the mixing of the 45.75-MHz pix IF carrier and the 41.25-MHz sound IF carrier in the video detector. If permitted to reach the CRT, the pix is degraded by a fine-grained interference pattern similar to that produced when the fine tuner is misadjusted. The interference is normally prevented by using one or more 4.5-MHz traps between the detector and the CRT, and by aligning the IFs with the sound IF marker no higher than 10% on the response curve. This also helps prevent intercarrier buzz in the sound.

If troubled with a beat pattern, check the resonance of the 4.5-MHz trap(s) by injecting a 4.5-MHz signal from an *accurate* signal generator to some point ahead of the tuned circuit(s) being checked. Connect an ac EVM or other suitable indicator at the CRT input. Tune the generator for the lowest dip and note the frequency. If necessary, tune the generator to *exactly* 4.5 MHz and, using a fairly strong signal, adjust coil slugs for a dip on the meter.

If an *accurate* generator is not available, carefully adjust the fine tuner for best pix and sound. While observing the pix, adjust the coils until the interference clears up. If the problem persists, check the IF alignment as in Sec. 13-30.

14-43 TROUBLESHOOTING COLOR SETS

Most of the procedures just described for a BW set also apply to color sets; the procedures that follow are peculiar to color sets only. In general, troubles associated with a color set are divided into three categories: *raster problems, BW pix problems,* and *color pix problems.* Thus the order of priorities in troubleshooting is, first, to obtain a normal *white* raster, then a BW pix, and, finally, the color pix.

Typical trouble symptoms are listed in Table 14-2. As an aid to troubleshooting, see the flow drawings, Figs. 14-18, through 14-21. Certain color problems associated with the video and matrixing output amplifiers are described at this time. Most color problems are covered in Chapter 16.

14-44 RASTER TROUBLES

Conditions for obtaining a raster on a color set are much different than for a BW set, and the number and kinds of defects that can adversely affect the raster or kill it are correspondingly greater. With a color CRT, there are three guns *all working together* to develop the raster. The proper operation of each gun is dependent on the condition of the gun, its operating voltages, and certain setup adjustments. Some of these factors affect the operation of each gun individually; others affect all three guns at the same time.

Bias is the most critical factor, and abnormal bias is the cause of many problems. Most are caused by defects in dc-coupled signal stages feeding the three guns, specifically the video and color output circuits. When a defect changes the bias of all three guns, the overall brightness is affected. When the bias of individual guns is affected, it upsets the color balance and the "whiteness" of the raster. For example, when a gun is either weak or overbiased (due to a defect in its associated circuit), the color contributed by that gun is either lost or weakened. Since it requires all three primary colors to produce *white,* the result is a *color-tinted* raster.

14-45 BIAS CHECKING

When bias troubles are suspected, measure the voltage between the grid and cathode of each of the three guns for both extreme settings of the brightness control. This should be done under no-signal conditions by switching off channel and with the service setup switch at the service position. When a bias problem exists, examine the schematic to determine, if possible, what defects could account for the condition. As described earlier for BW sets, check the voltage between grid and ground and from cathode to ground (on the gun with the problem) to determine which of the two circuits is at fault. Subsequent testing is normally restricted to that circuit only.

Where ac coupling is used between the amplifier and the troublesome gun, the coupling capacitor may be leaky or shorted. Where dc coupling is used, proceed as described in Sec. 14-28 to localize the problem. Typical troubles in the high-level stages are leaky or shorted capacitors and *open* or *shorted* transistors. Such shorts can in turn damage other components, which can often be found by inspection. Shorted spark gaps are not uncommon. Feel the video and chroma output transistors. Either an overheating transistor or a cold transistor can provide a good clue; the hot transistor being caused by internal leakage or too much forward bias, and, for the cold transistor, an indication that it is not conducting.

608 THE VIDEO AMPLIFIER

14-46 NO RASTER OR DIM RASTER

Troubleshooting for this condition usually begins by checking for HV and the condition of the CRT. This is followed by measuring all voltages at the CRT socket while adjusting the brightness control(s) and the R-G-B screen grid controls (see Sec. 15-71). For the possibility of severe *blooming,* try *backing off* the brightness control. In a brightly lighted room a very dim raster may be mistaken for *no raster*. Since brightness increases with reduced beam deflection, try switching the setup switch to the service position and a bright line may become visible. If no setup switch is provided, turn down the height control.

When excessive bias (more than 10 V with brightness control fully cw) is measured between the grid and cathode of one or more guns, take readings from grid(s) to ground and from cathode(s) to ground to determine which circuit is at fault. For symptoms of no raster or dim raster, either the cathode(s) will have too much positive voltage or the grid(s) not enough.

Start by checking the power output transistors, either video output or matrixing transistors, depending on the circuit. Either an *open* or a *shorted* transistor will change the dc voltages when direct coupling is used. For ac coupling, a leaky or shorted coupling capacitor will cause the same problems.

Starting at the CRT and working upstream, check input and output voltages at each dc-coupled transistor as described earlier (see Sec. 14-28 and Fig. 14-6e). Check for voltage variations as the brightness control is adjusted. Compare all readings with those shown on the schematic. Suspect the last stage checked that has abnormal voltages. As an alternative approach, start at the detector, working toward the CRT until a stage with abnormal voltages is located. Check the transistor in the suspected stage, its applied voltages, and associated components, as appropriate.

14-47 EXCESSIVE BRIGHTNESS

If unable to darken the screen with the brightness control(s), suspect either a grid-to-cathode short in the CRT or loss of CRT bias. Another symptom of a CRT short is little or no pix. When suspected, try tapping the neck of the CRT with the brightness control *turned down* as described in Sec. 14-27.

When the CRT bias measures less than about 30 V with the control turned down, and the troublesome circuit feeding either the grids or the cathodes has been identified, check voltages at all dc-coupled stages working toward the detector. The faulty stage is probably the last one checked having the abnormal voltages. For this problem (loss of CRT bias), look for a voltage condition *opposite* to that obtained when the CRT bias is *excessive*. When the faulty stage is located, check the transistor, its operating voltages, and associated components as appropriate. Typical causes of this complaint are a shorted clamp diode at the CRT input and transistors that are either *open* or *shorted*.

REMINDER: _____

An accidental short while probing, or when there is a shorted transistor or capacitor, may destroy a *number of transistors* between that point and the CRT.

14-48 ABL TROUBLESHOOTING

The automatic brightness limiter (Sec. 14-15) protects the CRT and other components from excessive beam current. A sensing voltage, proportional to the beam current, is applied to one of the video amplifier stages that is dc-coupled to the CRT. If beam current becomes excessive for any reason, voltage changes in the amplifier(s) cause an increase in CRT bias to reduce the current to a safe value. The ABL circuit usually operates in conjunction with the brightness control(s) and the LDR control function when provided.

ABL troubles may be indicated when there are symptoms of excessive brightness with blooming. Check for overcurrent sensing voltage at its source, the horizontal output–HV circuit. Create a temporary overcurrent by turning up the HV, brightness, and CRT screen controls to produce symptoms of severe blooming. Verify that the sensing voltage (either dc or the ac ripple component) changes as the brightness control is adjusted. Check for an *increase* in CRT bias (and a reduction in brightness) as the control is turned full on.

When there is no ABL action, verify the presence of sensing voltage at either the dc control amplifier in the video section or the ABL control stage if there is one. Check for voltage variations at the control stage and the amplifiers as the brightness control is adjusted.

NOTE:

The control stage is normally biased to cutoff and does not conduct until the sensing voltage reaches a critical level.

When the stage is suspect, check the transistor, its operating voltages (particularly the B to E bias which is critical), and all associated components as appropriate. After repairs have been completed, readjust the HV and other controls as necessary.

14-49 BLOOMING

This is a common problem with color sets because of the relatively high beam current and loading of the HV supply. In most cases, blooming can be cured by proper setup adjustments, usually by turning down the master brightness and CRT screen controls. Sometimes it is a combination of problems: low HV, borderline operation of the ABL circuit, poor regulation of the HV power supply, plus insufficient CRT bias. With most large-screen sets, some blooming is normal, and pix enlargement up to ¼ in. or so on all sides at full brightness is generally considered acceptable. In extreme cases, due to loading of the HV, the raster may be lost entirely. Blooming is usually accompanied by some defocusing. When the raster is lost, try backing off on the brightness control to see if it reappears.

For severe blooming, start by cutting back on the master brightness and the R-G-B screen controls. If the problem persists, measure the HV with no load (brightness control turned down), and again for full load with the control fully cw. With proper HV regulation, the voltage should not drop more than 1 kV or so with a 30-kV supply. The greater the drop the poorer the regulation. Troubleshoot the HV and regulator circuits as in Sec. 9-19. Adjust the HV to the amount specified on the schematic.

When the ABL circuit is suspect, check its operation and troubleshoot as described in Sec. 14-48.

14-50 ARCING

Arcing in any set can have serious consequences, especially a color set where it is quite common. Besides being a fire hazard, it can destroy expensive components such as power output transistors, flyback transformer, and even the CRT. Most color sets use spark gaps extensively, which offers some protection (see Sec. 14-9).

Learn to distinguish between the two kinds of arcing, LVPS arcing when the voltage is relatively low but the current is high, and HV arcing, which is the reverse, high voltage at low current. LV arcing can occur at almost any circuit point, usually to ground, when there is a buildup of conductive residue or a deterioration of insulation. HV arcing may occur in or around the flyback transformer, the CRT, and its associated components. Arcing is usually visible on close inspection in a darkened room. It also produces telltale sounds and smoke, which helps to pinpoint the trouble site. Corona is a type of HV discharge that is recognizable by its hissing sounds and the sweetish odor of ozone. HV arcing is most noticeable when the HV is maximum as the brightness is turned down. Measure the HV and adjust to the value specified on the schematic. Check the operation of the HV regulator (Sec. 9-21). For other particulars on arcing, see Sec. 9-15.

14-51 SERVICE SETUP SWITCH PROBLEMS

The slide-type switch sometimes becomes erratic, causing loss of vertical sweep, loss of pix, and other problems. Cleaning with a suitable solvent usually helps. Allow time for the solvent to evaporate before applying power. Internal LV arcing is fairly common and can trip the circuit breaker. If arcing persists, replace the switch.

14-52 COLOR-TINTED RASTER

Since a proper mixture of the three primary colors is required to develop a *white* raster, it is not surprising that with many sets sooner or later the color balance is upset and the raster takes on a grayish, off-white appearance, or color tint. In most cases, where the problem seems minor, only a touch-up of the service adjustments is required. The adjustments (as described in Sec. 15-118) are made in the following sequence:

1. Adjustment of the R-G-B screen controls, then the drive controls (gray-scale tracking).
2. CRT degaussing, if needed.
3. Purity adjustments, if needed.
4. Repeat step 1.

When setup adjustments alone do not correct the problem and the raster is a *distinct color,* the possibilities are that one or more of the CRT guns is either weak or dead, or there are improper operating voltages on one or more guns (bias or screen voltage). Switch off channel and with the screen controls turned up examine the raster, noting which color(s) is missing in the mixture. Visually check for heater glow of all three guns.

Check for a red, green, and blue raster as follows: with brightness on full, turn down the screen controls (or color cutoff controls where the set uses pre-CRT matrixing) cut off all three guns and darken the screen. Turn up the red screen or cutoff control to get a red field. Turn off the red gun. Repeat the procedure to obtain, if possible, a green field and a blue field. When one or two colors are weak or missing, suspect weak or dead guns or a voltage problem. Check the CRT, particularly the guns that are suspect. With one or two weak guns, tinting often changes during warm-up.

If the CRT checks normal, measure the screen voltage applied to all three guns, noting variations as the controls are adjusted. Compare with voltages shown on the schematic.

REMINDER: ⎯⎯⎯⎯⎯⎯⎯⎯⎯⎯⎯⎯⎯⎯⎯

With the pre-CRT matrixing system, the screen grids are tied together and supplied from a common B+ source. Beam currents of individual guns are controlled by *bias* variations with the controls in the three output amplifier circuits.

When the trouble still exists, suspect a bias problem. With the setup switch in the service position and the brightness control turned up, check for excessive bias on the troublesome gun and, if found, which circuit (grid or cathode) is suspect. Depending on which matrixing system is used, the fault may be with either the video *or* color stages. To isolate the defect, check for abnormal voltages as described in Sec. 14-28.

14-53 PIX TROUBLES

Whereas a normal white raster is prerequisite to a BW pix, a good BW pix, free of contamination, is required before you can expect good color. Raster problems are caused by a defective CRT, incorrect operating voltages, or adjustments; *pix* problems involve the *signals,* the Y signal or the chroma signals. Color pix problems described at this time are limited to those caused by the video amplifiers and matrix output circuits. For other color troubles, see Chapters 15 and 16.

14-54 NO PIX OR WEAK PIX

For loss of both BW and color pix, suspect video stages ahead of the chroma takeoff. If there is no BW pix but *some* color, look for trouble in the video section *downstream* from the chroma takeoff.

To localize the defect, follow the procedures described for a BW set (Sec. 14-35), signal tracing with a CRO or injecting a signal from an AF generator. The *delay line,* which is easy to locate and is approximately midway in the video section, is a good starting point. When scope checking, refer to the schematic for proper amplitude and polarity of the signal at each stage.

REMINDER:

At and beyond the stage where blanking pulses are introduced, the CRO display will consist of both video and blanking pulses.

When the faulty stage is located, check the transistor, its operating voltages, and associated components. An open, shorted, or leaky transistor is the most common defect, especially in the output stage. Feel power transistors as a check on their operating temperature. Suspect trouble when they are either too warm or too cool.

An open or shorted delay line will kill the pix and sometimes the raster. If an open coil is suspected, bridge with a wire to see if the raster and pix are restored. In late-model sets, a defect in the comb filter can weaken or kill the signal.

14-55 COLOR PIX (NO BW PIX)

This symptom indicates loss of the Y signal between the chroma takeoff point and the CRT. There is nothing to prevent the chroma signal from reaching the CRT, but because of its limited frequency range (0 to 0.5 MHz), the pix quality will be very poor. Troubleshooting procedures are the same as for no pix or weak pix (Sec. 14-54).

14-56 NO COLOR OR WEAK COLOR

If there is a normal BW pix, the trouble is probably in the chroma stages. Try adjusting the killer threshold control. If the BW pix is also weak (possibly with snow), suspect stages ahead of the detector or a weak reception problem. A strong signal is necessary to produce color in some cases. Other causes of color loss are described in Sec. 16-25.

14-57 BW PIX OK, POOR COLOR PIX

Poor color can mean a lot of things and for this symptom there are many possibilities, for example, weak color, wrong colors, poor color sync, things not considered at this time. Many of these possibilities can be eliminated, however, if there is a normal raster and a BW pix that is free of color contamination.

REMINDER:

Conditions involving the CRT, its operating voltages, and setup adjustments will affect both the raster and the BW pix. Troubles in these areas must be corrected *before* considering color problems.

This narrows the possibilities to things that have a bearing on the *chroma signals* (such as IF misalignment) and circuits that *are not dc-coupled* to the CRT, in particular, the R-G-B matrixing stages.

If one or more colors are weak or missing from the pix, decide which ones and look for trouble in the *corresponding* color stages that are *ac-coupled* to the CRT. Scope check input and output signals of suspected stage(s). When there is little or no amplification, isolate the defect by checking the transistor(s), operating voltages, and components. With the pre-CRT matrixing systems, if the BW pix is normal, chances are the output amplifiers are OK.

As an alternative to scope checking, signal injection may be used. With an AF signal fed to the R - Y amplifier, you should get a black and red bar pattern on the CRT. When fed to the G - Y amplifier, a black and green pattern should be obtained, and black and blue bars when fed to the B - Y amplifier. Relative gain can be judged by the intensity of the colors. For troubles in the upstream chroma stages, see Sec. 16-23.

14-58 TROUBLESHOOTING THE COMB FILTER

A defect in the comb filter can produce the same trouble symptoms as defects elsewhere in the video amplifier section: loss of either BW or color pix or both, loss of fine pix detail, poor color quality, and so on. When the comb filter is responsible for complete loss of pix, the initial signal tracing will normally localize the fault. For less drastic symptoms, scope tracing the signals *within* the circuit is usually required. In the absense of a CRO when the comb filter is suspect, try connecting a capacitor from the detector to the first video amplifier to see if the BW pix is improved or restored. For loss of color, connect the capacitor to the input of the first chroma stage.

Since the luminance-only signal is dependent on a properly operating chrominance-only circuit, the latter is normally checked *first*. If the CRO shows no luminance output (Q 401 collector for the circuit of Fig. 14-17), check for chroma-only output at the emitter of Q 402. If the signal at that point is OK, check for a defective Q 401. If necessary, check its operating voltages and all components in the luminance-only circuit.

If signal at the Q 402 output is missing or abnormal, but OK at the detector, suspect the chroma-only circuitry. Check for signal at the input and both outputs of Q 404. For loss of signal at both outputs, check the transistor, its operating voltages, and associated components. Check for signal at pins 1 and 2 of the DD. Check for the *delayed* composite signal at pins 3 and 4 of the DD, the base input of Q 402, and the emitter of Q 401. Check Q 402 and its operating voltages. Check for bias variations as the R 426 control is adjusted. When input signal to the DD checks OK, but there is little or no output, try a substitute DD after first determining there are no external defects. A dual-trace scope can be useful here for simultaneous viewing of the delayed and nondelayed composite video signals.

- **Alignment**

Unless adjusted properly, the benefits of a comb filter are lost and, in some cases, receiver performance is worse than without a filter. Alignment is critical and requires an accurate bar generator that is *phase locked* to the station signals, specifically the horizontal sync pulses. Lacking the proper equipment, an ordinary color bar generator can be used, but results are less than perfect.

For the circuit of Fig. 14-17, the chroma null amplitude control R 401 adjusts the strength of the delayed chroma-only signal applied to Q 401. The luminance null amplitude control R 426 is adjusted for maximum rejection of the luminance signal. The chroma null phase adjustment L 426 varies the amount of delayed chroma that is mixed with the nondelayed video. The luminance null phase adjustment L 427 determines the band of frequencies (3 to 4 MHz) that is applied to the DD. The controls R 426 and L 427 are normally not adjusted in the field. Adjustment procedures vary with different sets. For specific procedures, refer to the receiver service notes. When adjustments are completed, check BW and color pix for quality, detail, and freedom from beats.

14-59 CHECKING FREQUENCY RESPONSE (FIG. 14-3)

When loss or attenuation of the low or high video frequencies is suspected, a response check can be made as follows. Inject a signal from a suitable test generator to the input of the first amplifier stage. For high frequencies, use an RF generator with a small series blocking capacitor. For low frequencies, use an AF or function generator with a large blocking capacitor (at least 10 MF). As an indicator, connect a CRO or ac EVM to the signal input of the CRT. Adjust generator output below the level where overload may occur. Vary generator tuning, recording the output for different frequencies, and compare.

NOTE:

RF generators have less output than AF generators so it is difficult to compare the two.

14-60 DELAY LINE (DL) PROBLEMS

An open DL where dc-coupling is used between that point and the CRT usually results in loss of the raster. With ac-coupling, the Y signal will be lost. A grounded DL can also have the same effect. A receiver without a DL would have ghostlike symptoms where the color pix is displaced to the right of the BW pix objects. An improper replacement DL can produce the same effects, but to a lesser degree.

Termination resistors at each end of a DL are of particular importance. An open resistor can cause *ringing* and ghosting in the pix. Such multiple images are not tunable and are not to be confused with misconvergence, video ringing

614 THE VIDEO AMPLIFIER

in peaking circuits, an antenna mismatch, or multipath reception ghosts.

When checking a DL for leakage or the termination resistors, don't be misled by the shunting resistance of associated transistor junctions. Isolate the suspected component or take two readings, one with the ohmmeter leads reversed. When replacing a DL use only the *correct* replacement.

14-61 INTERFERENCE

A type of interference peculiar to a color set is the 920-kHz beat interference produced by the mixing of the 4.5-MHz sound IF carrier with the 3.58-MHz color subcarrier. It shows up as a fine-grained pattern on the pix, much like the wormy effect caused by 4.5-MHz beat interference as described in Sec. 14-42. Both kinds of interference occur when a fine tuner is misadjusted. It is normally prevented by the use of traps that keep the 4.5- and 3.58-MHz signals from reaching the CRT via the video amplifier stages. Many sets have two 3.58-MHz traps, one at the chroma takeoff and another in the video amplifier section. They are normally adjusted at the same time, using the procedure described in Sec. 14-42. The beat interference can be observed on the back porch of each horizontal sync pulse with a CRO connected at the detector output.

14-62 IC PROBLEMS

With some sets, the low-level video amplifier stages are contained in one or more ICs. Troubleshooting is limited to checking input and output signals and applied voltages, followed by substitution when the IC is suspect. Check all associated external components *prior to* substitution to avoid possible damage to the new unit.

14-63 INTERMITTENTS

Almost all the symptoms indicated in Table 14-1 can be intermittent in nature. As long as the symptoms are present, troubleshooting procedures are the same as when the symptoms are constant. There are two ways to approach such problems: wait for the symptoms to develop or try to *create* the symptoms (Sec. 5-13). For loss of raster or video, the most common problems, a dual-trace scope can be very useful. Simultaneously monitor the signal at widely separated points in the video amplifier section, gradually narrowing the problem to a single stage by scoping its input and output.

One common problem is an intermittent CRT (open or shorted elements) where the pix is lost and the screen cannot be darkened (see Sec. 14-27). Dirty contacts in a service setup switch also cause intermittent problems. Try wiggling the switch while observing the pix.

14-64 AIR CHECKING THE REPAIRED SET

After repairs are completed, allow the set to operate for a minimum of 30 min. Note any abnormal conditions that develop *during* or *after* warm-up. If not done previously, now is the time to make final touch-up adjustments of certain controls, such as master brightness, R-G-B drive controls, AGC threshold, and the color killer. Check BW pix for color fringing and if necessary, touch-up the purity and convergence adjusters as described in Sec. 15-118.

14-65 SUMMARY

1. The brightness of a raster is determined by the intensity of the CRT beam current. Beam current depends on the bias voltage, which is controlled by the brightness control(s). When turned fully cw, the near-zero bias results in high beam current and a bright raster. At the full ccw setting, the bias is maximum to cut off the CRT and produce a dark screen.

2. Other receiver functions besides the brightness control operate together to control the CRT bias and brightness level, for example, automatic brightness control with an LDR sensor, ABL as used in color sets, vertical and horizontal retrace blanking, and the signal itself. Amplitude variations of the video signal add to and subtract from the normal bias voltage to create brightness variations of the reproduced pix objects.

3. Three very common raster problems are no raster, dim raster, and excessive brightness. Factors affecting brightness are the condition of the CRT, the amount of HV, and other applied voltages. Bias is the most critical voltage, and abnormal bias is often caused by defects in video and chroma amplifiers that are dc-coupled to the CRT.

4. The brightness control for BW sets is usually in the CRT grid or cathode circuit. With color sets, the control is in a video amplifier stage that is dc-coupled to the CRT. Adjusting the control varies the bias and conductance of the control transistor. With dc-coupling, dc voltage variations are passed from one stage to the next up to the CRT input.

5. It is impossible to obtain good pix contrast with a weak signal even with normal bias. The pix will have a washed-out appearance where both bright and dark objects show up as gray. Turning up the control will produce white but no black. With the control turned down you get black but no whites.

6. Pix contrast is determined by the peak-to-peak amplitude of the composite video signal. Conditions causing loss of contrast include insufficient CRT bias, poor reception, and defects in the signal path between the antenna input terminals and the CRT, that is, the tuner, IF strip, detector, and video amplifiers. Excessive contrast is caused by too much signal.

7. The frequency response of the video amplifier(s) must be relatively flat from near zero to over 4 MHz. Poor low-frequency response reduces the contrast of large objects that are produced by frequencies in the 0- to 1-MHz range. Poor high-frequency response (in the 3- to 4-MHz range) causes loss of the fine pix detail.

8. For good *low-frequency* response, a video amplifier uses large values of coupling and bypass capacitors and/or dc coupling between stages. For optimum *high frequency* response, circuit impedances must be kept low by using fairly low value load resistors with each transistor. Distributed circuit capacities must be kept low. Peaking coils and video peaking are used in most sets to sharpen the pix.

9. Video peaking is desirable to sharpen or crispen a pix, but too much peaking can cause ringing. Video ringing causes ghostlike outlines on fine pix detail. Damping resistors are sometimes connected across peaking coils to reduce their Q to prevent this problem, and to broaden the PC response.

10. The video (or Y signal) in a color set has two functions: to develop the fine pix detail that is not reproduced in color, and to establish the brightness level of the various colors in the pix.

11. Until the comb filter became popular, all color receivers separated the chroma from the video at the first or second video amplifier stage following the detector. Separation was accomplished with a tuned coil broadly resonant to 3.58 MHz, the color subcarrier frequency. The takeoff coil and the first chroma stage(s) had a relatively narrow bandwidth to pass *only* the chroma signals extending from about 3- to 4 MHz.

12. With pre-CRT matrixing, each of the three matrix amplifiers has two inputs, a demodulated chroma signal (R - Y, G - Y, or B - Y), applied to the base of the transistor, and a BW signal applied to the emitter. There is no phase reversal between the base input and the collector output driving the CRT cathode.

13. A comb filter provides better separation of the luminance (BW) signals and the chroma signals. This results in an enhanced pix with sharper detail and eliminates beat interference on the pix. Beat interference is caused by the overlapping of the video and chroma signals that coexist in the IF bandpass.

14. Separation of the chroma and luminance signals in a comb filter is accomplished by phase cancellation. Each successive line of pix information is delayed and inverted, and then compared with the next line of information. The out-of-phase luminance signals cancel out, leaving only the chroma signal. To produce the luminance-only signal, the delayed chroma cancels the chroma in the nondelayed signal, leaving only the luminance information.

15. With ac coupling, the coupling capacitor(s) pass the signal but block the dc component, which is necessary to obtain the correct background brightness of each scene being televised. Ideally, this requires the dc level to be restored at the CRT input. A typical dc restorer consists of a diode that rectifies the signal peaks to develop the proper dc voltage. Direct-or dc-coupled stages do not require a restorer since the dc component was never lost.

16. Many sets have provision for automatically regulating the screen brightness in accordance with ambient room lighting conditions. An LDR (light-dependent resistor) located on the front of the set senses changes in the light falling on the screen. As the light increases, the LDR, via a dc control stage, causes the CRT bias to increase, reducing the beam current and the pix brightness. The reverse occurs with a decrease in room lighting.

17. Most color sets have protective spark gaps connected to certain terminals of the CRT socket. These devices prevent damage to circuit components in the event of arcing in the CRT.

18. The CRT must be biased to cutoff during vertical and horizontal retrace periods. This is accomplished with blanking pulses, samplings of the vertical and horizontal sweep circuits, that are applied directly, or via the video amplifiers, to the CRT. For greater effectiveness, most color sets use blanker amplifiers. Without blanking, the beam retrace becomes visible, for example, the vertical retrace lines often observed when the brightness is turned up.

19. There are several takeoffs in the video amplifier section, for example, the sync takeoff and AGC takeoff when keyed AGC is used. With a BW set the sound takeoff is either at the detector or a video amplifier stage. With a color receiver there is a chroma takeoff from either a video stage or from the comb filter when used.

20. With a color set, it is important that the luminance and chroma signals reach the CRT at the same time. When the two signals do not coincide, the pix objects produced in color will be displaced from those produced in BW, creating a ghostlike effect. A delay line is included in the video section to provide the necessary delay of the BW signal, by approximately 1 μs.

21. A color CRT can be damaged by excessive beam current. Excessive current results in excessive brightness, poor contrast and color, blooming, and defocusing. An ABL (automatic brightness limiter) circuit helps prevent these problems by sensing changes in beam current and increasing the bias as necessary.

22. Matrixing is the mixing of the luminance signal and the three primary colors (red, green, and blue) in the proper proportions to produce white and all complementary colors. There are two ways of doing this: (a) in the CRT (CRT matrixing) where three discrete color signals drive the three guns by different amounts, with the luminance signal applied *to all three guns,* and (b) by mixing the luminance and chroma signals in three matrixing amplifiers that individually drive the three guns. This is called pre-CRT matrixing and is currently the preferred method.

23. A white raster is required before you can obtain a BW pix that is free of color contamination. White light is produced by a mixture of 30% red, 59% green, and 11% blue. This requires a correspondingly different amount of beam intensity for each of the three guns. Individual beam currents are regulated by red, green, and blue screen voltage controls and, in some cases, drive controls.

24. A color-free BW pix must be obtained before you can get a good color pix. This requires the proper amount of signal drive for each of the three guns. This is the function of the R, G, and B drive controls. They are adjusted to obtain the whitest possible raster in a process called gray-scale tracking. This means uniform whiteness for all settings of the brightness control.

25. Most color receivers have a service setup switch. It is used as an aid in setting up the various CRT controls and adjusters. In the service position, the switch disables the vertical sweep to produce a thin horizontal line, and interrupts the signals so they will not affect the bias and beam currents of the three guns.

26. With a color CRT, there are three guns all working together to produce the raster, and raster problems occur when defects prevent normal operation of *any* or *all* of the guns. This includes a weak or dead gun or improper operating voltages, particularly screen and bias voltages. With dc coupling, practically any defect in the video amplifier stages between the detector and the CRT will create such problems.

27. When matrixing takes place in the CRT, the CRT is supplied with two kinds of signals, the demodulated R - Y, G - Y, B - Y signals (no luminance added), plus the separately applied luminance signal from the video amplifier. With pre-CRT matrixing, the chroma and luminance signals have already been combined, and the matrixed output signals driving the CRT are simply known as red, green, and blue signals.

REVIEW QUESTIONS

1. State three ways of biasing a BW CRT.

2. What change in pix brightness occurs by (a) decreasing the negative voltage on the CRT grid? (b) making the cathode more positive? Explain why in each case.

3. What is the required signal polarity when feeding (a) the grid of the CRT? (b) the cathode? Explain why, in each case. (c) What is the effect if signal polarity is reversed?

4. Explain how brightness is controlled in a color set.

5. (a) What change takes place in a pix when the signal amplitude is increased? Explain why.
(b) What control is used for this purpose?
(c) Where is it located in the circuit?

6. What is the function of the (a) CRT drive controls? (b) screen controls?

7. What would be the effect on a pix if a delay line were not used? Explain why.

8. A weak or missing luminance signal results in a color pix having very little detail. Explain why.

9. (a) Explain why vertical retrace lines sometimes appear when a brightness control is turned full on.
(b) Why is it more apt to occur on a weak station?
(c) What receiver circuits normally prevent this problem?

10. (a) Explain the interleaving of chroma and luminance information in the composite video signal. (b) What problems does it create? (c) Why are they combined in this manner?

11. Although both the grid and cathode of a CRT are supplied with dc *from the LVPS,* a dc restorer is still needed if there is a capacitor in the signal path from the detector. Explain why.

12. How is the frequency response of the video amplifier affected by (a) using a smaller value of coupling capacitor? (b) a larger coupling capacitor? (c) increasing the value of the load resistor?

Review Questions

13. (a) Why are peaking coils used?
(b) Why are they often shunted with a resistor?
(c) Such resistors shown on a schematic may appear missing in the chassis. Explain why.

14. The frequency response for a small-screen set can be considerably less than 4 MHz. Explain why.

15. Why is a comb filter delay device only required to pass frequencies between 3 and 4 MHz?

16. What prevents the chroma signal from reaching the CRT via the video amplifiers?

17. State two advantages of pre-CRT matrixing over CRT matrixing.

18. What is the purpose of a 3.58-MHz coil at the point of chroma takeoff?

19. Each transistor in a dc-coupled amplifier gets its bias from the previous stage. Explain.

20. Why are emitter followers used in the video section considering they provide no amplification?

21. Explain the purpose of a video peaking (crispening) control.

22. What is the purpose of retrace blanking? How does it work?

23. What signal(s) are present at the cathodes of a color CRT when (a) matrixing takes place in the CRT? (b) when the pre-CRT matrixing system is used?

24. What design considerations are built into a video amplifier to ensure good (a) low-frequency response? (b) high-frequency response?

25. (a) Explain the purpose of the 3.58-MHz trap in the video amplifier of a color set.
(b) Is such a trap required in a BW set when receiving color transmissions?

26. What is the purpose of a blanker, and why is it found mostly in color sets?

27. What is the expected result if the time constant of a blanking circuit is (a) too short?
(b) too long? Explain your answer.

28. Explain the need for dc restoration.

29. (a) State the purpose of a delay line.
(b) What would be the effect on a pix if a DL were not used?

30. With pre-CRT matrixing, what signal(s) are supplied to the video output amplifiers when receiving (a) BW? (b) color? (c) as above, when CRT matrixing is used?

31. (a) What control(s) simultaneously control the beam current of all three guns of a color CRT?
(b) What controls the current of each gun individually?

32. Explain why it is impossible to obtain a normal color pix if the Y signal is weak or missing.

33. What system of matrixing is used when signals are applied to (a) the grids and cathodes of the color CRT? (b) only the cathodes?

620 THE VIDEO AMPLIFIER

34. Explain the operation of a brightness control in a video amplifier stage several stages ahead of the CRT.

35. What makes it possible to interleave chroma and luminance signals within the same bandpass without seriously affecting each other?

36. Why is a 3.58-MHz chroma takeoff coil not required when a comb filter is used?

37. (a) What is the purpose of the drive controls?
(b) Some sets have three controls, others have only two. Explain.

38. What is affected when the drive controls are adjusted (a) the raster? (b) a BW pix? (c) a color pix? (d) each of these?

39. It is impossible to obtain good contrast with a weak signal. Fill in the missing words: With a weak signal, turning up the brightness control weakens the _____ areas of a pix, and turning down the control weakens the _____ pix objects.

40. What is meant by gray-scale-tracking and how is it accomplished?

41. (a) Explain the significance of the minus sign when indicating R - Y, G - Y, B - Y chroma signals.
(b) Do these signals exists at the CRT input for both systems of matrixing?

42. (a) What is the purpose of an ABL circuit?
(b) Explain the operation of a typical system.

43. What is the purpose of a master brightness control?
(b) How and when is it adjusted?

44. Why are chroma stages automatically disabled when receiving BW?

45. Why is the contrast control located in a stage downstream from the chroma takeoff?

46. What three controls are found in the video amplifier section of most color sets?

47. What is a comb filter and why is it used?

48. Loss of the Y signal results in a very poor color pix. Explain why.

49. Explain the purpose and operation of an automatic brightness control circuit using an LDR.

50. (a) What are the two matrixing systems in current use?
(b) In what ways do they differ?
(c) Briefly explain the operation of each system.

51. Explain the difference between a Y signal and a -Y signal.

52. Since a pix is produced by frequencies up to 4 MHz, why do the chroma stages only handle frequencies up to 0.5 MHz?

53. Explain the importance of 4.5-MHz traps in a color set.

54. Name three signal takeoffs in the video amplifier section (a) of a BW set? (b) in a color set?

55. Since sync pulses are not prevented from reaching the CRT, why don't they show up on a pix?

56. The HV applied to a CRT tends to drop as the brightness control is turned up. Explain. (b) How is the HV stabilized?

57. The larger the screen size, the higher the frequency that is needed for good pix resolution. Explain.

58. What are the effects on a pix when the video amplifier has (a) poor high-frequency response? (b) poor low-frequency response? (c) Where the low frequency response is poor the sync may also be affected. Explain why.

59. As the brightness control is turned up it takes a stronger signal to produce black in the pix. Explain why.

60. When a comb filter is not used, how is chroma kept out of the video amplifier stages and the luminance signal out of the chroma stages?

61. What is the function of the dc control stage in a multistage video amplifier?

62. Which signal is delayed in a comb filter (a) chrominance signal? (b) luminance signal? (c) composite video signal? (d) both (a) and (c)?

63. Explain how the chroma and luminance signals are mixed in the (a) CRT matrixing system; (b) the pre-CRT matrixing system.

TROUBLESHOOTING QUESTIONNAIRE

1. The G/C bias on a CRT with B+ applied to both grid and cathode is incorrect. Would you suspect (a) the grid circuit? (b) the cathode circuit? (c) either (a) or (b)? (d) How would you decide?

2. When a set has no raster and the HV and CRT check OK, how would you proceed?

3. A PNP video output transistor is capacitively coupled to the cathode of a BW CRT.
(a) State the probable effect on the raster if the capacitor becomes shorted.
(b) What if it is an NPN transistor? Explain your reasoning.

4. The brightness control is in the cathode circuit of a BW CRT. One end of the control is supplied with B+ through a resistor *R1*. The other end goes to ground via resistor *R2*. State the effect on the raster if (a) *R1* becomes open, (b) *R2* opens up. Explain why.

5. How would you determine if the ABL circuit is doing its job?

6. A BW set has no raster. It is restored with a jumper between the grid and cathode of the CRT.
(a) What trouble is indicated?
(b) How would you proceed?

7. A BW set has two 4.5-MHz traps, one in the detector and one in a video amplifier stage.

622 THE VIDEO AMPLIFIER

(a) Which one is the sound takeoff?
(b) Which one might you adjust if there are weak sound and a beat pattern on the screen?
(c) Interference but sound is normal?
State your reasoning.

8. The detector of a BW set is dc-coupled through a one-stage amplifier to the cathode of the CRT.
(a) Is the transistor an NPN or a PNP?
(b) What is the signal polarity at the detector output?
(c) State the effect if the transistor opens up.
(d) Develops a collector-to-emitter short.
State your reasoning.

9. A set troubled with intermittent loss of pix is being monitored at the input–output of a video amplifier stage, using a dual-trace scope. Where would you look if (a) signal is lost at both points along with the pix? (b) signal at the input remains but the output display is lost with the pix? (c) both input and output signals are present when the pix is lost?

10. State three possible causes of severe blooming in a color set.

11. What trouble is suspected if retrace lines appear when the brightness control is turned up? How would you proceed?

12. What condition is indicated when a color pix and BW pix are not in register?

13. (a) State the possible cause when the screen intermittently brightens up and cannot be darkened with the brightness control?
(b) How would you proceed?
(c) Why is the pix also lost at such times?

14. State the cause when a color pix of poor quality is obtained, but no BW pix when receiving BW or when the color control is turned down?

15. When a set has a normal raster but no BW or color pix, would you suspect the (a) video amplifier stages? (b) chroma stages? Explain.

16. (a) What are the symptoms of 920-kHz beat interference?
(b) How would you treat this problem?

17. HV arcing often occurs only when the brightness is turned down. Explain why.

18. A + voltage is applied to both the grid and cathode of a BW CRT. What is the effect of too much positive dc on (a) the grid? (b) the cathode?

19. What are the symptoms of a dead Y amplifier?

20. When one of the guns of a color CRT is dead, what is the effect on (a) the raster? (b) a BW pix? (c) a color pix? Explain why in each case.

21. A white raster cannot be obtained. State the possible causes when the raster is (a) yellow; (b) magenta; (c) red.

22. When there is color contamination on an off-channel raster, color outlines around objects of a BW pix, and a poor color pix, which condition would you treat first? Explain why.

23. Are drive controls adjusted while observing a (a) raster? (b) BW pix? (c) color pix? Explain why.

24. How would you check the CRT beam current(s)?

25. State the probable effect on the (a) raster, (b) BW pix, and (c) color pix if one of the output transistors of the pre-CRT matrixing system becomes open. Explain your reasoning.

26. The cathode of a BW CRT is supplied with $+100$ V. The brightness control in the grid circuit is fed from a $+90$-V source.
(a) What is the range of grid-to-cathode bias variation when adjusting the control?
(b) What happens to the raster if the ground end of the pot opens up?
(c) If the "hot" end of the pot becomes open?
(d) If a capacitor from the control arm to ground becomes shorted?

27. A set has a normal BW pix but no color. Would you suspect the (a) CRT? (b) video amplifiers? (c) matrix amplifiers? (d) all the above? (e) none of the above?

28. What system of matrixing is used where signals are applied to (a) the grids and cathodes of the CRT? (b) only the cathodes?

29. What trouble is indicated when a set has a normal BW pix but one color is missing from the color pix?

30. (a) How would you determine if the LDR brightness control circuit is operating? (b) What tests would you make if it is not?

31. (a) Would you suspect the CRT when there is a normal raster but one color is missing from the pix? Explain.
(b) How would you proceed?

32. Give two possible causes of a wide, dark, horizontal bar.

33. A set has ghost images that change with the fine tuner. Would you suspect video ringing or a multipath reception problem?

34. Give a number of possible causes of (a) no raster; (b) excessive brightness. How would you proceed in each case?

35. There is considerable color tinting when a set is first turned on. The raster gradually whitens as the set warms up. What would you suspect?

36. How would you distinguish between hum bars and the shading at the top of the pix due to retrace blanking problems?

37. What are the symptoms if the horizontal blanker is not doing its job?

38. What information is obtained by examining the hammerhead display?

624 THE VIDEO AMPLIFIER

39. What stages are suspect if there is a normal raster but, (a) no BW or color pix? (b) color pix but no BW pix? (c) BW pix but no color?

40. An NPN output transistor is directly coupled to the CRT cathode. How is the raster affected if the transistor becomes (a) open? (b) shorted?

41. A set has poor low-frequency response. What is the effect on (a) the pix? (b) the sync?

42. A receiver has three dc-coupled video amplifier stages. If normal voltages are measured at the first stage and the transistors in both the other stages check bad, where would you look for the cause? Why?

43. A set has unstable sync and pix smearing, both indicating poor low-frequency response. Would you check video stages ahead of or after the sync takeoff?

44. Would an open delay line cause (a) loss of BW pix? (b) video ringing? (c) loss of color?

45. Which gun(s) of the CRT are suspect when there is a (a) greenish-yellow raster? (b) a bluish-red raster? (c) a greenish-red raster?

46. What are the symptoms (a) of chroma getting into the luminance stages? (b) of luminance signal getting into the chroma stages?

47. (a) How can a defect in a video amplifier several stages ahead of the CRT cause a CRT bias problem?
(b) What test results would indicate such a problem?

48. What video amplifier stages are suspect when there is (a) a weak pix and poor sync? (b) a weak pix but normal sync? Explain.

49. A scope shows there is pulse clipping in a video amplifier stage.
(a) What condition is suspect if everything in the stage checks OK?
(b) How would you proceed?

50. An NPN video output transistor is dc-coupled to the grids of a color CRT. What would be the effect on the raster if the transistor stops conducting? Explain.

51. A set has several dc-coupled stages driving the CRT. The brightness control is in the first stage, and there is a CRT bias problem. How would you locate the faulty stage by (a) signal tracing? (b) with dc voltage tests? (c) If all voltages are abnormal except on the first transistor, which stage is defective?

52. When a pix has poor resolution and no scanning lines are in evidence on the off-channel raster, would you suspect (a) a narrowed IF bandpass? (b) poor focus? (c) poor high-frequency response in the video amplifier?

Chapter 15

PIX TUBES AND ASSOCIATED CIRCUITS

The CRT used in a TV receiver consists of one or more electron "guns" and a fluorescent screen. Electrons emitted from a heated cathode are directed at the screen, where they hit and energize phosphor particles, which in turn emit visible light. The electron beam(s) are deflected vertically and horizontally to produce the raster. The picture is produced by beam-current variations created by amplitude and frequency fluctuations of the video signal. The pix is developed sequentially, line by line, with the arrival of the video signal. Picture details appear at their proper locations on the screen because of the synchronization of the video with raster scanning. A pix is developed one element at a time, but because of persistence of vision and the retentivity of the screen phosphors, the screen appears to be constantly illuminated, and an entire pix is seen at all times. Because of the rapidity of the scanning there is no noticeable flicker.

15-1 THE BLACK-AND -WHITE (BW) CRT

A (BW) CRT has two main parts: the large *bowl* (sometimes called the *bell*), which includes the screen, and an "electron gun" in the neck of the tube. The CRT "envelope", which contains a very high vacuum, is made of heavy glass, varying in thickness from about $1/8''$ at the neck to approximately $1/4''$ at the faceplate. Except for the method of deflection, the tube is not too different from that used in an oscilloscope.

15-2 THE BOWL

At the front of the bowl is the phosphor-coated screen. The flare of the bowl, both inside and outside, is coated with a graphite compound called the *aquadag* (Fig. 2-2). The outer coating is grounded to the receiver chassis with spring contacts. High voltage (from the HV power supply) is applied to the inner aquadag coating and one of the gun elements via a button-type contact receptacle called the *ultor,* located on the side of the bowl. This structure of two conductive coatings with the glass "dielectric" in between acts as a capacitor to filter the HV.

15-3 DEFLECTION ANGLES

The beam of electrons developed by the gun is deflected by a *deflection yoke* mounted on the neck against the flare of the bowl. While scanning the raster, the beam forms an ever-changing angle between the yoke and various points on the screen. The deflection angle of a CRT is the angle from the yoke to any corner of the screen. If the beam is deflected beyond this amount, it will strike the edge of the flare, creating a shadow (neck shadow) on the screen. CRT deflection angles vary from around 50 degrees to 110° or more, depending on the tube. In general, the larger the screen and the shorter the neck, the greater the angle of deflection. Some modern large-screen, short-neck CRTs have a deflection angle of 114°. Deflection angles for different CRTs are specified in a tube manual.

15-4 IMPLOSION PROTECTION

Because of the high vacuum, a pix tube when struck from the front may *implode* inward with great destructive force. The danger is from flying glass, particularly if it's a large-screen color set with its thick faceplate. Early TV sets had a *safety glass* mounted on the cabinet in front of the screen. All modern CRTs have the safety glass attached to the faceplate. This is known as *integral implosion protection*.

15-5 THE SCREEN

A CRT is generally identified by the size of its screen, which, on a rectangular tube, is measured diagonally from corner to corner. Early sets used round tubes which didn't conform very well with the shape and aspect ratio of the transmitted pix. Tube sizes range from as small as one or two inches up to 30 or 40 inches. Currently, 25″ is the most popular screen size for a console receiver. Projection type sets have a larger screen but the CRT itself is relatively small. A recent innovation is a pocket-size receiver with a *flat* 2″ screen. The electron gun is parallel to the screen surface and the beam is deflected at right angles.

The phosphor coating on the inner side of a CRT faceplate is made to glow or fluoresce with a whitish light when bombarded with electrons. Its particles continue to glow for a short time after the impact, usually for the equivalent of two or three scanning lines, depending on the *retentivity* of the particular coating. The whiteness of the light depends on the kind of phosphor used. A bluish-white is most common. The amount of light emitted by the screen is determined by the intensity of the beam current, which is adjustable by means of the brightness control.

15-6 ION BURN PREVENTION

An electron beam is made up of electrons and negative *ions*. The ions, which are many times more massive than the electrons, serve no useful purpose, and are potentially destructive. If permitted to strike the screen, their impact can burn the phosphor, producing an *indelible blemish* on the face of the tube. An ion burn may appear as a round, brownish discoloration about the size of a half dollar at the center of the screen, or as two diagonally-crossed burn marks.

In early-model sets, ion burns were prevented by means of *ion traps*, small magnets critically positioned on the neck of the CRT. In modern CRTs, it's prevented by the "aluminized" screen, which has a thin film of aluminum deposited on the back of the phosphor coating. The film is "porous" to the small electrons, permitting them to reach the screen, while effectively blocking the passage of the larger ions. As a bonus, the aluminized film acts as a mirror-like reflector, producing a 50% increase in light output. Some of the "secondary electrons" that bounce off the screen are returned to the phosphor for added brightness.

15-7 THE ELECTRON GUN

The function of the electron gun is to develop and direct a beam of precisely controlled electrons at the phosphor-coated screen. The beam is controlled by a number of built-in elements and several external components mounted on the neck. Two means of control are employed: electrostatic, and magnetic. The electron gun of a BW CRT is shown in Fig. 15-1.

• **Electrostatic Control**

DC voltages are applied to the gun elements via the base connections on the neck and the HV ultor contact. The gun contains a control grid which is supplied with a negative voltage. Other elements are supplied with a positive voltage. Beam control depends on the amount of voltage and its polarity (unlike charges attract; like charges repel). This method is similar to that used in most small vacuum tubes, although the construction of the elements is different.

• **Magnetic Control**

Whether travelling in a vacuum or in a wire, an electron stream is surrounded by a circular magnetic field. When acted upon by some external magnetic force, this field, and the beam that produces it, can be deflected. This same principle is used to produce motion in an electric motor. The magnetic flux from components on the CRT gun can readily penetrate the glass neck to reach the beam. The flux may be either a "steady state" magnetic field from a *permanent* magnet, or a changing magnetic field from an electromagnet, as in the deflection yoke.

The elements that compose the electron gun are described in sections 15-8 through 15-12.

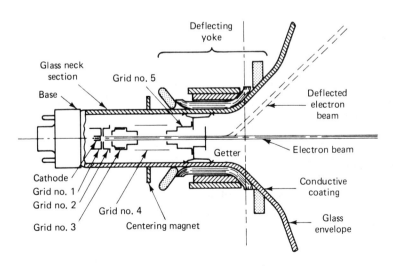

Fig. 15-1 The black-and-white CRT electron gun.

15-8 THE HEATER/CATHODE

Electrons are emitted from an indirectly heated cathode. Although they operate as *one,* the heater and the cathode are two separate elements. The cathode is a small cylindrical tube that encloses the heater. The forward end of the tube, which is closed off, is coated with an oxide material that is capable of producing quantities of electrons at low temperatures. The heater is a spiral-wound "filament" in close proximity within the cathode for maximum heat transfer. The two elements are electrically insulated from each other. The heater of most CRTs is energized from an AC source, usually between 5 and 6 volts. For a shortened warmup period, the CRT used in some small portables has a heater that draws very little current but at a higher voltage. Some sets have the "instant on" feature where the CRT is energized at all times at reduced voltage (see Sec. 6-3).

15-9 THE CONTROL GRID (GRID NO. 1)

Physically this is not a grid in the true sense, although its function is the same as its counterpart in a small vacuum tube. The control grid is a small stubby cylinder whose forward end is blocked off except for a tiny hole through which the electrons pass on their way to the screen. It is positioned immediately in front of the cathode. The grid is "biased" with a negative voltage controllable by means of the brightness control. Its function is to control the beam intensity, and thus the brightness of the screen. Adjusting the control cw reduces the bias, thereby increasing the beam current and hence the brightness. Beam current cutoff and a dark sceen occurs at between −30 and −100 V, with the control turned ccw. The control grid also helps *shape* the beam into a circular pattern. With a video signal applied between grid and cathode, the beam current is made to fluctuate creating a pix of varying degrees of brightness.

15-10 THE FIRST ANODE (GRID NO. 2)

This element is located after the control grid, where it serves two purposes: to accelerate the electrons, and to contribute to beam focus. In a color CRT it's known as the screen grid. Like Grid No. 1, this element is a truncated cylinder closed at its forward end except for a small hole. Supplied with a fairly high positive voltage, it attracts electrons. With the same attraction from all sides, the electrons pass through the opening rather than striking the element itself. The screen voltage for a typical BW CRT is about 150 V, and for a color CRT between 300 and 500V.

15-11 THE FOCUS ELECTRODE (GRID NO. 3)

Because of the mutual repulsion between electrons, the beam, upon leaving the cathode, tends to fan out. Unless this tendency is overcome, it would cause blurred images where the beam "splashes" onto the screen. To obtain *sharp* images, the beam must be brought to a "pinpoint" where it strikes its target (Fig. 2-2). This can be accomplished with either magnetic or electrostatic fields. Early CRTs used magnetic focusing, with either a PM or an electromagnet on the tube neck. All modern CRTs use electrostatic focus; suitable voltages are applied to the built-in electrodes.

Figure 15-1 shows the arrangement of elements which constitutes an *electronic lens system.* The lens system includes Grid No. 3 and two additional elements, Grid No. 4 and Grid No. 5.

In general, their construction is similar to the grids previously described. Focusing takes place because of the difference in potential between the various elements. Grid No. 4 and Grid No. 5 are connected internally to the inner aquadag which is supplied with HV. Some BW CRTs operate with B+ on the focus grid, adjustable with a focus potentiometer. Most modern tubes are the so-called "self-focusing" type, in which G3 connects either to ground or to some fixed voltage. A choice is sometimes provided with two or more terminal connections.

As the beam leaves the cathode and is attracted by the positively charged electrodes, it continues to diverge. As it passes through G4, the interaction of the electrostatic fields forces the electrons to converge into focus at the surface of the screen.

An important consideration regarding focus is the length of beam travel; in modern wide-angle tubes having nearly flat screens, it poses a problem. This problem has, however, largely been overcome by the use of specially designed yokes that maintain good focus over most of the screen area.

15-12 GRIDS NO. 4 AND NO. 5

Sometimes referred to collectively as the *second anode*, these electrodes are the most forward elements in the gun. Connected internally to the inner aquadag, they are supplied with high voltage to accelerate the electron beam and contribute to the focusing action.

15-13 BASE/SOCKET CONNECTIONS

Except for the HV, all electrode connections emerge at the base of the neck. As with most vacuum tubes, different CRT types have a variety of basing and socket configurations. Socket connections are shown on schematics and in tube manuals. As an aid to installing the socket correctly, most CRTs have a plastic "post" at the center of the base which is "keyed" to the socket. Base pins and socket connections are numbered, reading cw either from the keyway or from a large gap between contacts.

15-14 CRT OPERATING VOLTAGES

There's a considerable variation between voltage requirements for the different CRTs, depending on tube type and screen size. As might be expected, the larger the screen, the higher the voltages in most cases. Operating voltages for the various CRTs are listed in a tube manual.

15-15 BEAM CURRENT

Although the HV applied to a CRT is in the thousands of volts, the beam *current* is relatively low, on the order of 100 μA maximum for a typical medium-size tube. The path of the electrons after emission from the cathode is as follows: through the gun elements to the screen surface; then, by bouncing or leakage, to the inner aquadag; from the ultor contact to the HV rectifier; from the HV power supply to ground or to the LVPS; then back to the CRT cathode. Any interruption of beam current, such as from loss of HV or a bad CRT, will of course result in a dark screen.

15-16 SPARK GAPS

Some large-screen BW sets and practically all modern color sets use protective spark gaps at one or more of the CRT socket contacts. They

630 PIX TUBES AND ASSOCIATED CIRCUITS

protect the CRT and associated components against destructive high-impulse voltages. Such gaps are often built into the CRT socket. In lieu of "gaps", some sets use special S.G. capacitors or small neon lamps.

15-17 EXTERNAL GUN COMPONENTS

There are two items mounted on the neck of a BW CRT that are essential to its operation. They are the *deflection yoke* and the *centering ring*. On large-screen CRTs there may also be a couple of small magnets used as pincushion adjusters.

15-18 THE DEFLECTION YOKE

The function of the yoke is to produce a *raster* by deflecting the beam in two directions, vertical and horizontal. Unless deflected, the beam would produce only a high-intensity spot at the center of the screen. Loss of vertical deflection *only*, results in a thin bright horizontal line across the center of the screen. Similarly, a *vertical* line would indicate loss of *horizontal* sweep, but this seldom occurs since most horizontal-sweep problems also result in loss of HV and a dark screen.

A typical yoke for a BW set is shown in Fig. 15-2. It consists of two pairs of coils: one pair for vertical deflection and one for horizontal. Depending on the set, each pair of coils may be connected either in series or in parallel. The coils, which are wound on a circular ferrite core,

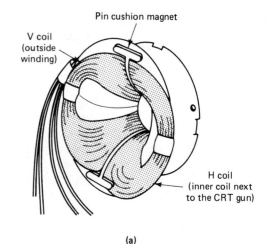

(a)

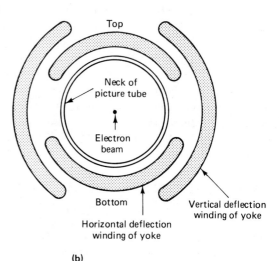

(b)

Fig. 15-2 The deflection yoke.

are oriented as shown in Fig. 15-2b, with the vertical windings in a horizontal plane and the horizontal coils in a vertical plane. Greater deflection is required in a horizontal direction, so the horizontal coils are on the inside, closer to the electron beam. Fairly heavy wire is used because of the relatively high currents. The yoke is mounted on the neck of the CRT against the flare of the bowl. One coil of the vertical yoke is on the left side of the neck, the other on the right. One of the horizontal coils is above the neck, and the other below it. The yoke can be moved in two directions: back and forth on the neck, and by turning. Wing nuts are usually provided for locking it into position. Beam deflection is the result of magnetic fields developed by the yoke coils when supplied with currents from the vertical and horizontal sweep circuits.

15-19 THE RASTER CENTERING RING

This device (Fig. 15-3) consists of two flat washer-like "ring" magnets fastened together in a manner that enables them to be independently rotated. Small projecting tabs are provided for their adjustment. The unit is mounted on the CRT gun immediately behind the yoke. The magnetic flux from the rings passes through the glass neck to react with the magnetic fields developed by the electron beam. The strength of the magnetic flux developed by the rings, its polarity, and orientation relative to the beam can be altered by rotating the rings relative to each other. This causes the beam, and therefore the raster, to be shifted in any desired direction.

15-20 PINCUSHION ADJUSTERS

Large-screen CRTs are prone to *pincushion distortion* (see Sec. 7-14, Sec. 9-6, and Fig. 2-2), where the four sides of the raster bow inward, or the opposite effect, a barrel-shaped raster where they bow outward. With a BW set, correction is made with two small flat permanent magnets mounted on each side of the CRT, as shown in Fig. 15-2a. The raster edges are straightened by bending the magnet supports to move the magnets closer to or further away from the CRT gun.

15-21 BEAM DEFLECTION

Beam deflection in a CRT follows the same basic principles as those that produce motion in

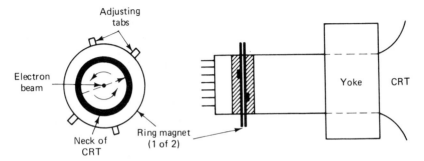

Fig. 15-3 Typical raster-centering ring. Centering magnets provide a fixed amount of offset to the electron beam.

an electric motor. A brief review of fundamentals is in order at this time.

Current flowing through a wire (or the beam current in a CRT) produces a circular magnetic field, as shown in Fig. 15-4a. The direction of the flux lines depends on the direction of current flow. In the example, the current is considered to be flowing away from us which results in a ccw magnetic field. The same is true for a CRT beam when viewed from the rear.

When current flows through a *coil,* the magnetic fields around each turn of wire combine as shown in Fig. 15-4b. We now have an *electromagnet,* where all the flux lines leave one end of the coil, the north pole, and enter the opposite end, the south pole. Within the coil, the flux lines go from south to north as indicated.

When a current-carrying wire (or CRT beam) comes under the influence of an external magnetic field (such as the coils of a deflection yoke), the flux lines interact as shown in Fig. 15-4c, to produce motion. Above the wire (or electron beam) the flux lines are in the same direction and can be considered to be *compressed.* Below the wire (or beam) in the example, the flux lines are in opposite directions and tend to cancel out. Analogous to the result of high-and low-pressure points in a hydraulic system, the wire (or beam) is forced downward. Reversal of the flux lines from either the wire or the coil produces the opposite effect: the motion is upward. The *amount* of deflection is proportional to the strength of the current and the density of the flux, and the motion is always at right angles to the lines of force.

Figure 15-4d shows two sets of yoke coils energized from the vertical and horizontal sweep circuits. In this example each pair of coils is connected *series aiding,* i.e. the magnetic fields developed by opposite coils are in the same direction. These fields penetrate the gun of the CRT to interact with the flux developed by the beam current. With sweep current flowing through the *vertical* coils only, the beam would be deflected downward as in Fig. 15-4c and e. If the sweep current and yoke flux is reversed, the beam will be forced upward as in (f). The result would be a thin bright vertical line on the screen.

When only the *horizontal* coils are energized, as in Fig. 15-4g and h, the motion is *sideways,* producing a *horizontal* line on the screen. In normal set operation, both sets of coils are continuously energized, causing the beam to move in two directions at the same time.

15-22 RASTER SCANNING

During the active trace of each horizontal line, the sawtooth current in the horizontal yoke coils is deflecting the beam at a uniform rate from left to right across the screen. At the same time, an increasing current in the vertical coils is forcing the beam downward. Hence, each horizontal trace slopes slightly downward to the right. At the completion of each horizontal line a horizontal sync pulse triggers a reversal, forcing the beam back to the left side of the screen. During this period the screen is blanked out and the horizontal retrace cannot be seen. With the beam at the left side of the screen another reversal takes place, the screen becomes unblanked, and the increasing current produces another horizontal line. This continues until the beam reaches the bottom of the screen, at which point the vertical yoke current is maximum until a vertical sync pulse arrives to initiate vertical retrace.

As the current through the vertical coils starts to decrease, the decaying flux causes the beam to move upward as the screen is darkened by a blanking pulse. As might be expected, the beam does not go directly to the top of the screen. Because the retrace time is long compared to the duration of each horizontal trace, the beam moves upward in a zig-zag manner under the

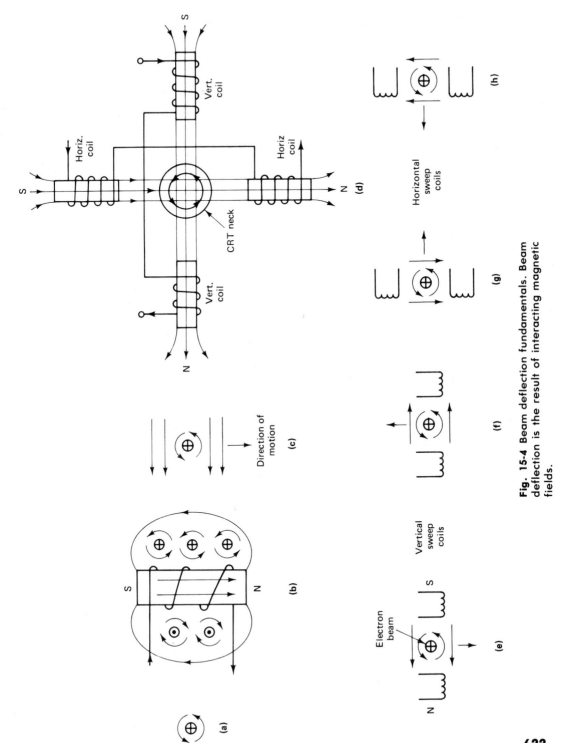

Fig. 15-4 Beam deflection fundamentals. Beam deflection is the result of interacting magnetic fields.

influence of both vertical and horizontal sweep circuits. (see Fig. 1–5).

REMINDER: _____

Horizontal trace and retrace continues uninterrupted during each vertical retrace interval. Under normal conditions, the screen is blanked out and vertical retrace, like horizontal, is not visible to the observer.

As the current through the vertical yoke drops to zero, the cycle begins over and the beam starts downward to trace out another *field*.

REMINDER: _____

The field repetition rate is 60 fields per second, and it requires two such fields to make up a *frame* (a complete pix), so that pix occur at a rate of 30 per second. Alternate fields are *interlaced*, as described in Sec. 1–7. Raster scanning is synchronized with the video signal as received.

15-23 DEFLECTION SENSITIVITY

Deflection sensitivity is an indication of how much sweep power is required to deflect the beam by a given amount. The horizontal sweep requirements are considerably greater than the vertical, because of the shape of the screen, and the larger the screen the more power is needed for both. The factors involved are the following.

- **HV and Beam Stiffness**

An increase in HV applied to the CRT produces a "stiff" beam with reduced deflection. It is sometimes called the "sticking potential". With an increase in HV the beam is accelerated to a greater degree and is therefore under the influence of the yoke fields for shorter periods of time. The reverse is also true: a *reduction* in HV produces *greater* sweep and an increase in raster height and width, but at the expense of brightness and focus.

- **Yoke Location**

The normal position on the neck is snug against the flare of the bowl. With the yoke moved back closer to the electron source, more deflection is obtained, but this results in *neck shadows* where the beam strikes the edges of the flare. With the yoke backed off slightly, shadows may be observed only at the corners of the screen. If it is too far back, the increased shading may be observed as a *round* raster. Most yokes are wound with "wrap-around" coils for closer contact with the CRT bowl. In color sets, the yoke is *not* in close contact with the tube flare.

- **Neck Diameter**

The smaller the diameter of the neck the closer is the beam to the yoke, and hence the greater is the deflection for a given amount of sweep power. For this reason, the neck size of a CRT used in a small portable may be as little as $3/4''$. At the other extreme, a color CRT, with its wider neck, requires considerable sweep power.

15-24 DEFLECTION ANGLES

A yoke is designed to produce a deflection angle that satisfies the requirements of the CRT it is used with. For example a 110° CRT requires a yoke capable of at least 110° of sweep without distortion, neck shadows, or impaired focus. In general, large-screen, short-neck CRTs require wide-angle yokes to match.

15-25 THE SIGNAL INPUT CIRCUIT

A *raster* is a necessary prerequisite to obtaining a pix. To obtain a *pix*, however, the CRT must be supplied with a *signal* also. Depending on the set, the video signal may be fed to either the control grid (G1) or the cathode. Since the voltage

The Video Output Stage

The signal from the output amplifier is coupled to the CRT either *directly* or via a coupling capacitor. Cathode drive is most common. Picture contrast depends on the amplitude of the video signal, which ranges from about 30V p-p for a small CRT to 100V or more for a large tube. A negative-going signal is required when feeding the grid, and a positive-going signal when feeding the cathode.

Brightness Control

This is a CRT bias control. Variation of the bias changes the intensity of the beam current and therefore the brightness of the pix. With most BW sets, the control is in either the grid or the cathode circuit of the CRT. With DC coupling, bias depends also on the operating conditions of the video amplifier(s). Bias, beam current, and screen brightness are constantly changing in accordance with variations in the signal amplitude. For example, if normal setting of the brightness control establishes the "no-signal" bias at −40V, then when a 100V p-p signal is received, the grid voltage fluctuates between +10V and −90V. The negative reading corresponds to beam-current cutoff and black pix objects, and the positive figure to high beam current and the brightest objects in a pix.

DC Restorer

The video signal has both AC and DC components. The DC component, which originates at the camera, establishes the average background illumination for each scene being televised. With direct-coupled video amplifiers in the receiver, the DC component is preserved and there's no problem. If, however, one or more coupling capacitors are used between the detector and the CRT, the DC level is lost and must be restored. For this, most sets use a diode at the CRT input which develops the DC by rectifying the signal at the pulse level which has a constant amplitude.

Retrace Blanking

Vertical and horizontal blanking pulses are used to darken the screen during vertical and horizontal rectrace intervals. Most BW sets feed the pulses directly to the CRT, at either the grid, the cathode, or the screen grid. The pulses are supplied by the sweep circuits. Amplitude and polarity is consistent with that of the video signal at the point of injection.

15-26 PRODUCING BAR AND LINE PATTERNS

A pix is produced when the video signal creates variations of beam current during the scanning of each horizontal line. The *size* of pix objects depends on the *rate of change*. *Slow* variations (from *low* video frequencies) produce large objects; rapid variations (from *high* video frequencies) produce small objects. The smallest pix detail is determined by the highest frequency transmitted and processed by the receiver, 4.2 MHz. On a 25″ screen for example, this is equivalent to some 300,000 pix elements. The amount of detail that can actually be *resolved*, however, depends on *screen size*. Naturally there is a limit to the detail that can be observed on a small screen, so such sets can get by with a restricted bandpass.

Picture development can be demonstrated by *bar patterns,* produced by injecting signals from a generator at the CRT input or at some previous amplifier stage. Several bar patterns are shown in Fig. 15-5. Such displays have practical uses in servicing: as an indicator when signal tracing

636 PIX TUBES AND ASSOCIATED CIRCUITS

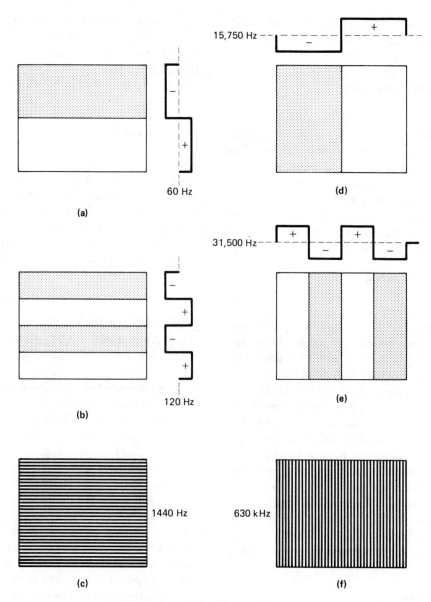

Fig. 15-5 Generating bar and line patterns.

for a weak or dead stage; for checking linearity; and so forth.

The *number of bars* or lines displayed depends on the *frequency* of the generator signal. Low frequencies produce *few* bars, *high* frequencies *many* bars. When used in signal tracing, the *number of bars* is of no importance. The rough indicator of signal levels is the *darkness,* or contrast of the bars, where black bars correspond with a strong signal and gray bars to a weak signal. For *linearity* checking, good linearity is present when all the bars are uniformly spaced. When bars are crowded or stretched at certain points, *pix objects* will also be affected for those areas of the screen.

Vertical bars are produced when the generator frequency is equal to or higher than the horizontal sweep frequency of 15,750 Hz. Frequencies lower than 15,750 Hz produce *horizontal* bars. Bars appear *stationary* when the generator frequency is an *exact multiple* of the sweep frequency. *Odd* multiples of 60 Hz, the vertical sweep frequency, result in horizontal bars that *roll*. The direction of the roll is indicative of whether the frequency is higher or lower than 60 Hz. Odd multiples of 15,750 Hz, or higher frequencies, result in *diagonal* bars: for example, the fine-line patterns often observed on a pix as caused by CB and other types of interference. Generally, diagonal patterns are constantly shifting because they are not synchronized with the horizontal sweep.

The number of horizontal bars produced by a generator signal is equal to the generator frequency divided by the vertical sweep frequency. For example, one bar for 60 Hz, two bars for 120 Hz, and so on. Figure 15-5a shows how one bar (one light and one dark area) is developed from a 60 Hz signal, where one cycle corresponds with the vertical sweep rate. While the beam is scanning the upper portion of the screen, the negative half cycle biases the CRT to cutoff to darken the screen. While scanning the lower half of the screen, the positive half cycle results in high beam current and the bright area. The signal here feeds the CRT grid. For cathode feed, conditions would be reversed. Where a square or rectangular waveform is used as in this example, the change is abrupt. If the signal is a *sine wave,* the change is gradual with the bar gradually fading from black to white, as with a *hum bar,* caused by poor filtering of the LVPS. In Fig. 15-5b, two alternately dark and light bars are produced by two cycles of a 120 Hz signal in the time it takes to complete one vertical scanning. Fig. 15-5c shows a pattern of approximately 24 bars or lines where the frequency is 1440 Hz. If the bars aren't stationary, the pattern can be synchronized with the vertical hold control.

The number of *vertical* bars produced by a generator (or other) signal is equal to the signal frequency divided by the horizontal sweep frequency: for example, one bar for 15,750 Hz, two bars for 31,500 Hz, and so on. Fig. 15-5d shows how one cycle of a 15,750 Hz voltage produces one dark and one light area in the time required to complete one horizontal trace. Since the same thing is repeated for each line, it appears as a vertical bar. In e, we see two such bars where the frequency is twice the sweep rate. In f, we have approximately 40 bars or lines from a frequency of 630 KHz. If the bars aren't stationary they can be synched with the horizontal hold control.

A *crosshatch generator* simultaneously produces both vertical and horizontal bar patterns.

15-27 THE PICTURE

A pix is produced in the same way as the bar patterns described above, i.e. by *intensity modulation* of the CRT beam. The difference is that for *bars,* frequency and amplitude of the generator signal is *constant,* whereas a *video signal* is continually changing in both frequency and amplitude. CRT bias, beam current, and screen brightness are determined by the setting of the brightness control. When a signal is received, amplitude variations of the signal create changes in grid/cathode voltage (bias), beam current, and brightness to produce the light and dark areas of the pix. For bright objects, the bias is reduced; for black objects the CRT is biased to cutoff. Gray objects are produced by biases between these two extremes. Pix *contrast* is determined by the *average* amplitude of the signal. The *size* of pix objects (horizontally), is determined by *frequency,* the same as for bar patterns, low frequencies for large objects and high frequencies for fine detail.

The Composite Video Signal

The signal as it appears at the input of a BW CRT is shown in Fig. 15-6a. It consists of video information and blanking and sync pulses. The pulses are *recurrent;* the *video* is constantly changing. In producing a pix, the raster is scanned *twice*, 262½ lines for each interlaced *field,* for a total of 525 lines per *frame* (a *complete* picture). The field rate is 60/second, and so the frame rate is 30 per second. The signal, as shown in Fig. 15-6a, represents four "lines" of pix information, the last two lines of pix at the bottom of the screen, and the top two lines where scanning of the next field commences. Not all 525 lines contain pix information; a few lines at both top and bottom of the screen are blanked out, and some are "used up" in vertical retrace. At point *A* on the drawing, when the scanning of a field is completed, the screen becomes blanked out by the vertical blanking pulse. The beam continues to move downward until point *B*, where vertical retrace is initiated by the vertical sync pulses. (Figures 1-6 and 10-4). From B to C the CRT beam is moving upward during vertical retrace. At *C*, the screen becomes unblanked and the beam is moving downward, as the first line of pix information for the next field is traced out.

For *color,* the only difference is the presence of a *color burst* on the "back porch" of each horizontal sync pulse (see Fig. 1-6c), and the chroma signals that are interleaved with the video.

Two important characteristics of the composite video signal are *amplitude* and *polarity*. Amplitude depends on receiving conditions and proper operation of all stages between the antenna input and the CRT. The requirements vary for different types of CRTs and screen sizes, from about 30 V p-p for a small screen to as much as 150V p-p or more for a large tube. Ideally, signal strength at the blanking level should be high enough to achieve beam-current cutoff.

Beam current is reduced and the screen darkened with *increases* in signal level, and *polarity* is determined by which element of the CRT is being driven. When feeding the grid, the signal must be negative-going; for cathode drive, positive-going. Typical CRO displays of the composite signal are shown in Fig. 15-6b and c.

Blanking Pulses

As shown in Fig. 15-6, *horizontal blanking pulses* interrupt the video at the completion of each line of pix information. Their purpose is to overbias the CRT and darken the screen during horizontal retrace periods. The duration is about 57 μseconds each, the average time required. Maximum *permissible* time is 63.5 μseconds.

The vertical blanking pulse darkens the screen during *vertical* retrace. Its duration is much longer than the horizontal, about 1200 μseconds. Insufficient blanking creates problems such as the visible retrace lines described earlier.

Sync Pulses

The purpose of sync pulses is to initiate retrace and to synchronize raster scanning with the video information. As can be seen in Fig. 15-6, a horizontal sync pulse "rides" on each horizontal blanking pulse. Sync pulses represent about ⅓ of the total p-p amplitude, and extend *beyond cutoff* into the so-called "blacker-than-black" region, where they are not normally seen on the screen.

As described earlier (Sec. 7-10), the vertical blanking pulse is *serrated*. Six of the serrations are vertical sync pulses which are integrated into a single pulse for triggering the vertical sweep oscillator, to initiate retrace. Other pulses, the *equalizing* and *horizontal* sync pulses, are also present (Fig. 10-5). Their purpose is to maintain horizontal sync during vertical retrace.

Sync pulses are not prevented from reaching

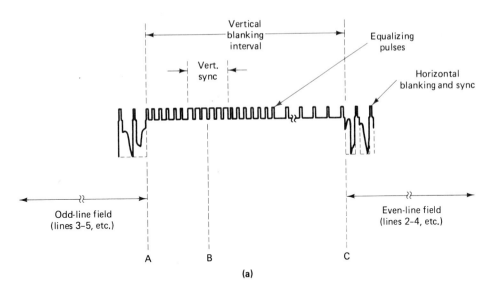

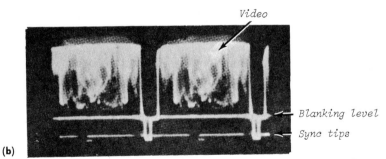

Fig. 15-6 (a) The composite video signal; (b) CRO photo—polarity reversed from (a) with pulses downward; CRO sweep frequency is 7,875 Hz to show two lines of video with horizontal sync pulses; (c) with CRO sweep set for 30 Hz to show two complete fields of video and the (compressed) vertical blanking and sync pulses.

the CRT, where they serve no purpose, but neither do they create problems since they occur only when the screen is blanked out. However, blanking and sync pulses *can* be observed by misadjusting the hold controls and turning up the brightness. This "hammerhead" display can be useful in trouble diagnosis.

- **The Video Signal**

The video, which is only part of the composite signal, has three characteristics: amplitude, polarity, and frequency.

Amplitude: This is adjustable with the contrast control to obtain the desired amount of pix contrast. Good contrast requires a signal strong enough to drive the CRT between two extremes, high beam current for white objects, and cutoff for black objects. Besides a strong signal, the setting of the brightness control is an important consideration. Suppose for example that for normal operation the brightness control establishes the average bias at -30 V. With a signal voltage of approximately 55 V p-p, the grid is made to vary between -50 V (CRT cutoff at the blanking level) and -5 V (the lowest signal levels, which produce the white areas of the pix). The pix produced will have good contrast, because the beam current variations create brightness changes from white to black.

Let us see why it's impossible to obtain good contrast with a *weak* signal. With a signal of only 20 V p-p, for example, the tube cannot be driven to cutoff, so normally black objects will appear dark gray. At the lowest signal levels, the minimum grid voltage will be -20 V, the beam current will be low, the objects that should be white will appear light gray. With a weak signal, adjusting the brightness control doesn't help either. If the control is turned down for a no-signal bias of -40 V, the weak signal will now drive the CRT to cutoff to produce black objects, but the low beam current will make normally white areas dark gray. With the brightness control turned *up*, we may have the condition where the average bias is -20 V. With the higher beam current, white objects may appear normal, but since the grid never goes beyond -30 V, the beam current doesn't decrease as it should and there will be no black areas in the pix; normally black objects will appear light gray.

Polarity: As stated earlier, the signal must be negative-going when fed to the CRT grid, positive-going when driving the cathode. Polarity is established at the video detector and may change one or more times before it reaches the CRT. When polarity is reversed at the CRT input it results in a *negative pix:* white objects show up *black,* and black objects appear *white.*

Frequency: The frequency of video is constantly changing in accordance with televised images.

15-28 DEVELOPING THE PIX

Abrupt changes in a pix, from white to black, or black to white, are created when the CRT is alternately turned on and off. For gray areas of a pix, the change is more gradual. As each bright area of the pix is illuminated, the phosphors continue to glow for some time after beam impact, usually for the duration of about two horizontal lines, depending on the type of phosphor. The eyes of the viewer retain the image, and as with movies, an illusion of smooth motion is created due to persistence of vision.

As explained earlier, high video frequencies produce rapid interruptions or changes in beam current to develop the small objects in a pix, and low frequencies produce slow variations to create large objects, the same as for bar patterns. In other words, the *rate of change* determines the "size" of pix objects on a horizontal plane: the more rapid the change, the greater the number of fine pix details that can be traced out in the scanning of each horizontal line.

Figure 2-3 shows how one line of a pix is created by variations of the video signal. As the beam moves from left to right across the screen, variations in the amplitude of the video signal vary the beam current to produce changes in pix tone from white, through various shades of gray, to black, as each line is traced out in sync with its counterpart in the station camera. In the example, the horizontal sync pulse at the extreme left has triggered horizontal retrace, which lasted for the duration of the blanking pulse; the horizontal sweep voltage has dropped to zero; the screen has become unblanked; and we're ready to start tracing another line of pix information. The horizontal sweep voltage starts to increase, moving the beam to the right, with the arrival of the video information which will modulate the beam current. Starting at the left, the amplitude is low for a brief period, creating high beam current and the white pix background. Then the video increases, almost abruptly, to the blanking level and the beam is cut off while scanning the dark area of the hair. Now come the rapid up-and-down fluctuations for the light and dark areas around the eye, a longer period between the eyes, then more rapid fluctuations for the other eye, followed by an increase, almost to the blanking level, for the hair again. Finally the voltage tapers off to produce the light area on the right side of the pix. At this point, another horizontal blanking pulse arrives to darken the screen, the sync pulse initiates retrace, the decreasing sweep voltage forces the beam back to the left side, and we're ready to scan the next line as the vertical sweep continues to deflect the beam toward the bottom of the screen.

- **Frequency and Pix Resolution**

The maximum number of elements that make up a TV pix is around 300,000. This figure is arrived at as follows: The maximum number of *vertical* elements is about 500, equivalent to the number of active scanning lines per frame. This is true, regardless of screen size. With a pix aspect ratio of 4 to 3, this means 4/3 × 500, or 664 elements per line, and, for a complete frame, 664 × 500, or a little over 300,000 per pix. This compares with around 500,000 elements for one frame of 35 MM motion pix film, and 125,000 for 16 MM. At a scanning rate of 30 frames/second, this represents some 9,000,000 (300,000 × 30) elements per second. It takes two square-wave cycles of video to reproduce each pix element, and therefore a frequency of about 4,500,000/second, or 4.5 MHz. Allowing for imperfections in the system, such as imperfect interlacing, a maximum frequency of 4.2 MHz is established for the finest pix detail.

The number of pix elements (fine detail) that can *actually* be resolved depends on a number of factors such as the screen size, the diameter and shape of the electron beam, focus, pix smear, and freedom from ringing and other interference. For the maximum permissible frequency of 4.2 MHz, optimum fine detail can be obtained on a large screen; but on a small tube, because of its average beam diameter of 0.05″ and the closeness of its scanning lines, an upper frequency limit of around 2 or 3 MHz is usually adequate.

For a better understanding of resolution, consider the typical test pattern of yesteryear shown in Fig. 15-7a. Such patterns are useful in evaluating receiver performance and making adjustments. In b, note the relative sizes of pix detail and corresponding frequencies, and in c, frequencies for different portions of the vertical "wedge". Note the limitations on resolution for frequencies higher than 4 MHz at the narrow part of the wedge. The horizontal wedge in d provides a check on vertical resolution. Line spacing at the left corresponds with scanning lines on a large-screen CRT and at the right with those on a small screen.

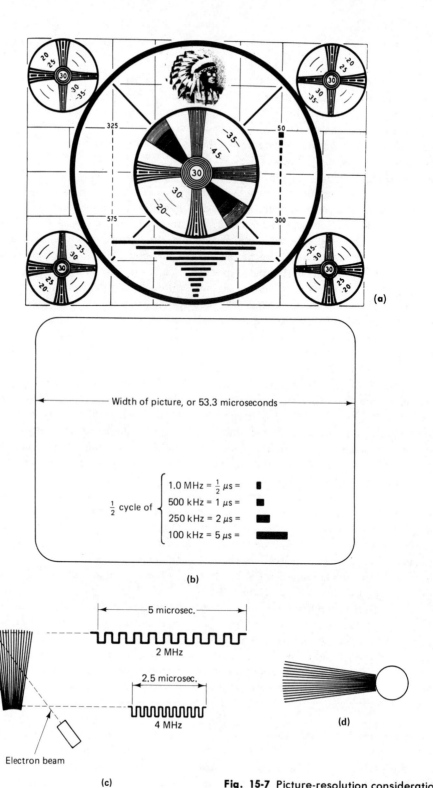

Fig. 15-7 Picture-resolution considerations.

15-29 COLOR

The understanding of how a color pix is produced, and the troubleshooting of color sets, requires a clear grasp of color principles.

As explained earlier, there are three *primary colors,* red, green, and blue, and a mixture of these in various proportions produces "secondary" or "complementary" colors, and in fact all the colors present in nature, including white, which is not actually a *color.* Visible white light is a mixture of *all* colors, including the primaries and their complements. Because light is being produced by the TV screen, colors are said to be *additive* and recognizable by *reflection;* this situation is different from the mixing of paint pigments and dyes, where the mixtures are "subtractive", as determined by the colors that are *absorbed.*

By definition, a primary color is one that when mixed with another produces a third distinctive color, its complementary. Every color has three characteristics; *Hue, saturation,* and *brightness.* Hue is *color,* the characteristic that distinguishes one from another; for example red, yellow, purple, etc. Saturation is the strength or vividness of the color, depending on the amount of white light in the mixture; for example, deep red is highly saturated, whereas *pink* is diluted, although both are of *the same hue.* Brightness means what it says, the lightness or darkness of a color which is not to be confused with saturation.

Table 15-1 shows how different colors may be added to produce other colors. For example, an equal mixture of red and green produces yellow. Equal proportions of red and blue produce magenta (purple). An equal mixture of green and blue produces cyan (aquamarine or turquoise). With *unequal proportions* of the primaries we obtain other colors. For example, red and green: where red is predominant we obtain a reddish yellow (orange); where green is predominant we get a greenish yellow. Equal

Table 15 - 1 Color Mixing

Color Mixture	Color Obtained
R + G =	Yellow
R + B =	Purple
G + B =	Cyan
R + G + B =	White or gray
R + yellow =	Reddish yellow (orange)
R + purple =	Reddish purple (magenta) or brown
R + cyan =	Deep violet
G + yellow =	Yellowish green
G + purple =	Bluish green (turquoise)
G + cyan =	Greenish blue
B + yellow =	Bluish green
B + purple =	Deep purple
B + cyan =	Light purple
R + white =	Pink
G + white =	Pale green
B + white =	Light blue

proportions of the three primaries produce white. Different shades of white (off-white or gray) are produced by dull primaries of different brightness. Black of course is the complete absence of light.

An interesting characteristic of the human eye is its inability to discern color on very small objects, and to see a mixture of hues as *one color.* For example if we project fully saturated primaries onto a screen as shown in Fig. 15-8, so they overlap, we see not only the primaries, but their three complementary colors yellow, magenta, and cyan, in addition to white. Because color can only be recognized in relatively large objects the finest details of a TV color pix, are viewed as black and white.

A TV station generates two kinds of signals, fully saturated chroma signals and luminance signals, which are transmitted separately. At the receiver, these signals are recombined to reproduce the original colors seen by the camera.

644 PIX TUBES AND ASSOCIATED CIRCUITS

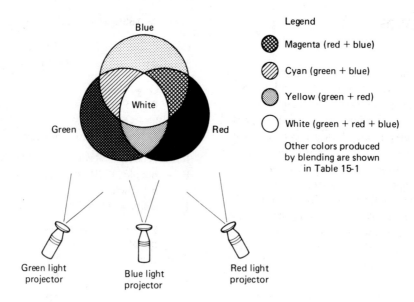

Fig. 15-8 Color mixing.

The screen of the color CRT contains red, green, and blue phosphors, which, when hit by electron beams of varying intensities, provide us with colors running the full gamut of the visible spectrum.

15-30 THE COLOR PIX TUBE

Unlike a BW CRT, a color pix tube has *three* electron guns, a *shadow mask* or *grill,* and a screen coated with three different color phosphors corresponding with the three primary colors, red, green, and blue. There are two basic types of tubes, the *delta* or *shadow-mask* type CRT, and the *in-line* type. In construction and operation the two types are quite different. In a delta CRT the electron guns are arranged in a triangular fashion and their three beams pass through openings in a shadow mask to illuminate corresponding color *dots* on the screen (see Fig. 15-9). In an in-line CRT (Fig. 15-14), the guns are grouped horizontally in line, and their beams go through slots in an *aperture grill* to illuminate color *stripes* on the screen. In both types the beams are precisely controlled to strike only their corresponding color phosphors, and no others. Both types are equally capable of producing a good color pix.

Color pix tubes differ from their BW counterparts in other ways also. Because of the very high vacuum, they are more ruggedly built, with thicker glass, around $1/8''$ at the neck to $1/2''$ or so at the faceplate, and with a special *leaded* glass as protection against harmful X-rays.

A color CRT is less efficient than a BW tube, giving less light output from its screen phosphors. Also, its deflection sensitivity is considerably less, so greater sweep power is required. There are two reasons for this; the higher voltage used (which increases the "sticking potential"), and the greater diameter of the tube neck, which places the beams further from the magnetic influence of the yoke and other external devices that control beam aiming. Still another difference is the amount of beam *current,*

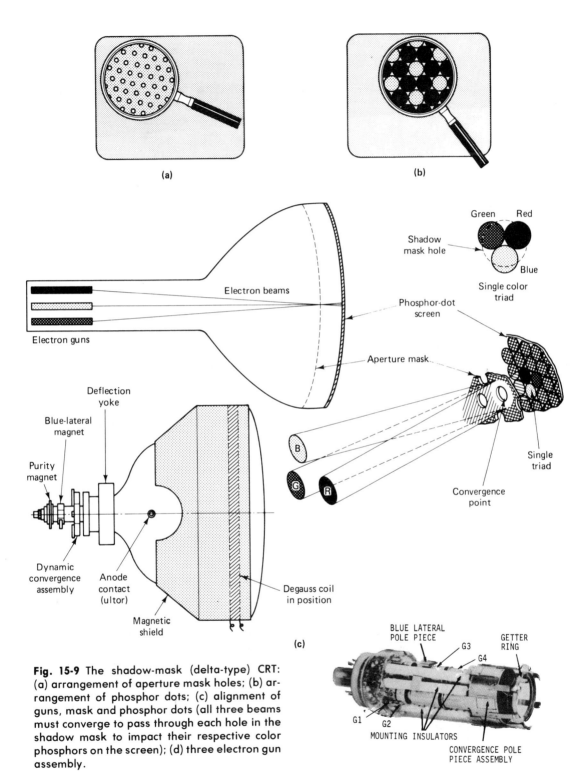

Fig. 15-9 The shadow-mask (delta-type) CRT: (a) arrangement of aperture mask holes; (b) arrangement of phosphor dots; (c) alignment of guns, mask and phosphor dots (all three beams must converge to pass through each hole in the shadow mask to impact their respective color phosphors on the screen); (d) three electron gun assembly.

which because of *three* guns, and higher operating voltages, is around 1 MA compared to 100 μA for a typical BW tube.

15-31 THE DELTA (SHADOW-MASK) CRT

The delta CRT is the ''original'' color tube. Except for certain improvements such as increased screen size its basic operation remains unchanged. Although still widely used, this type, largely because of greater difficulties in beam control (convergence), is rapidly being replaced in newer sets with in-line type tubes. Construction and operation of the delta CRT is described in the following paragraphs.

15-32 THE SCREEN

The red, green, and blue phosphors are deposited on the screen in the form of *dots*. The dots are geometrically arranged in triangular groups of three called *triads* (Fig. 15-9). Each dot is slightly smaller than the diameter of an electron beam, roughly 0.05 in. There are many thousands of dots, which can be observed by close inspection of a lighted screen. A 19″ CRT has approximately 250,000 triads, the equivalent of about 750,000 dots; a 21″ tube has about 360,000 triads (over a million tiny dots). Each dot is separated from its neighbors by a few thousands of an inch. The phosphors are uniformly deposited over the entire screen surface, which is backed with an aluminum coating for increased light output. In Fig. 15-9, note the relative locations of the dots; blue is at the bottom apex of each triad.

As an experiment, examine the dot pattern on a large delta CRT using a magnifying glass. Note the absence of color when the set is turned off, and the different colors when the phosphors are *excited*. While watching a pix, observe the color variations as they occur and the absence of color for black areas, and note how all three primaries appear as *white* when viewed from a distance.

• **The Black-Matrix Tube**

For increased brightness and pix contrast, most modern color tubes have a black area around each color dot. This permits the use of larger dots, for a corresponding increase in light output. The blackened areas also help prevent color contamination due to beam ''splashover'' onto adjacent color dots.

15-33 THE SHADOW MASK

As described in Sec. 2-8, this is a perforated metal plate mounted about ½″ behind the faceplate. It has many thousands of tiny holes, one for each triad on the screen surface. Its positioning is *very precise,* with each hole directly behind the center of each triad. The holes are slightly smaller than the diameter of the three electron beams. In operation, the beams of all three guns are made to converge and cross over in each hole, and, upon emerging, to strike their respective color dots on the screen, a seemingly impossible task, considering the close tolerances involved and the fact that the beams are being constantly deflected across the face of the tube.

External components are mounted on the neck of the CRT to control the aiming and direction of the three beams precisely, so that they will pass through the holes and hit the corresponding color dots. The shadow mask enables each beam to strike its own color dots without overlapping onto adjacent dots (see Fig. 15-9). Note the relative sizes of the beams, the aperture holes, and the phosphor dots; note also the different beam angles and how they strike at different points as determined by the spacing between the dots. Because of beam crossover, the blue gun is uppermost in the CRT neck whereas the blue phosphor dots are at the bottom of each triad.

15-34 THE ELECTRON GUNS

There are three *identical* guns, each named in accordance with one of the three primary colors. They are arranged in an equilateral triangle at the rear of the tube neck, with the blue gun at the top as shown in Fig. 15-9. Gun structure and operation is not very different from that in a BW CRT. Each gun operates in conjunction with its corresponding color dots on the screen.

15-35 INTERNAL GUN ELEMENTS

Each gun consists of a heater, a cathode, a control grid (G1), a screen grid (G2), a focus grid (G3), and G4, the second anode; see the callouts in Fig. 15-9. One thing unique to a color CRT is the magnetic "pole piece" assembly shown in the photo. It is attached to the three G4 grids. It consists of sectionalized plates whose purpose is to concentrate magnetic fields from an external *convergence assembly* to each of the three beams as they emerge from the guns. Another magnetic pole piece assembly (not visible in photo) is attached to G3 of the blue gun. It concentrates the magnetic field from an external *blue lateral adjuster magnet* to the beam of the blue gun only.

15-36 BASE/SOCKET CONNECTIONS

A typical delta CRT has 12 base connections. The heaters of the three guns are internally connected in parallel. The second anode (G4) is connected internally to the convergence pole piece assembly, and to the HV via the inner aquadag coating. The three (G3) focus grids are connected internally and to a common base pin. The G1 and G2 grids of each gun all have separate base pins. A tube manual shows the base/socket connections for the different tube types.

15-37 OPERATING VOLTAGES

Except for the heater, all operating voltages are considerably higher than for a BW CRT. The HV for example is usually around 25 KV depending on tube type and screen size. For a projection type CRT it may go as high as 80 KV. Typical focus voltage is between 5 KV and 7 KV. Controls are provided for adjusting most of the applied voltages. In early TV sets, the 6.3 V for the CRT heaters was supplied from a filament transformer or a winding on the LV power transformer. Most modern sets obtain the heater voltage from a winding on the flyback transformer. For operating voltages of a particular CRT refer to a tube manual.

15-38 BEAM CURRENT

The beam current of a color CRT is at least three times greater than for a BW tube. At full brightness it is around 300-400 μA per gun, for a total of about 1 MA. The amount of current is determined by the conditions of the guns and by the applied voltages. Excessive beam current can weaken or destroy a CRT, so most color sets incorporate an ABL circuit as protection against overload.

15-39 HV CONTROL AND THE X-RAY HAZARD

High voltage in excess of 20 KV or so produces potentially harmful X-ray emission. This is normally prevented by enclosing HV components in a shielded compartment. However, the HV specified for each set should never be exceeded. Most sets have a high-voltage adjustment and an HV regulating circuit. Spark gaps are provided as described earlier to protect the CRT and other components against HV arcing.

15-40 EXTERNAL COMPONENTS

Precise control of the CRT beams is accomplished in two ways: by proper dc *voltages* applied to the gun elements, and by *magnetic fields* developed by external components described in the following paragraphs.

15-41 THE DEFLECTION YOKE

Except for the *three* beams that are deflected as *one*, conditions for raster scanning are the same as for a BW set. About the only difference is the yoke, which is generally much larger because of the greater sweep-power requirements. (Because of the greater neck diameter, the *deflection sensitivity* of a color CRT is reduced because the yoke flux is farther from the electron beams). All of the CRTs now in use are of the short-neck, wide-angle variety, with a yoke to match. As in a BW set, provision is made for adjusting the yoke in two ways: *rotation* (raster tilt adjustment), and fore-and-aft. In a BW set the yoke is positioned tight against the flare of the bowl; in a color set, normal position is slightly back of the flare. This is a *color purity* adjustment. Once made, both yoke adjustments are locked, usually with wing-nuts. Besides its effect on purity, yoke positioning is a critical factor in the convergence of the three beams. Because of this, at least one make of set had the yoke *bonded* to the CRT, after adjustments were made at the factory.

15-42 THE DEGAUSSING COIL

For the three beams to strike their respective color dots, they must be made to converge within each hole of the shadow mask. Understandably, all convergence adjustments are extremely critical, and any external magnetic fields will adversely affect the adjustments and the pix. As explained earlier, the steel shadow mask and parts of the chassis and cabinet tend to become magnetized, by electrical appliances such as vacuum cleaners and even by the earth's magnetic field, weak as it is. To offset such external fields, older sets used small adjustable "rim magnets" positioned around the CRT faceplate inside the cabinet. All modern color sets use one or more *degaussing coils* positioned up front around the CRT (Fig. 15-9). The coil is periodically and momentarily energized from the ac power line, to maintain the shadow mask and other parts in a *demagnetized state*. Some early sets had a pushbutton switch for manually energizing the coil when required. All modern sets employ automatic degaussing (ADG) (see Section 6-3). Typical circuits are shown in Fig.

6-4. *Each time* the set is turned on, the coil is momentarily and automatically energized.

In addition to a degaussing coil, a metal shield is sometimes placed around the bowl of the CRT (Fig. 15-9).

15-43 THE PURITY RING

Although used for a different purpose, this device is essentially the same as the ring magnet assembly used for *raster centering* on a BW CRT (see Sec. 15-19 and Fig. 15-3). Positioned on the CRT gun at the gap between grids 1 and 2, it is adjusted (by rotating the two magnetized "washers") to obtain proper *beam landing* and good single-field color purity.

15-44 THE BLUE LATERAL ADJUSTER

This adjustable permanent magnet is mounted on the tube neck directly over the internal pole pieces attached to G3 of the blue gun. As one of the several *static convergence* adjusters it is used to adjust the beam of the blue gun in a horizontal direction. Early model sets used a small magnetized rod, adjustable by twisting. With modern sets it is usually a thumbscrew adjustment as part of the purity ring assembly (Fig. 15-22), which is secured to the CRT neck by some form of clamp arrangement.

15-45 CONVERGENCE AND BEAM LANDING

A normal pix is obtained *only* when the beam of each gun strikes its corresponding color dots on the screen, and this occurs only when all three beams converge in each and every hole in the shadow mask. This represents the ideal condition of *color purity,* in which it's possible to view a "pure" raster for each of the three primary colors. When the beams are properly *converged,* the three color fields *almost* overlap; they are displaced from each other by an amount equal to the spacing between the dots. Improper adjustments of purity and/or convergence result in color *fringing* on a BW pix and in color *bleeding* around the images of a color pix.

Figure 15-9 shows an expanded view of a single triad, with each beam landing on its proper color dot. Under this normal condition, and assuming that each dot is illuminated in the proper amount, the triad, from a distance, would appear white. On close inspection, the three colors are easily distinguishable.

Because of the "flatness" (lack of curvature) of a CRT screen, the beams travel a greater distance to the edges, as compared to the center areas of the screen. For a BW tube this creates no problem except some minor defocussing and insignificant variations in linearity of the sweep. In a color CRT, there are serious problems to overcome. As the beams sweep toward the four edges of the screen, and the distance travelled increases, they tend to converge *before* reaching the mask openings (Fig. 15-10). As they fan out, some of the beams miss the apertures and/or are unable to strike their correct phosphor dots. This problem increases with the size of the screen, and, because of the relative height and width of the raster, is greatest in a horizontal direction.

A further problem is that the deflection sensitivity of the three beams varies as different areas of the screen are being scanned. Unlike in a BW CRT where the beam travels through the center of the yoke, the three guns of a color CRT are not concentric with either the tube neck or the yoke, and are off center in different directions.

Proper beam landing and convergence is ob-

650 PIX TUBES AND ASSOCIATED CIRCUITS

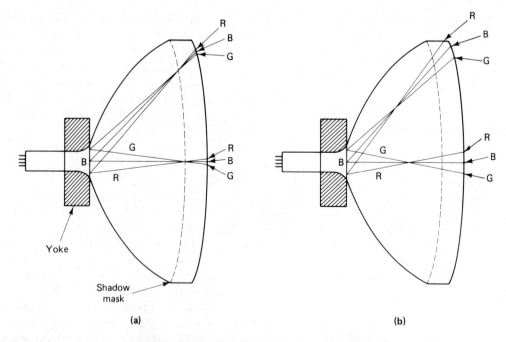

Fig. 15-10 Convergence and beam-landing problems. Perfect convergence is necessary for the beams to strike their proper phosphors. (CRT geometry and beam deflection exaggerated for clarity): (a) normal beam crossover, convergence, and landing; (b) misconvergence (beam crossover too soon).

tained with a *convergence assembly* which is mounted on the tube neck.

15-46 THE CONVERGENCE ASSEMBLY

This unit (Fig. 15-11 and 15-12) has three protruding "members", one for each of the three guns. It is positioned on the CRT neck directly over the internal convergence pole pieces described earlier. Each of the three identical arms of the assembly has two parts: an adjustable permanent magnet, and a two-coil electromagnet. Magnetic flux from each of the three sections is induced into the corresponding, built-in pole pieces of each associated gun to help direct its beam (Fig. 15-12). Each section is identified (red, green, or blue) in accordance with the gun it controls.

15-47 STATIC CONVERGENCE

Static convergence is the convergence of the three beams for proper beam landing over the *center* portions of the screen. Each of the three sections of the assembly has a small adjustable permanent magnet, which, when twisted, varies the flux density between the internal pole pieces of the corresponding gun. Beam movement is at right angles to the flux lines, and as each adjuster is turned its beam is forced to move *diagonally*, in-and-out in line with the center of

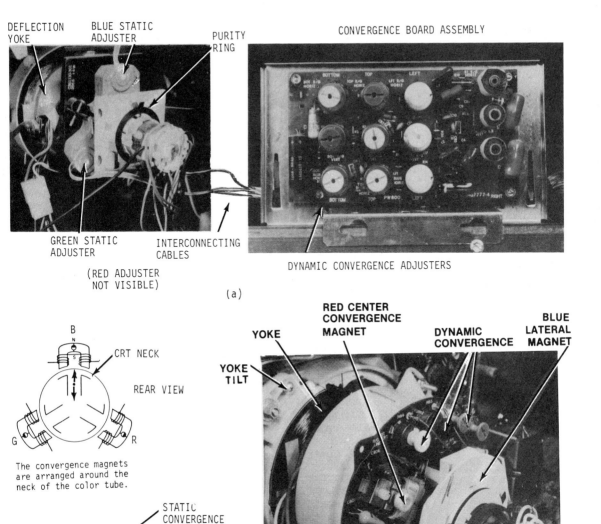

Fig. 15-11 Delta CRT convergence assemblies and adjusters: (a) early type system with extended convergence panel; (b) here the dynamic convergence adjusters and circuitry are part of the CRT neck assembly.

651

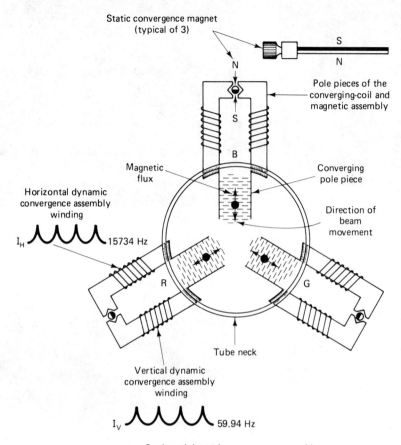

Fig. 15-12 Delta convergence-assembly functional drawing. Vertical coils are energized from the vertical sweep circuit; horizontal coils from the horizontal sweep circuit. (Parabolic waveform voltages are applied to all three convergence units.)

the CRT neck (Fig. 15-12). For static convergence, older sets used adjustable magnetic bars positioned over each gun.

15-48 THE BLUE LATERAL ADJUSTER FUNCTIONS

This is another static-convergence adjuster. It is required because of the position of the blue gun relative to the yoke coils, which is different from that of the other two guns. It is an adjustable permanent magnet that produces a *vertical* magnetic field to move the beam of the blue gun *sideways*. In early sets it consisted of a small magnetized rod secured to the tube neck with a spring. In late-model sets it may consist of magnets controlled by rotatable plastic gear-driven assembly as shown in Fig. 15-22. It is positioned on the neck between the purity ring and the convergence assembly. All four static convergence adjusters are adjusted for proper beam landing while observing either a crosshatch/dot pattern or a BW pix.

15-49 DYNAMIC CONVERGENCE

As shown in Fig. 15-10, convergence problems increase with the angle of beam deflection, and are greatest at the four edges of the screen, where the beams travel the greatest distance. Correction is made by causing the beams to speed up (sweep at a faster rate) as the deflection angle increases (*both vertically and horizontally*), and slowing down the sweep while scanning the center areas of the screen. This is done by creating variations of flux density across the built-in pole pieces to change the angle at which the beams enter the yoke, in accordance with the ever-changing angle of deflection.

The variation of flux density at the pole pieces, and hence the beam "aiming" (prior to deflection), is accomplished by feeding the coils of the convergence assembly with samples of current from the vertical and horizontal sweep circuits. Since *non-linear* changes are required, a specially shaped *parabolic waveform* is used (see Fig. 15-12). Each of the three members of the convergence assembly has two coils. One coil is energized from the vertical sweep circuit and the other from the horizontal sweep circuit (see Figs. 7-15 & 9-6). Thus, the strength of the flux at any instant is determined by the relative amounts of currents in the coils, which either add to or subtract from the magnetic fields produced by the static convergence adjusters.

With the beams aimed at the center of the screen, the "troughs" of the parabolic waveforms result in minimum current and low flux density; the beams move outward away from the center of the tube neck, reducing the influence the yoke flux has on the beams; and the sweep is slowed down. As beam deflection approaches the four edges of the screen, the peaks of the waveform produce maximum current through the convergence coils; the maximum flux forces the beams inward toward the center of the neck; and the angles at which the beams enter the yoke change, to speed up the deflection.

Note that the sweep is maximum when the beams are directed to the four *corners* of the screen, which corresponds with the peaks of the parabolic waveforms, and the sweep rate is lowest during the troughs of the waveforms when the beams are striking the center of the screen.

Each of the three beams requires a different amount of correction for each of the four sides of the screen. This differentiation is accomplished by controlling the amplitude and shape (tilt) of the parabolic waveforms feeding the three members of the convergence assembly. This process calls for considerable circuitry, which differs for each receiver. Besides resistors, capacitors, diodes, and coils, there are up to 12 controls, each with a distinctive functional name. (see Table 15-2). In early sets, the waveshaping circuits and the controls were contained on a *convergence board* (Fig. 15-11) connected to the convergence assembly by an extended cable. The unit is removable to facilitate making adjustments while observing the screen. In late-model sets the controls are often on a PCB to which the CRT socket is attached.

15-50 PINCUSHION CORRECTION

With rectangular, large-screen CRTs, the four sides of the raster tend to bow inward producing a pincushion effect. In BW sets, correction is accomplished by means of bar magnets; in color sets it is done by changing the amplitude of the vertical and horizontal sweep voltages while scanning different areas of the screen. Vertical deflection is increased while the beam is at the

Table 15-2 CRT operating controls.

CONTROL	FUNCTION	STAGE OR CIRCUIT CONTROLLED
Brightness	Panel control. CRT bias control for pix brightness.	DC-coupled video amp. stage.
Master brightness (Kine Bias)	To establish range & limits of user brightness control.	CRT grid/cathode circuits or in video amp.
Screen controls	For individual adjustment of R-G-B screen grid voltages.	B+ or B boost fed to CRT screen grids.
Master screen	In lieu of individual screen controls.	As above.
Drive controls	To adjust amount of signal drive to each gun.	CRT grid/cathode input circuits.
Purity ring	To obtain pure R-G-B color fields.	On CRT neck.
Static convergence adjusters	Blue lateral and R-G-B static adjusters. For beam convergence at center of screen.	Part of convergence assembly on CRT neck.
Dynamic convergence controls	For beam convergence at edges of screen.	With convergence circuitry on convergence board, PCB, or subchassis.
Pincushion adjusters	To align raster edges with cabinet masking bezel.	Vert. and horiz. sweep circuits feeding yoke.
Raster centering	To adjust vertical and horiz. centering of raster.	DC fed to vert. and horiz. yoke coils.
Yoke adjustments	For raster tilt, and longitudinal adjustment of yoke for purity "fireball".	Mechanical adjustments on yoke support.
Focus	To adjust beam focus.	HV power supply.
HV	Sets HV to specified value.	HV power supply.

top and bottom of the screen for the middle of each horizontal trace. Similarly, the left and right sides are filled out by increasing the horizontal deflection in the middle of each vertical trace. Correction is made by introducing a parabolic waveform into the vertical and horizontal yoke coils, as explained in Sec. 7-14 and 9-6. Overcorrection results in a barrel-shaped raster.

15-51 CRT OPERATING CONTROLS

There are many controls and adjusters, as listed in Table 15-2, that directly and indirectly affect the operation of a delta-type CRT. Some are user (panel) controls; others are service-type controls used in the CRT set-up procedures to

15-52 THE IN-LINE PIX TUBE

In-line type CRTs are rapidly replacing the delta-type tube in most late-model color sets because of fewer problems in converging the beams. There are three main differences between an in-line and a delta CRT: instead of a triangular grouping, the three guns are arranged *in-line horizontally;* the screen phosphors are *stripes* instead of dots; and the mask (called the aperture grill) has vertical *slots* instead of holes. There are two types in current use: the "domestic" three-gun in-line tube, and an imported "single-gun" tube called the *trinitron.*

15-53 THE 3-GUN IN-LINE CRT

Physical and functional characteristics of this type CRT (see Fig. 15-14), and how it differs

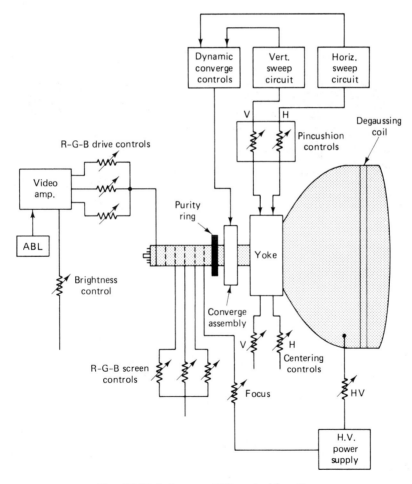

Fig. 15-13 Delta-type CRT control functions.

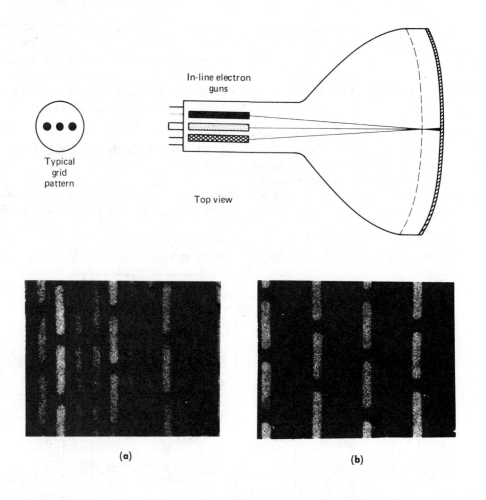

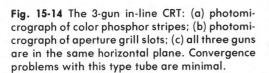

Fig. 15-14 The 3-gun in-line CRT: (a) photomicrograph of color phosphor stripes; (b) photomicrograph of aperture grill slots; (c) all three guns are in the same horizontal plane. Convergence problems with this type tube are minimal.

from a delta-type tube, are explained in the following paragraphs.

15-54 THE SCREEN

Figure 15-14 shows an enlarged section of the screen where the three-color phosphors are deposited as *vertical stripes*. The stripes, which are arranged in groups of three (one for each of the three primary colors), are only a few thousands of an inch wide, (depending on screen size), which is slightly *smaller* than the diameter of the electron beams. Each stripe is surrounded by a black area for increased brightness and to reduce *color bleeding*.

15-55 THE APERTURE GRILL

This is equivalent to, and serves the same purpose as, the shadow mask in a delta-type CRT. Instead of holes, the grill is *slotted*. The slots correspond with the phosphor stripes on the screen. The grill is positioned about ½" back of the screen with each slot directly behind a green phosphor stripe.

15-56 THE ELECTRON GUNS

Except that the three guns are positioned side-by-side (in-line) (see Fig. 15-14) they are basically similar to the guns in a delta CRT. There are some differences, however. For example, in some in-line tubes the three control grids (G1) are internally connected together as are the three screen grids (G2), and the signals are applied *only to the cathodes*. The common control grids connect either to ground or B+. Other in-line tubes have separate control grids and screen grids, the same as a delta CRT.

Another difference is in the method used for convergence. Some early in-line tubes used electromagnetic convergence much like that used with delta tubes. Such tubes had built-in pole pieces to transfer the magnetic flux from external convergence units to the red and blue beams (see Fig. 15-15)—more on this later. The 3-gun in-line CRTs used in most modern sets have no built-in convergence elements. Convergence is accomplished with ring magnets and by "wedging" the yoke (Fig. 15-15b).

Except that the three beams are in a horizontal plane, operation is not too different from that of a delta-type tube. On leaving the guns, the three beams (during the scanning process) converge in the vertical slots of the aperture grill to strike their corresponding color stripes on the screen. Note that the *green* beam is not bent, but always travels *in a straight line*, and that the distance travelled by *each* of the three beams during vertical deflection *is the same*, which greatly simplifies any problems of vertical misconvergence. Because the beams are wider than the grill slots, each beam actually goes through *two* slots to simultaneously illuminate two phosphor stripes of the same color.

External Gun Components: Besides the degaussing coil and the deflection yoke, there are few external components because of the relative simplicity of convergence systems required for in-line guns.

- **Color Purity**

To develop pure color fields, some in-line tubes use a conventional purity ring. Many late-model sets have a special kind of ring magnet assembly (Fig. 15-15b).

- **Convergence**

Some tubes use ring magnets; some use electromagnets. The unit shown in Fig. 15-15b is

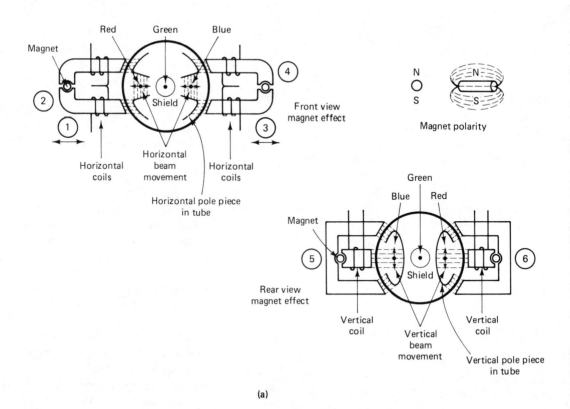

Fig. 15-15 (partial)

vergence, the flux produced by the vertical section moves the two beams in a vertical direction as the raster is scanned vertically. Note the small PM adjusters used for adjusting vertical and horizontal *static* convergence.

• **Typical In-Line CRT Circuitry**

Sample Circuit B at the back of this book uses an in-line CRT. Note the internally connected control grids and screen grids; also the protective spark gaps built into the CRT socket. The HV at the CRT ultor is approximately 30.5 KV and is not adjustable. Focus voltage which is adjustable is about 7 KV. Screen grid voltage (about 700 V) is adjustable by the master screen control R 1781. The average cathode bias and raster brightness is controlled by the front panel brightness control and the brightness range control via the video amplifiers. Individual CRT beam currents are adjustable with the drive controls and CRT cutoff controls that individually bias the three CRT cathodes. Pre-CRT matrixing is used with the video and chroma signals matrixed in the three chroma output stages that separately drive the CRT cathodes.

Reminder: _____

With this system, no signals are fed to the control grids.

Fig. 15-15 (a) and (b) In-line CRT purity and convergence assemblies; (c) few adjustments are required for converging the in-line guns.

used to adjust both purity and static convergence. It consists of 6 magnetized rings that are individually adjusted by turning cw or ccw. The two rings at the front are for purity, the other four are for static convergence. With in-line tubes there is no need for a blue lateral adjustment. With this arrangement, *dynamic convergence* is achieved by inserting small wedges between the yoke and the CRT, as shown.

Instead of ring magnets, some older in-line tubes used an electromagnet convergence assembly similar in some respects to that used with a delta-type CRT. Simplified drawings of the two-section assembly are shown in Fig. 15-15a, along with a photo of a similar unit in position. As in the delta CRT system, the coils are energized from the vertical and horizontal sweep circuits. The assembly consists of two sections: one for vertical convergence, the other for horizontal convergence. The convergence coils are connected in series with the coils on the deflection yoke. In the horizontal section, sawtooth current through the coils produces a magnetic field that is induced into built-in pole pieces in the tube to move the red and blue beams *horizontally* by the proper amounts according to the changing angles of deflection. The green beam is not affected. For vertical con-

15-57 THE TRINITRON

This is a "so-called" *single-gun* pix tube. Once the electrons are emitted from three separate in-line cathodes, beam aiming, focusing, and acceleration is accomplished with gun elements *common to all three beams*. This type CRT has a set of four built-in horizontal convergence plates (see Figs. 15-16 and 15-17a). The plates are

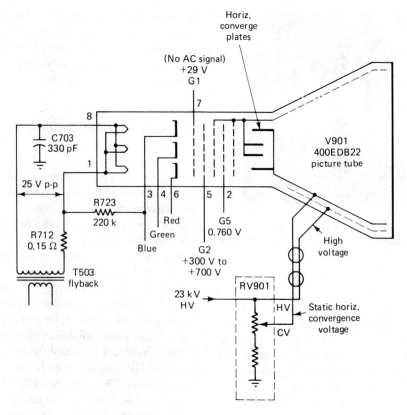

Fig. 15-16 The single-gun trinitron pix tube.

mounted upright in the same horizontal plane as the three beams. The aperture grill of a trinitron is slightly different from other in-line tubes in that it is constructed in *one piece*. Figure 15-17b shows the beam aiming of the three types of CRTs for comparison.

In Fig. 15-16, note that HV is applied to the convergence plates, adjustable by means of a *static horizontal convergence control*. Variations in voltage bend the red and blue beams for crossover in the grill openings as shown. This is not required for *vertical* convergence. Dynamic horizontal convergence is sometimes accomplished by superimposing a high-amplitude parabolic waveform voltage from the horizontal sweep on the HV dc applied to the convergence plates. Since a trinitron has only one control grid and screen grid the beam currents must be individually varied with controls in the three cathode circuits.

15-58 EXTERNAL GUN COMPONENTS

Besides the yoke, the only component on the CRT neck is a small coil (called a *twist coil*) positioned near the rear of the tube with its windings concentric with the tube neck (Fig. 15-17c). It serves the same function as the purity ring on other type tubes. The coil is energized with DC. Beam-landing adjustments are made by rotating the coil which moves the red and blue beams as shown in Fig. 15-17d. Note that the *green* beam is not affected. Vertical static con-

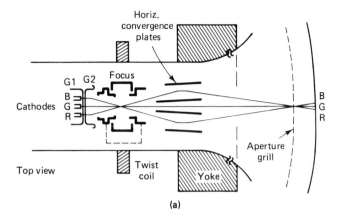
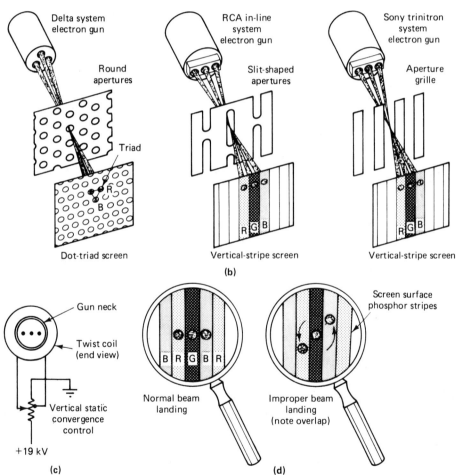

Fig. 15-17 Trinitron single-gun elements and beam control.

From:
Alvin A. Liff, *Color and Black & White Television Theory and Servicing*, ©1979, p. 294. Reprinted by permission of Prentice-Hall, Inc., Englewood Cliffs, NJ.

vergence is adjusted with a control that varies the amount and direction of the dc current flowing in the coil.

15-59 SIGNAL INPUT CIRCUITS

For a BW set, the video signal is applied to either the grid or the cathode, usually the latter. A color set requires two kinds of signals: chroma, and luminance (the Y signal). In a delta-type CRT in which the two signals are matrixed (mixed) in the tube itself, the Y signal is fed to the three cathodes and the chroma (R-Y, G-Y, and B-Y signals) are fed to the grids (Fig. 14-15). With pre-CRT matrixing, the mixing occurs in stages ahead of the CRT (for both delta and in-line tubes), and the red, green, and blue signals feed the cathodes.

15-60 DEVELOPING THE WHITE RASTER

Each of the three guns of a color CRT contributes to developing the raster, and proper operations depends on the condition of each gun and the applied voltages. The beam currents of the three guns are controlled individually, usually by varying the screen-grid voltages. If all three screen-grid controls are turned down, the screen will go dark. Assuming that proper set-up adjustments have been made for color purity (beam landing), and that the brightness control is turned up, if the red screen grid control is turned up there will be beam current from the red gun and we will see a red raster. With the red control turned down and the green control turned up we get a green raster. When only the blue control is turned up we get a blue raster.

As explained earlier, white is a mixture of the three primary colors in the proportions of 30% red, 59% green, and 11% blue. Obtaining these proportions depends on the relative adjustments of the three screen controls. There are many shades of white, and the "whiteness" of the raster is determined by the *brightness* of the three colors; high-intensity red, green, and blue produce the *whitest white,* and *dull* colors produce the various shades of off-white, i.e. *gray*. Black of course is the complete absence of brightness, as when the set is turned off.

Raster whiteness varies somewhat with different sets, depending on the nature of the screen phosphors. Many viewers prefer a slightly bluish tint. As a set ages, the emission of one or more guns may drop off, and the screen may take on a tint of some color; for example, yellowish if the blue gun is weak, or greenish if both the red and blue are weak. Whereas white in its various shades is produced by a 30-59-11% proportionate mix of the three primaries, changing the mix makes it possible to obtain practically any hue present in nature.

15-61 DEVELOPING A BW PIX

A *white raster* is a necessary prerequisite to obtaining a BW pix that is free of color contamination. When receiving a BW telecast, the chroma stages of the receiver are disabled and only the BW (luminance or Y signal) reaches the CRT. *Each* of the three guns of the CRT contributes to developing the pix, and each gun must be driven with the same proportionate signal level as in the R-G-B mixture required to produce a white *raster*. Under these conditions all three guns are driven to cutoff at the same time to produce black, and their beam currents are increased in the proper proportions to produce the various shades of white in the pix.

- **Gray-Scale Tracking**

Too much or too little signal drive to one or more of the guns results in variations of whites and grays that change with the pix content and the setting of the brightness control, even with a normal off-channel raster. In some cases, certain areas of the pix may become color-tinted, changing with the scene. The proper amount of signal drive to each of the three guns is established with R, G, and B *drive controls* connected between the video output stage(s) and the three CRT inputs. In a process called gray-scale tracking they are adjusted to obtain a uniform level of whiteness for all settings of the brightness control.

Ideally, there should be no visible color when viewing a BW pix. Color outlines around different pix objects indicate improper beam landing and the need for purity and/or convergence adjustments. In some sets, however, a small amount of misconvergence at the extreme edges of the screen is considered normal and acceptable.

15-62 PRODUCING A COLOR PIX

A color-free *BW pix* is a necessary prerequisite to obtaining a good color pix. When receiving a color telecast, the chroma stages are enabled and, when CRT matrixing is used, the CRT is driven by two kinds of signals: the luminance (Y) signal, and the chroma signals. The Y signal *simultaneously* varies the beam currents of all three guns. The demodulated chroma signals create beam current variations of the three guns *individually*. At the station, phase-modulated chroma signals are produced that correspond with the colors seen by the camera at any given instant. At the receiver, the signals are demodulated to produce R-Y, G-Y, and B-Y *color difference* signals that drive the three guns of the CRT. The relative amplitudes and polarities of these signals create beam current variations to produce the colors seen by the camera. The original *brightness* of the colors is provided by the Y signal.

When the chroma signals drive two of the guns to *cutoff,* the green and blue guns for example, only the red gun will be operating, to produce red in the pix. With the red and green guns driven to cutoff, only blue is reproduced; with the red and blue guns at cutoff, only green images. With two guns conducting (red and green for example), we get yellow. Although both the red and green phosphors are illuminated, they are not separated by the eye at a normal viewing distance, and so we see the mixture as a different hue, yellow. Other colors are produced in the same manner in accordance with the *relative* amplitudes and polarities of the three chroma signals driving the guns (see Table 15-1).

This can be more easily understood by the examples shown in Fig. 15-18. In (a), the signals feeding the three guns are in-phase, and during the positive half cycles all guns are driven to cutoff to produce a dark screen (black pix objects). In (b), the three signals are in-phase and negative-going, and all guns are conducting equally to produce some shade of white in the pix. In (c), the positive half cycles are driving the red and green guns to cutoff while the blue gun is conducting to produce blue pix detail at that instant of the scanning. In (d), the red and blue guns are cutoff while the green gun is conducting to produce green objects. In (e), two guns (red and blue) are conducting and the green gun is cut off. With the two phosphors illuminated by the same amount we see it as magenta. In (f), the blue and green guns are conducting equally to produce cyan. In (g), the red and green guns are conducting equally to produce yellow. In h, as in g, the blue gun is cut off and the red and green guns are conducting *but by different amounts,* because of the phase difference. With a predominance of red, we get a reddish yellow, (orange).

The intensity (degree of saturation) of all colors is determined by the amplitude of the chroma signal and by the setting of the front panel *color control.* Variations in *hue* are obtained by using the hue (or tint) control, which varies

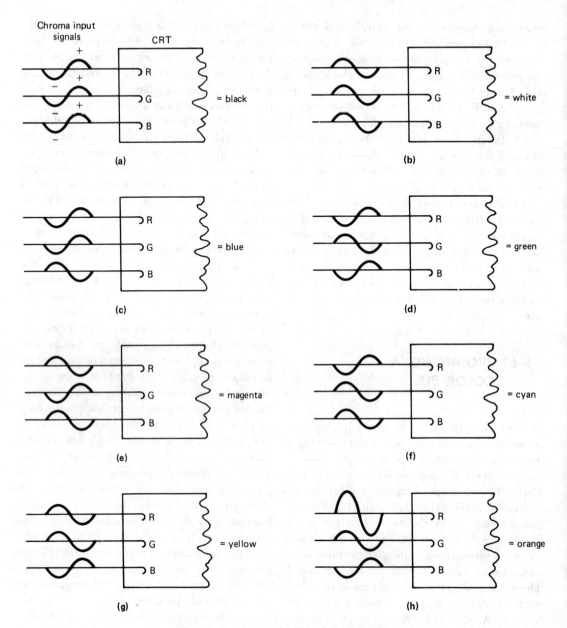

Fig. 15-18 Color reproduction: (a) all guns cut off; (b) all guns conducting equally; (c) only blue gun conducting; (d) only green gun conducting; (e) red and blue guns conducting, green gun cut off; (f) blue and green guns conducting, red gun cut off; (g) red and green guns conducting, blue gun cut off; (h) red gun conducting heavily, low conduction of green gun, blue gun cut off.

the phase relationship between the color difference signals and the color subcarrier. The brightness of colors is determined by the amplitude of the (Y) signal. *Manual control* of brightness is done by means of the brightness control.

With the *pre-CRT* matrixing system (Fig. 14-16), color brightness levels are established in matrixing amplifiers *ahead of the CRT,* and the CRT is driven by only one kind of signal: the R, G, and B signals from the output amplifiers.

- **Chroma Demodulation, Phasing, and Matrixing**

Both the luminance and the chroma *IF* signals must be *demodulated* to extract the modulation frequencies needed to drive the CRT. Two (sometimes three) demodulators are used for the chroma. At the demodulator inputs the signals exist as *phase-modulated* sidebands of the color subcarrier. At the demodulator outputs there are two (or three) signals whose amplitudes vary in accordance with the phase variations of the input signals, which correspond with the colors being seen by the camera. These signals represent *fully saturated* colors, and until matrixed with the brightness component (the Y signal) would produce harsh and unnatural-looking colors. Also, since chroma frequencies only extend to 1.5 MHz, the absence of frequencies between 1.5 MHz and 4.2 MHz (as provided by the Y signal) results in an almost unrecognizable pix of no detail.

To provide the fine detail and restore the original brightness of the colors, the Y signal and the chroma are recombined in the demodulators, the output amplifiers, or the CRT, depending on receiver design.

- **The Hue Control**

As a *phase adjustment,* this control provides the viewer with some control over color. It is normally adjusted for some recognizable tint such as skin color (flesh tones); once it is set, *all* colors should be correct. When turned fullly cw, pix takes on a greenish appearance; when fully ccw, magenta.

15-63 PRODUCING COLOR-BAR PATTERNS

A color-bar pattern consists of a number of vertical bars, each bar of a different hue and displayed in a particular sequence. It is used for checking equipment performance and making adjustments at the TV station, and is an almost indispensable aid in troubleshooting color receivers. Such patterns are sometimes transmitted, but rarely. For receiver servicing, a color-bar generator (Fig. 3-8, Sec. 3-12) is used. By analyzing the color bars you can quickly determine such things as: the proper operation of all chroma stages; whether the signals reaching the CRT guns are correct and, if not, which circuit is responsible; color sync conditions, as an aid in making adjustments; and the correctness of the operation of the ACC and the hue control.

There are two types of color-bar generators in common use; the seven-bar *NTSC generator,* and the 10-bar *gated rainbow generator.* The latter type is more popular. The waveforms generated by each and the patterns they produce are shown in Fig. 15-19. Some understanding of how the bar signals are generated and how they produce the patterns can be helpful.

The generation of a color bar pattern depends on two basic considerations: 1) the phase difference between two signals of slightly different frequencies is constantly changing; 2) phase variations of the chroma signal produce different colors on the screen. A typical bar generator puts out an amplitude-modulated carrier equivalent to the frequency of channel 3 or 4. The modulating signal is produced by a

666 PIX TUBES AND ASSOCIATED CIRCUITS

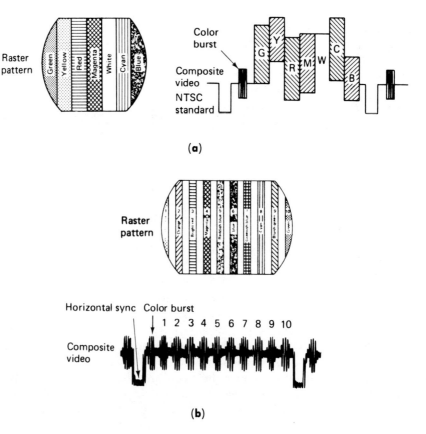

Fig. 15-19 Producing color-bar patterns. A bar pattern is the fastest way to diagnose color problems.

From:
Alvin A. Liff, *Color and Black & White Television Theory and Servicing*, ©1979, p. 563. Reprinted by permission of Prentice-Hall, Inc., Englewood Cliffs, NJ.

crystal-controlled color-subcarrier oscillator whose frequency differs from that of the subcarrier oscillator in the receiver by the frequency of the horizontal sweep (15, 734 Hz). The actual frequency of the receiver oscillator is 3.579545 MHz (not 3.58 MHz). Thus the generator subcarrier frequency is 3.563811 MHz. During the scanning of each horizontal line, the relative phase between the two signals changes by 360°. Such a signal applied to the CRT produces a constantly changing color for each horizontal line. For convenience, the colors are broken up and appear as bars of a number of specific hues. This interruption of color is the result of keying the generator signal at a 188.81 KHz rate.

In Fig. 15-19a, note the signal developed by an NTSC type generator during each horizontal scan period. This particular generator produces seven *bursts,* besides the *color burst* used for synchronization. Each burst is of a specific amplitude and phase, to develop the colors as shown, including white. The signal is sometimes referred to as the "crankshaft waveform", which it resembles.

The signal developed by a gated rainbow generator is shown in Fig. 15-19b. Twelve bars are produced, 10 of which develop the 10 corresponding color bars as indicated. With ten bursts, and each horizontal trace represented by 360°, each burst is 30° different in phase from the adjacent bursts, and therefore produces a different color. The horizontal sync pulse and color burst occur during the first 60°. The colors are numbered 1 through 10 starting with yellow-orange on the left and ending with green at the extreme right side of the screen. Note the specific phase angles that develop each color, for example, 90°, 180°, and 300°, for the three primaries, red, blue and green.

Besides differing in phase, each color developed by the station signal has a different level of saturation. Thus the crankshaft signal produced by an NTSC generator develops *true* colors. By comparison, a gated rainbow generator produces color bars of uniform intensity because each of the bursts have the same amplitude, as indicated in the figure.

In a properly operating receiver, adjustment of the hue control normally produces a phase shift of about 60°, moving all bars to the left or the right by one bar.

The uses of a color-bar pattern are described in Chapter 16.

15-64 PROJECTION-TYPE RECEIVERS

Because of the screen-size limitations of conventional sets, projection receivers are staging a comeback. Currently there are two basic types. A somewhat elaborate (and expensive) system uses three CRTs, one for each of the primary colors, projection lenses, and a silver screen. Projection is from the front (RW). It is used mostly for viewing by large audiences. The second type, more suited to domestic needs, is self-contained, and uses a single CRT which projects *from behind*. A typical late-model receiver of this type is described as follows. Except for the optical system, it is not too different from a "conventional" receiver. The main difference is in the CRT and its supply circuits.

Because of light losses, the system requires a CRT capable of producing high-intensity images on its screen. Such a tube requires a rather high HV, typically around 35 KV, for a beam current of 2 MA or so, double that of a conventional CRT. Because of the radiation hazard from X-rays, special shielding is required.

The essentials of a typical optical system are shown in Fig. 15-20a. Most of the items are mounted on a removable board. The CRT, which is a 13″ in-line type, is located at one end of an enclosed "tunnel". At the other end of the tunnel is a 7″ lens assembly. Light from the CRT (which is mounted at an angle) passes through the lens assembly and the enlarged image is reflected from an adjustable "lower" mirror, to a large "upper" mirror, then onto the rear of the viewing screen.

The large *flat* rectangular screen measures about 45″ diagonally. It has two different surfaces, a smooth side facing outward and a *grooved* side facing inward toward the optical system. The inner surface is called a *fresnel lens* and consists of closely-spaced concentric grooves (Fig. 15-20b) which duplicate the function of a rounded high-power conventional lens. The image is formed on a surface in front of the grooves.

Shortcomings of projection TV are: less brightness and pix sharpness compared to a direct-view TV, and the need for viewing the pix almost straight-on. The widely-spaced scanning lines can also be objectionable unless viewed from a distance.

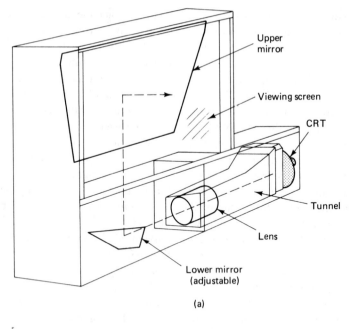

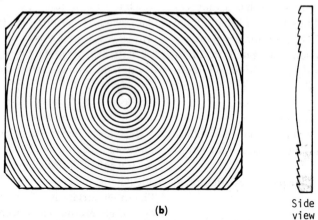

Fig. 15-20 Projection receiver (G.E.). With this system, projection is from behind the screen: (a) the precision mirrors direct the light from the projection lens onto the viewing screen; (b) the surface of the Fresnel lens appears to be a number of concentric grooves.

15-65 TROUBLESHOOTING THE CRT AND ASSOCIATED CIRCUITS

The experienced technician is able to recognize the more common symptoms of a defective CRT or improper set-up adjustments. He also knows that he can be wrong in his diagnosis and that there's a big difference between *suspecting* a component and *proving* it's defective. Since most of the symptoms of a bad CRT are often caused by troubles *elsewhere* in the receiver, and because the CRT is the most expensive item in the set, it must never be condemned and replaced without first performing a number of essential tests.

15-66 USE OF A CRT TEST JIG

Sometimes, when making house calls, the problem appears unrelated to the CRT or its associated components, and it's expedient to leave the CRT and the cabinet and take only the chassis to the shop. At the shop, the receiver is operated in conjunction with a CRT Test Jig, which, as described in Sec. 3-20 (see Fig. 3-15), consists of a substitute CRT, complete with its own sweep, convergence components, and HV. After returning and installing the chassis, certain touch-up adjustments are required that could not be made at the shop.

15-67 ANALYZING THE SYMPTOMS

Except for sound, all trouble analysis is based on what we see or don't see on the pix tube screen, and time spent in careful observation and evaluation of conditions usually pays off. Although the CRT is the key element, as important as its condition are the applied voltages, the associated components, and the set-up adjustments.

Decisions made as a result of the preliminary analysis determine the course of action and what tests are to be made, at least in the beginning. Table 15-3 lists most of the trouble symptoms relating to the CRT and its operating conditions. Troubleshooting procedures are shown in the flow drawings of Fig. 15-21. For additional symptoms and procedures refer to Chapter 14.

15-68 INSPECTION

With the set off, make a preliminary inspection for the *obvious*, such as broken wires, disconnected connectors, improper positioning of components on the CRT neck, etc. With the set on, check for CRT heater glow, signs of arcing, the odor of burning insulation, etc. If there's no raster, try resetting the circuit breaker if tripped.

15-69 FIRST-IMPRESSION SUSPECTS

Bad pix tube? Improper adjustments? Circuit or component troubles? Although it's too soon to make a *firm* judgment, we must start somewhere. Based on the observed symptoms, and depending on whether we seem to have a *raster* or a *pix* problem, decide which tests to make first. For example, if there's no raster one might logically start by testing the CRT or checking for HV. If both are OK, the next step would be to check *all* CRT operating voltages. On the other hand, if it's a pix problem and the raster is normal, none of these tests are called for and we'd begin by checking signal voltages. When signals also check normal, it could very well be an adjustment problem, particularly if it's a color set. Frequently the initial tests are a matter of personal preference based on symptoms, circumstances, and past experience.

Table 15-3 Trouble Symptoms Chart: CRT and associated circuits (BW & Color Sets)

RASTER PROBLEMS (Reminder: Raster troubles also show up on a *pix*.)

Trouble Condition	Probable Cause	Remarks/Procedures
No raster/dim raster	Defective CRT. Excessive bias. Weak or missing HV or other applied voltages.	Check for heater glow. Test CRT. Measure all operating voltages. For a weak CRT, try a filament booster and/or rejuvenate. Check for excessive bias between grid and cathode with brightness control turned up. If a color CRT, turn up master brightness and screen controls. A weak CRT will often brighten during warm-up. Don't evaluate brightness in complete darkness or with light shining on the screen. If dirty, clean faceplate.
Excessive brightness (unable to darken screen)	Loss of CRT bias, or grid-to-cathode short.	To obtain beam current cut-off control grid must be at least -30 V relative to the cathode. Check for too much positive voltage or not enough negative on the grid, that is, if the cathode is either too negative or not enough positive. Check brightness control circuit and video amplifiers. If control ineffective and loss of pix, suspect short in CRT.
Poor focus	Weak or gassy CRT. Loss of focus voltage or insufficient range of focus control.	Evaluate focus on the raster, not by the quality of a pix.
Raster distortion	Pincushion adjustments. Speaker or other magnetic influences too close to the CRT. Raster bending due to poor LVPS filtering. Keystone-shaped raster caused by a defective yoke.	Action is determined by the nature of the problem.
White flashes on screen	HV arcing. CRT shorting out.	Locate source of arcing. May be all but invisible corona at some hidden site. Observe in darkened room.

Table 15-3 (cont'd)

RASTER PROBLEMS (Reminder: Raster troubles also show up on a *pix*.)

Trouble Condition	Probable Cause	Remarks/Procedures
Blooming	Loss of CRT bias, excessive screen voltage, or poor HV regulation.	See Sec. 14-30 and Table 14-1.
Pix inverted or transposed left to right	Yoke is mounted upside-down, or turned; or the coil leads are reversed.	If transposed, lettering on a pix will be backward.
Small raster	Low B voltage often due to low line voltage. Excessive HV. May be trouble in both vertical and horizontal sweep circuits or misadjusted size controls.	If necessary, monitor for line voltage fluctuations over 24 hr. period. If low voltage, don't compensate by adjusting width and height controls.
Arcing in CRT	Loss of CRT vacuum. Foreign particles in CRT.	No visible heater glow with the tube neck getting warm indicates air in the tube. New tubes sometimes arc then clear up with no harm done. If sustained arcing try tapping neck with the tube face down.
No pix, or loss of contrast (raster OK)	No signal or weak signal at CRT input(s).	Signal trace all stages up to the CRT input.

COLOR-RECEIVER SYMPTOMS

Color-contaminated raster	Defective CRT. Improper operating voltages. Improper set-up adjustments.	Check and compare intensities of all three color fields. Adjust purity and gray-scale tracking. If a purity problem test all three guns and their operating voltages. *Thoroughly degauss the CRT.*
Greenish blemish around edges of faceplate	Indelible staining caused by chemical reactions.	Generally not too objectionable when viewing a color pix. As with an ion blemish, CRT replacement is the only option.
Color contamination on a BW pix (normal raster)	Convergence problem, due to improper set-up adjustments, defective convergence compo-	Degauss CRT and repeat purity and gray-scale tracking adjustments. Note which

671

Table 15-3 (cont'd)

RASTER PROBLEMS (Reminder: Raster troubles also show up on a *pix*.).

Trouble Condition	Probable Cause	Remarks/Procedures
	nents or circuit problems, or a defective or magnetized CRT.	colors are not converged and in which direction. For center of screen, adjust the PM static convergence adjusters. For edges of the screen make dynamic convergence adjustments according to the type of CRT.
No BW pix and poor color pix (normal raster)	Weak or missing luminance (Y) signal at CRT inputs.	Pix is produced by the chroma signals only and with no Y signal colors are *harsh* and vivid with loss of all fine pix detail.
No color or poor color (BW pix OK)	Chroma signals weak or missing at CRT inputs. Defect in color sync stages.	See chapter 16.

15-70 SYMPTOMS AND TROUBLESHOOTING PROCEDURES

Most of the procedures described at this time are generally applicable to both BW and color sets. Troubles peculiar to color sets are described later.

15-71 NO RASTER OR DIM RASTER (SOUND OK)

Start by turning up the brightness control. If a color set, also turn up the three screen controls. Visually check for CRT heater glow at the rear of the tube (*three* heaters if a color CRT). If no heater glow check the following possibilities: burned-out heater(s), break in heater circuit wiring, defective CRT socket, poor socket-to-tube contacts (which are often intermittent), loss of heater supply voltage, or loss of CRT vacuum (air in the tube). When there's air in the CRT, there may be no heater glow even when the filament shows continuity and the voltage is OK. However the neck will still get warm and there may be arcing within the gun(s).

REMINDER:

Many sets obtain CRT heater power from the flyback transformer. Check for HV. If OK, the CRT heater voltage is probably normal. Other causes of no raster or loss of brightness are: excessive bias; and CRT operating voltages that are weak or missing.

• **Check HV**

Connect ground lead of HV probe to chassis. Touch probe tip to HV ultor contact on the

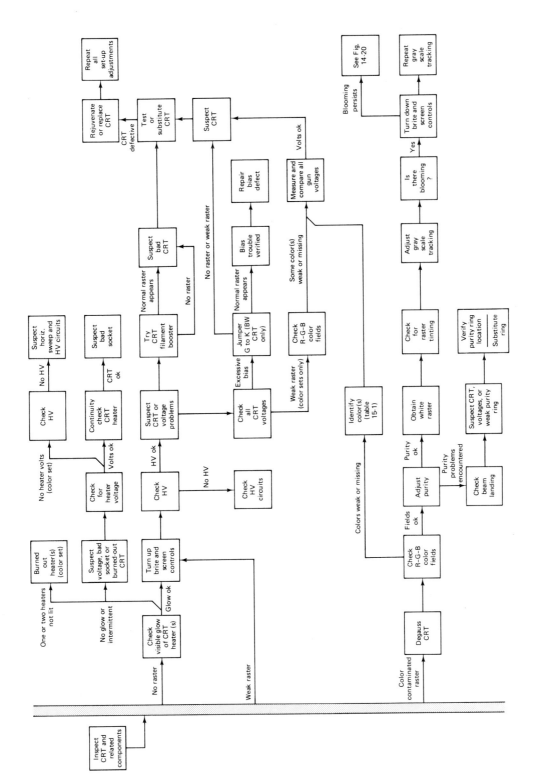

Fig. 15-21(a) Troubleshooting flow diagram.

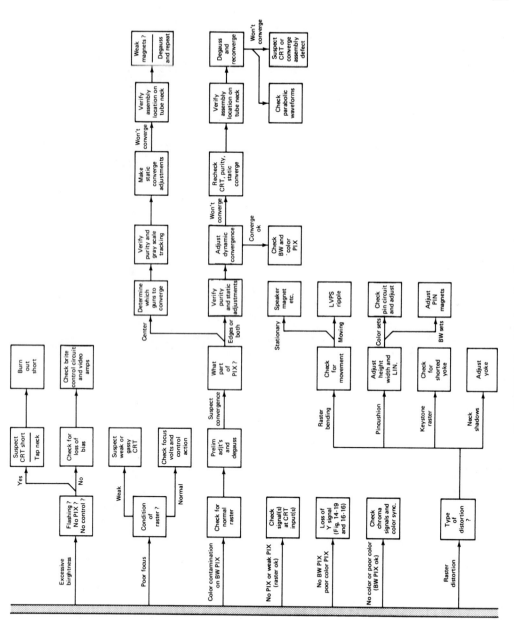

Fig. 15-21(b) Troubleshooting flow diagram.

15-71 No Raster or Dim Raster (Sound OK)

CRT and measure voltage. If rubber suction cup is used, push probe between rubber and the tube bowl until contact is made. Take proper precautions to avoid shock. If necessary, adjust HV to value specified in service data.

If no HV or low HV, remove HV lead from ultor contact and recheck for voltage on lead under "no load" conditions. If normal voltage is restored, suspect: poor HV regulation; an internal HV short in the CRT; or the CRT is operating with a high positive bias. If no HV with HV lead removed from the CRT, check the HV and horizontal sweep circuits per Chapters 9 and 10.

- **Arc Testing the HV**

When a probe is not available, the presence or absence of HV can be determined by a "spark test". After a little practice, a relative estimate on the *amount* of voltage can also be made. Spark testing, although practiced by many, does entail certain *risks*, and for that reason should not be practiced indiscriminantly or under all circumstances. Besides the shock hazard, HV surges can sometimes damage or destroy components. Also, any sustained overload can overheat components in the HV/beam current circuit, such as the HV rectifier and the flyback transformer.

An alternative and "safer" procedure in lieu of a probe is to use a small neon lamp attached to an insulated rod. When it is held near a flyback transformer or HV lead, the intensity of its glow provides a rough indicator of HV.

The procedure for spark testing HV is as follows: With the set *off*, disconnect the HV lead from the CRT ultor contact. Extend the lead away from the chassis and clear of all components. Connect a jumper wire to chassis ground. Turn the set on. Grasping the ground wire in one hand, bring it close to the end of the HV lead, noting the intensity of the spark, if any.

CAUTION: ———————

Avoid creating a sustained arc and never ground the HV directly. The length and intensity of the arc indicate the amount of HV, which varies with the type of set and the screen size. With a 21" BW set, for example, a $\frac{3}{8}$" arc might be considered normal. For a large-screen color set it's *much greater,* often up to 2" or more, so heed the precautions. A good spark to ground usually gives off a loud crackling sound.

When there's little or no HV dc, check for HV *AC* from the flyback transformer to the HV rectifier.

CAUTION: ———————

Never attempt to measure this voltage or the meter will be damaged; and don't make arc tests to ground as described above. Testing should be limited to drawing a *corona arc* by holding a well-insulated screwdriver (or the equivalent) *near* the contact (but never close to the flyback). Unlike the HV dc, a corona arc has a "soft", fat appearance and emits a faint hissing sound. Normal length of a corona arc is somewhat less than for the HV dc. If no corona or weak corona, look for trouble in the horizontal sweep circuit.

- **Arc Testing for Beam Current**

When there's no raster but HV is OK up to the CRT, chances are there is loss of beam current. With set *off,* position the HV lead close to, but not touching the ultor contact. Turn set on. With the brightness control fully cw there should be a high intensity "beam current" spark between the lead and the CRT. Change the distance between the lead and the ultor as appropriate.

CAUTION: ———————

Avoid personal contact with the lead, especially for a color set with HV in the 25KV range.

676 PIX TUBES AND ASSOCIATED CIRCUITS

If no arc is obtained (indicating lack of beam current), suspect a defective CRT, excessive CRT bias, or loss of one of the applied voltages. If there is an arc, see if it changes (as it should) when the brightness control is adjusted. The beam current arc should increase with the control cw and disappear with the control fully ccw (because the tube is biased to cutoff). When these changes don't occur, suspect a CRT bias problem, a HV short in the CRT, or a shorted HV rectifier. For a BW CRT, try a jumper grid-to-cathode. If arc increases, check for excessive bias. For a color CRT, a comparison beam-current check should be made for all three guns by alternately adjusting the screen controls.

As an aid in localizing a problem to either the CRT or its operating conditions, or as a temporary expedient to restore set operation, it's sometimes helpful to connect two similar sets back-to-back, using the chassis of one set to drive the CRT of the other set, or simply to "borrow" the HV.

CAUTION: _____

Make sure the CRTs and the chassis are compatible, i.e., the CRT socket connections and the applied voltages. To avoid shocks, discharge the CRT by grounding the ultor before connecting or disconnecting the HV lead.

15-72 CHECKING CRT OPERATING VOLTAGES

At least one of the applied voltages is normally checked when there is a raster problem. Depending on the symptoms, check for the presence or absence of voltages, the amount of the voltages *as specified on the schematic,* and the polarity. Except for the HV, all voltages are checked at the CRT socket connections.

REMINDER: _____

Loss of a voltage can be due to a shorted CRT. When suspected, remove the socket and recheck voltage(s) under no-load conditions. Typical operating voltages are listed in Table 15-4.

• **Bias Voltage**

As with any vacuum tube, bias is measured between grid and cathode. Consider the symptoms. For loss of brightness, the cause may be too much bias with brightness control turned up. For excessive brightness, check for too little bias with control turned ccw. Excessive brightness where the control has little or no effect is often caused by an intermittent CRT short. Try tapping the neck. Incorrect bias is due to one of three things: incorrect cathode voltage, incorrect grid voltage, or an internal short or leakage in the tube. Compare voltage readings with those shown on the schematic. For a color CRT, bias should be checked for each of the three guns, and compared.

• **Screen Grid Voltage**

Loss of screen voltage will kill the raster; low screen voltage results in loss of brightness. For a color CRT, check screen volts for each gun while adjusting the screen controls. Compare with schematic. Where there is a wide variation, look for a shorted or leaky capacitor or a defective dropping resistor.

REMINDER: _____

The voltage will measure high if the tube isn't drawing current.

• **Focus Voltage**

The focus voltage for a BW tube is not too critical and many sets provide terminals for different voltage values. Select the voltage that

Table 15-4 Typical CRT operating voltages.

CRT Element/Voltage	Typical Voltage	Remarks
Heater (filament(s) not lit)	6.3 Vac (60 Hz from LVPS or 15,734 Hz from H.O.T.)	Check for heater voltage, burned out heater(s) or bad socket or connections.
HV	10-15 kV for BW, 25 kV for color	Measure at ultor contact. Should vary with HV adjuster
Focus	0-300 V for BW, 1.5 kV-5 kV for color	Should vary with focus control
Screen grid(s)	100-250 V for BW, 300-400 V for color	If color CRT, should vary with screen grid controls
Cathode(s)	0-100V for BW, 100-300 V for color	Should vary with brightness control (also KINE BIAS and drive controls if color CRT)
Control grid(s)	0-40 V for BW, 100-200 V for color	
Bias (measured grid-to-cathode) (brightness control fully ccw)	0-50 V for BW, 0-150 V for color	Grid voltage(s) measured in reference to cathode(s). Grid(s) are *negative* and vary with brightness control.

gives the best focus on the scanning lines, for normal setting of the brightness control. In some cases best focus is with the grid grounded.

A color CRT operates with a focus voltage in the 5KV to 7KV range. Note the variation as the control is adjusted. Good focus is hard to obtain with a weak CRT. When focus is best at one extreme of the control, check for a defective focus rectifier or a high-megohm dropping resistor that's changed in value.

15-73 SMALL RASTER

For loss of both height and width, the most common cause is low *B* voltage. Sometimes the cause is low line voltage. If raster slowly fills out during warm-up, look for a defective LV rectifier or filter or other cause of poor LVPS regulation.

REMINDER: ─────────────────────────

Excessive HV will also reduce raster size. If a color set, check for misadjusted HV or troubles in the HV regulator or ABL circuits.

15-74 RASTER BRIGHTENS DURING WARM-UP

The usual cause is a weak CRT.

15-75 DISTORTED RASTER

For pincushion or barrel effect, adjust PM magnet on BW set if provided, and pincushion adjustments if it's a color set, until raster edges are parallel with the mask opening. A keystone (trapezoidal) raster is usually caused by an open or shorted yoke coil.

15-76 EXCESSIVE BRIGHTNESS

If screen cannot be darkened with brightness control turned down, suspect loss of CRT bias or excessive screen voltage. For color sets, try the master brightness control. If loss of pix, suspect a shorted CRT.

15-77 NECK SHADOWS

For a BW set, position yoke tight against the bowl of the CRT. A yoke with a too-narrow deflection angle will also cause this problem. For a color set, adjust the yoke tilt, and raster centering.

15-78 TILTED RASTER

Turn yoke either cw or ccw until raster edges are parallel with mask opening and the scanning lines are horizontal. Check for neck shadows, then lock yoke into position.

15-79 RASTER NOT CENTERED

Not to be confused with insufficient height or width. For BW set, adjust centering ring on CRT neck. For color sets, adjust vertical and horizontal centering controls.

15-80 BLACK HORIZONTAL BARS

There are several possibilities: *sound bars* which cover the entire screen and fluctuate with the sound; *hum bars*, either one or two wide bars that move slowly up through the pix; *loss of height* (black border at top or bottom of pix or both)—*vertical centering* is off; vertical blanking bars which obscure part of the raster at either top or bottom of the screen (vertical retrace lines may or may not be observed).

15-81 RIPPLE ON RASTER EDGES

Left and right sides of raster have an undulating scalloped appearance, due to PS ripple voltage in the horizontal sweep circuits. Usually accompanied by hum bars. The waviness also appears on a pix, but is not to be confused with horizontal *pulling*—a sync problem.

15-82 POOR FOCUS

Scanning lines may be fuzzy and indistinct or not seen at all. Not to be confused with troubles causing loss of *pix* detail. Typical causes: weak or gassy CRT, improper adjustment of focus control, or a defect in focus circuit.

15-83 BLOOMING

Caused by poor HV regulation, or insufficient CRT bias; if a color set, may be improper operation of ABL circuit, or the screen controls may be set too high.

15-84 BRIGHT HORIZONTAL FLASHES ON SCREEN

Often caused by poor connections. If raster brightens and cannot be darkened, check for an intermittent short in the CRT. May also be internal arcing in the tube or corona arcing in or around the flyback transformer.

15-85 ARCING IN CRT

Sporadic arcing in a *new* CRT is quite common but it usually clears up after a short time. Sustained arcing in a tube if there's no raster often indicates loss of vacuum. Another symptom is no visible heater glow, and sometimes a bluish glow around the gun elements. Arcing in a tube that's still functioning can often be cleared up by "flashing" with several hundred volts of DC (see Sec. 15-112).

15-86 ION BURNS

This is a rare problem since the advent of aluminized screens. Such blemishes are indelible and the only recourse where pix quality is impaired is to replace the tube.

15-87 ARCING EXTERNAL TO THE CRT

One common problem is HV arcing due to poor grounding contacts with the outer aquadag. Such arcing, which is usually visible in a darkened room, causes flashes on the screen. Tighten contacts or relocate contact with the aquadag. *Corona arcing* is recognized by the sweetish odor of ozone and its faint hissing sound. It tends to emanate from sharp HV points around the flyback transformer, the HV rectifier, and the ultor contact. Round off any sharp points and increase spacing where possible. In mild cases, try HV acrylic spray. For arcing around the ultor contact, clean the glass and trim edges of the rubber section cup, or replace if badly deteriorated. Remove all dust and grime. Arcing in the *yoke* or *flyback* is fairly common, and usually calls for replacement. If necessary, remove yoke from CRT neck and make visual inspection with set turned on and the yoke still connected. Turn down brightness control to avoid burning the screen. To pinpoint out-of-sight arcing in or around the HV cage, try listening through a length of rubber tubing as a "stethescope". Check for high line voltage. If a color set, adjust HV to value specified and check for improper operation of the HV regulator or ABL circuits.

15-88 CRT SOCKET DEFECTS

Socket problems, which are fairly common, are of two kinds, and both are internal, making diagnosis difficult and repairs all but impossible. One problem is wires broken loose from the contacts; the other problem is broken or "sprung" contacts. Both types of defects tend to be intermittent and show up when the socket and/or connecting wires are moved. Heater connections to the CRT give the most trouble because of the relatively high heater current. Check for fluctuations in the heater glow.

When an open circuit is suspected, partially remove socket for access to the CRT base pins. Check voltages or signals at the exposed pins as appropriate. If necessary, make continuity check between connections on the chassis and the corresponding tube pin(s). When connections on the chassis are not immediately accessible, use a sharp needle to puncture the wire insulation near the socket. If an intermittent condition, wiggle the wires and socket while observing the meter.

In some cases, sprung contacts can be tightened with a sharp-pointed tool. For dirty or tarnished contacts, apply contact cleaner to the base pins and the socket holes. Remove and install socket several times until good contact is made. Make sure the tube pins are properly aligned. Unless repairs are considered permanent, the socket should be replaced.

15-89 YOKE DEFECTS

Because of their low impedance, yokes used in solid-state receivers have fewer breakdowns than those used in tube-type receivers. Yoke troubles include arcing, and shorted, open, and grounded windings. For suspected arcing, remove yoke from the CRT and examine with the set operating, as described earlier. An ohmmeter check will detect an open coil, where trouble symptom is loss of deflection. "Ringing" the coils is the best test for *shorted* windings (indicated by a keystone raster). For further details on yoke testing, see Sec. 7-21, and 3-18.

15-90 RASTER WIDTH, HORIZONTAL LINEARITY, OR CENTERING PROBLEMS

See Chapter 9.

15-91 RASTER HEIGHT, VERTICAL LINEARITY, OR CENTERING PROBLEMS

See Chapter 7.

15-92 NO PIX HIGHLIGHTS

A dim raster where white pix objects have a silvery, milky appearance that changes as the brightness control is turned up; also poor focus and loss of pix detail. Condition may improve during warm-up. Check for a weak CRT.

15-93 SCREEN GOES DARK

If raster *abruptly* cuts out, suspect intermittent troubles affecting any of the CRT operating voltages. Monitor all voltages including the HV. Try wiggling the tube socket. Where raster *gradually* fades out, look for CRT heater problems.

15-94 RASTER SLOW COMING ON

A delayed raster that suddenly appears may be caused by a horizontal oscillator that is slow in starting up, an intermittent open or short in the CRT, or intermittent "thermal" troubles affecting CRT operating voltages. Low B+ is another possibility, and this results in a small raster. When raster *gradually* appears, the CRT is probably weak.

15-95 PIX INVERTED AND/ OR TRANSPOSED

This sometimes occurs after replacing a yoke. Pix will be inverted if the yoke is upside-down or the leads to the vertical coils reversed. Pix will be transposed left to right if horizontal coil leads are reversed. (This becomes apparent from lettering that is backward). Pix will be inverted *and* transposed if the yoke is turned by 180°.

15-96 SYMPTOMS AND TROUBLE-SHOOTING PROCEDURES (COLOR SETS ONLY)

Problems affecting both BW and color sets were described previously. The *following* troubles are those resulting from improper operation of the color CRT, as caused by a defective pix tube, abnormal voltage conditions, or incorrect set-up adjustments. For pix troubles caused by problems with the video and chroma *signal stages,* see Chapter 16.

15-97 RASTER IS RED, GREEN, OR BLUE

This indicates two of the guns are not working. Possible causes are: defective CRT; loss of screen grid voltage on two guns, or the screen controls are turned down; or, two of the guns are biased to cutoff, as with an *in-line* CRT where the three cathodes are driven separately. Test the CRT and check operating voltages.

15-98 RASTER IS YELLOW, MAGENTA, OR CYAN

One of the guns is not working. Determine which color is missing. A yellow raster indicates the blue gun is not conducting. A magenta raster means loss of the green; a cyan raster, loss of the red. Check for a defective CRT, loss of screen voltage on the troublesome gun, the screen control being turned down, or excessive bias on the one gun.

15-99 COLOR-TINTED RASTER

One or two guns are conducting excessively or not enough. Possible causes are: defective CRT; misadjusted screen controls; improper bias on one or two guns; or a purity problem. Check CRT, adjust screen controls, and adjust the purity ring.

15-100 GRAY-SCALE TRACKING PROBLEM

Raster changes from a color tint, through gray, to white as the brightness control is varied. Adjust the drive and screen-grid controls. If conditions change during warm-up, suspect a weak CRT.

15-101 JUMBLED WHIRLS OF COLOR

A pattern of ever-changing colors that covers the entire screen. Trouble may clear up after warm-up. Probably the degaussing coil is energized permanently or for too long. If considerable AC voltage exists across coil after warm-up, check the ADG components.

15-102 NEED TO FREQUENTLY DEGAUSS CRT

ADG circuit not operating properly; or set is exposed to strong magnetic fields, as from a vacuum cleaner, hair dryer, or other appliance. Check the ADG circuit and positioning of the degaussing coil.

15-103 COLOR-SPOTTED RASTER

Check for misconvergence. If condition changes during warm-up suspect a loose or warped shadow mask. Try a substitute CRT.

15-104 COLOR CONTAMINATION

Color impurities over portions of screen, usually near the edges. Suspect misadjusted purity. Adjust yoke and purity rings (or twist coil) to obtain a pure single-color field.

15-105 COLOR OUTLINES ON BW PIX

Verify purity is OK. If color outlines are present only near center area of screen, adjust *static convergence* adjusters. If color on edges of screen, adjust *dynamic convergence*.

15-106 NO BW PIX RECEPTION

Normal raster. Color pix is poor. Check for loss of the luminance signal at the CRT inputs.

15-107 NO COLOR PIX (BW PIX OK)

Check for loss of chroma signals at CRT inputs. For other possibilities see Chapter 16.

15-108 BLOOMING

Raster/pix blooming is more common in color sets that in BW sets, because of the higher beam current. Usual cause is that the master brightness or screen controls are set too high. Readjust, and if trouble persists check operation of the HV regulator and ABL circuits.

15-109 PIX TUBE DEFECTS AND TROUBLE SYMPTOMS

Most of the trouble symptoms created by a defective CRT can also be caused by misadjustments or troubles *elsewhere* in the set, particularly those that affect the CRT operating voltages. *All* such possibilities should be investigated before condemning a tube that's only suspect. If, after verifying all applied voltages, it's still a prime suspect, the CRT should still be tested (in the set) before considering replacement. When its condition is questionable, testing the CRT may be the *first step* in the troubleshooting procedure.

The most common CRT defects are: burned-out heater(s); shorts or opens (often intermittent); weak or "dead" CRT (loss of emission); gas or loss of vacuum (glass is cracked); leakage between elements; loose particles that can lodge between gun elements; and missing screen phosphors.

A defect peculiar to *color sets* is a loose or distorted shadow mask or grill, but this is rare.

Most CRT defects result in either a weak raster or no raster. For example, there may be no beam current with a burned-out heater, loss of vacuum, or complete loss of emission. The same is true for an internal open circuit (and sometimes a short). Low emission due to a weak gun results in a dim raster, and a silvery effect on bright pix objects that changes with the brightness control. Internal arcing or intermittently open or shorted elements produce flashes on the screen. A warped shadow mask or grill makes it impossible to maintain good convergence.

15-110 PIX TUBE TESTING

Tests should be made for *each* of the conditions described above; and this is where a good CRT checker (as described in Sec. 3-19) can be a real time saver, particularly for leakage and shorts, for gas, and for comparing emissions of the three guns of a color CRT. Where a proper CRT checker is not available, a CRT adapter can be used with a "regular" tube tester. In lieu of a checker, alternative methods for testing a CRT are as follows:

- **Filament Voltage Booster**

CRT "brighteners" have long been used to extend the life of a weak CRT. They are also useful in testing when loss of emission is suspected. If an increase in heater voltage noticeably increases raster brightness then the CRT is weak. CRT boosters (Sec. 3-20) are available in different types, and to match different CRT basing configurations. A BW booster cannot be used with a color CRT. As a temporary expedient in lieu of a booster, any convenient source of voltage between 7 and 8 V can be used.

- **Ohmmeter Testing for Leakage/Shorts**

Turn set *off* and remove the CRT socket. Using the highest ohms range, check for leakage/shorts between each base pin and all other pins. All readings should be *infinity* except for the heater which is only a few ohms.

- **Neon Bulb Leakage Test**

Inter-electrode leakage sometimes occurs only when the CRT is heated. A "dynamic" test for leakage can be made with a neon bulb and 100V or so from some convenient DC source. Heater voltage can be obtained from the CRT socket. Probe between all base pins. A bright glow on the neon bulb indicates a direct short; a dim glow, leakage.

15-111 PIX TUBE REJUVENATION

The life of a weak CRT can often be extended by one of two methods: with a filament booster as described above; or by a process called "flashing" which strips away any *cathode buildup*, exposing a fresh surface for increased electron emission. CRT rejuvenators (Fig. 3-14, Sec. 3-19) are available for performing this function where several hundred volts is momentarily applied to the CRT control grid(s). CRT testing and rejuvenation can often be performed by one instrument. Lacking such an instrument, the emission of a weak CRT can often be restored using the following procedure.

Remove the CRT socket. Feed heater voltage from the socket to the two heater pins of the CRT. Ground the cathode pin. Connect one end of a test lead to the control grid (G1).

NOTE: ─────────────────────────────

For convenience and to reduce the chance of shorts, use a socket of the proper type from a discarded chassis.

Obtain a source of several hundred volts (+) DC (the amount is not critical). Such a voltage point can often be found on the set being worked on. Turn on the set and wait a few seconds for the heater(s) to warm up. *Momentarily* touch the grid lead to the source of B + . A flash may be observed in the neck of the tube. *Caution:* Do not make contact for more than a split second or the tube will be destroyed. After flashing once, remove connections, restore the CRT socket and observe results, if any. If no improvement, repeat the procedure, as often as necessary. In cases where all the coating has been stripped from the cathode, the emission cannot be restored and the tube must be replaced.

For a color CRT, flash *only* the gun(s) that are known to be weak. If the emission falls off shortly after flashing a tube, it may help to install a filament booster.

WARNING: ─────────────────────────────

A tube should be flashed only as a last resort. Repeated flashings can further weaken or even destroy the tube.

15-112 FLASHING TO REMOVE SHORTS

A short or leakage between any two elements can often be cleared up by flashing with several hundred volts of dc. Turn set *off* and remove the CRT socket. With the heater *not lit,* momentarily apply the voltage across the appropriate base pins (avoid the heater pins). If trouble persists, turn the set (or the tube) face down on a padded surface and tap the neck with a blunt, nonmetallic object. Foreign particles may become dislodged and fall into the bowl, where they will do no harm.

15-113 REPLACING PIX TUBES

The following procedures are of a general nature. Since procedures vary somewhat for different receivers, refer to service literature for specific details.

15-114 REPLACING A BW CRT

• **Removal**

With most BW sets the CRT is attached to the chassis.

    1. Remove chassis (with CRT) from the cabinet after turning the set off.

    2. Disconnect the HV lead from the ultor contact. Discharge the CRT (more than once) by grounding the ultor with a jumper wire.

    3. Remove the CRT socket and centering ring magnet.

    4. Remove the hold-down band securing the front of the tube to the chassis.

    5. Prepare a clean padded surface on which to deposit the tube when removed.

    6. Slowly remove the CRT by pulling it *straight out* until it clears the yoke. Unless it is held *horizontally,* the neck may be broken by the snug-fitting yoke. Avoid hitting the chassis. *Caution:* Do not lift or carry the tube *by the neck.*

    7. Carry tube and place face down on the previously prepared surface.

• **Installation**

    1. Position chassis to receive the new CRT.

    2. Install CRT by slowly inserting its neck into the yoke. Avoid hitting the chassis. Any up, down, or sideways movement may snap the neck.

    3. Position front of tube into support saddle making sure all cushions and pads are in place. Attach hold-down band and tighten. Do not overtighten band.

    4. Connect HV lead to the ultor. Install centering ring on tube neck. Attach the CRT socket. Verify good contact between the grounding springs and the aquadag coating.

    5. Turn set on and make set-up adjustments as described in Sec. 15-117.

    6. Remove dust from inside of cabinet. Clean inside and outside of safety glass and face of the CRT. For plastic safety glass avoid using cleansing powders that may scratch the surface. Make final inspection for dust and fingerprints on safety glass and the CRT screen.

    7. Install chassis and CRT into cabinet.

15-115 REPLACING A COLOR CRT

With most color sets the CRT is attached to the cabinet rather than the chassis.

• **Removal**

    1. Turn set off and disconnect HV lead from the CRT ultor contact. Discharge the CRT (more than once) by grounding the ultor with a jumper wire.

    2. Remove CRT socket and disconnect all

wires and cables connecting to components on the tube neck.

3. If necessary, remove chassis from cabinet.

4. To facilitate installation, note position of all components on the tube neck. Carefully remove all these components from the neck. Note position of the degaussing coil and remove.

5. Turn set face down on a pad or blanket.

6. Carefully remove all CRT mounting hardware. Take *special care* not to strike the tube with metal tools.

7. Prepare a clean padded surface to deposit tube when removed.

8. Again discharge the CRT by grounding the ultor as before. Grasping the *faceplate* and the *bowl*, carefully lift the tube from the cabinet. *Important:* Do not lift or carry the tube *by the neck*. While supporting the tube with one hand *under the faceplate*, carry it to the previously prepared surface and place it face down. *Reminder:* A CRT implosion can cause serious injury from flying glass.

- **Installation**

1. Remove dust from inside cabinet. If cabinet has a safety glass, clean both inside and outside. Make careful visual inspection to be sure there are no fingerprints or dust on the inside surface of the glass.

2. Prepare new tube for installation. If necessary, remove any support band or other hardware from the old tube and install on new tube. Make sure any cushions or pads are properly in place. Do not overtighten support band. Note which way the tube is to be oriented relative to the ultor contact. If cabinet has a safety glass, clean the faceplate of the CRT.

3. Discharge the tube as described earlier. *Reminder:* A tube can retain a charge for surprisingly long periods, and to touch the ultor while handling can have serious consequences.

4. Carry tube to the cabinet. Unless tube has built-in implosion protection avoid touching the faceplate. Avoid touching glass around the ultor (fingerprints can create HV leakage). Carefully place tube face down into cabinet making sure the ultor is facing in the right direction.

5. Install mounting brackets and other hardware previously removed. Make sure all pads and cushions are in place. If secured with a band, do not overtighten. Place cabinet upright and verify that tube is properly positioned in the mask opening. Check for any foreign particles that may have been trapped, in view, in front of the faceplate.

6. Install degaussing coil. Install all components on the CRT neck. Install metal shield over bulb as applicable. Install chassis if previously removed. Reconnect all wires and connectors to the degaussing coil, and to the components on the neck. Connect HV lead to the ultor contact. Attach grounding spring(s) for good contact with the aquadag coating.

7. Orient and restore all neck components to their approximate original positions. Turn set on and verify that raster and pix are obtained. Make CRT set-up adjustments per Sec. 15-121 or Sec. 15-123 as appropriate.

15-116 PIX TUBE DISPOSAL

Unless a defective CRT is used as a "blank" for a trade-in credit, it must be "safed" before discarding. This is done by breaking the seal at the end of the gun to allow air to seep in *slowly* and neutralize the vacuum. Breakage of the bowl or faceplate can result in a *dangerous implosion*. A suggested safing procedure is as follows. Goggles and protective clothing is advisable.

1. Place tube face down in a carton or other container. Cover with a heavy blanket or the equivalent leaving only the base of the neck exposed.

2. If there's a composition keyway guide at the base, break it off to reveal the stem of the vacuum seal. Using pliers, snap off the seal. Air will enter with a slight hissing sound.

The tube is now safe. Avoid breathing any air-borne particles of the screen phosphor that may escape. Be careful in handling broken glass. Skin cuts or abrasions can become infected by the phosphors.

15-117 BW RECEIVER SET-UP ADJUSTMENTS

Certain adjustments are usually required as a receiver ages, during or after making repairs, and in particular, after replacing a pix tube. As appropriate to the circumstances, make the following adjustments. Most of this information is applicable to both BW and color sets.

1. After the CRT warms up, turn up the brightness and observe the off-channel raster.

2. If there are *neck shadows*, move yoke tight against the CRT and lock into position.

3. *Raster tilt:* Loosen and twist yoke until the four sides of the raster are aligned parallel with the mask opening. Lock the adjustments.

4. *Raster centering:* Center the raster vertically and horizontally using the centering ring magnets on the CRT neck. By rotating the two magnets relative to each other using the extended tabs, the raster can be moved in any direction. Make sure no neck shadows appear.

5. *Focus:* Adjust for the sharpest scanning lines. *Reminder:* Unless properly focussed, the fine pix detail will be lost.

6. *Pincushion correction:* Position the corrector magnets (when provided) until the four sides of the raster are straight, and parallel with the mask opening.

7. *Height and vertical linearity:* Because of interaction, these two controls are usually adjusted alternately. Adjust height control to more than fill the screen vertically. Adjust linearity for uniform spacing of the scanning lines, and/or by using a bar/crosshatch pattern. Switch on a channel to make sure the top and/or bottom of a pix is not cut off, and that there's no black showing.

8. *Width:* Adjust to more than fill the screen horizontally. Check that none of the pix is cut off, or that there's black showing on the sides.

9. *Horizontal linearity:* Adjust for uniform spacing of lines on a crosshatch pattern and proper proportions of pix objects.

10. Switch to an active channel and adjust brightness and contrast.

11. *Vertical hold:* Adjust vertical sync. Check vertical locking on all channels. Readjust as necessary. If poor vertical sync (pix roll), try readjusting the height and vertical linearity controls.

12. *Horizontal hold:* Pix should lock at the approximate mid-setting of the control. If it doesn't, adjust the *horizontal frequency coil* if provided. Check for good horizontal sync on all channels.

13. *AGC:* Turn adjustment (to increase pix contrast) until pix becomes unstable and loses sync. Back off the adjustment until stabilized. Check all channels.

14. *Channel selection:* Check for reception of all channels. Make the necessary tuner adjustments if fine tuner must be constantly readjusted.

15. *AFT:* If set has automatic fine tuning, reception should be the same for all channels with the AFT switched on or off. Adjust as required.

16. *Buzz control:* When this control is provided, check all channels for buzz in the sound, especially when lettering appears on the screen. If necessary, adjust for minimum buzz without distortion.

15-118 COLOR RECEIVER SET-UP ADJUSTMENTS (GENERAL)

Because of their critical nature, purity and convergence adjustments are made *after* all circuit defects, if any, have been corrected, and after making the preliminary adjustments described in Sec. 15-119. For good color, most receivers, especially older sets using delta CRTs, can benefit by a periodic touch-up of purity and convergence. Such adjustments are always required after replacing a CRT.

15-119 PRELIMINARY ADJUSTMENTS

In addition to the adjustments described for a BW set (Sec. 15-117) make adjustments and perform procedures as follows. Some of the adjustments are different from those of a BW set. Raster centering, for example, is adjusted with potentiometer type controls.

- **External Gun Components**

Verify that all components on the CRT neck are correctly positioned relative to their corresponding internal gun elements. Their exact locations depend on the type of CRT. This information is generally included with the receiver service data. Component locations for a typical delta tube are shown in Fig. 15-22.

- **Master Brightness Control**

When provided, make a tentative adjustment to establish the "range" of the front panel brightness control, between a dark screen, and maximum brightness with little or no blooming.

- **HV Adjustment**

If provided, measure and compare with value specified on schematic, and adjust as necessary.

- **Degaussing**

This should always be done prior to making purity and convergence adjustments, because ADG circuits are not always 100% effective in demagnetizing the CRT, the chassis, and certain cabinet fixtures. A typical service-type degausser is described in Sec. 3-20. The procedure is as follows. 1. With degausser coil energized from a 120 V ac outlet, hold it flat against the CRT faceplate. It doesn't matter whether the set is on or off. Move coil slowly over the surface of the screen and around the edges of the cabinet. 2. Slowly back away from the set. When about 6 or 8′ from the screen, turn the coil 90°. With coil edgewise to the set, switch off the power. Never switch the power on or off when in close proximity to the set as this may defeat the operation.

15-120 PURITY AND CONVERGENCE

These two separate but interrelated functions are equally important to obtaining a good quality pix. They are also the most critical of all the set-up adjustments. The adjustments establish conditions for proper beam landing at all points on the screen, the object being to obtain a *white* raster and a BW pix free of color contamination: both requirements for good color. Because misadjustment of purity and convergence can produce the same trouble symptoms as circuit defects, and vice versa, these adjustments should not be attempted hastily and without a good understanding of the procedures. Some symptoms that may indicate the need for such adjustments include: loss of brightness, a color-tinted raster, color outlines on a BW pix, a blurred pix with loss of fine detail, color contamination over portions of the screen, and severe blooming.

Purity and convergence set-up procedures vary somewhat for different receivers and are different for each of the three types of CRTs. Procedures that follow are of a general nature. For detailed specifics, refer to the receiver service literature.

REMINDER:

Such adjustments are made *after* making all the preliminary tests and adjustments described earlier.

Controls and adjusters used in these procedures are listed in Table 15-5 along with their functions. Note that some of the controls are used with only one type of CRT; others are applicable to all three types.

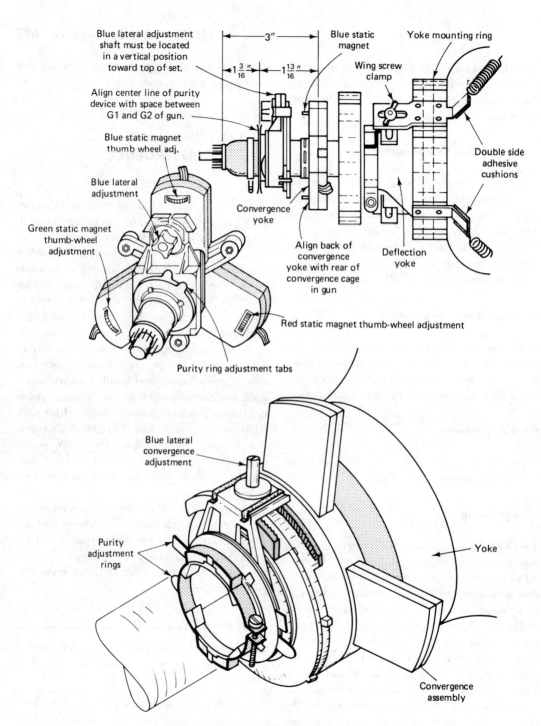

Fig. 15-22 Delta CRT external component locations. To be effective, all external components must be positioned directly over the corresponding gun elements.

Table 15-5 Controls/Adjusters and their Functions

	Delta CRT	3-gun, in-line CRT	Trinitron
Deflection yoke	Raster tilt and field purity (fireball) adjustments	Same as for delta CRT except adjustments are made for a green field	Same as for delta CRT but for a green field.
Purity ring magnets	To obtain pure single-color fields	Same as for delta CRT	Uses twist coil.
Twist coil			Same function as delta purity ring.
Static convergence adjustments	For beam convergence at center of screen (see Table 15-2)	Same as for delta CRT	Same as for delta CRT but not usually provided.
Dynamic convergence adjustments	Beam convergence at edges of screen (see Table 15-2)	May employ wedges for convergence at screen edges	Usually horizontal convergence only.
R-G-B screen controls	To obtain white raster	Same as for delta CRT if provided	As for delta CRT but uses grid/cathode controls.
Master screen control	To adjust for optimum beam current	Usually not provided	Usually not provided.
Drive controls	For gray-scale tracking	Same as for delta CRT	Same as for delta CRT and to obtain white raster.

15-121 DELTA CRT SET-UP PROCEDURES

There are four main divisions to the adjustment procedures. These adjustments, which should be performed in the order indicated, are as follows: 1) *purity,* to obtain a white raster; 2) *static convergence,* for beam convergence at center portions of the screen; 3) *gray scale tracking,* to maintain a white raster; and 4) *dynamic convergence,* for beam convergence at the edges of the screen.

In most cases (except where a CRT is replaced) only a touch-up of the purity and static adjusters is required. Dynamic convergence (which involves the use of special equipment) can be tedious and time-consuming and is performed only when absolutely necessary. In older sets, a small amount of misconvergence at the extreme edges of the screen is considered normal and acceptable.

Convergence adjusters used with a delta CRT are shown in Table 15-6 along with their functions. The purity and static convergence adjusters are located on the neck of the CRT. With most delta-type sets, the dynamic adjusters (and their associated circuitry) are located on a removable *convergence board* (see Sec. 15-46 and Fig. 15-11). Convergence adjustments are made in conjunction with other adjusters listed in Table 15-5.

Table 15-6 Delta CRT Controls and Adjusters

Static convergence adjusters	Function
Red gun static adjuster	An adjustable permanent magnet on the convergence assembly (Fig. 15-12). Moves red gun beam *diagonally* for convergence at center of screen.
Green gun static adjuster	As above, but for the *green* gun.
Blue gun static adjuster	As above, but moves the blue gun beam *vertically*
Blue lateral adjuster	This PM adjuster moves the blue beam in a *horizontal* direction.
Dynamic convergence controls	(Note: Many early sets had different names for the dynamic controls, such as *shape, tilt,* and *amplitude.*).

Vertical convergence controls	Function
Top RG vert lines (amplitude)	Varies amount of voltage fed to the red and green coils to converge lines at top center.
Bottom RG vert lines (tilt)	Varies amount of sawtooth component of parabolic waveform to converge lines at bottom center
Top RG hor lines (differential tilt)	Controls polarity of sawtooth current fed to red and green coils to converge lines at top center
Bottom RG hor lines (differential amplitude)	A "bridge arm" control to vary the parabolic currents divided between the red and green coils, to converge lines at bottom center.
Top blue hor lines (amplitude)	Controls amplitude of voltage fed to the blue vertical coil to converge center horizontal lines at both top and bottom.
Bottom blue hor lines (tilt)	Controls sawtooth current flowing through the blue coils to converge lines at top and bottom.

Horizontal convergence controls	Function
Right RG vert lines (differential)	Controls current through the red and green coils to converge lines at the right side of the screen.
Left RG vert lines (tilt)	Controls amount of current through the red and green coils to converge lines at left side.
Right RG hor lines (amplitude)	Shapes and controls amount of current through the red and green coils to converge at right side.
Left RG hor lines (differential tilt)	A bridge balance control to vary amount of currents through the red and green coils to converge lines at the left side.

Table 15-6 (cont'd)

Right blue hor lines (amplitude)	Controls amplitude of current through the blue coils to converge lines at right side.
Left blue hor lines (tilt)	Shapes and controls the current through the blue coils to converge lines on the left side.

15-122 ADJUSTMENT PROCEDURES

Turn on the set, then set up a large mirror in front of the screen. Before making any adjustments, allow a five or ten minute warmup period for the CRT and all voltages to stabilize. *Note:* Some interaction can be expected for all set-up adjustments. Where good purity cannot be obtained, for example, the cause may be severe misconvergence of the three beams. Similarly, the inability to obtain good convergence can be caused by improper purity adjustments. Static and dynamic convergence adjusters also interact with each other in the same manner. For such problems, alternately adjust as often as necessary.

Because of the influence of the earth's magnetic field, it is desirable to adjust the receiver at its normally used location and facing in the same direction as previously.

• **Purity**

Purity adjustments are made to obtain proper beam landing over the entire screen area. The procedures are as follows:

1. Switch to a blank channel. Disable the green and blue guns by one of two methods, . . . by turning down the screen controls, or with a *gun-killing switch box* (Sec. 3-20) connected between the control grids and ground to overbias the guns to cutoff. Turn up the red-screen control.

2. Loosen the deflection yoke locking screws and move yoke as far back as possible toward the base of the tube. A red blotch should be observed near the center of the screen.

3. Adjust the purity ring by moving the tabs on the ring magnets relative to each other to obtain a bright red "fireball" at the center of the screen. Rotate the assembly as necessary to center the fireball. Don't be concerned about neck shadows.

4. Slide the yoke forward on the tube neck to obtain a full-sized uniform red raster without neck shadows. Avoid turning the yoke to prevent changes in raster tilt. Secure the yoke locking adjustments.

5. Inspect raster for impurities and, if necessary, make touch-up adjustments to the purity ring.

The purity ring establishes beam landing at the *center* of the screen, and the yoke positioning determines color purity at the raster *edges*. Proper beam landing can be determined by close inspection with a magnifying lens. If necessary, repeat steps 3 and 4 for different settings of the brightness and red-screen controls.

If unable to obtain a pure, uniformly red raster, check the following possibilities: the need for a more thorough degaussing; severe misadjustment of convergence; purity ring improperly positioned on the tube neck, or weak ring magnets; misadjusted drive controls in the case of some sets; a loose or warped shadow mask in the CRT; an inoperative red gun, where either the gun is defective, or it is not conducting because of voltage problems. *Note:* It is possible to obtain a red raster even when the red gun is not functioning. This will occur if the purity ring is misadjusted so that the beam from the green or blue gun illuminates the red phosphor dots. Similarly, a green or blue raster can be obtained when *their* guns are not operating.

6. Disable the red gun (using the killer switch or turning down the red-screen control). Turn up the green-screen control to obtain a green raster. Check for color impurities. Repeat for blue raster. Compare the brightness levels for the

three single-color rasters. When one or more colors is weak or missing, suspect a defective CRT or abnormal operating voltages.

• Producing a White Raster

REMINDER: _____

White is produced by a proper mixture of red, green, and blue, as determined by the relative intensities of the three beams and the efficiencies of the screen phosphors.

1. If receiver is provided with a service set-up switch (Sec. 14-18), move switch to the service position to obtain a thin horizontal trace for all beams. Turn down all three screen controls to darken the screen. Slowly turn up the red control until a red trace becomes *barely visible* at normal setting of the brightness control. Turn up the green and blue controls to produce a barely visible green and blue trace. Return the service set-up switch to the normal position.
2. Where the receiver has no set-up switch, simply adjust the three screen controls to obtain the whitest possible raster. Check for adequate brightness with freedom from blooming as the brightness control is turned up. *Note:* For loss of brightness, either turn up the master brightness control if provided, or advance all three screen controls then readjust in the proper proportions. For excessive blooming, adjust the master brightness and/or screen controls to a lower setting. Other causes of blooming are poor HV regulation, or an inoperative ABL circuit.

• Gray-scale Tracking

This adjustment is made to obtain the whitest whites under varying *signal conditions*. It is accomplished by adjusting the amount of signal drive to each of the three guns, using the drive controls. Where DC coupling is used to the CRT inputs, the drive controls also affect the bias of each gun and can create problems of *raster tinting*.

1. Tune to an active station and adjust brightness and contrast controls for normal viewing. Turn down color control to view pix in black and white.
2. Adjust the two or three drive controls to obtain the brightest pix highlights.
3. Check for color tinting as brightness control is adjusted over its full range. If blue tinting occurs, reduce setting of the blue drive control. For green or red tinting, back off on the appropriate control. *Reminder:* The screen "tone" (shade of white) varies with different receivers, and is influenced by ambient room lighting. With many sets, a slightly bluish tint is considered normal.

For abnormal raster tinting: when the drive control(s) is ineffective, check for a weak gun, or insufficient signal drive to one or more guns.

• Static Convergence

When properly converged, the three beams of the CRT pass through each hole in the aperture mask to illuminate the corresponding color dots on the screen. When misconverged, color fringing is observed around BW pix objects, and the colors of a color pix are not in perfect register. Static convergence is the convergence of the beams over the center area of the screen; adjustments are made with the four PM-type adjusters mounted on the convergence assembly (Fig. 15-11). *Reminder:* The purity ring simultaneously moves *all three beams*. Static adjusters control the beams *individually*. Adjustments are made using one of two methods: with a dot/crosshatch generator, or by observing a BW pix. The generator method is faster and more accurate.

A dot/crosshatch generator (see Sec. 3-12 and Fig. 3-8) produces two kinds of patterns: a display of uniformly spaced white dots, or a crosshatch of vertical and horizontal lines. Either pattern is suitable for making convergence adjustments. Some examples of normal and abnormal convergence are shown in Fig. 15-23.

The generator signal is fed to the receiver; the dot or crosshatch pattern is produced by all

15-122 Adjustment Procedures   693

blue dots or lines are displaced and are seen in *color*. Figure 15-24 shows the direction of beam movements as the adjusters are turned. Procedures are as follows:

1. Feed generator signal to receiver input and observe pattern produced on the screen.

2. If red dots or lines are visible in the center area of the screen, adjust the red static adjuster to move the dots diagonally until they overlap the dots or lines produced by the green and blue beams. If all three beams are now converged the dots or bars will appear white. Do not be concerned at this time with misconvergence at the edges of the screen. *Reminder:* Don't confuse the generator dot pattern with the screen phosphor dots.

3. If green dots or lines are visible, adjust the green static adjuster until they merge.

4. If blue dots or lines are visible *above* or *below* the merged pattern, adjust the blue static adjuster, causing the dots or lines to move *vertically* until they overlap the red/green gun patterns.

5. If blue dots or lines are observed to the *left* or *right* of the merged pattern, adjust the blue lateral adjuster to obtain overlap and the absence of color.

Repeat all steps as often as necessary, then recheck purity and gray-scale tracking making touch-up adjustments if required. If a generator is not available, an alternate procedure is as follows.

1. Tune in a station and turn down color control to view pix in black and white.

2. If red, green, or blue outlines are observed around pix objects in center of the screen, adjust the appropriate adjusters until no color is observed. Double check for different channels and for different settings of the brightness/contrast controls. *Note:* A problem with this method is the pix which may be constantly moving; if possible select a stationary pix. *Reminder:* The red and green adjusters move their beams *diagonally;* the blue adjuster moves the blue beam *vertically,* and the blue lateral adjuster moves it *sideways.*

When one or more adjusters are ineffective

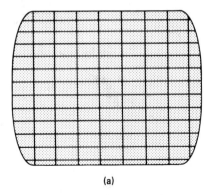
(a)

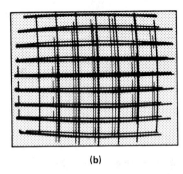
(b)

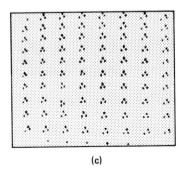
(c)

Fig. 15-23 Normal and abnormal convergence patterns: (a) pattern for a properly converged set; all lines (or dots) appear *white* with no colors visible; (b) serious misconvergence using a cross-hatch pattern; lines of all three colors are visible; (c) misconverged dot pattern; dots of all three colors are visible.

three of the CRT guns. When properly converged, the dots or lines produced by each gun are made to precisely *overlap,* and appear as *white.* When misconverged, the red, green, and

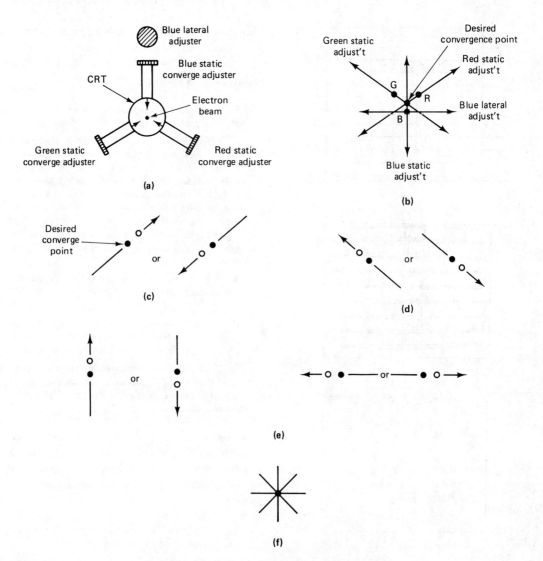

Fig. 15-24 Static convergence: (a) static convergence and blue lateral adjusters (rear view of CRT); (b) three misconverged dots of generator dot pattern (viewed facing the screen); arrows indicate direction of beam movement (and dots) as adjusters are adjusted; (c) red beam misconvergence; (d) green beam misconvergence; (e) blue beam misconvergence; (f) all beams converged.

and perfect center convergence cannot be obtained, check for improper location of the convergence assembly on the tube neck, or a weak adjuster magnet. Try interchanging adjusters.

- **Dynamic Convergence**

As opposed to static convergence, this is convergence at the four *edges* of the screen where the problems of beam landing are greatest. For the adjustment procedure either a white dot or crosshatch pattern may be used. Making adjustments while viewing a pix (as described for static convergence) is impractical. All dynamic convergence controls (Table 15-6) are grouped together, either on the chassis, on a CRT socket PCB, or on a detachable convergence board (Fig. 15-11). Before commencing, it's a good idea to mark the setting of each control so they can be readily restored to their original settings if necessary.

The following procedures are typical for most sets using a delta CRT. For specifics peculiar to a given set refer to the receiver service data. Normally, all vertical controls are adjusted first, then the horizontal controls, usually for the red and green beams, followed by convergence of the blue beam. Due to interaction, all adjustments must be repeated at least two or three times. To speed up the operation start by correcting the most obvious or serious problems before going through the entire procedure.

REMINDER: _____

With older sets, some misconvergence can be expected at the *extreme* edges of the screen and may be disregarded unless it seriously degrades the pix.

Vertical Dynamic Convergence

1. Feed dot/bar generator to receiver input and observe convergence conditions in a vertical direction. Disable the blue gun using either a gun-killer switch or the blue-screen control.

2. If set has a removable convergence board, remove and position it for easy access to the controls. Identify all vertical convergence controls.

3. Select three checkpoints along the vertical center line as shown in Fig. 15-25a.

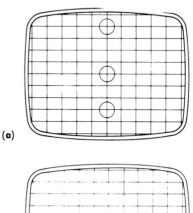

(a)

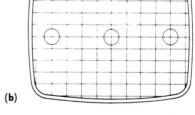

(b)

Fig. 15-25 Dynamic convergence: (a) vertical checkpoints; (b) horizontal checkpoints.

4. Adjust the *top RG vert lines* control (the first dynamic control listed in Table 15-6) to converge the red and green vertical lines at the top checkpoint. *Note:* RG indicates the control simultaneously affects both the red and green beams.

5. Adjust the *bottom RG vert lines* control to converge the red and green vertical lines at the bottom checkpoint.

6. Adjust the *top RG hor lines* control to converge the red and green horizontal lines at the top checkpoint.

7. Adjust the *bottom RG hor lines* control to converge the red and green horizontal lines at the bottom checkpoint. Repeat steps 4 through 7 as often as necessary until they no longer interact and the desired results are obtained.

696 PIX TUBES AND ASSOCIATED CIRCUITS

 8. Disable the green gun and turn the blue gun on.

 9. Adjust the *top blue hor lines* control to converge the blue and red horizontal lines at the top checkpoint.

 10. Adjust the *bottom blue hor lines* control to converge the blue and red horizontal lines at the bottom checkpoint. Repeat steps 9 and 10. Recheck center convergence and repeat steps 4 through 10.

Horizontal Dynamic Convergence

 1. Identify all horizontal convergence controls. Turn the red and green guns on and disable the blue gun. Select three checkpoints along the horizontal center line as shown in Fig. 15-25b.

 2. Adjust the *right RG vert lines* coil to converge the red and green vertical lines at the right checkpoint.

 3. Adjust the *left RG vert lines* control to converge the red and green vertical lines at the left checkpoint.

 4. Adjust the *right RG hor lines* to converge the red and green horizontal lines at the right checkpoint.

 5. Adjust the *left RG hor lines* control to converge the red and green horizontal lines at the left checkpoint. Repeat steps 2 through 5.

 6. Turn the green gun off and the blue gun on.

 7. Adjust the *right blue hor lines* coil to converge the blue and red lines at the right checkpoint.

 8. Adjust the *left blue hor lines* control to converge the blue and red lines at the left checkpoint.

 9. Repeat steps 2 through 8. Readjust all static adjusters as necessary to correct any convergence areas at the center of the screen. Disconnect generator and tune in a station. Check for any color fringing on a BW pix. *Caution:* After converging, do not change the setting of any preliminary adjustments, particularly the vertical and horizontal size, linearity, and centering controls.

When there's a problem in obtaining good dynamic convergence, check out the following possibilities: a loose or warped CRT shadow mask; purity or static-convergence problems; convergence assembly defective, or improperly positioned on the tube neck; coils in convergence assembly not energized with parabolic waveform voltage; the deflection yoke improperly located. If one or more controls are ineffective, the identity of the control(s) provides a clue to defective circuit(s).

15-123 IN-LINE CRT AND TRINITRON SET-UP PROCEDURES

Establishing and maintaining good convergence with an in-line CRT is much simpler than for a delta tube with its numerous interacting adjusters, and more nearly perfect convergence can be expected. The following set-up procedures are of a general nature, but typical. For specific step-by-step procedures for a given set, refer to the manufacturer's service notes.

As with a delta CRT, there are four operations to perform: purity, gray-scale tracking, static convergence, and dynamic convergence. In these procedures it is assumed that there are no circuit or component defects and that all *preliminary* set-up adjustments have been made (including a thorough degaussing), as described in Sec. 15-119.

• **Purity and Gray-Scale Tracking**

These adjustments are not too different from those described earlier for a delta CRT. Depending on the type of tube, purity adjustments are made with either a conventional purity ring, a six-section PM ring adjuster, or, in the case of a trinitron, a twist coil. In most cases, purity is adjusted using a *green* raster instead of the red raster of the delta tube set-up. After disabling the red and blue guns, and retracting the deflection yoke, the purity ring (or twist coil) is ad-

15-123 In-Line CRT and Trinitron Set-Up

justed for a green blotch at the center of the screen.

REMINDER: ─────────────────────

Only the red and blue beams are affected since the green beam travels in a straight line.

With a trinitron, the adjustment is made by turning the twist coil on the tube neck. The yoke is then moved forward to obtain a full pure-color raster. After checking the purity of the other two color fields, the yoke is locked into position.

The next step is to develop a white raster by adjusting the screen-grid controls if provided, or the controls in the cathode circuits if the CRT has internally connected grids. Gray-scale tracking is essentially the same as for a delta CRT, as are the procedures if unable to obtain good purity and tracking.

• **Convergence**

With in-line tubes there are only a few adjustments; this fact greatly simplifies the set-up procedure. Vertical convergence is no problem because the three guns are equi-distant from the top and bottom of the screen and because the screen phosphors are arranged vertically. Provisions must be made however for adjusting *horizontal* convergence. Procedures vary according to the type of tube.

Converging Three-gun in-line CRTs In some early-model in-line tubes, an electromagnetic convergence yoke is located on the tube neck. Except that it affects *only* the red and blue beams, its operation is similar to that of the convergence assembly used with a delta tube. Two typical units are shown in Fig. 15-15. Most modern tubes use the PM-type unit shown in Fig. 15-15b.

Static Convergence For the convergence assemblies shown in Fig. 15-15c, static convergence adjustments are made with small rotary PM type adjusters, much as with a delta CRT. The procedure is as follows: (see Fig. 15-26).

1. Adjust the red adjuster to converge the horizontal RG lines of a crosshatch pattern at the center of the screen.

2. Adjust the blue adjuster to converge the horizontal BG lines at the center.

3. Adjust the red side adjuster to converge the vertical RG lines at the center.

4. Adjust the blue side adjuster to converge the vertical BG lines at the center.

5. Repeat the entire procedure at least twice.

For the configuration shown in Fig. 15-15a, proceed as follows:

1. Release lock screws holding the two sliding magnet assemblies.

2. Converge the vertical BG lines at the center and both sides of the screen by sliding the magnet assembly sideways.

3. Adjust blue magnet No. 2 to converge the vertical BG lines at the center and both sides.

4. Converge the vertical RG lines at the same checkpoints by sliding the magnet assembly.

5. Adjust red magnet No. 4 to converge the vertical RG lines.

6. Adjust blue magnet No. 5 on the front assembly to converge the horizontal BG lines at center of screen.

7. Converge the horizontal BG lines at top and bottom by selecting different blue-coil taps at the top of the assembly using the flexible lead provided.

8. Adjust red magent No. 6 on the front assembly to converge the horizontal RG lines at the center of the screen.

9. Converge the horizontal RG lines at top and bottom by selecting the proper taps on the red coil.

10. Repeat all adjustments and lock the two screws holding the sliding assemblies into position.

698 PIX TUBES AND ASSOCIATED CIRCUITS

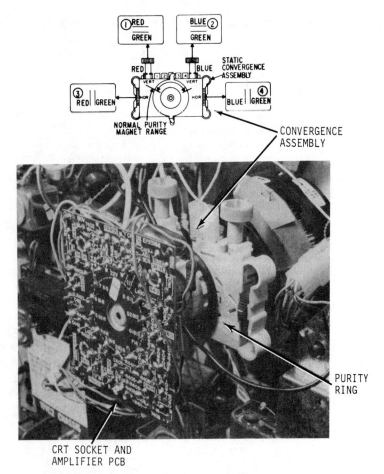

Fig. 15-26 Electromagnetic convergence of in-line CRT.

To adjust static convergence for tubes employing a six-part PM unit (Fig. 15-15b), proceed as shown in Fig. 15-27.

Dynamic Convergence Most in-line tubes except trinitron are adjusted using small *wedges* inserted between the deflection yoke and the bowl of the CRT as shown in Fig. 15-15b. Insert wedges as indicated to tilt the yoke in the desired direction for converging the crosshatch pattern lines at the four edges of the screen. With a *bonded yoke*, no adjustment is necessary. If unable to obtain good convergence, check the same possibilities as for a delta CRT.

Trinitron Convergence No provision is made for adjusting *vertical* convergence on small-screen trinitron tubes, but large CRTs have a *vertical static-convergence control* to vary the amount and polarity of DC current flowing through the twist coil to create a magnetic influence on the red and blue beams. Horizontal convergence is accomplished *electrostatically* by varying the dc voltage applied to the built-in horizontal con-

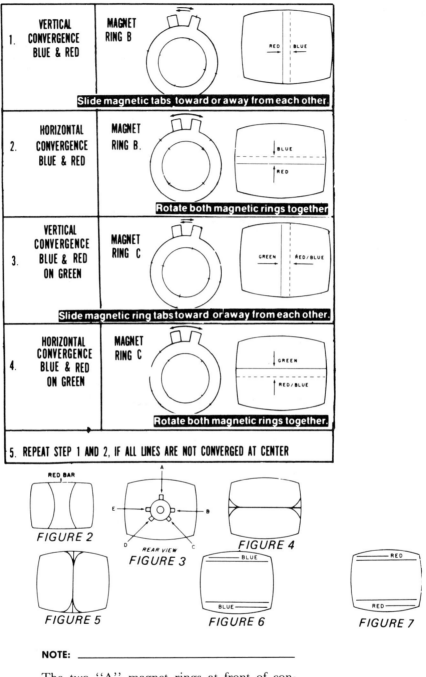

NOTE:

The two "A" magnet rings at front of convergence assembly are purity adjustments (see Fig. 15-15).

Fig. 15-27 (partial)

A. A vertical tilt of the yoke upward or downward will shift the red and blue fields of the raster as indicated below:

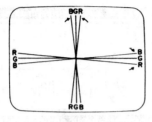 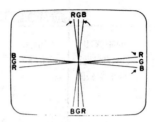

To correct, place wedge at top of yoke.　　To correct, place wedge at bottom of yoke.

B. A horizontal tilt of the yoke will shift the red and blue fields of the raster as indicated below:

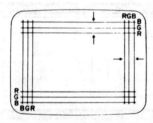 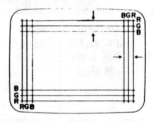

To correct, place wedge at right side of yoke (from rear of CRT).　　To correct, place wedge at left side of yoke (from rear of CRT).

(b)

Fig. 15-27 Converging in-line tubes with magnet rings and wedges: (a) static convergence—green is the center gun, converge the red and blue, then converge red and blue on green; (b) yoke wedging for dynamic convergence.

vergence plates. Adjustment procedures are as follows:

1. Verify that all preliminary set-up adjustments have been made, the CRT has been thoroughly degaussed, and that purity and gray-scale tracking is OK.

2. Display a crosshatch pattern on the screen as explained earlier.

3. Where a vertical static-convergence control is provided, adjust to converge the RB horizontal lines with the green lines at the center of the screen.

4. Adjust the horizontal static-convergence control to converge the RB vertical lines with the

green lines at the left and right sides of the screen.

5. Recheck purity and gray-scale tracking, making touch-up adjustments as necessary.

REMINDER:

The only provision for *dynamic* convergence is the parabolic voltage applied to the horizontal deflection plates in some cases, which may be adjusted with a *tilt* control. If provided, adjust control to converge lines at the extreme edges of the screen.

If unable to obtain good convergence, check for the following: a defective CRT; low voltage or loss of voltage on the internal convergence plates; loss of horizontal parabolic sweep voltage at the plates; improper purity adjustments.

15-124 PURITY AND CONVERGENCE PROBLEMS (GENERAL)

When unable to obtain good purity and convergence, after checking possible defects suggested earlier, proceed as follows: For purity problems with delta and 3-gun in-line tubes, check location of purity ring on the tube neck. If a weak purity ring magnet is suspected try substitution followed by the purity adjustments described earlier. For a trinitron, check continuity of the twist coil and check the applied DC voltage as the vertical static-convergence control is adjusted. The control should produce voltage variations of both positive and negative polarities. If purity problems persist, verify condition of the CRT and make a *thorough* and prolonged degaussing of the CRT including all sides of the cabinet.

For static convergence problems with delta and in-line tubes, verify proper location of convergence assembly on the tube neck. If a weak adjuster magnet is suspected, remove all three and compare, using a pocket compass. Try interchanging magnets on the three guns. For vertical static-convergence problems with a trinitron, check the twist coil and its applied voltage as for purity problems. For horizontal static-convergence problems, check the HV DC applied to the horizontal convergence plates while adjusting the horizontal static-convergence control. Compare voltage with that specified for the receiver. When a parabolic waveform is superimposed on the DC, scope-check for proper amplitude and waveshape, and, if necessary, check components in the waveshaping circuit from the output of the horizontal sweep circuit.

For dynamic convergence problems with a delta CRT, verify proper position of the convergence assembly on the tube neck. Repeat degaussing procedure, and static-convergence adjustments. Determine which is at fault, vertical or horizontal convergence, and check continuity of the corresponding coils on the convergence assembly (see Figures 15-11, and 15-22). Frequently a break occurs at one of the terminals due to stress and movement of the extended connecting wires. Scope-check the parabolic waveforms applied to the coils. When applied voltages are weak, missing, or improperly shaped, check the applicable sweep circuit source and inputs/outputs of the convergence board. Check all components of the suspected circuit on the convergence board, control pots, the slug-tuned coils, diodes, resistors, and capacitors. For dynamic-convergence problems with in-line and trinitron tubes, check out the possibilities described earlier.

15-125 TROUBLESHOOTING PROJECTION RECEIVERS

With few exceptions, troubleshooting is the same as for a direct-view receiver. Because of the high voltages applied to the CRT, arcing

702 PIX TUBES AND ASSOCIATED CIRCUITS

can sometimes be a problem, and the HV should never exceed the value specified in the service literature. Take extra safety precautions when measuring or working around the HV.

Ordinarily, the optical system requires no servicing in the field although adjustments are needed after replacing a CRT. Dust can be a problem if it enters the optical (tunnel) compartment. Remove all dust from inside the cabinet with a vacuum cleaner *before* gaining access to the optical components. When necessary, remove dust from lenses, mirrors, and the face of the CRT with a soft brush. Dirt accumulations can be removed with mild soap and water.

When servicing on the bench, the pix may be viewed by setting up a mirror in front of the CRT, or by projecting onto a white wall or movie screen to obtain a full-size pix. Avoid looking directly at the high intensity CRT screen. When replacing a CRT remember that there are two kinds of focusing: the conventional *electrical* focus control, and the *optical* focus as determined by the relative positioning of the CRT and the lens system. Check focus by observing the screen phosphor dots while switched off channel. To focus, first adjust the focus control, then, if necessary, the position of the CRT and the optical components.

15-126 INTERMITTENTS

Usually the best approach to servicing intermittents of any kind is through trying to create the symptoms, rather than through waiting for them to show up. For intermittent loss of raster or pix, monitor all CRT applied voltages and signals at the CRT terminals, as appropriate to the symptoms.

REMINDER: _____

Troubles in dc-coupled video or chroma output stages will upset both the signal inputs and the dc operating voltages, the CRT bias in particular.

One common problem associated with the CRT is intermittent contacts between the socket and the base pins, or an intermittent break in the socket connecting wires, often internally. Try wiggling the wires and twisting the socket while observing the pix. Clean socket contacts and base pins as necessary. Sprung or dirty CRT heater contacts cause fluctuations in the visible heater glow.

Intermittent opens or shorts in a CRT are very common, especially between a grid and a cathode. The usual symptom of a grid-to-cathode short is loss of pix, full brightness, and an ineffective brightness control. Try tapping the CRT neck with the control turned down. Internal leakage or shorts can often be corrected by *flashing* (see Sec. 15-112).

Intermittent HV arcing in and around the CRT is a common problem. Check for leakage around the ultor contact, and sparking at the aquadag grounding springs.

15-127 AIR-CHECKING THE REPAIRED SET

After repairs and/or set-up adjustments are completed, air-check the receiver, allowing a reasonable amount of time for any abnormal or intermittent conditions to develop. Critically evaluate pix quality on all channels, making any last-minute touch-up adjustments that may be needed. If a color set, avoid changing adjustments that may upset convergence. Make sure that all clamps holding the yoke and other components on the tube neck have been tightened, and that all adjusters have been locked in position.

15-128 SUMMARY

1. A BW pix tube consists of an electron gun and a phosphor-coated screen. Electrons striking the screen produce visible light. As the raster is being scanned, amplitude variations of the video signal vary the CRT bias to produce variations of beam current and a pix with brightness variations between white and black.

2. After being emitted from a heated cathode, the beam of electrons is attracted to the distant screen by a high voltage applied to the inner aquadag coating on the bowl of the tube. To reach the screen, the beam passes through small holes in a number of elements in the gun: a control grid that controls the electron flow; a screen grid that accelerates the electrons; and a focus grid that focuses the diverging beam to a pinpoint as it strikes the screen.

3. A pix is reproduced sequentially one element at a time, and line by line, during the scanning process. Because of the retentivity (afterglow) of the screen phosphors, the ability of the eye and brain to retain an image, and the rapid scanning rate, a total pix is observed with no noticeable flicker.

4. The deflection angle of a CRT is the angle formed from the center of the deflection yoke to diagonally opposite corners of the screen. Early type CRTs had a long neck and a narrow deflection angle of about 50°. Modern tubes have a short neck and a wide angle up to 114°. The yoke must be designed with a deflection angle to match the CRT.

5. Early model sets had an external "safety glass" in front of the screen as protection against implosion due to the high vacuum. Modern CRTs have built-in implosion protection. To prevent breakage, common-sense precautions must be observed when working around and handling a CRT.

6. A color CRT has *three* electron guns and three different screen phosphors corresponding with the three primary colors, red, green, and blue. The phosphors are deposited on the screen in groups of three. The three electron beams are precisely controlled to strike only their corresponding color phosphors.

7. Raster centering on a BW CRT is accomplished with a *centering ring* which consists of two flat washer-like magnets mounted on the tube neck. Rotating the magnets changes the strength of the magnetic flux which penetrates the glass to react against the electron beam, moving it in any desired direction.

8. The brightness control controls the CRT bias, thus the beam current, and thus the pix brightness. As with any vacuum tube, bias may be varied by changing either the cathode or the control-grid voltage. In most BW sets the control is at the CRT input. In color sets, the control is in a video-amplifier stage that is dc-coupled to the CRT.

9. The video signal may feed either the grid or cathode of the CRT. For grid drive, the signal has a negative-going polarity where the peaks of the signal drive the tube to cutoff to produce black in the pix. For cathode drive, a positive-going signal accomplishes the same thing. In a color set, the signals may simultaneously drive the grids *and* cathodes.

10. Good pix contrast can be obtained only when the signal is strong enough to drive the CRT between the extremes of cutoff and maximum brightness. With a *weak* signal, the pix will be either too light or too dark depending on the setting of the brightness control.

11. A signal of constant frequency applied to the CRT produces a bar pattern. The higher the frequency the greater the number of bars. If the frequency is an even multiple of the vertical sweep frequency, the bars will be horizontal; if an even multiple of the horizontal sweep frequency, the bars will be vertical. For odd-multiple frequencies the bars will slope and cannot be synched.

12. A pix is produced by variations of beam current as the raster is being scanned. Rapid variations produce small objects; slow variations produce wide objects. Producing a large pix object requires slow variations repeated for a great many scanning lines. Video signal variations are between 30 Hz and 4.2 MHz.

13. A large-screen set requires a video bandwidth of at least 4 MHz. to produce all the fine pix detail. On a small screen, good resolution is possible with a bandwidth of only 2 or 3 MHz, because extremely small pix details are too close together to be observed.

14. Obtaining optimum resolution of the fine pix detail requires a bright raster, *perfect* beam focus, good high-frequency response of the video stages, and freedom from ghosting or ringing problems.

15. Every color has three distinctive characteristics: *Hue,* its identity which distinguishes it from other colors; *saturation,* the depth or intensity of the color; and *brightness.*

16. A color receiver makes use of the three primary colors: red, green, and blue, from which all other colors can be derived. When only the red gun of the CRT is conducting to excite the red phosphors on the screen, red pix objects are produced. Similarly, the green and blue guns produce the greens and blues in a pix. With two or three guns simultaneously conducting by different amounts we view the mixture as another distinct color. Appropriate proportions of red, green, and blue produce a mixture seen as white.

17. The eye cannot discern color on extremely small objects; therefore the finest pix detail is reproduced only in black and white. Hence, video signals require a bandwidth of 4.2 MHz, but chroma signals need only 1.5 MHz.

18. A color CRT is driven by two kinds of signal: the video (luminance or Y signal,) and the chroma signals. When receiving black and white, the chroma stages are disabled and the three guns are simultaneously driven with the Y signal to produce the pix in black and white. When receiving color, chroma signals drive the guns individually to produce the various colors in the pix. The Y signal provides brightness information for the colors and is responsible for producing the fine detail.

19. There are three basic types of color CRTs; the delta or shadow-mask type, the three-gun in-line tube, and the trinitron "single-gun" in-line tube. In a delta CRT the three guns are arranged in a triangular formation. In in-line tubes the guns are positioned horizontally in line. In a trinitron, electrons emitted from three separate cathodes travel together through a single gun structure.

20. In a delta CRT, phosphor dots of the three primary colors are deposited on the screen in groups of three called triads. Behind the screen is a perforated metal plate called a shadow mask. Each hole in the mask is directly behind a triad. To reach the screen phosphors, each of the three beams must pass through and cross over within the mask openings. If they don't, some of the beams may not get through, or will excite the wrong colors.

21. In in-line tubes, the phosphors are arranged as vertical *stripes* of the three primary colors. Instead of a shadow mask, there's an *aperture grill* which serves the same purpose. The grill has slots, precisely positioned behind the screen phosphors.

22. Most color sets employ *spark gaps* as protection against damage to the CRT and associated components in the event of HV arcing. The spark gaps are usually built into the CRT sockets and connect between the various gun terminals and ground. Internal arcing in a CRT is quite common, but generally, if it is only sporadic and of brief duration, no damage is done.

23. High voltages in excess of 16 KV or so give off harmful X-rays. As protection against this hazard, HV circuits are contained in a shielded compartment. Never adjust the HV to a value greater than specified. With some sets, failure of HV regulator circuits or the ABL circuit can increase the HV beyond safe limits.

24. Despite the HV applied to a CRT, the beam current is relatively low, around 100 μA or so for a BW tube and about 1 MA total for the three guns of a color tube, even at maximum brightness. The path of the current is from cathode to screen, to the inner aquadag coating, from the ultor contact through the HV rectifier(s), then through the flyback transformer, then via the LVPS back to the cathode.

25. A delta CRT has a *convergence assembly* mounted on the tube neck between the yoke and the base. The assembly has three identical parts. Each part consists of an adjustable permanent magnet and a pair of coils. The coils are energized from the vertical and horizontal sweep circuits. Magnetic flux from each unit is concentrated between internal CRT pole pieces to influence the individual beams. The object is to obtain proper beam landing at all points on the screen.

26. To obtain a color-free BW pix and a good color pix, the three beams of the CRT must be made to converge in the shadow mask or aperture grill openings, and, upon emerging, to strike only their corresponding color phosphors. This is accomplished by proper adjustment of a *purity ring* and the convergence assembly.

27. A purity ring is essentially the same as the magnetic centering ring used with a BW CRT. It is mounted on the neck of a color tube to the rear of the convergence assembly. When it is properly adjusted, the three beams will illuminate their corresponding screen phosphors.

28. A shadow mask or aperture grill tends to become magnetized by the earth's magnetic field and other influences. This upsets the purity and convergence adjustments, and so adversely affects pix quality. Most color sets have a built-in *degaussing coil,* a large-diameter coil inside the cabinet close to the screen. Each time the set is turned on, the coil is momentarily energized with 60 Hz ac to demagnetize the CRT.

External service-type degaussing coils are used when the internal ADG system is not completely effective.

29. With the ever-changing angle of deflection, electron beam(s) travel a greater distance to the edges of the screen than to the center. This creates problems of convergence. Continuous and automatic correction is made by feeding the coils in the con vergence assembly with a parabolic waveform voltage to create a constantly changing influence on the three beams. PM adjusters are used for static convergence at the center of the screen and adjustments to the parabolic voltage take care of the screen edges (dynamic convergence).

30. Vertical convergence is no problem with in-line tubes because the distance of beam travel to the top and bottom of the screen is the same for all three beams. For static convergence, modern in-line tubes have a six-section, multi-purpose ring magnet assembly mounted on the tube neck. Dynamic convergence is accomplished by tilting the yoke using small wedges.

31. In addition to purity, convergence, and pincushion adjusters, there are other controls associated with the operation of a color CRT; brightness controls; R-G-B screen-grid controls to adjust individual beam current of the three guns; and R-G-B drive controls to adjust the signal drive to each gun. Other adjustments include yoke positioning adjustments, HV, raster centering, and focus.

32. A color-bar generator is a useful piece of test equipment that produces a color-bar pattern on the pix tube screen. A typical bar pattern consists of 10 vertical bars of different hues. Troubles involving the CRT, adjustments, and defects in the chroma stages show up as missing or misplaced color bars.

33. Making CRT set-up adjustments is an important part of servicing a color set. There are five main steps which are generally performed in the sequence as follows: 1) making all *preliminary* adjustments such as height, linearity, focus, and the like (this includes a thorough degaussing of the CRT); 2) making purity adjustments to obtain a single-color field; 3) obtaining a white raster and making gray-scale adjustments for white highlights in the pix with no background tinting; 4) making static-convergence adjustments for beam convergence at the center of the screen (this may be done with a dot/crosshatch generator or by observing a BW pix); 5) making dynamic-convergence adjustments to obtain good convergence at the edges of the screen (either a dot or crosshatch pattern should be used for this operation).

34. A normal raster is prerequisite to obtaining a good pix, and there are many things that can affect the raster: HV, focus, CRT-bias and other operating voltages, the sweep circuits, and the condition of the CRT. For a color CRT, the set-up adjustments are of particular importance.

35. Optimum pix resolution can be obtained only if there's a bright raster in *perfect focus*. Adjust focus off-channel by observing the scanning lines. Severe defocusing is indicated if the lines cannot be seen at all. If focus is not uniform over entire screen surface, adjust for best focus at the center.

36. Breakage of the faceplate of a pix tube, especially a large-screen color CRT, can cause injury from flying glass. Handle with care, and never discard a tube without first making it safe. To neutralize the vacuum, place the tube face down in a large carton and break the exhaust stem.

REVIEW QUESTIONS

1. State three main differences between a BW and a color CRT.
2. How is raster centering accomplished on: a) BW set? b) a color set?
3. Why is beam control so much more critical for a color set than for a BW set?
4. a) What is meant by the deflection angle of a CRT? b) Name four design features of a CRT that affect this angle.
5. What is meant by ion bombardment of the screen and what is the end result?
6. At which extreme setting of the brightness control is the CRT beam current the highest? Explain why.
7. What control(s) of a color set control the beam currents or a) all three guns? b) each gun individually?
8. Explain the beam focusing action in a) delta-type CRTs; b) in-line type CRTs.
9. The beams in a trinitron CRT are accelerated by four positively charged elements. a) Name them. b) What "AC" voltage is also applied to one of these elements? c) Why?
10. Explain the purpose of the shadow mask or aperture grill in a color CRT.
11. a) Compare the construction of the three guns in a color CRT. b) What identifies each of the three guns to a specific color?
12. Since a screen is coated with three different color phosphors, how is it possible to obtain pure rasters for each of the three primary colors?
13. Why does a CRT stop operating when it loses its vacuum?
14. a) Explain the pincushion and barrel-shaped raster. b) Why is it generally a problem only with large-screen CRTs? c) How is the problem overcome on BW sets? d) On color sets?
15. What colors are produced by equal mixtures of a) red and green? b) red and blue? c) green and blue? d) What mixtures produce the following: pink, orange, cyan, magenta, gold? e) How does pink compare with red as to hue and saturation? f) Does a change in brightness affect the hue of a color? Explain.
16. All modern color sets have a built-in degaussing coil. Explain its function.
17. a) What is meant by beam convergence? b) How is it accomplished in a delta-type

CRT? c) In an in-line CRT? d) What is the result of misconvergence on the raster, BW pix, and color pix?

18. a) Explain the difference between static and dynamic convergence. b) Why are convergence problems greatest with large-screen sets and at the edges of a pix? c) Is the problem greatest with a delta or an in-line type CRT? Why?

19. The larger and "flatter" the screen, the more difficult it is to maintain good focus and convergence, especially at the edges. Explain why.

20. A delta CRT has four adjusters for converging the beam at the center of the screen. Name and describe each of them.

21. Describe the purity ring and explain its function.

22. a) Dynamic convergence of a delta CRT requires a *parabolic* waveform. Explain why. b) Where do these voltages originate?

23. Some in-line CRTs have an electromagnetic *twist coil* positioned on the neck; other sets use four adjustable ring magnets for the same purpose. What is their function?

24. What makes it possible to obtain a white raster and white pix objects on a color CRT when there are no white phosphors on the screen?

25. Explain the differences between the two types of in-line CRTs.

26. Why is vertical misconvergence not a problem with in-line type CRTs?

27. a) What is meant by "beam landing"? b) How can it be checked?

28. What color raster do you get when the following guns are not operating? a) red gun? b) green gun? c) blue gun? d) R and B guns? e) R and G guns? f) B and G guns?

29. a) Why must the crossover point of the three beams of a color CRT be precisely within the openings in a shadow mask or aperture grill? b) What is the effect if the crossover is too soon?

30. What two kinds of voltage are applied to the convergence plates of a trinitron CRT? Explain the purpose of each.

31. On close inspection, what is the appearance of the color dots or stripes on a large-screen CRT a) with set turned off? b) while observing the raster? c) A BW pix? d) A color pix? Explain your answers.

32. When properly adjusted, where do the three beams of a color CRT converge a) at the mask apertures? b) slightly ahead of the mask? c) at the screen surface?

33. a) What is the operating bias of a CRT when $+120$ V is measured on the cathode and $+70$ V on the grid? b) How much cathode voltage is required to increase the bias to -100 V?

34. Why is the finest pix detail not reproduced in color?

35. Which of the three beams of an in-line CRT is not affected by the purity and convergence adjusters? Explain.

36. a) What is meant by a "pure" raster?) b) Give an example of an impure raster.

37. What component is used to adjust raster purity on a) a delta CRT? b) an in-line CRT?

38. What is the function of a control sometimes used to vary the HV applied to the convergence plates of an in-line CRT?

39. a) Why does a beam tend to spread out after leaving the cathode? b) Name two ways of bringing the beam back into focus.

40. Although attracted by the positive charge on the grids and anodes, most of the beam electrons pass through the openings in the elements rather than striking them. Explain why.

41. How is deflection sensitivity affected by an increase in HV? Explain why.

42. Why is beam convergence a greater problem at the edges of a screen than at the center?

43. a) How does a color set "recognize" whether a BW or color program is being received? b) What action occurs in each case?

44. As screen size is increased, the time required to scan from left to right becomes a) longer? b) shorter? c) the same? Explain your answer.

45. The typical color set requires more HV than a BW set. Explain why.

46. A large-screen CRT requires an RF/IF bandpass of 4.2 MHz but for a small screen, considerably less. Explain why.

47. Explain the differences in the waveforms and bar patterns produced by a) an NTSC color bar generator b) a keyed rainbow generator.

48. Each of the three beams of a color CRT simultaneously impacts more than one phosphor dot or stripe at any given instant. Explain why.

49. Which of the three characteristics of a video signal determines the horizontal size of reproduced pix elements a) amplitude? b) frequency? c) polarity?

50. Why are the three guns of a color CRT seldom driven with equal amounts of signal?

51. Explain the basic operation of a typical projection type receiver.

52. Explain how a color bar pattern is produced by a color bar generator.

53. What is a service set-up switch and how is it used?

54. Which of the three guns of a color CRT are made to conduct to produce a) BW pix? b) yellow pix objects? c) magenta-colored objects?

55. Name three things that influence the strength of the magnetic flux developed by a convergence assembly.

56. What changes occur as the beams of a color CRT are deflected toward the edges of the screen? a) The *amount* of deflection, b) the *rate* of deflection? or, c) the beam current intensity? Explain.

57. What determines a) vertical resolution of a pix? b) horizontal resolution?

710 PIX TUBES AND ASSOCIATED CIRCUITS

58. Do low video frequencies produce a) tall pix objects? b) wide objects? c) Both?

59. How many bars are produced when a CRT is fed with the following frequencies: a) 393.75 KHz? b) 126.3 KHz? c) 12 KHz? d) In each case state whether the bars are vertical, horizontal, or diagonal.

60. Why does a pix on an in-line CRT appear to be made up of both vertical and horizontal scanning lines?

61. Four different signals are supplied to a color CRT. Name them.

62. With many color sets, loss of HV coincides with loss of CRT heater glow. Explain why.

63. Explain purpose of the two yoke positioning adjustments.

64. A delta CRT has two sets of built-in magnetic pole pieces. Explain the location and purpose of each.

65. What method is used for dynamic convergence of an in-line CRT?

66. Is white light a mixture of a) all colors of the spectrum? b) red green and blue? c) both a and b? Explain your answer.

67. Complete reconvergence of a delta CRT can be very time consuming. Explain why.

68. Why is a blue lateral adjuster not required with an in-line CRT?

69. All three beams of a delta CRT must be deflected by different amounts. Explain why.

70. Fill in the missing words: The _____ on the neck of a color CRT simultaneously moves all three beams. The _____ adjusters control each beam individually.

TROUBLESHOOTING QUESTIONNAIRE

1. State probable cause where there's no visible heater glow, or raster, but the neck of a CRT gets warm.

2. A color CRT heater has continuity but it doesn't light up. Voltage checks show a loss of HV, and no AC heater voltage across the CRT filament contacts. What would you suspect? Hint: All B+ voltages check normal.

3. What symptoms indicate the need to adjust the following controls: a) purity, b) static convergence, c) dynamic convergence, d) drive controls, e) screen controls?

4. Flesh tones are normally observed while adjusting the tint control. Explain why.

5. A set has a poor color pix and there are color fringes on a BW pix. How would you proceed?

6. What adjustment would you make when there are blue outlines a) above or below pix objects at the center of the screen? b) on either side of pix objects?

7. What are white-dot and crosshatch patterns? For what purpose are they used?

8. A person in a televised scene undergoes a color change when moving from one side of the screen to the other. What trouble is indicated?

9. Small changes in purity and convergence are observed when a set is relocated in a room. Give two possible causes and what you would do in each case.

10. What is the normal sequence for making the following set-up adjustments? static convergence; purity; raster size and linearity; pincushion; dynamic convergence; focus; HV; R,G,B drive; raster centering.

11. What troubleshooting procedures make use of a color-bar generator or a vectorscope?

12. Name the tests that can be performed by a good quality CRT tester.

13. Explain the difference between color fringing on pix objects and tinting of the raster.

14. Why is color impurity and misconvergence more noticeable on a BW pix than a color pix?

15. How would you determine by observing a raster and BW pix whether there's a problem with color purity or misconvergence?

16. What would you suspect when there's good color purity and convergence but poor color when viewing a color pix?

17. Give possible causes where difficulty is experienced in obtaining a) good color purity b) good convergence c) how would you proceed?

18. a) How would you proceed when there are sounds of arcing from inside a HV cage? b) What are the most likely arcing sites? c) How does HV arcing differ from corona discharge?

19. a) Name 10 things to check and adjustments to be made *prior* to making purity and convergence adjustments b) What if adjustments are made in the *reverse* order?

20. A BW set has no raster. Connecting a jumper between control grid and cathode of the CRT restores the raster. a) State probable cause. b) What if the raster is not restored?

21. The signal from the collector of an NPN video output transistor is coupled to the CRT cathode via a coupling capacitor. State probable effect on the raster and pix if the capacitor becomes a) open, b) shorted. Explain.

22. When an exact CRT replacement is not available, name 6 things to consider when choosing a substitute.

23. A set having normal HV but no raster is being checked by arcing between the HV lead and the CRT ultor contact. What is suspect if, a) there's no arc with brightness control fully CW? b) there's a high intensity arc that doesn't vary with the control?

24. Pix brightness, contrast, and color should be evaluated under conditions of normal ambient room lighting. What error in diagnosis may be made if viewing is done in a) a completely dark room? b) a brightly lit room?

25. Explain a procedure for "safing" a CRT before discarding it.

26. Give three indications that a CRT has lost its vacuum.

27. State five possible causes of an impure color field.

28. What color field is normally used in making purity adjustments on a) a delta CRT? b) an in-line CRT? c) If good purity is obtained with one color why are further adjustments for the other two colors generally unnecessary?

29. The position of the yoke on the CRT gun is important in obtaining good purity over the entire screen. Explain why.

30. State three possibilities where a CRT cannot be 100% converged.

31. How does misconvergence show up on a) a white-dot pattern? b) a crosshatch pattern? c) a BW pix? d) a color pix? e) How do the dot and crosshatch patterns appear when convergence is OK?

32. A CRT can retain an HV charge for very long periods. Standard safety practice calls for discharging a tube under *all circumstances* before handling. a) Explain the danger. b) How would you discharge a tube?

33. Are the four static convergence adjusters of a delta CRT adjusted while observing a) a raster? b) a BW pix? c) a color pix?

34. It's almost impossible to make dynamic convergence adjustments while observing a pix. a) Explain why. b) How should it be done?

35. Even when brightness and/or focus is improved, the HV of a color set should not be adjusted higher than specified. Explain why.

36. a) What symptoms indicate that an ADG coil is energized at all times? b) How would you verify your suspicions?

37. a) Explain how it's possible to obtain a bright raster of a particular color even though the electron gun for that color is not operating. How would you check and evaluate conditions by adjusting b) the three screen controls? c) the purity ring?

38. A set has poor pix and color. How would you quickly determine if it has a convergence problem or if the CRT is magnetized?

39. a) For a BW set, why is the yoke normally positioned firmly against the CRT bowl? b) Is this also true for a color set?

40. Why is it customary to degauss a CRT prior to making set-up adjustments, even when a set has a built-in ADG system?

41. Explain procedure for adjusting raster tilt.

42. a) What controls are used for gray-scale tracking? b) Explain the procedure.

43. When intermittently the screen brightens, the pix is lost, and the brightness control becomes inoperative, what would you suspect? Hint: Normal operation is restored when the set is jarred.

44. Name 6 things to look for when evaluating a raster.

45. Give a logical step-by-step troubleshooting procedure for each of the following trouble symptoms: a) no raster; b) dim raster with poor focus; c) severe blooming; d) color-tinted raster; e) raster OK but color outlines on BW pix objects.

46. An abnormal dc voltage reading is obtained at the cathode input of a CRT. State possible causes if the video amplifier is a) capacitively coupled to the CRT; b) directly coupled.

47. Where should the following items be positioned on the neck of a delta color CRT? a) purity ring; b) deflection yoke; c) convergence assembly; d) blue lateral adjuster.

48. What two adjustments would you make for neck shadows on a BW pix tube?

49. Intermittent arcing is observed in the gun of a newly installed CRT. What would you do?

50. Give three possibilities where the CRT heater glow fluctuates or goes out.

51. The keyed plastic post at the base of a CRT is broken off, and there's no heater glow after connecting the socket. a) State probable cause. b) How would you correctly orient the socket? c) Does loss of the post mean the CRT has lost its vacuum?

52. Why should a CRT never be discarded without first relieving the vacuum?

53. After adjusting the screen controls to obtain a white raster, severe blooming is observed. What would you do?

54. a) What trouble is indicated when there's a change in raster tinting as the brightness control is adjusted? b) What controls would you adjust to correct the problem? c) What would you suspect if the trouble persists?

55. Describe procedure for adjusting purity on, a) a delta CRT, b) an in-line CRT.

56. Give four symptoms that indicate a CRT has lost its vacuum.

57. When a thin bright line is observed indicating loss of sweep, what's the first thing you would do? Explain why.

58. For a dim raster, which would you suspect, low HV or a weak CRT, if there's also: a) severe blooming? b) No evidence of blooming? State your reasoning.

59. What trouble symptoms normally call for adjusting a) the R-G-B static convergence adjusters? b) The blue lateral adjuster? c) Which way do the beams move? Explain the procedure when viewing a pix.

60. Which is normally adjusted first, purity or convergence? Explain why.

61. When checking for single-field color purity, a) what controls are used to disable the guns individually? b) Give another method.

62. State probable cause of the following symptoms after replacing a yoke: a) inverted pix; b) pix transposed left to right; c) pix inverted and transposed.

63. If a CRT is installed "upside down" would the pix be inverted? Explain.

714 PIX TUBES AND ASSOCIATED CIRCUITS

64. Name four possible causes when a CRT requires frequent adjustment of purity and/or convergence. How would you proceed to determine which cause is acting on a particular tube?

65. What is the effect if a yoke is turned 180° from its normal position? b) 90°?

66. Does misconvergence show up on a) the raster? b) a BW pix? c) a color pix? d) all of the above?

67. What results can be expected from misadjustment of a) the R-G-B screen controls? b) the drive controls?

68. What is the effect on raster and pix of a color set if the three guns are a) not conducting in the correct proportions? b) not being driven by signals of the correct proportionate levels?

69. Describe procedure for making purity and convergence adjustments on a trinitron.

70. What adjustment establishes color purity at a) the center of a screen? b) the edges of the screen?

71. What results can be expected if all three screen controls of a color set are a) turned full on? b) turned too far ccw?

72. a) Explain why "flashing" a CRT sometimes restores the cathode emission. b) What other condition can be corrected by flashing?

73. Give six precautions to observe when replacing a CRT.

74. Problem: no raster. When HV lead is removed from the CRT, a strong arc can be obtained between the lead and the ultor contact, but intensity of the arc doesn't change with the brightness control. State two possible causes.

75. What would you suspect if the raster on a BW CRT a) slowly brightens over a period of time? b) starts off small then slowly fills the screen? Explain your answers.

76. Give three uses for a crosshatch generator.

77. When a black edge is showing at the top of the raster a) It could be a height problem, b) poor vertical linearity c) a vertical centering problem d) vertical blanking, or possibly all of the above? e) How would you proceed?

78. What would you check for a raster centering problem on a, a) BW set? b) a color set?

79. Improper positioning of components on a CRT gun is a common cause where set-up adjustments have little or no effect. Explain.

80. There's a calculated risk in sparking the HV on a solid state receiver. Explain.

81. What problems can be caused by a) excessive HV? b) Insufficient HV?

82. State several CRT voltage problems that can cause a weak raster or no raster.

Chapter 16

THE CHROMA SECTION

16-1 SIGNAL DEVELOPMENT AT THE STATION

To fully understand the nature of chroma signals and how they're processed in a *receiver* requires a good understanding of how chroma and Y signals are developed *at the station*. Chapter 1 gives a brief overview of what takes place at the station; a review of that chapter is suggested before proceeding with the more detailed descriptions contained in the following paragraphs.

• **The Color Camera**

The *camera* is the source of the signals that produce a pix on a receiver screen either in BW or in color. A typical camera contains three camera tubes (CRTs). Each tube develops a raster, all scanned in unison, at the same rate and in the same manner as at the receiver. By the use of color filters, each tube "sees" only one of the three primary colors, red, green, or blue, and the amount of its color contained in the complementary colors. Each tube develops an output corresponding with the intensity of the color being viewed at that instant.

• **Developing the Luminance (Y) Signal**

The luminance signal conveys only *brightness* information. This is the signal used by a BW receiver. It is also used by a color receiver to reproduce fine pix details and supply brightness information for all colors. There is no output from any of the three camera tubes when viewing a black object, but each tube is able to recognize the amount of *white light* in its particular color.

REMINDER: _____

White is a combination of *all* colors. Hence, the three tubes produce *simultaneous outputs* while white is being viewed. The relative amplitudes of the three outputs correspond to the amount of each color that goes to make up white, which is 30% red, 59% green, and 11% blue. The outputs of the three tubes are fed to an "adder" (*Y* matrix) where they are combined to form the luminance signal (see Fig. 1-7). Some cameras

715

contain a fourth CRT where the Y signal is developed separately. The matrix also develops I and Q chroma signals from camera outputs in the following proportions:

$$I = 0.60 \text{ R}, 0.28 \text{ G, and } 0.32 \text{ B}$$
$$Q = 0.21 \text{ R}, 0.52 \text{ G, and } 0.31 \text{ B}$$

- **Developing the Chrominance Signal**

The limited band of frequencies allocated to each channel imposes serious limitations on the amount of information that can be transmitted, and, for compatibility, the NTSC system requires that all color information be transmitted within the same bandpass as for BW transmissions. Hence the interleaving of video and chroma as described in Sec. 14-13. Another part of the solution is the transmitting of only two of the three primary colors. It has been stated that all colors can be derived from the three primaries: red, green, and blue. Actually, it takes only two, red and blue, from which green can be extracted at the receiver.

The two color signals are produced by modulating a 3.58 MHz *subcarrier* with the matrixed outputs from the red, green, and blue cameras. The resulting chroma sideband frequencies are then used to modulate the main station transmitter along with the luminance, synchronizing, and sound signals. To prevent interference on BW receivers, the color subcarrier itself is not transmitted.

Because there are two colors to be transmitted, two chroma modulators, designated *I* and *Q*, are required. In addition to its corresponding color signal, each modulator is supplied with the 3.58 MHz unmodulated subcarrier as shown in Fig. 1-7. Although coming from the same source, each subcarrier is of a different *phase*. It is this phase difference that enables the two color signals to retain their identities when *combined* at the modulator outputs. This system, called the I and Q system, uses a phase difference of 90°. Phase shifting (delay) networks are used to make one subcarrier lag the other by the required amount. One network delays the subcarrier applied to the *I* modulator by 57°. With an additional delay of 90°, the *Q* modulator operates with a subcarrier of 147°. The terms I and Q mean *in-phase*, and *quadrature*, respectively. With the two subcarriers separated by 90°, the chroma contained in the sidebands at the modulator outputs is also separated by the same amount. A sampling of the subcarrier is transmitted as a *color burst*, and used by the receiver as a phase reference for color synchronization.

Each modulator has an output that varies in *amplitude* in accordance with the applied chroma information. Because the two outputs are 90° out of phase with each other, when combined, the vector sum of the two outputs varies in *both phase and amplitude*. The *amplitude* at any given instant represents the degree of *saturation* of the color, and the *phase* (relative to the subcarrier) corresponds with *hue*.

REMINDER: _____

Every color has three distinct characteristics, hue, saturation, and brightness.

From this we see that each color of the visible spectrum corresponds *with a specific phase angle* between zero and 360°, as shown in the phasor diagram of Figures 16-1 and 16-8. The angles for yellow, magenta and cyan are mid-way between the primaries from which they are derived. The modulation vectors are also shown for comparison. Note the I and Q axes, separated by 90° with phase angles of 57° and 147° respectively. By intent, the I axis is close to the red vector and the Q axis close to the blue vector. Note that the three primaries are 120° apart, and their complementaries of opposite polarity. Instead of I and Q, stations once used R−Y (red minus Y) and B−Y axes which are also shown on the diagram. These axes are at 90° and 180° respectively but still close to the angles for red and blue. Relative amplitudes are indicated on the vectors.

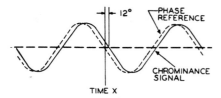

YELLOW

When the secondary color yellow is transmitted, the chrominance signal may be represented by a vector which lags the phase-reference vector by 12 degrees. The magnitude or length of the vector for fully saturated yellow is 0.45.

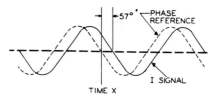

I

During testing, an I signal may be encountered. This signal may be represented by a vector which lags the phase-reference vector by 57 degrees. The fact that the I vector has no prescribed magnitude is indicated on the phase diagram by the dotted portion of the vector.

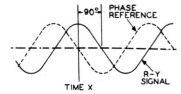

R-Y

During testing, an R - Y signal may be encountered. This signal may be represented by a vector which lags the phase-reference vector by 90 degrees. The fact that the R - Y vector has no prescribed magnitude is indicated on the phase diagram by the dotted portion of the vector.

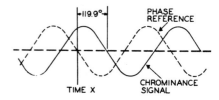

MAGENTA

When the secondary color magenta is transmitted, the chrominance signal may be represented by a vector which lags the phase-reference vector by 119.9 degrees. The magnitude or length of the vector for fully saturated magenta is 0.59.

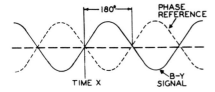

B-Y

During testing, a B - Y signal may be encountered. This signal may be represented by a vector which lags the phase-reference vector by 180 degrees. The fact that the B - Y vector has no prescribed magnitude is indicated on the phase diagram by the dotted portion of the vector.

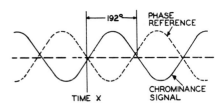

BLUE

When blue is transmitted, the chrominance signal may be represented by a vector which lags the phase-reference vector by 192 degrees. The magnitude or length of the vector for fully saturated blue is 0.45.

Fig. 16-1 (partial)

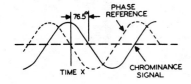

RED

When red is transmitted, the chrominance signal may be represented by a vector which lags the phase-reference vector by 76.5 degrees. The magnitude or length of the vector for fully saturated red is 0.63.

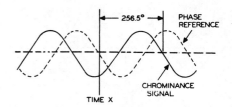

CYAN

When the secondary color cyan is transmitted, the chrominance signal may be represented by a vector which lags the phase-reference vector by 256.5 degrees. The magnitude or length of the vector for fully saturated cyan is 0.63.

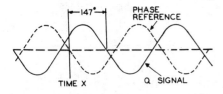

Q

During testing, a Q signal may be encountered. This signal may be represented by a vector which lags the phase-reference vector by 147 degrees. The fact that the Q vector has no prescribed magnitude is indicated on the phase diagram by the dotted portion of the vector.

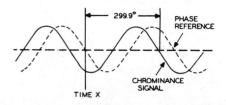

GREEN

When green is transmitted, the chrominance signal may be represented by a vector which lags the phase-reference vector by 299.9 degrees. The magnitude or length of the vector for fully saturated green is 0.59.

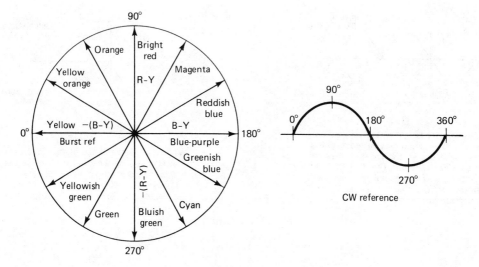

Fig. 16-1 Phase angles and chroma vectors.

Note that chroma angles zero through 360° correspond with one complete cycle of the subcarrier reference signal.

Development of the Y, I, and Q signals by the station can be summarized as follows: Each of the three camera tubes recognizes but one of the three primaries: red, green, or blue, plus their brightness levels, in the pix being scanned. Outputs of the three tubes are matrixed together in the proper proportions to develop the luminance (Y) signal. They are also combined (in different proportions) to form the I and Q signals that are used to modulate two subcarrier signals differing in phase by 90°. With the Y component removed, the I and Q signals represent fully saturated colors. The two modulator outputs are added vectorially to form the chrominance (chroma) signal. The chroma signal varies in both amplitude and phase. Amplitude variations are in accordance with amplitude levels at the modulator inputs, and correspond with color intensity or saturation. Instantaneous phase angles between the chroma and the color subcarriers establish the hue of the color. Thus, every color transmitted relates to a particular phase angle.

At the receiver, all functions are performed in reverse order: all colors are reproduced by demodulating the chroma sidebands using a locally generated subcarrier locked in sync with the station.

16-2 THE RECEIVER

At the *station,* the camera outputs were converted into two kinds of signals, a luminance (Y) signal, and a chrominance signal, that were processed separately. The Y signal establishes the brightness level of all colors viewed on the receiver screen, the fine pix detail that is *not* reproduced in color, and all video information to develop a BW pix. The chroma signal provides the color information for hue and saturation. The chroma signal was developed by modulating (encoding) two out-of-phase subcarriers with fully saturated red and blue signals from the cameras. At the modulator outputs, the two encoded signals, each retaining its original identity, were combined into one, then recombined with the Y signal to modulate the transmitter.

At the *receiver,* the process is reversed by: separating the chrominance from the luminance signal; processing each through separate channels; demodulating (decoding) the chroma into *color difference signals;* then recombining the chroma with the Y signal to drive the three guns of the CRT. Demodulation requires a locally generated subcarrier, which, for color synchronization, must be precisely controlled in both frequency and phase. Control makes use of the color burst which is a sampling of the subcarrier generator at the station.

Figure 16-2 represents a block diagram of a typical chroma section of a receiver. As indicated, there are three main functional areas: the *signal stages,* the *color-sync stages,* and what might be called the *chroma-switching stages.* The signal stages process the received chroma signals from the point of *chroma takeoff* to the inputs of the CRT. The color-sync stages maintain color synchronization with corresponding circuits at the station. The switching stages enable the chroma stages when receiving color and disable them when receiving BW.

Some sets have all chroma-related circuits on a PC module board. In late-model sets, all functions are performed by one or two ICs. Study the block diagram, noting the names of the various blocks and their interconnections, and other particulars such as the tuned circuits, all resonant to the subcarrier frequency, 3.58 MHz. Modern sets, particularly those using ICs, tend to dispense with *adjustable* tuned circuits. Other things to note are: alternate connections between certain blocks depending on receiver design; alternate locations of the tint (hue) control; and the fact that several stages are gated or keyed by the horizontal sweep circuit. The fol-

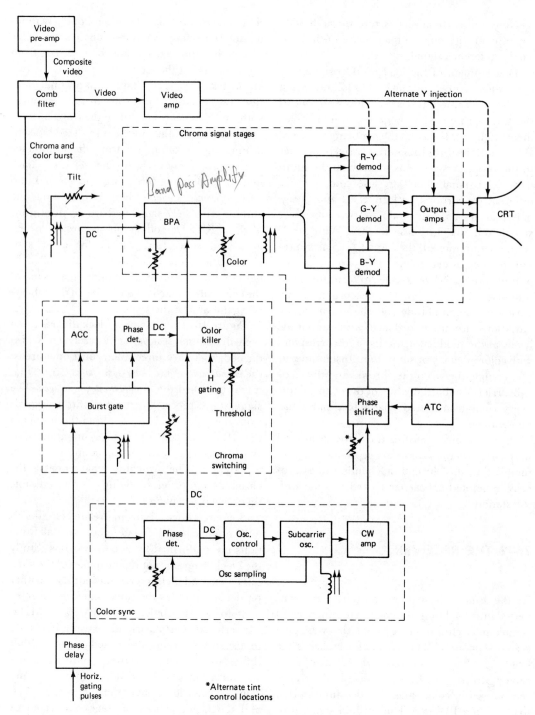

Fig. 16-2 Chroma processing stages.

lowing paragraphs represent a brief overview of functions described in Chapter 2. Detailed descriptions of each stage are given in upcoming sections of this chapter.

16-3 THE CHROMA SIGNAL STAGES

As shown in Fig. 16-2, this includes those stages through which the received chroma signals must pass en route to the CRT, namely the comb filter or other tuned filter at the point of chroma takeoff; the bandpass amplifier (BPA); the chroma demodulators; and the chroma-output or matrixing stage. Note that the luminance signal, sometimes called the video or Y signal, takes a separate path, via the video-amplifier stages, to be recombined with the chroma signal in the demodulators, the output amplifiers, or the CRT, depending on receiver design.

• **Signal Input**

The composite video signal (BW video and chroma information) feeds the BPA from a *chroma takeoff point* in the video amplifier, usually the first stage. Separation of the chroma from the video is accomplished in one of two ways, with a *comb filter* (as described in Sec. 14-13), or with a *tuned filter* having a narrow (1 or 2 MHz) bandwidth. The object is to pass only the chroma IF signals up to 1.5 MHz *maximum* while rejecting video information which extends to approximately 4 MHz. The *color burst* may also be taken from the chroma takeoff point to feed the burst gate.

• **IF Amplification**

In the video IF strip the chroma information was interleaved with the video and they still coexist up to the point of chroma takeoff. At this point, the chroma still exists as high frequency I and Q sidebands of the 3.58 MHz subcarrier, which have yet to be demodulated. Thus the BPA is actually another IF amplifier, but designed to pass chroma information *only*. This requirement is met by limiting the bandpass to either 1 or 2 MHz, depending on the set. From the output of the BPA, the amplified and still combined I and Q sidebands are simultaneously fed to the demodulators as shown in Fig. 16-2.

Note the circuits that interface with the BPA: the color-killer, for example, which functions as an electronic switch to enable the BPA when receiving color, and to disable it when receiving BW. The purpose is to prevent background "noise" from reaching the CRT via the chroma stages when the images were transmitted in BW. Such interference shows up as colored snow (confetti) on a BW pix.

Another input to the BPA comes from the automatic-color-control circuit (ACC), which is incorporated in most late model sets. The ACC automatically controls the gain of the BPA, and therefore color saturation, in accordance with the strength of the received signal, using the burst signal as a reference.

In some cases, the BPA is gated by horizontal sweep pulses to prevent unwanted signals from reaching the CRT during retrace periods.

With most sets the color control is found in the BPA. Since color saturation is a function of signal amplitude, it provides the user with a means of controlling chroma gain, and therefore the intensity of the reproduced color on the screen. Some sets also have the tint/hue control in the BPA.

• **Demodulation (chroma sampling)**

This is the means of recovering the original red, green, and blue signals used to modulate the subcarrier at the station. It requires a locally generated subcarrier to take the place of the subcarrier that was not transmitted. The subcarrier is developed by a crystal-controlled oscillator operating at approximately 3.58 MHz. The configuration shown in Fig. 16-2 uses two demodulators, one to demodulate the I (or R-Y) signal, the other to demodulate the Q (or B-Y) signal. The subcarrier oscillator shown here has two outputs. The frequency of each output is the same, but they differ in phase by some amount determined by the type of demodulator system in use. Other systems are described in Sec. 16-12. The relative phase angle of the two outputs is determined by phase-shifting networks as indicated. One oscillator output of a particular phase angle is injected into the R-Y demodulator to produce an R-Y output. The other oscillator signal of a different phase feeds the B-Y demodulator to produce a B-Y output. Although each demodulator is supplied with the same chroma inputs, each "recognizes" only one signal, according to the phase of the oscillator signal it receives. The "-Y" designations of the demodulator outputs signify that they are *fully saturated* color signals, since they are not yet combined with the brightness component.

The *green* color difference signal, which was not transmitted as such, is obtained (in the Fig. 16-2 configuration) by combining proportionate amounts of R-Y and B-Y signals as indicated. In other systems, green is recovered further downstream.

- **Tint Considerations**

The tint or hue of a color is determined by the phase relationship between the chroma signal and its corresponding subcarrier. Since hue can be changed by slight variations of one or the other, the tint control may be found in any one of several stages as shown in Fig. 16-2: the oscillator, the BPA, the oscillator phase detector, or the burst gate. Most sets now have provision for automatic tint control (ATC).

- **Matrixing the Y and Chroma Signals**

Until recombined with the luminance (Y) signal, the demodulated chroma signals represent *fully saturated* colors, and as such would produce harsh and unnatural-looking colors. Also, if the Y signal is not present for any reason, the result is loss of fine pix detail (no pix at all when receiving a BW telecast). Matrixing is the recombining of the chroma and luminance signals to restore the original brightness levels to the colors. It can take place in one of three areas: the demodulators, the output-amplifier stages, or the CRT. Mixing the signals ahead of the CRT is called pre-CRT matrixing; mixing them in the CRT is called CRT matrixing. (see Figs. 14-15 and 14-16). With CRT matrixing (used mostly on older sets) the Y signal is simultaneously applied to the three CRT cathodes, and the chroma signals individually to the three grids. With pre-CRT matrixing, the Y signal is applied to one input of the demodulator or output-amplifier transistors, the chroma to the other inputs. Regardless of which system is used, signal polarities must be opposite and correct for the transistor or CRT element being driven, so they both work together to either increase or decrease current flow.

With CRT matrixing, if the color-difference signals are sufficiently strong, they may drive the CRT *direct* with no additional amplification; otherwise, one or more amplifiers are required as is the case with pre-CRT matrixing. Such amplifiers are known by various names according to the functions they perform: color-difference amplifiers, matrix amplifiers, or simply video-output amplifiers. Three amplifiers are required with three discrete outputs to drive the three CRT guns individually. With CRT matrixing, the signals feeding the guns are R-Y, G-Y, and B-Y signals. With pre-CRT matrixing, they're called red, green, and blue signals, in accordance with the guns they drive.

Many early model sets used *passive* components for matrixing: numerous resistors connected in a voltage-dividing configuration to mix the signals in their proper proportions as required by the three guns.

- **Developing a BW Pix**

All three guns must be operating properly to produce a white raster and a BW pix. The three guns, however, are driven by only the Y signal, because the chroma is inhibited by the disabling action of the killer stage. The same conditions apply whether receiving a BW telecast or reproducing BW images on a color pix. To produce white objects all three guns must be conducting by approximately the same amount. Slight variations of beam current from one or more guns result in an off-white or in extreme cases a color tint. Black areas on the screen occur when all three guns are driven to cutoff. Different degrees of shading are caused by simultaneous variations of beam current from all three guns.

- **Developing a Color Pix**

A color pix is produced by varying the beam current of each gun *individually*. The variations are created by changes in both the amplitude and the polarity of the chroma signals driving the three guns. With CRT matrixing, the drive signals are the R-Y, G-Y, and B-Y color-difference signals. With pre-CRT matrixing, they are the "pure" red, green, and blue signals. The hue of a color being produced at any given instant is determined by the relative amplitudes and polarities of the signals feeding the three guns. For example, if the red gun is made to conduct while the green and blue guns are being driven to cutoff, only the red phosphors on the screen will be illuminated. If the blue and green guns are made to conduct while the red gun is cut off, both the green and blue phosphors will be illuminated (as can be observed on close inspection) but from a distance we interpret the mixture as cyan. In this example, if the blue signal is stronger than the green signal then the observed hue would tend to be more bluish. Conversely, if the green signal is the stronger of the two, we see a more greenish cyan. Other examples are shown in Fig. 15-18.

REMINDER: _____

Signal *polarity* determines whether or not a particular gun will conduct; when a gun is conducting, *amplitude* determines the intensity of conduction. When a signal is driving a CRT cathode, it must have a positive-going polarity to cause conduction; for grid drive, it must be negative-going.

Needless to say, if one or more primary color signals is not produced *that color and all its secondary colors* will be missing from the screen.

16-4 COLOR SYNC

The hue of a color is determined by the phase relationship between the oscillator-subcarrier and the chroma-sideband signals. This requires a *stable* oscillator that is locked into the station oscillator in both frequency and phase. Sync is maintained by comparing a sampling of the generated oscillator signal with the received color-burst signal. When there's a difference in either frequency or phase, a DC "error" voltage is developed which forces the oscillator back into step with its counterpart at the station. The action is similar to that of a horizontal AFC circuit, although color sync must not be confused with sweep synchronization.

With normal color sync, all colors are present in a pix, properly located, and of the correct hue. In addition, the tint control has a normal range, with correct skin tones obtained at its approximate mid-setting. Loss of color sync shows up in various ways. If oscillator *frequency* is off by a small amount the result is a rainbow pattern of horizontal bars called the "barberpole effect". The greater the frequency error the greater the number of bars. If oscillator frequency is an exact multiple of the burst frequency, the bars will be stationary, otherwise they will be in motion. When oscillator frequency is correct but of the wrong *phase,* the colors may all be present but of the wrong hue (a frequency error more than ¼ Hz results in a noticeable change in tint). That is, the colors that make up the pix will be *misplaced*. When color sync stability is borderline, there may be color lock on some channels but not on others, because of variations in burst amplitude. If the oscillator's frequency and phase are correct but there's the wrong phase from one of its outputs, colors developed by the corresponding demodulator will be incorrect. Phase errors also affect the range and setting of the tint color.

For *frequency* stability, the oscillator is crystal-controlled. Proper *phasing* is maintained with an *automatic frequency and phase control* circuit (AFPC), usually abbreviated to APC.

The color-sync stages, as shown in Fig. 16-2, include the oscillator, an oscillator-phase detector, and the burst gate. The burst stage also works in conjunction with the color-killer stage as explained earlier for chroma switching. Both operations make use of the received color burst.

- **The Color Burst**

The color burst is an 8-to-10 cycle sampling of the station subcarrier which was not transmitted. The burst signal is inserted on the back porch of each horizontal-blanking pulse, immediately after the sync pulse. It serves two purposes: operating the killer stage, and controlling the frequency and phase of the receiver-subcarrier oscillator.

- **Controlling the Oscillator**

From the burst takeoff, at the same point as chroma takeoff or in a later BPA stage, the chroma and burst signal feed the burst-gate input. The purpose of the burst stage is to amplify the burst signal while rejecting all other signals. This is done by keying the stage with strong pulses from the horizontal sweep circuit. The burst stage is keyed to conduct only for brief periods at the start of each horizontal retrace period, so only the burst signal gets through.

The amplified burst signal is supplied to the phase detector along with a sampling of the oscillator output. Any difference in frequency or phase results in a DC error voltage that forces the oscillator back into sync. Thus the sync becomes stabilized for each horizontal trace and retrace period. There is no need for control during *vertical* retrace.

Loss of the burst signal or improper operation of the phase detector will obviously result in complete loss of color sync.

16-5 CHROMA SWITCHING

Most sets have a color-killer stage which automatically enables one of the chroma stages when receiving color and automatically disables the chroma when receiving a BW transmission. The object is to prevent "white noise" from reaching the CRT via the chroma stages when receiving BW. However, this automatic function is not entirely necessary since turning down the color control accomplishes the same thing. The stages involved in the switching operation are shown in Fig. 16-2, and switching is accomplished with DC voltages developed in a phase detector and in the killer stage. Note the alternate connections used by different sets. Some make use of the oscillator-phase detector, others use a separate killer-phase detector as shown, and a few sets drive the killer to conduct or not conduct with the burst signal instead of DC. In some cases, the killer, like the burst stage, is gated with horizontal sweep pulses.

When receiving color, the killer is biased to cutoff, either by the burst signal or by a DC control voltage developed in the appropriate phase detector. The non-conducting killer in turn forward-biases a BPA or other chroma stage into conduction, allowing the chroma signals to reach the CRT. When receiving a BW transmission, the *absence* of the burst turns the killer transistor on, biasing the chroma stage into cutoff. Most sets have a color *threshold control* to establish the level at which the killer operates. With the control turned down (ccw), for most signal levels the BPA will be operating at all times and there will be colored snow (confetti) on a BW pix. At the other extreme (fully cw) the chroma is disabled and no color is produced on the pix.

In some sets, the subcarrier oscillator is periodically shock-excited into oscillations by the color burst. With this arrangement, no killer stage is required because with no burst the oscillator doesn't develop a signal and there can be no output from the demodulators.

A weak or missing color burst is a common cause of loss of color, for without the burst the chroma stages are continuously disabled.

- **Controls and Adjusters**

Controls associated with the chroma stages are shown in Table 16-1.

16-6 THE CHROMA CIRCUITS

A detailed description of all stage functions including typical circuits follows.

16-7 BANDPASS AMPLIFIER (BPA)

The BPA, sometimes called a *chroma IF amplifier,* has two main functions: to separate chroma signals from the video signal, and to amplify the chroma at IF frequencies. The BPA is a tuned amplifier with anywhere from two to five stages, depending on whether the set uses high-level or low-level demodulation. With low-level demodulation, one or two stages is typical with most of the chroma amplification following the demodulators. With high-level demodulation, most of the chroma gain is obtained prior to demodulation. Included in the BPA is the front-panel *color control* which controls the strength of the chroma signal and therefore the intensity of the colors on a pix. Most late-model sets provide ACC to automatically vary the gain of the BPA in accordance with the strength of the received signal. Some sets also have the tint (hue) control in the BPA. The signal input to the BPA is from a chroma takeoff point in the video amplifier, or, in late-model sets, from a comb filter. The amplified output of the BPA is fed in parallel to each demodulator, where the I and Q sidebands are *sampled* for the chroma information they contain.

- **Chroma Signals and Bandpass Considerations**

The signal being amplified in the BPA is the same signal produced by the chroma modulators at the station, namely, the I and Q sidebands of a 3.58 MHz subcarrier. Although coexisting and treated as one signal, they retain their separate identities because of the phase difference between their two carriers. The I signal, containing frequencies from 0-1.5 MHz, corresponds with a subcarrier phase of 57°. The Q signal, containing frequencies from 0-0.5 MHz is in quadrature, with a subcarrier phase of 147°. Although I signals up to 1.5 MHz are transmitted, they are not fully utilized in modern receivers. For practical and economic reasons, only frequencies from 0-0.5 MHz are made use of, the same as for the Q sidebands.

- **Broad-Band and Narrow-Band Color**

Early-model sets demodulated on the I and Q axes, the same as at the station, thus utilizing the full 1.5 MHz of the I signal. This process, however, required a 2 MHz bandwidth for the BPA. Since I and Q demodulation is now obsolete, present day receivers limit the bandpass to +/- 0.5 MHz, a total of *1 MHz.* Demodulation makes use of different axes, as determined by receiver design.

Table 16-1 Controls and Adjusters. (Note: Some sets do not have all of the controls listed.)

Control or Adjuster	Location	Purpose
Color control	Front panel. BPA circuit.	To control chroma-signal amplitude for desired color intensity.
Tint control	Front panel. May be in burst-gate, BPA, phase-detector, or subcarrier-oscillator circuit.	A phase control to obtain desired hues.
Tint range	Service adjustment.	To establish range of front panel tint control.
Killer threshold	Rear panel. Color-killer circuit. A service adjustment.	To establish point of killer turn-on by burst signal.
APC control	Service adjustment. Subcarrier-oscillator or phase-detector circuit.	AFPC color-sync control.
ACC control	Service adjustment. ACC circuit.	To establish ACC operating threshold.
ATC control	ATC circuit.	To obtain desired skin tones.
Drive controls	Service adjustments. In chroma-output stages or CRT inputs.	For gray-scale tracking.
BPA tuned circuits	Input and output of BPA.	To obtain desired chroma-bandpass response.
Tilt control	Input of BPA.	As above.
4.5 MHz traps	In video amplifier and BPA.	To prevent beat interference on pix.
3.58 MHz traps	In demodulator circuits.	To prevent interference on pix.
Burst-tuned circuit	At burst-gate output.	For maximum output of burst gate and to eliminate video and chroma signals.
Reactance-control tuned circuit	In AFPC reactance-control circuit, when used.	For proper operation of reactance control of subcarrier oscillator.
Oscillator tuned circuit	In subcarrier-oscillator circuit.	To establish the oscillator frequency of 3.58 MHz.
Comb filter adjustments	In comb filter circuit.	For optimum separation of luminance and chroma signals.

16-7 Bandpass Amplifier

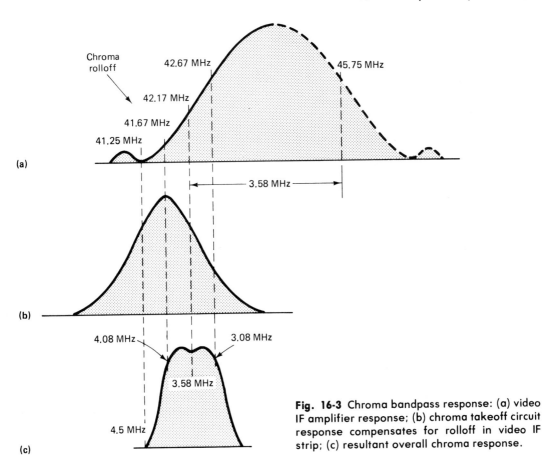

Fig. 16-3 Chroma bandpass response: (a) video IF amplifier response; (b) chroma takeoff circuit response compensates for rolloff in video IF strip; (c) resultant overall chroma response.

- **High-Frequency Roll-off Compensation**

High-frequency chroma signals tend to be attenuated in the video IF strip because of the dropoff at one extreme of the overall response (See Fig. 16-3a). This dropoff is compensated for at the input to the BPA with a *tilt control* and/or proper alignment of the chroma takeoff coil, as shown in Fig. 16-3b. Note how the slope or *tilt* of the curve is opposite to that of the video IF response. For proper color, this adjustment is *extremely critical*. Figure 16-3c shows an idealized overall response for the BPA, which is essentially "flat" from 3.08 MHz to 4.08 MHz. Although frequencies lower than 3.08 MHz and higher than 4.08MHz receive *some* amplification, the dropoff is fairly abrupt. In practice, the curve may have a slight dip or peak at the center frequency of 3.58 MHz. It's important that the two "markers" representing the I and Q sideband limits be at the same level for equal amplification of both sidebands. A BPA bandpass that's too wide results in weak color, and in interference from video reaching the CRT. A narrowed bandpass can cause excessive color, poor color quality, and sometimes oscillations.

- **Tint Control**

Hue can be controlled by changing the phase of the chroma signal, *or* that of the subcarrier. A typical tint control has a range of about 60° and

728 THE CHROMA SECTION

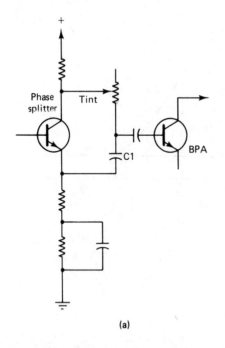

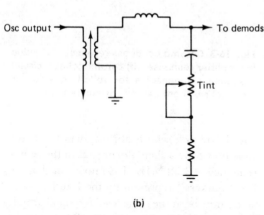

Fig. 16-4 Tint control circuits.

can change the average color of a pix from green at one extreme to magenta at the other extreme with normal colors at its approximate mid-setting. Adjustment is usually made by judging *skin tones*. Two typical tint-control circuits are shown in Fig. 16-4. In Fig. 16-4 a, one of the BPA stages is a phase splitter. The signal is fed to the next stage from both its collector and emitter, which are 180° out of phase with each other. The relative phase of the signal, which is limited to 90° or less by capacitor C1, is controlled by the pot.

In Fig. 16-4b, the tint control forms part of a phase-splitting network at one of the outputs of the subcarrier oscillator. When a tint control has a limited range, but all colors are accounted for, the usual cause is AFPC misalignment, and a touch-up of the tuned circuits in the burst-phase detector and oscillator circuits will often correct the problem.

16-8 TYPICAL BPA CIRCUIT

A five-stage BPA is shown in Fig. 16-5. Typically, there are only two *adjustable* tuned circuits: the chroma-takeoff coil L1 at the input, and the double-tuned output transformer that feeds the demodulators. The coil L2 is broadly resonant to 3.58 MHz. The input coil is peaked to approximately 4 MHz, the output transformer to 3.58 MHz. The bandwith of the takeoff coil L1 is established by the shunting resistors R1 and R2. The takeoff coil, because of the narrow bandpass, removes most of the low video frequencies and helps to compensate for high-frequency rolloff as described earlier. In lieu of the takeoff coil, most late-model sets now use a comb filter for better separation of video from chroma. In this circuit, burst takeoff is from the second stage, Q2.

The gain of the first stage is controlled by ACC action. Manual control of color intensity (by dc) is by means of the color control at the input to the third stage. Maximum gain is with the pot control arm at *ground* potential. On/off control of Q5 is accomplished by means of an input from the killer stage. Q5 is also cut off during horizontal retrace by negative-going blanking pulses applied to the transistor. The object is to prevent the burst signal from reaching the CRT.

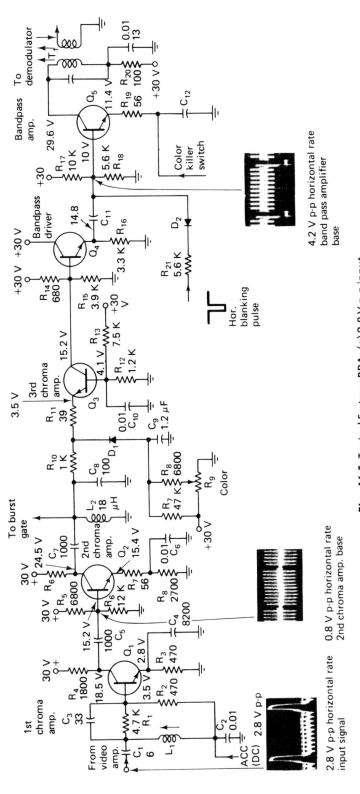

Fig. 16-5 Typical 5-stage BPA: (a) 2.8 V p-p input signal; (b) 0.8 V p-p 2nd chroma amp. base; (c) horizontal blanking pulse; (d) 4.2 V p-p bandpass amplifier base.

From:
Alvin A. Liff, *Color and Black & White Television Theory and Servicing*, ©1979, p. 571. Reprinted by permission of Prentice-Hall, Inc., Englewood Cliffs, NJ.

Some receivers have a 4.5 MHz trap in the BPA to reduce the chance of 920 KHz interference that results when the 4.5 MHz sound IF beats with the 3.58 MHz subcarrier. As might be expected, loss of signal in the BPA results in loss of all color. A *weak* signal reduces color intensity.

16-9 AUTOMATIC COLOR CONTROL (ACC)

This automatic function is provided on most sets to eliminate the need for frequent adjustment of the color control. It operates by regulating the gain of the BPA in accordance with the strength of the received signal. It can be compared to *AGC* which controls the gain of the tuner and video IF strip. Thus, the gain of the BPA (and the strength of the color) increases if there's a decrease in the received signal, and decreases if there's an increase in the received signal. The signal-level reference is the color burst. As with AGC, a DC voltage is developed that's proportional to the strength of the received signal. It is used as a variable bias to vary the gain of the controlled stage, in this case, the BPA. Either forward or reverse bias may be used depending on the type of transistor. ACC *delay* is sometimes used to avoid reducing gain on very weak signals.

Figure 16-6 shows one of several possible circuit arrangements in common use. The amplified color burst from the output of the burst gate is rectified with the ACC detector diode D1 and applied as a + voltage to the ACC amplifier Q1. The transistor, operating as a DC amplifier, reverses the polarity and applies a negative reverse ACC bias to the base of the BPA transistor Q2. ACC action is delayed by diode D2 until the burst is strong enough to overcome the normal bias on Q2. In some circuits, an ACC "threshold" control is provided in lieu of delayed ACC. In this circuit, the amplified DC from the output of Q1 is also used to control the killer stage. In some cases, the DC control voltage developed by the ACC amplifier is applied to the BPA *via the killer,* which functions as an intermediate control stage when turned on. Some sets do both, controlling the BPA directly *and* with the color-killer. Since the DC developed by a subcarrier-oscillator phase detector is a product of the burst signal, the DC control voltage is sometimes taken from that stage.

As an adjunct to ACC, receivers equipped to intercept and process the VIR signal, transmitted by most stations, are automatically controlled in accordance with the quality of colors viewed by the cameras.

Trouble symptoms associated with the ACC circuit include weak color, loss of color, excessive color, or loss of ACC action.

16-10 CHROMA-SWITCHING CIRCUITS

The presence of a color burst, as sensed by the burst gate, indicates reception of a color transmission. Absence of the burst indicates that a BW program is being received. If permitted to reach the CRT, noise generated in the chroma stages shows up as colored snow (confetti) on a BW pix. This is prevented by disabling one of the chroma stages when there's no burst. As shown in Fig. 16-2, the switching circuits include the burst gate, a phase detector, and the color-killer. Any chroma-signal stage may be switched: the BPA, the demodulators, or the color-output stages. As explained earlier, the sensing of a color burst causes a phase detector to develop a DC voltage. This DC voltage is used to control the subcarrier oscillator, the color-killer, and sometimes the ACC. During color reception, it turns off the killer which in turn enables the controlled chroma stage. If there's no burst, the killer is turned on and the chroma stage thus switched off.

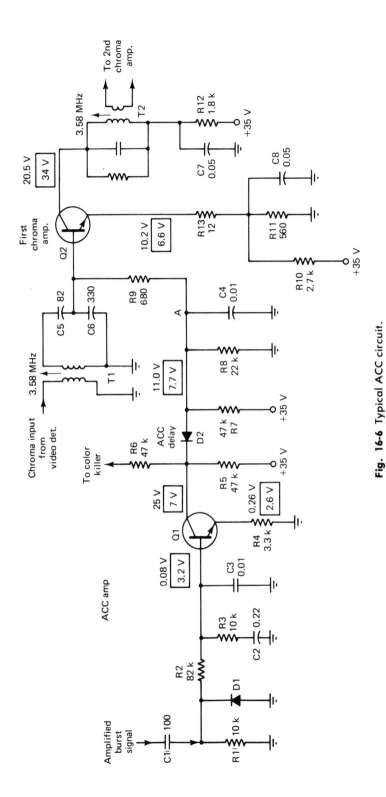

Fig. 16-6 Typical ACC circuit.

732 THE CHROMA SECTION

• The Burst Gate

Sometimes called a *burst amplifier* (its true function), this stage produces a strong color burst that serves two purposes: color sync and chroma-switching. The signal at the burst-gate input consists of chroma IF sidebands and the color burst. With the burst gate operating properly, only an amplified burst signal is present at its output. The chroma is eliminated in two ways: by limiting the bandpass to about 0.5 MHz, and by gating (keying) the transistor to conduct for the duration of the burst signal only, which occurs during horizontal retrace. Gating is accomplished by high-amplitude horizontal-sweep pulses that are shaped and *delayed* to coincide with the arrival of the burst signal that comes later. This timing is *critical,* as becomes evident by loss of color sync when a horizontal-hold control is even slightly misadjusted.

A typical burst gate has a double-tuned transformer at its output that couples the amplified color burst to the subcarrier-oscillator phase detector as shown in Fig. 16-2. It is tuned to 3.58 MHz. Adjustment is fairly critical since a strong color burst is essential for good color sync, for chroma-switching, and, in some cases, for strong color, especially in a weak-signal area.

• The Color Killer

A color killer can be likened to an electronic switch. It is made to conduct (turned on) during BW reception when there is no color burst to develop a DC control voltage. Under these conditions the killer develops a DC voltage that over-biases the controlled chroma stage to cutoff. During color reception the presence of a burst signal results in a DC control voltage that turns the killer transistor off and thus the chroma stage on. The DC control voltage comes from rectification of the burst signal in one of three circuits: the oscillator-phase detector, a killer-phase detector, or a simple diode rectifier. The amplitude of the DC is proportional to the strength of the color burst. The operation of a typical phase detector (discriminator) is described in Sec. 8-6. In some cases, the killer is controlled by the output of an ACC amplifier as described earlier. Like the burst gate, the killer is keyed by horizontal-sweep pulses to conduct only during the burst interval. Unless they were gated, strong noise impulses might trigger the chroma stage into conduction when only BW was being transmitted.

Most early sets and some current models have a *killer threshold control.* This service adjustment establishes the operating point where the killer is activated by the presence or absence of the color burst. Its adjustment is quite critical, especially in weak-signal areas. Normal setting is at a point where colored snow barely disappears on a blank channel. At this setting there should be strong color on the weakest channels and no snow on a BW pix.

A set having a *burst-driven oscillator,* which develops an output *only* when a color burst is received, doesn't require a killer stage. When receiving BW, the oscillator is inoperative, and with no CW input, there is no color output from the demodulators.

A typical chroma-switching circuit is shown in Fig. 16-7.

16-11 THE CHROMA DEMODULATORS

All colors are derived from just two signals developed at the station. These are the I and Q sideband signals produced by modulating the 3.58 MHz subcarrier with the camera outputs. Each sideband by itself represents certain colors; other colors are represented by information contained in *both* sidebands. Every color corresponds with a specific phase angle, i.e., the phase difference between the chroma information and the subcarrier (see Fig. 16-8). Although the phase of the subcarrier is fixed, the

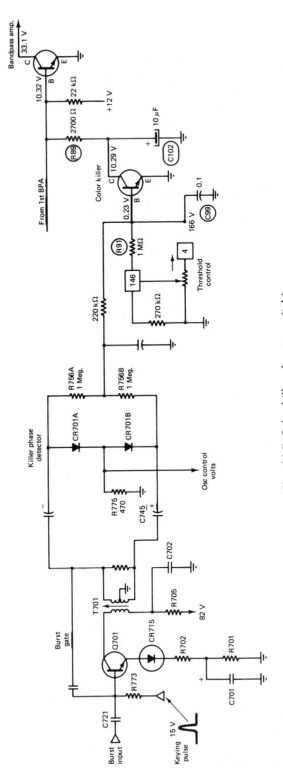

Fig. 16-7 Color-killer chroma-switching.

734 THE CHROMA SECTION

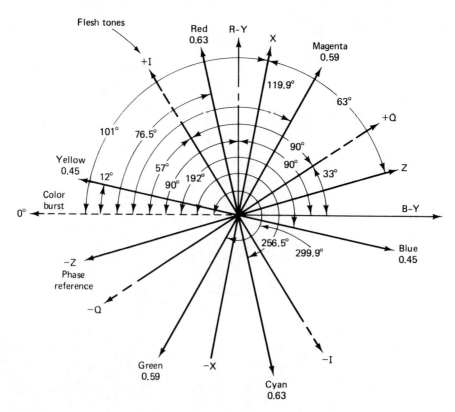

Fig. 16-8 Color phase diagram. Demodulation axes are chosen to emphasize certain colors.

phase angles representing the various hues are constantly changing according to the colors viewed by the cameras. Figure 16-8 also shows the phase angles for the I and Q axes, which are 57° and 147° respectively, a separation of 90°. Each color modulator at the station has an output that varies in *amplitude* in accordance with the applied chroma information. The *vector sum* of the two modulator outputs varies in both amplitude *and phase*. Hue is determined by the instantaneous phase angle of the vector, saturation by the instantaneous amplitude.

Operations at the receiver are performed in a reverse order and manner from those at the station. Here the color information contained in the I and Q sidebands is recovered and used to drive the CRT to produce the original colors viewed by the cameras. Demodulation requires a subcarrier to replace the subcarrier that was not transmitted. A receiver may have either two or three demodulators, each requiring two inputs: the I and Q chroma, and a subcarrier signal. The phase of the subcarrier is different for each demodulator.

Early-model receivers demodulated on the I and Q axes, the same as used at the station, but this method is no longer used. Receivers may demodulate on any number of axes (different phase angles) depending on receiver design—specifically, on the type of demodulator (several are in common use). The chief difference among them is the angle of the two axes. The phase angles used are established by the phased outputs of the subcarrier oscillator. In addition to hue, each phase angle vector (Fig. 16-8) corresponds to a certain amplitude. This is a con-

16-11 The Chroma Demodulators

sideration in receiver design where a particular demodulator may be used to compensate for variations in the light-producing efficiencies of the CRT screen phosphors used. In Fig. 16-8, the color red, for example, corresponds with an angle of 76.5°, about mid-way between the +I and R-Y axes.

- **Demodulator Input Signals**

Each demodulator is supplied with two signals: the I and Q chroma signal, and a locally generated subcarrier signal. The 3.58 MHz signal developed by the subcarrier oscillator is a CW (unmodulated) signal of constant amplitude and phase. Delay networks at the oscillator outputs produce the necessary phase shifts required by each demodulator. Phase angles vary according to the type of demodulator system in use.

The chroma input signal is the same I and Q modulation sidebands developed by the station modulators. Each sideband represents certain colors. Colors associated with the I sideband are red, and all its secondary colors, namely yellow, magenta, and variations of these. Colors associated with the Q sideband are blue, the secondary colors cyan and magenta, and their variations.

REMINDER:

Green was not transmitted as such since it can be reproduced by a mixture of red and blue. As stated earlier, each color corresponds to a specific phase angle. This is the angle *relative to the subcarrier,* and not to be confused with the phase difference between the two demodulation axes.

- **Demodulator Output Signals**

Each demodulator produces an output for two different angles, zero degrees and 180°. At whichever of these angles is used, the maximum output voltage of an R-Y demodulator represents the color red. Other angles, corresponding with all the red-derived hues, results in voltages of varying amplitudes. The same is true for the B-Y demodulator where the highest voltage, at either 0° or 180°, corresponds to blue, and other blue-related hues are represented by somewhat lower voltages. Colors requiring a mixture of both red and blue require outputs from *both* the demodulators.

The color produced on a CRT screen at any given instant is determined by the polarity and relative amplitudes of the chroma-signal drive to each of the three guns. These R-G-B drive signals are a by-product of the color-difference signals developed by the demodulators. Prior to demodulation, hue is determined by the phase of the signal, color saturation by its amplitude variations. Once demodulated, phase is no longer a factor, and colors are determined by amplitude and polarity.

Primary colors are produced when there's output from only one demodulator at any given time. Secondary colors require an output from two or three demodulators. For example, with an output from the R-Y demodulator only, only the red gun of the CRT is driven to conduction and we see red in the pix. Similarly, either green or blue is produced when there's an output from either the G-Y or B-Y demodulators, respectively. With an output from *two* demodulators, the R-Y and B-Y demodulators for example, both the red and blue guns are made to conduct to produce a secondary color. Assuming the same light-producing efficiency for the red and blue phosphors, equal outputs of the two demodulators and equal drive to the two guns produce magenta. Unequal outputs with R-Y the stronger of the two result in a reddish magenta. Where the B-Y output is stronger we get a bluish magenta. Similarly, other secondary colors are produced when there are simultaneous outputs from any two or three demodulators.

From the above example, we see that hue and saturation are *both* determined by the amount of voltage developed by the demodulators, conducting separately or in combination.

THE CHROMA SECTION

• **Sampling the Chroma**

Each demodulator produces an output by "sampling" the chroma signal, at intervals coinciding with each change in phase. Conduction occurs *briefly* at such times when the chroma signal applied to one input of the transistor is in phase with the CW subcarrier signal applied to the other input. Since the two inputs are 180° apart, the two signals must have opposite polarities. An NPN transistor, for example, must have a positive-going signal at the base input and a negative-going signal at the emitter. The opposite is true for a PNP transistor. An example is shown in Fig. 16-9a, for several sequential time periods. Note the opposite polarities of the two input signals; note also that the output signal is reversed 180° from the chroma input at the base. The signals are shown as a series of pulses. Time period 1: the strong positive-going pulse at the base is aided by a strong negative-going pulse at the emitter; they turn on the transistor and develop a strong negative-going output at the collector. Time period 2: With a negative-going pulse at the base the transistor is turned off; there is no output. Time period 3: the weak chroma pulse produces a correspondingly weak output. Time period 4: No chroma, therefore no output. Time period 5: Strong CW signal opposed by weak negative-going chroma; result is weak output. Time period 6: Strong positive-going CW signal resulting in transistor turn-off and no output. To summarize, the changing phase relationship between the two input signals determines the net bias on the transistor, the amount of conduction if any, and therefore the output. As a further aid in understanding the concept, see Fig. 16-9b which shows chroma sampling for two demodulators, identified as X and Z. The phase angles for this type of demodulator are 102° and 166° with a phase difference of about 64°. Note that the subcarrier applied to the Z demodulator is lagging the subcarrier at the X demodulator by that amount, which also coincides with the angular difference between the conduction periods of the two transistors, represented by the pairs of broken vertical lines. Note the relative amplitudes of the output pulses that coincide with the positive-going pulses of the CW signals. At time period 1, for example, a strong chroma signal produces a strong output from the X demodulator, but only a weak output from the Z demodulator some 64° later. The outputs at other time periods vary with chroma amplitude for whichever demodulator is conducting and by what amount. Since the same chroma is applied in parallel to both demodulators only one chroma wavetrain is shown. Which color is produced by the various pulses depends on the phase angle of the chroma relative to the subcarrier at that point in time (see color phase diagram, Fig. 16-8).

The outputs of demodulators described thus far are identified in accordance with the type of demodulator system in use, for example: R-Y; X or Z: These are color-difference signals, because the Y (brightness) component has yet to be added. Once the Y and chroma have been combined (matrixed) the signals are identified simply as red, green, and blue signals.

• **Matrixing**

The Y signal may be recombined with the chroma in the demodulators, the matrix/output amplifiers, or the CRT itself. If the recombination takes place ahead of the CRT the process is called pre-CRT matrixing. CRT matrixing is seldom used in late-model sets, as explained earlier. Although the Y signal has no effect on *hue*, without it all colors will appear harsh and unnatural and there will be no fine detail in the pix.

REMINDER:

The chroma signals can produce pix images only for frequencies up to 1.5 MHz. The Y signal frequencies up to about 4.1 MHz are responsible for the fine detail. The finest details

16-11 The Chroma Demodulators

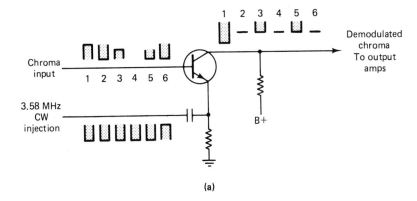

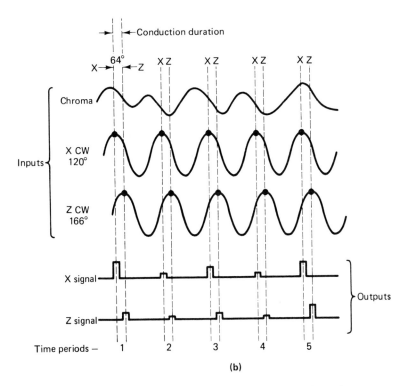

Fig. 16-9 Chroma-sampling: (a) typical demodulator.

are not produced in color because we see them only as variations of B and W.

When matrixing takes place in the demodulators, each demodulator will be supplied with *three* inputs, the chroma, the subcarrier, and the Y signal. After the reinsertion of the Y signal, the outputs are no longer called color-difference signals. They now exist as the true red, green, and blue signals required to drive the CRT and produce a correct pix.

738 THE CHROMA SECTION

• Developing the Green (or Blue) Signal

Because green is not transmitted, it must be developed at the receiver by combining proportionate amounts of the red and blue signals. These proportions are: 0.51 (R-Y) and 0.19 (B-Y). With a separate G-Y demodulator, a G-Y output is developed by sampling both the R-Y and B-Y input signals. When only two demodulators are used, green is obtained by sampling the outputs of R-Y and B-Y amplifiers following the demodulators.

Some early model sets used a R-Y demodulator and a G-Y demodulator to produce a B-Y signal in the same manner as described above. Such an arrangement is possible since any color can be developed simply by selecting the appropriate phase for the subcarrier injection.

• High- and Low-Level Demodulation

High-level demodulation is being used if the demodulators are preceded by several stages of chroma amplifiers. If the output of the demodulators is sufficiently strong the CRT may be driven *directly*, without any additional amplification. Most sets, however, have at least one power output stage, since a typical CRT requires between 80 and 100 V of signal drive. Most modern sets employ high-level demodulation.

By comparison, low-level demodulation takes place at low signal levels, usually after one or two BPA stages. With this system, additional amplifiers are always required after demodulation.

16-12 DEMODULATOR SYSTEMS

It is possible to demodulate on any two axes, representing two different phase angles, as established by the phased outputs of the subcarrier oscillator. Current receivers use phase angles from as low as 50° to as much as 130°. Figure 16-10 shows six different demodulator systems. Some use two demodulators, some use three. Other differences are how the green signal is developed and where matrixing with the Y signal takes place. The *main difference* that distinguishes one system from another, however, is the choice of phase angles. Regardless of the system used, the results are the same.

The choice of a demodulator system is a design consideration based on such things as the relative sensitivities of screen phosphors and which colors therefore need emphasizing, and production costs. A demodulator can be designed to correspond with any desired hue. To obtain a strong red for example, requires an R-Y demodulator supplied with the same subcarrier phase that corresponds with the color red at the station, namely, 90°. Similarly, to emphasize other colors, using appropriate phase angles, provided by the oscillator.

Each of the systems shown in Fig. 16-10 is identified by the type of signal at the demodulator outputs; for example, an R-Y demodulator produces an R-Y output, an X demodulator develops an X-designated output, a green demodulator produces a *G* output, and so on. Phasor diagrams are shown with each system to indicate the phase angles in each case. Such angles are considered approximate and vary somewhat with different receivers. Note that systems using two demodulators sample the chroma twice for every 360° rotation that corresponds with a complete cycle of the subcarrier. With three demodulators, sampling occurs three times for each cycle.

• I and Q Demodulator

Figure 16-10a, shows an I and Q demodulator. Although currently obsolete, it is included here for purposes of comparison. The symbol *I* refers to an in-phase subcarrier of zero degrees, and Q is identified with the quadrature phase of 90°. Note that the phase angles actually used, 57° for

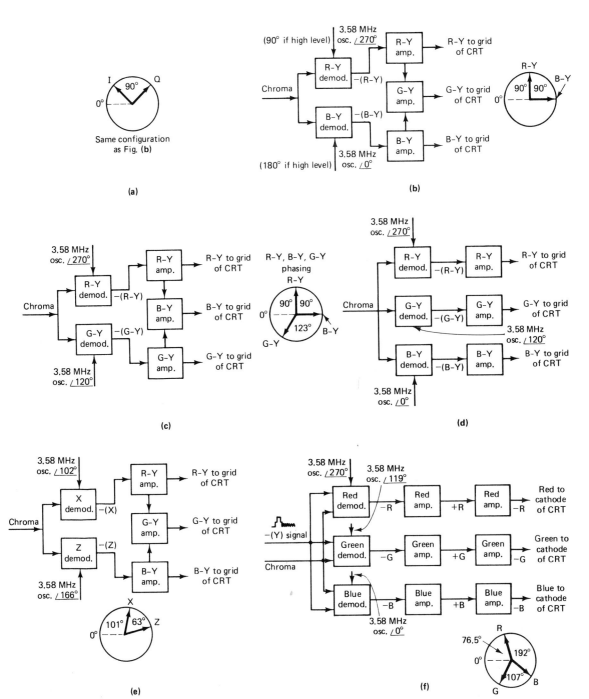

Fig. 16-10 Chroma-demodulator systems: (a) I and Q phasing; (b) R-Y, B-Y phasing; (c) R-Y, G-Y phasing; (d) R-Y, G-Y, B-Y phasing; (e) X-Z phasing; (f) R, B, G phasing.

From:
Alvin A. Liff, *Color and Black & White Television Theory and Servicing*, ©1979, p. 631. Reprinted by permission of Prentice-Hall, Inc., Englewood Cliffs, NJ.

the *I* sideband, and 147° for the Q sideband, a separation of 90° is the same as at the station.

- **R-Y/B-Y Demodulator**

Figure 16-10b shows an R-Y/B-Y demodulator system. Like the I and Q system, the angles here are also in quadrature. Two sets of angles are indicated: zero and 270° for low-level demodulation, and 90 and 180° for high-level demodulation. As can be seen from Fig. 16-8, the R-Y axis represents colors ranging from red to bluish green. The B-Y axis represents colors from blue to greenish yellow.

REMINDER: _____

A demodulator output is maximum when the phase of the chroma is the same as the phase of the applied subcarrier; correspondingly lower outputs for other phase angles represent the various secondary colors. This means that at zero or 270° we get maximum R-Y output from the R-Y demodulator, and at 90 or 180° we get maximum B-Y output from the B-Y demodulator. These two signals of course represent *fully saturated* colors; the brightness component has yet to be added.

With this system, a G-Y signal is obtained by combining samples of the demodulated R-Y and B-Y signals in a G-Y output amplifier as shown.

- **R-Y/G-Y Demodulator**

Figure 16-10c shows an R-Y/G-Y system that is sometimes used. Although the circuitry is essentially the same as in (b), conditions are reversed. Here the *R-Y* signal is demodulated by using a phase angle of 120°. The B-Y signal is obtained by combining samples of R-Y and G-Y signals in a B-Y amplifier.

- **R-Y/G-Y/B-Y Demodulator**

In Fig. 16-10d, *three* demodulators are used, each injected with a different subcarrier phase. Using an angle of 120° that corresponds with green, the G-Y signal is obtained direct from the chroma input signal.

- **X/Z Demodulator (Fig. 16-10e)**

For reasons of simplicity and economy, this two-demodulator system is used in many receivers. The terms X and Z are arbitrarily chosen to differentiate from other systems. It is a *non-quadrature* system using phase angles *greater* than 90°, although they are separated by only 63°. Angles around 102° and 166° are chosen to reduce crosstalk interference between the two demodulators. In Fig. 16-8 note that the X angle is only 11° from the color represented by the R-Y axis (yellow), and the Z axis is close to the green-yellow vector. The G-Y signal is obtained by combining the two outputs in a G-Y amplifier as in (b).

- **RGB Demodulator (Fig. 16-10f)**

In this system, matrixing can take place either in the output amplifiers or in the demodulators (the latter is shown in the example). Like the chroma, the *Y* signal is fed in parallel to each of the three demodulators. Thus the demodulator outputs can no longer be called color-difference signals. They are the red, green, and blue signals required by the CRT. In the other examples shown, where the CRT is driven with ''-Y'' signals, matrixing takes place in the CRT. In the example (f), each demodulator is followed by a driver amplifier and a power output stage. Since both the chroma *and the Y signals* make use of all stages shown, a defect at any point will have an effect on both color *and BW* pix reproduction.

16-13 TYPICAL DEMODULATOR CIRCUITS

The X demodulator of a typical X/Z system is shown in Fig. 16-11a. The Z demodulator (not shown) is identical to the X demodulator except for a different phase of the subcarrier injection voltage. Chroma is supplied to both demodulators via the tuned transformer T1. The subcarrier, at *zero degrees* is injected into the emitter. The two signals have approximately equal amplitudes. With both the base and emitter grounded, the transistor is operating class *B* with a zero no-signal bias.

As with all color demodulators, the transistor is switched on and off by the subcarrier signal, and conduction occurs for only brief periods during the peaks of its negative half-cycles. Some 63° later the Z demodulator is made to conduct to sample the chroma in the same manner. The demodulator output which appears across the load resistor R4 is in the form of short-duration pulses at a 3.58 MHz rate. The output level varies in accordance with both the phase and amplitude of the applied chroma signal. Capacitor C1 by-passes any residual 3.58 MHz subcarrier signal to ground.

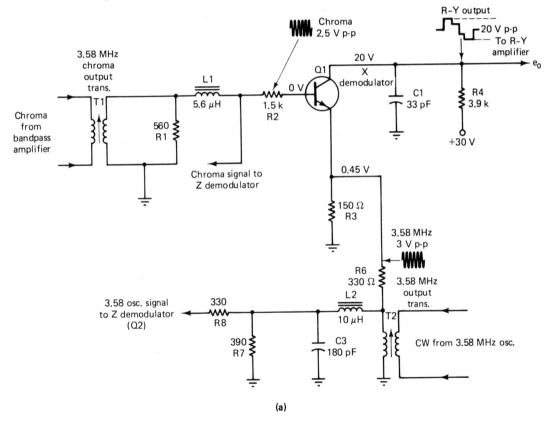

Fig. 16-11 (partial)

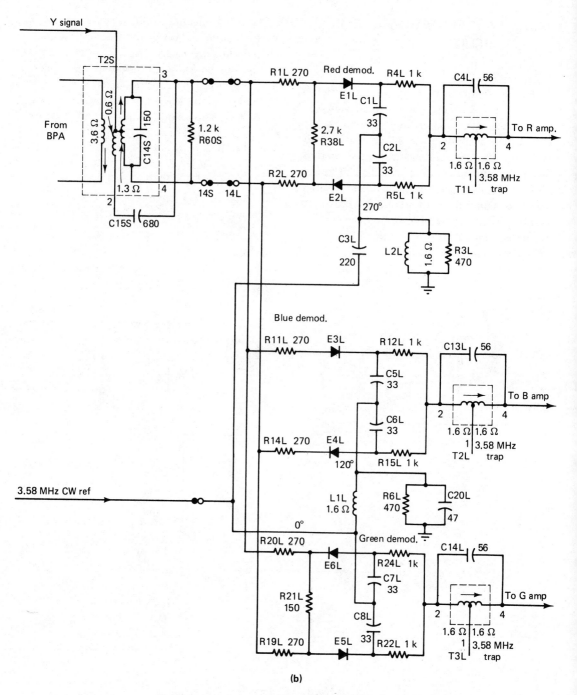

Fig. 16-11 Typical demodulator circuits. Note: The Z demodulator (not shown) is identical to the X demodulator.

The Diode Demodulator

Instead of transistors, many sets use diodes to demodulate the chroma signal. A typical circuit, as shown in Fig. 16-11b, uses six matched diodes, two for each demodulator. It is sometimes called a "balanced demodulator" system because most of the 3.58 MHz subcarrier is cancelled out by "push-pull" action of the diodes. This circuit uses the RGB system; matrixing takes place in the demodulators as in Fig. 16-10f. Note the names of the demodulators as distinguished from color-difference type demodulators. Except for the difference in subcarrier phase, each of the three demodulators is identical, so only one need be described at this time.

A dual-diode type color demodulator is actually the same as any *phase detector* such as used for horizontal AFC, color AFPC, and the like. As in other applications, the phase of two signals is compared, in this case, the chroma and subcarrier. In this circuit, the chroma from the BPA is applied to each diode from opposite ends of the BPA output transformer, and the subcarrier signal is injected at the junction of two small capacitors, C1L and C2L. In some circuits the two signals are applied in an opposite manner, with two subcarrier inputs and one chroma input. Results are the same in either case.

Whenever the chroma phase is the same as the subcarrier phase, the two diodes conduct, rectifying the peak levels of the chroma to produce an output that is passed on to the next stage. As peak rectifiers, the diodes conduct only briefly, to sample the chroma during alternate half-cycles of the subcarrier. Maximum output from the red demodulator (which is supplied with a subcarrier at 270°), is obtained when the phase of the chroma is either 270 or 90°. At zero or 180° there will be no output. At all other angles, the output will be somewhere in between.

Matrixing is accomplished by injecting the *Y* signal at the center tap of the BPA output transformer where it combines with the chroma input to each demodulator. Note the bridged-T type 3.58 MHz traps at the output of each demodulator to remove any residual subcarrier signal, which might produce 920-KHz beat interference.

16-14 AUTOMATIC TINT CONTROL (ATC)

Due to errors in transmission, the tint (hue) of colors tends to change from one station to the next, and also during scene changes, especially when cameras are switched going to and from commericals. Some changes are subtle and can be ignored; other changes may be severe, such as a face suddenly becoming green or purple. Clearly there is a need for a front-panel tint control. ATC, now provided on most sets, helps to overcome the inconvenience of having to make periodic adjustments. (Don't confuse ATC with *ACC,* a completely different function.)

Numerous ATC circuits have been devised. Currently, ATC is accomplished either by widening the demodulator phase angle (the difference between the applied subcarrier phases), or by altering the phase of the chroma between the BPA and the demodulators, or by doing both. A representative circuit of the former method is shown in Fig. 16-12.

Except for the added components, this is a typical X-Z demodulator. Transistors Q3 and Q4 are part of a phase-shifting network. They are enabled/disabled by an ATC "defeat switch." The subcarrier reference signal is injected into the X demodulator emitter at an angle of 70° and into the Z demodulator emitter at 170° (a difference of 100°), with the switch at OFF and Q3 and Q4 disabled. When the switch is closed, Q3 and Q4 are made to conduct, connecting phase-shifting components L2, R7, R8, and C2 into the circuit. They widen the phase angle to 130° by reducing the R-Y angle to less than 90° and increasing the B-Y angle to more

744 THE CHROMA SECTION

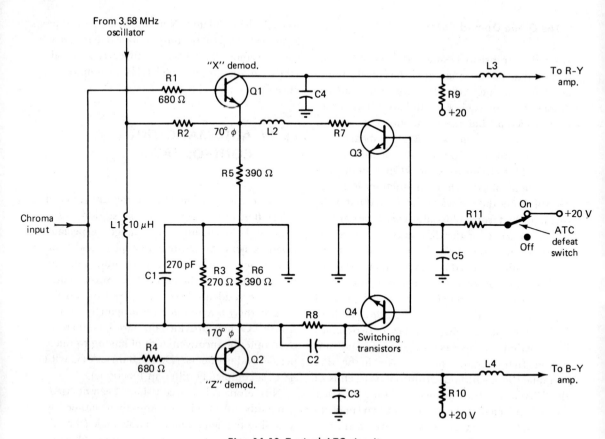

Fig. 16-12 Typical ATC circuit.

than 180°. The effect is to increase the orange output of the X demodulator and shift the Z output toward blue. The net result is an improvement in flesh tones, but at the cost of some sacrifice to the fidelity of other colors. In Fig. 16-8, note that flesh tones correspond with an angle of about 57°.

For sets equipped with VIR sensing, tint is automatically controlled in other ways.

16-15 THE CHROMA OUTPUT CIRCUITS

The output from demodulators is quite low, seldom more than 5 V p-p, and additional amplification is required to boost the level to 100 V or so, the amount of drive required by a typical CRT. In most cases, each demodulator is followed by a *driver amplifier* with a gain of about × 10, then a power output stage. With CRT matrixing, the output stages are called R-Y, G-Y, and B-Y amplifiers. With pre-CRT matrixing, they're called R-G-B output amplifiers. Block diagrams of both systems are shown in Figures 14-15 and 14-16. With CRT matrixing, the demodulated color-difference signals individually drive the CRT control grids, and the three cathodes are driven in parallel by the Y signal. With pre-CRT matrixing, matrixing takes place in either the demodulators or the output amplifiers and the *combined signals* drive either the grids or cathodes (but not both).

16-15 The Chroma Output Circuits

The circuit of a typical chroma output section is shown in Fig. 16-13. Here the three output transistors have fairly high gains, so no driver stages are required. As in most sets, direct coupling is used between the demodulators and the CRT, eliminating the need for DC restorers at the CRT inputs. Pre-CRT matrixing is used with the Y signal injected at the emitters of the three transistors. The demodulator is an IC with four inputs and three outputs. Note the signal levels and polarities at the different circuit points, and the three drive controls for regulating the drive to each of the three guns. For the waveforms indicated, the red and blue beams are driven to cutoff with only the green gun conducting to illuminate the green phosphors. Although the three output circuits are identical, a slightly reduced signal to the green amplifier, as compared to the other two, reduces the relative drive to the green gun. This is to compensate for the increased light-producing efficiency of the green phosphors used in this set. In other sets, the drive levels may be the same, or all different.

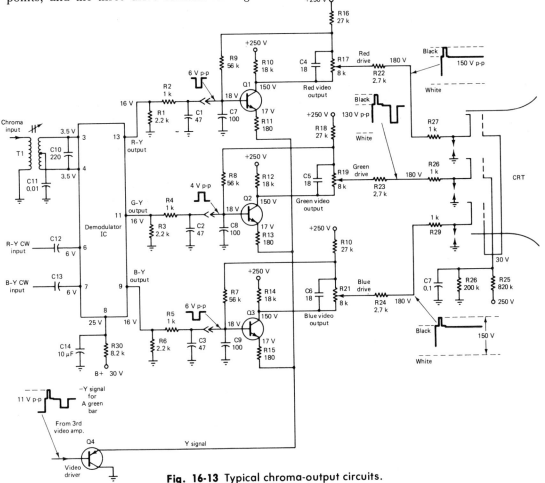

Fig. 16-13 Typical chroma-output circuits.

From:
Alvin A. Liff, *Color and Black & White Television Theory and Servicing*, ©1979, p. 646. Reprinted by permission of Prentice-Hall, Inc., Englewood Cliffs, NJ.

THE CHROMA SECTION

• Developing a BW and Color Pix

The following is a brief summary of the chroma *signal stage* functions and how they contribute to the development of a pix.

At the chroma-takeoff point in the video amplifier, the chroma and video signals separate, each taking a different path to the CRT. The chroma signal, consisting of the I and Q sidebands of the color subcarrier, is amplified at IF frequencies in a multi-stage bandpass amplifier (BPA). Because the bandpass is only 1 MHz, only the chroma signals are amplified and passed on to the next stage, the demodulator(s). Here the color-difference signals are recovered from the two sideband signals. Either two or three demodulators may be used. If only two, usually the green signal is developed by combining R-Y and B-Y signals in the demodulator or the following stage. In some cases, blue is developed by combining R-Y and G-Y signals.

The color-difference signals represent saturated colors and must be matrixed with the Y signal to restore the original brightness information to the colors. Matrixing may occur in either the demodulators, the following output stages, or the CRT itself.

Only the Y signal is used to produce a BW pix; when a color set is receiving BW, the chroma stages are disabled to prevent noise from getting through and producing colored snow. *Each of the three guns* is simultaneously driven by the Y signal to develop a pix in BW, and their relative drive amplitudes are important.

REMINDER: _____

White is a combination of the three primaries in their proper proportions.

Both the Y signal and the three chroma signals are needed to produce a color pix. As explained earlier, with CRT matrixing, the Y signal is applied to the three guns in parallel, and the three color-difference signals to the three guns individually. With pre-CRT matrixing, matrixing takes place at some point prior to the CRT input and the combined signals drive either the grids or cathodes of the CRT. With this arrangement, the chroma output circuits must be operating to produce both a color *and BW* pix.

All colors can be derived from the three primaries (red, green, and blue); the screen of a color CRT is coated with phosphors of all three. The three phosphors emit light, either singly or in combination, when excited by electrons from the three guns. *Which* phosphors are illuminated depends on the amplitude and polarity of the three color-drive signals at any given instant. For example, if the red signal causes conduction of the red gun while the green and blue guns are driven to cutoff, then pix images will be reproduced in red. Similarly, we obtain green if only the green gun is conducting and blue images when only the blue gun is conducting. If two or three guns are conducting simultaneously, all three phosphors are excited in proportion to the relative strengths of the three beam currents (Fig. 15-18). Because the phosphor particles are too small to be seen individually, from a distance we observe them as a mixture, *a different color,* according to the relative intensities of the light emitted from the three phosphors.

Which colors are produced, and their intensities, at any given instant during the scanning of a pix, is established by the demodulators, where the chroma is sampled. The hue of the color is determined by the instantaneous phase angle between the ever-changing chroma and the subcarrier, which has a constant phase. The strength of a color varies with the signal amplitude. The brightness of the reproduced color is dependent on the amplitude of the Y signal at that time.

16-16 THE SUBCARRIER OSCILLATOR

A 3.58 MHz color subcarrier is required to demodulate the chroma IF sidebands. Because the station subcarrier is not transmitted, an equivalent subcarrier must be generated at the receiver. The signal developed by the subcarrier oscillator is an unmodulated (CW) AC voltage of constant amplitude. Its *frequency* and *phase* must be *precisely* controlled. For *frequency stability,* all subcarrier oscillators are *crystal* controlled. Some form of AFPC/APC circuit is used to control the oscillator *phase,* using the received color burst as a reference.

Three circuits commonly used for generating the subcarrier are 1) the basic oscillator, 2) the burst-driven oscillator, and 3) the burst-driven crystal circuit.

• **The Basic Oscillator**

A typical circuit is shown in Fig. 16-14a. As in any oscillator, some form of positive feedback is required to develop and sustain oscillations. Here, part of the transistor output is fed back (in phase) via C727 to reinforce the signal at the base. Some circuits rely on the inter-electrode capacity of the transistor for feedback. The voltage-dividing resistors R727 and R728 provide a small amount of forward bias for oscillator "start up". The component Y701 in the base circuit is an oscillating type crystal for stabilizing the frequency. A crystal can be considered a high-Q "tank circuit" having a natural resonant frequency: in this application, 3.58 MHz. The oscillator frequency is adjustable by means of the tuned output transformer T702. When the oscillator frequency is close to that of the crystal (within about 200 Hz), the crystal takes over to control the frequency.

In this circuit, the oscillator *phase* is controlled by a phase detector in conjunction with *varactor* diode CR725. Note the CW amplifier transistor used to boost the oscillator output to the level required by the demodulators, usually between 5 and 10 V p-p.

• **Oscillating Crystals**

The "crystal" used to control an oscillator consists of a thin wafer of quartz material sandwiched between two flat spring-loaded plates. Conversely to crystals used in some phono cartridges and microphones, this type is made to vibrate mechanically when an AC voltage of the proper frequency is applied to its two terminals (the "piezoelectric effect"). The resonant frequency of a crystal depends mainly on its thickness: the thinner the crystal the higher the frequency. A subcarrier-oscillator crystal is ground with great accuracy to oscillate at precisely 3.579545 MHz. Its frequency can be varied by a small amount by shunting with a variable capacitor or the equivalent, such as a varactor, or a FET reactance-control circuit. Without a crystal, an oscillator would be "free-running" and subject to *frequency drift.*

• **The Burst-Driven Oscillator (Fig. 16-14b)**

Sometimes called an *injection-locked oscillator,* this circuit is crystal-controlled only when a color burst is received. The burst signal from the burst gate causes the crystal to "ring", controlling the oscillator in both frequency and phase. In this circuit the ringing voltage is amplified before being applied to the oscillator. With this configuration, no additional AFPC circuits are required. One serious shortcoming is that sync stability is dependent on the strength of the color burst, and color lock-in can be unstable in a weak signal area, especially where noise is present.

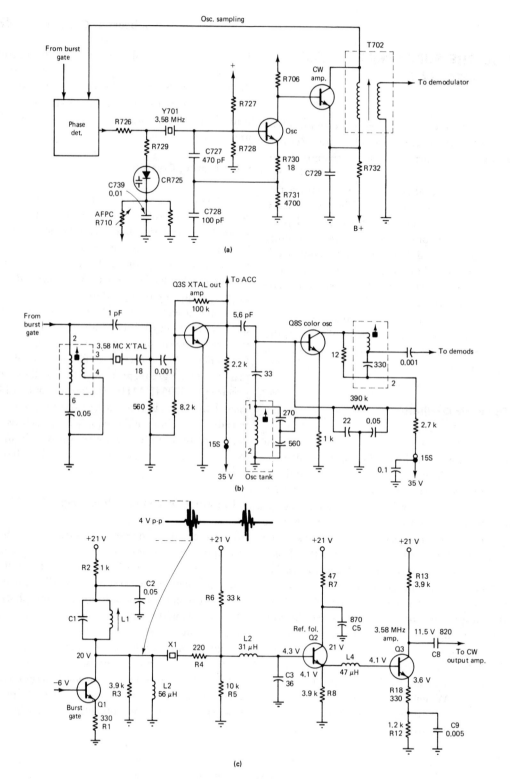

Fig. 16-14 Subcarrier-generating circuits.

• The Crystal-Ringing Circuit (Fig. 16-14c)

Unlike other systems, this circuit *contains no oscillator*. As in Fig. 16-14 b, the color burst from the burst gate causes the crystal to ring at the subcarrier frequency. The ac voltage thus generated is then amplified by two or three stages to serve as a subcarrier for the demodulators. The high "Q" of the crystal enables it to ring continuously between color bursts. Generally, sets using this circuit don't require a killer stage, because there is no color burst when receiving BW and therefore no subcarrier to feed the demodulators. With no subcarrier, the demodulators are disabled and no noise or other interference can get through to the CRT. With this circuit, a weak or missing burst will create problems, but loss of color sync is not possible because there is no free-running oscillator to cause frequency drift.

16-17 SUBCARRIER PHASE-SHIFTING NETWORKS

A subcarrier oscillator has either two or three outputs, depending on the number of demodulators used, and each output is of a different phase. One output may come directly from the oscillator at zero phase (relative to the color burst), while phase-shifting networks condition the other output(s), or a delay may be introduced into all outputs, depending on the demodulator requirements.

Phase-shifting networks make use of coils, transformers, capacitors, and resistors, and sometimes a transistor. In some cases an oscillator-output transformer is used: the opposite ends of the secondary winding are 180° out of phase with each other; if one end is grounded, the other end may be either in phase or 180° out of phase with the oscillator, depending on the direction of the windings. Coils and capacitors are of course reactive and can make one voltage either lag or lead the other, by amounts depending on the components' values. For example, the larger a coil and the smaller a capacitor the greater the delay. Component values are quite critical, since the slightest phase error has an adverse effect on hue which may or may not be correctable with the tint control. Typical delay circuits are shown in Figs. 16-11, 16-12, and 16-14.

• Tint Control

As explained earlier, hue can be controlled in a number of ways: by changing the phase of the chroma signal, the burst signal, or the color subcarrier; and some sets combine the phase-shifting networks with the tint control. In Fig. 16-4, for example, a pot is used to control the effective reactance of a capacitor, and therefore the phase angle at one of the oscillator outputs.

Another method of changing phase is with a resonant circuit. At resonance, the circuit is not reactive and no phase shift occurs. Off resonance, there's either a leading or a lagging phase shift depending whether above or below resonance. The tint control varies the tuning, usually with a pot in series with a small fixed capacitor.

16-18 OSCILLATOR CONTROL CIRCUITS

Normal color sync has been achieved when all colors on a pix are present and of the correct hue. It requires precise synchronization between the subcarrier oscillator and the received color burst, which is a sampling of the subcarrier generated at the station. The subcarrier must be correct in both *frequency* and *phase*. An error in frequency results in a "barberpole" of colored bars. A phase error causes the wrong colors to be reproduced, and alters the "range" of the tint control. If there are no oscillator defects, *fre-*

750 THE CHROMA SECTION

quency is adequately controlled by the crystal. *Phase* control is somewhat more involved. Several methods of phase control (AFPC) are described below.

- **Phase Detector/Varactor Control**

This system, as shown in Fig. 16-14a, uses a *varactor diode* (see Sec. 5-9) to control the oscillator. The varactor gets its DC control voltage from a conventional phase detector. Note the feedback loop, where a sampling of the oscillator is applied to the phase detector for comparison with the color burst. When the two are in phase (the normal condition for color sync), no DC "error" voltage is developed and no change is initiated. When the oscillator either lags or leads the burst signal by some amount, a positive or negative voltage is developed to force the oscillator back into step.

The varactor, which simulates a variable capacitor, is shunted across the oscillator input circuit. The amount of simulated capacity depends on the amount of applied DC. The varactor is normally reverse-biased from B+ by the voltage-dividing resistors, as shown. One of the resistors, called the AFPC control, is adjustable. Its setting establishes color sync when the DC is zero. In some sets this pot serves as a tint control, as it indirectly controls the phase.

If the oscillator lags the burst by more than 90°, a negative DC is developed, reducing the varactor capacity and hence making the oscillator speed up. Conversely, a lag of less than 90° develops a positive DC and an increase in varactor capacity, to make the oscillator slow down. The greater the phase error, the greater the DC correction voltage of either polarity.

- **Phase Detector/FET Control**

Instead of a varactor, some sets use a FET, operating as a simulated capacitive reactance. When the drain current is made to lead the AC drain voltage by 90°, the oscillator "sees" the FET as a capacitor. The DC error voltage from the phase detector controls the amount of voltage delay, and therefore the amount of simulated capacitive reactance. With the FET connected across the oscillator input, this action results in the necessary phase correction.

- **Burst-Injection Oscillator Control See Fig. 16-14 b)**

This circuit doesn't require a phase detector, because the oscillator phase is directly controlled by the applied color burst after it has been amplified. This is a "brute force" system and any attenuation of the burst signal has an adverse effect on color-sync stability.

- **Crystal-Ringing AFPC (Fig. 16-14 c)**

As in the circuit shown in Fig. 16-14b, no phase detector or varactor is required. With this circuit, color sync is automatically correct at all times, because the color burst itself takes the place of the usual oscillator-generated subcarrier. This circuit was used mostly in older-model sets; as in the previous circuit (b), stable color sync is dependent on a strong signal.

For optimum performance, the tuned circuits of all AFPC systems must be accurately and critically aligned.

16-19 CHROMA ICS

Practically all modern sets use one or two ICs to perform all chroma functions. Figure 16-15 shows a 24-pin IC that contains all the chroma circuits except the output stages. The output stages are not included because they require power transistors and other large heat-producing components. The IC contains most of the circuitry for the following stages: BPAs, demodulators, chroma pre-amps, burst gate, killer, oscillator, and AFPC. Large components such as controls and tuned circuits are external

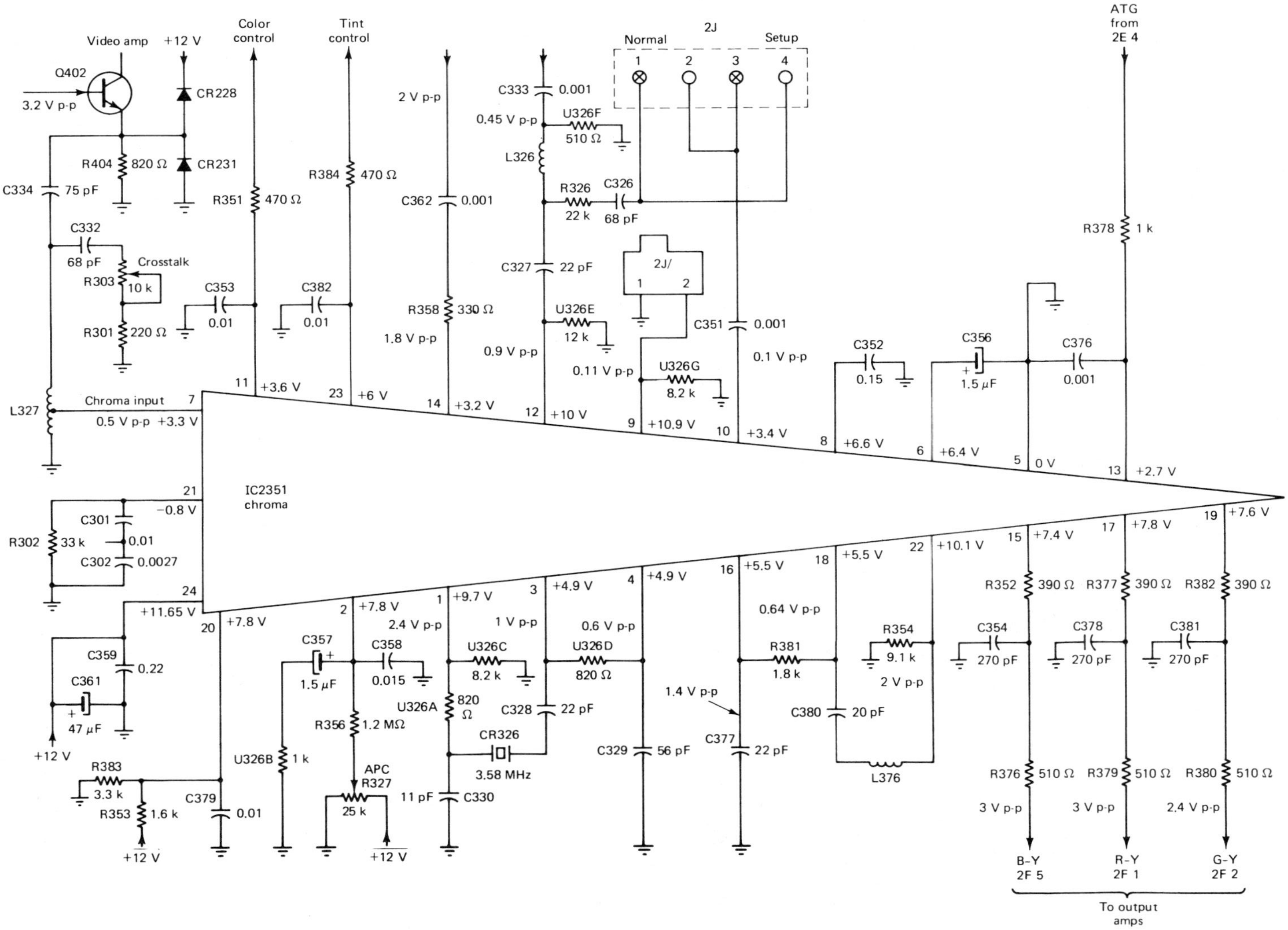

Fig. 16-15 Typical chroma IC.

752 THE CHROMA SECTION

to the IC. There is only one input to the IC, the composite signal from the chroma takeoff. There are three outputs, the R-Y, G-Y, B-Y signals that drive the RGB output stages.

Note the absence of the numerous tuned circuits normally associated with chroma stages. Other things to note are: the chroma takeoff coil, which is not adjustable; the "crosstalk" control, which adjusts for bandpass tilt at the BPA input; the APC control, which sets the bias of the oscillator-control varactor; coil L326, in one of the BPA stages—it is the only *adjustable* tuned circuit; and the crystal. The components connected to pins 16, 18, and 22 comprise the phase-shifting networks at the oscillator output. The color and tint control (not shown) operate as *dc* controls for their respective functions.

- **Another Typical Chroma Processing Circuit**

Sample Circuit D at the back of this book also uses an IC for most of the chroma functions. Note the functional blocks within the IC (IC 300) and their associated components external to the IC. The chroma takeoff (at the video input to the signal module) feeds composite signal to the BPA via pin 15 of the IC. Instead of a comb filter, a fixed-tuned filter separates the chroma and video. Unlike the previous circuit, there is no crosstalk control. From the BPA, the amplified *chroma only signal* goes to a color gain amplifier, then via phase shifting networks to the demodulators. The color control varies the signal amplitude with DC applied to the color gain amplifier via pin 20 of the IC. The tint control which connects to pin 2 of connector PG 6 (at lower left of the diagram) controls the hue by changing the DC applied to the collector of Q 350. The demodulators are also supplied with two phases of the subcarrier oscillator. The three outputs of the demodulators then feed the RGB output amplifiers (where matrixing takes place), and from there to the CRT cathodes.

For control of the color killer and the subcarrier oscillator, the composite signal (at TP 30) is amplified then applied to the burst gate along with a delayed horizontal gating pulse. The burst signal at the burst gate output feeds the color gain amplifier which is controlled by the color killer to prevent snow on the screen during BW reception. The APC control block samples an output from the subcarrier oscillator and is controlled by DC via the APC control R 320, to stabilize the oscillator in both frequency and phase relative to the incoming color burst. The 3.58 MHz crystal connects to pin 6 of the IC.

16-20 TROUBLESHOOTING THE CHROMA SECTION

Despite the great number of interrelated circuits, the variety of trouble symptoms, and the many possible causes of each kind of trouble, this section of the set is no more difficult to service than other sections—perhaps even less so, everything considered. In fact, most color-related troubles can be diagnosed by adjusting controls, by observation, and by accurate interpretation of what appears on the TV screen. After the preliminary diagnosis, tests are made to localize the fault, followed by further tests to pinpoint the defect. For many color problems there is no actual defect and troubles may be cleared up with proper adjustments and alignment of the APC circuits. For sets using ICs, troubleshooting is limited to checking input/output signals, DC voltages, and external components. With ICs, a proper schematic is a *must*.

16-21 SYMPTOMS ANALYSIS

Time spent in evaluating symptoms greatly contributes to troubleshooting efficiency; the CRT screen provides the clues. A good *raster* and a good *BW pix* are prerequisite to a good color pix, so *each* must be established, and any problems corrected *in that order*. While assessing the observed symptoms, note the response as the following controls are adjusted: the fine tuner, the color and tint controls, and the killer threshold.

• **Colored Snow**

For loss of color, a common problem, check for the presence of colored snow (confetti) when tuned to receive color,....and the absence of colored snow on a raster and BW pix. Snow is caused by "noise" generated in the front end of the receiver, or background interference picked up by the antenna. It is most bothersome in weak signal areas where the signal-to-noise ratio is low. For loss of pix on a BW set, the presence of snow indicates an uninterrupted signal path from the antenna input to the CRT. Excessive snow is associated with a weak signal or with AGC or tuner troubles. Little or no snow indicates trouble in the IF stages, the detector, or the video amplifier.

Noise reaching the CRT via the chroma stages produces *colored snow*. To prevent such interference on a BW pix, the chroma signal path is interrupted by the color-killer. A malfunctioning killer stage is a common cause of no color pix.

For loss of pix, one of the first steps is to check for snow. The absence of colored snow (assuming a normal BW pix) usually means the chroma signal path is interrupted. The three most likely possibilities are: there is a defect in the BPA; the BPA is disabled by the killer or ACC circuits; or the subcarrier-oscillator signal is missing. Colored snow should be *sharp* and *grainy;* if fuzzy and blurred, suspect BPS misalignment.

Many color problems, including loss of color, can be caused by a *weak signal,* even when BW reception is acceptable. Check the antenna system and try a substitute. When in doubt, use a bar generator for comparison. Adjust the AGC threshold as appropriate. Trouble symptoms, and suggested procedures for each, are shown in the flow drawings of Fig. 16-16 and Table 16-2.

Table 16-2 Color Troubleshooting Chart

Trouble Symptom	Probable Cause	Remarks/Procedures
No color (BW pix ok)	Weak signal. Dead BPA or subcarrier oscillator. Defect involving *both* demodulators or output stages. BPA turned off by defect in ACC or killer stage. Killer threshold control turned down. Loss of color burst, if burst-driven oscillator. Severe misalignment. Defective comb filter.	Scope-trace station signal or bar signal through BPA. Check for oscillator injection to demodulators. Disable killer and ACC. Turn up color and threshold controls. Check for a dead oscillator. Check BPA and video IF alignment. Check CW oscillator amplifier and phasing networks.
Weak color	Weak BPA stage. Weak oscillator signal. Other causes as above.	Check for weak stages rather than a dead stage.

Table 16-2 (cont'd)

Trouble Symptom	Probable Cause	Remarks/Procedures
Excessive color	Defect in killer, ACC or BPA stages. Video IF or BPA misalignment.	Defeat killer and/or ACC. If normal color restored, troubleshoot these stages. Check alignment.
No BW or color pix (raster ok)	Loss of signal between antenna input and the chroma takeoff. AGC defect or AGC threshold turned down.	Turn up AGC threshold control. Try clamping AGC bus. Signal trace tuner, IF, detector, and first video-amplifier stages. Check for snow.
Colored snow (on raster and BW pix)	Killer-circuit defect. Killer threshold control misadjusted. Defective phase detector.	Adjust threshold and troubleshoot killer stage. Check DC output of phase detector.
Color background on BW pix. No color pix. Raster ok	Dead oscillator or CW amplifier.	Check for oscillator signal.
No BW pix. Harsh color pix with no detail	Loss of Y signal between chroma takeoff and the CRT, or matrix amplifier. Defective comb filter.	Signal-trace video stages downstream from chroma takeoff or comb filter.
Colors missing (raster and BW pix ok)	One demodulator or one output amplifier defective. No oscillator injection to one demodulator. Phase-shift problems.	Determine which colors are missing and troubleshoot appropriate demodulator or output amplifier.
Some colors weak	One demodulator or output amplifier is weak. Low oscillator injection to one demodulator.	(as above)
Wrong Colors (raster and BW pix ok)	Defects in demodulators or output amplifiers. Wrong oscillator phase. Misalignment.	Determine which color(s) are wrong and check appropriate demodulator(s) or amplifiers. Check IF and BPA alignment. Check phase-shifting networks.
Erratic color changes	Unstable purity/convergence conditions, or defec-	If only color pix affected, check for oscillator drift

Table 16-2 (cont'd)

Trouble Symptom	Probable Cause	Remarks/Procedures
	tive CRT. Video or chroma at burst-gate output. Demodulator troubles. ATC circuit defects. Loss of keying pulses to burst gate.	or unstable APC circuits. Check demodulators and phased outputs of oscillator. Check ATC functions. Check gating pulses.
Poor flesh tones	Weak red gun of CRT. Insufficient red drive to CRT. Weak R-Y signal, from demodulator or output amplifier. Misalignment. CRT set-up misadjustments.	Check CRT and set-up adjustments. Check for weak R-Y signal. Check alignment of video IF strip, BPA, and APC circuits.
Loss of color sync	Defect in burst gate, APC circuits, or oscillator. Weak signal. APC misalignment. Loss of burst signal.	If a frequency problem, isolate oscillator and check oscillator tuning. If not an oscillator defect, suspect oscillator-control circuit. If a phase problem, check color burst, phase detector and oscillator control circuit. Perform APC alignment.
Loose (weak) color sync	As above.	As above
Smeared colors	Gassy CRT. Misalignment of video IF, BPA, or APC.	Check for weak or gassy CRT. Check alignment.
Hum bars (color pix only)	Poor filtering of DC feeding the chroma stages.	Determine which colors affected and check filters of corresponding DC power sources.
No automatic color control	Burst gate, ACC rectifier, or ACC circuit defect.	Check burst signal and ACC/killer DC voltages. Try variable bias supply.
920 KHz beat interference on color pix	Misalignment of video IF, BPA or trap circuits.	Adjust traps, and, if necessary, align the IF and BPA circuits.
Intermittent color (raster ok)	All color stages are suspect.	Monitor all chroma and chroma-related signals. Realign APC circuits.

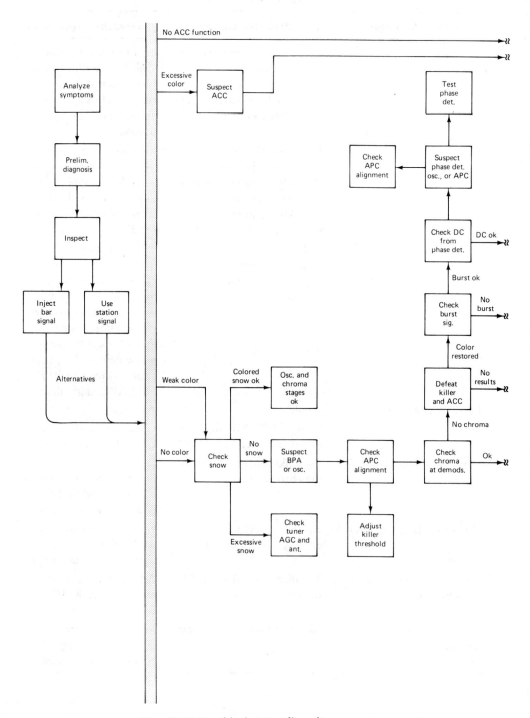

Fig. 16-16 Troubleshooting flow diagram.

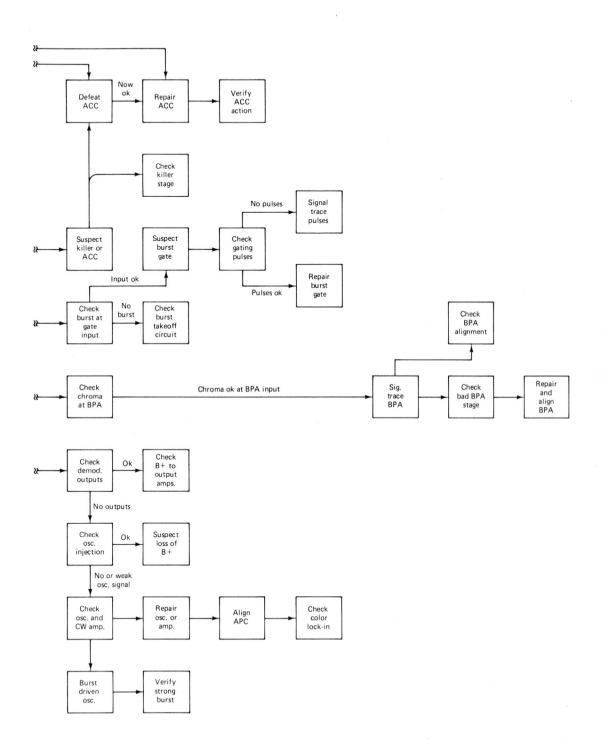

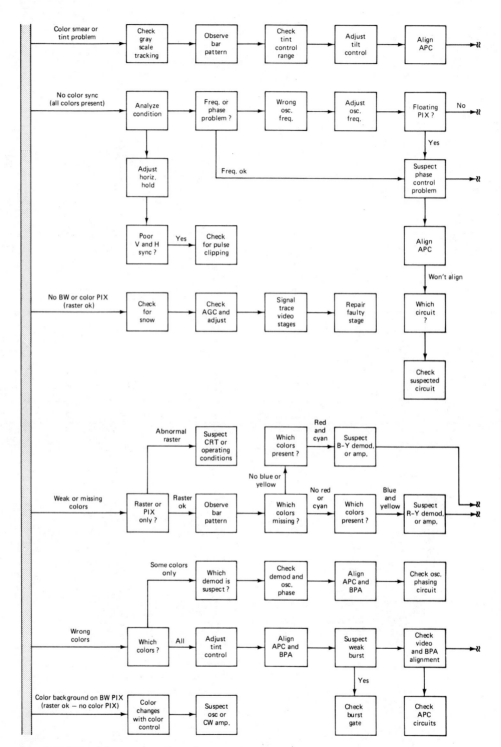

Fig. 16-16 Troubleshooting flow diagram (continued).

758

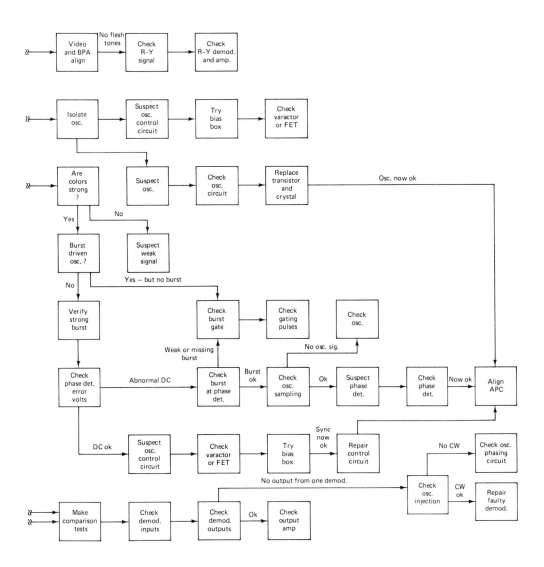

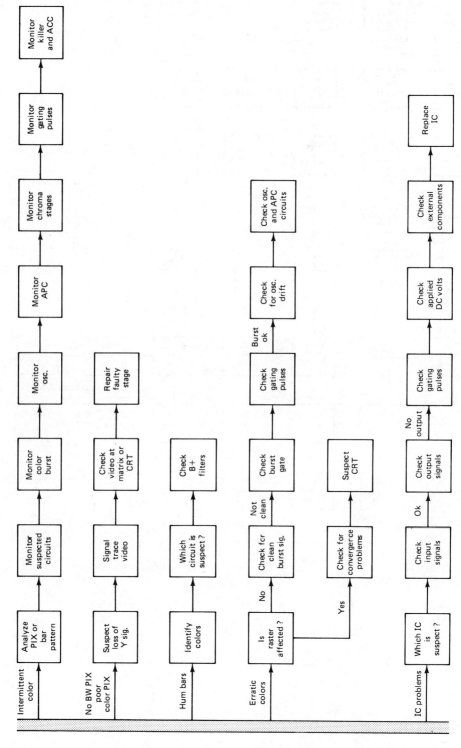

Fig. 16-16 Troubleshooting flow diagram (continued).

16–22 INSPECTION

With set *off,* inspect for obvious defects, poor soldering, cracked PCBs, loose edge connectors, and so on. Examine the schematic, noting all chroma-related stages and ICs. Identify them in the chassis. With set *on,* check for overheating components or power transistors, and high-wattage resistors that run cold.

16–23 LOCALIZING THE PROBLEM

Localizing troubles in the chroma section in the shortest possible time without being sidetracked by too many false leads or misinterpreted test results can be a real challenge. Of greatest importance is an *accurate diagnosis of the symptoms* and what stages could be responsible; next is deciding on what tests to make, and which comes first. Of course no single troubleshooting approach is suitable for all circumstances; it depends on the symptoms, the equipment on hand, and the individual preferences of the technician.

For problems with color, the first step is to evaluate the raster. Check for brightness, perfect focus, and complete absence of color tinting. Evaluating a BW pix is the next step. Check for contrast, resolution, and the absence of background tinting or color outlines around images. Correct any abnormal conditions *before* considering problems with color reception.

- **Signal-Tracing Methods**

The approach to a particular color problem is determined mainly by the symptom(s). Loss of color for example may call for scope tracing the chroma *signal stages* between the point of chroma takeoff and the CRT. For loss of signal in the video stages, any one of several alternate methods may be used. The signal may be either the *station signal,* or a signal provided by a generator. The preferred method makes use of a *color-bar generator* and a CRO. Lacking this equipment and depending on the nature of the tests, other types of generators and indicators may be used with varying degrees of success.

Loss of color sync, as another example, usually calls for scoping the generated subcarrier signal and the color burst, in conjunction with DC voltage checking. Another approach here, and one with some merit, is to go through the APC alignment procedure. (See Sec. 16–37). If a certain adjustment cannot be performed as expected, it can pinpoint the stage or circuit responsible. Most older sets can benefit by an APC touch-up, anyway.

- **Using the Station Signal**

When chroma is lost or attenuated in one of the "signal" stages (BPA, demodulators, or output amplifiers), its progress can be traced with a CRO, AC EVM, or other form of signal tracer. The received color burst can be checked at the input and output of the burst gate in the same manner.

- **Using the Color-Bar Generator**

This is by far the fastest and most accurate method for localizing *and* isolating most defects in the chroma section. Its most common uses include: scope-tracing for weak or missing chroma signals; troubleshooting abnormal color problems, such as wrong colors and missing colors, color-sync problems including APC alignment, and chroma-phase errors; checking relative amplitudes of the three color signals at the CRT input; and checking the range of the tint control.

As described in Sec. 15–63, a color-bar generator produces a phase-modulated RF signal which is usually fed to the receiver antenna terminals. As with a station signal it passes through and is processed by all signal stages up to the CRT. The bar signal, as viewed by a CRO at any stage, is a "comb" pattern consisting of a horizontal sync pulse, a color burst, and ten phase-related pulses. Each pulse corresponds with a particular hue. Scope-tracing the bar signal can proceed either *upstream,* starting at the CRT and working toward the chroma takeoff until a normal signal is found, or downstream from the chroma input toward the CRT until an abnormal display appears. Normal waveforms for various circuit points are shown on most schematics.

Assuming an uninterrupted path through the chroma stages, a color-bar pattern such as that shown in Fig. 15-19 appears on the pix-tube screen. Each bar of a different hue corresponds with one of the 10 pulses of the bar signal, which are separated by 15°. With a rainbow-type generator, the colors range from yellow on the left to green on the right. Bar positions will change with the setting of the tint control. The bar pattern is used in checking colors actually reproduced by the CRT.

In the absence of color, the bars appear in BW and in various shades of gray.

16-24 TROUBLE SYMPTOMS AND STAGE CHECKING

Suggested step-by-step procedures for most chroma trouble symptoms are shown in Fig. 16-16. Further details for each condition are contained in the following paragraphs.

16-25 NO COLOR OR WEAK COLOR

With the color control turned up, check for colored snow on a blank channel or on a BW pix. Strong colored snow (confetti) having a sharp grainy texture indicates that the chroma signal stages (BPA, demodulators, and output amplifiers) are able to pass a signal. It also indicates that the oscillator is working. Excessive snow usually indicates trouble at the front end: the antenna, tuner, or AGC. Colored snow on a BW pix means the chroma stages are conducting at all times and not being turned off by the color-killer. Try adjusting the killer threshold control.

Check BW reception. Assuming a normal BW pix, if there is no color or weak color, all chroma stages are suspect, including the oscillator, the killer, and the alignment of the AFPC. Check the APC alignment (for procedures see Sec. 16-37). Check for chroma signal at the demodulator inputs. If chroma is weak or missing at the demodulator inputs, disable the killer by adjusting the threshold control or by applying an override bias using a bias box. If this restores the color, suspect a faulty killer stage or loss of the color burst.

With the CRO, check for a strong color burst at the burst-gate output. Verify that it's really a burst and not a false indication of interference. An inferior CRO can't always distinguish between the two. Verify that the burst is "clean", with no video or chroma getting through the gate. If there is no burst at the output, check for burst at the input, and if necessary as far back as the burst takeoff point in the BPA. If there's a good input burst but a weak or missing burst at the output, check for a weak or missing gating pulse coming from the horizontal sweep circuit. If burst is OK, check input/output DC volts at the killer transistor. If OK, suspect the ACC circuit. Apply an override bias to see if the BPA can be turned on.

When disabling the killer and ACC *doesn't* restore the color, signal-trace the BPA, and repair if a bad stage is located. If suspected, check for BPA misalignment, but don't attempt alignment without good cause. Alignment procedures are described in Sec. 16-42.

When the CRO shows strong chroma at the demodulator inputs, check for output at each demodulator.

REMINDER: ─────────────

One demodulator not working results in loss of *some* colors only. Loss of *all* color can only be caused by a defect involving *all* demodulators. Also bear in mind that the demodulators cannot pass chroma signals (or snow-producing white noise) unless supplied with a strong CW subcarrier from the oscillator.

If there is little or no demodulator output, check transistor conduction by making DC voltage checks across emitter or collector circuit resistors. If there is no conduction, check for oscillator voltage at each demodulator. If there is no oscillator injection, suspect a dead oscillator (or CW amplifier if there is one).

REMINDER: ─────────────

Assuming that the chroma stages are ok, a change in colored snow on a raster or BW pix when the color control is adjusted indicates that the oscillator must be working. If there is no change, the oscillator is suspect. If it's a burst-driven oscillator, verify injection of a strong burst signal.

If there is no color or weak color, and the demodulator outputs are normal, check for loss of chroma in the matrix or output amplifiers. As with the demodulators, a defect in *one* amplifier only, would not cause loss of *all* color. Look for a "common" trouble like loss of *B* voltage.

REMINDER: ─────────────

In DC coupled amplifiers, a defect will have an adverse effect on the *raster*.

16-26 EXCESSIVE COLOR

This trouble can usually be traced to the killer or ACC circuits. Try backing off on the killer threshold control. See if normal color can be restored by disabling the killer and/or ACC using override bias. If there is no improvement, check alignment of the BPA (insufficient bandwidth), and, if necessary, the *video* IF response.

16-27 LOSS OF ACC FUNCTION

An ACC defect can produce numerous symptoms besides loss of color control. Try defeating the ACC as described for excessive color. Check the color burst and/or the DC control voltage developed by the phase detector or ACC rectifier. See if it varies with the fine tuner. If phase detector is suspect, see Sec. 16-36.

16-28 NO BW PIX, POOR COLOR PIX

If color pix is harsh and completely lacking in detail, suspect loss of the Y signal somewhere between the point of chroma takeoff and the CRT. Check for snow with the color control turned down. Normal amounts of snow indicate that the video amplifier is ok; no snow indicates a dead amplifier stage. Check the comb filter if there is one. Signal-trace the video up to where it's matrixed with the chroma. With CRT matrixing this will be the CRT input; with pre-CRT matrixing, it's at either the demodulators or the output amplifiers. Trouble between the point of injection and the CRT is unlikely because a defect in only *one* demodulator or amplifier will not prevent passage of the Y signal.

16-29 NO PIX (BW OR COLOR)

Turn up contrast and color controls. If there is no snow, and the set uses pre-CRT matrixing, suspect a defect in the demodulators or output amplifiers between the point of video injection and the CRT. If these circuits are ok, check for signal loss ahead of the chroma takeoff (see Chapter 14).

16-30 WEAK OR MISSING COLORS

Check the raster. If there are missing colors and a color-tinted raster, suspect one or more defective CRT guns, abnormal CRT operating voltages, or improper set-up adjustments (see Chapter 15). If normal raster and colors are missing from the pix only, determine which colors are observed and which are missing. For this, a color-bar pattern is preferable to a broadcast pix. If blues and yellows are observed, but no red or cyan, suspect the R-Y (X) demodulator or corresponding amplifier. If red and cyan are observed, but no blue or yellow, suspect the B-Y (Z) demodulator or corresponding amplifier. Scope-check the station signal or bar signal at the faulty stage, making comparisons with the response at the trouble-free stage.

No output from one demodulator usually means loss of oscillator injection to that stage. For this problem, suspect the corresponding phase-shifting network. When no defect is discovered, suspect a drastic phase shift. Examine the color-bar pattern (or vectorscope display) while adjusting the tint control. The range of a tint control is typically about 60° maximum. Check the APC alignment.

16-31 WRONG COLORS

Observe bar pattern while adjusting the tint control. Suspect a demodulator defect or an oscillator-phase problem. When only *some* colors are wrong, determine which colors and which demodulator or phase-shift network is suspect.

Check the appropriate circuit; if no defect is found, check the APC alignment.

When all colors are wrong, there's probably a phase problem with either the oscillator, the BPA, or the burst gate. Check color-sync stability by noting the lock-in range as the fine tuner is adjusted. If sync is marginal, check for a weak color burst at the gate output, or questionable APC action. Check oscillator stability and the operation of the phase-detector and oscillator-control circuits, as appropriate (see Sec. 16-36). Check all phase-shifting components. Perform APC alignment. If no improvement, check alignment of the BPA, including adjustment of tilt control if provided.

16-32 ERRATIC COLORS

Check for changes in the off-channel raster that may be caused by a CRT defect such as a loose or warped shadow mask or aperture grill. Check for color around objects of a BW pix that changes during warmup. This can be caused by convergence-circuit defects, unstable purity or convergence adjusters, or expansion/contraction of the yoke coils. For erratic color changes on a color pix only, check for chroma or video at the burst-gate output. Such signals can upset normal operation of the APC circuit, resulting in random shifts in oscillator phase and in the hues that are developed. Where the burst output is not "clean", suspect the gate transistor, improper operating voltages, a weak gating pulse, or trouble in the pulse delay circuit.

16-33 COLOR SMEAR

Check for perfect focus. Where color edges are not sharply defined, check the quality and resolution of a BW pix. Where both BW and color pix are affected, suspect video ringing or mismatching at the antenna input. Check for smearing on a color-bar pattern. Where BW pix is normal, color smearing is sometimes caused by misalignment of the video IF strip, more often by BPA misalignment. Sometimes a slight touch-up of the chroma takeoff coil or the tilt control is all that's needed.

16-34 TINT PROBLEMS

When the quality of a color pix is only fair, first verify that there is a good raster with no color tinting, and normal gray scale tracking. Verify convergence on a BW pix. Observe a color pattern to determine which colors are weak or predominant. Adjust tint control. If control range is limited, check for oscillator phase errors or misalignment of the APC or BPA. If required, make a critical alignment of both. If tint control covers the normal range from green to magenta but with poor skin tones in-between, check for a weak R-Y signal.

REMINDER: ───────────────

Skin tones require a large amount of red, some green, and some gray from the *Y* signal.

16-35 NO COLOR PIX; COLOR BACKGROUND TO BW PIX

Assuming a normal white raster, suspect loss of the CW oscillator signal. With a dead oscillator or CW amplifier, no colored snow is observed, and BW snow doesn't change with the color control.

16-36 LOSS OF COLOR SYNC

Loss of color sync has no effect on a BW pix. In color reception the usual symptom is the *barber-pole effect,* where all colors are present, but broken up into horizontal bands. The BW pix in the background is unaffected. Loss of sync occurs when the receiver subcarrier oscillator is not locked in frequency and phase with its counterpart at the station. Color sync makes use of the color burst which is a sampling of the station subcarrier oscillator. With a phase-detector-controlled oscillator, a DC error voltage is developed to lock the oscillator in sync. A burst-driven oscillator is automatically locked in step with the burst, without the need for intervening control circuits. Before the oscillator can be *phase-locked,* it must be operating at or close to the correct frequency of 3.58 MHz. For frequency stability the oscillator is crystal controlled.

There are many possible causes of lack of color sync, involving the oscillator, the burst gate, and in most cases the phase-detector and oscillator-control circuit, as shown in Fig. 16-2. Start by tuning in a station or a bar pattern from a generator. Critically adjust the horizontal hold. If vertical and horizontal sync are unstable, check for possible pulse-clipping in the video amplifier, which can also attenuate the color burst. Check for the color burst on the hammerhead display (Fig. 10-8).

Next, determine whether loss of sync is due to a *frequency* problem or a *phase* problem. If frequency is correct (or nearly correct) you get two or three bands of color moving slowly through the pix. If the frequency error is great, the bands become greater in number. The slope of the bands is an indication of whether the oscillator is running fast or slow. If it is off frequency, adjust the tuned circuit to determine if oscillator is *capable* of operating at the correct frequency. When it is properly adjusted, the pix will either lock in sync or slowly "float" across the screen. If this condition cannot be achieved, the trouble is with the oscillator or some interfacing circuit such as the phase detector, varactor or FET control, CW amplifier, or phase-shifting networks. If necessary, each in turn must be isolated from the oscillator as testing progresses. The usual cause is the control circuit, or an abnormal error voltage developed by the phase detector.

Disconnect or ground out the control circuit as applicable and try readjusting the oscillator coil. If still unable to obtain the correct frequency, suspect a defect in the oscillator circuit itself. Typical troubles are a bad transistor, a bad oscillator coil, or a bad crystal. The crystal is checked by substitution. If the correct frequency can be restored, perform a complete APC alignment.

A phase-locking problem is indicated when a floating pix can be obtained, but it won't lock or remain locked. First verify that there is a strong signal (judge by the intensity of the colors). A weak signal means a weak color burst which cannot do its job.

Check for a clean strong burst at the gate output. If OK, check for DC error voltage developed by the phase detector. Voltage should be close to zero with a floating pix, and should change when the fine tuner is adjusted. Where there's an unbalance as indicated by a positive or negative voltage, troubleshoot the circuit as described in Sec. 8-14. If the error voltage is ok, suspect the varactor or FET control circuit, whichever is used. Test the varactor or FET, preferably by substitution. If color sync is restored, perform an APC alignment and check for lock-in by adjusting the fine tuner.

As stated earlier, an alternative to this step-by-step procedure for loss of color sync is to attempt an APC alignment. When a particular adjustment can't be made it provides a clue to the faulty circuit.

16-37 APC ALIGNMENT PROCEDURES

Alignment of the APC circuits is extremely critical and the slightest misadjustments can create numerous problems, such as: loose, or complete loss of, color sync; no color, or weak colors; wrong or missing colors; reduced range of the tint control; and inability to obtain good flesh tones. Most sets have three or four adjustments as follows:

- **The Burst Transformer:** Tuned circuit at the output of the burst gate, made to resonate at 3.58 MHz for maximum amplification of the color burst.

- **Reactance Coil:** connected in a FET-type control circuit when used; tuned to 3.58 MHz.

- **Varactor Control Pot:** used with varactor control circuit to adjust the oscillator frequency.

- **Oscillator Coil:** used to adjust the oscillator frequency.

- **Oscillator Output Transformer:** used to obtain maximum output of the oscillator signal to drive the demodulators.

The following APC alignment procedure is applicable to most receivers. For specific details and variations, refer to the receiver service literature.

1. Tune in the best possible pix from a color telecast; or inject a signal from a color-bar generator, and observe the bar pattern. Turn up the color control and set the tint control to mid-position.

2. Kill the burst signal by connecting a jumper wire between base and emitter of the burst-gate transistor. Connect an EVM on a low DC range to one end of a load resistor at the phase-detector output. A positive or negative reading between 10 and 20 V can be expected. This reading represents the oscillator signal after rectification by one of the diodes. When little or no DC is measured, the oscillator is probably dead.

16-37 APC Alignment Procedures

3. If there's a reading, adjust the oscillator coil (or varactor pot if provided) for maximum DC on the meter. A "floating" bar pattern or pix should be observed on the screen. This indicates that the oscillator, although not locked in sync (because of the loss of burst), is capable of operating at the proper frequency. If this "float-by" display cannot be obtained, the oscillator circuit either has a defect or is being forced off-frequency by a control circuit defect or an abnormal error voltage from the phase detector. The greater the number of color bands the greater the frequency error. For crystal control, the free-running frequency must be within +/- 200 KHz of 3.58 MHz.

4. Remove jumper from the burst-gate transistor. Restoration of the burst should result in an increased reading on the EVM. Adjust the burst transformer for maximum DC. Where removal of the jumper failed to increase the reading, check for a weak or missing color burst.

REMINDER: _____

There are two conditions that can upset the color sync: a weak or missing burst, even though the oscillator is on frequency; and the oscillator's being off-frequency, even though there's a strong burst.

5. Kill the DC error voltage by grounding the junction of the phase-detector split-load resistors. The oscillator is now free-running and subject to drift. Adjust the reactance coil or oscillator pot, as applicable, to obtain a floating pix with the color bands nearly vertical.

6. Remove the jumper to restore the DC error voltage. Color should lock abruptly into sync. Check lock-in stability by adjusting the fine tuner. Assuming some oscillator drift, colors should lock at two or more color bands. If oscillator frequency must be *exact* to obtain lock-in, the sync is not very good. If removing the jumper causes an abrupt change in frequency, the phase detector is probably at fault.

7. If good color sync is still not possible, check the DC error voltage at the junction of the phase-detector load resistors. It should be zero for a floating pix. If not, there's an imbalance of the two diode circuits. The most common causes are: slight leakage of one or both coupling capacitors; load resistors not equal in value; leakage in one or both diodes. The resistors and diodes must be replaced as *matched pairs*.

8. Where the error voltage appears normal and changes with the fine tuner, but there is still no color sync, the varactor or FET control circuit is probably at fault. Check the applied operating voltages; then check the varactor or FET, preferably by substitution.

9. If set has an adjustable oscillator-output transformer, connect the EVM to one of the demodulators (or an equivalent test point isolated from the oscillator). Adjust the coil for maximum reading. See if reading can be increased further with a touch-up of the oscillator coil or potentiometer.

10. If necessary, go back and repeat all adjustments at least one more time. In addition to color sync, check the general quality of the reproduced color pix or bar pattern. Check the range of the tint control. If normal skin tones are not obtained at mid-setting, make a slight touch-up of the burst transformer.

REMINDER: _____

When a tuned circuit won't adjust properly, there are two possibilities: a defect in the circuit, and a defect in the coil itself. Check for shorted turns, and for a broken or missing slug. When any circuit repairs are made subsequent to alignment, the APC procedure should be repeated.

Except for the fact that it has one tuned circuit fewer, the procedures for a burst-driven oscillator system are essentially the same as those given above for a phase-detector-type APC system.

THE CHROMA SECTION

16-38 HUM BARS

AC ripple voltage getting into the video stages produces the familiar BW horizontal bars. AC in the chroma stages shows up in different ways—bands of color on both a BW and color pix, or colors that change in intensity—depending on what stages are involved. To help localize the problem, adjust the color and threshold controls and note whether the symptoms appear on a raster or only when a signal is being received. The color of the bars also provides a good clue. If bars are red or cyan, check for ripple on the DC feeding the R-Y demodulator or amplifiers. For blue or yellow bars, investigate the B-Y demodulator or amplifiers. Use CRO to locate AC superimposed on the station or bar signal. As you would for BW hum bars, check for filters that are open or have lost capacity.

NOTE:

On color sets, hum bars drift slowly upward because the vertical sweep rate is slightly less than the power-line frequency.

16-39 TROUBLESHOOTING IC CIRCUITS

With most sets now using ICs in the chroma section, troubleshooting is limited to checking input/output signals, DC voltages at the pin connections, and associated external components. Analyze the symptoms; if there is more than one IC decide which is the more likely suspect according to the symptoms. Where chroma is lost between input and output the IC is suspect *but not the only possibility*. Before replacement, check for presence of gating pulses of the proper amplitude. If there's a problem with the BW pix, check the *Y* signal input if pre-CRT matrixing is used. Check applied DC voltages and external components as appropriate.

REMINDER:

An external defect can ruin an IC replacement if such a fault exists and is overlooked. Use extreme caution when probing or making measurements. Be sure the set is turned off before making and breaking connections, including test instruments, and when replacing components. Use mini-clips at the IC pins to avoid shorts; if possible, use connecting points other than the IC pins. For other hints see Sec. 5-9.

When the signal appears normal up to the power output stages, the IC is probably OK.

16-40 INTERMITTENT CHROMA TROUBLES

To localize the problem, monitor all signals relating to the problem at various circuit points while attempting to create the abnormal condition (see Sec. 5-13). Check points may include the following: chroma and burst input; burst-gate output; demodulator inputs for both chroma and CW signals; gating pulses at the burst gate and killer; horizontal blanking input to the BPA or other chroma stage; and the oscillator output. For chroma-switching problems, monitor the DC biasing voltage at the input and the output of the killer, the ACC, and the controlled BPA stage. When trouble has been localized, avoid disturbances that may restore operation before the defect can be identified.

16-41 KILLER THRESHOLD ADJUSTMENT

Adjusting the threshold control can provide useful clues in troubleshooting most chroma problems. Unfortunately, many modern sets have dispensed with this control. Tentative adjustments are usually made a) when the symptoms are: weak color or loss of color, improper ACC action, colored snow off channel and on a BW pix, and b) as a means of enabling/disabling the killer when signal-tracing the BPA section. The control establishes the point at which the killer transistor either conducts or doesn't conduct, to enable the BPA when receiving color and to disable the chroma when receiving BW. If the control is set too high (ccw on most sets), a color pix is not affected, but color fringing will be observed on a BW pix because the BPA is turned on at all times, even when not receiving a color burst. If the control is set too low, color is weakened or lost. The adjustment is correct when a) there's strong color on all channels, and b) there's no *colored* snow off channel and when receiving BW. An ineffective threshold control usually means a killer-circuit defect or loss of the burst signal.

Although critical, especially under weak signal conditions, the adjustment is simple, and is made after all repairs have been completed. Switch off-channel to obtain a snowy raster. Turn control all the way up, then adjust it back to the point where colored snow barely disappears. Check for strong color on all channels; if necessary, make a compromise touch-up in favor of color even if it causes some color contamination to be observed on certain channels.

16-42 BANDPASS AMPLIFIER ALIGNMENT

Good color depends on the proper alignment of the video IF strip and the BPA, because both contribute to the amplification of all chroma signals contained in the I and Q sidebands of the color subcarrier. As with the IF strip, alignment of the BPA tuned circuits is extremely critical, so much so that the slightest misadjustment can seriously degrade the color. Troubles caused by BPA misalignment include: wrong colors; weak or excessive color; poor color registration; blurring and smearing of colors, with a loss of fine pix detail; difficulty obtaining proper hues, especially skin tones; phase errors; non-uniform amplification for all colors; and a critical fine tuner.

Normal overall response curves for the video IF strip and the BPA are shown in Fig. 16-3 for comparison. Markers representing the color subcarrier and its sidebands are located on the low-frequency slope of the IF curve. They receive unequal amplification: the 42.67 MHz sideband is near the top of the curve, where the gain is high; the 41.67 MHz sideband is near the bottom where the gain is less. Compensation for this rolloff is made in the alignment of the BPA, as shown in Fig. 16-3b. This curve represents the alignment of the chroma-takeoff coil at the input of the BPA. The result is the overall response shown in Fig. 16-3c, where both sidebands receive equal amplification. The bandwidth of the overall response is approximately 1 MHz, from 3.08 MHz to 4.08 MHz, centered at 3.58 MHz, the color-subcarrier frequency. With a comb filter it may be somewhat wider to obtain color on finer detail. Early-model sets with "wide band" color had a bandpass of 2 MHz. An ideal response is a flat-topped curve; in practice there's usually a slight dip in the center. Unlike the IF response, bandwidth is the frequency difference between the two markers near the top of the curve, not at the 50% level.

The frequencies shown on the video IF response are produced in the *tuner* by heterodyning the RF carriers with the tuner oscillator. The corresponding lower-frequency markers shown in Fig. 16-3b and c are developed in the *video detector* by beating the chroma signals against the 45.75 MHz pix carrier.

The BPA of a typical receiver has two tuned circuits: a single tuned coil at the input, and a double-tuned transformer at the output. The input coil in conjunction with a *tilt control* develops the curve of Fig. 16-3b. The output transformer establishes the bandpass. In those sets which use a comb filter to separate video from chroma, there is no need for the input coil and tilt control. Some late-model sets use non-adjustable tuned circuits in the BPA; others employ low- and high-pass filters.

Misalignment of the video IF strip and/or the BPA can be suspected when: no circuit defects can be found; the quality of color is still poor after making repairs; or there's evidence that the adjustments have been tampered with. Generally, the condition of alignment should be evaluated before making actual adjustments, since it may not be required. In some cases however it may be expedient to "try" each adjustment, especially since many sets can benefit from at least a touch-up alignment.

- **BPA Alignment Considerations**

Of the different methods of alignment, the sweep method, as used with the video IF strip, is the fastest and most accurate. Start by turning up the color control and the killer threshold. Defeat the AGC, by connecting an override bias of the correct amount and polarity to the AGC bus. Defeat the ACC, if provided, by switching it off or applying an override bias.

16-43 VSM ALIGNMENT

VSM (video sweep modulation) is the preferred method for aligning the BPA. As for sweep alignment of the video IF strip (Sec. 13-41), it requires a sweep generator, a marker generator, and a CRO to obtain a response curve. Such a curve represents the bandpass characteristics of the amplifier and the relative amount of amplification for different frequencies. Alignment using this method is comparatively simple; the most difficult part is obtaining the curve and making sure it truly reflects conditions that actually exist. (See Sec. 13-31 and Fig. 13-17).

The curve obtained with VSM includes the effects of the video IF strip as well as those of the BPA; therefore, the former affects the latter, and troubles in the video section cannot be corrected in the BPA. The equipment hookup is not very different from that shown in Fig. 12-23. The sweep signal may be injected at the antenna terminals, for an overall response, or at the video detector if only a BPA response is desired. For VSM, an additional item is required, an *RF modulator*. The modulator is supplied with three different signals; the sweep signal, one or more marker signals, and an unmodulated 45.75 MHz signal. The modulator output connects to the antenna terminals, the mixer stage, or the video detector, as appropriate. The CRO connects to the BPA output or the demodulator inputs, whichever is convenient. The color control sometimes makes a good test point. Since the signal at this point has yet to be detected, a demodulator probe is required with the CRO.

When fed to the mixer, the sweep generator is tuned to approximately 44 MHz. When feeding the BPA, the dial is set for 3.58 MHz. A sweep of 0-5 MHz is used. In the modulator, the 45.75 MHz CW carrier is modulated by the swept signal, which, after going through the receiver stages, traces out the response on the CRO. The marker signals identify specific points on the curve, in particular the 3.08 MHz and 4.08 MHz bandpass limits. Modern sweep equipment can speed up and simplify alignment; but more modest equipment, with the proper techniques, can achieve the same results. A suggested procedure is as follows: Examine the overall response noting the location of the 3.08 MHz and 4.08 MHz markers which *must* be at the *same height* for equal amplification of the two sidebands. A flat top, or a slight rise or dip at the center, is not important, but one marker higher than the other results in fuzzy colors and loss of pix detail. Generally there's no need for an external 3.58 MHz marker, because in most cases leakage from the receiver-subcarrier oscillator marks the center of the curve. Where the response isn't right, try adjusting the double-tuned output transformer. If this doesn't help, adjust the chroma takeoff coil (and the tilt control if there is one). The object is to obtain the conditions shown in Fig. 16-3b, to compensate for video IF rolloff. There is no need to observe *this* curve, however; all that's needed is the end result. If necessary, repeat the adjustment of the output transformer.

REMINDER: _____

When adjusting, it is the curve that moves. The markers remain fixed.

Some sets have a 4.5 MHz sound trap in the BPA. If this coil isn't tuned properly it causes a wormy, 920-KHz beat pattern on a color pix. Normal location of the 4.5 MHz sound marker (Fig. 16-3c) is at the base line where it receives no amplification. A quick and effective way to adjust the trap is: misadjust the fine tuner until the beat pattern becomes visible on the pix tube screen; then adjust the trap to reduce or eliminate the beat.

If a normal BPA response cannot be obtained, check the video IF response as described in Sec. 13-30, particularly conditions on the low-frequency side of the curve. For good color the slope must be reasonably straight with marker locations as shown in Fig. 13-3. The 41.67 MHz and 42.67 MHz markers *must be* equi-distant from the 42.17 MHz subcarrier marker. The 42.67 MHz marker should be near the top but *not beyond* the bend of the curve. The 41.67 MHz marker should be about 10% up from the base line. If all markers are too low on the curve, the result is weak color; if too high, the result can be excessive color and other problems. Make adjustments as necessary to obtain the proper response.

If there is a normal video response, recheck the response of the BPA, making adjustments as previously described. If there's still a problem, and the BW pix is less than perfect, check the video detector and any video amplifiers ahead of the chroma takeoff. Look for conditions that might have an adverse effect on the frequency response.

• **Other Alignment Methods**

If the equipment (Fig. 3-9) is available, the *bar sweep* method of alignment can be fast and accurate. It is also possible to align the BPA with an accurate RF generator and EVM, as described in Sec. 13-40. Feed the generator signal of 3.58 MHz to the video-detector test point and connect the meter (with a demodulator probe) to the demodulator inputs. Adjust the takeoff coil, tilt control, and output transformer for a maximum meter reading. Tune the generator to 4.08 MHz and readjust the takeoff coil and tilt control for maximum indication at this frequency. Compare voltage readings for 3.08 MHz and 4.08 MHz. If not equal, touch up all adjustments until they are.

Another method is by observing a color pix or, preferably, a bar pattern. Tune set to obtain the best possible color. Peak the BPA output transformer to obtain the strongest color. Adjust the chroma-takeoff coil and tilt control to obtain *uniform color* across the color bars or pix objects while eliminating any fuzziness at the edges.

REMINDER: _____

Any circuit repairs subsequent to alignment should be followed by a touch-up of the BPA adjustments, the tilt control in particular.

16-44 AIR-CHECKING THE REPAIRED SET

Air-checking a set after all *known* repairs have been made is an important part of TV servicing, because many relatively minor imperfections may be overlooked while troubleshooting the main problem. Also, some troubles show up only after a set has been operating for a time. During this "burn in" check, periodically examine the pix with a critical eye. Check for such things as: strength of color, color quality, resolution of both a BW and a color pix, focus, color sync stability, raster tinting, absence of color on a BW pix, and the range of the tint control. Now is the time to make any last-minute adjustments to the killer threshold and the AGC; any fuzziness in the color pix calls for a touch-up of the tilt control in the BPA. Verify that all CRT set-up adjustments are locked or tightened, so that they can't change when set is jarred or transported.

16-45 SUMMARY

1. Each CRT in a color camera has an optical filter that enables it to "see" only one of the three primary colors (red, green, or blue) plus the amount of that color contained in its complementary colors. The voltage output of each tube corresponds with the intensity of the color being viewed. The camera outputs are fed to a matrix where all three are proportioned to produce a luminance (Y) signal and I and Q chroma signals.

2. The I and Q signals from the color camera are used to modulate two subcarrier signals having a phase separation of 90°. The modulator outputs are combined to form a chrominance signal that varies in both amplitude and phase. Amplitude corresponds with color intensity. Hue is determined by the instantaneous phase angle between the chroma and the subcarriers. Each color thus becomes associated with a particular phase angle.

3. The I and Q signals are transmitted as sidebands of the 3.58 MHz subcarrier within the same 4 MHz bandpass as the Y signal that supplies brightness information for the colors. This is called interleaving. Although combined, the I and Q signals still retain their identities and don't interact since each relates only to a particular subcarrier phase.

4. At the receiver, demodulation of the chroma information is accomplished with the aid of a locally generated subcarrier. For accurate reproduction of the pix colors, it's imperative that the subcarrier be locked in phase with its counterpart at the station. The locking is made possible by the transmission of 8-to-10-cycle bursts of the subcarrier; this "color burst" is used as a phase reference. The method is called "color synchronization."

5. A color phase (phasor) diagram (Fig. 16-8) provides a convenient method for depicting the various colors and their corresponding phase angles and relative amplitudes, and typical modulation and demodulation axes commonly used. *Reminder:* the sweep from 0° to 360° corresponds with one complete cycle of the subcarrier reference signal. At the receiver, the chroma is "sampled," once for each cycle of the subcarrier, to produce hue-related voltages corresponding with the phase angles indicated on the diagram.

6. Some sets have three demodulators, one for each primary color. If only two demodulators are used, the third color (usually green or blue) is produced from a mixture of the other two.

7. IF signals consisting of the I and Q sidebands of the color subcarrier are fed to each demodulator in parallel. The demodulators are also supplied with a CW signal from the subcarrier oscillator. The oscillator phase is different for each demodulator. Demodulation takes place as each demodulator is made not to conduct, or to conduct by different amounts according to the phase difference between the chroma and oscillator inputs, which correspond with the color being scanned by the camera at that instant. Without the oscillator signal there can be no conduction and therefore no color pix.

8. The demodulator outputs are the same-color difference signals produced at the station. Each of these signals, R-Y, G-Y, and B-Y, corresponds with one of the primary colors. Other colors are produced by proportionate mixtures of these signals which drive the three guns of the CRT. White is produced by a proportionate mixture of all three. Zero output simultaneously from each demodulator results in black pix objects.

9. A color difference signal represents a fully saturated color containing no brightness information. Brightness levels are provided by the Y signal, the same signal that produces a BW pix. The Y signal is combined (matrixed) with the chroma in either the demodulators, the output amplifiers, or the CRT itself. After matrixing, the signals are called simply the red, green, and blue signals.

10. Two kinds of pictures are produced on the screen of a color CRT, a pix in BW and a color pix. The two superimposed pictures must be in perfect register. The BW pix is produced from the video (Y) signal; the color pix, from the chroma information. The finest pix *detail* is produced only in BW.

11. Colors on a color pix are developed as each gun is made to conduct by different amounts. This is determined by the amplitude and polarity of the three drive signals at any point in time. To develop red, the red gun is made to conduct while the other two are cut off; similarly for blue and green. Secondary hues are produced by two or three guns conducting at the same time.

12. To produce images in color at the proper times relative to the scanning of the raster, each demodulator must be made to conduct at precisely the right time, by the process known as color synchronization. Demodulator conduction is governed by the applied signal from the subcarrier oscillator; therefore that signal must be synchronized with its counterpart at the station, by the use of the transmitted color burst.

13. During each horizontal retrace period, a comparison is made in both frequency and phase between the color burst and the generated oscillator signal. If there's a difference, a DC "error voltage" is developed to force the oscillator into sync with the station. Some sets use a "burst-driven" oscillator, eliminating the need for extra sync stages.

14. From a "takeoff point" in the video amplifier, the composite signal (video, chroma, color burst, sync, and blanking pulses) is applied to the BPA and the burst gate. The BPA amplifies the chroma signals prior to demodulation. Most of the video is rejected, because of the BPA's narrow (1 MHz) bandwidth. The BPA may also be gated (by horizontal sweep pulses) to ensure rejection of the burst and sync pulses.

15. The function of the burst gate is to amplify the burst signal while rejecting all other

signals that might upset the color sync. A "clean" burst output is obtained by keying the gate (with horizontal sweep pulses) to conduct only for the duration of the color burst. Loss of the burst signal results in loss of color synchronization, and in some cases total loss of color. A weak burst results in "loose" color sync.

16. The color burst serves another purpose: to enable the chroma stages when receiving color, and to disable them when receiving BW. Chroma switching is accomplished by the color killer. When receiving color, the burst signal turns the killer off, and the killer in turn switches on the BPA. When receiving BW, conditions are reversed; the absence of the burst causes the killer to conduct, developing an overbias to turn off the BPA. The switch point for killer conduction is set by a killer threshold control.

17. Good color depends on the uniform amplification of all chroma frequencies. The dropoff on the low-frequency slope of the video IF response tends to attenuate the high chroma frequencies, those around 4.08 MHz in particular. Compensation is made by peaking the BPA input circuits to this approximate frequency. Slight misadjustments of the input tuned circuit or its associated tilt control cause changes in hue, and fuzzy indistinct color images.

18. The tint control is a phase control for producing changes in hue. Normal hues should occur at the approximate mid-setting of the control. Easily-recognized skin tones are used when adjusting for the correct hues. Since hue is determined by the phase difference between the chroma signal and the subcarrier oscillator, abnormal operation of any of these stages will change the color. When color shift is beyond the range of the tint control, it can often be restored with a proper APC alignment. The circuits involved are the burst gate, phase detector, subcarrier oscillator, and oscillator control.

19. The station modulators develop I and Q signals separated in phase by 90°, the phase angle between the two oscillator signals. Different receivers use different phase angles for *demodulation,* the angle established between the two or three oscillator outputs. Most receivers use either R-Y, G-Y demodulation, or X/Z demodulation. I and Q demodulation is seldom used. Figure 16-8 shows the different phase angles.

20. Fine pix detail is produced by high video or chroma frequencies. Although the station sends out chroma up to 1.5 MHz, most modern sets make use only of frequencies up to 0.5 MHz (hence the 1-MHz bandpass limits of the BPA), for economy and because color cannot be observed on very small pix details anyway.

21. At the station, the hue of a color viewed by the cameras determines the phase of the chroma developed relative to that of the subcarrier oscillator. At the receiver, the difference in phase between the received chroma and the locally generated subcarrier determines the color that's reproduced by the CRT.

22. The output of an R-Y or X demodulator corresponds to the color red and its complementary colors. The output of a B-Y or Z demodulator corresponds to blue and its complementaries. The output of a G-Y demodulator (if there is one) corresponds to green and its complementaries. An output from only one demodulator produces only a primary color. Other colors are developed by a combined output of two or three demodulators in the proper proportions. The amplitude of a demodulator output corresponds with the strength or intensity of the color. Hue is determined by the phase difference between the chroma and that of the applied oscillator signal.

23. Since *all* chroma signals are processed by the BPA, a BPA defect affects *all colors*. A defect in one demodulator or one output amplifier usually affects only one primary color and its complementaries.

REVIEW QUESTIONS

1. Why do some sets have power transistors in the chroma output stages and others not?

2.. What system of APC is used where loss of the burst signal results in a) loss of all color? b) loss of color sync only? Explain your answer.

3. Explain why APC misalignment can cause: a) loss of color; b) weak color; c) wrong colors; d) loss of color sync.

4. Identify the tuned circuits in: a) the chroma signal stages; b) the color sync stages. c) State the purpose of each.

5. What characteristic of the demodulator output signals determines a) the hue of a color? b) the intensity of a color?

6. What is the function of the 3.58 MHz subcarrier signal at a) the station? b) the receiver?

7. Where does matrixing of the Y signal take place when the signals driving the CRT are a) color-difference signals? b) red, green, and blue signals?

8. What type of demodulation is used when there are three outputs from the subcarrier oscillator, all differently phased?

9. Why is a color demodulator sometimes called a synchronous detector?

10. How is the *frequency* of the subcarrier oscillator maintained? b) How is it kept in *phase* with the color burst?

11. How do the demodulators develop outputs for producing a) one or more primary colors? b) all other colors?

12. Fill in the missing words. The wrong phase from one of the oscillator outputs causes _____ colors to be wrong. If the oscillator is not phased with the burst signal the result is _____.

13. a) How does the BPA reject most of the video signal? b) Why is the BPA sometimes gated?

14. a) How does the burst gate reject all signals except the color burst? b) What problems are created if video and chroma get through?

15. Some sets have two phase detectors in the chroma section. Explain their functions.

776 THE CHROMA SECTION

16. Explain varactor control of a subcarrier oscillator.
17. Describe the signal developed by a subcarrier oscillator.
18. Explain the uses of a color-bar generator in troubleshooting.
19. Describe the barberpole pattern and the condition it represents.
20. Explain purpose and operation of the color killer.
21. State two ways in which the Y signal contributes to developing a good color pix.
22. Explain why a good raster is prerequisite to a normal BW pix and why both are essential to obtain a good color pix.
23. What stages are dependent on a strong color burst? Explain why in each case.
24. At what times is the BPA disabled? Explain how and why.
25. What is meant by a color-difference signal? Name the color-difference signals.
26. Many sets require a different amount of signal drive to each of the three CRT guns. Explain why.
27. Explain the purpose of a 4.5 MHz trap in the BPA and 3.58 MHz traps in the chroma demodulators.
28. What is meant by high- and low-level demodulation?
29. The 3.08 MHz and 4.08 MHz markers on the overall BPA response *must be* at the *same level* near the top of the curve. Explain why.
30. Explain how proper alignment of the BPA compensates for the chroma rolloff in the video IF strip.
31. a) Explain the 920-KHz beat pattern often seen on a pix: its appearance, how it's developed, and how it's minimized with proper receiver design. b) Why is the interference peculiar to color sets?
32. Explain why a dead subcarrier oscillator results in loss of all color.

TROUBLESHOOTING QUESTIONNAIRE

1. What symptoms indicate loss of the Y signal? Explain.
2. What would you suspect if a CRO shows strong video and chroma at the output of the burst gate?
3. For loss of color, it's customary to defeat the killer. Explain why.
4. Why must the ACC be disabled when performing BPA alignment? Briefly explain the alignment procedure.
5. What is meant by APC alignment? Explain step by step how it is done.
6. State the procedure for adjusting the killer threshold control.

7. a) What trouble is indicated when colored snow is observed off-channel and on a BW pix? b) How would you proceed?

8. Give three possible causes of an unbalanced phase detector.

9. For poor color pix quality and a color-tinted raster, would you suspect a) the CRT? b) improper set-up adjustments? c) APC alignment? or d) a demodulator defect?

10. What problem is indicated during APC alignment if a) a floating pix cannot be obtained with the oscillator frequency adjustment? b) pix float-by is obtained but no color lock-in?

11. What adjustment(s) would you try where there's a good BW pix but color images are blurred and indistinct?

12. Strong colors usually indicate a strong color burst. Explain why.

13. When observing the color bands while adjusting the subcarrier oscillator, how can you tell a), when you're getting *close* to the correct frequency? b) when the frequency is *correct*?

14. What troubles can be expected from loss of gating pulses to a) the burst gate? b) the killer? c) the BPA?

15. With some sets the subcarrier oscillator stops functioning when switched off-channel or when receiving BW. Explain.

16. A CRO shows normal chroma at the demodulator inputs. State possible causes for lack of output from: a) one demodulator; b) both demodulators.

17. Explain the procedure for signal-tracing all chroma stages with a bar signal.

18. How would you proceed when normal green and magenta can be obtained at the extremes of the tint control, but skin colors are poor?

19. A set has loss of color; color is restored when the killer is defeated. How would you proceed?

20. a) What circuits are suspect when there's loss of all color? b) certain colors only?

21. Where would you check for PS ripple when there's a hum bar a) on the raster, BW, and color pix? b) on color pix only? c) on BW pix only?

22. When using an inexpensive CRO, how would you determine whether the signal at the burst-gate output is truly a burst signal, or merely interference?

23. How would you proceed when the oscillator won't generate the correct frequency?

24. Explain a method for adjusting 4.5 MHz traps without the aid of instruments.

25. What problems are expected when the killer a) won't conduct? b) is conducting at all times? c) State the corrective procedure in each case.

26. a) A weak oscillator can result in weak color. Explain why. b) How would you measure the strength of oscillations?

27. What stages are suspect for the following trouble symptoms: a) weak or missing color; b) some colors missing; c) poor color sync?

28. Under what conditons would you suspect misalignment of a) the BPA? b) the APC circuits?

29. Give several possible causes of weak color.

30. State several possible causes of loss of signal in the BPA.

31. One approach, when there's loss of color sync or there are weak or wrong colors, is to attempt an APC alignment. Explain why this can localize the defect.

32. What pix colors are lost, besides the primaries, when there's a dead R-Y demodulator or output amplifier? b) defective G-Y stages? c) defective B-Y stages?

Chapter 17
THE SOUND SECTION

Each TV station is assigned a 6-MHz band of frequencies within either the VHF or UHF spectrum, and the pix and sound signals are transmitted separately within this allocation. Within its assigned channel, each station develops a specific pix RF carrier and a sound RF carrier (see Table 1-1). The two carriers are separated by exactly 4.5 MHz. Pix and sound information is contained in the sidebands of the two carriers. The sound signal occupies a relatively narrow bandwidth (50 KHz) compared to the 4 MHz required for the pix and synchronizing signals (See Fig. 1-3). Both sidebands are transmitted.

17-1 THE FM SOUND SIGNAL

The TV sound carrier is *frequency modulated*. FM has several advantages over AM: the capability of higher fidelity (up to 15 KHz or more); less chance of noise pickup and other types of AM interference; and fewer problems with crosstalk between the sound and video signals.

As described in Sec. 1-23, the amplitude of an FM signal is *constant*, but the RF carrier is continuously *changing in frequency*, in accordance with the audio modulation. See Fig. 1-8. The *volume* of sound corresponds with the *amount of frequency deviation* each side of the carrier frequency. The higher the sound level the greater the deviation, up to a maximum of 50 KHz. For 100% modulation, the signal varies +/− 25 KHz above and below the assigned carrier frequency. For lower volume levels the frequency deviation is correspondingly less.

The *rate* of frequency variations corresponds with the frequency of the audio. Low audio frequencies (the bass tones) produces a slow change in the carrier frequency. High (treble) frequencies correspond to a rapid change in frequency. The range of TV sound is approximately 40 Hz to 15 KHz. A sweep generator produces an FM signal, but the sweep rate is fixed at 60 Hz.

17-2 RECEIVER FUNCTIONS

At the receiver, the pix and sound RF carriers are converted into IF carriers; a pix IF carrier of 45.75 MHz and a sound IF carrier of 41.25 MHz. In the video detector, whose main function is to demodulate the video signal, the two IF carriers are combined to produce a new (and lower) sound IF carrier of 4.5 MHz, the frequency difference between the two carriers.

780 THE SOUND SECTION

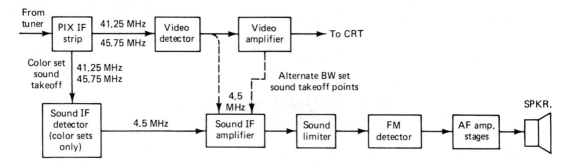

Fig. 17-1 Sound-section block diagram.

The block diagram of Fig. 17-1 shows the sound stages of a typical receiver. Except for the difference in sound takeoff and an additional detector stage, the sound section of a color set is essentially the same as that of a BW receiver. The complete sound section of a typical color set is shown in Fig. 17-2.

In a BW set, the sound takeoff may be anywhere between the video detector and the CRT input. In most sets, the takeoff is at the detector output. When the takeoff is from the video amplifier, the additional sound amplification means that fewer stages are required in the sound section. One problem, however, is incomplete separation of video from sound signals, which shows up as an annoying buzz in the sound. This is called *intercarrier* or *sync* buzz and is caused by the 60-Hz vertical blanking and sync pulses and by high-amplitude low-frequency video information. It is most noticeable when large blocks of lettering appear on the screen.

In a color set, if the 4.5 MHz sound IF is permitted to beat with the 3.58 MHz color subcarrier, a 920-KHz beat pattern is produced which gives the pix a fine-grained "wormy" appearance. To minimize this problem, the sound takeoff for a color set is ahead of the video detector, usually at the last video IF stage as shown in Fig. 17-1. This necessitates an extra detector stage for the sound strip, a *sound IF detector*, for producing the 4.5 MHz sound IF from a mixture of the 45.75 MHz and 41.25 MHz IF carriers, a process done in the video detector of a BW set.

Once developed, the 4.5 MHz sound IF signal is amplified by one or more tuned sound IF stages. These stages may be preceded or followed by one or more *sound limiter* stages whose function is to discriminate against any AM-type signals, either video or "noise" bursts, that may degrade the sound. The amplified sound IF signal is then demodulated to recover the audio sidebands from the frequency-modulated IF carrier. Following the FM detector are one or more audio amplifier stages that drive the speaker.

In most late-model sets, processing of the sound signal is accomplished by a single IC.

• **Preemphasis and Deemphasis**

This is a method for improving the signal-to-noise ratio for high audio frequencies from about 5 KHz and beyond. At the station, the high frequencies receive more amplification than the lows. At the receiver, the high's are deemphasized to their normal levels. In the process, background noise which is high-pitched, as most is, is also reduced. In effect, the action is much like that of a conventional tone control, except that the constants are fixed.

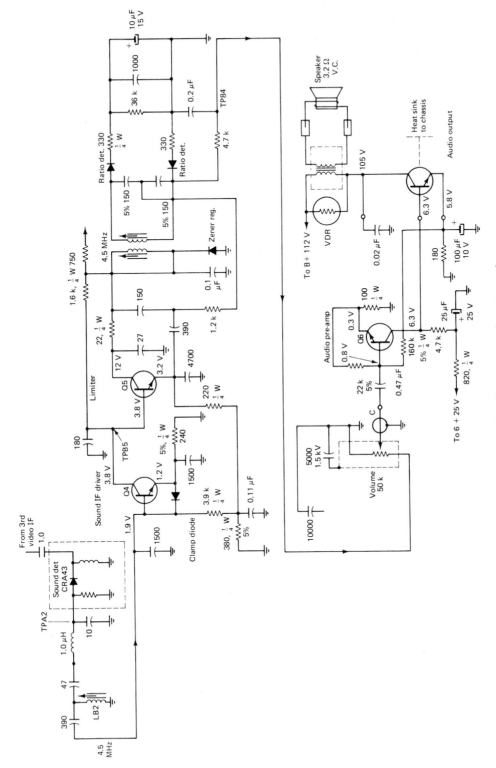

Fig. 17-2 Typical color-set sound-section schematic.

17-3 THE SOUND TAKEOFF

This is the point at which the sound and video became separated; the video is sent to the CRT, the sound signal to the speaker. As previously stated the point of sound takeoff is different for BW and color sets.

• **BW Sets**

The sound takeoff (at the video detector or amplifier) interfaces with the input of the first IF amplifier or limiter stage. The takeoff circuit is composed of one or more tuned circuits, resonant to 4.5 MHz. The tuned circuit(s) accept the 4.5 MHz sound IF signal while rejecting the video of the composite signal. That is, they prevent video from getting into the sound section and at the same time keep the sound out of the video stages.

REMINDER: _____

4.5 MHz is beyond the upper limit of the video frequencies that make up the pix.

Several takeoff circuits are shown in Fig. 17-3. Coupling may be inductive, capacitive, or both.

Figure 17-3a uses a single tuned coil and a small coupling capacitor. The capacitor value is usually less than 5 pF, to aid in rejecting the video and other interference. The circuit shown in 17-3b uses a single-tuned transformer. In 17-3c, a double-tuned transformer is employed. Coil L1 is a tank that feeds the sound strip; L2 is an absorption trap.

• **Color Sets**

Because the sound takeoff is in the video IF strip, the takeoff coil(s) are tuned to 41.25 MHz. Typical circuits are similar to those shown in Fig. 17-3. Both 41.25 MHz traps and

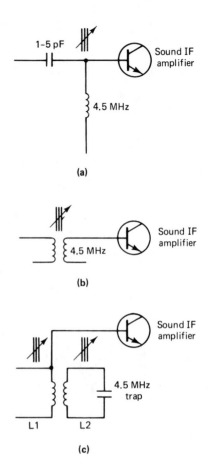

Fig. 17-3 Sound takeoff circuits.

4.5 MHz traps are often used to prevent intercarrier buzz and 920-KHz beat interference.

17-4 THE SOUND IF DETECTOR

This stage, which is peculiar to color sets, is located between the sound takeoff, at the last video IF stage, and the sound IF amplifiers. In this stage (as in the video detector of a BW set) the 41.25-MHz and 45.75-MHz IF carriers are combined to produce the 4.5-MHz sound IF carrier. Any video that is present at the detector output is rejected by a 4.5 MHz tuned circuit somewhere in the subsequent IF stages. The detector may be either a diode or a transistor; a diode is more common. This detector *does not demodulate;* the signal at its output is the same FM signal as at its input except for the difference in frequency.

17-5 THE SOUND IF AMPLIFIER

Like the video IF strip which it resembles, the sound IF section is a tuned IF amplifier whose function is to amplify the 4.5-MHz FM sound signal prior to detection. Differences that distinguish it from the video If strip are: fewer stages, the absence of AGC control, the resonance of its tuned circuits, and the narrow bandpass. Except when it is contained in an IC, in a typical set it may have one or two stages. The tuned circuits are peaked at 4.5 MHz. Consistent with the maximum allowable sweep of 50 KHz, the bandpass is between 50 KHz and 100 KHz. A bandpass exceeding 50 KHz is desirable to make the fine tuner less critical for the user, and less susceptible to tuner-oscillator drift. The wider the bandpass the less the stage gain; if *less* than 50 KHz it can cause amplitude distortion on strong signals.

REMINDER: ⎯⎯⎯⎯⎯⎯⎯⎯⎯⎯⎯⎯⎯⎯⎯⎯⎯⎯⎯

An incorrect oscillator frequency develops the wrong pix and sound IF carriers. However, the frequency difference between the two remains *constant* at 4.5 MHz. For a narrowed IF response or when the tuned circuits aren't precisely peaked to 4.5 MHz, the result is weak, and sometimes distorted, sound.

To prevent oscillations, each IF stage is usually *neutralized,* with a small capacitor to provide out-of-phase feedback from output to input. Ferrite beads may also be used on critical leads.

17-6 SOUND IF LIMITERS

Only the *FM* signal should be permitted to reach the sound detector. Signals that vary in *amplitude* are prevented from reaching the speaker by some form of AM rejection. This prevention is the function of a limiter stage, whose job is to reject amplitude-modulated signals: the video signal; blanking and sync pulses that cause intercarrier buzz; and strong static-producing noise impulses.

Although diodes are sometimes used, most sets employ transistors for this purpose, and in some cases one of the IF amplifiers functions as a combined amplifier/limiter. To operate as a limiter, a transistor is biased for class C with reduced collector voltage so it will saturate at relatively low signal levels. When overdriven into saturation, any amplitude variations greater than the IF carrier are effectively removed by clipping action. This does not distort the sound, which is encoded in the frequency. A typical limiter provides little or no stage gain.

17-7 FM DETECTORS

The purpose of this detector (which is not to be confused with the IF detector previously described) is to demodulate the FM sound signal, i.e., to convert the IF frequency variations into audio (AF) signals capable of producing sound. Whereas a simple diode circuit will detect an *AM* signal (e.g. the video detector), an *FM* detector is considerably more complex. There are several types of detectors, described as follows.

784 THE SOUND SECTION

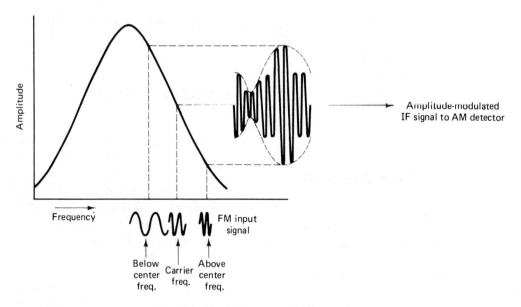

Fig. 17-4 Slope detection of FM signals.

• Slope Detection

An FM signal can be heard on an AM receiver when the IFs are slightly mistuned. This is called slope detection, a system no longer used but mentioned here for purposes of illustration. Figure 17-4 shows a typical IF response curve. With the tuned circuits properly peaked to the correct frequency, equal amounts of output are produced by an FM signal that sweeps the same amount above and below resonance. If the circuit is mistuned, operation extends to one slope or the other as shown in the drawing. Note that the output now varies in amplitude with the IF frequency variations of the input signal; maximum output occurs when the frequency deviation is small, reduced output at an increased amount of sweep. The output variations correspond with amplitude changes of the audio and are not to be confused with *audio frequencies*, which are a function of the sweep *rate*. A conventional AM detector is still required to recover the audio. The system is no longer used because non-linearity of the response slope results in unequal outputs above and below resonance, creating distortion.

• The Foster-Seely Discriminator (Fig. 17-5a)

This basic detector used by many sets in the past is rapidly becoming obsolete. Circuit-wise, and in operation, it is essentially the same as the phase detector used for horizontal AFC, and color AFPC. The IF signal is supplied to the two discriminator diodes in two ways: inductively, to obtain positive and negative-going signals from opposite ends of a transformer, and "directly", from a transformer center-tap. The diodes conduct alternately and by different amounts with the changing phase relationships between the two applied signals. Depending on its conduction, each diode develops an IR drop across its associated load resistor. At the carrier frequency (the "center" or "resting" frequency between frequency deviations), the

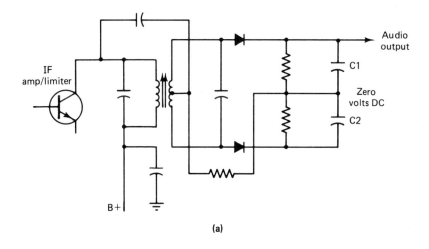

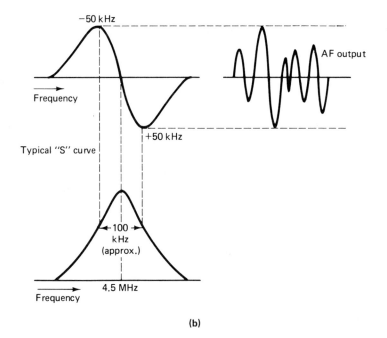

Fig. 17-5 Basic FM-detector operation. An FM detector converts frequency variations into amplitude changes at an audio rate: (a) typical Foster-Seely discriminator; (b) IF bandpass response.

diodes conduct equally, producing equal and opposite voltages across the series-connected load resistors. This corresponds with zero audio output. As the FM signal sweeps above and below resonance, the diodes conduct by different amounts producing a net output that changes in amplitude with frequency changes of the incoming signal.

This type of detector has a high output, but unfortunately does not discriminate against AM interference. For this reason it must be preceded by one or two limiter stages.

• **The "S" Curve**

The S-shaped curve of Fig. 17-5b shows graphically the conditions for FM detection.

The lower curve represents the overall response of a typical sound IF system. In this example the bandpass is twice the amount needed for the maximum allowable sweep of 50 KHz. The ''S'' curve shows the positive and negative-going outputs from the detector. The *crossover point* (where conduction shifts from one diode to the other) corresponds with 4.5 MHz, the resonant peak of the IF response. The frequencies represented by the two peaks of the S curve correspond with the upper and lower bandpass limits, in this case, 4.45 MHz and 4.55 MHz. The useful operating range is the linear portion of the curve between these two frequencies. The actual amount of the curve that is used varies with the amount by which the incoming signal is being swept above and below resonance. For low volume levels only a small portion of the curve is used, close to the crossover point. At high volume levels the sweep may extend up to +/− 25 KHz above and below 4.5 MHz. Unless the two peaks are equidistant from the crossover point and the response is reasonably linear, distortion will occur at high volume. Sweep alignment of the FM detector, as described later under troubleshooting, makes use of an S curve as shown in the drawing.

• **The Ratio Detector**

This type of detector has long been used; it is still used by many sets, with the exception of those employing ICs. It develops only half the output obtained from a discriminator, but this disadvantage is offset by its ability to reject AM-type interference. Although it is theoretically unnecessary, most sets also use a limiter stage for optimum rejection of noise and buzz. There are numerous variations of the ratio detector, but basically there are two types, the *balanced* and *unbalanced* detector.

The Balanced Ratio Detector. This circuit shown in Fig. 17-6a is similar in some respects to the discriminator-type detector shown in Fig. 17-5. The main differences are: the two diodes are connected series-*aiding;* a large capacitor (5-50 Mfd) is connected across the load resistors; and the audio output is taken from the junction of two small capacitors instead of from one end of the load resistors. As in the discriminator, the diodes conduct according to the signal voltage applied to each at any given instant. The signal voltages depend in turn on the ever-changing phase relationship between signals of opposite polarity from the secondary of the input transformer, and an in-phase signal applied in parallel to both diodes via capacitors C1 and C2. This in-phase signal comes from a small *tertiary coil* (L3) inductively coupled to the transformer primary. The direction of winding is such that this voltage has the same phase as the primary voltage. Note the paths of the electron flow of the two diodes: From D1 to C1, L3, L1, and back to the diode. From D2 to L2, L3, C2, and back to D2.

In effect, the signal-input transformer converts FM to AM, and the *amplitude-modulated* IF

17-7 FM Detectors

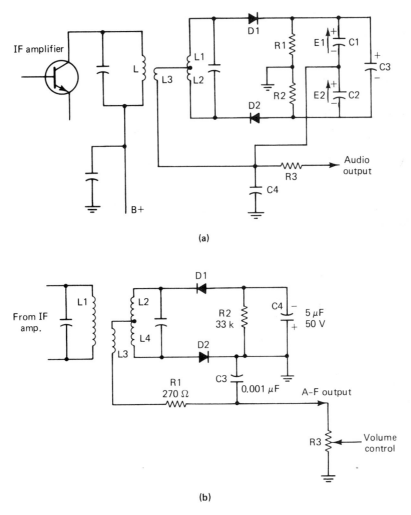

Fig. 17-6 Ratio detectors. A ratio detector is insensitive to AM type noise and signals: (a) balanced ratio detector; (b) unbalanced ratio detector.

signal is then rectified (detected) by the diodes to develop an audio signal. For an unmodulated incoming signal, and at such times as the signal passes through the center carrier "resting" frequency, the transformer can be considered "resistive", and the two diodes are supplied with equal and opposite voltages. Under these conditions they conduct simultaneously and by equal amounts to charge the two capacitors C1 and C2. The next result of the two oppositely charged capacitors is zero AF output.

When the signal sweeps *above resonance* (higher than 4.5 MHz), the transformer secondary becomes *inductive* and the current *lags* the induced voltage by approximately 90°. This changes the relative voltage applied to the diodes causing one to conduct more than the other placing a greater charge on its associated capacitor.

When the signal sweeps *below resonance* (lower than 4.5 MHz), the circuit becomes *capacitive*, and the secondary current *leads* the induced voltage by some angle. The two voltages, which are added vectorially, results in greater conduction of the other diode to place a stronger charge on its capacitor. Thus the audio, taken from the junction of the two capacitors is made to swing both positive and negative at an audio rate in accordance with the sweep of the FM signal.

AM rejection is accomplished with the *stabilizing capacitor* C3. This capacitor is charged by the sum of the voltages developed across capacitors C1 and C2, and the split load resistors R1/R2. This serves as a *stabilizing voltage* to establish the signal level at which the diodes conduct. To prevent the DC from fluctuating with the audio and any noise bursts, the capacitor value must be fairly large (with a long time constant) for low reactance to the lowest audio frequencies. Note that the audio output is dependent not on the *amount* of total voltage across C4 and the series-connected capacitors, but on the *ratio* of the voltages between C1 and C2, which is constantly changing.

The network comprising R3 and C4 at the AF output is for deemphasis of the high audio frequencies to counteract their pre-emphasis at the station.

The Unbalanced Ratio Detector. Because of its simplicity, this ratio detector, as shown in Fig. 17-6b is the more popular of the two circuits. Note how it differs from the balanced-type detector. There are no series-connected capacitors or resistors; the audio is taken from the center tap of the input transformer; and one side of the circuit goes to ground to provide a common return path for the diode currents. The resistor R2 sometimes consists of two resistors to provide a center tap for alignment purposes.

Note the path of the diode currents: For D1: to C4, via ground to R3, R1, through L3 and L2, and back to the diode. For D2: through L4 and L3, R1, R3, and via ground back to D2.

Circuit operation is not too different from other ratio detectors. Transformer action is the same: the vector addition of in-phase and out-of-phase voltages is fed to the diodes as *amplitude* variations of the FM signal. The diodes are thus made to conduct by different amounts, depending whether the incoming signal is being swept above or below the resonance of the transformer, and by how much. Capacitor C3 at the audio output is charged by diode conduction. Although the diodes are connected series-aiding, the C3 charging currents are in *opposite* directions, and the net amount of voltage developed across the capacitor, and its polarity, at any given instant is determined by which diode is conducting the most. The voltage across C3 thus varies in amplitude with frequency variations of the signal. The volume control at the AF output is part of the deemphasis network.

The stabilizing capacitor C4 performs the same function as described earlier: to provide a "stabilizing bias" for the diodes, and to "absorb" any AM fluctuations. The capacitor is an electrolytic type connected with the polarity as shown. A common cause of intercarrier buzz is its becoming open or drying up and thereby losing capacity.

• **The Quadrature Detector**

This circuit, as used in most sound ICs, is currently the most popular FM detector. Except for the absence of a "buzz" control, its operation is similar to the "gated beam" detectors found in many tube-type sets. The circuit has two signal inputs: the FM signal from the sound IF stage, and a sampling of the signal that has undergone a phase shift. Quadrature (90°) phase shifting is accomplished with a special "ceramic filter". Demodulation takes place by comparing the phase of the two signals: the regular IF input signal, and the phase-shifted signal which is tuned by the circuit's only resonant circuit, a quadrature (quad) coil. Computer-type logic circuits are used for this operation, and an output is obtained when the two signal voltages are of the same polarity. The audio output is in the form of pulses whose width varies at an audio rate in accordance with the sweep deviations of the FM signal. Usually the volume level is controlled by varying the DC power applied to the AF voltage amplifier.

17-8 THE AUDIO AMPLIFIERS

Sound plays a secondary role to what is observed on the TV screen, and the requirements for most sets are not too stringent. The audio power output requirement, for example, is minimal, seldom exceeding 1 or 2 watts, and averaging around $\frac{1}{2}$ watt for comfortable listening. Frequency response too can be somewhat limited, for all but the more expensive consoles, provided that the tone quality is pleasing to the average listener and that there's no distortion. With most sets, hi-fi sound is *impossible,* in view of the small and inexpensive speaker generally used and the lack of proper baffling.

The typical TV receiver has three audio stages, each named according to its function: an AF *voltage* amplifier, a *driver,* and a *power output* stage (see Fig. 17-7). Older sets used discrete components; in modern sets, all or most functions are performed by one or two ICs.

17-9 THE AF VOLTAGE AMPLIFIER

With a gain factor of about × 10, the function of this stage is to boost the relatively low signal voltage provided by the detector, which is seldom more than about 1 V p-p. This may be enough to operate a pair of high-impedance headphones but is much too weak to drive a speaker. Some sets have as many as three such voltage amplifiers, depending on the amount of detector output and the drive requirements of the following stages. In ICs, there may be several transistors connected in pairs, in the Darlington configuration.

Direct (DC) coupling is usually employed between stages, for economy and to reduce the attenuation of low (bass) frequencies. Voltage amplifiers usually operate class A in the conventional common-emitter hookup, for maximum gain with minimum distortion. Negative feedback is often used for stability, to improve the frequency response, and to raise the input impedance of an amplifier. This stage usually contains the volume and tone controls. As with any high-gain amplifier, signal input leads, including the "hot" leads of a "conventional" volume control, use shielded wire to prevent hum pickup. Shielding is not needed, however, on the "single wire" volume control used with most ICs, where gain is controlled with *DC*.

17-10 DRIVER AMPLIFER

An output transistor operating class B draws some current from the previous stage. This power (about 3 MW) is furnished by the driver stage, which uses a medium-*power* transistor as distinguished from the preceding *voltage* amplifier. A class A output stage generally doesn't require a driver. A driver transistor usually operates class A-B (midway between A and B), as an emitter-follower with the output taken from the emitter. Although contributing little or no *voltage* gain, this operating mode provides a high input impedance, and a low output impedance to match the next stage, a requirement for maximum power transfer.

790 THE SOUND SECTION

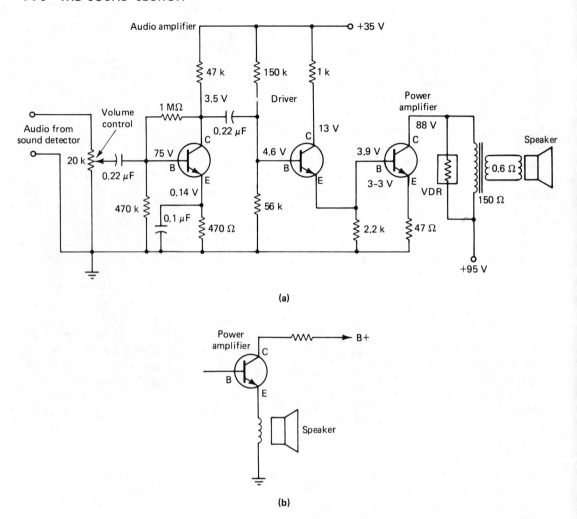

Fig. 17-7 Typical single-ended AF amplifier.

17-11 THE OUTPUT STAGE

A speaker is an electromagnetic device that requires audio *power* for its operation. Because the voice coil is a very low impedance (typically between 8 and 32 Ohms), this power must consist mostly of *current,* at low voltage.

REMINDER:

Power (in watts) is the product of voltage and current ($W = I \times E$), so increasing either voltage or current means reducing the other. Once the AF signal level has been amplified sufficiently, it's the job of the output stage to develop and transfer this power to the speaker. The high audio current is obtained in one of two ways:

17-11 The Output Stage

with a step-down transformer, or by using transistors operating as emitter followers.

There are two types of output circuits in common use: "single ended" circuits (using one transistor), and circuits with two transistors operating in "push-pull". For sets with ICs, the output transistor may be either external to the IC, or contained in the IC along with other AF stages. At present, this is possible only if the audio power is low (about 1.5 watts or less), so that a medium-power transistor can be used without a heat sink. Amplifiers of greater power use high-power transistors that require an efficient heat sink, which is not practical in an IC.

• Single-Ended Output Stage

The typical circuit, shown in Fig. 17-7a, uses a transistor operating class A in a common-emitter configuration. Bias is developed in two ways: by the drop across the emitter resistor, and from the previous directly-coupled stage. Note the VDR (voltage-dependent resistor) which protects the transistor and transformer from transient pulses. Some negative feedback is provided by the un-bypassed emitter resistor.

The output transformer (O.T.) has a high step-down ratio to convert the transistor output power from high-voltage/low-current to low-voltage/high-current. The transformer secondary provides a 1:1 impedance match with the low-impedance speaker. To prevent damage to the transformer and transistor from high-amplitude "noise spikes", a zener diode and/or a VDR may be connected as shown.

The circuit shown in Fig. 17-7b doesn't require a transformer. Here the speaker is driven *directly* from the low-impedance emitter circuit of a transistor operating as an emitter follower. With this arrangement, a shorted speaker can damage or destroy the transistor, and a shorted transistor can burn out the speaker.

• Push-Pull Output Circuits

Currently, most sets have two transistors in the output stage, operating class B. Since current flows for only part of a cycle (compared to class A where there's current flowing at all times), this arrangement is more efficient than a single-ended output, providing more undistorted power with less heat dissipation. In push-pull operation, the transistors conduct alternately on alternate half cycles of the applied signal. This requires signals of opposite polarities. In the past, they were obtained from a center-tapped input-coupling transformer (there's a 180° phase difference between the voltages at opposite ends of a transformer). Current practice is to use a transistor as a *phase inverter;* signals 180° apart are obtained from the emitter and collector circuits. As stated earlier, transistors operating class B require a certain amount of *drive power* which must be furnished by the previous stage.

Figure 17-8 shows two push-pull circuits commonly used. The circuit in 17-8a drives the two transistors with a phase inverter/splitter. Equal voltages of opposite polarity are developed across the equal-value load resistors R1/R2 and are capacity-coupled to the output transistors. A center-tapped output transformer is required. The output transistors are biased class B (to cutoff) and don't conduct in the absence of a signal. While the positive-going signal at the collector of Q1 is driving Q2 into conduction, the negative-going signal at the Q1 emitter keeps Q3 cut off; then the process is reversed and Q3 is made to conduct while Q2 is cut off. Because of degeneration in the Q1 emitter circuit, Q1 doesn't contribute to the amplification. With current flowing in opposite directions through the two halves of the output transformer, both half-cycles of the signal are induced as AC into the secondary winding to drive the speaker. Any even-harmonic distortion is cancelled out by bucking action, and an emitter by-pass capacitor is not required if the Q2/Q3 circuits are properly balanced.

- **Complementary-Symmetry Circuits**

These popular circuits, found in most hi-fi equipment, are currently used in many TV receivers. They have low distortion, and with their low output impedance can directly drive the speaker with no need for an output transformer. There are several variations of these circuits, one of which is shown in Fig. 17-8b. Such circuits are also used extensively in vertical sweep circuits to drive the yoke (see Sec. 7-13 and Fig. 7-13 and 14).

The basic complementary-symmetry circuit of Fig. 17-8 uses two *different* type transistors (a PNP and an NPN) operating class *B*. Their collector/emitter circuits are connected *in series*, between B+ and ground. Operating as a "pair", one transistor "complements" the other. The applied *B* voltage is split between the two transistors and the DC measured at their emitters is approximately one-half the source voltage. From the junction of the emitters, the audio output is fed via a large-value coupling capacitor to the speaker. To pass the relatively high speaker current (especially for the low frequencies) in this low-impedance circuit requires a typical capacity of between 100 and 500 mfd. The resistor across the speaker serves as a protective load in case the speaker is disconnected while the set is on.

Transistor Q1 is a conventional AF *voltage* amplifier, and Q2 a driver amplifier, both operating in the common-emitter mode. Negative feedback from the output stage is supplied to Q1 via the voltage-dividing network of R1, R2, and C1. The amplified signal from the collector of Q2 feeds the base of Q3 directly, and the base of Q4 via the diode D1 and R3. Transistor Q1 gets its DC operating voltage via the B/E junction of Q2. Note the absence of a phase inverter.

Unlike a transformer-fed output stage, the signal *polarity* feeding each output transistor is the *same,* and push-pull action occurs because the transistors are *different types.* With a positive-going signal at the base of Q3, its forward bias is increased and the transistor is turned on. At the same time, the same polarity of signal driving Q4 reduces its bias, turning the transistor off. With a reversal of signal polarity, the opposite is true: Q3 is turned off and Q4 conducts. Thus the two transistors are made to conduct alternately, but for *less than 50%* of each cycle. For minimum distortion, the duration of conduction, which is established by the bias, is quite critical.

Because each transistor conducts for less than a full half-cycle, a problem is created by a "gap" in the output before one transistor takes over after the other leaves off. This is called *crossover* or "notch" distortion, a kind of distortion that is particularly annoying at low volume levels. In this circuit (Fig. 17-8b), the distortion is minimized by offsetting the bias of Q4 with the diode D1 and resistor R3. With silicon-type transistors, two diodes are often used for *bias stabilization* to reduce distortion and prevent *thermal runaway*. With germanium transistors, a VDR (voltage dependent resistor) or a zener diode may be used for this purpose. Besides optimum bias, minimum distortion requires that the two transistors be operated as a *matched pair*.

- **Another Circuit Configuration**

One variety as shown in Fig. 7-14 is called a *Quasi-complementary* circuit. Note how it differs from the circuit just described. For example, the two output transistors are of the same type, and therefore must be preceded by a phase inverter to drive them with opposite-going polarities. With identical transistors, both positive and

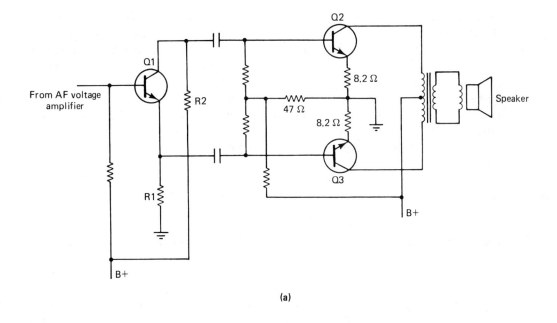

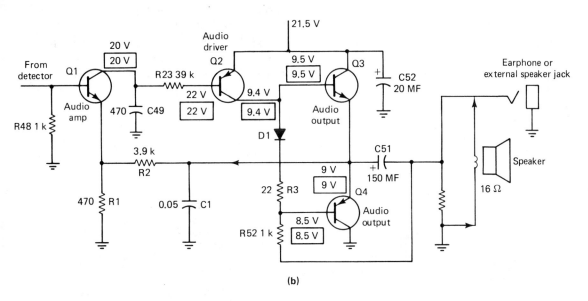

Fig. 17-8 Push-pull output amplifiers operate class B for greater power and efficiency: (a) phase inverter/splitter driving a p-p output stage; (b) complementary symmetry output circuit.

negative DC power sources are required. Under no-signal conditions, with the transistors conducting equally, the DC voltage at their junction is approximately *zero*. This makes it possible to feed the speaker *directly* from the junction without the need of a large coupling capacitor. With this arrangement, however, the speaker can be damaged by an unbalance of the transistor currents, and removing or shorting the speaker with the set on can destroy the transistors. Some output circuits have a protective fuse, rated at about 1 A.

17-12 THE SPEAKER

The speaker converts audio currents into sound waves, to faithfully reproduce the sounds originating at the station. It is an electromagnetic device, operating on the same "motor principle" that deflects the beam of a CRT (see Fig. 15-4). A basic speaker such as shown in Fig. 17-9 consists of the following parts: a PM-type magnet attached to a metal *frame;* a paper *cone* (to move the air); a *voice coil* attached to the apex of the cone; and a flexible *cone support*. The voice coil is wound on a paper form cemented to the cone. Connections to the coil are via two flexible wires attached to "eyelets" on the side of the cone. The magnet is fitted with *pole pieces,* providing a small circular "air gap" within which the voice coil is free to move. The voice coil is centered and supported within the gap by the flexible cone support. The outer rim of the cone is flexible and cemented to the speaker frame. Audio currents flowing through the coil produce a magnetic field that reacts against the flux of the PM magnet to move the cone in and out with each AC reversal, to create sound waves. The volume of the sound varies with the amplitude of the current, the flux produced, and the excursion (distance of back-and-forth movement) of the cone. The rate of movement varies with the frequency of the AF currents. A good speaker has a frequency response from about 50 Hz to 15 KHz and up.

Speakers are rated according to their power-handling ability, efficiency, and frequency range. The inexpensive speaker found in most

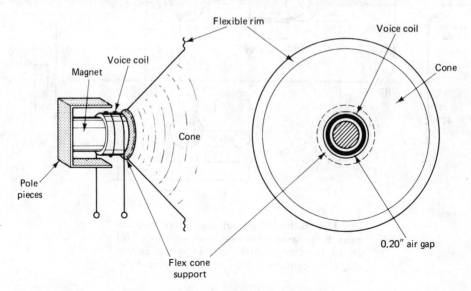

Fig. 17-9 Speaker construction details.

17-14 Negative Feedback

receivers is rated at about 5 watts, with a limited frequency response. For maximum *undistorted* power output, the speaker voice-coil impedance must be matched to the AF output stage. The impedance of speakers varies from as low as 3.2 ohms to as high as 40 ohms in some cases. Typical *impedances* are 4, 8, 16, and 32 ohms. The *ohmic resistance* of a voice coil is considerably less than these figures.

- **Multiple Speakers**

To improve the quality of sound, many console-type receivers use two or more speakers. One of them may be a "woofer" for the bass tones, the other a "tweeter" for the highs. A *coaxial* speaker has two cones for better response at both low and high frequencies. Two or more speakers must be *phased,* i.e. their connections "polarized" so that their cones move in and out together. If they were working in "opposition", the quality of sound would be degraded.

- **Crossover Networks**

Ideally, low frequencies should drive a woofer, with a tweeter receiving the highs. In hi-fi systems, special L/C/R *crossover networks* are used to divide the frequencies. Most TV sets simply connect a capacitor in series with the tweeter, for low reactance to the highs and increased opposition to the low frequencies. The capacitor is non-polarized and usually consists of two electrolytics connected back-to-back.

- **The Earphone Jack**

Many sets have a jack into which headphones may be plugged for private listening. As shown in Fig. 17-8b, the jack has an extra contact for disabling the speaker. In some cases, there's a low-value "load" resistor to prevent damage to the output transistors. A phone jack can easily be installed in any set. With a high-impedance headset, connection is best made at some upstream point in the audio amplifier, such as the hot side of the volume control.

17-13 VOLUME AND TONE CONTROLS

The typical volume control is a potentiometer connected at the input of the AF voltage amplifier where it functions as a variable voltage-divider for the audio signal. Its value is fairly high (between 10 and 50K ohms) to minimize circuit loading. Its resistance element is usually *tapered* for non-linear control depending on the circuit. In some cases, the volume control is one section of the tone control and/or the receiver turn-on switch. Sometimes the volume control has a *fixed tap* for emphasizing low frequencies at low volume-level settings of the control.

When ICs are used, the volume control, as explained earlier, controls the signal *indirectly* by varying the DC supplied to one of the IC transistors. With this arrangement, the control may be remotely connected without the need for shielded wires.

Hi-fi systems are somewhat complex with separate bass and treble controls. In most TV sets, the tone control is simply a potentiometer and capacitor connected in series between some high-level point in the AF amplifier and ground. The capacitor, depending on its value, bypasses some of the high frequencies to ground giving the effect of an increase in bass response. The pot provides a means of varying the capacitive reactance, and thus the overall tone.

17-14 NEGATIVE FEEDBACK

Negative (or "inverse") feedback is an inexpensive way of reducing distortion and improving the quality of sound developed by an AF amplifier. It may be accomplished in several ways; usually it is achieved by taking some of the signal from the output of a stage and feeding it back, *out of phase,* to the input of the stage, or to some earlier stage. This is *opposite* in effect to "positive" or regenerative feedback, where the feedback is *in-phase,* producing oscillations. Positive feedback tends to increase signal levels; negative feedback results in a *decrease in stage gain.* This decrease, however, is usually of secondary importance to the improved response. Any amplitude or frequency distortion in the amplifier between the output and input of the feedback "loop" is reduced by *cancellation.* The greater the amount of feedback (within limits), the better the response, but at the expense of a proportionate reduction in gain.

Negative feedback can also be used to increase the impedance of the base input circuit of a transistor, or to reduce any positive-feedback instability in a circuit. An example of negative feedback is shown in Fig. 17–8b, where the feedback loop encompasses three stages.

17–15 SOUND–SECTION ICs

The sound functions on modern sets are performed by ICs. Figure 17–10 and Sample Circuit C at the back of this book show a typical sound module where a single IC contains all the functional stages including the power output stage. From the sound takeoff in the video IF strip, the 41.25 MHz and 45.75 MHz IF carriers are combined in the sound IF detector to produce the 4.5 MHz sound IF carrier which is tuned by coil L 114 and fed to the IC. In the IC, the signal goes through an IF amplifier/limiter and a low-pass filter to feed the quadrature-type FM detector. The detector is tuned by the external coil L 188. From the detector the audio signal is amplified, then coupled externally through capacitor C 182 to the speaker. The volume control is part of a voltage-dividing network for DC control of the audio amplifier via the DC volume-control block. The applied *B* voltage is regulated within the IC.

17–16 TROUBLESHOOTING THE SOUND SECTION

With the possible exception of the AF output stage, the sound section gives very little trouble, and when it does troubleshooting is fairly simple and straight-forward—in fact, no more difficult than for the corresponding stages of a radio receiver. This is because there are relatively few possible trouble symptoms, no interrelated stages, and no difficulties in tracing circuits. With ICs, of course, there's the need to make numerous tests before considering replacement.

In most cases, troubles can be quickly localized to one of two areas: the FM detector and the upstream IF stages; or the audio stages up to and including the speaker. The volume control is approximately mid-way and makes a convenient test point for isolating the problem to one section or the other. Generally, the only test instruments required are an EVM and some means of signal-tracing. Suggested troubleshooting procedures are presented in the flow drawings of Fig. 17–11 and Table 17–1.

17–17 INSPECTION AND SYMPTOMS ANALYSIS

As you would for troubles elsewhere in the set, make a preliminary inspection for obvious defects in the sound stages, first with the set off, then with power applied. Watch for potentially destructive conditions, like arcing, and overheating resistors and transistors, particularly in the power output stage. (*continued on page 803.*)

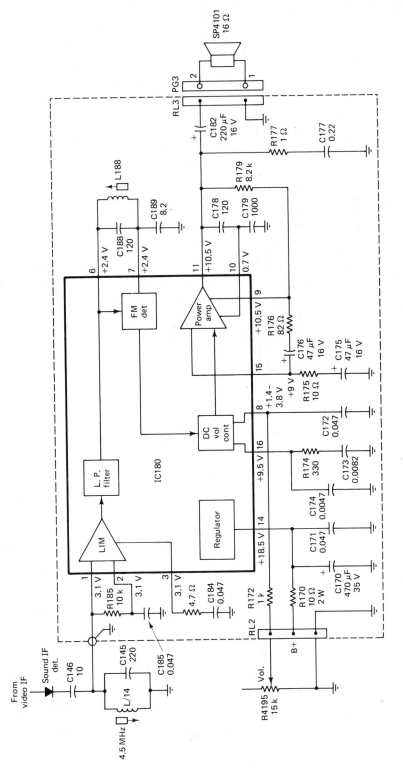

Fig. 17-10 Typical sound-section IC.

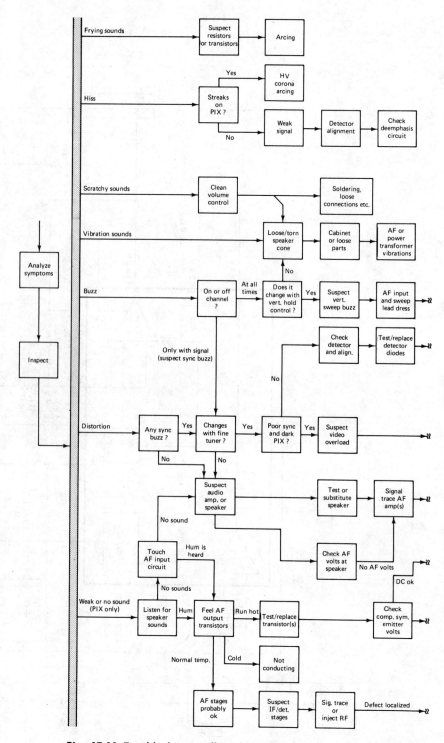

Fig. 17-11 Troubleshooting flow drawing.

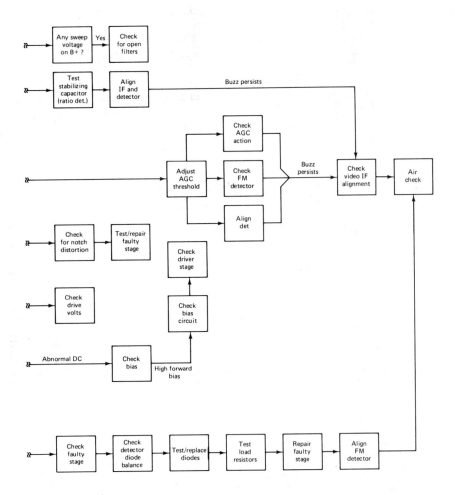

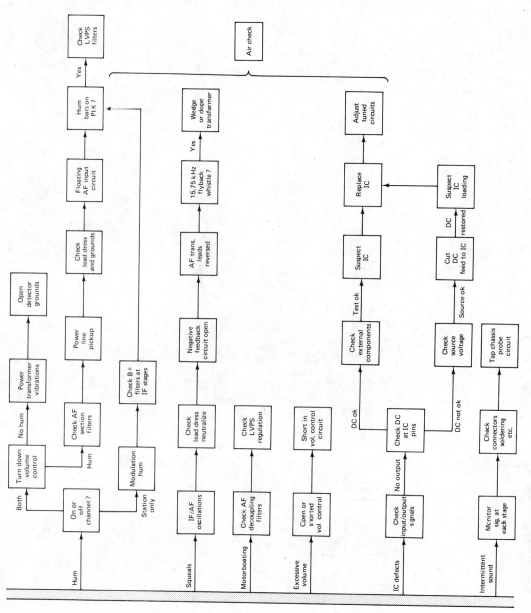

Fig. 17-11 Troubleshooting flow drawing (continued).

Table 17-1 TV Sound Trouble Chart

Trouble Symptom	Probable Cause	Remarks/Procedures
No sound, no pix	Dead stage ahead of sound takeoff . . . tuner, video IF strip, and video detector or amplifier if a BW set.	In some cases, a bad stage ahead of the sound takeoff can cause loss of pix but *some* sound.
No sound, pix ok	Trouble in sound section after sound takeoff. . . . sound IF detector (if a color set), sound IF stage, limiter, FM detector, AF voltage amplifier, power output stage, or speaker.	Listen for faint hum or noise from speaker; If some sound, trouble is probably ahead of the volume control. Check for hum when volume control is touched. Localize trouble by signal tracing or injection.
Weak sound	Loss of gain in one of the above sound stages; either a defect or misalignment. Leaky transistors or capacitors. Resistors changed value. Shorted output transformer.	As above. If no defect, check the alignment.
Intermittent sound	Intermittent loss of signal in one of the sound stages. Poor soldering, loose connections, etc.	Monitor for signal cutting out at various stages. Tap and probe as appropriate.
Distorted (garbled) sound	FM detector defect or misalignment, audio amplifier defect, or bad speaker. Video overload. Output stage notch distortion.	Suspect detector defect or misalignment if changes with fine tuner. Check AF stages.
Sync buzz	FM detector defect or misalignment. Limiter stage defect. Video overload. Video IF misalignment.	Buzz and distortion usually go together, and changes with fine tuner. Adjust AGC threshold. Check other possibilities listed.
Vertical sweep buzz	Coupling between AF stages and vertical output circuit or yoke.	Sweep buzz changes with adjustment of vertical hold control. Check means of coupling, . . . inductive . . . poor lead dressing, open decoupling filters, etc.
Squeals or oscillation	Positive feedback in IF or AF stages.	Determine means of feedback and correct. Check lead dress.
Motorboating	Open or dried-up decoupling capacitors.	Try bridging all suspected capacitors with suitable substitutes (make and break connections only when set is OFF).

Table 17-1 (cont'd)

Trouble Symptom	Probable Cause	Remarks/Procedures
Excessive volume (little or no control)	Volume control open at ground end.	Check control
Hum	Open/dried up LV filter capacitors, or APF circuit defect. Open ground at AF input circuit. Induction pickup from power transformer or power line. Leakage in power switch/volume control.	For hum on- and off-channel, check filtering of B+ feeding AF stages. For modulation hum (only on a station) check filtering at IF/detector stages.
Vibration sounds	Loose or torn speaker cone. Vibrating transformer cores (loose power trans. core causes hum; buzz from vertical sweep trans; station sounds from AF output trans.).	Repair or replace speaker. Wedge or dope loose laminations. Inspect for loose cabinet or chassis parts and secure.
Scratchy noises Frying sounds	Speaker voice coil rubbing. Dirty volume control or tuner. Poor soldering. Loose connections. Burned resistor (carbonized). Corroded connections in coils or transformers.	Repair or replace speaker. Clean dirty controls. Tighten connections. Check coils/transformers for resistance variations.
Hiss or sizzling sounds	Detector defect or alignment. Weak signal pickup. De-emphasis or tone control defects. HV corona arcing.	Repair and/or align detector. Listen at HV cage for corona arcing. See if it changes with brightness control.
IC Sound problems	Defective IC or external components.	Replace IC only after checking signal input/output, voltages, and external components.
Popping sounds	Speaker may be bottoming out at high volume. Arcing.	Check for HV arcing, and flashovers in the CRT.
Static or other interference	Pickup from nearby electrical equipment. Poor AM rejection by limiter stage or FM detector.	Determine if pickup is by antenna or the AC power line. If set is ok there may be no easy remedy.
Overheating transistors	Transistor has excessive leakage. Too much forward bias. Excessive signal drive. Inefficient heat sink.	Check/replace transistor. Check for cause of wrong bias. Inspect heat sink.

17-17 Inspection and Symptom Analysis **803**

Table 17-1 (cont'd)

Trouble Symptom	Probable Cause	Remarks/Procedures
Poor tone quality	Multiple speakers improperly phased. AF amplifier defect. Overheating output transistors.	Check speaker phasing. Check frequency response of AF amplifier.
High-pitched 15.75 KHz whistle (not audible to some people)	Horizontal-sweep oscillator squegging. Loose flyback core or windings.	Sound emanates from transformer, not the speaker.

A list of sound trouble symptoms follows. For some of them, the fault may or may not be with the sound section. When the sound *only* is affected, the trouble is obviously in the sound section. When raster and/or pix are also affected, look elsewhere. Although defects in other circuits can affect the sound, it's seldom the other way around. Of course it's always possible to have two or more *unrelated* troubles in separate sections of a set.

• **Weak Sound Or No Sound**

Listen for a faint background hum at the speaker with the volume turned up. If there's none, suspect the audio stages or the speaker. If there's some hum or noise, look for trouble between the sound takeoff and the output of the FM detector. If there's some sound, but loss of pix, it's natural to check the video section downstream from the sound takeoff; but don't be misled. Certain defects in the video IF strip *ahead* of the sound takeoff will often let some sound through, leading to an incorrect diagnosis. For loss of both pix and sound there is usually a *common cause:* the antenna system, AGC, loss of *B* voltage, the tuner, the video IF strip; or in a BW set, the video detector or amplifier. Start by adjusting the AGC control.

• **Distortion**

This is not to be confused with static sounds or other kinds of interference. Distortion may be caused by a circuit *defect,* by the speaker, or by misalignment of the FM detector. The distortion may be the same in each case, so certain tests must be made to determine the nature of the problem.

• **Background Noises**

This includes such things as hum, buzz, mechanical (vibration) sounds; hiss; howls, squeals or motorboating, and scratchy, staticy sounds, each with a different cause. With some, the pix is also affected; with others, only the sound.

Hum is caused by poor filtering of the LVPS or inductive/capacitive pickup by the audio amplifier. There are two kinds of buzz, intercarrier (sync) buzz, and vertical sawtooth buzz. With different circuits responsible, it's necessary to distinguish between the two. Sawtooth buzz changes with the vertical hold control. With sync buzz there are several possibilities: video overload as caused by loss of AGC (where the pix is also affected); trouble with the FM detector (either a defect or misalignment); or video IF misalignment. Vibration sounds may be caused by a bad speaker, loose baffles, cabinet resonances, transformer laminations, etc. Hissing sounds may be caused by HV corona arcing (which also shows up on the pix), arcing, or a defect in the deemphasis network.

804 THE SOUND SECTION

High-pitched squealing may be caused by IF oscillations; for lower-pitched howling sounds suspect the audio stages. The "putt-putt" sounds of motorboating usually indicate an open decoupling or filter capacitor. For staticy sounds there are numerous possibilities: the speaker, arcing, dirty volume control or tuner contacts, loose connections, and so on. Most of the above symptoms can be intermittent in nature.

17-18 LOCALIZING THE PROBLEM

As with any problem, the troubleshooting approach depends on such things as the trouble symptoms encountered, the available test equipment, and the inclinations and experience of the technician. If weak or missing sound, determine if the speaker and AF amplifiers are working by listening for residual background hum. Try touching the input of the first audio amplifier with a screwdriver or the equivalent. A good hum can normally be expected, due to pickup from AC power wiring. Avoid shorting or grounding closely-spaced contacts such as IC pin connections. Where possible, make contact with the same circuit at some alternate point such as the "hot" side of the volume control (unless it's a DC controlling type). If there's hum, see if it varies with the control. Where there's no hum, try a substitute speaker. Use a speaker whose impedance is close to the original and avoid opening or shorting the speaker leads with the set on. In some cases this can cause damage to the output transistor(s), by removing or changing the load. Handle the speaker with care to avoid puncturing the cone. If still no hum, check for loss of DC power in the AF section. Check for an open in the normally closed contacts of the earphone jack if there is one. Check for wires broken loose from the volume control, a common problem. If an abnormally loud hum, suspect the ground lead. If still no hum, suspect trouble in the audio stages. If normal hum, the audio section is probably ok and the trouble is somewhere upstream from the detector.

• **Locating the Bad Stage**

Either trace the *station signal,* stage-by-stage, or inject a signal from a generator. For signal-tracing, use a CRO, an aural tracer, or an AC EVM as appropriate. An aural tracer is desirable since it provides a "listening check" on sound quality. Under "make do" circumstances, a pair of headphones is suitable, or the audio section of any radio, stereo, or whatever. A demodulator probe is required when signal-tracing IF stages. For signal injection, use an AF generator for the audio stages and a modulated RF generator (tuned to 4.5 MHz) in the IF section. A "noise" generator can also be used. Signal-tracing or injection may begin at the speaker and work upstream toward the sound takeoff, or may go in the opposite direction. If you desire, follow the progress of a generator signal rather than the station signal. For a complaint of *weak* sound, be aware of the relative signal levels to expect at various points, e.g. around 1 V p-p at the detector output, 2 to 5 V p-p at the first AF output, approximately 5 V p-p from the power output stage, and slightly less across the speaker terminals. Actual readings of course depend on the strength of the received or injected signal and the setting of the volume control; therefore they are seldom shown on a schematic.

REMINDER: ─────────────────

Sound limiters and phase inverters provide little or no amplification.

If no signal at the IF input, suspect the sound-takeoff coil, or, if it's a color set, a bad IF detector. If sound is *distorted,* listen carefully at both high and low volume levels. If speaker is suspect, try substitution. If there's also *buzz* that changes with the fine tuner, suspect a detector defect or its alignment. Test detector diodes and other components. For a quadrature-type detector, touch-up the quad coil while listening to a station. If necessary, perform a *complete* alignment of the sound IF and detector stages.

If distortion persists, and there is no buzz, check the quality of station sound (with an aural tracer or the equivalent) at various points in the AF amplifiers. If you desire, inject a sine wave from an AF generator and examine the waveform (for distortion) at different check points. Try different signal levels as appropriate.

Background sounds are distinctive and easily identified, but finding the cause can be a problem. For *hum,* both on- and off-channel, suspect poor filtering of the DC power supplying the AF amplifiers. If it occurs only when a station is received (modulation hum), check for ripple on the DC feeding the IF/limiter stages. There may or may not be hum bars on the pix. Avoid bridging filters with the set on.

Loose connections and poor soldering are common causes of crackling, static-like sounds. Clean noisy controls and tuner contacts as necessary. Probe circuits in the sound section using an *insulated* probe. Gently stress PCBs. Try tapping. Resolder all suspicious connections, particularly around large terminals and lugs where cracks tend to occur. For intermittent problems, monitor signals and DC voltages at various check points as appropriate. Other background sounds are localized as described earlier.

With ICs, of course, no testing can be done on the IC itself. Signal-trace inputs and outputs as previously described, and check the applied DC voltages and external components. When the voltages and signal input are OK, but the output signal is abnormal or missing, the IC is suspect and should be replaced.

17-19 ISOLATING THE DEFECT

After localizing a problem to a single stage, the next step is to pinpoint the *cause.* Normal procedure is to check the transistor and any diodes in the troublesome circuit, transistor operating voltages, and, if necessary, all other components in the stage, in that order. Generally, as with other sections of the set, *alignment* is considered only as a last step when no defect can be found. For troubleshooting purposes, the sound section can be considered as 5 functional areas; the sound IF detector, the IF/limiter stages, the FM detector, the audio amplifiers, and the speaker. In the following procedures, each is treated separately.

17-20 TROUBLESHOOTING THE SOUND IF DETECTOR

This stage (found only in color sets) gives very little trouble, because of the weak signals at this point in the set, because of the few components, and because of the absence of DC power (except when a transistor is used). If a CRO (with demodulator probe) shows normal signal input but little or no 4.5 MHz output, check the diode (or transistor). Check transistor voltages, where applicable. When replacing a diode, make sure of the correct polarity. If troubles persist, suspect the tuned circuit, either a defect or misalignment.

17-21 TROUBLESHOOTING THE IF/LIMITER STAGES

Except in rare cases of transistor breakdown, these stages too are almost trouble-free. Possible trouble symptoms are: no sound, weak sound, oscillations, or sync buzz, and other kinds of AM interference, if the limiter isn't doing its job. Check the transistors, their operating voltages, and other components as appropriate. For loss of gain where no defect can be found, suspect misalignment. For oscillations, check for an open neutralizing capacitor and check the lead dress.

17-22 TROUBLESHOOTING FM DETECTORS

Trouble symptoms commonly associated with the detector are no sound or weak sound, and buzz and distortion (which usually go together). For a weak or dead stage, check the diodes (or transistor) and the alignment. For buzz and distortion the possibilities are: an imbalance in the bridge circuit (defective diode, transformer, or load resistor); an ineffective limiter stage; an open or dried-up stabilizing capacitor in a ratio-detector circuit; and video overdrive, or misalignment of the video IF strip.

For a discriminator-type circuit, perform the tests described in Sec. 8-14. In particular, check and compare the DC voltages developed across the two load resistors; they should be equal and of opposite polarity. If the circuit is unbalanced, check and compare the load resistors and the conductance of the diodes. Defective diodes should be replaced as a *matched pair*. The presence of DC voltage indicates that the signal is reaching the detector and is being rectified by the diodes. If there is little or no DC, look for the cause at the detector input or at some point upstream, including misalignment of the tuned circuits.

Most of the foregoing tests also apply to a ratio detector. For AM rejection, this circuit is dependent on the large value *stabilizing capacitor* across its load resistor(s). An open or dried-up capacitor is a common cause of buzz and distortion. Try bridging with a substitute, making sure of the correct polarity. If the capacitor checks OK and overload is suspected, try detuning the sound takeoff coil and/or adjusting the AGC threshold control.

REMINDER: _____

Insufficient AGC is a common cause of overdrive.

If reducing the signal clears up the buzz and distortion, suspect video IF misalignment; perhaps the 41.25 MHz marker is too high on the curve, where it receives too much amplification. If pix quality is good, an alternative to realignment in some cases is to use a smaller coupling capacitor at the sound takeoff. Try a 1pF capacitor or a "gimmick" consisting of two insulated wires twisted together. If trouble persists, try adding a 41.25 MHz trap in the video IF strip ahead of the sound takeoff.

17-23 ALIGNMENT

Common trouble symptoms that *may* indicate misalignment of the sound section are: loss of sound, weak sound, and distorted garbled sound usually accompanied by buzz. Compared to those used for video IF alignment, the procedures are simpler and considerably less time-consuming. With separate, discrete sound stages there are several tuned circuits that may require adjusting: the sound takeoff, one or more sound IF/limiter circuits, and either a detector transformer or quadrature coil. With *ICs* there's usually only the sound takeoff and a quad coil. The sound IF/limiter tuned circuits are simply peaked to 4.5 MHz. Alignment of the *detector* coils is extremely critical and must be done with great care to prevent buzz and distortion.

17-23 Alignment

There are three methods for aligning the sound section: alignment by "ear"; alignment with an unmodulated RF generator and a CRO or EVM; and sweep alignment. Alignment by ear is usually limited to slight touch-up adjustments, and the procedure is simple. Tune to a station and adjust the fine tuner for a good quality pix. Peak the sound takeoff coil and all IF/limiter tuned circuits, including the primary of the discriminator or ratio detector transformer, for *loudest* sound. Don't touch the fine tuner; maximum sound doesn't coincide with the best pix. Adjust the detector-transformer secondary coil, or quad coil, as applicable, for the clearest undistorted sound.

For alignment with instruments, proceed as follows.

- **Preliminary Steps**

1. Disconnect antenna, short out the terminals, and tune to a blank channel to prevent interference pickup and misleading indications.

2. If circumstances warrant, disable the HV to reduce the chance of shock.

3. Connect signal generator (RF or sweep generator as applicable) through a small blocking capacitor to some point ahead of the sound takeoff. Tune generator for a moderately strong signal at precisely 4.5 MHz. Connect the indicator (CRO or EVM) to the appropriate point at the detector output as specified in the following procedures.

- **Alignment Using RF Generator and EVM**

Connect meter, on a low DC range, to the junction of the detector load resistors. If it is a ratio detector with only one resistor, parallel it with two 100k-ohm resistors in series to provide a temporary junction point. Tune generator each side of 4.5 MHz noting changes in meter reading. Peak all tuned circuits (except the detector-transformer secondary coil or quad coil) for maximum indication at 4.5 MHz. When calibration accuracy of generator is in doubt, first peak the coils using a *station* signal, using it as an accurate standard. Readjust the generator as appropriate.

Move meter connection to the ungrounded end of the detector load. Adjust the transformer secondary coil for a *minimum* (null) indication on the meter. Switch to a lower meter range, reducing the generator output as necessary. Touch-up the adjustment for the lowest possible indication, which is normally close to zero. Check results on a station. If there is distortion or buzz at the best pix setting of the fine tuner, make a further touch-up of the adjustment. If the problem persists, recheck the circuits for defects.

- **Sweep Alignment**

Sweep alignment, the preferred method for aligning the video IF strip, is seldom used for the sound section because of the equipment requirements and the time required for their set-up. As shown in Fig. 12-23 and described in Sec. 12-31, the equipment consists of a sweep and marker generator, and a CRO. One advantage of this method, however, is that it provides a visual check on alignment *conditions* before making any adjustments. Accuracy of course is dependent on the marker generator, and unless it's close to 100% accurate at 4.5 MHz, the expected results will not be obtained. Alignment procedures are as follows.

Connect sweep generator to some point ahead of the sound takeoff. Adjust generator for a sweep of about 200 KHz, centered at 4.5 MHz. Connect the CRO to the detector output (usually the hot end of the volume control). An "S" curve should be obtained as shown in Fig. 17-5b and described in Sec. 17-7. Adjust generator tuning to center the curve on the scope. Adjust CRO controls to obtain the proper curve proportions, as indicated.

Loosely couple the marker signal to the same point as sweep generator injection. A marker blip should appear on the S curve. Adjust marker-generator output until the blip is barely visible and doesn't distort the curve. Adjust the detector-transformer secondary coil until the marker falls at the crossover point on the S curve.

Adjust the sound takeoff coil, all sound IF and limiter coils, and the primary of the detector transformer, for maximum amplitude of the response. Alternately, the coils may be peaked with the marker generator and an EVM as described earlier, or the CRO may be connected at the detector input, with a demodulator probe. With this method, a bandpass curve is obtained as shown in Fig. 17-5b.

Adjusting the secondary of the detector transformer is the final step in the alignment procedure. The coil is adjusted until the two peaks of the S curve (Fig. 17-5) are equidistant from the center crossover frequency of 4.5 MHz, and the trace between the peaks is fairly straight and linear. Disconnect equipment and check results on a station.

- **IC Alignment**

Tune to a strong station and adjust quadrature coil for best sound. Reduce signal strength until background *hiss* is heard. Adjust sound-takeoff coil for maximum sound. Touch-up the quad coil for best sound quality without buzz.

17-24 TROUBLESHOOTING THE AF STAGES

Assuming that the trouble has been localized (at least tentatively) to a single AF stage, and the problem is loss of sound, start by checking the transistor and its operating voltages. If it's the AF *voltage* amplifier, or driver, the usual defects are: a bad transistor; an open, shorted, or leaky coupling or decoupling capacitor; and a defective resistor, particularly in the biasing network. When direct coupling is used, a defect in this stage will upset the bias and operating conditions of the following stage(s). Besides weak or missing sound, audio *distortion* can also be caused by these same defects.

- **Distorted Sound (garbled, muffled, or tinny)**

Distortion and loss of volume in an audio amplifier usually go hand in hand, both caused by the same defect. The usual causes are: a bad transistor; improper operating voltages, particularly the bias; a wrong-value load resistor or impedance; and signal overdrive or speaker overload. With complementary symmetry output circuits, two common problems are crossover (notch) distortion, and thermal overload. As stated earlier, locating the stage where the signal becomes distorted can best be done with an AF generator and CRO, a distortion analyzer, or an aural tracer. The stage where distortion occurs however is not necessarily at fault; it may be overdriven from some defect upstream. Distortion with sync buzz is usually caused by the FM detector.

The characteristic operating curves of Fig. 17-12 illustrate the importance of correct bias and signal drive. Fig. 17-12a shows the conditions for normal operation of a typical NPN transistor operating class A. The no-signal bias of +0.6 V locates the operating point mid-way on the linear portion of the curve. With a sine-wave input there's an undistorted sine-wave output.

Figure 17-12b shows the effect of too much forward (+) bias. The positive-going half-cycles drive the transistor into saturation, clipping or compressing the upper portions of the signal. In 17-12c, there is insufficient forward bias, so the negative-going half-cycles drive the transistor

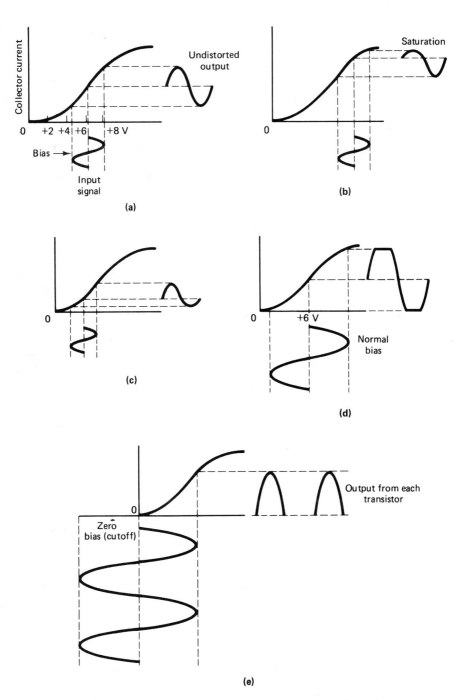

Fig. 17-12 Output transistor operating conditions: (a) normal class A operation; (b) insufficient forward bias; (c) excessive forward bias; improper bias or signal overdrive creates distortion in an AF amplifier; (d) signal overdrive; (e) class B.

toward cutoff, distorting the lower parts of the output signal. In 17-12d, the no-signal bias is correct but the transistor is overdriven. Note the flattening of *both* upper and lower portions of the output signal as the transistor is over-driven alternately to saturation and to cutoff.

The conditions for class *B* are similar to those of Fig. 17-12c. To prevent distortion, two transistors must be operated in push-pull to provide an output for both positive and negative half-cycles of the signal. Class *B* (as compared to class A) operation is much more efficient since there is practically no current flow in the absence of a signal.

Another kind of distortion is due to a poor frequency response. For loss of bass response, check for an open or dried-up electrolytic-type coupling capacitor. Loss of high frequencies may be caused by a leaky/shorted capacitor in the tone-control circuit or a by-pass from some AF signal point to ground. If necessary, the frequency response may be checked with either a square wave or a sine wave from an AF generator. A square-wave signal at some low frequency is injected into the input of the amplifier, and the output signal is observed on a CRO. A square wave is rich in harmonics, so a good square-wave output indicates a good response. A distorted square wave indicates a poor response.

With a sine wave signal, the frequency of the generator is varied over the full audio range from about 50 Hz to 15 KHz or more, while the output *level* is monitored with an indicator such as a CRO or AC EVM. Some dropoff can be expected at the two extremes. Repeat the test for both extreme settings of the tone control.

- **Troubleshooting the Output Stage**

Because of the relatively high peak levels of signals, and the high DC currents, this stage is where most breakdowns occur. The most common defect is a leaky or shorted transistor. When this occurs it can cause secondary problems such as burned emitter- or collector-circuit resistors, a damaged output transformer or speaker, or a blown "fuse resistor" if there is one. Sometimes breakdowns occur in reverse, i.e., a circuit short, a leaky/shorted coupling capacitor, or some upstream defect will overload the transistor(s). In this case, one or more junctions will check leaky or *open*. As a rule, when a bad transistor or other component is found, check further for the *cause*.

Checking the operating temperature of an output transistor (by touching) can indicate the nature of a problem. When a transistor overheats (thermal runaway), it indicates excessive current due to: leakage in the transistor, an inefficient heat sink, too much forward bias, an open or shorted output coupling capacitor (electrolytic), or an open or shorted speaker. When an output transistor doesn't warm up, it means there's little or no conduction. Check for an open transistor junction, insufficient forward bias, loss of B voltage, or an open emitter/collector circuit.

- **Troubleshooting a Complementary-Symmetry Circuit**

In addition to the foregoing problems, this circuit has its own unique troubles. One of the first tests to make regardless of the symptoms is a check on the DC voltage at the common output junction of the two transistors. If the circuit uses two identical transistors and is powered from two DC sources of opposite polarity, the voltage measured at their junctions should be *zero*. A voltage greater than one or two volts of either polarity indicates a circuit imbalance, *which must be corrected*. If the circuit uses two different transistors and one DC power source, the voltage at their junctions should measure one-half the supply voltage, within one or two volts. If not, check for the cause of the imbalance. Also, check and compare the operating temperatures of the two transistors. They should be equally warm to the touch. If one runs hotter than the other, it, or both of them, will eventually fail unless the cause is found and corrected. Although a leaky/shorted transistor is common, it isn't the only possible cause.

Initial testing of class B output stages should be performed under no-signal "quiescent" conditions (volume control turned down), so that the current drain is low. When a transistor overheats, it's a good idea to reduce the power while investigating the cause. Use a *variac* in the AC power line or a 10 watt 50-100 ohm resistor in the DC supply circuit. Another precaution is to insert a low-candlepower car or Christmas-tree lamp in the DC line, or in place of a burned-out fuse, to restore operation. A lamp provides a visual, relative indication of current. An ammeter would indicate actual amounts but at the risk of a burn-out. Bridging an open *fuse resistor* in the emitter circuit, however, is not practical; it would upset the bias. Normal no-signal "idling" current is typically about 5 to 10 MA. It is best tested by measuring the IR drop across an emitter resistor and applying Ohm's law.

If there is no overheating, it's safe to turn up the volume. With a signal, the current may increase to 1 A or more. The stronger the signal the greater the current. At high volume, monitor the temperature of the transistors and watch for overheating emitter resistors. If the circuit has no fuse, it's a good idea to add one (at the DC input).

Make an "in-circuit" test of each transistor. If it is warranted, repeat the test after either removing the transistor or breaking two of the three connections. Defective power transistors should be replaced as a *matched pair*. Use new mica heat-conducting insulators with liberal amounts of thermal grease applied to *both* sides.

Check the DC operating voltages of each transistor, including the driver. Make three bias checks, B to ground, E to ground, and, most important, B to E (the true bias). Compare readings with those shown on schematic. Reminder: Abnormal voltages may be the *cause,* if a transistor is conducting too much or not enough, or the *result* of a defective transistor.

REMINDER: _____

To prevent notch distortion, the bias of one output transistor is usually offset slightly from the other. Abnormal bias readings may be caused by: a bad transistor, a defective resistor in the biasing voltage divider network (including any VDR or bias-stabilizing diodes), a leaky coupling capacitor, or (where direct coupling is used) a defect in the previous stage. To check for notch distortion (which is most noticeable at low volume levels) scope-check the sine wave injected from an audio generator as it appears across the speaker terminals. There should be no flattening of the peaks or gaps between cycles. Also check and compare the two output signals individually from B to E. If necessary, adjust the bias offset. If distortion persists, check the negative-feedback circuit.

If the output stage checks OK, investigate the driver or phase inverter. Check the latter for equal and opposite signals at the emitter and collector. An unbalanced output can be caused by a leaky/shorted coupling capacitor or by one of the load resistors' having changed value.

• **Checking Power Output**

If there's a suspected loss of power output it may be checked as follows. Note the impedance of the speaker and connect a load resistor of the same value in its place. The wattage rating should be at least 5 W. Connect a CRO and an AC EVM, calibrated for RMS, across the resistor. Feed a 400-Hz sine wave from an AF generator through a blocking capacitor to the input of the voltage amplifier. Slowly increase the generator output until the scope indicates distortion. Clipping of one half-cycle indicates an imbalance of the output transistors. Clipping of both half-cycles indicates signal overdrive. Reduce signal level until distortion clears up. Measure the RMS AC voltage and compute the power output; $W = E^2/R$. Compare with the expected power output.

17-25 SPEAKER TESTING AND REPAIRS

Repair shops use "test speakers" for substitution when the condition of the receiver speaker is in doubt, and for substitution when the original speaker is left in the customer's home. When substituting, use a speaker of the correct impedance and avoid shorting or opening the speaker connections when the set is operating. Failing to heed these precautions can result in damage to the AF output transistors.

Trouble symptoms that *may* be caused by a bad speaker are: no sound; weak sound; intermittent sound; poor sound quality; distortion; scratchy noises; and rattling, vibration-type sounds. Typical speaker defects include: a punctured or torn cone; a loose cone (at the rim where it's cemented to the frame); a flexible inner cone support that's become unglued; an open voice coil; a distorted or off-center voice coil rubbing against the pole pieces; iron filings or other foreign material in the magnetic air gap; a break in the flexible leads connecting to the voice coil; or a weak magnet (which is rare).

For loss of sound, verify the presence of AC signal voltage across the voice coil terminals. If signal is fed through an electrolytic coupling capacitor, check for signal attenuation (particularly at low frequencies) at the speaker end of the capacitor. Don't mistake an open in the earphone jack for an open voice coil.

A suspected speaker of course can be checked by substitution. Actual speaker testing depends on the symptoms. In most cases, a defective speaker is replaced rather than repaired. For loss of sound, make an ohmmeter check of the voice coil. A reading between 3 and 20 ohms is typical (not to be confused with its impedance). A decided click should be heard when the meter connection is made and again when it's broken. A 1.5 V battery can also be used for the "click test", but prolonged contact can damage the voice coil. If speaker checks open or intermittent, look for a break in the flexible wires between the voice coil and the terminals. This problem can also cause the speaker to cut out intermittently.

Mechanical buzzing sounds, especially at low frequencies, are often caused by a puncture, crack, or tear in the cone, or the cone's having become unglued at the rim, or similar problems with the inner flexible cone support. To repair a damaged cone, use rubber cement or a small patch of cotton-backed surgical tape, never hard cement or scotch tape that may come loose and vibrate. To recement the inner cone support membrane, the voice coil must be centered in the air gap until the cement sets. Centering can be done with shims made from thin strips of film. For access it may be necessary to remove the felt dust cover.

Fuzzy, scratchy sounds can be caused by the voice coil's rubbing against the pole pieces, by dust or filings in the air gap, or by a loose connection. Dust can be removed with compressed air; filings pose a greater problem. To check for a rubbing voice coil, grasp the outer edges of the cone and gently move it in and out, while listening for a scraping noise. Another method is to drive the speaker with a low-level low-frequency signal. A quick fix can often be made by distorting the frame. Using heavy pliers, grasp edge of frame and twist. Check for rubbing and if no improvement, repeat the procedure at several points of the frame until the voice coil is centered.

A small low-wattage speaker should never be used with a high-power amplifier. With a limited excursion, it will "bottom out" on high-level bass tones. At high volume levels it will soon be destroyed.

Two or more speakers must be properly phased, i.e. their cones moving in and out at the same time. Reversing the connections to one speaker degrades the quality of sound. When polarity is in doubt, momentarily connect a 1.5 V battery to each speaker noting which way their cones move. Make the appropriate connections so that they operate in phase.

17-26 TROUBLESHOOTING SOUND ICs

Troubleshooting procedures are much the same as described earlier for ICs in other sections of a set; check input/output signals, applied DC voltages, and external components, in that order.

REMINDER:

Missing or abnormal DC voltages at IC pins may be caused by either an external defect *or* a defective IC. For complete loss of sound, try a substitute speaker. Where DC control over volume is used, check for DC voltage variations as the control is adjusted. Check for broken wires to the front-panel controls. With most ICs there are only two tuned circuits, the input (4.5 MHz) coil, and a ratio detector or quad coil. For weak sound, distortion, and/or sync buzz, touch-up these adjustments while listening to a station.

If all DC voltages, the input signal, and external components check OK, but there's abnormal or no sound output, replace the IC.

17-27 INTERMITTENT SOUND

This is a fairly common problem. Causes include: loose or broken connections, poor soldering, or cracks in PCB wiring; a defective volume control or speaker. Intermittent frying sounds are often caused by a charred resistor, or by corrosion in a coil, the tuned-circuit coils, or an audio output transformer. Intermittently open capacitors are very common. Twist and probe the pigtails as appropriate. Apply strain and tapping to PCBs, probe wiring with insulated probe, and check for intermittent edge connectors. For elusive problems, monitor signal (by aural checking) at various circuit check points.

17-28 AIR-CHECKING THE REPAIRED SET

During this very important phase of TV repair, don't neglect the sound. This is the time to catch any previously overlooked problems or troubles that are slow in developing. Make sure there's no sync buzz or other interference when the fine tuner is adjusted for best pix. Listen closely for the scratchy, fuzzy sounds of a bad speaker at both low and high volume levels. Monitor the temperature of AF output transistors. If two transistors, make sure they're equally warm to the touch. Indications of thermal runaway call for additional troubleshooting.

814 THE SOUND SECTION

17-29 SUMMARY

1. To prevent interference with adjacent stations, and because of the great many stations and the limited expanse of the broadcast band, an AM radio station must operate within a maximum allocation of 10 KHz. This limits the highest permissible modulating frequency to 5 KHz. Free of that limitation, the FM sound of *TV* can go up to 15 KHz or more, providing much better fidelity. FM also has the advantage of less interference from AM-type noise impulses, and less crosstalk from the video.

2. In FM, the RF carrier frequency is continuously changing in accordance with the AF modulation. The loudest sounds produce a maximum frequency deviation of 25 KHz above and below the assigned carrier frequency. For weaker sounds, the amount of sweep is reduced proportionately. The *rate of sweep* corresponds with the moduating frequency: low audio frequencies produce a slow sweep, higher frequencies, a more rapid sweep, to a maximum of 15 KHz.

3. At the receiver, the sound signal goes through several frequency changes. In the tuner, the beat between the pix and sound RF carriers (which are 4.5 MHz apart) produces a sound IF frequency of 41.25 MHz. In the video detector of a BW set the 41.25 MHz sound carrier beats with a 45.75 MHz pix IF carrier to produce a new and lower sound IF of 4.5 MHz, the difference between the two. In a color set, the 4.5 MHz beat is developed ahead of the video detector, in a *sound IF detector* stage. Regardless of the alignment of the receiver's tuned circuits, the sound IF is *fixed,* as established at the station.

4. Most noise and interference is AM, and steps are taken to prevent such signals from reaching the speaker. A particular problem is intercarrier ''sync'' buzz, caused by vertical-blanking and sync pulses, and high-amplitude low-frequency video signals. Most receivers use a detector that discriminates against AM, preceded by one or more limiter stages. To further reduce the problem, the video IF strip is aligned with the 41.25 MHz sound IF marker low on the response curve where it receives very little amplification. Reminder: A ''beat'' signal takes on the characteristics of the weaker of the two signals being heterodyned.

5. To improve the signal-to-noise ratio, high audio frequencies are overemphasized at the station (preemphasis). In restoring them to their normal level (deemphasis), the receiver also reduces high-frequency ''white'' noise proportionately.

6. The sound takeoff is the point where the sound IF carrier separates from the video. In a BW set, the 4.5-MHz sound takeoff may be anywhere between the output of the video detector and the input of the CRT. In a color set, the 41.25-MHz sound takeoff is at the last video IF stage; following the sound takeoff, the pix and sound IF carriers are combined in a *sound IF detector* to produce the 4.5 MHz beat. From here on, the sound stages for BW and color sets are identical.

7. To allow for tuner-oscillator drift, and to make fine tuning easier, a typical sound IF strip has a bandpass of about 100 KHz, twice the requirements for the FM sound signal which has a maximum deviation of +/− 25 KHz. With the IF frequency *fixed* at

17-29 Summary

4.5 MHz, misalignment of the tuned circuits creates no change except to reduce the sound level.

8. An FM detector first converts the FM signal into an AM signal at IF frequencies, then demodulates it (usually with diodes) to recover the audio sidebands. The AF range is from about 50 Hz to 15 KHz. Three popular types of FM detectors are: the discriminator, the ratio detector, and the quadrature detector.

9. A discriminator-type detector develops a positive or negative DC voltage across two split-load resistors. It changes with frequency variations above and below the carrier frequency. At the 4.5 MHz carrier frequency the equal and opposite voltages across the two resistors cancel. A frequency shift above resonance develops the positive half-cycle of the AF output signal; frequency deviations below resonance produce the negative half-cycles. This type of detector has high output, but no AM-rejection capability.

10. The AF output of a ratio detector depends on the ratio between the output voltages developed by the two diodes conducting alternately by different amounts. The AF output is taken from the electrical center of the "bridge" circuit. AM rejection is obtained with a large-value capacitor (stabilizing capacitor) connected across the diode load resistor(s). This is an electrolytic-type capacitor which tends to dry out and lose capacity, causing problems of sync buzz.

11. Most modern sets use ICs in the sound section. The detector may be either a ratio or a quadrature detector. Usually there are only two tuned circuits, one for the sound takeoff and one for the detector. They are normally adjusted "by ear" for the loudest, clearest sound without buzz or distortion. Volume is usually controlled by variations of the DC applied to the AF amplifier stage. In some cases the AF output stage is built in; in other cases it's external to the IC.

12. The typical sound section has two AF amplifiers, a voltage amplifier to bring up the signal level, and a power output amplifier to develop the *power* to drive the speaker. The output stage may be "single ended", or two transistors operating push-pull. For class *B* operation the output stage is preceded by a semi-power driver stage.

13. Some sets have a "step down" output transformer to match the output transistor(s) to the low-impedance speaker. An impedance mismatch results in distortion and loss of power. Output transistors operating as emitter followers have a low output impedance and there is no need for a matching transformer. Coupling to the speaker may be *direct*, or via a large-value coupling capacitor.

14. Many sets use some form of *complementary symmetry* AF output stage. Two transistors are used, operating push-pull, class *B*. One type of circuit uses two identical transistors supplied from two power sources of opposite polarities. The output transistors are driven by the two outputs of a phase inverter. Another (and more popular) circuit uses a PNP and an NPN transistor, using a single power source, and driven in parallel (the same polarity) from the driver amplifier.

15. Two problems peculiar to complementary-type output circuits are crossover (notch) distortion, and thermal runaway of one or both transistors. Notch distortion is the result of a "gap" in the output where both transistors are cut off. This is corrected

816 THE SOUND SECTION

for by a slight offset of the bias of one transistor. Thermal runaway (overheating) occurs when there's too much forward bias; typical causes are transistor leakage and defective biasing resistors.

16. For greater sound power and better quality many sets use two or more speakers. One speaker functions as a "woofer" for the low bass tones, and the other as a "tweeter" for the high frequencies. The signals may be split according to frequency by means of a crossover network. Two or more speakers must be connected "in phase" so that their cones move in and out in unison.

17. Negative feedback is an inexpensive way to reduce distortion in an AF amplifier. A sampling of the output signal is fed back (out of phase) to some previous stage via an RC voltage-dividing network. Any distortion present at the input is reduced by cancellation. Negative feedback creates an unavoidable loss of gain: the greater the feedback the less the gain.

18. Many sets are provided with an earphone jack for quiet listening. When the phones are plugged in the speaker is disconnected and sometimes replaced with a low-value load resistor. Reminder: Never short or disconnect a speaker with the set operating. Depending on the circuit, such an action can damage or destroy the output transistor(s).

REVIEW QUESTIONS

1. What is the function in a color set of a) a sound IF detector? b) an FM detector?

2. How is video kept out of the sound section, and how are sound signals prevented from reaching the CRT?

3. Explain negative feedback in an audio amplifier, why it is used, and how and why stage gain is affected.

4. Why is the AF signal amplified, then reduced to a low level, to feed the speaker?

5. What can be used in place of an aural tracer for a listening check of the sound signal?

6. What is meant by sync buzz and sawtooth buzz and how would you distinguish between the two?

7. State the causes of thermal runaway in an AF output stage and how it can be recognized.

8. a) What is meant by crossover or notch distortion in a push-pull output amplifier? b) What steps are taken to prevent this condition?

9. Describe the changes that take place in the sound signal in the tuner, sound IF detector, and FM detector of a color set.

10. Compare, and explain the differences between, a discriminator and a ratio detector.

Review Questions

11. a) Why is a phase inverter required to drive a complementary-symmetry output circuit using two identical transistors? b) For proper operation, the voltage measured at the emitters of the transistors should be *zero*. Explain why.

12. a) When a complementary circuit uses two different transistors, they are driven in parallel with the same signal polarity. Explain why. b) What voltage should be measured at the transistor emitter terminals?

13. The S curve obtained during sweep alignment of the FM detector shows one peak stronger than the other and the trace between them is badly curved. Explain why this produces distorted sound.

14. Intercarrier sync buzz is most noticeable when large block lettering appears on the screen. Explain why.

15. Why is the sound takeoff of a color set ahead of the video detector?

16. Explain how a limiter stage prevents AM-type signals from reaching the detector.

17. What is meant by DC control of volume and where is it used?

18. a) Even with two transistors, a class *B* output stage is more efficient than class *A*. Explain why. b) Why are *two* transistors required?

19. What is meant by bias stabilization of a complementary-symmetry circuit, and why is it necessary?

20. Explain the need for: a) an AF voltage amplifier; b) an AF power amplifier.

21. For a BW set, explain the pros and cons of sound takeoff either at the video-detector output or after the video amplifier.

22. a) Why is it desirable to split the signal feeding a woofer and a tweeter? b) What type of circuit is used? c) Why should the two speakers be phased?

23. A DC-type volume control does not require a shielded lead. Explain why.

24. a) Why is a push-pull output stage preceded by a driver stage? b) How does the driver differ from a regular voltage amplifier?

25. Why are most AF output transistors connected in the emitter-follower mode?

26. What is the purpose of the *tap* on a volume control?

27. Why do most AF amplifiers use DC coupling?

28. Does a misaligned sound IF section alter the frequency of the 4.5 MHz carrier? Explain.

29. a) What sound stage is most susceptible to hum pickup? Explain why. b) What precautions are taken to prevent this problem?

30. What are the advantages of FM over AM?

31. a) Why does a speaker have such a low impedance? b) Give three typical impedance values. c) What results can be expected when a speaker is not properly matched to the AF output stage?

32. Explain how the *S* curve is developed during sweep alignment of the FM detector.

33. Fill in the missing words. In FM, the volume of sound corresponds with _____ _____. The frequency of the audio corresponds with _____.

TROUBLESHOOTING QUESTIONNAIRE

1. State possible causes and a general troubleshooting procedure for the following: 1) loss of sound; 2) distorted sound with buzz; 3) distorted sound but no buzz; 4) buzz without distortion; 5) intermittent sound.

2. What is the effect if the electrolytic coupling capacitor feeding the speaker a) loses capacity? b) becomes open?

3. Explain a procedure for phasing two speakers.

4. Give several possible causes of hum.

5. Why must push-pull output transistors be replaced as a matched pair?

6. State several possible causes when the tuned circuits of an FM detector cannot be aligned.

7. Explain a procedure for aligning the sound section without the aid of instruments.

8. When the accuracy of a signal generator is in doubt it is better to align the sound section on the station signal. Explain why.

9. Name four things to consider when replacing a speaker.

10. Give possible causes and procedure for the following background sounds or noises: a) scratchy sounds; b) vibrations; c) buzz; d) squeals; motorboating; e) hiss.

11. What voltage indication can be expected at the junction of the split load resistors of an FM detector when the two diode circuits are a) balanced? b) unbalanced?

12. What is the effect of an open stabilizing capacitor in a ratio-detector circuit?

13. What sound problem can be expected when the 41.25 MHz sound marker is located too high on the video IF response? Explain why.

14. Explain why opening or shorting the speaker connections in some cases can damage the AF output transistor(s).

15. What would you suspect for loss of sound on a color set if there's a normal signal at the sound takeoff, but no signal at the input of the sound IF transistor?

16. What would you suspect when an AF output transistor runs a) too hot? b) cold?

17. What problem can be expected when a PNP output transistor is inadvertently installed in place of an NPN type? Explain why.

18. A CRO shows a gap in the AF output from a complementary-output circuit. State probable cause and how you would proceed.

19. Because of their low cost, defective TV speakers are usually replaced rather than repaired. For what kinds of defects and under what circumstances might a repair be attempted?

20. The mica insulator is inadvertently left out when replacing an AF output transistor. What problems can be created by this omission?

21. Explain the importance of: a) the grease film on an output-transistor insulator; b) the heat sink.

22. a) Explain why shielded wire must be used for connecting a voltage-dividing volume control. b) Why is this unnecessary for DC control of volume, as used with ICs?

23. What is the expected result if a negative-feedback loop opens up? Explain.

24. Even slight leakage in an audio coupling capacitor will cause severe distortion. Explain why.

25. Explain why a shorted coupling capacitor feeding the speaker can destroy both the output transistor(s) and the speaker.

26. A shorted winding in an AF output transformer can be hard to detect. What trouble symptoms does it create, and how would you proceed when you suspect it?

27. A set has distorted sound. With an AF generator feeding the volume control, a CRO shows a distorted sine wave at the output of the first AF stage. State several possible causes.

28. Give procedure for localizing the cause when there's hum: a) even on a blank channel; b) only when a station is received.

29. Where would you check for hum that a) varies with the setting of the volume control? b) is not affected by the control?

30. Fill in the missing words: A defective video detector causes loss of _____ on a BW set, and loss of _____ on a color set.

31. a) Explain why a slight touch-up of a sound-detector adjustment will often clear up problems of distortion and sync buzz. b) How would you proceed if such a touch-up doesn't help?

32. What problems can be expected if one of the diodes of a FM detector is connected with reverse polarity?

33. Fill in the missing words: In an AF output stage, a leaky/shorted transistor can overload and damage _____ . A defective resistor in turn can be caused by a shorted _____ . Therefore, when one or the other is found, check further for a *cause.*

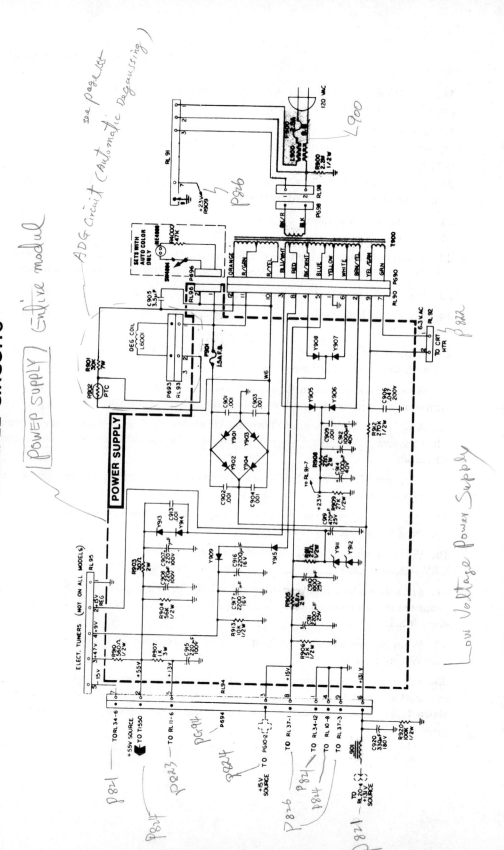

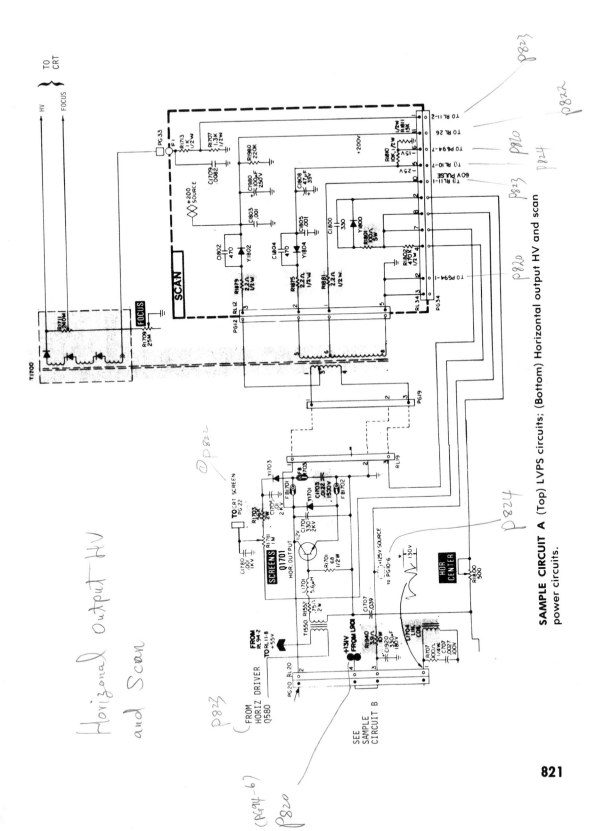

SAMPLE CIRCUIT A (Top) LVPS circuits; (Bottom) Horizontal output HV and scan power circuits.

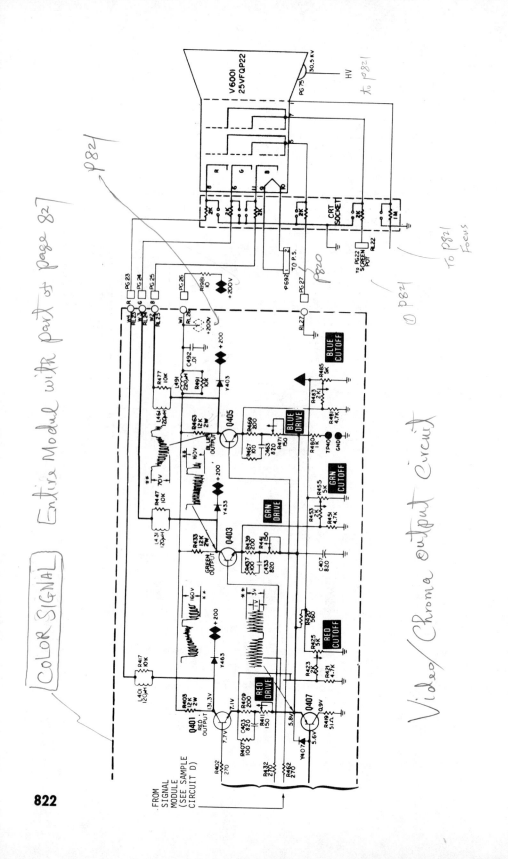

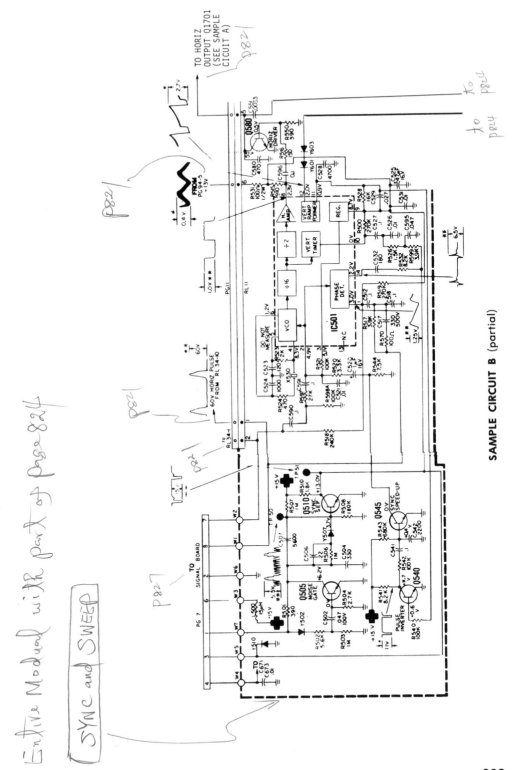

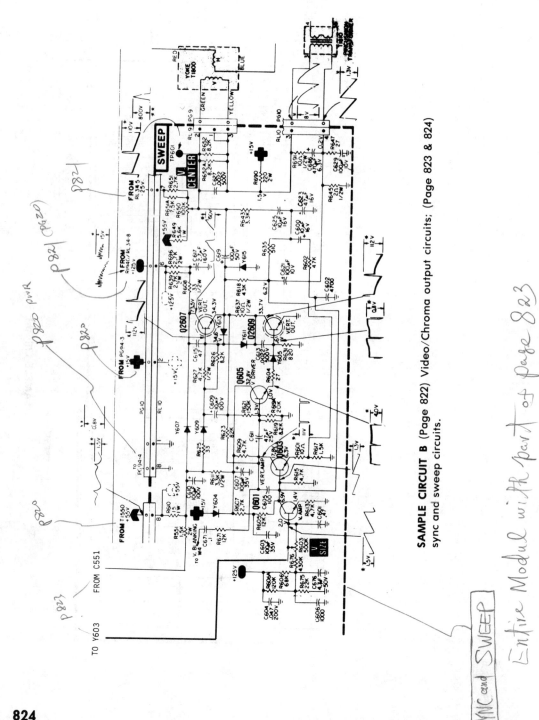

SAMPLE CIRCUIT B (Page 822) Video/Chroma output circuits; (Page 823 & 824) sync and sweep circuits.

SAMPLE CIRCUIT C (partial)

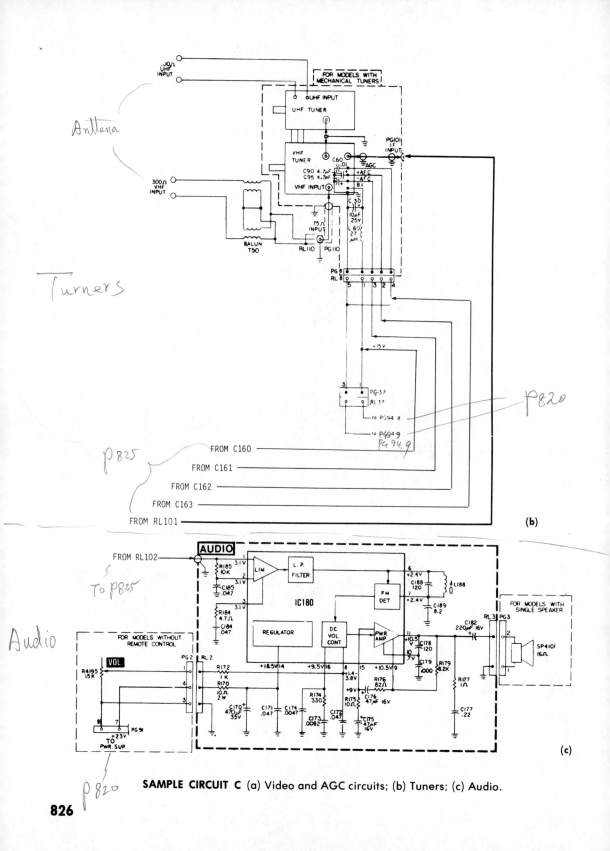

SAMPLE CIRCUIT C (a) Video and AGC circuits; (b) Tuners; (c) Audio.

SAMPLE CIRCUIT D Video and chroma circuits.

827

INDEX

A

ABL (automatic brightness limiter) • 295, 574, 576, 648
ABL troubleshooting • 603, 608—609
Absorption trap (see traps)
ACC (automatic color control) • 33, 721, 730
ACC troubleshooting • 730, 763
AC (alternating current) • 153—154
AC coupled stages • 555, 612
ADG (automatic degaussing) • 35, 155, 648—649
ADG troubleshooting • 172
Adder • 16, 571, 715
Adjacent channels • 408, 413, 492, 495—496, 520
AF (audio frequency) • 779—780, 789
AF amplifier • 27, 28, 789, 791—792
AF amplifier troubleshooting • 808—811
AFC (automatic frequency control) • 25, 200, 250—251, 258, 723
AFC troubleshooting • 271, 341
    alignment • 272—273
AFPC (automatic frequency and phase control) • 29, 32, 85, 724, 748—750
    (see color sync)
AFPC troubleshooting • 765—767
    alignment • 761, 765—767
AFPC stages • 720
AFT (automatic fine tuning) • 28, 402, 410, 422, 424
AFT troubleshooting • 463, 468—469
    alignment • 468—470

AGC (automatic gain control) • 26, 351, 502—503
    amplified • 359, 362, 384
    delayed • 365, 367, 381
    forward AGC • 354, 355, 374, 380, 414, 458
    gated (keyed) • 32, 353, 356—359
    reverse AGC • 354, 355, 374, 380, 414, 458
    threshold control • 27, 337, 338, 340—343, 356—358, 510
AGC troubleshooting • 337, 370, 387, 516, 523
    adjustment • 337, 343, 375, 381—382, 386, 458, 753
AGC keyer (see Keyer)
Air check • 141, 179, 236, 274, 314, 387, 476, 541, 614, 702, 772, 813
Alignment (tuned circuits) • 140
    defined • 140
    of tanks • 489
    of traps • 412
    purpose of • 406—409
Alignment procedures (general) • 459, 462, 527, 540
Alignment methods • 60, 459—462, 527—540, 770—771
AM (amplitude modulation) • 17, 503
Ambient room lighting • 562, 602
Amplifiers (general) • 203
Antenna input circuits • 402—403
Antenna orientation • 495
Anti-hunt (AFC) • 257, 272
Anti-pincushion circuits • 291
APF (active power filter) • 152, 161—162, 178

APF troubleshooting • 175—176
Aquadag • 296, 299, 562, 625, 629, 647
Arc gate • 299, 564
Arcing • 90, 123, 168, 176—177, 230, 299, 307—308, 314, 564, 603, 609, 679
Arc testing HV • 313—314, 679
Arc testing CRT • 675—676
ATC (automatic tint control) • 722, 743
ATC troubleshooting • 765
Audio amplifiers (see AF amp)
Audio distortion (see sound)
Audio muting (see muting)
Automatic brightness control (ABC) • 562, 574
ABC troubleshooting • 602
Automatic brightness limiter (see ABL)
Automatic color control (see ACC)
Automatic degaussing (see ADG)
Automatic fine tuning (see AFT)
Automatic frequency control (see AFC)
Automatic gain control (see AGC)
Automatic phase control (APC) (see AFPC)
Automatic shutdown (see shutdown mode)
Automatic tint control (see ATC)
Autotransfers • 71—72, 123, 155, 172, 207
Avalanche (see zener) • 65, 125, 165

B

B boost • 248, 283—284, 295, 298
B voltage • 152—153
Back porch • 12, 16

828

INDEX 829

Balanced circuits • 32, 208, 424, 567, 786
Balanced modulators • 14—15
Balun (see tuners)
Bandpass considerations • 2, 5, 27, 408—409, 459, 488—489, 495, 779
Bandpass amplifier (see BPA)
Bandswitching (see tuners)
Bandwidth • 6
    BPA • 727
    RF/mixer • 460—461
    sound IF • 783
    video IF • 488, 493, 495, 532
Bar patterns • 49, 337, 519, 635—637, 723
Battery operated sets • 167, 178—179
Battery testing • 74, 179
BCD logic (binary coded decimal) • 429, 435
Beam (CRT) • 626—629
Beam current (see pix tube)
Beam current cutoff • 562, 635
Beam deflection • 23, 201—204, 632, 638
Beam intensity • 626
Beam landing • 638, 649
Beam retrace (see retrace)
Beam synchronization • 322
Beam trace (see beam deflection)
Beats (see heterodyne)
Beat frequencies (see interference)
Beta • 66, 130, 228
Bias •
    automatic • 26, 324
    computation • 111
    control • 229
    CRT (see pix tube)
    cutoff • 367
    excitation • 322, 324
    fixed • 324
    forward • 353, 357
    measurements • 111, 342
    reverse • 324, 358
    self • 324
    stabilization • 210, 232, 792
    transistor • 354
Bias box • 337, 372, 380
Bias clamp • 337, 372
Bifilar transformer • 385, 495, 523
Bimetal strip (CB) • 154
Black • 306, 635
Blacker than black • 337, 638
Blanker • 33, 289, 576
Blanker troubleshooting • 231
Blanking • 10, 214, 289, 338, 635, 638
Bleeder current • 162

Bleeder resistance • 162
Block diagrams • 24, 30, 152, 192, 246, 322, 352, 403, 454, 488, 568—569, 720, 780
Blocking oscillator • 195, 198, 250
Blocking oscillator troubleshooting • 225—226
Blue (see RGB colors) • 735
    from red and green • 738
    lateral adjuster • 647, 649, 652
Boost voltage (see B boost)
BPA (bandpass amplifier) • 29, 33, 493, 721, 725, 728
    bandpass requirements • 725
    rolloff compensation • 728, 769, 777
BPA troubleshooting • 769
    alignment • 769—771
Bridge rectifier • 159—160
Bridged-T filter/trap • 496—497, 503, 743
Brightness • 28
    automatic control of (see auto. bright. control)
    master control • 573—574
    of color • 13, 567, 643
    of raster • 28, 567
Broadband color (see wide band)
Buffer • 203, 259
Burst amplifier (see burst gate)
Burst gate • 32, 722, 724, 730, 732
Burst gate troubleshooting • 762
    alignment (see AFPC)
Burst signal (color burst) • 32
Bus •
    AGC • 337, 351, 373—374, 380, 385
    DC • 179
Butterfly wings pattern • 205
B-Y (see color difference signals) • 15

C

Camera • 7, 13, 715
Capacitance • 405
    defined • 405
Capacitors • 405, 120
    defects • 117, 120
    functions • 120
    testing (see component testing)
    types of • 161
Carbonized leakage path • 137, 299
Carrier waves • 25, 404, 490
Cascade • 165, 328, 365, 567
Cascode • 414
Case histories • 140
Cathode ray tube (CRT) (see pix tube)

CB (see circuit breaker)
Centering • 25, 216, 233
    horizontal • 292
    vertical • 205
Centering ring • 26, 204, 216, 631
Channel frequency allocations • 2—5
Channel number display (see tuners)
Channel selector (see tuners)
Cheater cord • 154
Choke coils (see filters)
Chroma (color) •
    amplifier • 744—745
    control • 36
    demodulators • 719, 721, 732, 738, 740—741, 743
    developing, at station • 715, 716, 719
    modulators • 716, 719
    rolloff compensation • 506, 727
    sampling • 716, 721, 722, 736, 738
    sidebands • 15, 424, 568
    signal • 13, 568, 569, 725
    switching • 724, 730, 732
    takeoff • 561, 571, 578, 582
Chroma troubleshooting (general) • 752—772
Chrominance (see color)
Circuit breaker (CB) • 154, 170
Circuit disturbance • 122, 136—139
Circuit tracing • 84
Clamping • 95, 337, 375, 454, 555
Clipping (pulse) • 336—337, 342, 518
Coaxial cable • 403, 405, 422
Coils • 122
Color • 643
    automatic control of (see ACC)
    bar pattern • 15, 59, 665, 762, 764—765
    bar signals • 59, 665, 762
    burst • 12, 16, 518, 569, 638, 721, 724
    camera • 8, 13
    characteristics • 643
    complementary • 12, 34—35, 643, 715, 719, 644
    contamination • 583
    control • 36
    demodulation (defined) • 721—722
    demodulators (see chroma)
    difference amplifiers • 31, 578, 722
    difference signals • 15, 35, 554, 578, 580, 582, 663, 719, 722

830 INDEX

Color, (*Contd.*)
   field • 681
   hue (tint) • 31, 722
   killer • 29, 721, 724, 732
   matrixing • 578, 580, 582,
      612, 665, 716, 719, 722, 736
   mixing • 12—13, 31, 643—644
   mixtures • 718, 743, 746
   narrow band • 492—493, 533,
      725
   phase • 15, 663, 665, 667, 716
   phase detector (AFPC) • 32,
      743
   phosphors (see pix tubes)
   primaries • 716, 746
   purity • 34, 649
   reproduction • 663
   saturation • 12—13, 31, 665,
      719
   sidebands • 492, 534, 571, 665
   signals • 662, 750
   skin (flesh) tones • 107, 665,
      728, 744, 765
   subcarrier • 32, 411, 492,
      716, 719, 747
   sync • 16, 32, 569, 719, 723
   transmission • 719
   wide band • 492—493, 533,
      725, 769
Color characteristics • 12—13
Color controls • 36
Color demodulators • 31, 735
Color demodulator troubleshooting
   • 763—764
Color difference signals • 15
Color killer (see killer stage)
Color receiver schematic • 820—827
Color sync (AFPC) • 322, 569, 723
Color tint, control of • 36
Colpitts oscillator • 251, 416, 422
Comb filter • 569, 571, 583, 721, 728
Comb filter troubleshooting • 612
   alignment • 613
Comparator (see PLL, tuners)
Compatability • 7
Complementary colors (see colors)
Complementary symmetry amplifier
   • 208, 231, 792
Complementary amplifier troubleshooting • 231, 810—811
Components •
   damage to • 109, 135—136,
      266, 308, 368, 385, 412,
      457, 600, 608, 683
   defects • 117, 119—124, 138
   exact replacements • 118,
      313—314
   replacing • 133, 136—137,
      386, 463

Components, (*Contd.*)
   testing • 64—65, 88, 109,
      114, 117—118, 123,
      126—128, 163, 172, 230,
      312—313
Composite video signal •
   11—12, 505, 508, 638
Confetti • 365, 721, 725, 753
Contact cleaner • 455
Contact cleaning • 455—456
Continuous wave (CW) • 31, 48,
   404, 410
Contrast, control of • 28, 637
Controls, BW sets • 28, 367
Controls, color sets • 496, 582
Convergence • 34, 215
   defined • 649—650
   dynamic • 215, 235, 653, 695,
      698
   horizontal • 652—653, 696
   static • 215, 649—650,
      692—693, 698
   vertical • 192—193, 653,
      695—696, 698
Convergence (Delta CRT) • 647
   adjusters • 647, 649, 652—653
Convergence troubleshooting
   (Delta) • 687, 701—702
   adjustments • 692—693,
      695—696
Convergence (in-line CRT) • 701
   adjusters • 654, 658—659
Convergence troubleshooting
   (in-line) • 696, 701—702
   adjustments • 697
Convergence (Trinitron) • 696
   adjusters • 660
Convergence troubleshooting
   (Trinitron) • 701—702
   adjustments • 697—698,
      700—701
Core saturation • 167
Core slugs • 123, 407, 415, 464, 467
Corona discharge • 90, 299
Corona, troubleshooting for •
   307, 314, 610, 675, 679
Countdown system • 195, 200,
   251, 260
Countdown system troubleshooting
   • 235, 269
Coupling capacitor • 217, 225,
   555, 557, 782
Coupling methods • 408, 493, 555
Crankshaft waveform • 666
Crispening (see video peaking)
Critical components (see substitution)
Crosshatch pattern • 59
Crossover network (AF) • 795

CRO (cathode ray oscilloscope)
   (see test equipment)
Crosstalk (cross-modulation) • 228,
   365
CRT (see pix tube)
Crystal • 747
Crystal oscillator • 32, 427, 435,
   721, 724, 747
Crystal oscillator troubleshooting •
   749—751
Crystal ringing amplifier • 749
Crystal ringing, troubleshooting
   • 747, 750
Current cutoff bias • 323
Current measurement •
   by computation • 45, 116, 178
   direct • 45, 116, 178
   precautions • 45, 116
CW (see continuous wave)

D

Daisy-petal display • 63—64
Damped oscillations • 67, 287, 506
Damper • 283
Darlington transistor • 129, 203,
   228, 362, 789
DC (direct current) •
   control stage • 562, 573
   coupling (see direct coupling)
   restorer • 33, 635
Decoupling • 152, 162, 365, 557
Deemphasis • 17, 780, 788
Defeat switch • 402, 424
Deflection yoke (see yoke)
Degaussing • 35, 72, 155, 648—649
Degeneration • 559
Delay device (see comb filter)
Delayed AGC (see AGC)
Delay line (DL) • 569, 572, 573,
   611, 613
Delta CRT • 644, 654
Demodulators (see color demods) • 31
Detectors • 25, 27, 32
Detent (see tuners)
Dial calibration (see tuners,
   UHF) • 466
Dial lights • 151, 155, 168, 173
Differential amplifier • 506, 508, 582
Differentiating network • 214,
   255, 321, 328, 330, 343
Digital logic circuits (see tuners)
Digital readout • 54—55
Diodes • 123—126, 165
Diode testing • 15, 123—128, 524
Direct coupling (see coupling
   methods) • 33, 91, 113, 152,
   203—204, 225, 259, 274, 362,
   505, 555, 573, 789

INDEX

Direct coupled amplifier • 554, 789—791
DC coupled amplifier troubleshooting • 113, 602, 763
DC restorer • 555, 33
Discriminator • 424, 784, 786
Discriminator troubleshooting • alignment • 806—808
Distortion
    crossover • 232, 792
    raster • 195, 214, 216, 230, 249, 289
    sound • 792, 803—805, 808
    waveform • 195, 249
Distortion, testing for (audio) • 808, 810
Distributed capacity • 405—406
Driver amplifier • 203, 259

E

Earth's magnetic field (see degaussing)
Earphone jack • 795
Edge purity (see convergence)
Electrolytic capacitor • 161
Electromagnet • 632, 650, 659, 790
Electron beam • 628
Electron emission • 628
Electron gun (see pix tubes)
Electronic switch • 62, 732
Electronic tuner (see tuners, electronic)
Electrostatic deflection (CRO) • 50
Electrostatic focus • 627
Elevator transformer (balun) (see tuners)
Emitter follower • 203
Equalizing pulses • 330, 338, 638
Error voltage • 251—252, 255, 258, 266, 271—272, 424, 426, 428, 723
Evaluation • 41, 381, 526, 761
Eye characteristics • 643, 646

F

Fail safe (see overvoltage protection)
Feedback • 208, 493, 497
    negative • 194, 196, 203, 208, 228, 414, 559, 783, 791, 795—796
    positive • 195, 203, 208, 414, 497, 522
Feedback loop • 195, 207—208, 228, 231, 256, 272, 427
Feedthrough capacitor (see tuners)

Ferrite beads • 283, 503, 783
FET (field effect transistor) • 66, 130, 355—356, 510
FET troubleshooting • 113, 130—131, 457
    precautions • 131, 457
Field, scanning • 9—10, 634, 638
Filters • 161, 255, 328, 330, 415, 428, 433
Fine detail • 32, 558,—559, 606
Fine tuning (see tuners) • 28, 410—411
Fireball • 691
Fire hazards • 157
Floating pix • 271, 273
Flyback circuit • 282, 301
Flyback circuit troubleshooting • 312—313
Flyback transformer (H.O.T.) • 24, 283—284
Flyback troubleshooting (see component testing)
FM (frequency modulation) • 17, 491, 529, 779
    FM theory • 57, 779, 784, 786
FM detectors • 780, 783
Focus • 307, 606
    control of • 28
Foldover • 193, 267, 306
Forward AGC (see AGC)
Frame, repetition rate • 9—10, 634, 638
Free-running (see oscillators)
Frequency compensation, video • 505, 557—558
Frequency conversion (see tuners)
Frequency deviation (see FM) • 17, 57
Frequency divider (see tuners)
Frequency drift • 462
Frequency modulation (see FM)
Frequency response •
    chroma • 771
    sound • 794
    video • 25, 555—556, 606
Frequency response checking • 167, 196, 526—527, 606, 613, 810
Frequency synthesis (see tuners)
Fringe area • 386, 540
Front end (see tuner)
Full wave rectifier • 159—160
Full wave voltage doubler • 159
Fuse • 154
    link • 154, 167, 173
Fuse blowing • 179, 233
Fused circuits • 154, 210
Fusible resistor (fusistor) • 119, 156, 167—168, 172

G

Gain, of stages • 94, 259, 353, 365, 488, 554, 559, 565, 606, 789
Gated AGC (see keyer)
Gating pulses • 284, 357, 384, 721
Ghosts • 412, 506, 573, 605
Gimmick capacitor • 406, 414, 458
Graticule • 62, 64
Gray scale tracking • 34—35, 579, 662—663, 681, 692, 696—697
Green (see RGB colors) •
    from red and blue • 722, 735, 738
Grounds • 157, 176
Gun (CRT) (see electron gun)
G-Y (see color difference signals) • 31

H

Hairpin coil (see UHF tuner) •
Half wave rectifier • 156—157
Half wave voltage doubler • 157—158
Hammerhead display • 338—339, 518, 605, 640
Harmonics • 404, 431, 497, 505
Hartley oscillator • 249, 251
Haystack response • 409, 493, 531, 533
Heat dissipation • 162
Heat sink • 136, 312, 791
Heater, CRT •
    brightener • 72
    circuit fusing (see fuse link)
    instant-on • 155
    power sources • 155, 294
Heterodyne, defined • 404
Heterodyne beats • 16, 404, 411, 414, 489, 491, 496—497, 520, 743, 771, 780
High frequency rolloff (see chroma, BPA)
High Voltage (HV) (see PS, HV)
Hold down circuit • 283
Horizontal deflection
    beam retrace • 9—10
    beam trace • 9—10
    blanking • 247, 289
    countdown system • 247, 251—258
    driver • 247
    oscillator • 247, 249
    scanning lines • 9—10
    SCR system (see SCR)

Horizontal deflection, (*Contd.*)
    sweep amplifier • 281—282
    sync • 251, 518
    yoke • 282, 630
Horizontal dynamic convergence • 248
Horizontal output systems • 247, 281—282
Horizontal output troubleshooting • 300, 311
Horizontal output transformer (H.O.T.) (see flyback)
Horizontal sweep dependent circuits (see sweep)
Horizontal sweep oscillators • 249
Horizontal oscillator troubleshooting • 268
Horizontal sync • 200, 247, 255
Horizontal sync troubleshooting • 267, 271, 273
Hot chassis • 88, 157, 412
    hazard • 157
    isolation, testing for • 173
Hue • 13, 15, 643, 734
Hum modulation • 521, 603, 605
Hum voltage • 178
    testing for • 178
HV (high voltage) (see P.S.)
Hybrid sets • 151, 173

I

I and Q • 31, 492, 725, 734
    demodulation • 31
    modulation • 716
    phase angles • 716, 725, 734
    signals • 5—6, 492, 716, 721
ICs • 27, 369, 508, 510, 750, 752, 796
IC troubleshooting • 113, 116, 131—133, 224, 235, 382, 519—520, 614, 768, 813
    precautions • 131, 382, 768
Idling current • 229
IF (intermediate frequency) • 404, 487
IF amplifiers (general) • 510
IF traps (see traps) •
Image frequency • 409, 413
Impedance matching • 412, 493, 503, 789—791, 795
Implosion (see pix tube)
Impulse noise • 321, 324, 327, 783, 786
Inductance defined • 122, 405
Induction • 123, 424
Infrared (see remote control)
Infrared system troubleshooting • 474—475

In-line CRT (see pix tubes)
Insertion loss • 173, 272, 573
Inspection (see troubleshooting, general)
Instant-on • 155, 173
Insulation breakdown (see arcing)
Integrated circuit (IC) •
    troubleshooting (see IC)
Integrator network • 25, 192, 200—201, 321, 328, 343
Intensity marker (see markers)
Intensity modulation • 26, 637
Intercarrier sound • 487
Interference (external) • 341, 408—409, 412—413, 495—496, 520—521
Interference (internal) • 60, 341, 495, 502, 520—521
Interference remedies • 495, 606, 614
Interlace scanning • 9—10, 228, 634, 638
Interleaving • 15—16, 571
Interlock • 154, 170
Intermittents, troubleshooting • 137—138, 179, 224—225, 236, 267, 344, 387, 475—476, 541, 614, 702, 768, 813
Internal resistance • 153, 164
Ion bombardment • 626—627
    ion burn • 626—627, 679
Ion trapping • 626—627
Inverse feedback (see negative feedback)
Inverter • 203
IR drop • 153, 274, 311, 574
Isolating defects • 224

J-K-L

Job descriptions • 1
    bench technician • 1, 83—84
    outside technician • 1
Keyer AGC (see AGC keyer)
Keyer • 32, 356—358, 374
Keyer troubleshooting • 383—384
Keyboard (see electronic tuner)
Killer, color • 29
Killer stage troubleshooting • 768—769
Lead dress • 137, 178, 458, 463, 522
Leakage •
    capacitors • 120—121
    diode/transistor junctions • 124, 132
    HV (see corona)
LDR (light dependent resistor) • 119, 562, 574

LED (light emitting diode) • 28, 55, 126, 402, 430, 435
Light • 643
    ambient lighting • 562
    color mixing • 643
    white light • 643
Lightning damage • 412, 457
Limiter stage (see sound)
Line filter • 154
Line voltage problems • 172
Linearity • 28
Linearity troubleshooting • 226, 267
Lissajous patterns • 64, 267
Load • 296
Load resistors • 271, 505, 508, 791
Loading • 274, 310, 312, 495, 522, 795
Localizing troubles • 223
Low or high line voltage • 172
Low voltage power supply (LVPS) (see PS, LV)
Luminance • 13—14, 554, 567—568

M

Magnets • 627, 631, 645, 650, 652
Magnetic deflection • 627
Magnetic flux • 627, 631—632, 650, 653
Markers (sweep alignment) • 58, 409, 460—461, 490—492, 528, 530—534, 538—539, 769—771
Master brightness control • 654
Matched components • 255—256, 271, 424, 469, 767, 792, 806
Matrix • 736, 745
Matrixing (see color)
Mechanical defects • 467
Memory (see tuners)
Microprocessor (see PLL)
Misalignment • 606, 753
Misconvergence • 59
Mistracking • 490
Mixer (VHF tuner) • 414
    bandpass requirements • 404
    tuned circuits • 405—411
Mixer (UHF basic tuner) • 410, 422
Mixer troubleshooting (UHF) • 465—466
    alignment • 458—460, 462, 466
Mixer troubleshooting (VHF) • 456—458
    alignment • 458—459
Modulation
    AM • 2, 15
    defined • 15
    FM • 2, 17—18
Modulation, at station • 2, 15, 17

INDEX **833**

Modulator • 14—16
Modules • 133, 497
Module troubleshooting • 133—134
Monochrome • 7, 23, 567
Mosfet (see FET)
Multiple •
    defects • 85—86, 340, 387
    symptoms • 85—86, 137, 140, 168, 583
Multiplex • 18
Multivibrator (MV) • 195—198

N

Narrow band color (see color)
Negative feedback (see feedback)
Negative going signal (see signal polarity)
Negative pix • 26, 98, 337, 340, 518, 554, 640
Neon bulb • 122, 303, 563, 630, 675, 683
Neutralizing • 414, 458, 502, 521—522, 783
Noise • 324, 327, 341—342, 365
Noise circuits • 154, 324, 327, 340
Noise immunity • 24, 257, 321, 324, 340, 342, 357—358, 786—788
Noisy controls • 119—120
Notch filter (bridged T trap) • 497
NPN transistor • 127—129
NTSC (National Television Systems Committee) • 7, 16, 571, 716
Null (trap alignment) • 612

O

Off-set bias • 210
Ohm's Law • 162—163
Oscillation • 414, 497, 506, 520
Oscillation, troubleshooting for • 458, 462, 521—522, 527
Oscillating crystal (see crystal)
Oscillators, free running • 193, 201, 226—227, 248, 321, 747
Oscillators, general • 192, 195, 251, 410, 414—415
Oscillator troubleshooting, general • 92, 462—463
    Substitution for • 92, 100, 457, 462, 467
Oscillator control circuits • 257—258, 272
Oscillator control troubleshooting (general) • 765—766, 268—273
Oscillator drift • 410, 456, 747
Oscillator drift, troubleshooting • 456, 462, 464

Oscillator, horizontal (see horiz. sweep)
Oscillator injection • 414, 416, 457, 466
Oscillator, subcarrier (see subcarrier osc.)
Oscillator, UHF tuners • 410
Oscillator, VHF tuners • 404, 410
Oscillator troubleshooting (tuners) • 262—464
    alignment • 463—464
Oscillator tracking • 410, 415, 466
Oscillator, vertical (see vert. sweep)
Oscilloscope (CRO) • 50, 52
    uses • 530
Overall alignment (see VSM) • 535, 770—771
Overall response curves • 535, 770—771
Overcoupling • 408, 495, 503, 536
Overheating components • 90—91, 118, 172, 176, 267, 516
Overload •
    HV power supply • 311—314
    LVPS • 115, 175
    signal circuits • 227, 516, 518, 523, 810
Overload (PS), troubleshooting for • 175—176, 227, 267, 308
Overload (signal circuits) troubleshooting • 810
Overload distortion • 810
Overload protection • 167
Overvoltage protection • 167, 295, 297—298, 308
Override bias • 762—763
Ozone • 299, 307

P

Parabolic waveform • 235, 653
Parallel resonant circuit • 406—407
PCBs (printed circuit boards) • 28, 84, 133—134
PCB troubleshooting • 90, 134—136
Peaking •
    IF • 506
    video • 505, 557—559
Permanent magnet (PM) (see magnets)
Persistance of vision • 625
Phase •
    angles • 15, 424, 667, 716, 719, 725, 734, 736, 738, 740
    defined • 716—718
    detectors • 32, 251, 255—256, 730, 732, 750
    network • 716, 722, 728

Phase, (*Contd.*)
    splitter/inverter • 255, 322, 728, 791
    subcarrier • 716, 725
Phase inverter • 210, 341
Phase locked loop (see PLL)
Phasing • 16, 72
Phasor diagram • 716, 734, 738
Phosphors (see CRT screen) • 626—627, 646, 657, 723
Picture (pix) • 27, 34—35, 553, 579, 611—613, 637, 640
Pix (general)
    brightness • 608, 637
    contrast • 553, 637
    development • 26, 35, 579—580, 640—641, 662—663
    elements • 7, 635, 641
    fine detail • 7, 635, 637, 641, 665, 736—737
    focus • 676—677
    quality • 611, 641, 771—772
    signal • 637—638
Pix signals (see video signals)
Pix transmission • 18
Pix tubes, types • 26, 33—34
Pix tube (BW) • 25
    beam control • 628
    beam current • 628—629
    bias • 553, 559, 562—563, 607—608, 611, 628, 640
    deflection angle • 626
    drive signal • 553, 567, 635
    electron gun • 25, 625, 627
    elements • 25, 26, 628
    emission • 683
    external components • 630—631, 634
    implosion • 87, 626, 685
    operation of • 627—629, 637—638
    operating voltages • 628—629, 676
    pix development • 637, 640—641
    safety glass • 625—626
    screen phosphors • 25, 580, 625—626
    ultor • 299, 303, 308, 625
Pix tube (color) •
    aperture grille • 34, 644, 655, 657
    beam control • 574
    beam current • 574, 646—648
    bias • 553, 559, 562, 607—608, 611
    black matrix • 646
    convergence (delta) • 653

834 INDEX

Pix tube (color), (*Contd.*)
    convergence (in-line) • 659
    convergence (Trinitron) • 659—661
    degaussing • 648—649
    delta • 33, 644, 646
    drive signal • 553, 578, 635
    electron guns • 32, 646—647, 657
    elements • 647, 657, 659—660
    external components • 646, 648—649, 657
    focus • 628—629, 676, 678
    in-line • 33—34, 644, 655, 657
    magnetic shield, 649
    matrixing • 611, 662, 665
    operation of (delta) • 35, 646
    operation of (in-line) • 33, 657
    operation of (see Trinitron below)
    operating voltages • 111, 647
    phosphor dots • 32, 644, 646—647, 655
    phosphor stripes • 644, 655, 657
    producing BW pix • 723, 746
    producing color pix • 35, 663, 723, 735, 746
    producing RGB rasters • 649, 687—691
    producing white raster • 34, 607, 662, 692
    projection CRT • 626, 667
    purity • 649, 657
    shadow mask • 644, 646
    Trinitron • 34, 655, 659—660
Pix tube troubleshooting • 601, 610, 669—702
    convergence • 687—697, 701
    defects • 682
    degaussing • 681, 687
    flashing • 679, 683—684
    handling precautions • 87—88, 684—685
    purity, 689—691, 696—697, 701
    rejuvenation • 68, 683
    replacement • 684—685
    set-up adjustments • 86, 609—610, 678, 681, 686—690, 696
    testing • 610, 682—683
Piezoelectric effect • 433
Pincushion • 33, 192, 214, 217, 248, 289, 631, 653—654
Pincushion, troubleshooting • 192, 214, 234
    adjustments • 234, 289—291

PLL (phase locked loop) • 426—428
PNP transistor • 127—129
Polarized line plug • 173
Positive feedback (see feedback)
Positive going signal (see signal polarity)
Potentiometers • 119—120, 166
Power amplifiers • 204—205, 281—283, 300, 744—745, 789—791
Power consumption • 156, 173
Power dissipation • 162—163
Power line isolation (see hot chassis)
Power shutdown • 168
Power supply (HV) • 24, 248
    ABL (see automatic brightness limiter) • 648
    control of • 297, 648
    current requirements (see pix tubes)
    flyback transformer • 283—285
    flyback drive power • 282
    focus voltage • 248, 295, 298, 314
    rectifiers • 126, 296
    regulator • 248, 295—297
    scan rectification (see Scan rect.)
    shutdown • 168, 248, 295
    tripler • 296, 314
    X-ray radiation (see X-ray)
Power supply (HV) troubleshooting • 92, 126, 300, 302—303, 314 609—610, 675
Power supply (LVPS) • 23, 151, 153
    ADG • 154—155
    B+ (see B voltage)
    bridge rectifier • 159—160
    circuit breaker (CB) • 153—154
    current requirements, dc • 151, 153
    DC voltage distribution • 152—153, 162
    DC voltage requirements • 151—153, 162
    filament transformer • 154—155
    filters, APF • 152, 161
    filters, L/C, R/C • 152, 161
    fuses • 153
    fuse link • 154—155, 167, 173
    fusible resistor • 156, 167
    hot chassis • 88
    interlock • 154
    internal resistance • 163—164
    isolation transformer • 157, 168, 172

Power supply (LVSP), (*Contd.*)
    line voltage compensation • 72, 153, 167
    load • 153, 164—165
    overload protection • 153, 167
    power transformer • 155
    rectifier circuits • 157—160
    ripple • 156, 162, 178
    transformerless PS (hot chassis) • 88, 153, 155—156, 168
    voltage dividers • 162
    voltage doubler • 153, 157
    voltage regulation • 23, 152, 163—167
    voltage tripler • 158—159
Power supply (LV) troubleshooting • 92, 114—115, 168, 179
Power transfer • 204—206, 282, 789—794
Power transistors • 117, 126—127, 204
Power transformer • 155
Power transformer troubleshooting • 172
Preamp • 365, 369, 414, 497, 510
Preemphasis • 17, 780
Prescaler (see tuners)
Primary colors (see color)
Printed circuit boards (see PCBs)
Probes • 47, 73, 97, 517, 527
Programmable divider (see tuners)
Projection receivers • 667
Projection set troubleshooting • 701—702
PTC (positive temperature coefficient) • 156
Pulses •
    color burst (see color) •
    differentiated • 214, 330
    equalizing • 330
    horizontal sync • 24, 202, 254—257, 330—331
    integrated • 200—201, 328—329
    serrated • 328—331
    vertical sync • 328
Pulse clipping/compression • 98
Push pull • 567, 791
Purity • 34
    adjustment • 687
    magnet rings • 649
    of raster • 649

Q-R

Q • 287, 495, 558, 749
Quadrature • 716, 740
Quadrature detector • 788—789
Quadrature detector troubleshooting • alignment • 806—808

INDEX 835

Quasi-complementary symmetry • 208, 210, 217, 793—794
Rabbit ears • 337, 403
Radio frequency (see RF)
Radiation •
    silicon rectifier • 178
    X-ray • 87
Rainbow pattern (see color bars)
Raster • 9—10, 26, 586
    blanking • 214
    blooming • 678, 682
    brightness • 607, 678
    centering (BW) • 216, 678
    centering (horizontal), color • 26
    centering (vertical), color • 26
    color • 681
    contamination • 610—611
    defined (see scanning)
    development • 34, 607
    distortion • 678, 680
    focus • 678
    height • 193—194
    linearity • 195, 226
    purity • 610
    tilt • 678
    tinting • 607, 610, 681, 692
    white • 607, 610
    width • 292, 306
Ratio detector • 786—787
Ratio detector troubleshooting • 806
    alignment • 806—808
RC time constant • 250
Reactance • 406
Rectifiers (see power supply)
Red, green, blue (RGB colors) • 32
    beam convergence (see convergence)
    color amplifier (see color)
    color field (see color)
    drive control (see controls)
    gun (see pix tube)
    in color mixture (see color)
    in white (see color mixing)
    phase angle • 735
    purity (see pix tube)
    signal • 737
    screen control (see controls)
    screen phosphor • 32
Regeneration • 414, 497, 522
Regulator (see voltage regulators) • 163, 165
Rejuvenation (see pix tube troubleshooting)
Relay • 433
Remote control • 36, 431
    controlled functions • 431, 435
    memory • 433
    receiver • 431

Remote control, (*Contd.*)
    receiver turn-on • 433
    systems of transmission • 431
    transmitter • 431, 433, 435
Remote control systems • 36, 179, 431, 433
Remote control troubleshooting • 179, 472—475
Repairs (general) • 455, 812
Replacing components • 614
Resistors • 119
Resistor defects • 118—119
Resonance • 405—407, 527—528, 787—788
Response curves • 57, 409, 411, 459, 489—492, 496, 528, 530—531, 769—771
Restorer, dc. (see DC restorer)
Retrace •
    beam • 9, 564, 638
    blanking • 214, 564, 576
    lines • 214, 555, 564, 603
    time required (horizontal) • 202
    time required (vertical) • 10, 202
Retrace blanking pulses • 564—565, 567
Reverse AGC (see AGC)
RF (radio frequency) • 2, 17, 401—411
RF amplifier (UHF) • 421
    signal/noise ratio • 365, 412
    tuned circuits (see tuner, UHF)
RF amplifier (UHF) troubleshooting • 457
    alignment • 466
RF amplifier (VHF) • 365, 413
RF amplifier (VHF) troubleshooting • 458
    alignment • 458—462
Ringing • 123, 207, 230, 259, 283, 287, 506, 524, 573, 605, 747
Roadmapping • 84
Rolloff (see chroma)
R-Y • (see color difference signals) • 15, 31

S

Safety • (general)—86—88, 92, 283, 684—685
    shock hazards • 87, 157, 274, 302, 675
Safety capacitor • 313—314
S curve • 785—786
Sam's Photofact (see service data)
Saturable reactor • 215, 234, 291, 297

Saturation • 215
    color • 643
    transformer • 167, 172
    transistor • 324, 342, 354, 367
Sawtooth • 193—194, 196, 205, 283
Scanning • 9—10, 193, 200—202
Scan rectification • 23, 151, 161, 168, 178, 248, 284, 292, 294—295
    scan P.S. troubleshooting • 313
Schematic diagrams • 84
SCR (silicon controlled rectifier) • 126, 166—167, 287
SCR, horizontal sweep system • 247, 260, 284
SCR sweep troubleshooting • 309
Secondary PS (see scan rectification) •
Selectivity • 404, 408, 413, 489
Self created problems • 137
Semiconductors • 123
Sensitivity • 337, 351, 365, 457, 634
Series peaking • 505—506
Series resonance • 406—407
Serrations • 328, 330, 638
Service adjustments • 36
Service set-up switch • 86, 203, 580, 610
Servicing data (software) • 84
Shadow mask (see delta CRT)
Shielding • 402, 455, 502, 505, 524, 648, 667
Shielded wires • 522, 789
Shock excitation • 49, 207, 230, 250, 506, 725
Shock hazards (see safety)
Shorted coil windings • 312—313, 523, 535, 680
Short circuits • 120
Shunt • 45
Shunt peaking • 505—506
Shunt regulator (see LVPS)
Shutdown mode • 88, 168, 236, 266, 295, 303, 308, 311
Sidebands • 5, 408, 492
    chroma • 404, 487, 491, 721
    sound • 25, 404, 487
    video • 25, 404, 487, 491
Signals, general • 1, 16, 554
Signal control, with dc • 559, 562, 582, 789
Signal injection • 47, 91—94, 224, 454, 456, 518—519, 604, 804
Signal levels • 553, 580, 582, 638, 640
Signal overdrive • 94, 458
Signal path • 580, 582, 746, 753
Signal polarity • 26, 322, 323, 554, 640

836 INDEX

Signal polarity, (Contd.)
    negative-going • 26, 554, 635
    positive-going • 26, 554, 635
Signal reflections • 412
Signal snatching • 92, 99
Signal to noise ratio • 365, 412—414, 540—541, 780
Signal takeoff points • 321, 323, 336, 338, 357, 424, 489, 495, 503, 554—555, 569, 721, 780
Signal tracing • 91, 95—99, 223, 338, 517, 604, 761
Silicon controlled rectifier (see SCR)
Silicon grease • 229, 312
Sine wave stabilization • 249—250
Single-ended amplifiers • 204—205
Skin tones (flesh tones) (see color)
Slope detection • 784
Snow (BW) • 340, 365, 375, 436, 458
Snow (colored) (see confetti)
Software (see servicing data)
Soldering/desoldering • 135—137, 455
Sound •
    bars • 153
    bass response • 789
    distorted (see distortion, sound)
    FM detector • 780, 783
    IF amplifier • 27, 783
    limiter • 783
    signal • 17, 404
    sound IF detector • 780, 782
    takeoff • 780, 782
    traps • 496
    treble response • 810
Sound detectors • 27, 496
    discriminators • 784—785
    quadrature • 788—789
    ratio • 786—788
Sound stages, troubleshooting • 796—813
Spark gap • 298—299, 314, 412, 564, 578, 603, 629—630
Speaker • 28, 790, 794
    testing and repairs • 812
Spectroflex • 13
Spectrum • 431, 492, 572
Spot killer • 562—563, 567
Spot size • 628, 642, 645, 649
Spurious signals • 460, 533
Square waves (see waveforms) • 167, 196, 282
Stage gain (see gain)
Stagger tuning • 493, 495, 522, 536
Standing wave ratio (SWR) • 412
Star (silent tuning at random), P.L.L. • 431

Static convergence • 649, 692
Sticking potential (CRT) • 634, 644
Subcarrier • 15, 32
    amplifier • 748—749
    frequency • 747
    frequency control of • 747—750
    function of • 747
    generation of • 747
    injection • 721—722, 732—743
    oscillator • 747—750
    phase • 747
    phase shifting network • 749
    suppression at station • 32
Subcarrier control circuits • 749—750
Substitution •
    antenna • 381, 454
    components • 118, 122, 124
    critical components • 271, 283, 307, 314
    CRT • 676
    precautions (see components, replacing)
    receiver oscillators • 462
    tuners • 382, 454
    voltages • 308, 380, 676
Surge resistor • 161
Swamping (loading) resistors • 495
Sweep (see beam deflection)
Sweep alignment • 460, 529—530, 535
Sweep dependent circuits (horizontal) • 247
Sweep dependent circuits (vertical) • 192, 210
Sweep derived PS (see scan rect.)
Sweep deviation (see FM)
Sweep frequencies • 193
Sweep synchronization • 9—10, 200
SWR (see standing wave ratio)
Symptom analysis • 84—85, 370, 436, 510, 669, 752, 796
Sync, beam deflection • 9, 192, 200
Sync, color • 16, 32
Synchronization (see sync)
Sync pulses • 10—11, 271, 638
Sync separator • 321—322, 338, 340
Sync (sweep) troubleshooting • 332 341—344
Synchronous detector • 31, 506, 508
Synchronous detector troubleshooting • 524—525, 582
    alignment • 525, 540

T

Takeoff points (see signal takeoffs)
Tank coils • 489, 495, 747
Taps, on coils • 495, 565
TC (time constant) • 194, 324, 327—328, 353, 359, 565

Tertiary winding • 284
Test equipment •
    bias box • 69
    CR analyzer • 53, 67
    CRT brightener • 72
    CRT gun killer switch • 63—64, 72
    CRT tester/rejuvenator • 53, 68
    CRT test jig • 69—71, 669
    curve tracer • 66
    degaussing coil • 72
    demodulator probe • 47, 49, 52, 73, 97
    diode/transistor tester • 64, 129—130
    field strength meter (FSM) •72
    flyback/yoke ringer • 66—67, 123
    frequency counter • 60
    grid dip oscillator (GDO) • 52—53
    HV probe • 47,73
    isolation transformer • 71, 157, 168, 172
    kits • 43
    meters • 45
    oscilloscope • 50—52
    power supply • 69
    R/C sub box • 69, 118—119
    signal generators • 48, 57
    signal tracers • 49
    subber • 69, 71
    tube checker • 53
    variac • 72, 172
    vectorscope • 63
    video analyzers • 60
Test equipment troubleshooting • 74—76
Test pattern • 11, 641—642
Test point (TP) • 500, 504, 804—805
Test signals (VIR/VITS) • 16—17
Thermal runaway • 117, 179, 282, 792, 810—811
Thermistor • 119, 156, 231
Threshold controls • 29, 327, 732, 762
Thyristor • 126, 166, 287
Tilt •
    BPA tuning control • 727, 764, 769
    parabolic waveform • 235, 653
    yoke adjustment • 678, 680, 686—689
Time constant (TC) • 194
Time delay • 573
Tint (hue) • 749, 765
Tone controls • 28
Tools • 455

INDEX 837

Toroid • 154
Totem pole • 208
Tough dogs • 140
Trace, beam (see beam deflection)
Tracking •
   gray scale (see gray scale tracking)
   tuner oscillator (see oscillator)
Transconductance • 64
Transducer (see remote controls)
Transformers • 71—72, 122
Transformerless PS (see PS, LV) • 88, 119
Transistors • (See NPN, PNP)
Transistors, troubleshooting • 111, 113, 129—130
Transistor operating configurations • 203, 228, 259, 328
Transistor operating voltages • 111, 113
Transistor testing • 65, 126—130
Transmission line • 412
Transposed antenna leads • 476
Trapezoid • 205
Traps (wave traps) • 407, 412—413, 489, 495, 569
Triac (see diodes) • 166
Triad • 33
Triggering • 62, 126, 201, 203, 226—227, 328, 638
Trimmer capacitor • 407
Trimpot • 166, 210, 232
Trinitron (see pix tubes-color)
Tripler (see PS, HV, LVPS)
Troubleshooting, general • 84—86, 90, 137
Troubleshooting, by substitution • 70
Troubleshooting, trouble symptoms (general) • 84, 116, 171—177, 371—372, 437—441
   pix problems • 511—513, 587—591, 670—672, 753—755
   raster problems • 220—222, 261, 263, 301—302
   sound problems • 801—803
   sync problems, color sync • 753—755, 765—767
   sync problems, raster scanning • 261—263, 332—344
Tuned circuits • 404—405, 409, 413, 493, 496, 510
Tuned circuits, troubleshooting • 523, 535, 767
Tuned circuits, alignment (general) • 606
Tuned circuit fundamentals • 406—408
Tuners (see channel selectors)

Tuners, general • 25, 28
AFT • 402, 410, 422, 424, 457, 464
AGC • 453
antenna impedance matching • 412, 506, 606, 764
balun • 412
bandpass requirements • 401, 404
band switching • 409—410, 418—419
detents • 402, 415—416, 468
electronic (varactor control) • 418—419
feed-through capacitors • 403
fine tuner • 410—411, 415, 463
frequency conversion • 403, 404
input/output signals • 403—405
mixer stage • 403—404
oscillator stage • 404, 410, 414—415, 464
PLL • 429—431
RF stage • 403—404, 409, 413
shielding • 402
tuning methods • 409
UHF channels • 401, 409
UHF/VHF changeover switching • 405, 415
varactor control of • 402, 409, 415, 418
VHF channels • 401
Tuners, troubleshooting (general) • 436—476
   alignment • 455, 457, 459—462, 464—465
   repairs • 455
   substitution • 382, 454, 468
Tuner types • 415
Tuner, (UHF continuous) • 409, 418, 421
   (see tuners, general)
   hairpin coil • 421, 466
   tuning ratio • 421
   mixer stage • 421
   operation of • 405
   oscillator (see osc, UHF) • 421
   RF stage (see tuners, UHF)
   tuning capacitor • 421
   tuned lines • 421—422
Tuners, continuous, troubleshooting • 465—467
   alignment • 466
Tuners, drum/turret • 415—416
   (see tuners, general)
Tuner, (drum/turret) troubleshooting • 462—463
   alignment • 460—461, 464

Tuners (electronic), (PLL) • 426, 435
   bandswitching • 418, 430, 436
   bandswitching voltage • 418, 429—430, 436, 458
   channel readout display • 430, 436
   comparator • 200, 427—429, 436
   frequency dividers • 427—428, 436
   frequency synthesizing • 426, 428
   keyboard control • 401, 418, 426, 429—430
   memory • 431
   microprocessor • 426—430
   muting • 431, 436
   PLL theory • 426—428
   prescaler • 429
   programmable divider • 427—429
   reference oscillator • 427—429
   remote control • 426, 431, 433
   star system • 431
   tuning voltages • 418, 422, 426—429, 431, 436, 458, 463
   up-down search • 431
   varactor control of • 428
   VCO (voltage controlled oscillator) • 427—431, 436
Tuner (PLL) troubleshooting • 470—472
Tuners, varactor controlled • 435
   (see tuners, general)
Tuners (varactor) troubleshooting • 465—467
   channels, preset adjustments • 419
Tuners, wafer/switch • 415
   (see tuners, general)
Tuners, wafer/switch, troubleshooting • 415
   alignment • 464—465
Tuners, replacement of • 468
Tuning, defined • 408—409
Tuning methods • 409, 462
Tuning voltage (see varactor tuners)
Tuning wand • 461, 535
Tweeter (see speaker)
Twist coil • 216, 660

U-V

UHF (ultra high frequencies) • 2
UHF channel allocations • 3—5
UHF tuners (see tuners)
Ultor (see pix tube)
Up-down search (see PLL tuner)

838 INDEX

User controls • 28
Varactor • 33, 125
Varicap (see varactor)
VCO (voltage controlled oscillator) • 200
VDR (voltage dependent resistor) • 119, 156, 206, 791
Vectors • 64, 716
Vertical blanking • 192, 214, 231
Vertical controls • 193
Vertical retrace • 200, 214, 231
Vertical sweep • 23, 191—192
Vertical sweep oscillator • 191—195
Vertical oscillator troubleshooting • 225
Vertical sweep output circuit • 192, 203—205, 210
Vertical output troubleshooting • 229
Vertical sync • 200
Vestigial sidebands • 2, 17, 492
VHF (very high frequencies) • 2, 25, 401
VHF bands •
   high channels • 409
   low channels • 409
Video • 25
   amplifier • 25
   detector • 506
   frequencies • 488, 554, 606, 641
   IF amplifier • 25
   overload • 336—337, 605—606
   peaking • 505—506, 510, 605
   sidebands • 487—488
   signal • 14, 404, 638
   waveforms • 639—640
Video amplifier • 25, 554, 582
Video amplifier troubleshooting • 583—614
Video detectors • 25, 503
Video detector troubleshooting • 523—524
Video IF amplifier • 25, 405
   frequencies • 488

Video IF troubleshooting • 510—541
   alignment • 525—540
Video signal • 554, 638, 640
Vidicon • 8
VIR (vertical interval reference) • 16—17, 730, 744
VITS (vertical interval test signal) • 16
Visual inspection • 90—91
Volatile memory • 441
Voltage •
   AC • 153—155
   amplifier (see amplifiers, general)
   computation (see IR drop)
   controlled oscillator (see VCO)
   DC • 110, 152
   distribution • 152, 162
   divider (AC) • 357
   divider (DC) • 162, 354, 367
   drop (IR) • 116, 119
   focus, CRT • 314
   HV (see power supply, HV)
   levels • 110
   load voltage • 153
   measuring • 110—111
   multipliers • 158—159, 296
   no-load voltage • 153, 156
   pdc • 152, 503
   p-p • 47, 283, 565—568
   regulation • 153, 163—165, 368
   regulator (HV) • 163—166, 386
   regulator (LV) • 163—166, 177—178, 386
   regulating transformer (VRT) • 167, 172
   ripple • 120, 178, 223, 603
   RMS • 47, 167
Volume control • 789, 795
Voltage controlled oscillator (VCO) • 200
VRT (voltage regulating transformer) • 167, 172

VSM (video sweep modulation) • alignment • 770

W

Wattage (power) •
   computation • 811—812
   measurement • 811
   power consumption • 173
   resistor rating • 119
   speaker rating • 794—795
Waveform • 205
   analysis • 109
   distortion • 109
   linearity • 194
   shaping • 194, 328, 653
Wavetrap (see traps)
White • 643
   from color mixture • 12, 643, 746
   highlights • 35, 662, 680
   'noise • 365
   raster • 662
Wide band color (see color)
Width control • 29, 247, 292
Woofer (see speaker)

X-Y-Z

X-ray (see safety) • 648
   hazard • 87, 248, 283, 648
   protection • 87, 283, 648, 667
X-Z demodulator • 31, 736
Y (luminance) amplifier (see video amp) • 568
Y signal (video, color set) • 15, 29, 554, 715
Yoke • 23, 25, 204—206, 282—283, 626, 630—631, 634, 648
   coupling • 206, 282
Yoke, troubleshooting • 207, 230, 680
   adjustments • 678, 680, 686—689
Zener diode • 125, 165
Zero beat (see null)